A Meeting of Minds: In Recognition of the Contributions of Randall J. Cohrs

A Meeting of Minds: In Recognition of the Contributions of Randall J. Cohrs

Editors

Charles Grose
Ravi Mahalingam
Joel Rovnak

MDPI • Basel • Beijing • Wuhan • Barcelona • Belgrade • Manchester • Tokyo • Cluj • Tianjin

Editors

Charles Grose
University of Iowa
Iowa City, IA, USA

Ravi Mahalingam
University of Colorado
School of Medicine
Anschutz Medical Campus
Aurora, CO, USA

Joel Rovnak
Colorado State University
Fort Collins, CO, USA

Editorial Office
MDPI
St. Alban-Anlage 66
4052 Basel, Switzerland

This is a reprint of articles from the Special Issue published online in the open access journal *Viruses* (ISSN 1999-4915) (available at: https://www.mdpi.com/journal/viruses/special_issues/Randy_Cohrs).

For citation purposes, cite each article independently as indicated on the article page online and as indicated below:

LastName, A.A.; LastName, B.B.; LastName, C.C. Article Title. *Journal Name* **Year**, *Volume Number*, Page Range.

ISBN 978-3-0365-5259-0 (Hbk)
ISBN 978-3-0365-5260-6 (PDF)

Cover image courtesy of Joel Rovnak

Contents

About the Editors

Charles Grose

Dr. Grose is currently a Professor of Pediatric Infectious Diseases at the University of Iowa. He received his first NIH grant as an assistant professor in 1978, with the goal of defining the major antigenic determinants of varicella zoster virus. His laboratory has continued to study the molecular biology and immunology of varicella zoster virus with NIH funding up to the present time. He is a member of the editorial boards of the Journal of Virology, Virology and Viruses. Dr. Cohrs was a co-investigator on his current NIH grant, to study aspects relating to the attenuation of the varicella vaccine virus. His publication record includes an h-index of 52.

Ravi Mahalingam

Dr. Ravi Mahalingam is a Professor of Neurology at the University of Colorado School of Medicine. He also serves as a visiting professor at the Sri Ramachandra Medical College & Research Institute, Chennai, India. Ravi completed his master's in Chemistry at the Indian Institute of Technology in Mumbai, India in 1976, and his PhD in Biochemistry and Molecular Biology in 1980 at Southern Illinois University, Carbondale, Illinois, USA. His first post-doctoral fellowship was in Microbiology in Denver, Colorado, with Prof. Donald Cummings, and in 1986 he joined Don Gilden as a postdoctoral fellow and rose to the rank of full professor in the Department of Neurology in 2005. During this time, Randy and Ravi became close colleagues. Ravi was the first to use PCR to identify latent varicella zoster virus (VZV) DNA in human ganglia and later show the expression of the VZV ORF63 protein. He developed the unique animal model of primary VZV infection, latency and reactivation using simian varicella virus (SVV) infection in non-human primates. In collaboration with Prof. Wayne Gray, Ravi sequenced the complete SVV genome in 2001. His laboratory currently focuses on the multi-organ disease produced by varicella virus reactivation.

Joel Rovnak

Dr. Rovnak is an Associate Professor in the Department of Microbiology, Immunology and Pathology in the College of Veterinary Medicine and Bioscience at Colorado State University in Fort Collins, Colorado. His research is focused on the role of viral proteins in the regulation of host gene expression in infected cells. This includes investigations of viral oncogenesis using retroviral models, bovine leukemia virus, human T cell leukemia virus, and walleye dermal sarcoma virus, and the metabolic reprogramming of cells using flaviviral models, dengue virus and Zika virus. His work is specifically focused on the Mediator complex and, particularly, cyclin-dependent kinases 8 and 19, as direct targets in the virus control of host transcription. It was in this context that Dr. Rovnak initiated collaborations with Dr. Cohrs on the epigenetic analyses of cells infected by retroviruses, flaviviruses and herpesviruses. He also collaborated with Dr. Cohrs for 12 years in the establishment and performance of two scientific associations, the Colorado Alphaherpesvirus Latency Society (CALS) and the Rocky Mountain Virology Association (RMVA), which continue to serve and promote effective science communication.

Preface to "A Meeting of Minds: In Recognition of the Contributions of Randall J. Cohrs"

Randy Cohrs was a major contributor to the field of alphaherpesvirus latency, especially of varicella zoster virus. His work extended from his deep compassion for afflicted patients and their families and took him to the cutting edge of the molecular mechanisms of virus transcription control. He helped to introduce practicable therapies and vaccines to the clinic. He coupled his work with patient samples to in vitro systems and advanced next-generation sequencing techniques to bring clarity to the control of alphaherpesvirus gene expression in disease.

Randy also worked outside the herpesvirus field, both to share information and to learn from the work of others on various virus models. He wanted to know what others thought and he went to extraordinary measures to make it possible for them to share their work. While particularly focused on students (of all ages and disciplines), he trained everyone he met in the art of communication. He practiced that and forced others to practice it as well. To that end, Randy's contributions to science extend well beyond his chosen field. Whether it was the International Herpes Workshop, the Rocky Mountain Branch of the American Society for Microbiology, the Rocky Mountain Virology Meeting, or, his favorite, the Colorado Alphaherpesvirus Latency Society, he dedicated himself fully to scientific communication. The results have come to play in many fields as a result of his introduction of diverse persons and personalities in convivial and, for him, joyous interactions. He was a magnanimous, kind and charitable man who broke down all the barriers to scientific communication as often as he possibly could.

Charles Grose, Ravi Mahalingam, and Joel Rovnak
Editors

viruses MDPI

Editorial

The Enduring Legacy of Randall Cohrs: A Meeting of the Minds in the Rocky Mountains

Charles Grose [1,*], Joel Rovnak [2] and Ravi Mahalingam [3]

1 Virology Laboratory, Department of Pediatrics, University of Iowa, Iowa City, IA 52242, USA
2 Department of Microbiology, Immunology and Pathology, Colorado State University, Fort Colins, CO 80523, USA; joel.rovnak@colostate.edu
3 Department of Neurology, University of Colorado School of Medicine, Anschutz Medical Campus, Aurora, CO 80045, USA; ravi.mahalingam@cuanschutz.edu
* Correspondence: charles-grose@uiowa.edu

Randall Cohrs established the Colorado Alphaherpesvirus Latency Society (CALS) in 2011. The main function of CALS was to support an annual meeting of investigators interested in herpesvirus latency and reactivation. Randall Cohrs died on 30 July 2021. In memory of Randy, we have assembled a Festschrift of articles written by virologists who attended CALS, as well as the Rocky Mountain Virology Association and the International Herpesvirus Workshop. Randy received his PhD degree from Southern Illinois University; the title of his doctoral thesis in the department of microbiology was "The molecular biology of *Herpesvirus sylvilagus*". Thereafter, he moved to Colorado and later accepted a faculty position in the Department of Neurology at the University of Colorado School of Medicine. As a member of the Don Gilden laboratory, he devoted his career toward the study of transcription, latency and pathogenesis of varicella-zoster virus. In mid-career, he developed an interest in organizing scientific meetings. Based on knowledge acquired from that responsibility, he established CALS in 2011, specifically to bring together what he called a disjointed assemblage of virologists with shared interests in viral latency. CALS was a success. In this Festschrift, his friends and colleagues have submitted their contributions. The authors come from virology research centers in several countries, including the United States, the United Kingdom, the Netherlands, Germany, France, Switzerland, Israel, India, China and Japan. The viruses include, besides varicella-zoster virus and herpes simplex virus, cytomegalovirus, dengue virus, and coronavirus

Citation: Grose, C.; Rovnak, J.; Mahalingam, R. The Enduring Legacy of Randall Cohrs: A Meeting of the Minds in the Rocky Mountains. *Viruses* **2022**, *14*, 915. https://doi.org/10.3390/v14050915

Received: 26 April 2022
Accepted: 27 April 2022
Published: 28 April 2022

Publisher's Note: MDPI stays neutral with regard to jurisdictional claims in published maps and institutional affiliations.

1. Introduction to Festschrift

Randall Cohrs was the president of the Colorado Alphaherpesvirus Latency Society (CALS). The main purpose of CALS was to share scientific insights and foster collaborative research efforts among the gathered virologists. After his unexpected death in 2021, we have assembled a Festschrift in memory of Randy and his numerous achievements. Scientists who attended CALS, as well as Randy's friends and colleagues from other virology meetings, were invited to contribute an article.

2. The Life of Randall Cohrs

Randy was a child of the American tall grass prairie. This massive ecosystem encompasses most of the states of North Dakota, South Dakota, Nebraska, Kansas, Iowa as well as sections of Illinois west and south of Chicago. The prairies formed 10,000 years ago and were once home to millions of bison. Long after the bison disappeared, Randy was raised in Illinois. The land is dotted with small villages and medium-sized towns, interspersed with thousands of acres of farmland. In turn, the farmland is crisscrossed by thousands of miles of train tracks. Even as a child, Randy always wanted to know more about how things worked. Randy enjoyed high school, where he enrolled in all the science classes; but he also won a letter for high achievement in cross-country running. Like all teenagers, he also had a variety of summer and after-school jobs, such as a short-order cook, a groundskeeper and even a cemetery caretaker. However, he never passed up a chance to go fishing.

After high school, he spent both his undergraduate and microbiology graduate college years at Southern Illinois University (SIU) in Carbondale, Illinois. He always described these years as fulfilling. He obtained his Ph.D. degree under the mentorship of Professor Hassan Rouhandeh on 20 March 1985. The title of his doctoral thesis was "The molecular biology of *Herpesvirus sylvilagus*". Experimental infection with *Herpesvirus sylvilagus* produces clinical and histopathologic changes in its natural host, the cottontail rabbit (Sylvilagus genus), that are similar to those observed in humans acutely infected with Epstein–Barr virus (infectious mononucleosis). Randy's most important Ph.D. research publication was a paper in the *Journal of Virology*, entitled *Herpesvirus sylvilagus: Polypeptides of virions and nucleocapsids* [1]. As expected, his thesis was one of the thickest on the shelf in the library (Figure 1). Randy was especially proud when colleagues at his former department asked him to give a special seminar for the SIU graduate students.

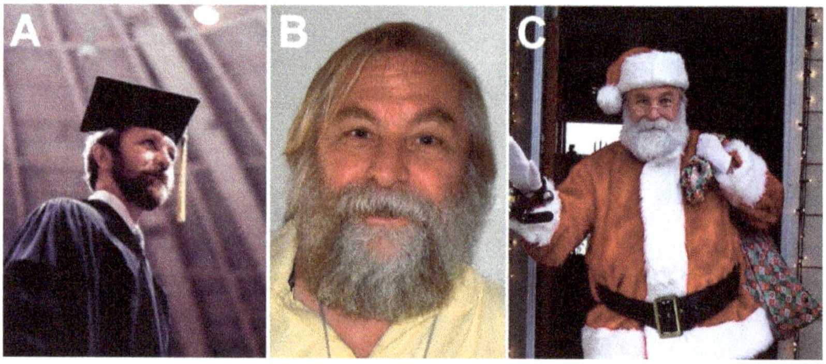

Figure 1. Photographs of Randall Cohrs during his career. (**A**). Graduation ceremony at Southern Illinois University in 1985. (**B**). Taking a break at a virology meeting in 2008. (**C**). Randy dressed as Santa Claus for a favorite children's event in Lakewood, Colorado, in 2013.

Randy moved to Colorado soon after graduation in order to be closer to his family who had moved there while Randy was in college. He accepted a post-doctoral position at the AMC Cancer Research Center in Lakewood, Colorado, with Professor Opendra Sharma, to study the role of vitamin E and the stability of RNA in cancer cells [2]. Soon afterward, he met Terri Carpenter; Randy and Terri were married on 3 January 1987, and they remained inseparable ever after. In 1989, when Don Gilden, Abbas Vafai and Ravi Mahalingam were looking for a RNA expert to investigate varicella zoster virus transcription, Randy Cohrs was their first choice. Randy was successfully recruited to a research-track faculty position in the Department of Neurology at the University of Colorado Health Sciences Center [3]. Therefore, except for the 4 years at the AMC Cancer Center, Randy would devote the rest of his scientific career to the study of herpesviruses (Figure 1). Randy was a very hard worker and was always focused on his next paper, grant or collaboration. His passion to understand basic biological processes was infectious. He also had insatiable thirst for other scientific fields such as astronomy, higher physics, and mathematics. Further, Randy easily engaged with the younger generation, infusing them with a curiosity for science. One of his frequently cited first-authored papers that drew considerable attention at virology meetings was a collaborative study funded by the National Aeronautics and Space Administration entitled *Asymptomatic reactivation and shed of infectious varicella zoster virus in astronauts* [4]. Another important first-authored paper entitled *Comparison of virus transcription during lytic infection of the Oka parental and vaccine strains of varicella-zoster virus* determined that viral IE62 (ICP4 homolog) was not the sole determinant of attenuation in the live varicella vaccine strain Oka [5]. Randy was also a member of the team that discovered the varicella-zoster virus latency-associated transcript [6]. Each of the authors of the current article were fortunate to have written one or more articles with Randy [7–9].

At the seventh annual meeting of the Rocky Mountain Virology Club in 2007, Kathryn Holmes, from the University of Colorado, summoned Randy, Tony Schountz and Joel Rovnak from across the room to the business meeting, already in progress, and conferred upon them the administrative duties of procuring snacks and beverages for the next meeting. Two years later, Randy and Joel volunteered to run the meeting and Tony set up the website. At the same time, they established the Rocky Mountain Virology Association (RMVA), a 501(c)3 corporation. The RMVA could accept tax-free donations, but more importantly, RMVA could apply for support from the National Institutes of Health (NIH). From 2010 forward, the annual RMVA meeting held at the scenic Colorado State University Mountain Campus has been partially funded by the National Institute of Allergy and Infectious Diseases. With this NIH support, the organizers invited distinguished speakers from around the country. This change expanded the RMVA from a small regional meeting to one with national status. During these meetings, nothing excited Randy as much as making connections between investigators across a broad range of fields (Figure 2).

Figure 2. The Rocky Mountain Virology Association meeting in 2019. The photo shows several virologists at the meeting taking a hike in the mountains around the Colorado State University conference center during an afternoon recess. Randy, as usual, is at the center of an active discussion along the trail. Besides Randy, those in the picture include Laura Ashton, Carmen Ledesma-Feliciano, Christie Mayo, Amy MacNeill, Jasmine McCoy, Lee Fortunato, Sven Heinz, Joel Baines and Joel Rovnak. The meeting was held in late September when the golden leaves on the aspen trees are most beautiful.

3. Colorado Alphaherpesvirus Latency Society

For a brief moment in this commentary, we introduce some pop-psychology to explain Randy's enthusiasm for organizing virology meetings. There has been considerable research into the traits acquired by children in a family, based on their birth order. Almost everyone agrees that the first-born child is typically a high achiever. Randy was the middle child, having both an older and a younger brother. Randy had the archetypical persona of a middle child, including being a people-pleaser, thriving on friendships, having a large social circle, and playing the role of peacemaker. We believe that these characteristics gave him the ability, along with the valuable experience acquired from supervising the RMVA meeting, to establish the Colorado Alphaherpesvirus Latency Society (CALS) in 2011 (together with

Don Gilden). We consider CALS to be one of the crowning achievements of Randy's career. For nine consecutive years (2011–2019) until halted by the COVID-19 epidemic, CALS was held annually in mid-May at the Christiania Hotel in Vail, Colorado. CALS began as a meeting of 40 scientists interested in latency of alphaherpesviruses, but expanded to 70 when Don and Randy were able to secure rooms in the nearby Tivoli Lodge for additional participants. The rules were simple: submit an abstract, and give a 10 min talk followed by a 10 min discussion. The sessions over 2 days were divided into themes relevant to latency. Each CALS session had one guest speaker; the speaker at the 2019 CALS was the Nobel laureate Stanley Prusiner [10]. At night, the group meandered through the village to a restaurant. Randy always held his fireside chat on Friday evening after dinner. Randy carried out all these activities so many times with such a passion that his scientific influence was felt well beyond the varicella and herpes world.

4. Anecdotes

Each author will present below an anecdote about Randy, which provides more insight into his persona both inside and outside the world of science.

4.1. From Charles Grose

Both Randy and I lived similar childhoods. We talked about trains, not model toy trains but real trains. Both of us were raised on the prairie, where trainspotting was a popular game for children. When a train was first seen or when a train whistle was heard miles away, he walked to the tracks to be only a few feet away when the 200-ton locomotive rumbled past him. The engineers always waved. As his first scientific experiment, he placed penny coins on the tracks and waited to see which became the widest after being crushed by the locomotive. Randy's favorite railroad was the Soo line because of its weird name, a phonetic abbreviation of the Minneapolis, Saint Paul and Sault Sainte Marie railroad (French word Sault = Soo in English). When Randy moved to Colorado, he switched his allegiance to the scenic Denver and Rio Grande Railroad. Joel later told me that Randy liked Joel's office because Joel could hear sounds of passing trains from his office.

4.2. From Ravi Mahalingam

In the late 1970s, both Randy and I were graduate students at SIU in microbiology and chemistry, respectively. Later, we were colleagues working together on varicella-zoster virus latency and reactivation in the same department at the University of Colorado School of Medicine. Having had the privilege of interacting closely with Randy for the past 4 decades, I noticed that the five words that Randy used most often was "how can I help you". Whether it was a person that he met in the street for the first time or a longtime acquaintance, Randy was always finding ways to help them when there was a need. I can never forget how Randy carried my hand carriage with the broken wheel up the hill to my room in Regensburg, Germany, and when he rolled my 3-year-old son in his stroller throughout the museums in Washington, DC. Additionally, Randy had a knack of making friends independent of language and cultural barriers. I remember him spending more than an hour with a total stranger (who did not know English), whom he bumped into during his visit to India. Randy's passion for life and curiosity to learn will be missed.

4.3. From Joel Rovnak

Randy and I spent hours brain-storming transcription regulation and working out chromatin immune precipitation protocols. He was an early proponent of next-generation sequencing and did everything he could to promote the core sequencing capacity at the University of Colorado School of Medicine and to support investigators with the skills to analyze sequence data. To that end, we spent a lot of time selecting guest speakers from the sequencing and transcription fields for our virology meetings, more than one of whom lamented that they were not virologists and questioned why were we inviting them to a virology meeting. They came to the meetings and came back again and again. One of

Randy's greatest joys was introducing people who would never meet otherwise. He knew that they needed to meet and just had not realized it yet.

5. The Final Day

On Friday, 30 July 2021, Randy was working late in the laboratory as usual. Just before he left, he arranged meetings the next day with students and colleagues to practice talks for a meeting on Monday. While driving home he must have known something was amiss with his health, because he took care to pull over to the side of the highway out of traffic, one last act of consideration for others. The impact of his death extends far and wide. Randy is buried in Golden, Colorado, at the historic Golden Cemetery, Section 238. Graciously, the journal *VIRUSES* has given his friends and colleagues an opportunity to assemble a Festschrift by which to recognize and remember the legacy of Randy Cohrs.

Author Contributions: All three authors contributed substantially to the writing of this article about Randall Cohrs. All authors have read and agreed to the published version of the manuscript.

Funding: Research by Cohrs at the University of Colorado was supported by both the NIH program project and individual research grants; Mahalingam was a co-investigator on several of these grants. Cohrs was a co-investigator on NIAID research conference grants written by Rovnak at Colorado State University. Cohrs was a co-investigator on NIAID research grant AI153817 written by Grose at the University of Iowa.

Acknowledgments: We thank Terri Carpenter Cohrs, the wife of Randy, and his younger brother Daniel Cohrs for providing details about the life of Randy Cohrs.

Conflicts of Interest: The authors declare no conflict of interest.

References

1. Cohrs, R.; Rouhandeh, H. Herpesvirus sylvilagus I. Polypeptides of virions and nucleocapsids. *J. Virol.* **1982**, *41*, 1063–1072. [CrossRef] [PubMed]
2. Cohrs, R.J.; Goswami, B.B.; Sharma, O.K. Occurrence of 2-5A and RNA degradation in the chick oviduct during rapid estrogen withdrawal. *Biochemistry* **1988**, *27*, 3246–3252. [CrossRef] [PubMed]
3. Cohrs, R.J.; Srock, K.; Barbour, M.B.; Owens, G.; Mahalingam, R.; Devlin, M.E.; Wellish, M.; Gilden, D.H. Varicella-zoster virus (VZV) transcription during latency in human ganglia: Construction of a cDNA library from latently infected human trigeminal ganglia and detection of a VZV transcript. *J. Virol.* **1994**, *68*, 7900–7908. [CrossRef] [PubMed]
4. Cohrs, R.J.; Mehta, S.K.; Schmid, D.S.; Gilden, D.H.; Pierson, D.L. Asymptomatic reactivation and shed of infectious varicella zoster virus in astronauts. *J. Med. Virol.* **2008**, *80*, 1116–1122. [CrossRef] [PubMed]
5. Cohrs, R.J.; Gilden, D.H.; Gomi, Y.; Yamanishi, K.; Cohen, J.I. Comparison of virus transcription during lytic infection of the Oka parental and vaccine strains of Varicella-Zoster virus. *J. Virol.* **2006**, *80*, 2076–2082. [CrossRef] [PubMed]
6. Depledge, D.P.; Ouwendijk, W.J.D.; Sadaoka, T.; Braspenning, S.E.; Mori, Y.; Cohrs, R.J.; Verjans, G.; Breuer, J. A spliced latency-associated VZV transcript maps antisense to the viral transactivator gene 61. *Nat. Commun.* **2018**, *9*, 1167. [CrossRef] [PubMed]
7. Kennedy, P.G.; Rovnak, J.; Badani, H.; Cohrs, R.J. A comparison of herpes simplex virus type 1 and varicella-zoster virus latency and reactivation. *J. Gen. Virol.* **2015**, *96*, 1581–1602. [CrossRef] [PubMed]
8. Cohrs, R.J.; Gilden, D.H.; Mahalingam, R. Varicella zoster virus latency, neurological disease and experimental models: An update. *Front. Biosci.* **2004**, *9*, 751–762. [CrossRef] [PubMed]
9. Grose, C.; Yu, X.; Cohrs, R.J.; Carpenter, J.E.; Bowlin, J.L.; Gilden, D. Aberrant virion assembly and limited glycoprotein C production in varicella-zoster virus-infected neurons. *J. Virol.* **2013**, *87*, 9643–9648. [CrossRef] [PubMed]
10. Mangold, C.; Engel, E.; Ostler, J.; Wilson, A.; Cohrs, R.J. 2019 Colorado alphaherpesvirus latency society. *J. Neurovirol.* **2020**, *26*, 297–309. [CrossRef] [PubMed]

viruses

MDPI

Obituary

Meeting in Mind and a Smile on the Face: A Tribute to Dr. Randall J Cohrs

Jayasri Das Sarma [1,2]

[1] Department of Biological Science, Indian Institute of Science Education and Research—Kolkata, Mohanpur 741246, India; dassarmaj@iiserkol.ac.in
[2] Department of Ophthalmology, University of Pennsylvania, Philadelphia, PA 19104, USA

It is my privilege to have a mentor cum friend like Prof. Randall Cohrs, who played a significant role in mentoring me and moulding me into the scientist and mentor that I am today. I first met Randy at the 9th International Symposium on Neurovirology held in 2009 in Miami, FL, USA. Cohrs's then mentor and dear friend Dr. Donald H Gilden introduced me to him. Randy being a kind and generous human, we became friends in no time. Although by training and practice, he was a Herpes virologist and I was a Corona virologist, we often found ourselves discussing science over coffee. Our friendship made me realize that friendship knows no boundaries of age, ethnicity, language, and the branch of science you chose to pursue.

I had the chance to meet Randy again in 2010 when I was invited to deliver a grand round lecture at the Department of Neurology, University of Colorado by Dr. Gilden. Randy was a very compassionate human being; and this meeting further strengthened our friendship.

Randy was a mentor to me; he taught me an essential skill for a scientist in this era: networking and conducting meetings. In 2013, Randy encouraged me to apply for the Indo-US science and technology forum (IUSSTF) American Society for Microbiology (ASM) professorship award. We wrote the grant together, which was funded, and Randy served as my mentor for the program. This was our first venture into collaboration together and a very successful one, highlighted on the International Society of Neurovirology (ISNV) and Indo-US website (Figure 1). This collaboration gave me a sense of trust and confidence in my mentor and myself. I found myself discussing my challenges as a scientist with Randy, to which he would always have something thoughtful to add. Randy's mentorship helped me publish a few articles with him in journals like *PLoS ONE* [1] and *Clinical Immunology* [2] based on the Affymetrix microarray analyses combined with a multiplex protein array system for cytokines and chemokines, which for the first time highlighted the robust upregulation of various innate immune genes that are mainly involved in antiviral immune response, phagolysosome maturation, and MHC class-II expression in murine β-coronavirus infected chronically inflamed tissues. My wholehearted thanks to him on behalf of me and my student, for his immense help and contribution in this project which also led to the Ph.D. thesis of Dr. Kaushiki Biswas, her gratitude towards Dr. Cohrs is evident from her anecdote quoted below (Box 1).

Citation: Das Sarma, J. Meeting in Mind and a Smile on the Face: A Tribute to Dr. Randall J Cohrs. *Viruses* **2022**, *14*, 1124. https://doi.org/10.3390/v14061124

Academic Editors: Charles Grose, Ravi Mahalingam and Joel Rovnak

Received: 10 May 2022
Accepted: 16 May 2022
Published: 24 May 2022

Winter 2013
Vol. 13 No.1

ISNV

International Society for NeuroVirology

Editor: Dianne Langford, Ph.D. Associate Editor: Michael Nonnemacher, Ph.D.

Noteworthy News

Dr. Jayasri Das Sarma (pictured right), Associate Professor, Department of Biological Sciences, Indian Institute of Science Education and Research-Kolkata, India and Dr. Randall Cohrs (pictured left), Research Professor, University of Colorado, Denver School of Medicine have received the 2013 ASM-IUSSTF Indo-US Research Professorship Award to develop a bi-lateral research relationship between India and the US. They propose to establish a multiplex array analysis of differentially expressed host genes involved in innate immunity and demyelination in mice following mouse hepatitis virus infection. Evidence of molecular pathways driving microglial activation in (MHV) infection will advance the understanding of mechanisms by which viral-induced demyelination can occur through innate immune activation. This program is managed by the American Society for Microbiology (www.asm.org) and generously supported by the Indo-US Science and Technology Forum (www.indousstf.org).

Figure 1. ASM-IUSSTF Indo-US Research Professorship award. The ASM-IUSSTF Indo-US research Professorship award for the year 2013 was awarded to Dr. Jayasri Das Sarma, with Dr. Cohrs as her mentor. The news was highlighted in the ISNV newsletter.

Box 1. Anecdote from Dr. Kaushiki Biswas, Assistant Professor at Department of Biological Sciences, Presidency University, Kolkata, India

I had the privilege to meet Prof. Randall J Cohrs when I was a Ph.D. student, during the Indo-US symposium on Viral Infections of the Nervous System, 2014 at Gurgaon, India. I was working in the laboratory of Prof. Jayasri Das Sarma and was in my final year of Ph.D. As he was eager to interact with the young minds, an exclusive interaction session was arranged for us where we got the opportunity to have a direct meeting with him on a round table. Prof. Cohrs interacted with all the students and research scholars with curiosity and patience. He illuminated us with his profound knowledge of viruses and mentored us to a career in science. When I look back, I want to thank him for his thoughtful gesture, which also reflects a part of his persona. All throughout my PhD career his scientific publications on viruses acted as a reference and guiding path to my work. He had also provided his scientific insights into my PhD dissertation. I am proud to share a publication with him as a co-author [2].

In 2014, one of my BS-MS students, Mr. Jibin Sadasivan (Box 2), received the Du Pré grant, which was awarded to young MS researchers from the Multiple Sclerosis International Federation (MSIF), United Kingdom, to make short visits to established MS research centres outside of their own country. Randy was his host for the program. Jibin was indeed overwhelmed by Randy's mentorship style, reflected in his testimony.

Box 2. Anecdote from Mr. Jibin Sadasivan, Graduate Student at University of British Columbia, Vancouver, Canada

I met Randy in 2014 when I was doing a short internship in his lab at the University of Colorado. Moving to a new country for the first time wasn't easy, but Randy made sure that I had a smooth transition and that I did not feel left alone during my time there. He drove me to the lab every day and took me on short trips to Denver and Boulder on weekends. He made sure that I got souvenirs and pictures from the places we visited together. After my internship, we stayed in touch, and he continued to mentor me as I moved to grad school. Randy was a passionate mentor, a fantastic friend, and an inspiration to many of us, personally and professionally. I will be forever honoured and grateful to have had the opportunity to know him and work with him.

Prof. Gilden urged me to host the Indo-US bilateral symposium in India; this would facilitate cultural exchange in addition to science communication. With his support, we hosted a symposium in 2014 in New Delhi with Dr. Pankaj Seth (Indian Host), myself (Indian Co-Host) and Randy (US-Host), and Dr. Lynn Pullium (US Co-Host). The meeting received a great response. Dr. Randall Cohrs, Dr. Ravi Mahalingam, and others presented their work. Dr. Donald Gilden delivered a plenary lecture on temporal arteritis, which the audience received with overwhelming enthusiasm. My students got a chance to interact with Randy here for the first time, and they were overwhelmed with Randy's easy-going nature and the vital perspective that he gave them on their research projects. After this meeting, our group of scientists took a golden triangle tour to the cities of Agra, Jaipur, and Delhi to experience the culture of these Indian cities (Figure 2).

Figure 2. Meeting in mind and travel in heart. Virologists on a tour of the golden triangle after the Indo US bilateral symposium held in India in 2014. (**A**) Visit to the Taj Mahal. (**B**) Traditional Indian welcome of guests with garland. Besides Randy, the picture includes Dr. Stanley Perlman, Dr. Susan Wiess, Dr. Donald Gilden, Dr. Howard Lipton, Mr. Ed, Ms. Audry Gilden, Mrs. Pam Lipton, Dr. Kathryn Iacono, Dr. Jayasri Das Sarma, and Ms. Kimberly Dine.

Prof. Donald Gilden was also looking forward to the next Indo US bilateral symposium I hosted in 2016, but he could not attend the meeting due to his sad demise. Dr. Cohr continued to support my endeavours to conduct meetings. He was the Co-PI for the Bioanalytical tools and techniques workshop (BAW) 2018, held around the Indian festival of Colours, Holi (Figures 3 and 4). This festival celebrates the eternal and divine love of Radha Krishna and signifies the triumph of good over evil. Randy enjoyed playing Holi with colours, and "work hard, play hard" was his slogan for the students during this meet. The students thoroughly enjoyed his company during the festival and the meeting (Figure 5). Randy also urged me to apply for ASMCUE 2017, where I received the American Society for Microbiology Undergraduate Leadership Award. In 2019, he invited me to the 19th Rocky Mountain Virology Association Meeting, held at a captivating and picturesque Mountain Campus of Colorado State University at the Pingree Park. He gave me an opportunity to deliver the invited opening presentation on the protective role of Ifit2 during neurotropic virus infection, followed by a keynote lecture by Dr. W. Ian Lipkin. Randy was always up-to date with the current advances in Virology and was an enthusiastic for hosting people with upcoming and important studies.

Figure 3. The Bioanalytical tools and techniques workshop (BAW) 2018. Under the mentorship of Dr. Cohrs, we hosted the workshop, which received a great response. The picture depicts the entire group that attended BAW 2018, with Randy and me in the centre, along with Dr. Ravi, Dr. Seth, Dr. Brent, Dr. Patricia, and Dr. Sue.

Figure 4. Fun times at BAW 2018. (**A**) COVID biology lab (IISER Kolkata) group with Dr. Cohrs and Dr. Ravi Mahalingam. (**B**,**C**) Fun outing to a nearby waterfall during the workshop. (**C**) Randy, Dr. Palmer and Dr. Seth dressed in traditional Sikkim attire. (**D**) Randy and Dr. Seth's performance at the cultural program.

Figure 5. Soaked in colour. (**A**) Randy covered in Holi colours. (**B**) When a colleague becomes your dearest friend and brother; Dr. Ravi Mahalingam and Randy playing with Holi colours. (**C**) Randy enjoying Holi with the students during BAW 2018.

Randy was certainly a great leader. His leadership quality is evident from how he established and successfully conducted meetings for the Colorado Alpha Herpesvirus Latency Society (CALS) and led the European Society for Translational Medicine as the Vice-president and Rocky Mountain Virology Association (RMVA) as president.

In 2020, amidst the pandemic, we hosted the Indo US bilateral symposium in virtual mode on COVID biology, which hosted many experts from the field and led to interesting discussions on the outbreak and evolution of coronaviruses. He taught me that to be a good scientist, it is crucial to be a good mentor. The last meeting that I hosted with his guidance was the National Association of Biology Teachers (NABT) meeting in 2021, titled

"Tips for Teaching Science in a Pandemic: Ways to Lessen the Stress on both Students and Educators". This series was presented under the banner of NABT and supported by the Indian Institute of Science Education and Research (IISER Kolkata), the Indian Institute of Science (IISc Bangalore), the University of Colorado-School of Medicine and Northwestern Connecticut Community College, USA. Randy was very enthusiastic about having insights regarding the rich Indian tradition of learning and education that prevails since ancient times, The Gurukul System. The Gurukul System is a residential schooling system whose origin dates back to around 5000 BC in the Indian subcontinent. It was more prevalent during the Vedic age where students were taught various subjects and about how to live a cultured and disciplined life. Gurukul was actually the home of the teacher or Acharya and was the centre of learning where pupils resided until their education was completed. At the Gurukul, all were considered equal, and the guru (teacher) as well as shishya (student) resided in the same house, following a system of stages of life discussed in Hindu texts of the ancient and medieval eras. These four stages are: Brahmacharya (student), Gṛhastha (householder), Vanaprastha (forest walker/forest dweller), and Sannyasa (renunciate). The Asrama system is one important facet of the Dharma concept in Hinduism. Randy was overwhelmed with the practicality of this four stages of life.

In 2020, IUSSTF forum commenced a call-for-proposal for a collaborative effort involving scientists from India and US to explore the therapeutic targets to reduce SARS-CoV-2 infectivity. With the Principal Investigator from the US, Dr. Maria Nagel, I formed a group of scientists including Randy, who worked together to explore the efficacy of *Azadirachta indica* A. Juss (Neem) bark extract in restricting β-Coronaviral infection and replication. Randy played an active role in this project, and his experience in virology helped us throughout the project [3] (Figure 6). It is deeply saddening that we lost Randy. He was an inspiration and kept us all motivated.

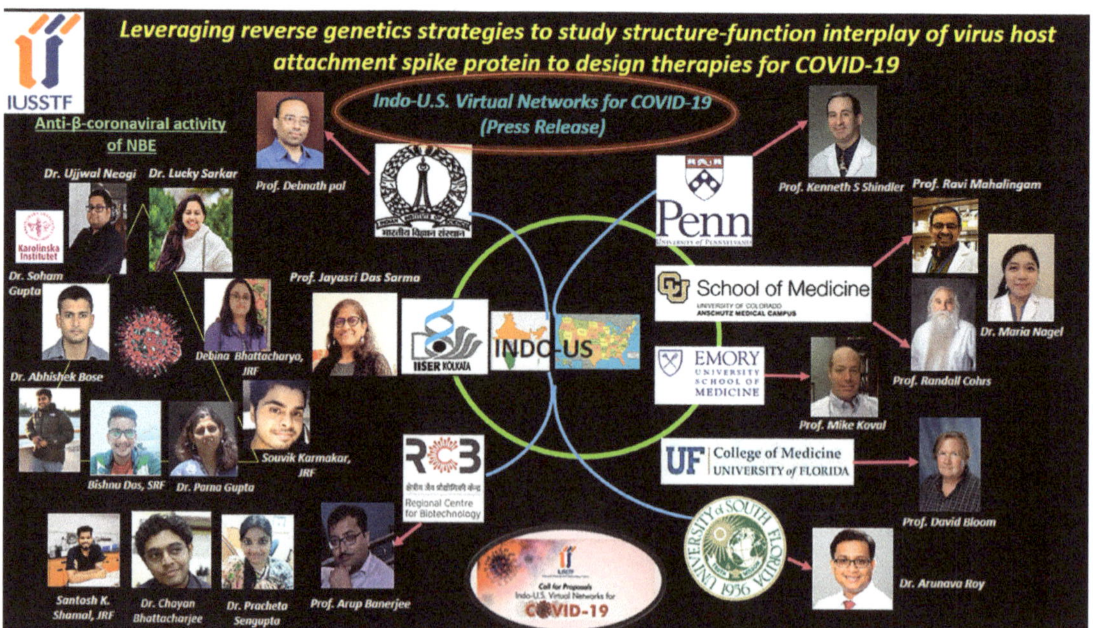

Figure 6. The Indo-US virtual network project. The picture depicts the entire team of scientists from India, US, and other parts of the world working together to understand the anti-β-coronaviral activity of NBE. Randy was a significant part of the group, and his absence will be felt forever.

The testimonies from all the students signify the friendly and warm personality of Dr. Cohrs (Box 3). His sad demise is a great loss for all of us. Randy was a genuine, kind, and compassionate person, and an inspiration to many. In his honour, at the BAW 2022 that I will be hosting, I am organizing a "Professor Randall J. Cohrs Memorial Lecture". I shall miss his guidance and presence, and I hope that he continues to support us in spirit. All my current students and alumni miss you Randy! However, we know that you are always with us.

Box 3. Anecdotes from Students and Alumni of my Lab who met Prof. Randall Cohrs

Dr. Lucky Sarkar, Post-Doctoral Fellow, Florida Research & Innovation Center, Cleveland Clinic, Florida, USA.

During my Ph.D. days, in February 2018, while working under Prof. Jayasri Das Sarma at IISER Kolkata, I had the privilege of interacting with Prof. Randall J. Cohrs in an International INDO-US Biological and Analytical Workshop, Gangtok, India. Before that, I had heard a lot about his work, personality, knowledge, and sense of humour from my mentor Prof. Das Sarma. Undoubtedly, he had a great love not only for Science but also for different cultures and traditions. He used to resemble our Bengali polymath and Nobel Laureate Rabindranath Tagore. He was always enthusiastic about meeting young research minds and take part in cultural programs like dance or music. He also encouraged me to participate in cultural programs like NABT and other national and international platforms. I was fortunate enough to work in an IUSSTF-funded virtual networking grant on COVID-19 since the pandemic struck the world. Since April 2020, we have been working hard on our project "*Azadirachta indica* A. Juss (Neem) bark extract and its Nimbin isomers restrict β-Coronaviral infection and replication", and drafting our manuscript together with other collaborators from India, USA, and also from Sweden. However, we suddenly lost such an eminent scientist and a humble soul one year before our paper was published in *Virology*, Elsevier, 2022. Dr. Cohrs has made one of the most tangible contributions to my research life. We miss you, Sir.

Dr. Abhishek Bose, Post-Doctoral Fellow, Department of Genetic Medicine at Weill Cornell Medicine, New York, USA.

During the PhD period, I was fortunate enough to get several chances to meet Dr. Randall Cohrs and it was my absolute privilege and honour to know him from a close distance. We met in the Indo -US science and technology forum (IUSSTF) conference and the BAW-bioanalytical methods workshop organized by Prof. Das Sarma in 2018. In the meeting, I not only got a chance to discuss Science but also interacted with Prof. Cohrs like a friend. Not only was he a renowned virologist and neurologist, but also one who would make us feel at home with his warm smile and jovial nature. He took active part in the cultural program organized as a part of the conference and also met and dressed like the local people of Sikkim. Later, when I joined a post-doctoral position in IUSSTF networking project on COVID-19 with Prof. Das Sarma, we enjoyed Prof. Cohrs's presence as a collaborator of the team, where he provided thoughtful scientific insights to steer the project at its inception. Also, he made an appearance in our spiritual forum "Kathanubhuti", where he listened to the glories of Indian cultural heritage and slokes from Srimad Bhagavat Gita with deep interest and shared his spiritual knowledge and love for the Hindu religion. His inclusive and kind nature will keep him alive bright as sunshine, among all of us, who remain inspired by him. We miss you Prof. Cohrs (fondly known as Randy), as a teacher, friend and philosopher, as someone I will always look up to. My prayers for you and for your loving family.

Funding: Dr. Cohrs was a co-investigator for the Indo-US Science and Technology Forum Virtual Network project for COVID-19. Our previous collaboration was supported by American society FOR Microbiolgy(ASM)- Indo-US Science and Technology Forum(IUSSTF) Professorship award.

Institutional Review Board Statement: Not applicable.

Acknowledgments: I thank all my current students and alumni for helping me write this obituary, and for collecting the wonderful pictures and memories for over a decade, to represent our long standing and fond memories with Cohrs.

Conflicts of Interest: The author declares no conflict of interest.

References

1. Chatterjee, D.; Addya, S.; Khan, R.S.; Kenyon, L.C.; Choe, A.; Cohrs, R.J.; Shindler, K.S.; Das Sarma, J. Mouse hepatitis virus infection upregulates genes involved in innate immune responses. *PLoS ONE* **2014**, *9*, e111351. [CrossRef] [PubMed]
2. Biswas, K.; Chatterjee, D.; Addya, S.; Khan, R.S.; Kenyon, L.C.; Choe, A.; Cohrs, R.J.; Shindler, K.S.; Das Sarma, J. Demyelinating strain of mouse hepatitis virus infection bridging innate and adaptive immune response in the induction of demyelination. *Clin. Immunol.* **2016**, *170*, 9–19. [CrossRef] [PubMed]
3. Sarkar, L.; Oko, L.; Gupta, S.; Bubak, A.N.; Das, B.; Gupta, P.; Safiriyu, A.A.; Singhal, C.; Neogi, U.; Bloom, D.; et al. Azadirachta indica A. Juss bark extract and its Nimbin isomers restrict β-coronaviral infection and replication. *Virology* **2022**, *569*, 13–28. [CrossRef] [PubMed]

viruses

MDPI

Article

Genetic Adaptation by Dengue Virus Serotype 2 to Enhance Infection of *Aedes aegypti* Mosquito Midguts

Steven M. Erb [1], Siritorn Butrapet [2], John T. Roehrig [2], Claire Y.-H. Huang [2] and Carol D. Blair [1,*]

[1] Center for Vector-Borne Infectious Diseases, Department of Microbiology, Immunology, and Pathology, Colorado State University, Fort Collins, CO 80523, USA; serb@rams.colostate.edu

[2] Division of Vector-Borne Diseases, Centers for Disease Control and Prevention, Fort Collins, CO 80521, USA; sjb7@cdc.gov (S.B.); jtr1@cdc.gov (J.T.R.); yxh0@cdc.gov (C.Y.-H.H.)

* Correspondence: carol.blair@colostate.edu

Abstract: Dengue viruses (DENVs), serotypes 1–4, are arthropod-borne viruses transmitted to humans by mosquitoes, primarily Aedes aegypti. The transmission cycle begins when *Ae. aegypti* ingest blood from a viremic human and the virus infects midgut epithelial cells. In studying viruses derived from the DENV2 infectious clone 30P-NBX, we found that when the virus was delivered to female *Ae. aegypti* in an infectious blood meal, the midgut infection rate (MIR) was very low. To determine if adaptive mutations in the DENV2 envelope (E) glycoprotein could be induced to increase the MIR, we serially passed 30P-NBX in *Ae. aegypti* midguts. After four passages, a single, non-conservative mutation in E protein domain II (DII) nucleotide position 1300 became dominant, resulting in replacement of positively-charged amino acid lysine (K) at position 122 with negatively-charged glutamic acid (E; K122E) and a significantly-enhanced MIR. Site directed mutagenesis experiments showed that reducing the positive charge of this surface-exposed region of the E protein DII correlated with improved *Ae. aegypti* midgut infection.

Keywords: dengue viruses; *Aedes aegypti*; adaptive mutation

Citation: Erb, S.M.; Butrapet, S.; Roehrig, J.T.; Huang, C.Y.-H.; Blair, C.D. Genetic Adaptation by Dengue Virus Serotype 2 to Enhance Infection of *Aedes aegypti* Mosquito Midguts. *Viruses* **2022**, *14*, 1569. https://doi.org/10.3390/v14071569

Academic Editors: Charles Grose and Ravi Mahalingam

Received: 14 June 2022
Accepted: 15 July 2022
Published: 19 July 2022

1. Introduction

Dengue viruses (DENVs) (*Flaviviridae, Flavivirus*) are the most medically important arboviruses infecting humans today. It has been estimated that 390 million DENV infections occur annually, of which 96 million cause frank dengue disease [1] ranging from self-resolving dengue fever to life-threatening dengue hemorrhagic fever and dengue shock syndrome (DHF/DSS).

DENVs are maintained in nature in a mosquito-to-human transmission cycle. The primary mosquito vector, *Aedes aegypti,* is ubiquitously distributed throughout the tropics and is increasingly expanding beyond the tropics [2,3]. Vaccine development has been complex and problematic [4] and no specific anti-viral treatment is available. Thus, control of transmission by mosquito vectors is an integral facet in dengue disease reduction. Multiple steps are required for the transmission of mosquito-borne viruses to vertebrate hosts. Mosquitoes are exposed to the virus when ingesting an infectious blood-meal from a viremic host. The crucial initial step in the transmission cycle is infection of and replication within midgut epithelial cells. Although putative midgut cell receptor proteins have been shown to bind DENVs, they have not been characterized [5,6], and receptor-binding domain(s) on DENV structural proteins are not well-defined. After amplification and spread in infected midgut cells, virus must disseminate into the hemocoel to infect and further amplify in secondary target tissues including the salivary glands. The transmission cycle is completed when a salivary gland-infected mosquito inoculates a new host with virus during probing to seek her next blood-meal. The 10-to 14-day interval between initial infection of the midgut epithelium and ability of the mosquito to transmit the virus in the saliva is the extrinsic incubation period [7,8]. Four genetically and antigenically distinct

DENV serotypes (DENV1-4) comprise the DENV serocomplex. Mature virus particles have a single-stranded positive-sense RNA genome (~11 kb) encapsidated in a protein core surrounded by a host-derived lipid bilayer anchoring 180 copies each of the virus-encoded membrane (M) protein and envelope (E) glycoprotein [9]. The E protein is arranged in an icosahedral scaffold of 90 homodimers that lie parallel to the virion surface. The flavivirus E protein, a Class II fusion protein, is responsible for host cell attachment and entry and virus envelope fusion with cellular endosomal membranes. The DENV2 E protein 2 Å crystal structure has been solved [10], revealing three distinct structural domains (DI, DII and DIII) that correlate with three previously described antigenic domains (C, B and A) defined by monoclonal antibody (MAb) mapping [11–14]. Domains I and II consist of linearly discontinuous amino acid sequences connected by four peptide strands that form a molecular hinge, which permits the translocation and insertion of the flavivirus-conserved DII fusion peptide into the host cell endosomal membrane to bring about virus-mediated cell membrane fusion [10]. DII also is involved in E protein homodimerization in the mature virion. DIII is an immunoglobulin-like structure connected to DI and has host-cell receptor binding properties [15].

Although it has been shown that different strains or genotypes of a single DENV serotype can exhibit differential specificity for infection of the same *Ae. aegypti* strain [16–18], to date there has been limited research focusing on DENV genetic determinants that influence infection of mosquitoes.

We have conducted studies to define the genetic determinants of the E gene/protein involved in infection of mammalian cells, using DENV2 strain 16681 derived from the infectious cDNA clone pD2/IC-30P-NBX (30P-NBX) [15,19,20]. We have also shown that DENV2 30P-NBX infects and replicates efficiently in various tissues outside the midgut of *Ae. aegypti* RexD strain mosquitoes after intrathoracic (IT) inoculation. However, infection of the midgut epithelium after introduction of virus by infectious blood-meal is inefficient [15]. The average midgut infection rate (MIR) for 30P-NBX in *Ae. aegypti* RexD in a large number of trials was 33.79%, compared to an average MIR of 78.74% for our standard laboratory strain DENV2 Jamaica 1409 (J1409) [21]. Given the importance of the E protein in determining DENV cell tropism/initiating infection, and that 30P-NBX has a relatively low infection rate after oral infectious challenge but not after IT inoculation, we hypothesized that mosquito midgut infection determinants are located in the E protein and that 16,681 had undergone mammalian cell culture-adaptive changes during multiple cell culture passages before preparation of the infectious cDNA clone that reduced its ability to infect midgut epithelial cells. Comparison of E protein AA sequences shows that there are nine differences between strains 16681 and J1409 and all except three of these, R120T, Q131L and T170I, are conservative. We noted that each time 30P-NBX was presented in a blood meal a small proportion of infectious clone-derived viruses efficiently initiated midgut infection, so in this study, we serially passaged 30P-NBX in *Ae. aegypti* RexD strain mosquitoes by oral infection to select for and identify E protein genetic adaptations conferring efficient midgut epithelial cell growth. We confirmed the importance of selected E protein domain II amino acid (AA) sequences in midgut infection by constructing and phenotyping 30P-NBX mutants.

2. Methods and Materials

2.1. Cell Culture and Viruses

Vero (African green monkey kidney epithelial) cells were grown at 37 °C in Dulbecco's modified Eagle's medium (DMEM) with 10% fetal bovine serum (FBS). C6/36 (*Aedes albopictus* larval) cells used for virus phenotyping experiments were grown in Ye-Lah medium [22] and C6/36 cells used for propagation of virus to infect mosquitoes were grown in Leibovitz L-15 infection medium containing 2% FBS, non-essential amino acids, penicillin-streptomycin, and L-glutamine, both at 28 °C. Both Vero cells and C6/36 cells are standard cell lines that have been used in our laboratories for a number of years. They were purchased from American Type Culture Collection (ATCC) (Vero CCL-81;

C6/36 CRL-1660), and expanded by culturing in the media indicated. Aliquots of the expanded cultures were preserved in liquid nitrogen. After a given cell line had been recovered from liquid N_2 and passed for experimental use no more than 20 times, it was discarded and replaced by a newly-recovered aliquot.

DENV2 strain 16681 was first isolated from the serum of a dengue hemorrhagic fever/shock syndrome (DHF/DSS) patient in Bangkok, Thailand, in 1964 [23]. The strain is a member of the DENV2 Asian 1 genotype and since isolation has been passaged multiple times in BS-C-1 cells, six times in LLC-MK2 cells, once in a rhesus macaque monkey, and twice in *Toxorhynchites amboinensis* mosquitoes [23,24]. The virus was additionally passaged once in Vero cells, twice in LLC-MK2 cells, and four times in C6/36 cells before construction of the infectious cDNA clone pD2/IC-30P-A [24]. The infectious cDNA clone was modified to facilitate introduction of site-specific mutations into the E gene and DENV2 derived from this infectious clone is termed 30P-NBX and exhibits the same phenotypes in mammalian and invertebrate cell cultures as parental strain 16681.

DENV2 strain Jamaica 1409 (J1409) was isolated from a human with dengue fever (DF) in 1983 and is a member of the American/Asian genotype. Strain J1409 was plaque-purified in LLC MK2 cells and passaged extensively (>25 times) in C6/36 cells [25–28] before construction of the infectious cDNA clone (J1409-ic) by Pierro and colleagues [27]. After transfection and virus recovery in C6/36 cells, infectious clone-derived virus J1409 was passaged once in C6/36 cells for use in this study.

To introduce selected mutations into the E gene of infectious cDNA clone 30P-NBX, the QuikChange® Lightning Site-directed Mutagenesis kit (Stratagene, Santa Clara, CA, USA) was used per the manufacturer's instructions. Engineered mutations for this study were targeted to DII of the E protein to produce mutants KK122/123EE and R120T. Resulting recombinant cDNA was amplified and used as a template for transcription of infectious DENV2 genomic RNA. In vitro transcription for transfection of Vero and C6/36 cells was performed as described previously [19,22,24] using the AmpliScribe™ T7 kit (Epicentre Technologies, Madison, WI). Positive-sense vRNA was transfected into Vero and C6/36 cells using a Bio-Rad Gene Pulser Xcell system (Bio-Rad, Hercules, CA, USA) as described previously [19]. Medium harvested from transfected Vero cells (day 12 post infection (PI), designated V-0) or C6/36 cells (day 14 PI, designated C-0) was centrifuged to remove cell debris, supplemented with 20% FBS, and stored at −80 °C. Aliquots of V-0 and C-0 were used to infect naïve Vero and C6/36 cells to produce V-1 and C-1 seeds, respectively. Genome cDNA of V-1 and C-1 seeds were fully sequenced to evaluate their genomic stability as described previously [19].

2.2. Assay of Virus Growth Kinetics in Cell Culture

Twelve- to fourteen-day growth curves were performed to investigate mutant virus growth kinetics in various cell types. Duplicate cell cultures were infected with each virus at a multiplicity of infection (MOI) of 0.001. Virus genomic equivalents were measured by RT- quantitative (q)PCR using 3′-NCR primers and probes and the iScript™ One-Step RT-PCR Kit (Bio-Rad, Hercules, CA, USA) as previously described [29], and samples were assayed for infectious virus by $TCID_{50}$ in C6/36 cells.

2.3. Infectious Virus Titration by 50% Tissue Culture Infectious Dose

Infectious virus end-point assays were performed in C6/36 cells with detection of virus antigen by ELISA as described previously [19]. C6/36 cells in 96-well plates were infected with 10-fold serial dilutions of virus and maintained for seven days at 28 °C with 5% CO_2. Cells were fixed with cold acetone and virus antigen was detected by ELISA using DENV2 polyclonal antiserum. Absorbance was read at 405 nm and 630 nm and the difference was expressed as the delta optical density (ΔOD). ΔOD values 2-fold or greater than the negative control were considered positive. Virus titers were calculated by the method of Reed and Muench [30].

2.4. *Aedes aegypti Mosquito Oral Infection by Artificial Blood-Meal*

Ae. aegypti RexD strain mosquitoes (RexD) are a laboratory colony originating from Rexville, Puerto Rico. *Ae. aegypti* Chetumal strain mosquitoes are a more recently-established laboratory colony originating from Chetumal, Mexico [17]. They exhibit significantly higher midgut infection and dissemination rates for DENV2 J1409 than RexD mosquitoes [8]. All mosquitoes were reared from eggs and maintained as adults at 28 °C, 80% relative humidity with a photocycle of 12 h light: 12 h dark. Adult female mosquitoes were maintained in one-pint cartons with organdy covering and given water and sugar until infection. To prepare DENV2 for infectious blood-meals, C6/36 cells were infected at a multiplicity of infection (MOI) of 0.001 and maintained for 12–14 days with a medium change at 7 days. Infected cells were scraped into the medium, which was mixed with an equal volume of defibrinated sheep blood and supplemented with ATP to a final concentration of 1 mM. Adult female mosquitoes 4–6 days post-emergence were starved for 24 h, deprived of water for 4 h, and exposed to the infectious blood-meal for 45 min using a 37 °C water-jacketed glass feeding device with a hog-gut membrane. Virus titers in the blood-meals for all experiments ranged from 6.2 to 9.2 $\log_{10}$ $TCID_{50}/_{mL}$. Previous studies with 30P-NBX and E protein mutant viruses showed no correlation between virus titers in this range and midgut infectivity (data not shown). Fully engorged mosquitoes were selected and maintained for 7 days, when midguts were dissected in PBS, fixed in 4% paraformaldehyde in PBS overnight, and analyzed for virus antigen via immunofluorescence assay (IFA) to determine midgut infection rates. Each oral infection experiment was repeated at least three times with at least 19 mosquito midguts analyzed per experiment. To investigate mosquito infection rate kinetics, assays of dissected mosquito midguts and head tissues were performed every two days for 14 days post blood-meal (PBM). Virus antigen was detected in midguts and head squashes to determine infection and dissemination rates by IFA. Experiments were repeated three times and 17–30 mosquitoes were dissected at each time-point PBM.

2.5. *DENV2 Passage in Aedes aegypti Mosquito Midguts*

Infectious virus derived from DENV2 strain 16,681 infectious cDNA clone 30P-NBX, 30P-NBX-derived DIII FG loop mutants 382VEPGΔ and VEP382RGD (382RGD) [15], and DENV2 strain J1409 infectious cDNA clone (J1409) [27] were serially passaged in *Ae. aegypti* RexD midguts and amplified in C6/36 cells. To start the passage experiment, virus was amplified in C6/36 cells and an aliquot of the cell-virus suspension was incorporated into an infectious blood-meal as described above. The remaining cell-virus suspension was stored at −80 °C for titration and sequencing of the DENV E gene. RexD mosquitoes were challenged with the infectious blood-meal and fully engorged mosquitoes were maintained for 10 days. Mosquito midguts were dissected and placed into 4% paraformaldehyde for IFA analysis to determine MIRs (19–36 mosquitoes) or pooled on dry ice for trituration (at least 20 mosquitoes). Midguts were triturated in L15 infection medium and filtered through a 0.2 μm membrane syringe filter. The filtrate was placed directly onto naïve C6/36 cells to start the next passage. Four passages were completed for each virus and the complete passage series was repeated once for 30P-NBX.

2.6. *Indirect Immunofluorescence Assay (IFA) of Mosquito Tissues*

Midgut and head squash IFAs were performed as described previously [31]. Virus antigen in midguts and head tissues was detected using flavivirus E protein DII group-reactive mouse MAb 4G2 (HB-112, ATCC, Manassas, VA, USA) in wash buffer (PBS, 0.05% Triton X-100) or PBS, respectively. Secondary antibody was ImmunoPure biotin-labeled goat anti-mouse IgG (Thermo Scientific, Waltham, MA, USA) with 0.005% Evan's Blue counter-stain, followed by streptavidin-fluorescein (GE Healthcare, United Kingdom). MIRs and head tissue infectivity (dissemination) rates were determined by dividing the number of virus antigen-positive midguts or head squashes by the total number analyzed. The relative infection intensity (RII) ratio is a quantitative measure of infection intensity in the midgut [31] in which positive midguts were scored on a scale of 0.5 to 4, where 0.5 denotes less than 25%, 1 denotes 25%, 2 denotes 50%, 3 denotes 75%, and 4 denotes 100% of the midgut surface area is positive for viral antigen. The RII ratio was determined by adding the infection intensity scores of all positive midguts in a treatment group and dividing by the total number of positive midguts. Student's t tests (p-value ≤ 0.05) were performed using Excel 2007 and chi-square analysis (p-value ≤ 0.05) was carried out using SAS 9.1.

2.7. Envelope Glycoprotein Gene Sequencing

DENV2 RNA was isolated from infected C6/36 cell cultures with the QIAamp Viral RNA Isolation Kit (Qiagen, Germantown, MD, USA), and the E gene was amplified using the Titan One-Step RT-PCR system (Roche, Indianapolis, IN, USA) per the manufacturer's instructions. PCR products were gel extracted using QIAquick Gel Extraction kit (Qiagen) and Sanger sequencing reactions were performed using ABI Prism BigDye Terminator v3.1 Cycle Sequencing Kit (Applied Biosystems, Carlsbad, CA, USA) at the Centers for Disease Control and Prevention, Fort Collins, CO. Sequences were analyzed using Lasergene Seqman (DNASTAR, Madison, WI, USA).

3. Results

3.1. Serial Passage of DENV2 in Aedes aegypti Mosquito Midguts

Infectious clone-derived DENV2 with four different E gene sequences were serially passed in *Ae. aegypti* mosquitoes to determine if adaptation to allow efficient midgut infection would result from the selection of virus genomes with enhancing mutations; the focus of our study was on the E gene due to its importance in initiating infection. DENV2 strain 16681 infectious cDNA clone-derived virus (30P-NBX); 30P-NBX-derived viruses with FG loop (DIII) deletion or mutation, 382VEPGΔ and VEP382RGD, which we had previously shown to decrease or not affect MIRs, respectively [15]; and DENV2 infectious clone J1409-ic-derived virus (J1409) were serially passaged by oral infection of *Ae. aegypti* RexD strain mosquitoes. Pre-passage virus MIRs of 30P-NBX and mutant virus VEP382RGD were similar, while mutant virus 382VEPGΔ had a significantly lower MIR and J1409 had a significantly higher MIR compared to 30P-NBX (Figure 1).

Each clone-derived virus was amplified in C6/36 cells, provided to RexD mosquitoes in an infectious blood-meal (SP0), and serially passaged four times (SP1-4) in mosquito midguts with amplification in C6/36 cells after each passage as described in Materials and Methods. Any increases in the rate of infection of midgut epithelial cells were assumed to result from mutations that occurred in the E gene during midgut infection since the E gene of 30P-NBX has been shown to be genetically stable during successive passages in C6/36 cells [29]. SP0 MIRs were similar to average MIRs previously determined for each virus (Table 1); 30P-NBX SP0 MIR fell within the range found in many previous determinations (data not shown). The MIRs of 30P-NBX and mutant VEP382RGD significantly increased after one passage in RexD midguts and in both cases, this increase correlated with a mixed nucleotide (nt) population of adenylic acid (A, parental) and guanylic acid (G, mutation) in consensus sequences at E gene position 1300. A transition from A to G causes a non-conservative AA change from lysine (K) to glutamic acid (E) in E protein DII at position 122 (K122E) (Table 1 and Figure 2). No other consistent nucleotide changes were seen in the remainder of E gene coding regions of genomes recovered from passaged virus. As revealed by consensus sequencing, four passages of 30P-NBX in mosquito midguts were required before E completely substituted K at position 122, while for mutant VEP382RGD, only two passages were required before this complete AA change occurred (Table 1). Two independent serial passage experiments were completed for 30P-NBX in RexD midguts and the second experiment yielded the same results (data not shown). In contrast, 382VEPGΔ and J1409 MIRs did not demonstrate any significant changes after four serial passages in mosquito midguts and no nt sequence changes were found in their E genes. Mutant 382VEPGΔ was lost after four passages and there was insufficient vRNA present in SP4 samples for sequencing. Our finding that a single AA mutation located in DII of the E protein appeared to significantly enhance mosquito midgut infection was unexpected.

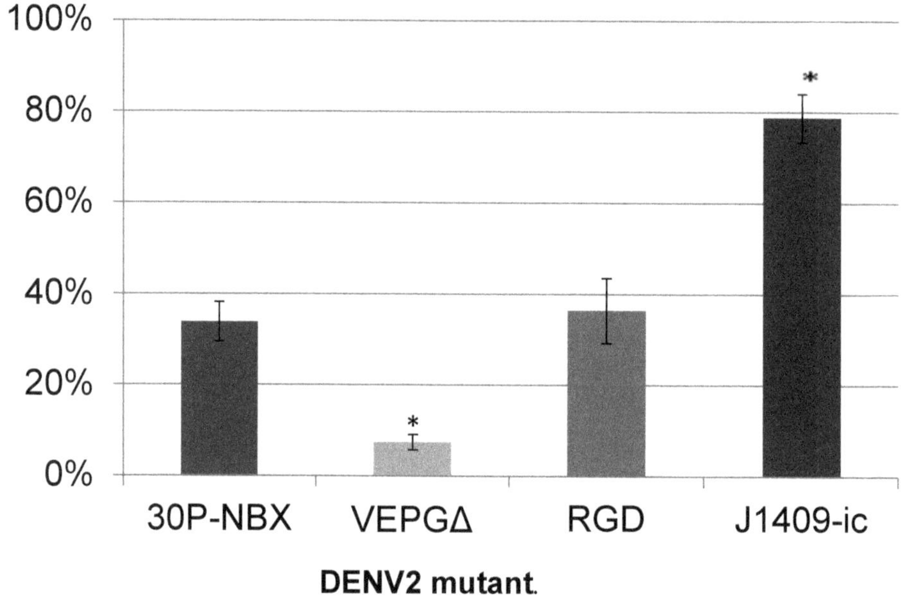

Figure 1. DENV2 MIRs in *Ae. aegypti* RexD mosquito midguts. RexD mosquitoes were orally challenged with each virus, maintained for seven days until midguts were dissected, and MIRs were determined by IFA. Data are the average of at least three experiments and significance was determined by comparison with 30P-NBX via student's *t* test (* p-value < 0.05).

Table 1. Midgut infection rates and genetic changes after serial passage of 30P-NBX, 382VEPGΔ, VEP382RGD, and J1409 in *Ae. aegypti* RexD mosquitoes.

	30P-NBX			382VEPGΔ			VEP382RGD			J1409		
Passage	Midgut Infection Rate (%) [a]	Confidence Interval (%)	Position 122 AA [b]	Midgut Infection Rate (%) [a]	Confidence Interval (%)	Position 122 AA [b]	Midgut Infection Rate (%) [a]	Confidence Interval (%)	Position 122 AA [b]	Midgut Infection Rate (%) [a]	Confidence Interval (%)	Position 122 AA [a]
0	6/34 (18)	7–34	K	1/24 (4)	1–20	K	11/25 (33)	20–51	K	17/19 (89)	68–97	K
1	13/19 (68)	46–86	K/E	5/31 (16)	7–33	K	25/30 (83)	66–93	K/E	15/28 (54)	36–71	K
2	22/36 (61)	45–75	E/K	4/34 (12)	5–27	K	27/29 (93)	78–98	E	22/35 (63)	46–77	K
3	15/34 (44)	29–61	E/K	1/27 (4)	0.9–18	K	12/21 (57)	36–76	E	24/31 (77)	60–89	K
4	18/19 (95)	75–99	E	0/35 (0)	0.07–10	NA [c]	26/30 (87)	70–95	E	25/29 (86)	69–94	K

[a] MIR determined by dissecting and staining midguts with E protein-specific fluorescent antibody at 7 days PBM. [b] AA present at DENV2 E protein position 122. Mixed consensus cDNA sequence determined for both strands. Codon for the first AA is present in greater amounts than the second AA, as determined by examination of the sequence chromatogram. [c] Not available. There was insufficient vRNA present for sequencing.

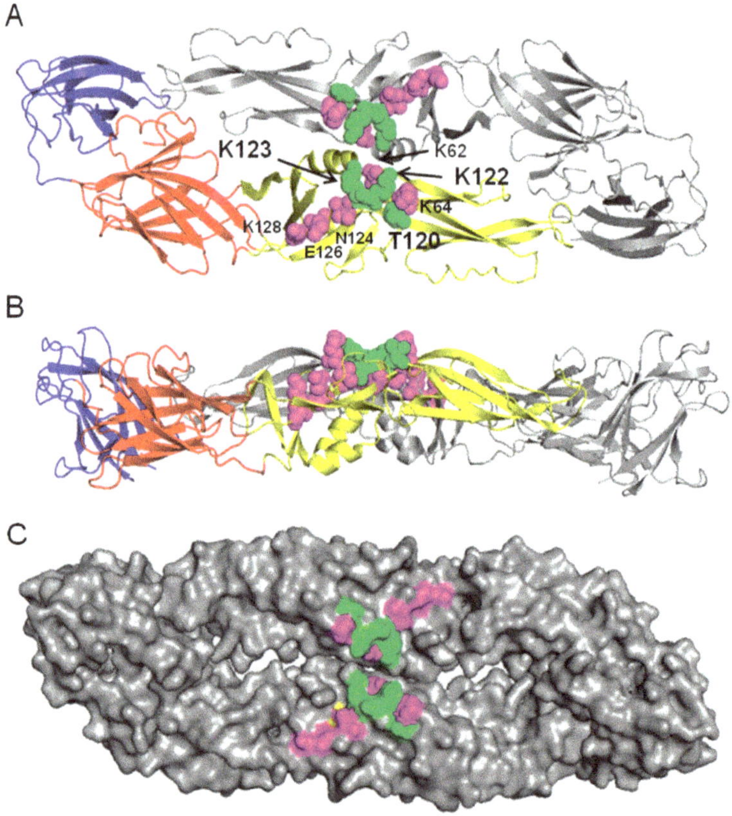

Figure 2. DENV2 E protein structure and location of relevant AAs in DII. (**A**) Top-down view of the DENV2 E protein homodimer with DI in red, DII in yellow, and DIII in blue in lower left monomer. AAs examined (T120, K122, and K123 in green) and discussed (K63, K64, N124, E126, and K128 in magenta) in this study are specified. Note, AA 120 is threonine in the published DENV2 E protein structure [10]. (**B**) Side-view of the DENV2 E protein homodimer. (**C**) Top-down view of the space-filling model of the DENV2 E protein homodimer to show surface-exposed AAs, which are colored the same as in A. Protein structures were obtained from the protein database bank (DENV2 E protein homodimer ID: 1oan) and were rendered in Polyview-3D.

3.2. *Verification of Adaptation-Mutant Phenotype by Site-Directed Mutagenesis to Construct E Protein DII Mutant Viruses*

The AA sequence of DENV2 16,681 (30P-NBX) E protein DII was aligned with those of other DENV2 genotypes, other DENV serotypes and other arthropod-borne flaviviruses. AA sequence alignments showed that there is considerable sequence conservation in DENV2, but sequence variability in other flavivirus DII AA 120 to 130 (Table 2). Comparison of E protein AA sequences revealed nine differences between strains 16,681 and J1409, only one of which occurs in this region. DENV2 strain 16,681 is the only flavivirus shown with positively-charged arginine (R) at position 120 All of the other DENV2 strains, including strain J1409, have uncharged threonine (T) at this position (Table 2). This difference is noteworthy due to the surface exposed location of T120 on the E protein and its close proximity to positively-charged KK122/123 (Figure 2). To determine if AA mutation K122E in DII of the DENV E protein was responsible for enhanced midgut infection, we introduced this and other potentially relevant mutations into the parental 30P-NBX genome by site-directed mutagenesis. After mutagenesis, RNA was transcribed in vitro and transfected into C6/36 cells (C-0) and virus produced was used to infect naïve C6/36 cells (C-1). Mutant KK122/123EE, which had previously been developed, was able to replicate after both transfection and one passage in C6/36 cells as evidenced by detectable viral antigen in cells. Consensus sequencing of the C-1 virus genome showed the expected full length genome sequence (Table 3), demonstrating that these mutations had no effect on infectivity of or replication in C6/36 cells. Mutant virus RNA also was transfected into Vero cells (V-0) and resulting virus was used to infect Vero cells (V-1). Full-length genome sequencing of KK122/123EE V-0 virus showed a partial reversion from E at position 123 to K, while E at position 122 remained unchanged. V-1 genome consensus sequencing revealed a complete reversion to K at position 123 while E at position 122 remained unchanged (Table 3). Recovered V-1 virus contained the mutation selected in *Ae. aegypti* midgut serial passage experiments with no other differences in the virus genome sequence compared to 30P-NBX (K122E), and was used as virus seed in all subsequent phenotypic studies. Virus recovery and genome sequencing showed that K123 was essential for replication in Vero cells, although either K123 or E123 was tolerated in C6/36 cells.

Interestingly, J1409 (T120, KK122/123) did not accrue any adaptive mutations during passage in mosquito midguts. Site-directed mutagenesis was used to introduce mutation R120T into the E protein of 30P-NBX (mutant designated R120T). Mutant R120T replicated in both C6/36 cells and Vero cells without alterations in genome sequence after both transfection and one virus passage (Table 3).

Growth kinetics of parental and mutant viruses K122E and KK122/123EE were analyzed by infecting duplicate cell cultures with each virus at a MOI of 0.001 and measuring virus genomic equivalents in medium by RT-qPCR or infectious virus by $TCID_{50}$, as described in Methods, every two days. Growth kinetics for both mutants were similar to 30P-NBX in C6/36 cells (Figure 3A), corroborating the transfection data and the ability of these viruses to replicate efficiently in this cell type. Interestingly, 30P-NBX consistently caused more cytopathic effects (CPE) in the form of syncytium formation, cell rounding, and cell detachment than K122E in the C6/36 cells routinely used to amplify virus for oral infection experiments (data not shown). In Vero cells grown at 37 °C, mutant K122E replicated similarly to 30P-NBX and had equivalent virus genome titers, in contrast to mutant KK122/123EE, which showed no virus RNA replication (Figure 3B). Sequencing virus RNA at the conclusion of the growth curves showed K122E was genetically stable, while there was not enough KK122/123EE viral RNA recovered from Vero cell medium for sequencing at the conclusion of the experiment. Growth kinetics of mutant R120T in Vero and C6/36 cells were determined by titrating infectious virus released into medium by $TCID_{50}$ (Figure 3C,D). R120T reached similar peak titers as 30P-NBX in both cell types, but peak titer was reached more rapidly than 30P-NBX in Vero cells (Figure 3D). Due to the apparent inability of mutant KK122/123EE to replicate at 37 °C, its temperature sensitivity was investigated by comparison of growth in Vero cells at 28 °C and 37 °C (Figure 3E). No apparent virus replication occurred at 37 °C up to day 2 PI in this experiment, as seen previously (Figure 3B). However, at 2 days PI, KK122/123EE began to replicate at a more rapid rate at 37 than at 28 °C, and this continued from days 4–8 PI. After day 8, the rate of virus replication became similar at the two temperatures and virus genome titers were similar by day 12 PI at both temperatures. Virus genome sequencing revealed no additional nucleotide changes in mutant KK122/123EE after replication at 28 °C, while mutant KK122/123EE grown at 37 °C partially or fully (results from duplicate cultures) reverted from E123 to K123. We attribute the difference from results shown in Figure 3B to a random mutation that changed E123 to K123 and was thereafter selected for more rapid growth. These results suggest that initial mutation is a stochastic event, followed by selection of mutants with a growth advantage. The results shown in Figure 3B,E suggested that lysines at both positions 122 and 123 of the E protein are not tolerated for replication at 37 °C. Interestingly, reversion of only AA 123 and not AA 122 occurred in the transfection, passage, and growth curve assays.

Table 2. E protein DII AA sequence alignment of DENV2 and other flaviviruses.

Virus	Strain	DENV2 E Glycoprotein AA Position																
		116	117	118	119	120	121	122	123			124	125	126	127	128	129	130
DENV2 Genotypes																		
Asian 1	**16681**	C	A	M	F	**R**	C	**K**	**K**	- [a]	- [a]	N	M	E	G	**K**	V	V
	PUO-218					T												
	M1					T											I	
Asian 2	New Guin C					T								K				
	PL046					T											I	
	CTD113					T								K				
Asian/Amer	Jamaica 1409					T												
	13382-Tizimin					T												
American	PR159					T											I	
	Ven2					T											I	
Cosmo	SL714					T											I	
	CAMR5					T											I	
Sylvatic	IC80-DAKAr578					T		L				K						
	P8-1407					T		L										
Other Mosquito-Borne Flaviviruses																		
DENV1	16007			K		K		V	T			K	L				I	
DENV3	PhMH-J1-97			K		Q		L	E			S	I					
DENV4	Thailand/1985			K		S		S	G			K	I	T		N	L	
YFV	Asibi			K				A				S		S	L	F	E	
JEV	Nakayama			K		S		T	S			K	A	I		R	T	I
MVEV	NG156			K		T		S	S			S	A	A		R	L	I
WNV	NY99			K		A		S	T			K	A	I		R	T	I
SLEV	Laderle			K					N			K	A	T			T	!
Tick-Borne Flaviviruses																		
TBEV	Neudoerfl		V	K	A	A		E	A	K	K	K	A	T		H		Y
POWV	LB			K		E		E	E	A	K	K	A	V		H		Y

Positively-charged AAs in DENV2 strain 16681 sequence are in bold. Blank spaces indicate AA is identical to that in DENV2 strain 16681. [a] Tick-borne flaviviruses have 2 AA inserted between positions 123–124 of mosquito-borne viruses.

Table 3. Transfection and infection by 30P-NBX E protein DII mutants in C6/36 or Vero cells.

Virus Mutant	C6/36 Cells		Vero Cells		
	C-0/C-1 [a] Virus Recovery	C-0/C-1/E Protein Sequence	V-0/V-1 [a] Virus Recovery	V-0 E Protein Sequence	V-1 E Protein Sequence
KK122/123EE	+/+	Unchanged [b]	+/+	K122 E [c], K123K/E part. rev. [c]	K122 E [c], E123K full rev. [c]
R120T	+/+	Unchanged [b]	+/+	Unchanged [b]	Unchanged [b]

[a] Transfection (C-0 or V-0) and recovered virus passage (C-1 or V-1) was considered positive if virus antigen was detected in cells by IFA. [b] Sequencing verified that recovered virus genomes contained the introduced mutations and had no additional changes in the genome. [c] Mutants with partial reversions (part. rev.) or full reversions (full rev.) in the E protein gene are specified.

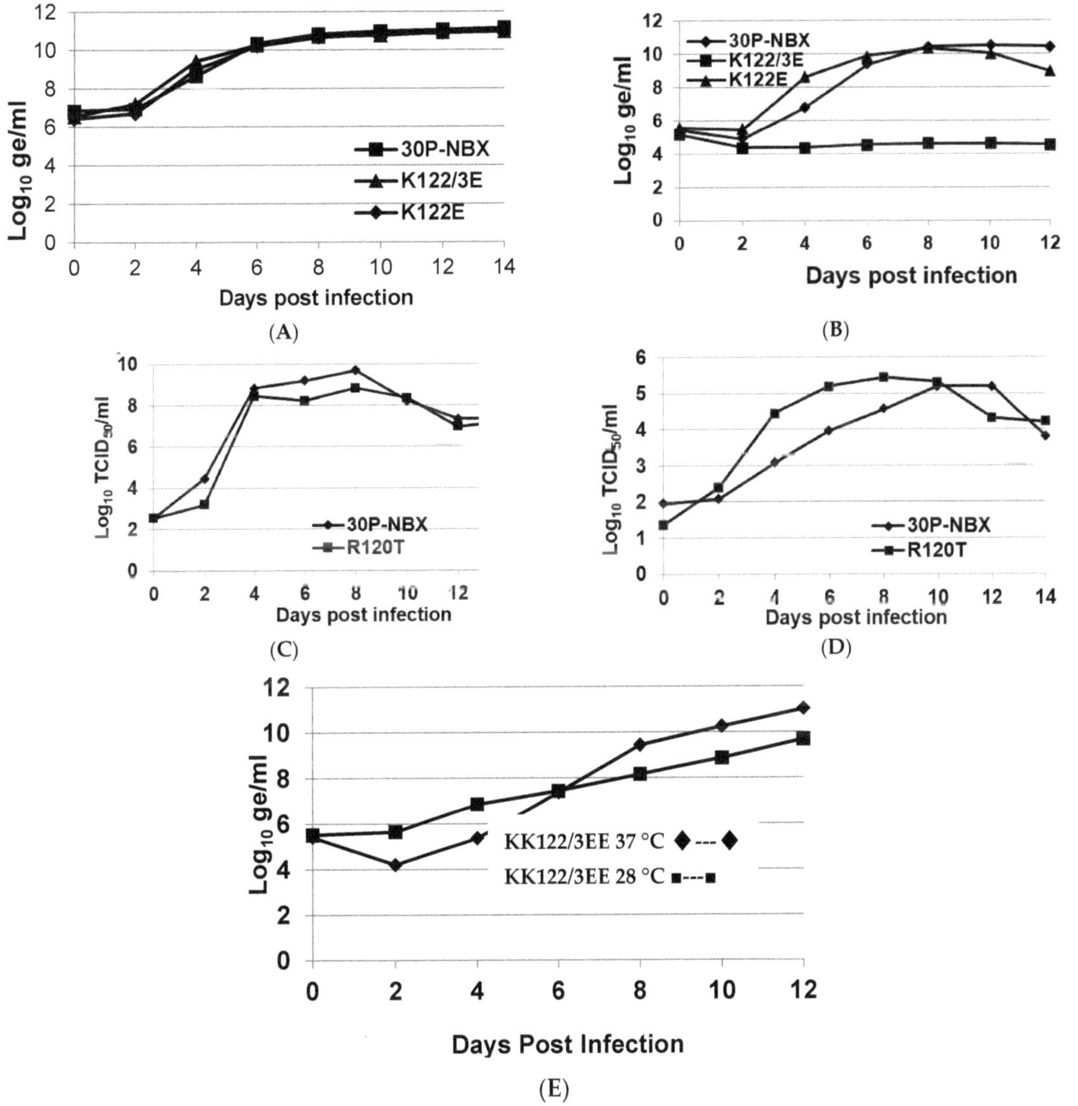

Figure 3. Virus growth kinetics in C6/36 and Vero cells. C6/36 cell cultures at 28 °C (**A**) and Vero cell cultures at 37 °C (**B**) were infected with DENV2s at an MOI of 0.001 and medium was sampled every two days for 14 or 12 days. Data presented are geometric mean titers (in $\log_{10}$ genome equivalents/mL) from duplicate flasks determined by RT-qPCR. C6/36 cell cultures at 28 °C (**C**) and Vero cell cultures at 37 °C (**D**) were infected with 30P-NBX and R120T at a MOI of 0.001 and infectious virus in medium was measured by $TCID_{50}$ every two days for 14 days. The data presented are geometric means (in $\log_{10}$ $TCID_{50}/_{mL}$) from duplicate flasks. (**E**) Duplicate Vero cell flasks were infected with mutant KK122/123EE and maintained at 37 °C or 28 °C. Titers were determined by RT-qPCR. Virus RNA was sequenced at the end of the growth curve to verify the status of engineered mutations. Mutant KK122/123EE retained the mutated sequence at 28 °C but the sequence partially or fully reverted to E123 at 37 °C.

3.3. Phenotypic Properties of E Protein DII Mutants in Ae. aegypti Mosquitoes

To verify that mutation K122E in E protein DII, which was selected during 30P-NBX serial passage in *Ae. aegypti* midguts, was responsible for enhanced midgut infection and replication, constructed mutants K122E as well as KK122/123EE and R120T were presented to adult female mosquitoes in infectious blood-meals and MIRs were determined at 7 DPM. Mutant viruses each had significantly higher MIRs [K122E (81.8%), KK122/123EE (81.9%) and R120T (79.6%)] than 30P-NBX (33.8%) in RexD mosquitoes. (Figure 4A) Complete genome sequences showed that the only difference in deduced AA sequences between K122E and 30P-NBX recovered from mosquitoes was at position 122, confirming that this mutation alone was responsible for the enhanced infection rates that developed during serial passage. Significantly increased MIRs of mutants KK122/123EE and R120T than parental 30P-NBX implicated reduction in surface-exposed, positively charged AAs in E protein DII in increased efficiency of midgut infection (see Figure 2). To determine if 30P-NBX MIRs were higher in a more susceptible *Ae. aegypti* strain, mosquitoes from the Chetumal colony were provided with infectious blood-meals containing 30P-NBX and mutants K122E and R120T. Although 30P-NBX had higher MIRs (38.85% compared to 33.79% in RexD), K122E and R120T also had significantly higher MIRs compared to 30P-NBX in Chetumal mosquitoes (Figure 4B). Four independent challenge experiments were conducted, each completed at least one month apart. Completing all experiments concurrently would have utilized mosquito eggs that were oviposited by the same parents. Also, all of the extrinsic environmental factors would have been similar for each of the repetitions, which might further bias our results. Conducting replicate experiments at different times reproduced our previously observed general experimental variation in MIRs by 30P-NBX and other DENV2 strains, as seen in Table 1.

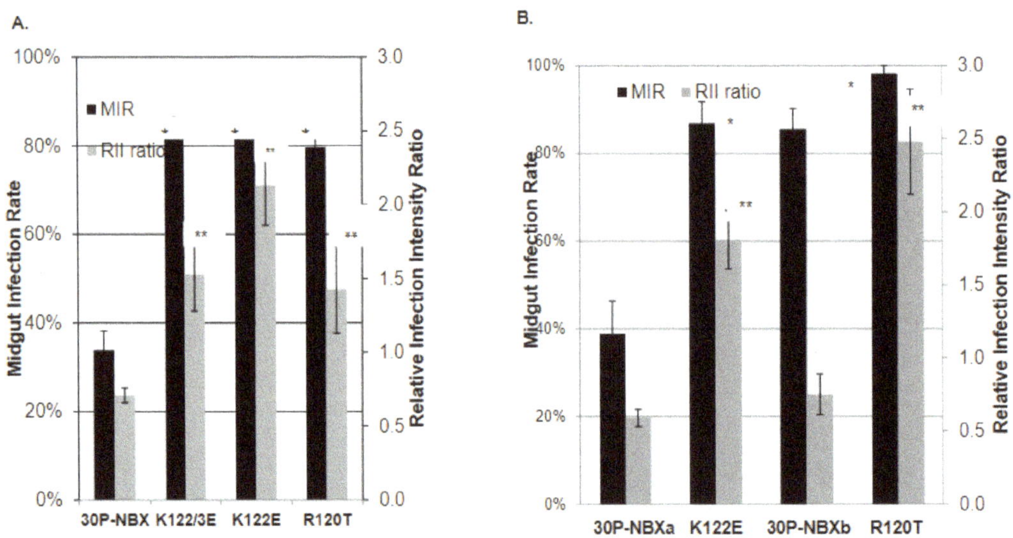

Figure 4. Virus MIRs and relative infection intensity (RII) ratios in Ae. aegypti RexD mosquito midguts (**A**) and Chetumal mosquito midguts. Mosquitoes were orally challenged with each virus, maintained for seven days until midguts were dissected, and MIRs and RII ratios were determined by IFA. In (**B**), 30p-NBXa and 30P-XNXb are data for internal controls for K122E and R120T, respectively. Data are the average of at least three experiments and significance of MIR (*) and RII ratio (**) differences were determined by comparison with 30P-NBX using student's *t* test (*p* value < 0.05).

IFA analysis of infected mosquito midguts to determine MIRs also showed that on average, in RexD infections by K122E, KK122/123EE, and R120T, DENV2 E antigen was detected over a significantly greater area of the entire midgut at all times PBM than in mosquitoes infected with 30P-NBX. To express this difference, a measurement of relative midgut infection intensity, the RII ratio, was developed as described in the Methods section and illustrated in Figure 5. Mutants consistently had higher RIIs than 30P-NBX (Figure 5) Our observations suggested that K122E initiates infection in a higher proportion of cells and spreads more rapidly and efficiently in the mosquito midgut than 30P-NBX.

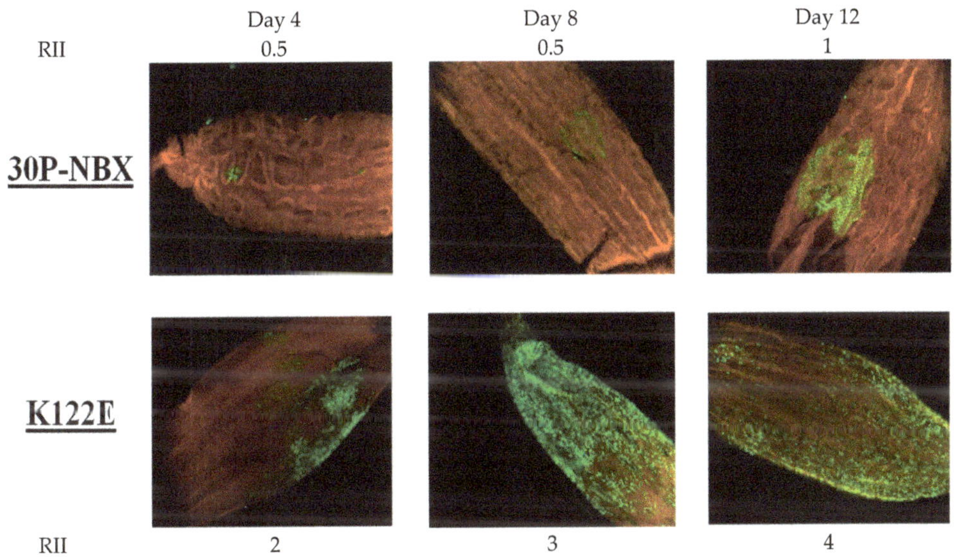

Figure 5. Demonstration of RII ratios in infected midguts from the time course experiment. RexD mosquitoes were orally challenged with 30P-NBX or K122E, midguts were dissected every other day for 14 days, and virus antigen was detected by IFA. Proportions of midguts displaying virus antigen (green) are representative of typical infection intensity seen for day 4 (**left**), day 8 (**middle**), and day 12 (**right**) PBM.

To determine if the more efficient and robust midgut infection by mutant K122E resulted in a higher infection rate in secondary mosquito tissues, implying a higher transmission rate, virus dissemination from the midgut was investigated. IFA analysis for DENV2 E antigen in head tissues from mosquitoes in which MIR and RII had been determined showed that both K122E and 30P-NBX began to escape the midgut at four days PBM. However, K122E had a significantly higher infection rate in head tissues than 30P-NBX by day six PBM, continuing until the end of the time course (observations not shown). Thus, mutant K122E both infects and disseminates from a higher proportion of mosquito midguts than 30P-NBX.

4. Discussion

In this study we demonstrated that single amino acid replacements/changes in DII of the DENV2 E protein of a virus with low MIR can result in significantly enhanced infection of *Ae. aegypti* mosquito midguts. Serial passage of 30P-NBX in RexD mosquitoes identified an adaptive mutation in DII of the E protein at position 122 from positively-charged lysine to negatively-charged glutamic acid (K122E) that correlated with increased infection rates in mosquito midguts. Incorporation of this mutation into the infectious clone recapitulated the results of the serial passage experiment, showing that this single AA mutation was solely responsible for the enhanced infectivity phenotype. We also showed that an alternative single mutation of spatially proximate positively-charged AA R120 to uncharged T significantly enhanced mosquito midgut infection compared to the parent virus. To our knowledge this is the first time mosquito infection determinants have been mapped to DII of the DENV E protein.

Time course experiments in mosquitoes showed that mutant K122E initiated infection in a significantly higher proportion of mosquitoes than 30P-NBX as early as two days post blood-meal (PBM), suggesting that early stage events such as attachment and/or entry were enhanced by the K122E mutation. The specific cellular receptor(s) for DENVs in either human or mosquito target tissues and receptor-binding domains on DENV structural proteins for these receptor(s) is/are unknown. It is possible that mutant K122E and 30P-NBX have different attachment affinities for a specific primary or ancillary receptor, thus contributing to differences in infectivity rates. In addition, mosquitoes were challenged with virus that was maintained at 37 °C and this higher temperature might have resulted in virion structural/conformational changes [32] that promoted enhanced receptor affinity in mutants K122E and R120T [33], as well as accelerated attachment and entry kinetics in the mosquito midgut that affected MIRs. Maintaining the blood-meal at a temperature of 28 °C or lower may help elucidate whether temperature differentially affects receptor-binding affinities or internalization of the two viruses.

DIII of the flavivirus E protein is widely accepted to have receptor binding properties and the FG loop (AAs 381–386) specifically was proposed to bind to mosquito cells. This was further suggested by the absence of this loop structure in the tick-borne viruses. Previously we showed that deletion of the FG loop AAs 382–385 (VEPG) attenuated virus infection in mosquito midguts as well as in Vero cells. Mutation of the FG loop AA sequence 382–384 from VEP to RGD did not significantly affect MIRs, suggesting that the FG loop structure itself and not the AA sequence is important for midgut infection [15]. Due to our focus on E protein functions required for midgut infection in this study, we included mutant viruses 382VEPGΔ and VEP382RGD in the serial passage experiments to determine if deletion or alteration of this DIII motif would place selective pressure on the E protein to acquire mutations that would compensate for these changes. Serial passage of 382VEPGΔ in mosquito midguts did not select for any enhancing adaptive mutations and the virus was lost between the third and fourth passages. Although this virus was able to infect mosquito midguts and secondary tissues in the first passage (albeit at a significantly lower rate than 30P-NBX), our inability to continuously passage 382VEPGΔ in midguts indicated that the presence of the FG loop is vital to the transmission cycle of the virus in vivo and no compensatory changes in other domains could rescue its loss. In contrast, mutant VEP382RGD acquired the K122E mutation more rapidly than 30P-NBX, after only two passages in mosquito midguts. This may suggest that the RGD substitutions imposed greater selective pressure for the K122E mutation. Additionally, multiple passages of wild type DENV2 strain 16681 in Vero cells resulted in a mixed K122K/E population (C.Y-H. Huang, unpublished data). These findings at first suggested that replacement of a positive charge at AA 122 was primarily a primate cell culture-adaptive mutation for DENV2, but the results of this study show the K122E mutation is also relevant to invertebrate systems. Whatever the selective pressure, it appears that reduction of the number of positively charged AAs in this region of DII constitutes a mutational hot spot in the E protein of DENV2 strain 16681.

The high MIR of J1409 may have precluded selective pressure for this DENV2 strain to accumulate mutations during passage in mosquito midguts. DENV2 E protein AA sequence alignments showed that strain 16681 has R120 while all other DENV2 strains, including J1409, have T120 (Table 3). The close proximity of AA120 to AA122 (Figure 2) suggests that replacement of a positively-charged AA at position 120 may have a similar effect to K122E and indeed, mutant R120T had a significantly higher MIR than 30P-NBX, showing that this AA substitution alone could also result in the increased midgut infectivity phenotype. Strain 16,681 has been extensively passaged in various mammalian and invertebrate systems since isolation and the positively-charged R120 could be the result of those passages. This is in agreement with findings that showed passage of DENV2 strain PUO-218 in cultured mammalian BHK-21 cells selected for a T120K change that resulted in higher binding capacity for glycosaminoglycans (GAGs) and reduced neurovirulence in mice [34]. The high MIR resulting from mutation R120T in 30P-NBX may explain why J1409 did not accrue any adaptive mutations in the E gene during midgut passage and similarly, the absence of R120 in other DENV2 strains may explain why E122 is not present in any natural isolates. A reduction in total surface-exposed, positively-charged AAs in this region of E protein DII appears to be a critical requirement for mosquito midgut infection, thus accounting for low MIRs for 30P-NBX, with R120/K122/K123 in its DII sequence, in even highly susceptible Chetumal strain mosquitoes. Reduced DII positive charge density is also tolerated in mammalian cells, as shown by the ability of K122E to replicate in Vero cells as efficiently as 30P-NBX. Nevertheless, engineering double mutations K122E and K123E while retaining R120T rendered the virus unstable in Vero cells at 37 °C (Figures 3B and 4). Interestingly, the double mutant was able to infect a significantly higher proportion of mosquito midguts than 30P-NBX, suggesting that KK122/123EE facilitates infection and is stable during replication in midgut cells at 28 °C.

Reduction of positive charge density on the E protein DII surface also could alter E protein monomer or homodimer stability. For example, two phenotypically distinct DENV2s with an E protein AA difference at position 62 were isolated from K562 and C6/36 cells inoculated with serum from the same DHF patient [35]. The virus isolated from C6/36 cells had E62 and could not bind to and infect B lymphocytes, whereas virus isolated from K562 cells had K62 and was able to bind to and efficiently infect B cells. The K562 cell-derived virus was capable of infecting C6/36 cells but with low efficiency. AA 62 is located in close proximity to AAs 122 and 123 at the homodimer interface (Figure 2) and Kinoshita et al. [35] speculated that K (compared to E) at this position causes high electrostatic repulsive effects with its sister AA on the opposite monomer, causing a loosening effect favorable to B cell binding. However, K122, K123, and R120 do not appear to have such close proximity to their equivalent AAs on the opposite monomer (Figure 2) so it is less likely that electrostatic repulsion affects these AAs to the same extent.

It also is possible that E protein DII mutations affect conformation and stability during virion assembly and maturation. DENV midgut infections generally start at a few foci of infection and spread laterally from each infected cell either by direct cell-to-cell transfer at the edges of a focus or by budding out of cells and diffusing short distances to infect local cells [8]. RII ratios in the midgut for mutant K122E were significantly higher than for 30P-NBX as early as day four PBM, showing that K122E causes an extremely productive infection that spreads rapidly and eventually encompasses the entire tissue. 30P-NBX midgut infections remain relatively restricted by comparison. Mutations on the exposed surface of DII may stabilize the E protein during assembly and maturation through the trans-Golgi pathway, helping to produce a higher infectious virus to particle ratio. Prestwood et al. [34] suggested that the N124D/K128E mutant virus produced a higher PFU to particle ratio than the parent virus in BHK-21 cells. Perhaps a similar phenomenon occurs in mosquito midgut cells.

Although the most potent and serotype-specific neutralizing anti-DENV monoclonal antibodies (MAbs) bind to epitopes on E protein DIII, both murine and human neutralizing MAbs have been mapped to DII epitopes [36,37]). The neutralizing MAb-binding region of DII is variable among the DENV serotypes and other flaviviruses (Table 3) and AA mutations in this region have been shown to create flavivirus MAb neutralization escape variants for DENV, JE, MVE, TBE, and YFV (summarized in [38]), suggesting that the AA sequence variability among flaviviruses in this region of the E protein may be reflective of enhanced immune pressure in the vertebrate host.

DENV2 are suggested to have evolved only recently from sylvatic DENV2 maintained in nature in a cycle between non-human primates and canopy-dwelling *Aedes* spp. Mosquitoes [39,40]. Phylogenetic analysis of DENV2 E proteins suggested that a L122K mutation (Table 3) was predicted to have accompanied DENV2 E protein evolution to the *Ae. aegypti*-human cycle, possibly due to immune selection [38–41]. Most of the AA mutations proposed to correlate with emergence of endemic/epidemic DENV1-4 from sylvatic progenitors were located in DIII, so it is unclear from the phylogenetic analysis whether the L122K change occurred independently or in combination with changes in DIII. If this mutation was necessary for the transition from the sylvatic to the human cycle, it could explain why no natural DENV2 endemic/epidemic isolates have mutations of K122.

The epidemiological implications of our findings are highlighted by the dissemination data from the time course experiment. Head tissue IFA analysis, used as a surrogate for transmission potential, showed that K122E disseminated from the mosquito midgut in a higher proportion of mosquitoes than in 30P-NBX. Even though this may correlate more with higher MIRs and/or higher proportion of infected midgut cells (RII) than with increased dissemination capacity over 30P-NBX virus, the ability of this mutant virus to disseminate from the midgut and infect secondary tissues is clearly greater than 30P-NBX. It will be interesting to see if mutant K122E can infect field-caught or genetically diverse laboratory strain *Ae. aegypti* mosquitoes as efficiently as the laboratory colonized strains used in this study. Increases in viral fitness that increase transmission rates by mosquitoes and produce higher viremia titers in humans can lead to genotype and strain displacements [42–45]. Given that the K122E mutation does not result in fitness costs for replication in mammalian cells, the epidemic potential of a virus that accumulates this point mutation would likely be high.

We have shown for the first time that single AA mutations in DII of the DENV2 E protein can significantly enhance infection of *Ae. aegypti* mosquitoes. Natural mosquito isolates of DENV2 should be monitored for variations in gene sequence at this surface-exposed region of DII and could provide biological markers for virus emergence in the future. Inclusion of arginine at position 120 in DII of live-attenuated DENV2 vaccine viruses with the predominant KK122-123 sequence in addition to attenuating mutations could reduce the transmission potential of vaccine viruses from vaccinees. Investigation of the contribution of this AA region to protective immunity, DENV transmission, and viral pathogenesis in mammals merits further attention.

Author Contributions: Conceptualization, S.M.E. and C.D.B.; methodology, S.M.E. and S.B.; investigation, S.M.E. and S.B.; resources, C.D.B., C.Y.-H.H. and J.T.R.; writing—original draft preparation, S.M.E.; writing—review and editing, C.D.B. and C.Y.-H.H.; supervision, C.D.B.; project administration, C.D.B. and J.T.R.; funding acquisition, C.D.B. and J.T.R. All authors have read and agreed to the published version of the manuscript.

Funding: This study was supported by the Centers for Disease Control and Prevention, and grants from the Pediatric Dengue Vaccine Initiative (PDVI TR-159A) and the College of Veterinary Medicine and Biomedical Sciences Research Council, Colorado State University.

Institutional Review Board Statement: Not applicable.

Informed Consent Statement: Not applicable.

Data Availability Statement: Not applicable.

Conflicts of Interest: The authors declare no conflict of interest.

Author Contributions: The findings and conclusions in this report are those of the authors and do not necessarily represent the official position of CDC.

References

1. Bhatt, S.; Gething, P.W.; Brady, O.J.; Messina, J.P.; Farlow, A.W.; Moyes, C.L.; Drake, J.M.; Brownstein, J.S.; Hoen, A.G.; Sankoh, O.; et al. The global distribution and burden of dengue. *Nature* **2013**, *496*, 504–507. [CrossRef] [PubMed]
2. MacKenzie, J.S.; Gubler, D.J.; Petersen, L.R. Emerging flaviviruses: The spread and resurgence of Japanese encephalitis, West Nile and dengue viruses. *Nat. Med.* **2004**, *10*, S98–S109. [CrossRef]
3. Kyle, J.L.; Harris, E. Global Spread and Persistence of Dengue. *Annu. Rev. Microbiol.* **2008**, *62*, 71–92. [CrossRef] [PubMed]
4. Normile, D. Safety concerns derail dengue vaccination program. *Science* **2017**, *358*, 1514–1515. [CrossRef] [PubMed]
5. Mercado-Curiel, R.F.; Esquinca-Avilés, H.A.; Tovar, R.; Díaz-Badillo, Á.; Camacho-Nuez, M.; Muñoz, M.D.L. The four serotypes of dengue recognize the same putative receptors in *Aedes aegypti* midgut and *Ae. albopictus* cells. *BMC Microbiol.* **2006**, *6*, 85. [CrossRef]

6. Mercado-Curiel, R.F.; Black, W.C.; Muñoz, M.D.L. A dengue receptor as possible genetic marker of vector competence in *Aedes aegypti*. *BMC Microbiol.* **2008**, *8*, 118. [CrossRef] [PubMed]
7. Richardson, J.; Black, W.; Salazar, M.I.; Molina-Cruz, A. Quantitative Analysis of Dengue-2 Virus RNA During the Extrinsic Incubation Period in Individual *Aedes aegypti*. *Am. J. Trop. Med. Hyg.* **2006**, *74*, 132–141. [CrossRef]
8. Salazar, M.I.; Richardson, J.H.; Sánchez-Vargas, I.; Olson, K.E.; Beaty, B.J. Dengue virus type 2: Replication and tropisms in orally infected *Aedes aegypti* mosquitoes. *BMC Microbiol.* **2007**, *7*, 9. [CrossRef]
9. Kuhn, R.J.; Zhang, W.; Rossmann, M.G.; Pletnev, S.V.; Corver, J.; Lenches, E.; Jones, C.T.; Mukhopadhyay, S.; Chipman, P.R.; Strauss, E.G.; et al. Structure of Dengue Virus: Implications for Flavivirus Organization, Maturation, and Fusion. *Cell* **2002**, *108*, 717–725. [CrossRef]
10. Modis, Y.; Ogata, S.; Clements, D.; Harrison, S.C. A ligand-binding pocket in the dengue virus envelope glycoprotein. *Proc. Natl. Acad. Sci. USA* **2003**, *100*, 6986–6991. [CrossRef]
11. Heinz, F.X. Epitope Mapping of Flavivirus Glycoproteins. *Adv. Virus Res.* **1986**, *31*, 103–168. [PubMed]
12. Mandl, C.W.; Guirakhoo, F.; Holzmann, H.; Heinz, F.X.; Kunz, C. Antigenic structure of the flavivirus envelope protein E at the molecular level, using tick-borne encephalitis virus as a model. *J. Virol.* **1989**, *63*, 564–571. [CrossRef] [PubMed]
13. Roehrig, J.; Bolin, R.A.; Kelly, R.G. Monoclonal Antibody Mapping of the Envelope Glycoprotein of the Dengue 2 Virus, Jamaica. *Virology* **1998**, *246*, 317–328. [CrossRef]
14. Roehrig, J.; Johnson, A.J.; Hunt, A.R.; Bolin, R.A.; Chu, M.C. Antibodies to dengue 2 virus E-glycoprotein synthetic peptides identify antigenic conformation. *Virology* **1990**, *177*, 668–675. [CrossRef]
15. Erb, S.M.; Butrapet, S.; Moss, K.J.; Luy, B.E.; Childers, T.; Calvert, A.E.; Silengo, S.J.; Roehrig, J.T.; Huang, C.Y.-H.; Blair, C.D. Domain-III FG loop of the dengue virus type 2 envelope protein is important for infection of mammalian cells and Aedes aegypti mosquitoes. *Virology* **2010**, *406*, 328–335. [CrossRef] [PubMed]
16. Fansiri, T.; Fontaine, A.; Diancourt, L.; Caro, V.; Thaisomboonsuk, B.; Richardson, J.H.; Jarman, R.G.; Ponlawat, A.; Lambrechts, L. Genetic Mapping of Specific Interactions between Aedes aegypti Mosquitoes and Dengue Viruses. *PLoS Genet.* **2013**, *9*, e1003621. [CrossRef]
17. Bennett, K.E.; Olson, K.E.; Muñoz, M.D.L.; Fernández Salas, I.; Farfán Ale, J.A.; Higgs, S.; Black, W.C.; Beaty, B.J. Variation in vector competence for dengue 2 virus among 24 collections of Aedes aegypti from Mexico and the United States. *Am. J. Trop. Med. Hyg.* **2002**, *67*, 85–92. [CrossRef]
18. Armstrong, P.M.; Rico-Hesse, R. Efficiency of Dengue Serotype 2 Virus Strains to Infect and Disseminate in *Aedes aegypti*. *Am. J. Trop. Med. Hyg.* **2003**, *68*, 539–544. [CrossRef]
19. Huang, C.Y.-H.; Butrapet, S.; Moss, K.J.; Childers, T.; Erb, S.M.; Calvert, A.E.; Silengo, S.J.; Kinney, R.M.; Blair, C.D.; Roehrig, J.T. The dengue virus type 2 envelope protein fusion peptide is essential for membrane fusion. *Virology* **2009**, *396*, 305–315. [CrossRef]
20. Butrapet, S.; Childers, T.; Moss, K.J.; Erb, S.M.; Luy, B.E.; Calvert, A.E.; Blair, C.D.; Roehrig, J.; Huang, C.Y.-H. Amino acid changes within the E protein hinge region that affect dengue virus type 2 infectivity and fusion. *Virology* **2011**, *413*, 118–127. [CrossRef]
21. Erb, S.M. Mutagenesis of the dengue virus envelope glycoprotein gene can significantly alter virus infectivity phenotypes in cultured cells and live mosquitoes. In *Microbiology, Immunology & Pathology*; Colorado State University: Fort Collins, CO, USA, 2011; p. 195.
22. Huang, C.Y.-H.; Butrapet, S.; Pierro, D.J.; Chang, G.J.J.; Hunt, A.R.; Bhamarapravati, N.; Gubler, D.J.; Kinney, R.M. Chimeric Dengue Type 2 (Vaccine Strain PDK-53)/Dengue Type 1 Virus as a Potential Candidate Dengue Type 1 Virus Vaccine. *J. Virol.* **2000**, *74*, 3020–3028. [CrossRef] [PubMed]
23. Halstead, S.B.; Simasthien, P. Observations related to the pathogenesis of dengue hemorrhagic fever. II. Antigenic and biologic properties of dengue viruses and their association with disease response in the host. *Yale J. Biol. Med.* **1970**, *42*, 276–292. [PubMed]
24. Kinney, R.M.; Butrapet, S.; Chang, G.-J.J.; Tsuchiya, K.R.; Roehrig, J.; Bhamarapravati, N.; Gubler, D.J. Construction of Infectious cDNA Clones for Dengue 2 Virus: Strain 16681 and Its Attenuated Vaccine Derivative, Strain PDK-53. *Virology* **1997**, *230*, 300–308. [CrossRef]
25. Deubel, V.; Kinney, R.M.; Trent, D.W. Nucleotide sequence and deduced amino acid sequence of the structural proteins of dengue type 2 virus, Jamaica genotype. *Virology* **1986**, *155*, 365–377. [CrossRef]
26. Deubel, V.; Kinney, R.M.; Trent, D.W. Nucleotide sequence and deduced amino acid sequence of the nonstructural proteins of dengue type 2 virus, Jamaica genotype: Comparative analysis of the full-length genome. *Virology* **1988**, *165*, 234–244. [CrossRef]
27. Pierro, D.J.; Salazar, M.I.; Beaty, B.J.; Olson, K.E. Infectious clone construction of dengue virus type 2, strain Jamaican 1409, and characterization of a conditional E6 mutation. *J. Gen. Virol.* **2006**, *87*, 2263–2268. [CrossRef]
28. Deubel, V.; Kinney, R.M.; Esposito, J.J.; Cropp, C.B.; Vorndam, A.V.; Monath, T.P.; Trent, D.W. Dengue 2 Virus Envelope Protein Expressed by a Recombinant Vaccinia Virus Fails to Protect Monkeys against Dengue. *J. Gen. Virol.* **1988**, *69*, 1921–1929. [CrossRef]
29. BButrapet, S.; Kinney, R.M.; Huang, C.Y.-H. Determining genetic stabilities of chimeric dengue vaccine candidates based on dengue 2 PDK-53 virus by sequencing and quantitative TaqMAMA. *J. Virol. Methods* **2006**, *131*, 1–9. [CrossRef]
30. Reed, L.J.; Muench, H. A Simple Method of Estimating Fifty Per Cent Endpoints. *Am. J. Epidemiol.* **1938**, *27*, 493–497. [CrossRef]
31. Brackney, D.E.; Olson, K.E.; Foy, B. The Effects of Midgut Serine Proteases on Dengue Virus Type 2 Infectivity of Aedes aegypti. *Am. J. Trop. Med. Hyg.* **2008**, *79*, 267–274. [CrossRef]
32. Zhang, X.; Sheng, J.; Plevka, P.; Kuhn, R.J.; Diamond, M.S.; Rossmann, M.G. Dengue structure differs at the temperatures of its human and mosquito hosts. *Proc. Natl. Acad. Sci. USA* **2013**, *110*, 6795–6799. [CrossRef] [PubMed]

33. Hung, J.-J.; Hsieh, M.-T.; Young, M.-J.; Kao, C.-L.; King, C.-C.; Chang, W. An External Loop Region of Domain III of Dengue Virus Type 2 Envelope Protein Is Involved in Serotype-Specific Binding to Mosquito but Not Mammalian Cells. *J. Virol.* **2004**, *78*, 378–388. [CrossRef] [PubMed]
34. Prestwood, T.R.; Prigozhin, D.M.; Sharar, K.L.; Zellweger, R.M.; Shresta, S. A Mouse-Passaged Dengue Virus Strain with Reduced Affinity for Heparan Sulfate Causes Severe Disease in Mice by Establishing Increased Systemic Viral Loads. *J. Virol.* **2008**, *82*, 8411–8421. [CrossRef] [PubMed]
35. Kinoshita, H.; Mathenge, E.G.M.; Hung, N.T.; Huong, V.T.Q.; Kumatori, A.; Yu, F.; Parquet, M.C.; Inoue, S.; Matias, R.R.; Natividad, F.F.; et al. Isolation and characterization of two phenotypically distinct dengue type-2 virus isolates from the same dengue hemorrhagic Fever patient. *Jpn. J. Infect. Dis.* **2009**, *62*, 343–350.
36. Roehrig, J.T. Antigenic structure of flavivirus proteins. *Adv. Virus Res.* **2003**, *59*, 141–175.
37. Wahala, W.; Kraus, A.A.; Haymore, L.B.; Accavitti-Loper, M.A.; de Silva, A.M. Dengue virus neutralization by human immune sera: Role of envelope protein domain III-reactive antibody. *Virology* **2009**, *392*, 103–113. [CrossRef]
38. Wang, E.; Ni, H.; Xu, R.; Barrett, A.D.T.; Watowich, S.J.; Gubler, D.J.; Weaver, S.C. Evolutionary Relationships of Endemic/Epidemic and Sylvatic Dengue Viruses. *J. Virol.* **2000**, *74*, 3227–3234. [CrossRef]
39. Twiddy, S.S.; Woelk, C.H.; Holmes, E.C. Phylogenetic evidence for adaptive evolution of dengue viruses in nature. *J. Gen. Virol.* **2002**, *83*, 1679–1689. [CrossRef]
40. Vasilakis, N.; Tesh, R.B.; Durbin, A.P.; Munoz-Jordan, J.L.; Da Rosa, A.P.A.T.; Weaver, S.C. Antigenic relationships between sylvatic and endemic dengue viruses. *Am. J. Trop. Med. Hyg.* **2008**, *79*, 128–132. [CrossRef]
41. Vasilakis, N.; Fokam, E.B.; Hanson, C.T.; Weinberg, E.; Sall, A.A.; Whitehead, S.S.; Hanley, K.A.; Weaver, S.C. Genetic and phenotypic characterization of sylvatic dengue virus type 2 strains. *Virology* **2008**, *377*, 296–307. [CrossRef]
42. Hanley, K.A.; Nelson, J.T.; Schirtzinger, E.E.; Whitehead, S.S.; Hanson, C.T. Superior infectivity for mosquito vectors contributes to competitive displacement among strains of dengue virus. *BMC Ecol.* **2008**, *8*, 1. [CrossRef] [PubMed]
43. Hang, V.T.T.; Holmes, E.; Veasna, D.; Quy, N.T.; Hien, T.T.; Quail, M.; Churcher, C.; Parkhill, J.; Cardosa, J.; Farrar, J.; et al. Emergence of the Asian 1 Genotype of Dengue Virus Serotype 2 in Viet Nam: In Vivo Fitness Advantage and Lineage Replacement in South-East Asia. *PLoS Negl. Trop. Dis.* **2010**, *4*, e757. [CrossRef] [PubMed]
44. Messer, W.B.; Gubler, D.J.; Harris, E.; Sivananthan, K.; De Silva, A.M. Emergence and Global Spread of a Dengue Serotype 3, Subtype III Virus. *Emerg. Infect. Dis.* **2003**, *9*, 800–809. [CrossRef] [PubMed]
45. Cologna, R.; Armstrong, P.M.; Rico-Hesse, R. Selection for Virulent Dengue Viruses Occurs in Humans and Mosquitoes. *J. Virol.* **2005**, *79*, 853–859. [CrossRef]

viruses MDPI

Article

Silent Reactivation of Varicella Zoster Virus in Pregnancy: Implications for Maintenance of Immunity to Varicella

Mirella Mourad [1], Michael Gershon [2], Satish K. Mehta [3], Brian E. Crucian [4], Nicole Hubbard [5], Jing Zhang [6] and Anne Gershon [7,*]

1 Department of Obstetrics and Gynecology, Columbia University Irving Medical Center, New York, NY 10032, USA; mjm2246@cumc.columbia.edu
2 Department of Pathology, Columbia University Vagelos College of P&S, New York, NY 10032, USA; mdg4@cumc.columbia.edu
3 JES Tech, Human Health and Performance Directorate, Houston, TX 77058, USA; satish.k.mehta@nasa.gov
4 National Aeronautics and Space Administration (NASA) Johnson Space Center, Human Health and Performance Directorate, Houston, TX 77058, USA; brian.crucian-1@nasa.gov
5 Infectious Disease Pathology, ProMedica Laboratory, Department of Pathology, University of Toledo College of Medicine and Life Sciences, Toledo, OH 43614, USA; nicole.hubbardmd@promedica.org
6 Department of Pathology, Columbia University Irving Medical Center, New York, NY 10032, USA; jz14@cumc.columbia.edu
7 Department of Pediatrics, Columbia University Irving Medical Center, New York, NY 10032, USA
* Correspondence: aag1@cumc.columbia.edu

Abstract: We encountered two cases of varicella occurring in newborn infants. Because the time between birth and the onset of the illness was much shorter than the varicella incubation period, the cases suggested that the infection was maternally acquired, despite the fact that neither mother experienced clinical zoster. Thus, we tested the hypothesis that VZV frequently reactivates asymptomatically in late pregnancy. The appearance of DNA-encoding VZV genes in saliva was used as an indicator of reactivation. Saliva was collected from 5 women in the first and 14 women in the third trimesters of pregnancy and analyzed at two different sites, at one using nested PCR and at the other using quantitative PCR (qPCR). No VZV DNA was detected at either site in the saliva of women during the first trimester; however, VZV DNA was detected in the majority of samples of saliva (11/12 examined by nested PCR; 7/10 examined by qPCR) during the third trimester. These observations suggest that VZV reactivation occurs commonly during the third trimester of pregnancy. It is possible that this phenomenon, which remains in most patients below the clinical threshold, provides an endogenous boost to immunity and, thus, is beneficial.

Keywords: VZV; reactivation; latency; saliva; salivary VZV DNA

Citation: Mourad, M.; Gershon, M.; Mehta, S.K.; Crucian, B.E.; Hubbard, N.; Zhang, J.; Gershon, A. Silent Reactivation of Varicella Zoster Virus in Pregnancy: Implications for Maintenance of Immunity to Varicella. *Viruses* **2022**, *14*, 1438. https://doi.org/10.3390/v14071438

Academic Editors: Charles Grose, Ravi Mahalingam and Joel Rovnak

Received: 11 May 2022
Accepted: 22 June 2022
Published: 30 June 2022

1. Introduction

Extreme stress, such as space flight, may cause varicella zoster virus (VZV) to reactivate from latency (in persons who have had varicella or have been vaccinated against it) without producing symptoms. This phenomenon was reported to occur in 50–65% of astronauts during and after short- and long-duration spaceflights [1,2] and in 17% of children who were immune to VZV and stressed by hospitalization in an intensive care unit (ICU) [3]. These reports employed the transient presence of VZV DNA in the saliva as an indication of viral reactivation. In part because the screening of saliva for the presence of VZV DNA is available only on a research basis, it is not clear what the incidence and prevalence are of asymptomatic reactivations of VZV in individuals who are immune to varicella. Although the frequency with which asymptomatic VZV reactivation occurs is unknown, published data imply that VZV DNA is rarely, if ever, detected in the saliva of healthy, young individuals [4].

We encountered two newborn infants who developed varicella just after birth, a timeframe that is much shorter than the incubation period for varicella. These observations suggested that VZV was passed from the mother to the fetus prior to delivery. Because neither mother displayed varicella or zoster, it was likely that asymptomatic reactivations of VZV occurred in late pregnancy and provided the virus that completed the varicella incubation period perinatally. Because the varicella in the newborn infants was mild, maternally derived immunity was also transferred and ameliorated the infection. These observations led us to ask whether the asymptomatic reactivation of VZV occurs frequently in pregnancy. To answer this question, we determined whether VZV DNA could be detected in the saliva of a group of pregnant women in the absence of symptoms, as has been reported in astronauts [1,2]. We also determined whether other herpesviruses might reactivate in pregnancy.

Case 1. This infant was born at Columbia Presbyterian Hospital in New York City in 2018. His mother, a 37-year-old healthy woman presented in active labor at 36 weeks of pregnancy. Her antenatal period was significant only for vaginal candidiasis. She had no history of febrile illnesses or rashes in the antenatal period and no chronic medical conditions. She recalled experiencing chickenpox in childhood and had not experienced clinical zoster. Her serum varicella IgG level was positive prior to pregnancy. Her male baby was delivered by Caesarean section for nonreassuring fetal heart rate with a rupture of membranes at delivery. The APGAR scores were 8 and 9; the infant's weight was 3295 g. The infant was hypoglycemic and hypothermic shortly after delivery. Ampicillin and gentamicin were administered after blood cultures were obtained. The postnatal examination revealed a mild, diffuse, pustular rash with scattered petechiae. Dermatology was consulted; the differential diagnosis included systemic bacterial or viral infection and transient neonatal pustulosis. On day one of life, bacterial and fungal blood and skin cultures were obtained and were found to be negative. VZV and herpes simplex virus (HSV) cultures from skin lesions were obtained and were found to be negative. Acyclovir 60 mg/kg/dose IV was administered since the infant was thought possibly to have a neonatal HSV infection. Direct fluorescent antigen (DFA) tests from the skin lesions then tested positive for VZV and negative for HSV. The polymerase chain reaction (PCR) tests obtained from skin lesions were positive for VZV and negative for HSV. A diagnosis of varicella was made in the infant, although there was no obvious exposure of the baby to VZV. A lumbar puncture was not performed. The diagnosis of varicella was made in the infant, and treatment with acyclovir was continued for a week. The infant's rash improved, and the baby appeared entirely well on a follow-up visit.

Case 2. This 39-week, 1-day-old male infant was born via Cesarian section due to failure to progress at Pro Medica Russell J. Ebeid Children's Hospital in Toledo, Ohio, in June 2021. The pregnancy was complicated by gestational hypertension. He presented with vesicular lesions on his right cheek in the first 24 h after birth. Subsequently, a mild rash was noted on his trunk that was thought to be erythema toxicum. Neonatal HSV infection was suspected, but a PCR test on a lesion on the cheek was negative for HSV DNA. It was positive, however, for VZV DNA (tested at ProMedica Laboratories, as well as in the laboratory at Columbia). As part of routine prenatal testing, his mother was found to have a positive IgG immunoassay for VZV IgG in her first trimester of pregnancy. She recalled experiencing varicella as a child. Except for the rash, the baby appeared well. New lesions developed on day four of life. By day six, all the lesions had crusted, and no further lesions developed. VZV IgG testing was positive on the infant. The mother was asymptomatic throughout the illness of her infant but reported that an aunt who babysat her daughter experienced shingles prior to the delivery. The infant was discharged and went home at 11 days of age.

2. Methods

We performed a prospective observational study in order to determine if we could identify evidence of the asymptomatic reactivation of VZV in pregnancy. Pregnant subjects from different trimesters in pregnancy who received prenatal care at our institution were approached and consented to participate in our study. The subjects were recruited from 3 general areas: obstetric ultrasound, the antepartum unit, and the labor and delivery unit. The subjects were excluded if they had any signs or symptoms of COVID-19 infection or other viral illness. Saliva was collected with commercial kits for this purpose obtained from Omnigene-oral Genotek (Ottowa, ONT, Canada), as described previously [5]. Nested PCR was employed at Columbia University for the detection of VZV DNA, as previously described. Real-time PCR was performed at NASA [1], as well as at Columbia University. This study was approved by the Institutional Review Board at Columbia University Irving Medical Center (AAAS6203), Institution Review Board at Johnson Space Center, NASA; Study ID: Pro2773, and the University of Toledo College of Medicine and Life Sciences.

3. Results

Saliva samples were obtained from 19 pregnant women. These women ranged in age from 25 to 46 years (mean 35). Their ethnic backgrounds were varied and included White (n = 9), Black (n = 4), Hispanic (n = 5), and Asian (n = 1). Ten patients provided a history of having experienced varicella in childhood; three did not, and six could not recall. Three remembered having received the live attenuated varicella vaccine and did not have a history of varicella. Most patients were healthy at baseline; however, two patients had a chronic medical condition: one had cystic fibrosis, and one had systemic lupus erythematosus. Samples were obtained from five women in the first trimester of pregnancy (weeks 9–12), and fourteen from women in the third trimester (weeks 25–40; average 34 weeks).

The saliva from the patients was independently assayed in two different laboratories: at Columbia and at NASA. From the Columbia laboratory, 11/12 (92%) third-trimester patients were found to have VZV DNA in their saliva by nested PCR. In the NASA laboratory, 7/10 (70%) were positive by real-time PCR. Their viral loads ranged from 508 to 1128 copies/mL with a mean + SE of 717 + 84 copies/mL saliva. Ten saliva samples were tested at both sites; the results agreed in 8/10 samples (80%). Two third-trimester patients had positive saliva samples at Columbia but negative tests at NASA. Saliva from 5/5 women in the first trimester were negative by qPCR at Columbia; 2/2 tested were negative at NASA. At NASA, no reactivation of HSV or cytomegalovirus (CMV) was identified in saliva from nine patients by qPCR.

In summary, at Columbia, of 12 women who were studied in weeks 25–40, 11/12 had positive saliva for VZV DNA (92%) using nested PCR. At NASA, of 10 patients studied in weeks 25–40, 7/10 (70%) were positive for VZV DNA by qPCR. In five women whose saliva was studied during the first trimester for VZV DNA, 0/5 were positive (Supplementary Materials).

4. Discussion

We became interested in whether the reactivation of VZV in pregnancy was common or not after observing two newborn infants with varicella. One reported infant was born at Columbia. The other was encountered 3 years later because the medical team responsible for the care of this infant had heard about the original Columbia patient and brought their patient to the attention of the Columbia physicians. The silent reactivation of VZV is of particular interest because it is a potential stimulant of maternal immunity to VZV. It is interesting that VZV was reactivated in the setting of decreased immunity in pregnancy, which is thought to occur so that the mother does not reject her fetus. Immunity to VZV, however, was not so diminished in the pregnant women that clinical zoster was able to occur. Reactivation was only identified in women during late pregnancy.

It is not fully understood why immunity to varicella persists for years, both in individuals who have experienced clinical varicella and also in those who have been vaccinated. Two theories exist [6,7]: One is that VZV may periodically and spontaneously reactivate from latency (with or without symptoms), leading to what is termed as internal boosting. The other is that periodic exposure to VZV in persons with varicella is required to maintain immunity, which is called exogenous boosting [8]. The demonstration of the asymptomatic reactivation of VZV from latency, therefore, has importance for vaccination programs, as well as being an interesting phenomenon in which newborn infants can, on rare occasions, develop mild varicella from their mothers.

The diagnosis of neonatal varicella in the first infant in this report led us to begin to determine whether the reactivation of VZV could be identified in late pregnancy, when an incipient mother presumably becomes at least somewhat immunocompromised. The second infant was born 3 years later, after the study was underway, at a different hospital. Both cases were demonstrated to be due to VZV by the PCR of skin specimens at Columbia. There is no question about the diagnosis of VZV in these infants. It was, at first, suspected that each baby might have a herpes simplex virus (HSV) infection, but HSV was not detected in either one by PCR, ruling out this possibility. The babies were treated appropriately with acyclovir and recovered. The course of their varicella was mild, as would also be expected, due to the presence in the infants of maternal antibodies to VZV (but not enough to prevent the clinical infection of the infants) and antiviral therapy.

When these young babies were diagnosed with varicella, it was clear that the VZV had to have come from their mothers, who themselves had no symptoms of VZV infection. The shortest possible incubation period of varicella is 7 days, so postnatal exposure was not possible. The phenomenon of varicella in 1- to 3-day old infants was described previously in three newborns many years ago [9]. In all the instances, the illness in the baby was mild. It was concluded that the mother had to be the source of the virus, but the mechanism by which this might have transpired was neither clear nor was it even discussed in older publications. We now postulate that the reactivation of VZV occurred in these mothers, who must have experienced an asymptomatic form of herpes zoster similar to that described in astronauts [1,2].

In the current study, we found that the silent reactivation of VZV during pregnancy was common. This was shown independently in two different laboratories. There was close agreement between the two laboratories for women in the third trimester (92% reactivation at Columbia vs. 70% at NASA). Nested PCR is more sensitive than qPCR, so this result was not unexpected. It is of interest that the saliva samples of five women tested in early pregnancy were negative for reactivation by qPCR at both laboratories.

In addition to the rare cases of clinical varicella in neonates who were born to healthy mothers that we described, other data suggest that VZV from healthy mothers can cross the placenta. About 20 years ago, VZV was reported in the cerebrospinal fluid (CSF) of two neonates with neurologic abnormalities and no history of rash. These infants were born to healthy mothers who completed normal pregnancies [10,11]. It was reasoned that the virus that infected the children might have been derived from a reactivation of maternal VZV during pregnancy leading to viremia, which was asymptomatic in the mothers but, nevertheless, crossed the placenta to enter the infants who, as a result, developed serious neurological abnormalities. More recently, the transfer of VZV DNA from 61 pregnant mothers to approximately one-third of their infants was reported in a study of sera obtained from pregnant women and their infants [12]. In these cases, however, there were no associated symptoms in the infants.

We found no evidence in our study on saliva of reactivation of HSV or cytomegalovirus (CMV) in the pregnant women we described. We expected that we might find reactivation of other herpesviruses, such CMV and HSV, which reactivate along with VZV in astronauts [2], but we did not.

About 50 years ago, Hope-Simpson proposed the hypothesis that VZV could reactivate from latency to cause zoster [6]. In addition, Hope-Simpson also proposed that VZV would periodically reactivate silently, liberating an amount of the virus that was too low to cause clinical disease but would, nevertheless, lead to episodes of asymptomatic viremia. His idea was that these asymptomatic reactivations would boost an individual's immunity to the virus. Despite this mechanism, however, immunity could still wane as a function of age, leading to the manifestation of clinical zoster [6]. Implicit in Hope-Simpson's hypothesis was that the silent reactivation of VZV could stimulate immunity to the virus. At the time of Hope-Simpson's publication, humoral immune responses were recognized, but cellular immunity to VZV had not yet been described. At that time, moreover, Hope-Simpson was unable to provide data to verify that his postulated asymptomatic reactivation of VZV actually did occur. Usually, in 70% of people over a lifetime, cellular immunity prevents VZV from reactivating and causing clinical zoster [13]. When cell-mediated immune function declines, however, as it does in the elderly, VZV may reactivate in neurons that project to the skin, causing symptomatic zoster with a classical skin rash [14]. This occurrence is well-known, but it is not evidence of the asymptomatic reactivation of VZV.

The periodic silent reactivation of VZV without symptoms that Hope-Simpson postulated could be a critical factor in the maintenance of long-term immunity to VZV. After primary infection, VZV becomes latent in peripheral neurons, including those of the enteric nervous system (ENS) [5,7]; moreover, reactivation in enteric neurons would be a strategic location to stimulate immunity because the gastrointestinal tract is the largest organ of the immune system in the body [15]. Since enteric neurons do not project to the skin, the reactivation of VZV in the gut can occur in the absence of skin lesions and, often, without provoking symptoms [5,16]. It is, thus, important to recognize and understand silent VZV reactivation because it may play a role in maintaining long-term immunity to VZV. Not only does wild-type VZV become latent after primary infection, but so also does the vOka strain of the live attenuated varicella vaccine [7,17].

Some epidemiological investigators assumed that exogenous exposure to circulating wild-type VZV is critical for the maintenance of immunity to VZV and built computer models to predict what would happen if VZV ceased to circulate and epidemics of varicella no longer occurred [8]. These models predicted that epidemics of zoster and a loss of immunity to varicella would be the inevitable long-term outcomes of decreased exposure to circulating wild-type VZV. The assumptions, however, that formed the basis of the dire predictions of the computer models did not include the possible effects of silent reactivation of VZV with the stimulation of immunity by endogenous exposure to VZV (endogenous boosting). Fortunately, enough time has passed since the licensure of the live attenuated VZV vaccine in the US, its nearly universal use, and the virtual disappearance of circulating VZV in the USA to ask whether the predicted epidemics of zoster have occurred. In fact, there is no convincing evidence that the incidence of zoster is greater in countries that routinely vaccinate against varicella than in those that do not [18]. Laboratory evidence of immunity to VZV in older adults is similar in countries that do or do not routinely use a varicella vaccine [19,20]. The incidence of zoster in cloistered nuns and monks who are not exposed to children and their epidemics of varicella, moreover, is not different from that in the general population [21]. The failure of the disastrous spread of zoster to occur suggests that endogenous boosting is important in maintaining immunity to VZV. It is, therefore, important to document instances of inapparent VZV reactivation. Such occurrences may be more frequent than is appreciated. Clearly, stress can reactivate VZV without symptoms, as has been reported in astronauts during and after space travel [1,2]. The reactivation of VZV without symptoms could be common due to the stresses of late pregnancy and delivery. Mothers are young and ought to have healthy, well-functioning immune systems. Such a group might be expected to have sufficient cellular immunity to prevent reactivated VZV from reaching the threshold necessary to manifest clinical zoster. In most cases, their immunity would also be sufficient to protect the fetus. The rare escape of reactivated VZV from immune surveillance might give rise to neonatal varicella, but in most instances, VZV reactivation during late pregnancy is probably beneficial to the mother and even to her child. The event would be silent and asymptomatic, but immunity to VZV, including protection against zoster, would receive a boost.

Supplementary Materials: The following supporting information can be downloaded at: https://www.mdpi.com/article/10.3390/v14071438/s1.

Author Contributions: Collected specimens, M.M. and N.H. Virologic analyses of saliva specimens, J.Z., S.K.M. and B.E.C. Conceptualization, M.G. and A.G., Wrote first draft of paper, M.G. and A.G. All authors have read and agreed to the published version of the manuscript.

Funding: This research was funded by National Institute of Health grant number R01 DK093094.

Institutional Review Board Statement: Approved by IRB at Columbia University and University of Toledo College of Medicine and Life Sciences.

Informed Consent Statement: Informed consent was obtained from all subjects involved in the study Signed by participants in accordance with Columbia University IRB and University of Toledo College of Medicine and Life Sciences.

Conflicts of Interest: The authors declare no conflict of interest.

References

1. Mehta, S.K.; Cohrs, R.J.; Forghani, B.; Zerbe, G.; Gilden, D.H.; Pierson, D.L. Stress-induced subclinical reactivation of varicella zoster virus in astronauts. *J. Med. Virol.* **2004**, *72*, 174–179. [CrossRef] [PubMed]
2. Mehta, S.K.; Laudenslager, M.L.; Stowe, R.P.; Crucian, B.E.; Sams, C.F.; Pierson, D.L. Multiple latent viruses reactivate in astronauts during Space Shuttle missions. *Brain Behav. Immun.* **2014**, *41*, 210–217. [CrossRef]
3. Papaevangelou, V.; Quinlivan, M.; Lockwood, J.; Papaloukas, O.; Sideri, G.; Critselis, E.; Papassotiriou, I.; Papadatos, J.; Breuer, J. Subclinical VZV reactivation in immunocompetent children hospitalized in the ICU associated with prolonged fever duration. *Clin. Microbiol. Infect.* **2013**, *19*, E245–E251. [CrossRef] [PubMed]
4. Birlea, M.; Cohrs, R.J.; Bos, N.; Mehta, S.K.; Pierson, D.L.; Gilden, D. Search for varicella zoster virus DNA in saliva of healthy individuals aged 20–59 years. *J. Med. Virol.* **2014**, *86*, 360–362. [CrossRef] [PubMed]
5. Gershon, A.A.; Chen, J.; Gershon, M.D. Use of saliva to identify varicella-zoster virus (VZV) infection of the gut. *Clin. Infect. Dis. Off. Publ. Infect. Dis. Soc. Am.* **2015**, *61*, 536–544. [CrossRef] [PubMed]
6. Hope-Simpson, R.E. The nature of herpes zoster: A long term study and a new hypothesis. *Proc. Roy. Soc. Med.* **1965**, *58*, 9–20. [CrossRef] [PubMed]
7. Gershon, A.A.; Breuer, J.; Cohen, J.I.; Cohrs, R.J.; Gershon, M.D.; Gilden, D.; Grose, C.; Hambleton, S.; Kennedy, P.G.; Oxman, M.N.; et al. Varicella zoster virus infection. *Nat. Rev. Dis. Primers* **2015**, *1*, 15016. [CrossRef] [PubMed]
8. Brisson, M.; Gay, N.; Edmunds, W.J.; Andrews, N.J. Exposure to varicella boosts immunity to herpes-zoster: Implications for mass vaccination against chickenpox. *Vaccine* **2002**, *20*, 2500–2507. [CrossRef]
9. Hyatt, H.W. Neonatal varicella. *J. Nat. Med. Assn.* **1967**, *59*, 32–34.
10. Mustonen, K.; Mustakangas, P.; Smeds, M.; Mannonen, L.; Uotila, L.; Vaheri, A.; Koskiniemi, M. Antibodies to varicella zoster virus in the cerebrospinal fluid of neonates with seizures. *Arch. Dis. Childhood. Fetal Neonatal Ed.* **1998**, *78*, F57–F61. [CrossRef]
11. Mustonen, K.; Mustakangas, P.; Uotila, L.; Muir, P.; Koskiniemi, M. Viral infections in neonates with seizures. *J. Perinat. Med.* **2003**, *31*, 75–80. [CrossRef] [PubMed]
12. Ranjan, V.; Mishra, A.; Kesarwani, A.; Mohan, K.V.; Lal, S.N.; Puliyel, J.; Upadhyay, P. Mother-to-Child Transfer of Reactivated Varicella-Zoster Virus DNA and Varicella-Zoster IgG in Pregnancy. *Viral Immunol.* **2020**, *33*, 72–76. [CrossRef] [PubMed]
13. Arvin, A.M. Humoral and cellular immunity to varicella-zoster virus: An overview. *J. Infect. Dis.* **2008**, *197* (Suppl. 2), S58–S60. [CrossRef] [PubMed]
14. Dorshkind, K.; Swain, S. Age-associated declines in immune system development and function: Causes, consequences, and reversal. *Curr. Opin. Immunol.* **2009**, *21*, 404–407. [CrossRef]
15. Pabst, R.; Russell, M.W.; Brandtzaeg, P. Tissue distribution of lymphocytes and plasma cells and the role of the gut. *Trends Immunol.* **2008**, *29*, 206–208. [CrossRef] [PubMed]
16. Naik, R.D.; Vaezi, M.F.; Gershon, A.A.; Higginbotham, T.; Chen, J.J.; Flores, E.; Holzman, M.; Patel, D.; Gershon, M.D. Association of Achalasia With Active Varicella Zoster Virus Infection of the Esophagus. *Gastroenterology* **2021**, *161*, 719–721.e2. [CrossRef]
17. Weinmann, S.; Chun, C.; Schmid, D.S.; Roberts, M.; Vandermeer, M.; Riedlinger, K.; Bialek, S.R.; Marin, M. Incidence and clinical characteristics of herpes zoster among children in the varicella vaccine era, 2005–2009. *J. Infect. Dis.* **2013**, *208*, 1859–1868. [CrossRef]
18. Harpaz, R. Do varicella vaccination programs change the epidemiology of herpes zoster? A comprehensive review, with focus on the United States. *Expert Rev. Vaccines* **2019**, *18*, 793–811. [CrossRef]

19. Carryn, S.; Cheuvart, B.; Povey, M.; Dagnew, A.F.; Harpaz, R.; van der Most, R.; Casabona, G. No Consistent Evidence of Decreased Exposure to Varicella-Zoster Virus Among Older Adults in Countries with Universal Varicella Vaccination. *J. Infect. Dis.* **2022**, *225*, 413–421. [CrossRef]
20. Gershon, A.A.; Gershon, M.D. Widespread Use of Varicella Vaccine Does Not Reduce Immunity to Zoster of Others. *J. Infect. Dis.* **2022**, *225*, 361–363. [CrossRef]
21. Gaillat, J.; Gajdos, V.; Launay, O.; Malvy, D.; Demoures, B.; Lewden, L.; Pinchinat, S.; Derrough, T.; Sana, C.; Caulin, E.; et al. Does Monastic Life Predispose to the Risk of Saint Anthony's Fire (Herpes Zoster)? *Clin. Infect. Dis. Off. Publ. Infect. Dis. Soc. Am.* **2011**, *53*, 405. [CrossRef] [PubMed]

viruses

MDPI

Communication

Variable Gene Expression in Human Ganglia Latently Infected with Varicella-Zoster Virus

Peter G. E. Kennedy [1,*] and Paul Montague [2]

1 Institute of Neuroscience and Psychology, University of Glasgow, Glasgow G61 1QH, UK
2 Institute of Infection, Immunity and Inflammation, College of Medical, Veterinary and Life Sciences, University of Glasgow, Glasgow G61 1QH, UK; paul.montague@glasgow.ac.uk
* Correspondence: peter.kennedy@glasgow.ac.uk

Abstract: Varicella-Zoster virus (VZV) is a pathogenic human herpes virus that causes varicella ("chicken pox") as a primary infection, following which it becomes latent in neuronal cells in human peripheral ganglia. It may then reactivate to cause herpes zoster ("shingles"). Defining the pattern of VZV gene expression during latency is an important issue, and four highly expressed VZV genes were first identified by Randall Cohrs in 1996 using cDNA libraries. Further studies from both his and other laboratories, including our own, have suggested that viral gene expression may be more widespread than previously thought, but a confounding factor has always been the possibility of viral reactivation after death in tissues obtained even at 24 h post-mortem. Recent important studies, which Randall Cohrs contributed to, have clarified this issue by studying human trigeminal ganglia at 6 h after death using RNA-Seq methodology when a novel spliced latency-associated VZV transcript (VLT) was found to be mapped antisense to the viral transactivator gene 61. Viral gene expression could be induced by a VLT-ORF 63 fusion transcript when VZV reactivated from latency. Prior detection by several groups of ORF63 in post-mortem-acquired TG is very likely to reflect detection of the VLT-ORF63 fusion and not canonical ORF63. The contributions to the VZV latency field by Randall Cohrs have been numerous and highly significant.

Keywords: varicella-zoster; virus; latency; gene expression; ganglion; autopsy

Citation: Kennedy, P.G.E.; Montague, P. Variable Gene Expression in Human Ganglia Latently Infected with Varicella-Zoster Virus. *Viruses* **2022**, *14*, 1250. https://doi.org/10.3390/v14061250

Academic Editors: Charles Grose, Ravi Mahalingam, Joel Rovnak and Oliver Schildgen

Received: 31 March 2022
Accepted: 8 June 2022
Published: 9 June 2022

1. Introduction

The issue of which viral genes are expressed during Varicella-Zoster virus (VZV) latency in human ganglia has been addressed by several groups over the last few decades. VZV is a ubiquitous pathogenic human herpes virus that causes varicella ("chicken pox") as a primary infection. Then, after a highly variable latent period in neurons in human trigeminal ganglia (TG), dorsal root ganglia (DRG) and peripheral autonomic ganglia, the virus may reactivate, either spontaneously or following various triggering factors to produce herpes zoster ("shingles"), which is a very painful vesicular rash occurring in a dermatomal distribution [1,2]. Herpes zoster may sometimes be followed by post-herpetic neuralgia (PHN), which causes severe pain in the affected dermatome that may prove highly resistant to all therapy [2]. It has been established that during the process of VZV latency the virus is located predominantly in neurons in the affected ganglia [3].

Professor Randall Cohrs of Colorado University made significant contributions to the investigation and our understanding of VZV transcription during human ganglionic latency. This question is important as knowledge of the extent of VZV gene transcription is essential for a better understanding of the mechanism of VZV latency—especially as the biological functions of some of the VZV genes are known [4]. In an early study, Cohrs and his colleagues [5] constructed a cDNA phage library from poly(A)+ RNA obtained from latently infected human ganglia obtained at autopsy within 24 h of death and detected VZV-specific inserts by PCR. Using this then-innovative technique they detected VZV gene 21-specific sequences in the ganglionic samples. This was followed up two years later [6] by the detection of VZV genes 21, 29, 62 and 63 from a human ganglia cDNA library, consistent with the previously reported detection of VZV genes 29 and 62 by another group using Northern blot analyses [7]. This was the first report of VZV gene 63 transcription during latency, and expression of this viral transcript has become a key hallmark of VZV latency. These four VZV latency-associated transcripts were also found to be present in post-mortem human TG in an extensive subsequent study in [8] using PCR in situ hybridisation technology, when it was also found that VZV gene 4 was also detectable in a few subjects. Three years later, in a collaboration between the laboratories of Dr Cohrs and the author, it was reported that both transcription and translation of VZV gene 66 could be demonstrated in latently infected human TG [9]. In view of this novel finding, it was suggested that the prevention of immediate-early (IE) 62 import to the nucleus by VZV 66-pk phosphorylation is one possible mechanism by which VZV latency is maintained.

One of the key problems with these types of study is that it is possible that viral reactivation may occur as a result of the death of the individual itself. This is an inevitable concern with any ganglionic specimen obtained at post-mortem around 24 h after death, which is the case with most of these studies. For this reason, it has long been suspected by some experts in the field that the extent of viral transcription detected in such studies may reflect possible viral reactivation as well as true viral latency. A clear indication of this possibility was provided by a subsequent study from Dr Cohr's laboratory. This study used a multiplex reverse transcription (RT)-PCR assay that allowed the rapid and sensitive detection of transcripts corresponding to all 68 unique varicella-zoster virus (VZV) open reading frames (ORFs) in five amplification reactions [10]. It was found that, while the detection of VZV gene 63 transcripts was the most frequently detected, a total of 10 viral transcripts were detected—including ORFs 4, 11, 29, 40, 41, 43, 57, 62, 63 and 68. This represented the expression of IE, early and late VZV transcripts, which was highly suggestive of some degree of VZV reactivation as well as genuine latency. The recent report of Depledge and colleagues [11] obtained post-mortem ganglia within about 6 h after the death of the individuals—a much earlier time than before, thereby avoiding the potential problem of possible viral reactivation having taken place. They detected only two viral transcripts using a highly sensitive enriched RNA-Seq method. As well as detecting VZV ORF 63, which was expected, these workers also detected a spliced latency-associated VZV transcript (VLT) that mapped antisense to the viral transactivator gene 61 [11]. These findings are considered further below. It is significant that Dr Cohrs was an author on this manuscript, reflecting his underlying desire to obtain the most accurate picture possible of a biological process.

In order to obtain further evidence of VZV reactivation, as well as latency, in post-mortem TG tissues obtained at 24 h after death, we carried out a comprehensive and systematic study of all 68 transcripts using a nested PCR stratagem.

2. Materials and Methods

2.1. Tissues and RNA Extraction

Four human ganglionic tissues (trigeminal ganglia—TG) were kindly donated and arbitrarily designated S24, S27, S35 and S36 by the Medical Research Council HIV Brain and Tissue Bank in Edinburgh, Scotland, with appropriate ethical clearance. All tissues were obtained at autopsy at around 24 h after death (and not earlier than this). To our knowledge, all four subjects—who were totally anonymised—were immunocompetent and had no recent history of herpes zoster. As depicted in Figure 1, the total cellular RNA was extracted from the four human trigeminal ganglia samples using the commercial reagent RNABee (ams Biotechnology), purified through a silica-membrane RNeasy spin column (Qiagen) and DNase-treated (Ambion). As a quality control check for the presence of genomic contamination in the RNA preps, standard end-point PCRs were performed on the four ganglionic cDNAs using β-actin intron spanning primers which yield different-sized genomic and cDNA PCR products (Figure 1).

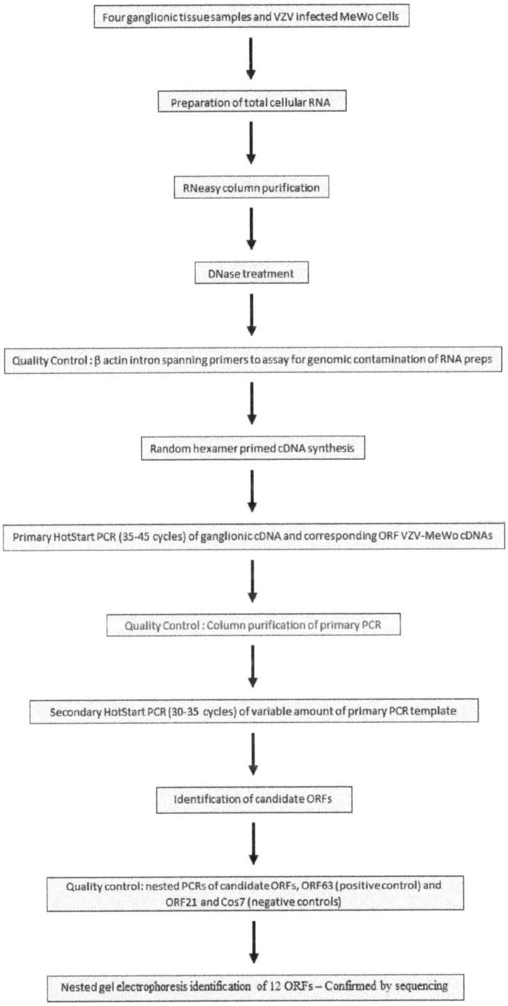

Figure 1. Flow chart depicting key stages in the nested PCR analysis to identify VZV genes expressed in ganglionic tissue.

2.2. Nested RT-PCR Assays

Nested PCR is an established and powerful technique for detecting very low abundance transcripts. A comprehensive and systematic Hot start-based nested PCR screen of the 68 ORFs catalogued at the inception of the study was undertaken. A Hot start approach minimizes non-specific amplification and primer dimer formation by suppressing enzymatic Taq activity until the first denaturation step has been reached. Due to the combined high cycle number employed, (65–80), the nested PCR process is also highly susceptible to template contamination sourced from a VZV laboratory environment. To minimize this potential contamination, nested PCRs were set up in a Class II tissue culture hood using dedicated molecular biology enzymes/chemicals and UV-irradiated plasticware and pipettes. Figure 1 highlights the key stages in a flow chart format of the nested PCR stratagem adopted to screen for VZV ORF genes expressed in ganglionic tissue. Random hexamer primed cDNAs were synthesized using the SuperScript III reverse transcription kit

(ThermoFisher Scientific, Waltham, MA, USA). Hot start primary PCRs (JumpStart RedTaq Reaction Mix Sigma) with the outer primers were routinely performed on 5 ng ganglionic cDNAs. cDNA prepared from VZV-infected MeWo cells provided a positive control for the amplification of correctly sized primary PCR products for each ORF (Figure 1). Following electrophoretic analysis of the primary PCR runs, only those ORF samples with a clean background indicative of the absence of non-specific priming were selected to assay for secondary amplification. Using the inner primer pairing, a Hot start (35–45) cycle was performed on an aliquot of the primary PCR product (JumpStart RedTaq Reaction Mix Sigma) and purified using a MSB SpinPCRapace column (Invitek). The synthesis of a correctly sized single secondary PCR product is determined in part by the combination of the primary PCR template load and cycle number. Accordingly, secondary Hot start PCRs comprised of a reduced cycle number (30–35) were tested on a variable amount of purified primary PCR product ranging from 2.5 to 25% (Figure 1). The predicted size differences between the primary and the smaller-sized secondary PCR products were confirmed by 2–3% agarose gel electrophoresis. For these workflow reasons, nested PCRs were initially performed on small batches of up to four ORFs. Positive-scoring secondary PCRs were repeated ($n \geq 5$). As a final quality control check for each latency candidate, a contemporaneous nested PCR was set up comprised of the candidate ORF, the latency hallmark gene ORF6 as a positive control, ORF21 encoding a nucleocapsid protein and Cos7 cDNA as a representative of a non-human non-neuronal cell type were included as negative controls for VZV gene expression (Figure 1).

3. Results

The outcome of this systemic nested PCR of 68 VZV ORFs screen revealed that 12 transcripts were detected in ganglionic tissue as evidenced by gel electrophoresis identification of a correctly sized single secondary PCR product and confirmed by sequencing (Eurofins UK). These are listed in Table 1, correlating ORF activity in each of the four ganglionic samples. Proof of concept of the nested PCR paradigm is illustrated in Figure 2 for ORF63, depicting the 386 bp primary product generated from VZV-infected MeWo cell cDNA and the smaller nested 326 bp PCR fragment. Nested ORF21, Cos7 and water controls were negative.

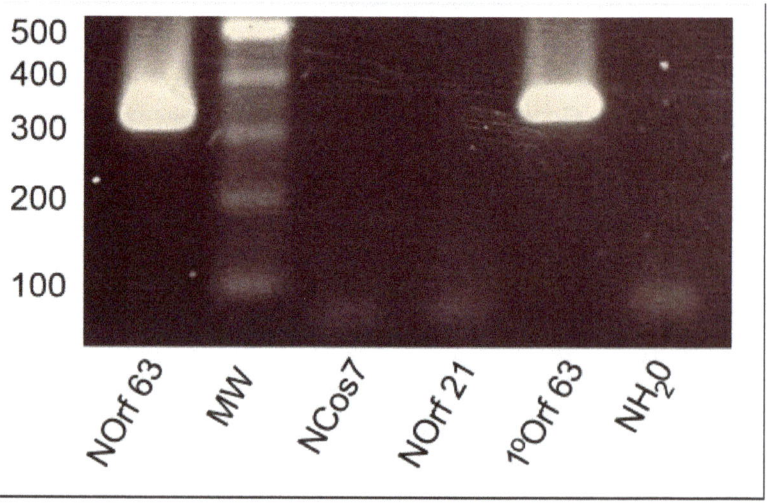

Figure 2. Gel analysis of nested PCR. Nested PCR of ORF 63. Primary PCR amplifies a 386 bp product which yields a 326 bp fragment following a secondary PCR nested run. Nested ORF21, nested Cos7 and nested water controls were negative.

Table 1. Activity profile of VZV transcripts detected in the current study.

		Donor Designated Ganglionic Samples			
Classic ORFs	**Proposed Function**	**S24**	**S27**	**S35**	**S36**
ORF4	IE Transcriptional activator	−	+	+	+
ORF21	E Replication in culture	−	−	−	−
ORF29	E Transcriptional modulator	+	−	+	+
ORF40	L Encapsidation	−	−	−	−
ORF62	IE Transcriptional activator	+	−	+	−
ORF63	IE Transcriptional activator	+	+	+	+
ORF66	E Protein kinase—Unknown	−	−	−	−
Nagel et al. [10]					
ORF11	Tegument protein	−	−	+	+
ORF41	Essential for growth in culture—Unknown	−	+	+	+
ORF43	Essential for growth in culture—Unknown	−	−	−	−
ORF57	Unknown	−	+	−	+
ORF68	Glycoprotein	−	−	−	−
Novel candidates					
ORF32	Substrate for ORF47	+	−	+	+
ORF37	L Glycoprotein	−	−	+	+
ORF42	Unknown	+	−	+	−
ORF58	Dispensable for replication—Unknown	−	−	+	−
ORF60	L Protein kinase—viral replication	−	−	+	+

This panel of 12 active genes was comprised of ORFs 4, 29, 62 and 63—all members of the initial grouping of "classic" latency genes [6,8]—and ORFs 11, 41, and 57, described in the Nagel study [10]. In addition, we report on the activity of a novel cohort comprised of ORFs 32, 37, 42, 58 and 60. However, the activity profile (Table 1) described here raises two main further issues. The first concerns the variation in ORF activity between the four ganglion samples, in that samples S24 and S27 had much less "genetic latency activity" than S35 and S36, while only ORF63—the hallmark latency gene—was active in all samples. The second issue concerns the failure to replicate key published findings. Most notably, we could not detect the classic ORFs 21, 40 and 66 in addition to the ORFs 43 and 68 identified by Nagel and colleagues [10].

4. Discussion

Since the pioneering work of Randall Cohrs in 1996 on latent VZV gene expression using cDNA libraries, our knowledge of this process has progressed considerably until the present time. It is noteworthy that Dr Cohrs was a co-author on several of these new analyses over the last 25 years—a testament to his commitment to defining the most accurate analyses of this process and not merely adhering strictly to fixed scientific dogma. It has become clear that latent VZV gene expression is even more restricted than had been previously thought.

In this overview of the subject, we also report some of the findings in our own laboratory—results which clearly demonstrate the widespread VZV gene expression that is detected when human ganglion samples are obtained many hours after death. An interpretation of such findings could be that both true viral latency and also viral reactivation as a consequence of the death of the individual had occurred—a view which is consistent with the study of Nagel et al. [10], where 10 VZV transcripts were detected using a multiplex RT-PCR method, though in neither this nor our own studies were primers targeting the VLT used. The detection of VZV transcripts corresponding to immediate early, early and late VZV genes is possibly suggestive of a degree of viral reactivation. However, it is not all certain that the virus is truly reactivating at these late stages, i.e., regaining the ability to replicate its genome and/or spread to new cells, since there is no definite evidence for this. An obvious difficulty is in relating the predicted function of the novel transcripts that we detected with the latent VZV condition. ORFs 37 and 60 are late genes encoding Glycoproteins H and L, respectively [12], while ORF 42 may be involved in the encapsidation process [13]. The candidacy of the other two transcripts is less of an issue in this regard, as ORF 32 encodes a small polypeptide with unknown function [14], and ORF 58 encodes a nuclear phosphoprotein [15]. At this juncture, it is uncertain whether these ORFs remain "latency candidates" that justify further experimentation or if they can only be classified as transcripts detected in latent tissue and are more likely to be representative of low-level VZV reactivation. Overall, the evidence suggests that the latter possibility is true.

Recent evidence has been based on analyses of human ganglia obtained as early as 6 h after death, which should preclude the possibility of viral reactivation, combined with the application of advanced gene sequencing methodology. Thus, Depledge et al. [11] reported their detection of only two VZV transcripts when they used a very sensitive and enriched RNA-Seq technique on autopsied TG obtained much earlier after death (~6 h) than had been the case in previous studies. As expected, they detected ORF 63 in these ganglia, and detection of this transcript has long been considered to be a hallmark of VZV latency [16]. However, they also detected a novel spliced latency-associated VZV transcript (VLT) that was found to be mapped antisense to the viral transactivator gene 61 [11]. No other VZV transcripts were detected in these tissues. This group also found that VLT is expressed in human TG neurons and encodes a protein which could be detected in both productively infected cells and herpes zoster skin lesions [11]. Moreover, their subsequent studies revealed that viral gene expression could be induced by a VLT-ORF 63 fusion transcript when VZV reactivated from latency [17]. The VLT-ORF63 transcripts appear to be key regulators of the transition of the virus from latency to reactivation. This important study provides the first meaningful investigation of why VLT expression is consistently detected and ORF63 expression intermittently detected in latently infected human ganglia harvested with short post-mortem intervals. The authors concluded that the prior detection by several groups of ORF63 in post-mortem-acquired TG reflected the detection of the VLT-ORF63 fusion rather than canonical ORF63 [17]. It is relevant that VZV small noncoding RNAs antisense to the VLT enhance replication, possibly by regulating its expression [18].

It makes sense at present to focus on the functionality of the VLT, which is clearly a very important novel latency transcript. The fact that the VLT and the VLT-ORF 63 fusion transcripts were the only ones detected so soon after death adds greatly to their significance. Regarding the data presented here, we think it likely that if TG are analysed around 24 h after death, then it is inevitable that a degree of viral reactivation will have already occurred, and this presumably explains why so many VZV transcripts representing all three classes of VZV genes are so frequently detected at these later post-mortem times.

Author Contributions: Both authors P.G.E.K. and P.M. contributed equally to the manuscript. All authors have read and agreed to the published version of the manuscript.

Funding: This research received no external funding.

Institutional Review Board Statement: Not applicable.

Informed Consent Statement: Not applicable.

Data Availability Statement: Not applicable.

Conflicts of Interest: The authors declare no conflict of interest.

References

1. Gershon, A.A.; Breuer, J.; Cohen, J.I.; Cohrs, R.J.; Gershon, M.D.; Gilden, D.; Grose, C.; Hambleton, S.; Kennedy, P.G.E.; Oxman, M.N.; et al. Varicella zoster virus infection. *Nat. Rev. Dis. Prim.* **2015**, *1*, 15016. [CrossRef] [PubMed]
2. Kennedy, P.G.E.; Gershon, A.A. Clinical features of Varicella-Zoster virus infection of the nervous system. *Viruses* **2018**, *10*, 609. [CrossRef] [PubMed]

3. Kennedy, P.G.E.; Grinfield, E.; Gow, J.W. Latent varicella-zoster virus is located predominantly in neurons in human trigeminal ganglia. *Proc. Natl. Acad. Sci. USA* **1998**, *95*, 4658–4662. [CrossRef] [PubMed]
4. Kennedy, P.G.E.; Mogensen, T.H.; Cohrs, R.J. Recent issues in Varicella-Zoster Virus latency. *Virology* **2021**, *13*, 2018. [CrossRef]
5. Cohrs, R.J.; Srock, K.; Barbour, M.B.; Owens, G.; Mahalingam, R.; Devlin, M.E.; Wellish, M.; Gilden, D.H. Varicella-zoster virus (VZV) transcription during latency in human ganglia: Construction of a cDNA library from latently infected human trigeminal ganglia and detection of a VZV transcript. *J. Virol.* **1994**, *68*, 7900–7908. [CrossRef] [PubMed]
6. Cohrs, R.J.; Barbour, M.; Gilden, D.H. Varicella-zoster virus (VZV) transcription during latency in human ganglia: Detection of transcripts mapping to genes 21, 29, 62, and 63 in a cDNA library enriched for VZV RNA. *J. Virol.* **1996**, *70*, 2789–2796. [CrossRef]
7. Meier, J.L.; Holman, R.P.; Croen, K.D.; Smialek, J.E.; Straus, S.E. Varicella-zoster virus transcription in human trigeminal ganglia. *Virology* **1993**, *193*, 193–200. [CrossRef] [PubMed]
8. Kennedy, P.G.E.; Grinfeld, E.; Bell, J.E. Varicella-zoster virus gene expression in latently infected andexplanted human ganglia. *J. Virol.* **2000**, *74*, 11893–11898. [CrossRef] [PubMed]
9. Cohrs, R.J.; Gilden, D.H.; Kinchington, P.R.; Grinfeld, E.; Kennedy, P.G.E. Varicella-Zoster Virus Gene 66Transcription and Translation in Latently Infected Human Ganglia. *J. Virol.* **2003**, *77*, 6660–6665. [CrossRef] [PubMed]
10. Nagel, M.A.; Choe, A.; Traktinskiy, I.; Cordery-Cotter, R.; Gilden, D.; Cohrs, R.J. Varicella-zoster virus transcriptome in latently infected human ganglia. *J. Virol.* **2011**, *85*, 2276–2287. [CrossRef] [PubMed]
11. Depledge, D.P.; Ouwendijk, W.; Sadaoka, T.; Braspenning, S.E.; Mori, Y.; Cohrs, R.J.; Verjans, G.M.G.M.; Breuer, J. A spliced latency associated VZV transcript maps antisense to the viral transactivator gene 61. *Nat. Commun.* **2018**, *9*, 1167. [CrossRef] [PubMed]
12. Storlie, J.; Maresova, L.; Jackson, W.; Grose, C. Comparative analyses of the 9 glycoprotein genes found in wild-type and vaccine strains of varicella-zoster virus. *J. Infect. Dis.* **2008**, *197* (Suppl. S2), S49–S53. [CrossRef] [PubMed]
13. Visalli, R.J.; Nicolosi, D.M.; Irven, K.L.; Goshorn, B.; Khan, T.; Visalli, M.A. The Varicella-zoster virus DNA encapsidation genes: Identification and characterization of the putative terminase subunits. *Virus Res.* **2007**, *129*, 200–211. [CrossRef] [PubMed]
14. Reddy, S.M.; Cox, E.; Iofin, I.; Soong, W.; Cohen, J.I. Varicella-zoster virus (VZV) ORF32 encodes a phosphoprotein that is posttranslationally modified by the VZV ORF47 protein kinase. *J. Virol.* **1998**, *72*, 8083–8088. [CrossRef] [PubMed]
15. Yoshii, H.; Sadaoka, K.; Matsuura, M.; Nagaike, K.; Takahashi, M.; Yamanishi, K.; Mori, Y. Varicella-zoster virus ORF 58 gene is dispensable for viral replication in cell culture. *Virol. J.* **2008**, *5*, 54. [CrossRef] [PubMed]
16. Kennedy, P.G.E.; Rovnak, J.; Badani, H.; Cohrs, R.J. A comparison of herpes simplex virus type 1 and varicella-zoster virus latency and reactivation. *J. Gen. Virol.* **2015**, *96 Pt 7*, 1581–1602. [CrossRef] [PubMed]
17. Ouwendijk, W.J.D.; Depledge, D.P.; Rajbhandari, L.; Lenac Roviš, T.; Jonjic, S.; Breuer, J.; Venkatesan, A.; Verjans, G.M.G.M.; Sadaoka, T. Varicella-zoster virus VLT-ORF63 fusion transcript induces broad viral gene expression during reactivation from neuronal latency. *Nat. Commun.* **2020**, *11*, 6324. [CrossRef] [PubMed]
18. Bisht, P.; Das, B.; Kinchington, P.R.; Goldstein, R.S. Varicella-Zoster Virus (VZV) Small Noncoding RNAs Antisense to the VZV Latency-Encoded Transcript VLT Enhance Viral Replication. *J. Virol.* **2020**, *94*, e00123-20. [CrossRef] [PubMed]

viruses

MDPI

Article

Simian Varicella Virus Pathogenesis in Skin during Varicella and Zoster

Ravi Mahalingam [1,*], Brittany Feia [1], Colin Coleman [2], Kusala Anupindi [1], Pratush Saravanan [1], Amalia Luthens [1], Amalia Bustillos [1], Arpita Das [3], Eileen de Haro [3], Lara Doyle-Meyers [3], Jayme Looper [4], Andrew N. Bubak [1], Christy S. Niemeyer [1], Brent Palmer [5], Maria A. Nagel [1,6] and Vicki Traina-Dorge [3]

1 Department of Neurology, University of Colorado School of Medicine, Anschutz Medical Campus, Aurora, CO 80045, USA; brittany.feia@cuanschutz.edu (B.F.); kusala.anupindi@ucdenver.edu (K.A.); psaravanan@berkeley.edu (P.S.); amalia.luthens@cuanschutz.edu (A.L.); amalia.bustillos@cuanschutz.edu (A.B.); andrew.bubak@cuanschutz.edu (A.N.B.); christy.niemeyer@cuanschutz.edu (C.S.N.); maria.nagel@cuanschutz.edu (M.A.N.)

2 Cell and Developmental Biology Program, School of Medicine, Oregon Health Sciences University, Portland, OR 97201, USA; colemaco@ohsu.edu

3 Division of Microbiology, Tulane National Primate Research Center, Tulane University, Covington, LA 70433, USA; adas@trudeauinstitute.org (A.D.); edeharo@tulane.edu (E.d.H.); ldoyle@tulane.edu (L.D.-M.); vtraina@tulane.edu (V.T.-D.)

4 Department of Veterinary Clinical Sciences, Louisiana State University, Baton Rouge, LA 70803, USA; jlooper@lsu.edu

5 Division of Allergy and Clinical Immunology, Department of Medicine, University of Colorado Anschutz Medical Campus, Aurora, CO 80045, USA; brent.palmer@cuanschutz.edu

6 Department of Ophthalmology, University of Colorado School of Medicine, Anschutz Medical Campus, Aurora, CO 80045, USA

* Correspondence: ravi.mahalingam@cuanschutz.edu; Tel.: +1-303-724-4324

Citation: Mahalingam, R.; Feia, B.; Coleman, C.; Anupindi, K.; Saravanan, P.; Luthens, A.; Bustillos, A.; Das, A.; de Haro, E.; Doyle-Meyers, L.; et al. Simian Varicella Virus Pathogenesis in Skin during Varicella and Zoster. *Viruses* **2022**, *14*, 1167. https://doi.org/10.3390/v14061167

Academic Editor: Jônatas Abrahão

Received: 30 March 2022
Accepted: 24 May 2022
Published: 27 May 2022

Abstract: Primary simian varicella virus (SVV) infection and reactivation in nonhuman primates is a valuable animal model in the study of varicella zoster virus disease [varicella (chickenpox) and herpes zoster (shingles)]. To understand SVV pathogenesis in skin, we inoculated 10 rhesus macaques with SVV, resulting in varicella rash. After the establishment of latency, eight of the monkeys were immunosuppressed using tacrolimus with or without irradiation and prednisone and two monkeys were not immunosuppressed. Zoster rash developed in all immunosuppressed monkeys and in one non-immunosuppressed monkey. Five monkeys had recurrent zoster. During varicella and zoster, SVV DNA in skin scrapings ranged from 50 to 10^7 copies/100 ng of total DNA and 2–127 copies/100 ng of total DNA, respectively. Detection of SVV DNA in blood during varicella was more frequent and abundant compared to that of zoster. During varicella and zoster, SVV antigens colocalized with neurons expressing β-III tubulin in epidermis, hair follicles, and sweat glands, suggesting axonal transport of the virus. Together, we have demonstrated that both SVV DNA and antigens can be detected in skin lesions during varicella and zoster, providing the basis for further studies on SVV skin pathogenesis, including immune responses and mechanisms of peripheral spread.

Keywords: simian varicella virus; varicella; zoster; skin pathogenesis

1. Introduction

Varicella zoster virus (VZV) typically causes childhood varicella (chickenpox), establishes latent infection in multiple ganglia along the neuraxis, and reactivates decades later to produce herpes zoster (shingles), predominantly in the elderly. Varicella presents as a disseminated vesicular/macular/papular skin rash; in contrast, zoster appears as a unilateral dermatomal-distribution vesicular rash corresponding to the ganglia associated with the affected dermatomes. Due to VZV's restricted host range, dissecting mechanisms through which VZV spreads to multiple organs, including skin, during primary infection and reactivation has been challenging. Using a SCID-hu mouse model of VZV pathogenesis, T lymphocytes have been shown to transport virus to skin from lymph nodes during primary infection [1]. However, it is unclear how VZV is transported from ganglia to multiple organs, including skin, during zoster. Earlier comparative studies of peripheral blood mononuclear cells (PBMCs) during varicella and zoster revealed that, viremia occurs at a much lower magnitude during zoster, despite abundant VZV DNA and virions detected in zoster vesicles [2–5]. Zoster skin lesions have been clinically categorized into four stages (erythematous, vesicular, pustular, and ulcerative), VZV spreads to the epidermis via sensory nerves during the erythematous stage and then to the dermis in the vesicular stage [6]. Further, virus antigens have been found mostly around the isthmus and less often near the bulb of the hair follicle [7]. Lymphocytes surrounding blood vessels have also been shown to contain VZV immediate early protein 63 (IE63), but not other structural proteins [8]. The selective detection of VZV antigens has also been reported following immunohistochemical analysis of biopsied skin samples from varicella and zoster [9]. Most studies related to VZV infection in skin have been performed on human skin biopsies because VZV is an exclusively human pathogen. Simian varicella virus (SVV) infection in non-human primates (NHP) has served as a useful model since its pathological and immunological features are similar if not identical to VZV infection in humans [10]. Upon experimental inoculation in rhesus macaques, SVV produces varicella, establishes latent infection, and can be reactivated via immunosuppression [11,12]. We previously showed that SVV antigens can be detected in sweat glands in zoster skin [13]. In this report, we extend our studies by analyzing skin scrapings and fixed skin samples (for SVV DNA and antigens, repectively), from the same rhesus macaques during both varicella and zoster.

2. Materials and Methods

2.1. Ethis Statement

Rhesus macaques that were used in this study were all housed at the Tulane National Primate Center (TNPRC) in Covington, LA. All animal housing, care, and research were performed in compliance with The Guide for the Care and Use of Laboratory Animals (National Research Council), the Animal Welfare Act guidelines, and the guidelines at the TNPRC, accredited by the Association of the Assessment and Accreditation of Laboratory Animal Care. Protocols were approved by the Institutional Animal Care and Use Committee of Tulane University (The Assurance Number A4499-01; Protocol Title: Varicella Virus Latency and Reactivation in the Nonhuman Primate, IACUC Protocol #PO177R3, 25 Oct 2021). All procedures were conducted by highly qualified veterinarians.

2.2. Monkeys

Ten male SVV-seronegative rhesus macaques were housed in the Tulane National Primate Center (TNPRC) in Covington, LA, and used for all experiments. The age, sex, and weight of each monkey are listed in Table 1. During acute infection and reactivation, monkeys from this study were housed in the same room, either in single or dual cages. During the immunsuppression and reactivation phase, treated monkeys, shown to be behaviorally compatable, were pair-housed with other treated monkeys. The two untreated monkeys, LT27 and LE91, were also pair-housed. A paraformaldehyde-fixed paraffin-embedded skin sample from a rhesus macaque that developed a fulminant SVV infection 53 days after irradiation (kind gift from From Dr. Steven Shipley) and an uninfected SVV-sero negative monkey (R110368) purchased from Primate biologicals (Bethesda, MD, USA) were used as positive and negative controls.

Table 1. List of monkeys used in this study.

Group	Monkey ID	Age [a]	Sex	Weight (Kg)	Primary Infection (dpi [b])	Reactivation	
						Treatment	Dpx [c]/mpi [d]
1	KI92	4.0	M	9.9	7	A [e]	84
	KG58	4.2	M	9.5	9	A	63
2	LB56	6.7	M	8.2	9	B [f]	45
	LT26	4.3	M	4.3	9	B	66
	LK67	4.7	M	8.1	9	B	42
	LR16	4.4	M	8.3	9	B	10
	LR42	4.4	M	8.0	9	B	56
	LR70	4.4	M	9.2	9	B	45
3	LT27	4.3	M	8.3	9	C [g]	3.8 [d]
	LE91	6.6	M	10.5	7	C	-
control	B321	3.0	M	5.9	D [h]	E [i]	56
control	R110368	5.0	M	U [j]	NA [k]	NA	NA

[a] Years. [b] Days post-inoculation. [c] Days post-immunosuppression. [d] months post inoculation, [e] A-Irradiation, tacrolimus, prednisone. [f] B-Tacrolimus. [g] C-Not immunosuppressed, [h] D-time of natural primary infection unknown, [i] E-Irradiation, [j] Unknown, [k] Not applicable.

2.3. *Acute SVV Infection and Establishment of Latency*

A deltaherpesvirus strain of wild type-SVV isolated from a naturally infected monkey (*Erythrocebus patas*) was propagated in rhesus fibroblasts (Frhl-2) (ATCC, Manassas, VA, USA) and a virus stock was prepared as described previously [14]. SVV-seronegative monkeys were inoculated intratracheally with 2.4–5 $\times$ 10^5 plaque forming units (pfu) of wild type-SVV-infected Frhl-2 cells. All monkeys were monitored following anesthesia by physical exam and blood collections every 3 to 7 days with skin scrapings and punch biopsy (4 mm) samples from skin collected upon the appearance of typical varicella rash lesions. All ten monkeys developed a typical varicella rash within 7–9 days post-inoculation (dpi) and monitoring continued until the establishment of latency confirmed by the absence of SVV DNA in PBMCs for two consecutive weeks.

2.4. *Immunosuppressive Regimens*

Eight months after primary infection, establishment of SVV latency in KI92 and KG58 (group 1, Figure 1) was confirmed (absence of detectable SVV DNA in two consecutive bleeds) and immunosuppresion phase was initiated with transport of the two monkeys in a van (3-h round trip) from the TNPRC in Covington, LA, USA to the School of Veterinary Medicine, Radiation Oncology facility at Louisiana State University in Baton Rouge, LA. They were anesthetized and exposed to a single dose of 200-cGy total body X-irradiation, then treated daily with oral tacrolimus (Prograf; 500 µg; 80 µg/kg of body weight/day) and prednisone (2 mg/kg/day) until virus reactivation was confirmed, and they were euthanized 2–3 months post-zoster (Figure 1). Two months after primary infection, establishment of latency was confirmed in group 2 animals: LB56, LT26, LK67, LR16, LR42, and LR70 (Figure 1). Immunosuppresion consisted of daily treatment with oral tacrolimus (Prograf; Sandoz Inc., Princeton, NJ, USA) 80–240 µg/kg of body weight/day) until SVV reactivation was confirmed, and they were euthanized 2–3 months post-zoster. Group 3 monkeys, LT27 and LE91, were infected with SVV but not immunosuppressed or transported (Figure 1). These monkeys were also monitored for viral reactivation. All animals were monitored by physical exams every 7 days, and blood samples were collected weekly until reactivation and collected again during euthanasia.

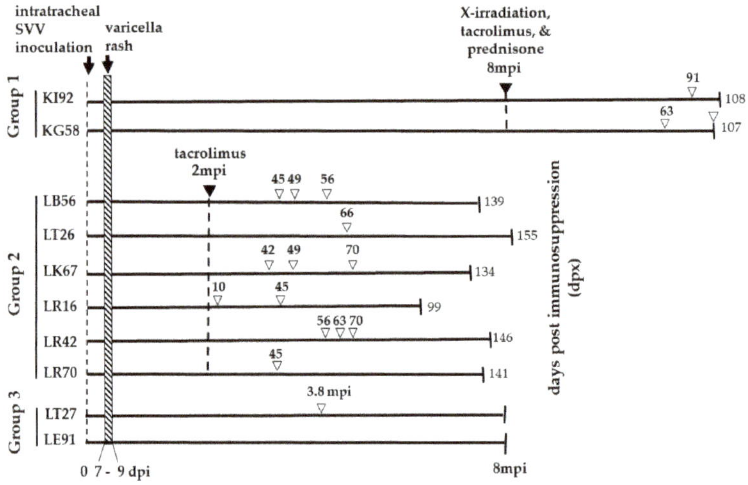

Figure 1. Experimental Design. The monkeys were divided into three groups. Two rhesus macaques (Group 1—KI92 and KG58) were intratracheally inoculated with 5×10^{5} pfu of SVV and developed varicella rash 7–9 days post-inoculation (dpi). Eight months' post-inoculation (mpi), both monkeys were exposed to X-irradiation (200 cGy) and daily oral treatment with tacrolimus (80 µg/kg/day) and prednisone (1 mg/kg/day) for the duration of the experiment. KI92 and KG58 developed zoster rash, 91- and 63-days post-immunosuppression treatment (dpx), respectively. KG58 had recurrent zoster at 107 dpx. KI92 and KG58 were euthanized 108 and 107 dpx, respectively. Six rhesus macaques (group 2—LB56, LT26, LK67, LR16, LR42, and LR70) were intratracheally inoculated with 2.4×10^{5} pfu SVV, and all developed varicella rash between 7–9 dpi. Two months later, all six monkeys were treated with tacrolimus (80 µg/kg/day) daily for the duration of the experiment. All six immunosuppressed monkeys developed a zoster rash at indicated dpx (numbers above, unfilled arrowheads). Monkeys LB56, LK67, LR16, and LR42 had recurrent zoster, as indicated by multiple unfilled arrowheads. All six monkeys were euthanized at the indicated dpx. Two rhesus macaques (group 3—LT27 and LE91) were intratracheally inoculated with 2.4×10^{5} pfu of SVV, and both developed varicella rash by nine dpi. Neither were immunosuppressed. One of the monkeys (LT27) developed zoster rash 3.8 mpi. Group 3 monkeys were euthanized 8 mpi. DNA extracted from skin scrapings obtained at the time of zoster from monkeys in all three groups were analyzed for the presence of SVV DNA by qPCR. Euthanasia is indicated by vertical lines on the right end of the lines. The numbers at the end of the lines indicate the time (dpx (groups 1–2) or mpi (group 3)) of euthanasia.

2.5. Determination of Anti-SVV Antibody Levels

Anti-SVV antibody titers in serum obtained from all monkeys prior to SVV inoculation and during the time of monitoring after inoculation until euthanasia were determined using a plaque reduction neutralization test (PRNT) as described previously [12].

2.6. Collection and Analysis of Blood Samples for SVV DNA by qPCR

Blood samples collected from monkeys at multiple times following SVV inoculation and immunosuppression were processed using a Ficoll gradient to isolate PBMCs and a commercial DNA extraction kit per manufacturer's instructions (Qiagen, Germantown, MD, USA), followed by quantitative PCR (qPCR) for SVV DNA in PBMCs as described previously [15].

2.7. DNA Extractions from Skin Scrapings and qPCR Analysis

During physical examinations, anesthetized monkeys demonstrating typical varicella lesions on the skin were identified and photographed. Single lesions were scraped using a scalpel and resuspended in 500 µL of PBS and frozen at −80 °C until DNA extraction. After thawing, cell suspensions were centrifuged at 20,000× *g* for 10 min at room temperature and the pellets were resuspended in 180 µL of ATL buffer (DNA extraction kit, Qiagen) and 20 µL proteinase-k and processed. The samples were incubated overnight at 56 °C to ensure optimal and complete lysis. DNA was extracted using Qiagen DNeasy Blood and Tissue kit according to the Quick-Start Protocol for tissues. In the final step, DNA was eluted using 70 µL of AE buffer and quantitated before qPCR using a Nanodrop spectrophotometer (ThermoFisher, Waltham, MA, USA). The DNA samples were analyzed by qPCR using primers specific for SVV open reading frame (ORF) 61 as described previously [11]. Briefly, limited dilutions of cloned SVV bacmid (containing 5000, 1000, 100, 50, 10, 5 and 1 copies of virus DNA in a background of 100 ng of salmon sperm DNA) was used in real-time qPCR using primers specific for SVV ORF61 generate a standard curve [16]. The number of copies of SVV DNA in the unknown samples were determined by comparing the C_t values. The C_T values for real-time PCR ranged from 33.8 (min) to 37.6 (max). The standard curve was linear between 1–5000 copies of SVV bacmid which was used in real-time qPCR. We performed 3 different PCR assays on each DNA extraction. A sample is considered positive only if two out of three independent PCRs are positive for SVV DNA.

2.8. Harvesting and Processing of Skin Samples for Immunohistochemistry

Typical papular/vesicular lesion of varicella or zoster rash in each monkey was identified. A 4-mm punch biopsy was obtained from an area different from the one used for DNA extraction. The tissues were fixed in 10% Zinc-formalin (Z-fix) (Analtech, Battle Creek, MI, USA), processed, and paraffin-embedded.

2.9. Immunohistochemistry

Formalin-fixed, paraffin-embedded (FFPE) skin sections (5 µm) were deparaffinized in xylene and ethanol for 15 min each. The sections were then rhydrated using graded ethanol washes and washed once with water. They were then subjected to antigen retrieval in citrate buffer (10 mM Sodium Citrate pH 6.0 and 0.05% Tween20). The citrate buffer was pre-heated in a steamer for 10 min, and the slides were submerged in the hot buffer and incubated on the benchtop for 5 min. The sections were then imunostained using the ImmPRESS kit along with Vector NovaRED substrate kit (Vector Laboratories, Burlingame, CA, USA) per the manufacturer's instructions. The primary antibody was either rabbit polyclonal antibody raised against SVV IE63 protein (1:7000 dilution), Rabbit polyclonal antibody raised against SVV nucleocapsid (1:25,000 dilution), or Rabbit polyclonal antibody raised against SVV glycoproteins H and L (gH + L; 1:5000) (a generous gift from Dr. Wayne Gray). Normal rabbit serum (at the same dilution as the primary antibody) was used as a control. Following the first staining, some of the sections were washed with PBS for 5 min and processed using ImmPRESS Horse Anti-Rabbit IgG polymer kit, peroxidase along with Vector Blue substrate Kit, Alkaline phosphatase (Vector Laboratories) as per the manufacturer's instructions. The primary antibody was a Mouse anti-β-III tubulin antibody (STEM CELL Technologies, Kent, WA, USA) at 1:500 dilution. Mouse IgG2a (BD Biosciences, Franklin Lakes, NJ, USA) at a dilution of 1:500 was used an isotype control. Positive controls consisted of skin sections from an acutely infected immunosuppressed rhesus macaque (B321) immunostained for SVV and β-III tubulin, which were observed under a microscope during substrate color reactions. Some of the sections were counter-stained using hematoxylin (1:10 dilution of stock) for 2 min. The slides were then mounted using glass coverslips with ProLong Gold Antifade Mountant (Life Technologies, Eugene, OR, USA) and imaged using an Olympus BX46 light mircoscope and CellSens Software (Olympus, Center Valley, PA, USA). Each staining was repeated at least three times to ensure reproducibility.

3. Results

3.1. Primary SVV Infection in Rhesus Macaques, Establishment of Latency, and Immunosuppression

Three groups of SVV-seronegative Indian rhesus macaques were used in this study (Table 1 and Figure 1). They were inoculated intratracheally with wild-type SVV as described in the methods. All 10 monkeys developed varicella rash between 7–9 dpi and resolved thereafter. Monkeys in group 1 were inoculated with 5×10^5 pfu/monkey whereas monkeys in groups 2 and 3 were inoculated with 2.4×10^5 pfu/monkey. The difficulty in using consistent amounts of virus in the inoculum is due to challenges in growing SVV at high titers and batch-to-batch variation. Thus, it is possible that the differences in infection are associated with differences in the quantity of virus inoculated, but this is unlikely given that all infected monkeys developed rash.

Earlier we demonstrated that in rhesus macaques [11,17], lack of viremia is a consistent marker for the establishment of latency. Eight month's post infection (mpi), establishment of latency was confirmed in the two monkeys in group 1 (KI92 and KG58). They were then immunosuppressed with a single dose of X-irradiation and a combination of tacrolimus and prednisone administered daily for the duration of the experiment and were monitored for reactivation. Two months after primary infection, establishment of latency was confirmed in the six monkeys in group 2 (LB56, LT26, LK67, LR16, LR42, and LR70) and the two monkeys in group 3 (LT27 and LE91). Group 2 monkeys were treated with tacrolimus daily for the duration of the experiment. Two monkeys in group 3 (LT27 and LE91) were not treated with immunosuppressants (Figure 1). All ten monkeys were monitored for reactivation.

3.2. Skin Rash during Varicella and Zoster

The extent of varicella rash was mild in three monkeys (KI92, LR42 and LE91), moderate in two monkeys (LB56, LR16) and extensive in five monkeys (KG58, LT26, LK67, LR70, and LT27) (Table 2). At different times following immunosuppression, monkeys in all three groups developed zoster. A representative image of skin rash during varicella and zoster in LR42 (group 2) is presented in Figure 2. In all SVV-infected monkeys, the extent of rash during varicella was more pronounced with multiple lesions spread throughout the body compared to zoster, which was minimal and more localized. Unlike zoster in humans, which occurs as a dermatomal rash, both immunosuppressed and non-immunosuppressed monkeys developed zoster rash in multiple regions of the body.

Table 2. Severity of varicella (7–9 dpi) rash and detection of SVV DNA in skin scrapings.

Group	Monkey ID	Severity [a]	SVV ORF 61 DNA (Copy #/100 ng)
1	KI92	2+	NA [b]
	KG58	4+	NA
2	LB56	3+	50
	LT26	4+	5×10^5
	LK67	4+	1×10^6
	LR16	3+	310
	LR42	2+	6×10^4
	LR70	4+	2×10^3
3	LT27	4+	1×10^7
	LE91	2+	4×10^5

[a] The type and severity of the rash is graded as macular, papular, vesicular, and hemorrhagic. Scoring parameters are based on number of lesions with 2+ = 6–10; 3+ = 11–20; 4+ = >21. [b] Not available.

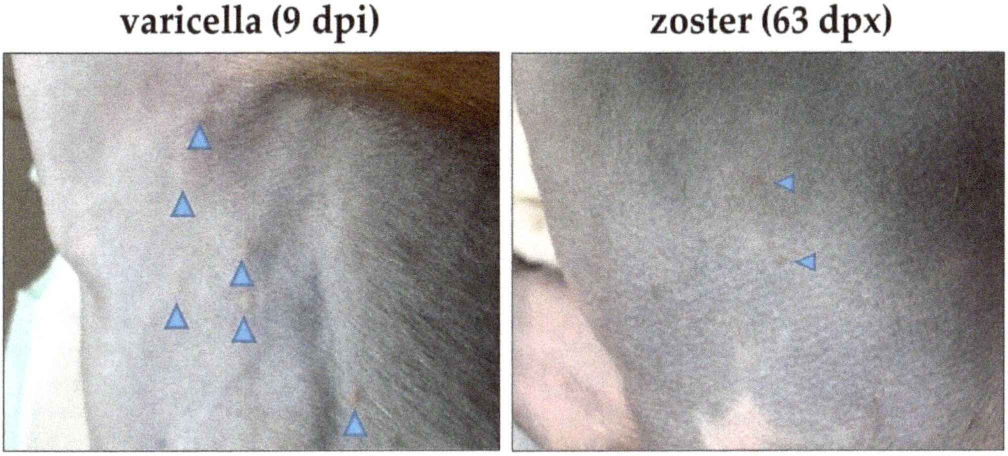

Figure 2. Skin rash after varicella and zoster in monkey LR42 (group 2). Varicella skin rash was seen in the torso nine days after SVV inoculation (dpi; **left image**). Zoster skin rash was seen on the torso 63 days after immunosuppression (dpx) with tacrolimus treatment (**right image**). The location of the vesicles is indicated by blue arrowheads.

In group 1, KI92 and KG58 developed zoster 91 and 63 dpx, respectively (Figure 1). In group 2, all monkeys developed zoster between 10–66 dpx. In group 3, LT27, one of two non-immunosuppressed monkeys also developed zoster 3.8 mpi, possibly due to unrelated reduction in SVV-specific cell mediated immunity, which has been documented [13]. The other control monkey (LE91) did not develop zoster. In several of our studies, we have observed that besides immunosuppression there is a possibility that transportation, single-cage housing, and repeated removal from cages for sample collection triggered SVV reactivation [12,13,15]. One of the two monkeys in group 1 (KG58) and four (LB56, LK67, LR16 and LR42) of the six monkeys in group 2 had recurrent zoster (rash appearing one week or more after initial zoster rash had cleared). Recurrent zoster in KG58 occurred at the time of necropsy (107 dpx).

3.3. Detection of SVV DNA in Skin Scrapings during Varicella

DNA extracted from skin scrapings obtained during varicella in groups 2 and 3 were analyzed using real-time qPCR for the presence of SVV DNA (Table 2). During varicella, SVV DNA copy numbers in skin scrapings ranged from 50 to 1×10^7 copies/100 ng. Although substantial amounts of SVV DNA were detected in most of the monkeys, the quantity of SVV DNA in skin, in general, did not always correlate with the extent of varicella rash.

3.4. Detection of SVV DNA in PBMCs during Varicella and Immunosuppression

DNA extracted from sequential blood samples obtained from monkeys in all 3 groups during primary infection and later during immunosuppression, were analyzed for viremia using real-time qPCR for SVV DNA. We performed three different PCR assays on each DNA sample. Any DNA sample is considered positive only if two out of three PCRs are positive for SVV DNA During varicella, viremia peaked at 7 dpi in all monkeys, except KI92 where 26 copies of SVV DNA/100 ng) were detected at 4 dpi (Table 3). In group 1, low level viremia was detected up to 64 dpi and was not detectable at later times after the establishment of latency (data not shown). In groups 2 and 3, viremia was absent in all by 42 dpi, confirming the establishment of latency. The highest quantity of SVV DNA among all 10 monkeys was detected in both skin scraping and PBMCs of LT27 (group 3) during varicella (Tables 2 and 3). It is possible that the detection of SVV DNA at 4 dpi could be due to input virus.

Table 3. Viremia during varicella.

Group	Monkey ID	SVV ORF 61 DNA Copy #/100 ng [a]								
		Days Post-Inoculation								
		4	7	9/10	11	14	21	29/30	42	64
1	KI92	26	4	6	20	0	0	0	NA [b]	1
	KG58	78	161	20	12	2	1	0	NA	3
2	LB56	0	95	0	NA	0	0	0	0	0
	LT26	21	103	9	NA	0	0	0	0	0
	LK67	0	103	31	NA	9	17	6	0	0
	LR16	7	36	8	NA	0	0	0	0	0
	LR42	7	282	40	NA	7	3	0	0	0
	LR70	21	128	16	NA	0	0	0	0	0
3	LT27	0	414	47	NA	0	0	0	0	0
	LE91	0	23	0	NA	0	0	0	0	0

[a] All samples done in triplicates and averaged. [b] Not available.

Unlike primary infection, detection of SVV DNA in PBMCs from groups 1 and 2 after treatment and immunosuppression was very sparse (Table 4). Low levels (2–10 copies/100 ng) of SVV DNA were detected three weeks after immunosuppression in five of the eight immunosuppressed monkeys (groups 1 and 2). In monkeys LK67 and LR42 (group 2), 3/12 and 3/13 sequential samples, respectively revealed low levels (3–10 copies/100 ng) of SVV DNA.

Table 4. Viremia during SVV reactivation.

Group	Monkey ID	SVV ORF 61 DNA Copy #/100 ng [a]												
		Days Post-Immunosuppression Treatment (dpx)												
		Pre	5/7	12/14	19/21	28	35	42	49	56	63	70	84	112
1	KI92	0	0	0	2	0	3	0	0	0	0	0	0	NA [b]
	KG58	0	0	0	0	0	0	0	0	0	0	0	0	NA
2	LB56	0	0	0	0	0	0	0	0	0	NA	0	0	0
	LT26	0	0	0	0	0	0	0	0	0	0	0	0	0
	LK67	0	0	0	8	3	0	0	6	0	NA	0	0	0
	LR16	0	0	0	3	0	0	0	0	0	NA	0	0	NA
	LR42	0	0	0	10	0	0	0	0	5	0	0	0	3
	LR70	0	0	0	0	0	0	0	0	0	NA	0	2	3
		Months post infection (mpi)												
		2	2.2	2.4	2.6	3	3.2	3.4	3.6	3.8	4.1	4.3	4.8	5.7
3	LT27 [c]	0	0	0	0	0	0	0	7	0	0	0	4	0
	LE91 [c]	0	0	0	6	0	0	0	0	0	0	0	0	6

[a] All samples done in triplicates and averaged. [b] Not available. [c] Not immunosuppressed.

3.5. *SVV-Specific Antibody Response during Varicella and Zoster*

Humoral response to SVV was measured using a plaque reduction (number of plaques) neutralization assay in all 3 groups of monkeys at multiple times following SVV inoculation and immunosuppression (Table 5). Anti-SVV antibody titers are expressed as the serum dilution that neutralized at least 80% of the SVV plaques compared to control cultures. In group 1, SVV-specific antibodies were detected 2 weeks post-inoculation and remained high during the establishment of latency until 70 dpi, when immunosuppression was initiated. At 14 dpx, both KI92 and KG58 showed decreases in their antibody levels that remained low even following zoster (KI92, 91 dpx and in KG58, 63 and 107 dpx; Figure 1). In group 2, all six monkeys had high antibody responses during acute infection (1:160 and 1:320) by 14 dpi that remained ≥1:80 at the start of immunosuppression. After immunosuppression, four of the monkeys (LB56, LT26, LK67, LR16) showed increased antibody levels around the time of zoster while two (LR42, LR70) remained stable (Table 5 and Figure 1). In group 3 after inoculation, LT27 had stable antibody titers (1:160–1:320) until they decreased to 1:80 at 3.8 mpi, when this monkey developed zoster and increased to 1:1280 at the time of euthanasia. In LE91, the SVV-specific antibody levels peaked at 1:80 at 14 dpi but declined, and without any zoster occurance, remained low through the course of the experiment.

Table 5. Neutralizing antibody titers during varicella and zoster.

Group	Monkey ID	Anti-SVV Antibody Titer [a]								
		PRE	Dpi [b]			Dpx [c]				Necropsy
			14	28	70	14	42	56	70	
1	KI92	<1:10	1:160	1:160	1:320	1:40	1:80	NA [d]	1:40	1:40
	KG58	<1:10	1:160	1:160	1:320	1:160	1:160	NA	1:40	1:80
2	LB56	<1:10	1:160	1:160	1:160	1:160	1:80	1:320	1:160	1:320
	LT26	<1:10	1:160	1:160	1:640	1:320	1:320	1:640	1:320	1:640
	LK67	<1:10	1:160	1:80	1:80	1:80	1:160	1:160	1:80	1:160
	LR16	<1:10	1:160	1:80	1:160	1:80	1:80	1:160	1:80	1:160
	LR42	<1:10	1:320	1:320	1:160	1:320	1:320	1:320	1:160	1:320
	LR70	<1:10	1:160	1:80	1:160	1:160	1:160	1:160	1:160	1:320
		Months post inoculation (mpi)								
		PRE	0.5	1.0	2.3	2.5	3.4	3.8	4.3	8.0
3	LT27 [e]	<1:10	1:320	1:320	1:160	1:320	1:160	1:80 [f]	1:80	1:1280
	LE91 [e]	<1:10	1:80	1:40	1:20	1:40	1:20	1:40	1:20	1:40

[a] Anti-SVV antibody titers are expressed as the serum dilution that neutralized at least 80% of the SVV plaques. compared to control cultures. [b] Days post-inoculation. [c] Days post-immunosuppression. [d] Not available. [e] Not immunosuppressed. [f] Time of zoster.

3.6. Detection of SVV DNA in Skin Scrapings during Zoster

Typical zoster lesions were carefully observed, photographed, scraped, DNA extracted, and analyzed by qPCR for the presence of SVV DNA. The samples were analyzed in triplicate and scored as positive only if two out of three PCRs contained SVV DNA. Details about the SVV DNA-positive skin scrapings obtained during immunosuppression are presented in Table 6. Seventeen skin scrapings collected from eight monkeys during zoster contained SVV DNA. Monkey KI92 (group 1) had the highest number copies (127 copies/100 ng) in skin scrapings obtained at 91 dpx. Copies of SVV DNA in the rest of the monkeys ranged from 2–38 copies/100 ng. Skin scraping from the non-immunosuppressed monkey LT27 (group 3) with zoster contained 10 copies/100 ng. While LR42 had SVV DNA-positive skin scrapings, as well as PBMCs at 56 dpx, and LT27 was viremic at 49 dpx and had positive SVV DNA-positive skin scrapings soon after. During the reactivation phase, only 2 out of 11 blood samples that contained SVV DNA correlated with the occurrence of skin rash, suggesting subclinical reaction.

3.7. Detection of SVV Antigens in Skin and Colocalization of SVV Antigens with a Neuronal Marker during Varicella and Zoster

Sections of zinc-formalin-fixed paraffin-embedded skin biopsies obtained before SVV inoculation, during varicella, during and zoster were analyzed by immunohistochemistry using rabbit polyclonal antibodies specific for SVV IE63 protein or SVV nucleocapsids and a mouse monoclonal antibody specific for βIII-tubulin (neuronal marker). Representative analysis of biopsied skin from monkey LK67 (group 2) before inoculation, nine dpi and 42 dpx are presented in Figure 3. Skin biopsy from monkey LK67 pre-inoculation did not show positive staining with rabbit anti SVV63 or mouse anti-βIII-tubulin antibodies (Figure 3A). SVV IE63 protein was detected in hair follicles (Figure 3B,C, arrows) in skin biopsies from LK67 at both nine dpi and 42 dpx. SVV IE63 protein was also detected in the epidermis and colocalized with βIII-tubulin in skin biopsy from LK67 at 42 dpx (Figure 3C, arrowhead). The concentrated βIII-tubulin colocalizing with SVV IE63 protein suggested hyperinnervation at the site of rash. Mixture of normal rabbit serum and isotype mouse IgG2a control antibodies did not show any positive staining with skin biopsies from LK67 at pre-inoculation, nine dpi, and 42 dpx (Figure 3D,E,F).

Table 6. Detection of SVV DNA in skin scrapings during zoster.

Group	Monkey ID	SVV ORF 61 DNA Copy #/100 ng [a]									
		Days Post-Reactivation (dpx)									
		10	42	45	49	56 [b]	63	66 [b]	70	91	107
1	KI92									127	
2	LB56			1	6	16, 9					
	LT26							1, 6			
	LK67		2		3				5		
	LR16	6		4							
	LR42					2	12		38		
	LR70			3							
		Months post inoculation (mpi)									
						3.8					
3	LT27 [c]					10					

[a] All samples done in triplicates and averaged. [b] Samples from two locations were analyzed. [c] Not immunosuppressed.

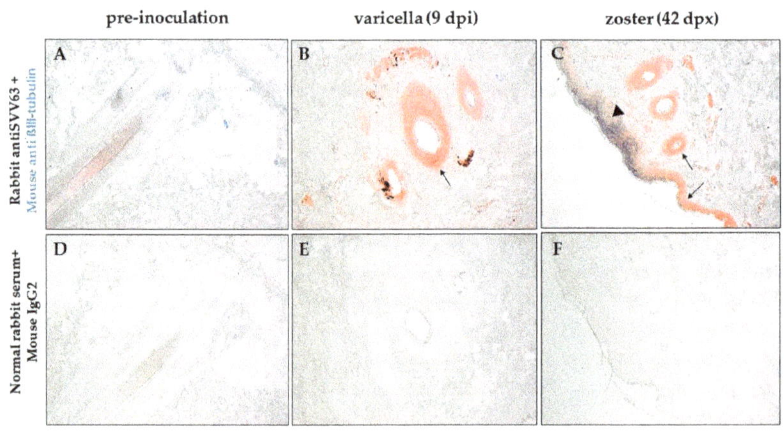

Figure 3. Immunohistochemical analysis of varicella and zoster skin rash from the same rhesus macaque. Sections of biopsied, zinc formalin-fixed, paraffin-embedded skin samples before inoculation with SVV (pre-inoculation), 9 dpi (days post-inoculation) and 42 dpx (days post-immunosuppression) from rhesus macaque LK67 (group 2) was analyzed using dual-color staining. Rabbit polyclonal antibodies against SVV IE63 protein (red) and mouse monoclonal antibody directed against human β-III tubulin (blue) were used as described in methods to identify the presence of virus antigen and its proximity to nerve endings. Normal rabbit serum and mouse anti IgG2a were used as negative controls (**D–F**). SVV IE63 protein was not detected in the pre-inoculation biopsy (**A**), but was found in hair follicles during varicella and zoster and in the epidermis during zoster (**B**, arrows). β-III tubulin was found to colocalize with SVV IE63 protein (**C**, arrowhead) in the epidermis during zoster. (Magnification, ×100).

A section of fixed skin obtained at necropsy from a rhesus macaque (B321) with acute SVV infection and generalized, fulminent varicella rash, after irradiation served as a positive control. The skin was analyzed using rabbit anti-SVV IE63 antibody and mouse monoclonal antibody specific for βIII-tubulin. SVV IE63 protein and βIII-tubulin were found to be colocalized in the epidermis (Figure 4A, inset), suggesting axonal spread of virus during varicella. Analysis of skin from monkey KI92 (group 1) at 91 dpx revealed colocalization of SVV and neuron-specific marker in sweat glands (Figure 4B, inset), also suggesting axonal transport of SVV to sweat glands and hair follicles during SVV reactivation. Analysis of an adjacent section of the same skin samples with rash from monkey KI92 using a mixture of rabbit polyclonal antibodies specific for SVV gH + L revealed the presence of the SVV antigens in the same sweat glands and hair follicle (Figure 4C). SVV gH + L were also seen in other areas of the skin including the epidermis. A mixture of normal rabbit serum and isotype mouse IgG2a did not show positive staining in another adjacent section of skin from monkey KI92 (Figure 4D). While SVV IE63 protein was absent in biopsied skin from the uninfected control rhesus macaque (R110368), hair follicles were positive for the neuronal marker βIII-tubulin (Figure 4E). SVV gH + L-specific antibodies did not show any positive staining in the skin section from the uninfected control monkey R110368 (Figure 4F).

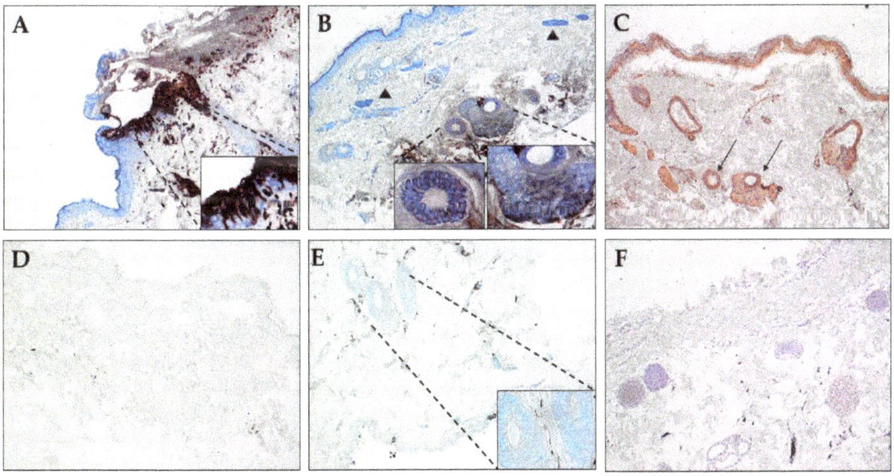

Figure 4. Colocalization of SVV antigen and βIII-tubulin in skin sweat glands during zoster. Sections of biopsied, zinc-formalion-fixed, paraffin-embedded skin samples from (**A**) an acutely infected immunosuppressed rhesus macaque (B321; SVV-positive control), (**B–D**) from rhesus macaque KI92 (group 1) 91 dpx (days post-immunosuppression), and (**E**,**F**) an uninfected SVV-seronegative rhesus macaque (R110368) were analyzed by immunohistochemistry. Rabbit polyclonal antibodies against SVV IE63 protein (red) (**A**,**B**,**E**), SVV gH + L, and (**C**) mouse monoclonal antibody directed against human β-III tubulin (blue; **A**,**B**,**E**) were used as described in methods. Nerve bundles containing βIII-tubulin are identified by arrowheads in panel B. Normal rabbit serum by itself (**F**) or with mouse anti IgG2a antibody (**D**) were used as negative controls. Colocalization of SVV IE63 protein with β-III tubulin can be seen in epidermis during acute infection (**A**, inset) and in sweat glands and hair follicles during reactivation (**B**, inset). SVV gH + L antigens can also be seen in sweat glands in an adjacent section (**C**, thin arrows). β-III tubulin, but not SVV IE63, protein can be seen in sweat glands and hair follicle in skin from an uninfected monkey (**E**, inset). Normal rabbit serum along with mouse anti IgG2a antibody on skin from zoster in monkey KI92 (**D**) or using rabbit anti SVV gH + L on skin section from the uninfected monkey (**F**) did not show any staining. Section in panel F was counterstained with hematoxylin. (Magnification, ×100, inset ×600).

Table 7. Detection of SVV DNA and antigens in skin of rhesus macaques during varicella and zoster.

Group	Monkey ID	Uninfected	Pre-infection	Acute Infection (dpi)	Reactivation (dpx/mpi [a])	Antiserum		
						SVV IE 63	SVV Nucleo-Capsid	ß-III Tubulin
	R110368	SVV- [b]		NR [c]	NR	-	-	+
1	KI92			NA [d]	91	+	+	+
	KG58			NA	107	+	+	+
2	LB56		SVV-			- [e]	-	+
				9		+	-	+
					45	+	+	+
					49	+	-	+
					56	+	-	+
	LK67		SVV-			-	-	+
				9		+	-	+
					42	+	+	+
					49	NA	NA	NA
					70	NA	NA	NA
	LR16		SVV-			-	-	+
				9		+	-	+
					10	+	+	+
					45	NA	NA	NA
	LR42		SVV-			-	-	+
				9		+	+	+
					56	+	-	+
					63	+	+	+
					70	+	+	+
	LR70		SVV-			-	-	+
				9		+	-	+
					45	+	+	+
3	LT27		SVV-			-	-	+
				9		+	+	-
					3.8 [a]	+	+	+

[a] Months post primary infection. [b] SVV- denotes sero-negative animal. [c] Not relevant. [d] Not available. [e] Background staining seen. +, denotes a positive antibody reaction.

Results of immunohistochemical analyses of skin sections from all three groups of monkeys before SVV inoculation, during varicella, and zoster using antibodies specific for SVV IE63, SVV nucleocapsid, and βIII-tubulin are presented in Table 7. Overall, the SVV IE-63 protein was detected more readily than the virus proteins within the nucleocapsid. As expected, the neuronal marker (βIII-tubulin), but not SVV-specific antigens, was detected in skin sections from the uninfected control monkey (R110368). The neuronal marker was found and colocalized with SVV IE63 protein in skin in 7/8 monkeys during varicella, and 7/8 monkeys during zoster, further confirming the neuronal route of virus transport during varicella, as well as zoster. Supplementary results (Figure S1) show the presence of SVVIE63 protein in close proximity to nerve bundles and colocalizing with βIII-positive cells in epidermis in skin from monkeys during varicella as well as zoster. Some of the robust staining in the epidermis may be due to cross-reactivity with beta tubulin class II which is present in epidermal keratinocytes.

In monkeys KI92 and KG58 (group 1), skin samples during varicella were not available, but zoster skin sections contained both SVV IE63 protein and proteins associated with SVV nucleocapsid. Detection of both SVV proteins (Table 7) and DNA (Table 6) in skin from KI92 during zoster suggested the presence of replicating virus, although skin samples for immunohistochemistry and SVV DNA qPCR were obtained from different areas of the skin. Biopsied skin during varicella rash (nine dpi) was obtained from five monkeys in group 2 (LB56, LK67, LR16, LR42, and LR70) and LT27 in group 3. SVV IE63 (an immediate-early gene) protein was detected in all six monkeys, whereas SVV nucleocapsid-specific antigens were found only in two (LR42 and LT27) of the six monkeys, likely due to higher level of expression of immediate-early genes. In two of the monkeys (LK67 (49 and 70 dpx) and LR16 (45 dpx)), biopsies were not available, but skin scrapings were found to contain SVV DNA (Table 6). During zoster, both SVV IE63 protein and proteins associated with nucleocapsids were detected in in 9/12 skin biopsies obtained from eight monkeys across all three groups. Taken together, our detection of SVV DNA in skin scrapings and SVV antigens in biopsies from all monkeys at various times, including during recurrent zoster, provided convincing evidence that SVV is associated with both varicella and zoster rash.

4. Discussion

Like VZV infection in humans, SVV causes varicella becomes latent, and reactivates to produce zoster in NHPs. Key findings in the report are summarized in Figure 5. Viremia and the extent of rash varied among the 10 monkeys (Table 2), probably due to the differences in their antiviral response. Peak viremia and varicella rash occurred at the same time (7–9 dpi) in all monkeys except one (KI92), in which peak viremia occurred at four dpi (Table 3). Extensive rash was seen during varicella (Figure 2), due to the hematogenous virus spread as indicated by high viremia (Table 3) [18]. There was not a complete correlation between the severity of varicella rash and SVV DNA in skin scrapings (Table 2). However, this observation was likley due to differences in sampling as scrapings and biopsies were obtained from from distinct single lesions, mostly from different areas of the body. Once establishment of latency is confirmed by the absence of detectable virus DNA in blood for two consecutive weeks, immunosuppression was initated. A limitation of this study is that animals were immunosuppressed with a combination of tacrolimus, irradiation, and steroids to trigger reactivation. In humans, reactivation occurs with a decline in cell mediated immunity (CMI) associated with aging and in the setting of immunosuppression such as seen in cancer and transplant patients treated with similar therapies (tacrolimus, irradiation, steroids). Because we could not indefinitely age the animals and wait for reactivation, our model immune suppresses to decrease VZV-specific CMI.

Monkeys in groups 1 and 2 were immunosuppressed at eight and two mpi, respectively, after viremia was absent for two weeks. Viremia following immunosuppression and during zoster was intermittent and much lower than viremia during primary infection, probably due to reduced hematogenous spread of virus and increased axonal spread following reactivation from individual ganglia (Table 4). Skin rash during zoster was also much less extensive compared to varicella (Figure 2). Lack of detectable viremia during zoster compared to varicella despite the presence of virus DNA in skin rash vesicles in humans have been documented [2]. As observed before, no significant changes in SVV-specific antibodies were observed in immunosuppressed monkeys during zoster, suggesting that humoral response is unlikely to play an important role in virus reactivation [12,13,15,19]. However, in one monkey (LT27, group 3) that was not immunosuppressed, we noticed a reduction in the virus-specific antibodies at the time of zoster, the reason for which remains unclear (Table 5). In the other non-immunosuppressed monkey (LE91-group 3), the antibody response after inoculation was low likely due to the relatively low-level viremia seen at seven dpi (Table 3).

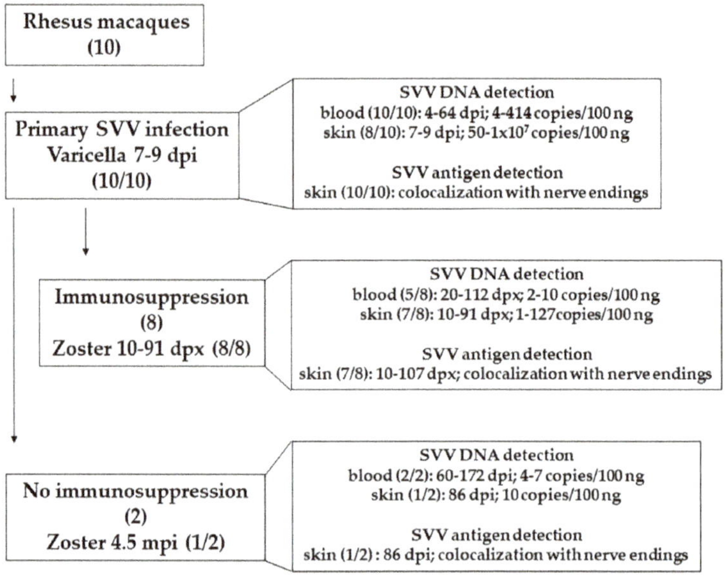

Figure 5. Summary of findings in this report. Ten rhesus macaques were inoculated with SVV. All of them developed varicella 7–9 dpi. SVV DNA (4–414 copies/100 ng) was detected in blood, 4–64 days post infection (dpi) in all 10 monkeys. SVV DNA (50–1 $\times$ 10^7 copies/100 ng) was detected in skin scrapings in 8 of 10 monkeys, 7–9 dpi. Analysis of biopsied skin samples, during varicella, revealed colocalization of SVV antigens with nerve endings in all 10 monkeys. Establishment of latency was confirmed by the absence of SVV DNA in blood. Eight of latently infected monkeys were immunosuppressed. Zoster rash developed in all monkeys 10–91 days post immunosuppression (dpx). SVV DNA (2–10 copies/100 ng) was detected in blood, 20–112 dpx, in 5 of 8 monkeys. During zoster (10–91 dpx), 1–127 copies of SVV DNA (per 100 ng) were detected in skin scrapings. In the affected skin, SVV antigens were found to be colocalized with nerve endings. In two monkeys were not immunosuppressed, SVV DNA (4–7 copies/100 ng) was detected 60–172 dpi. One of the two non-immunosuppressed monkey developed zoster, 4.5 mpi (months post infection). SVV DNA (10 copies/100 ng) and antigens was detected in skin scrapings during zoster (86 dpi). SVV antigen was found to be colocalized with nerve endings. Abundance of SVV DNA and antigens in skin was substantially higher during varicella compared to zoster. SVV infection and reactivation in rhesus macaques serves as an extremely useful model to study varicella and zoster in humans.

Our detection of SVV DNA in skin scrapings from all eight immunosuppressed monkeys at various times provided convincing evidence that presence of SVV corrlates with zoster rash. Recurrent zoster seen in seven out of eight immunosuppressed monkeys from groups 1 and 2 confirmed our earlier observation and what has been found in humans during VZV reactivation [13,20,21]. Monkey KI92 (group 1) had substantially higher copies of SVV DNA in skin scraping at 91 dpx (127 copies/100 ng) possibly due to the

Multiple modes of immunosuppression used, compared to only tacrolimus-treated monkeys in group 2 (Table 6). Detection of higher quantities of SVV DNA in zoster skin (91 dpx in KI92 and 70 dpx in LR42; Table 6) in the absence of viremia at these times (Table 4) further supports the notion of axonal spread of SVV during reactivation. Earlier, we documented that during varicella, SVV enters the ganglia before the appearance of skin rash [22]. Colocalization of SVV antigens with the neuronal marker in skin, along with high level of viremia, suggests either simultaneous hematogenous spread and axonal transport of virus to skin or spread of infection between keratinocytes during varicella. On the other hand, detection of minimal viremia, as well as SVV DNA in skin rash combined with colocalization of SVV antigen with the neuronal marker, supports the idea of virus being transported preferentially via the axonal route from the ganglia to the skin during reactivation and zoster.

While we detected SVV IE63 protein in all the varicella and zoster skin biopsies, proteins associated with SVV nucleocapsids were not detectable in two varicella and three zoster skin biopsies. Detection of some, but not other, VZV proteins in skin biopsies from varicella and zoster in humans has been reported [9]. Selective detection of VZV IE63 protein compared to glycoproteins in early zoster lesions has also been attributed to the reduced abundance of glycoproteins and preferential resistance of IE63 protein to proteolysis [23]. During zoster in humans, VZV gC has been found in axons and Schwann cells in skin biopsies [9], and VZV IE63 protein and glycoproteins have been seen in dermal nerves and perineurial dendrocytes [23]. SVV antigens were found to be colocalized with the neuronal marker in most, if not all, of the skin samples analyzed. Our detection of SVV antigens associated primarily with sweat glands and hair follicles was very similar to VZV antigens in human skin during zoster [6–8,23–25]. Together, our results show that compared to primary SVV infection (varicella), SVV causes a milder skin rash and less viremia during zoster. In addition, SVV DNA and antigens can be detected in both skin scrapings and biopsies. Huch et al. [26] showed that during zoster in humans, Langerhans and plasmacytoid dendritic cells are strongly associated with VZV antigen-positive cells in skin. The lack of availability of rhesus specific reagents have hampered the progress in this area. Our future studies will include identification of antibodies that cross react with NHP skin to understand the role of immune cells in controlling zoster rash, as well as neuronal subtypes that are involved in pathological manifestations, such as postherpetic neuralgia.

Supplementary Materials: The following supporting information can be downloaded at: https://www.mdpi.com/article/10.3390/v14061167/s1, Figure S1: Detection of SVV IE63 protein and βIII-tubulin in epidermis and nerve bundles in skin during varicella and zoster in rhesus macaques.

Author Contributions: Conceptualization, R.M., A.N.B., C.S.N., B.P., M.A.N. and V.T.-D.; methodology, R.M., B.F., L.D.-M., J.L. and V.T.-D.; data collection, R.M., B.F., C.C., K.A., P.S., A.L., A.B., A.D., E.d.H. and V.T.-D.; writing—original draft preparation, R.M., B.F., C.S.N., A.N.B., C.S.N., B.P., M.A.N. and V.T.-D.; writing—review and editing, R.M., B.F., C.C., K.A., P.S., A.L., A.B., A.D., E.d.H., L.D.-M., J.L., A.N.B., C.S.N., B.P., M.A.N. and V.T.-D.; funding acquisition, R.M., A.N.B., M.A.N. and V.T.-D. All authors have read and agreed to the published version of the manuscript.

Funding: This study was supported in part by Public Health Service grant P01 AG032958 (to R.M., A.N.B., B.P., M.A.N. and V.T.-D.) from the National Institutes of Health (NIH). This study was also supported in part with federal funds from the National Center for Research Resources and the Office of Research Infrastructure Programs (ORIP) of the NIH through grant P51 OD011104 to the Tulane National Primate Research Center (L.D.-M. and V.T.-D.).

Institutional Review Board Statement: The study was conducted in compliance with The Guide for the Care and Use of Laboratory Animals (National Research Council), the Animal Welfare Act guidelines, and the guidelines at the Tulane National Primate Research Center, accredited by the Association of the Assessment and Accreditation of Laboratory Animal Care. Protocols were approved by the Institutional Animal Care and Use Committee of Tulane University (The Assurance Number A4499-01; Protocol Title: Varicella Virus Latency and Reactivation in the Nonhuman Primate, IACUC Protocol #PO177R3, 25 October 2021).

Data Availability Statement: Not applicable.

Acknowledgments: We thank Anah Gilmore for technical help, Stanca Birlea for useful discussions, Steve Shipley for providing the control tissue samples, Wayne Gray for providing the SVV gH + L antibodies and the Tulane National Primate Research Center Veterinary Medicine staff for excellent animal care. We are grateful to Cathy Allen for preparing the manuscript and Teresa Mescher (Pubsciconsulting, Inc) for editorial assistance.

Conflicts of Interest: The authors declare no conflict of interest. The funders had no role in the design of the study; in the collection, analyses, or interpretation of data; in the writing of the manuscript, or in the decision to publish the results.

References

1. Ku, C.-C.; Zerboni, L.; Ito, H.; Graham, B.S.; Wallace, M.; Arvin, A.M. Varicella-Zoster Virus Transfer to Skin by T Cells and Modulation of Viral Replication by Epidermal Cell Interferon-α. *J. Exp. Med.* **2004**, *200*, 917–925. [CrossRef] [PubMed]
2. Kimura, H.; Kido, S.; Ozaki, T.; Tanaka, N.; Ito, Y.; Williams, R.K.; Morishima, T. Comparison of quantitations of viral load in varicella and zoster. *J. Clin. Microbiol.* **2000**, *38*, 2447–2449. [CrossRef] [PubMed]
3. Ishizaki, Y.; Tezuka, J.; Ohga, S.; Nomura, A.; Suga, N.; Kuromaru, R.; Kusuhara, K.; Mizuno, Y.; Kasuga, N.; Hara, T. Quantification of circulating varicella zoster virus-DNA for the early diagnosis of visceral varicella. *J. Infect.* **2003**, *47*, 133–138. [CrossRef]
4. Kronenberg, A.; Bossart, W.; Wuthrich, R.P.; Cao, C.; Lautenschlager, S.; Wiegand, N.D.; Mullhaupt, B.; Noll, G.; Mueller, N.J.; Speck, R.F. Retrospective analysis of varicella zoster virus (VZV) copy DNA numbers in plasma of immunocompetent patients with herpes zoster, of immunocompromised patients with disseminated VZV disease, and of asymptomatic solid organ transplant recipients. *Transpl. Infect. Dis.* **2005**, *7*, 116–121. [CrossRef] [PubMed]
5. Satyaprakash, A.K.; Tremaine, A.M.; Stelter, A.A.; Creed, R.; Ravanfar, P.; Mendoza, N.; Mehta, S.K.; Rady, P.L.; Pierson, D.L.; Tyring, S.K. Viremia in Acute Herpes Zoster. *J. Infect. Dis.* **2009**, *200*, 26–32. [CrossRef]
6. Muraki, R.; Baba, T.; Iwasaki, T.; Sata, T.; Kurata, T. Immunohistochemical study of skin lesions in herpes zoster. *Virchows Arch.* **1992**, *420*, 71–76. [CrossRef]
7. Muraki, R.; Iwasaki, T.; Sata, T.; Sato, Y.; Kurata, T. Hair follicle involvement in herpes zoster: Pathway of viral spread from ganglia to skin. *Virchows Arch.* **1996**, *428–428*, 275–280. [CrossRef]
8. Iwasaki, T.; Muraki, R.; Kasahara, T.; Sato, Y.; Sata, T.; Kurata, T. Pathway of viral spread in herpes zoster: Detection of the protein encoded by open reading frame 63 of varicella-zoster virus in biopsy specimens. *Arch. Virol. Suppl.* **2001**, 109–119. [CrossRef]
9. Annunziato, P.W.; Lungu, O.; Panagiotidis, C.; Zhang, J.H.; Silvers, D.N.; Gershon, A.A.; Silverstein, S.J. Varicella-Zoster Virus Proteins in Skin Lesions: Implications for a Novel Role of ORF29p in Chickenpox. *J. Virol.* **2000**, *74*, 2005–2010. [CrossRef]
10. Mahalingam, R.; Gershon, A.; Gershon, M.; Cohen, J.I.; Arvin, A.; Zerboni, L.; Zhu, H.; Gray, W.; Messaoudi, I.; Traina-Dorge, V. Current In Vivo Models of Varicella-Zoster Virus Neurotropism. *Viruses* **2019**, *11*, 502. [CrossRef]
11. Messaoudi, I.; Barron, A.; Wellish, M.; Engelmann, F.; Legasse, A.; Planer, S.; Gilden, D.; Nikolich-Zugich, J.; Mahalingam, R. Simian Varicella Virus Infection of Rhesus Macaques Recapitulates Essential Features of Varicella Zoster Virus Infection in Humans. *PLoS Pathog.* **2009**, *5*, e1000657. [CrossRef] [PubMed]
12. Traina-Dorge, V.; Doyle-Meyers, L.A.; Sanford, R.; Manfredo, J.; Blackmon, A.; Wellish, M.; James, S.; Alvarez, X.; Midkiff, C.; Palmer, B.E.; et al. Simian Varicella Virus Is Present in Macrophages, Dendritic Cells, and T Cells in Lymph Nodes of Rhesus Macaques after Experimental Reactivation. *J. Virol.* **2015**, *89*, 9817–9824. [CrossRef] [PubMed]
13. Traina-Dorge, V.; Palmer, B.E.; Coleman, C.; Hunter, M.; Frieman, A.; Gilmore, A.; Altrock, K.; Doyle-Meyers, L.; Nagel, M.A.; Mahalingam, R. Reactivation of Simian Varicella Virus in Rhesus Macaques after CD4 T Cell Depletion. *J. Virol.* **2019**, *93*, e01375-18. [CrossRef] [PubMed]
14. Mahalingam, R.; Clarke, P.; Wellish, M.; Dueland, A.N.; Soike, K.F.; Gilden, D.H.; Cohrs, R. Prevalence and distribution of latent simian varicella virus DNA in monkey ganglia. *Virology* **1992**, *188*, 193–197. [CrossRef]
15. Mahalingam, R.; Traina-Dorge, V.; Wellish, M.; Lorino, R.; Sanford, R.; Ribka, E.P.; Alleman, S.J.; Brazeau, E.; Gilden, D.H. Simian varicella virus reactivation in cynomolgus monkeys. *Virology* **2007**, *368*, 50–59. [CrossRef]
16. Mahalingam, R.; Kaufer, B.B.; Ouwendijk, W.J.D.; Verjans, G.M.G.M.; Coleman, C.; Hunter, M.; Palmer, B.E.; Clambey, E.; Nagel, M.A.; Traina-Dorge, V. Attenuation of simian varicella virus infection by enhanced green fluorescent protein in rhesus macaques. *J. Virol.* **2018**, *92*, e02253-17. [CrossRef]
17. Mahalingam, R.; Traina-Dorge, V.; Wellish, M.; Deharo, E.; Golive, A.; Messaoudi, I.; Gilden, D. Effect of Time Delay after Necropsy on Analysis of Simian Varicella-Zoster Virus Expression in Latently Infected Ganglia of Rhesus Macaques. *J. Virol.* **2010**, *84*, 12454–12457. [CrossRef]
18. Ouwendijk, W.J.D.; Mahalingam, R.; de Swart, R.L.; Haagmans, B.L.; van Amerongen, G.; Getu, S.; Gilden, D.; Osterhaus, A.D.M.E.; Verjans, G.M.G.M. T-Cell Tropism of Simian Varicella Virus during Primary Infection. *PLoS Pathog.* **2013**, *9*, e1003368. [CrossRef]
19. Mahalingam, R.; Traina-Dorge, V.; Wellish, M.; Deharo, E.; Singletary, M.L.; Ribka, E.P.; Sanford, R.; Gilden, N. Latent simian varicella virus reactivates in monkeys treated with tacrolimus with or without exposure to irradiation. *J. Neurovirol.* **2010**, *16*, 342–354. [CrossRef]
20. Yawn, B.P.; Wollan, P.C.; Kurland, M.J.; Sauver, J.L.S.; Saddier, P. Herpes Zoster Recurrences More Frequent Than Previously Reported. *Mayo Clin. Proc.* **2011**, *86*, 88–93. [CrossRef]
21. Lee, S.M.; Han, J.; Yang, C.M.; Choi, C.Y.; Khoramnia, R.; Chung, T.Y.; Lim, D.H. Chronic and recurrent herpes zoster ophthalmicus. *Medicina* **2021**, *57*, 999. [CrossRef] [PubMed]
22. Mahalingam, R.; Wellish, M.; Soike, K.; White, T.; Kleinschmidt-DeMasters, B.K.; Gilden, D.H. Simian Varicella Virus Infects Ganglia before Rash in Experimentally Infected Monkeys. *Virology* **2001**, *279*, 339–342. [CrossRef] [PubMed]
23. Nikkels, A.F.; Debrus, S.; Sadzot-Delvaux, C.; Piette, J.; Rentier, B.; Pierard, G.E. Localization of varicella-zoster virus nucleic acids and proteins in human skin. *Neurology* **1995**, *45*, S47–S49. [CrossRef] [PubMed]

24. Rinder, H.M.; Murphy, G.F. Eccrine duct involvement by herpes zoster. *Arch. Dermatol.* **1984**, *120*, 261–262. [CrossRef] [PubMed]
25. Sangueza, O.P.; Gordon, M.D.; White, C.R. Subtle clues to the diagnosis of the herpesvirus by light microscopy. Herpetic syringitis. *Am. J. Dermatopathol.* **1995**, *17*, 163–168. [PubMed]
26. Huch, J.H.; Cunningham, A.L.; Arvin, A.M.; Nasr, N.; Santegoets, S.J.A.M.; Slobedman, E.; Slobedman, B.; Abendroth, A. Impact of Varicella-Zoster Virus on Dendritic Cell Subsets in Human Skin during Natural Infection. *J. Virol.* **2010**, *84*, 4060–4072. [CrossRef] [PubMed]

viruses

MDPI

Communication

HLA-B*57:01 Complexed to a CD8 T-Cell Epitope from the HSV-2 ICP22 Protein Binds NK and T Cells through KIR3DL1

Kerry J. Laing [1,*], Victoria L. Campbell [1], Lichun Dong [1,†] and David M. Koelle [1,2,3,4,5]

1 Department of Medicine, University of Washington, Seattle, WA 98195, USA; vlc32@uw.edu (V.L.C.); lichun.dong@maxvax.cn (L.D.); dkoelle@medicine.washington.edu (D.M.K.)
2 Department of Laboratory Medicine and Pathology, University of Washington, Seattle, WA 98195, USA
3 Vaccine and Infectious Diseases Division, Fred Hutchinson Cancer Center, Seattle, WA 98109, USA
4 Department of Global Health, University of Washington, Seattle, WA 98195, USA
5 Benaroya Research Institute, Seattle, WA 98101, USA
* Correspondence: laingk@uw.edu
† Current address: MaxHealth Biotechnology, Chengdu 610041, China.

Abstract: HLA-B*57:01 is an HLA allelic variant associated with positive outcomes during viral infections through interactions with T cells and NK cells, but severe disease in persons treated with the anti-HIV-1 drug abacavir. The role of HLA-B*57:01 in the context of HSV infection is unknown. We identified an HLA-B*57:01-restricted CD8 T-cell epitope in the ICP22 (US1) protein of HSV-2. CD8 T cells reactive to the HSV-2 ICP22 epitope recognized the orthologous HSV-1 peptide, but not closely related peptides in human IFNL2 or IFNL3. Abacavir did not alter the CD8 T-cell recognition of the HSV or self-derived peptides. Unexpectedly, a tetramer of HSV-2 ICP22 epitope (228–236) and HLA-B*57:01 bound both CD8 T cells and NK cells. Tetramer specificity for KIR3DL1 was confirmed using KIR3DL1 overexpression on non-human primate cells lacking human KIR and studies with blocking anti-KIR3DL1 antibody. Interaction with KIR3DL1 was generalizable to donors lacking the *HLA-B*57:01* genotype or HSV seropositivity. These findings suggest a mechanism for the recognition of HSV infection by NK cells or KIR-expressing T cells via KIR3DL1.

Keywords: CD8 T cell; T-cell epitope; NK cell; KIR3DL1; HLA-B*57:01; herpes simplex virus

Citation: Laing, K.J.; Campbell, V.L.; Dong, L.; Koelle, D.M. HLA-B*57:01 Complexed to a CD8 T-Cell Epitope from the HSV-2 ICP22 Protein Binds NK and T Cells through KIR3DL1. *Viruses* **2022**, *14*, 1019. https://doi.org/10.3390/v14051019

Academic Editors: Charles Grose, Ravi Mahalingam and Joel Rovnak

Received: 17 April 2022
Accepted: 9 May 2022
Published: 11 May 2022

1. Introduction

Class I HLA molecules are fundamentally important in the immune control of viral infections by activating CD8 T cells and interacting with natural killer (NK) cells. The class I allelic variant HLA-B*57:01 is classified as a Bw4 family member and has been associated with improved disease outcomes for HIV/AIDS. The superior virologic control of HIV-1 is associated with the *HLA-B*57:01* genotype [1], HLA-B*57:01-restricted CD8 T-cell responses [2], and specific T-cell receptor (TCR) sequence attributes [3]. At the same time, the expression of the killer cell immunoglobulin-like receptor (KIR), KIR3DL1—an inhibitory receptor on NK cells—in *HLA-B*57:01* positive persons is associated with the elite control of HIV-1 and, potentially, increased NK function [4,5].

*HLA-B*57:01* is also associated with drug hypersensitivity reactions, particularly to abacavir, a medication used to treat HIV-1 [6]. Alterations in HLA-bound peptide conformation have been hypothesized as abacavir can bind in the peptide-binding groove of HLA-B*57:01. This may elicit inappropriate CD8 T-cell activity that manifests itself as severe hypersensitivity. The effect is dependent on the properties of the peptide C-terminus, whereby peptides ending with small hydrophobic residues, such as isoleucine, leucine, and valine, potentially become neo-epitopes in the presence of abacavir. Peptides ending in large hydrophobic residues, such as phenylalanine or tryptophan, may also lose binding, and, as a consequence, the recognition of pathogen peptides may be reduced [7,8].

While peptides complexed with HLA class I activate T cells through TCR engagement, they can also interact with immune cells via KIR. There are at least 17 known functional members of the polymorphic KIR family. These can be activating (S type, for short cytoplasmic tail, containing an immunoreceptor-tyrosine activating motif (ITAM)) or inhibitory (L type, long cytoplasmic tail, containing an inhibitory ITIM motif) [9]. KIR are expressed by both NK cells and a subset of T cells. Inhibitory KIR reduce the activation of NK cells in the presence of their appropriate HLA ligand. Along with other HLA allelic variants sharing the Bw4 determinant, HLA-B*57:01 appears to have specificity for KIR3DL1. Most persons possess one or two functional copies of either *KIR3DL1*, [9–11] or *KIR3DS1*, an activating variant of *KIR3DL1* that can occur at the same location in the KIR locus on human chromosome 19 [12]. *KIR3DL1* has many allelic variants that encode different protein sequences with differing expression levels on the cell surface, which impacts the strength of NK cell inhibition [9,13]. The specificity of the interaction with HLA-B*57:01 may depend on the amino acid residue 166 of KIR3DL1 (which is a leucine in the common allele *KIR3DL1*001*) [14,15] and can be blocked with the KIR3DL1-specific monoclonal antibody (mAb) DX9. KIR3DS1 has a contrasting amino acid (arginine) at the corresponding residue shown to modulate HLA-Bw4 binding [15]. The interaction of HLA-B*57:01 with KIR3DL1/DS1 residue 166 is further influenced by the HLA-B*57:01-bound peptide, specifically peptide position 8 (P8). For example, KIR3DL1 binding is abrogated if epitope P8 is a negatively charged (aspartic acid or glutamic acid) amino acid and KIR3DS1 binding is favored if epitope P8 is an aromatic amino acid (tryptophan, phenylalanine or tyrosine) [16], consistent with a structural role for peptides in modulating HLA-KIR interactions.

NK cells are large CD3-negative lymphocytes that are defined phenotypically by the cell surface expression of CD56 and/or CD16. Both markers vary in expression depending on the function and level of differentiation of the NK cells. The two major populations of NK cells, defined as $CD56^{bright}$ and $CD56^{dim}$, represent less differentiated, higher-cytokine-producing NK cells and more differentiated, more cytotoxic NK cells, respectively [17]. More prevalent in blood than $CD56^{bright}$ NK cells, $CD56^{dim}$ NK cells are more likely to express KIR.

NK cells have long been recognized as being functionally important for host defense against herpesvirus infections, including HSV species in particular. In 1989, Biron et al. reported severe herpes infections in a patient with phenotypic NK cell absence [18], later related to mutation in *GATA2* [19]. Since then, several other rare mutations have been reported in persons with quantitative or qualitative NK cell deficiencies, which make subjects susceptible to herpesviruses (reviewed in [20]). A lower level of activating receptors on $CD56^{dim}$ NK cells has been suggested as a potential correlate with HSV severity [21]. These reports emphasize the significance of defining how NK cells can sense HSV and mechanisms that could counteract HSV recognition.

During the study of HLA-B*57:01-restricted CD8 T cells that react to human herpesvirus-2, the major causative agent of genital herpes, we identified a viral peptide epitope that can activate CD8 T cells. Unexpectedly, a complex of this newly discovered HSV-2 epitope with HLA-B*57:01 also interacted with NK cells. We show that interaction with NK cells is mediated through KIR3DL1 binding, and that a sub-population of CD8 T cells, including from persons without HSV infection or HLA-B*57:01, can also bind to HLA-B*57:01-peptide via KIR3DL1.

2. Materials and Methods

2.1. Subjects and Specimens

Subjects seropositive for HSV-2 were recruited to the University of Washington Virology Research Clinic. The study was approved by the University of Washington Institutional Review Board (IRB STUDY00004400) and participants provided written informed consent. Class I HLA typing was performed at the Puget Sound Blood Center (now Bloodworks) or by PCR-based genotyping [22]. Peripheral blood mononuclear cells (PBMC) were isolated using Lymphoprep (Cosmo Bio USA, Carlsbad, CA, USA) and cryopreserved at

10×10^6 cells/mL/vial in 10% DMSO (Fisher, Hampton, NH, USA), 40% human serum (Valley Biomedical, Winchester, VA, USA), and 50% RPMI-1640 (Hyclone, Logan, UT, USA). Epstein–Barr virus (EBV) strain B95-8 was used to immortalize B lymphocyte continuous lines (B-LCL) from PBMC [23]. HSV-1 and HSV-2 seropositivity was assessed by serum Western blot [24].

Bulk T-cell lines were generated from HSV-2 reactive CD8 T cells, as previously described [25]. In brief, subject-specific monocyte-derived dendritic cells were generated from adherent PBMCs using IL-4 and GM-CSF [26] and were combined with an equal number (2×10^5) of UV-irradiated (Stratalinker XL1000, 180,000 μJ) HeLa cells that were uninfected (mock) or infected overnight with HSV-2 strain 186 (MOI 2.5). After incubation for 4 h at 37 °C/5% CO_2, 1×10^6 negative-selected (STEMCELL EasySep Human CD8 T Cell Isolation Kit) autologous CD8 T cells were added per well. T cells were harvested after 18–20 h of incubation and stained with anti–CD3-ECD (UCHT1, Beckman Coulter, Brea, CA, USA), anti–CD8-FITC (3B5, Thermo Fisher Scientific, Waltham, MA, USA), anti–CD137-APC (4B4-1, BD Biosciences, San Jose, CA, USA), and 7-AAD (BD Biosciences). Live (7-AAD negative), single CD3+ CD8+ CD137+ lymphocytes were sorted from HSV-2-exposed PBMC using a BD Biosciences FACS Aria II (University of Washington Cell Analysis Facility, Seattle, WA, USA).

Sorted cells were expanded by stimulation with 1.6 μg/mL phytohemagluttinin P (PHA, Remel, San Diego, CA, USA), in the presence of 2×10^6/mL γ-irradiated (3300 rads) allogeneic PBMC, for 2–3 weeks (37 °C, 5% CO_2) in a T-cell medium (TCM) containing RPMI-1640, 4% human serum, 4% defined fetal bovine serum (Hyclone, Logan, UT, USA), 2 mM L-glutamine (Hyclone), and 100 U/mL penicillin/streptomycin (Gibco). Natural IL-2 (nIL-2, 32 U/mL, Hemagen, Columbia, MD, USA) was added after 48 h and replenished with fresh TCM twice weekly. Polyclonal T-cell lines were further expanded using anti-CD3 as published [27] and the resultant T-cell lines were cryopreserved.

2.2. Antigens

Vero cells infected with HSV-2 (strain 186) stocks were harvested when cytopathic effect reached 80%, disrupted by sonication, and the cellular debris removed by low-speed centrifugation. The supernatant was aliquoted and stored at −80 °C. Each HSV-2 open reading frame (ORF) was cloned into Gateway™ pDONR221 (Invitrogen, Grand Island, NY, USA), as previously described [28,29]. HSV-2 ORFs were subcloned to pDEST103 [30] for the eukaryotic intracellular expression of eGFP-HSV-2 fusion proteins.

Targeted HSV-2 ICP22 (*US1*) peptides (8–10 amino acids) were designed using the algorithmic prediction of HLA binding [31] (Table 1). Sequence variants for an immunogenic HSV2-US1 (228–236) peptide RTRLGPRTW were identified for HSV-1 (HSV1-US1 (232–240), RaRLaPRTW), and for human IFNL3 (IFNL3 (76–84), RsRLfPRTW) and human IFNL2 (IFNL2 (76–84), hsRLfPRTW) using BLASTP (BLAST®, NCBI) with search limits set to HSV-1 or human. Each HSV and human peptide was predicted to be an avid binder to HLA-B*57:01 [31] (not shown). Peptides (Sigma, >70% pure) were reconstituted in DMSO.

2.3. Identification of HLA-B*57:01 Restricted CD8 T-Cell Responses to HSV-2 Proteins

Artificial APC (aAPC) expressing HLA-B*57:01 and individual HSV-2 ORFs were used to detect CD8 T-cell responses in CD8 T-cell lines enriched for HSV-2 reactivity. Briefly, HSV-2 ORFs cloned into pDEST103 (100 ng/well) and empty p103 negative control plasmids were co-transfected, in duplicate, with pcDNA3.1 constructs encoding HLA-B*57:01 (100 ng/well) into COS-7 cells as previously described [30]. After 48 h, T cells (1×10^5 in TCM) were added and co-cultures incubated for 24 h at 37 °C. PHA (1.6 μg/mL) were used as a positive control. Secreted IFN-γ in supernatants was measured by ELISA [30]. In brief, high binding ELISA plates (MICROLON™ 600) were coated overnight at 4 °C with mouse anti-human IFN-γ mAb (clone 2GI, Thermo Fisher, Waltham, MA, USA) and blocked with Tris-buffered saline (TBS) containing 0.1% bovine serum albumin (BSA) for 1 h at room temperature. Supernatants were added to blocked plates for 2 h at room temperature after washing with PBS-Tween (Dulbecco's phosphate-buffered saline (PBS, Corning) containing 0.1% Tween-20 (Fisher Scientific, Waltham, MA, USA)), after the incubation of the plates for 2 h at room temperature. After washing, detection antibody (mouse anti-human IFN-γ-Biotin conjugate, clone B133.5, Thermo Scientific, Waltham, MA, USA) was added for 2 h at room temperature, plates washed, and avidin-horse radish peroxidase (HRP) added for 30 min at room temperature. TMB substrate (KPL) was added to the washed plates and reactions stopped after 10 min with 1M phosphoric acid (Fisher Scientific). Results are reported as absorbance (OD_{450}) and determined using a Wallac Victor^3 plate reader (Perkin Elmer, Waltham, MA, USA). Responses to HSV-2 proteins were considered positive if both OD_{450} values were greater than 0.15 and at least 2-fold higher than negative controls.

Table 1. Peptides tested for reactivity with polyclonal HSV-2-reactive CD8 T-cell lines.

Peptide Name	Species	Gene/ORF	Protein	AA Position in Protein	Peptide Sequence
HSV2-US1 (36–45)	HSV-2	*US1*	ICP22	36–45	PSSSESEGKP
HSV2-US1 (50–59)	HSV-2	*US1*	ICP22	50–59	ESSSTESSED
HSV2-US1 (55–64)	HSV-2	*US1*	ICP22	55–64	ESSEDEAGDL
HSV2-US1 (105–114)	HSV-2	*US1*	ICP22	105–114	DASDGWLVDT
HSV2-US1 (158–167)	HSV-2	*US1*	ICP22	158–167	PASLPGIAHA
HSV2-US1 (228–236)	HSV-2	*US1*	ICP22	228–236	RTRLGPRTW
HSV2-US1 (295–304)	HSV-2	*US1*	ICP22	295–304	STSDDEISDA
HSV2-US1 (350–359)	HSV-2	*US1*	ICP22	350–359	WTSEEGSQPW
HSV2-US1 (366–375)	HSV-2	*US1*	ICP22	366–375	DTSSAERSGL
HSV1-US1 (232–240)	HSV-1	*US1*	ICP22	232–240	RaRLaPRTW *
IFNL3 (76–84)	Human	*IFNL3*	IFNL3 (IL28B)	76–84	RsRLfPRTW *
IFNL2 (76–84)	Human	*IFNL2*	IFNL2 (IL28A)	76–84	HsRLfPRTW *

* Variant amino acids different from peptide HSV2-US1 (228–236) are shown in lower case.

Epitope mapping in ICP22 was performed using autologous B-LCL as antigen-presenting cells. Equal numbers (1×10^5) of B-LCL and CD8 T cells were combined with individual peptides in TCM and incubated at 37 °C for 20–24 h. Peptides were screened at a final concentration of 1 μg/mL each peptide and <0.3% DMSO. For titration, a log10 dilution series was used between 1 μg/mL and 0.1 pg/mL. An equivalent concentration of DMSO served as a negative control, while PHA (1.6 μg/mL) was used as a positive control. Medium alone (MED) controls were additionally included in some experiments. To investigate if HLA-B*57:01 restricted T-cell reactivity to HSV2-US1 (228–236) would be influenced by abacavir, T cells were combined with B-LCL that were pre-treated with abacavir (10 μg/mL) or left untreated for 24 h [32]. HSV2-US1 (228–236) and related variant peptides (Table 1) were added at 1 μg/mL, and cultures incubated at 37 °C for 24 h. Peptide-driven T-cell activation was determined by measuring IFN-γ in cell supernatants by ELISA [33]. The HSV2-US1 (228–236) epitope was uploaded to the Immune Epitope Database (Epitope ID 9063706).

2.4. Detection of T-Cell and NK-Cell Binding by B*57:01 Tetramer Containing HSV2-US1 (228–236)

To confirm the HLA-restriction of the HSV2-US1 (228–236) RTRLGPRTW peptide, a custom PE-conjugated peptide-HLA-B*57:01 tetramer (ImmunAware, Hørsholm, Denmark), termed B57-RTR, was tested for binding to HSV-2-specific CD8 T-cell lines and PBMC. The specificity of B57-RTR tetramer for binding KIR3DL1 was assessed using COS-7 cells that overexpressed KIR3DL1. In brief, 100 ng of either pReceiver-M02:KIR3DL1 (EX-A1726-M02, Genecopoeia) or pReceiver-M02:eGFP (negative control, EX-EGFP-M02, Genecopoeia) expression plasmids were transfected into COS-7 cells in 96-well plates using 0.5 μL Fugene 6 (Promega, Madison, WI, USA) as per the manufacturer's instructions. After 48 h, trypsin-harvested cells were pre-incubated for 30 min at room temperature in FACS buffer (Dulbecco's phosphate-buffered saline (PBS), 1% BSA, 0.01% NaN_3) with 5 μg/mL mAb DX9 (Invitrogen, Waltham, MA, USA) or an isotype (mouse IgG1) control antibody

(Biolegend, San Diego, CA, USA), washed, and then stained with B57-RTR (or mock stained with FACS buffer) for 30 min at room temperature. Binding was evaluated using a BD Canto II flow cytometer (University of Washington Cell Analysis Facility, Seattle, WA, USA) and FlowJo Version 10 (for Mac, BD, Franklin Lakes, NJ, USA).

Tetramers were also used to stain T-cell lines (3×10^5) or PBMC (1×10^6) for 30 min at room temperature in the dark, after staining for 20 min with LiveDead (Near-IR, Invitrogen) and before staining for 30 min with anti-CD3-ECD (UCHT1, Beckman Coulter, Burea, CA, USA), anti-CD8-FITC (3B5, Thermofisher, Waltham, MA, USA), anti-CD56-APC (HCD56, Biolegend, San Diego, CA, USA), and anti-CD16-PacificBlue (3G8, Biolegend, San Diego, CA, USA) in FACS buffer. For some experiments, anti-KIR3DL1-AlexaFluor700 antibody (DX9; Biolegend, San Diego, CA, USA) was included prior to staining with tetramer. For other experiments, cells were pre-incubated for 30 min with the unconjugated DX9 antibody or the isotype control antibody (as above), before staining with B57-RTR. Stained cells were fixed in PBS containing 1% formaldehyde. Data were acquired using a four-laser BD Canto II flow cytometer (University of Washington Cell Analysis Facility, Seattle, WA, USA) and analyzed using FlowJo Version 10.

3. Results

3.1. HLA-B*57:01-Restricted CD8 T Cells Recognize HSV-2 Proteins

HSV-2 specific CD8 T cells were isolated from the PBMC of two subjects with the HLA-B*57:01 genotype by selecting activated (CD137^{high}) cells after HSV-2 antigen was cross-presented by monocyte-derived dendritic cells. The specific recognition of individual HSV-2 proteins presented by HLA-B*57:01 was determined by measuring IFN-γ responses to aAPC expressing only this HLA allele (Figure 1A). Several HSV-2 ORFs were antigenic. Both T-cell lines reacted to *US1* and *UL39*, while single subjects reacted to *UL7*, *US8*, and *UL46*. Responses to ICP22 (the protein product of ORF *US1*) were decoded using peptides predicted to bind HLA-B*57:01. This identified the amino acids 228–236, abbreviated RTR, as CD8 T-cell antigens for both subjects (Figure 1B,C). The peptide remained active at concentrations as low as 10 ng/mL (Figure 1C).

Since HLA-B*57:01 is associated with hypersensitivity reactions to abacavir, we assessed whether abacavir has any impact on CD8 T-cell activation by RTR, or by sequence-similar peptides from HSV-1 or human interferon lambda (Table 1). We observed the cross-recognition of the HSV-1 cognate peptide, but not the interferon-derived peptides. Abacavir did not alter peptide-driven responses (Figure 1D).

3.2. HLA-B*57:01:US1 Tetramer Binds to Both T Cells and NK Cells

We constructed a tetramer of RTR and HLA-B*57:01 to confirm the HLA-B*57:01 restriction of the T-cell response to ICP22 (Figure 1). Unexpectedly, the B57-RTR tetramer was bound by both CD3^{pos} T cells, and by CD3^{neg} cells in PBMC from subject 12. To determine the phenotype of the CD3^{neg} cells, we co-stained PBMC with NK cell markers and observed the binding of B57-RTR to CD3^{neg} NK cells that were positive for the NK markers CD56 and/or CD16 (Figure 2).

Since other HLA-B*57:01-peptide ligands can bind KIR3DL1 on NK cells [34], we examined if the B57-RTR tetramer could bind to KIR3DL1. Tetramer binding to KIR3DL1 was confirmed by flow cytometry using COS-7 cells transfected with KIR3DL1, while COS-7 cells transfected with negative control (EGFP) expression plasmid did not bind B57-RTR (Figure 2A). Furthermore, tetramer binding was blocked by pre-incubating transfected COS-7 cells with anti-KIR3DL1 mAb DX9, but not with an isotype control antibody, supporting a direct KIR3DL1 interaction with the HLA-B*57:01:US1 peptide complex.

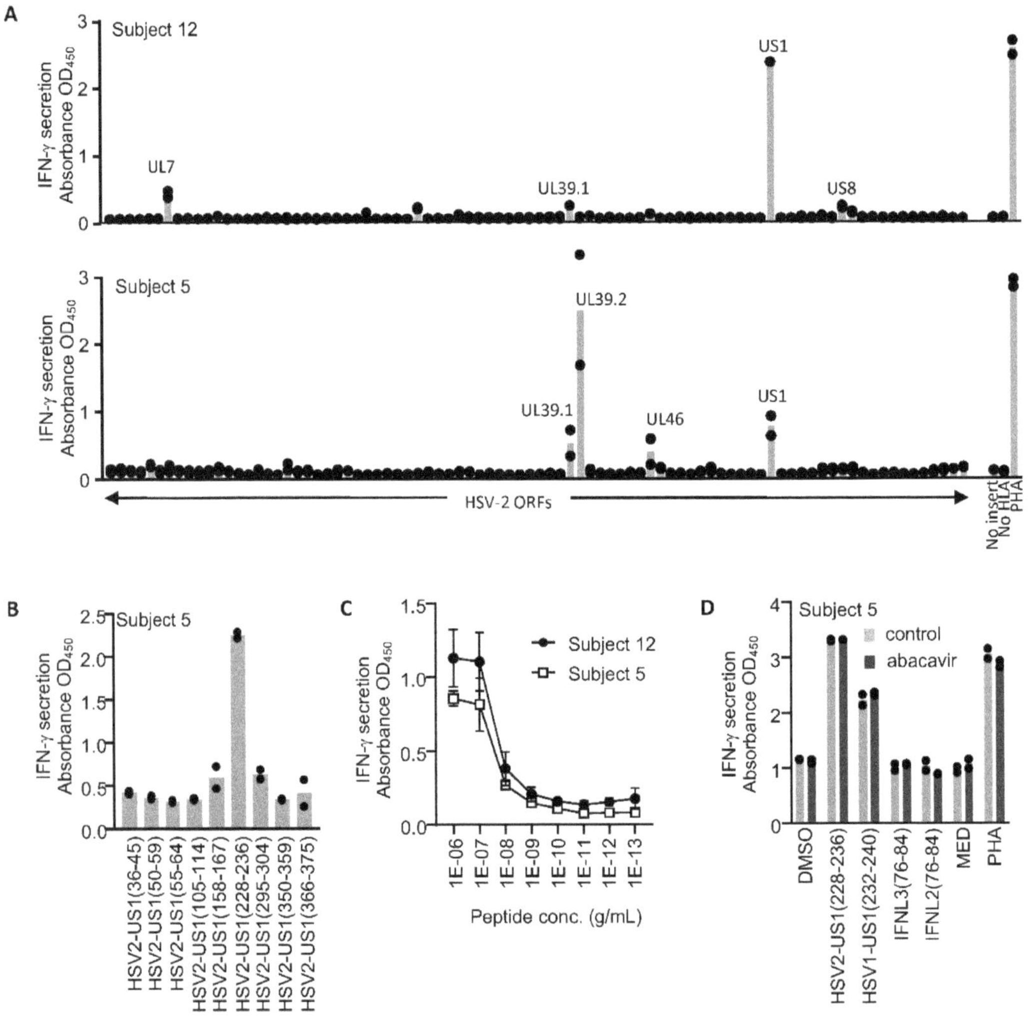

Figure 1. (**A**) T-cell lines generated from CD8 T cells activated (CD137^{high}) following exposure to HSV-2 cross-presented antigen were screened for their protein specificity. An expression library representing all HSV-2 ORFs was co-transfected with HLA-B*57:01 into COS-7 cells to generate artificial antigen-presenting cells (aAPC). HSV-2-specific CD8 T-cell lines were exposed to aAPC and activation measured by IFN-γ secretion. HSV-2 ORFs that generated responses are labeled. (**B**) An epitope at HSV-2 ICP22 (ORF US1) aa 228–236 (sequence RTRLGPRTW; abbreviated RTR) was detected by testing predicted HLA-B*57:01 binding peptides. (**C**) RTR peptide was active to 10 ng/mL for T-cell lines from both subjects. (**D**) Exposure to abacavir (10 µg/mL) did not alter the recognition of RTR or the homolog from HSV-1 (HSV1-US1 (232–240)), and did not reveal activation in the presence of peptides with similar sequences in human IFNL3 (76–84) RsRLfPRTW or IFNL2 (76–84) hsRLfPRTW. MED, medium alone; PHA, phytohemagluttinin P. Alternative amino acids are shown in lower case. All tests were performed in duplicate with raw data indicated and means shown as bars (**A**,**B**,**D**) or with means shown as symbols and standard deviations shown as error bars (**C**).

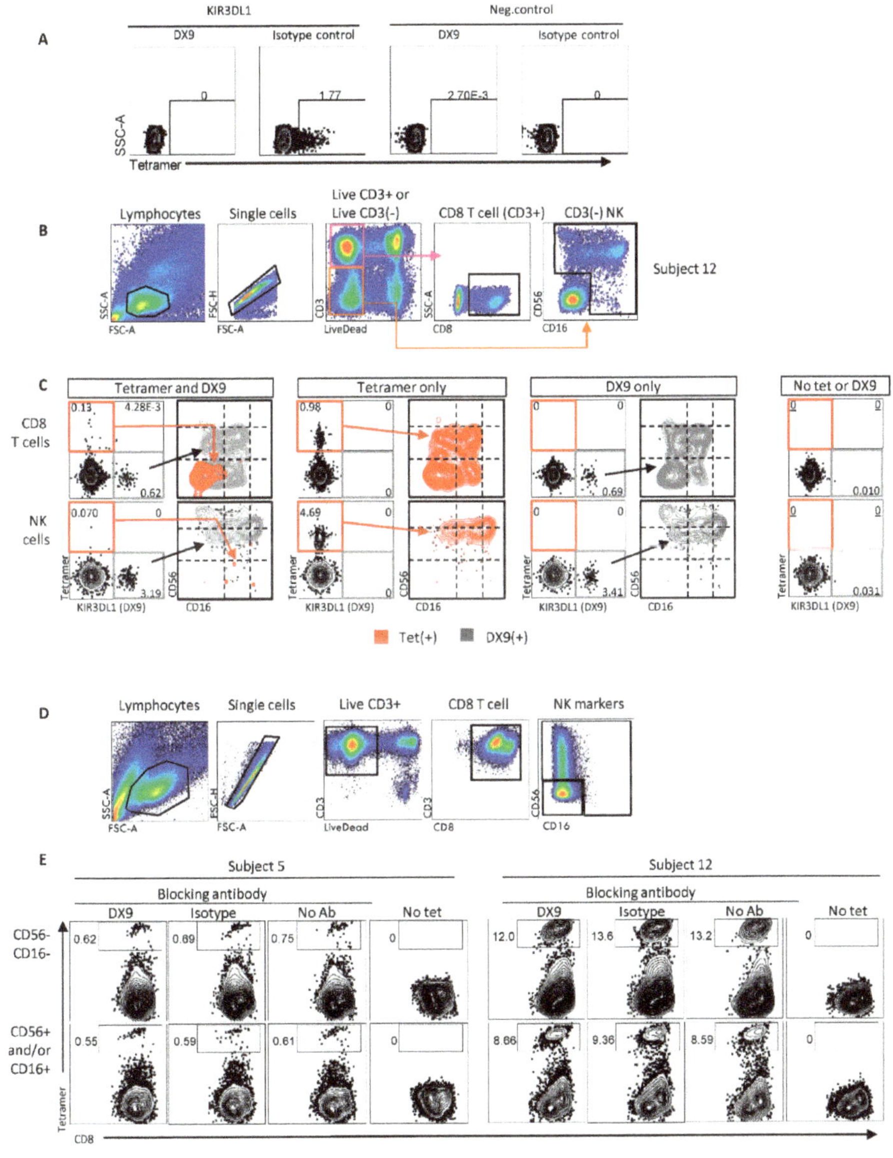

Figure 2. B57-RTR tetramer binds NK and T cells in a KIR3DL1-dependent and -independent manner. (**A**) B57-RTR tetramer binds COS-7 cells transfected with KIR3DL1, but not to cells transfected with a negative control expression plasmid. The interaction was blocked by anti-KIR3DL1 antibody DX9 but not an isotype control. (**B**) Gating scheme showing CD8 T cells (CD3^{pos}CD8^{pos}) and NK cells (CD3^{neg} and (CD56^{pos} and/or CD16^{pos})) subsets within PBMC lymphocytes. (**C**) B57-RPR tetramer binding of CD8 T cells and NK cells (from (**B**)) was compared in the presence or absence of DX9 fluorescently conjugated antibody. DX9 stained discrete subsets of both CD8 T and NK cells and largely inhibited tetramer binding. A modest frequency (0.13%) of CD8 T cells, but only a few NK cells, remained tetramer-positive in the presence of DX9. The CD56 and CD16 phenotypes for B57-RTR tetramer positive (red gate) and DX9-positive events (gray gate) are shown. Note: Tetramer or DX9 gated events are plotted using contour plots with outliers shown. In these overlay graphs, event frequencies

are scaled within the population applied to a grey DX9 or red tetramer layer, and are thus are meaningful for comparing phenotypes but are not quantitative. "Tet" and "tetramer" refer to fluorescent B57-RTR tetramer. (**D**) Gating scheme to separate HSV-2-specific CD8 T-cell lines into CD56^{neg} and CD56^{pos} populations. (**E**) Both HSV-2-specific CD8 T-cell line subsets (from (**D**)) bound B57-RTR tetramer, but binding was not reduced by DX9, consistent with tetramer binding through TCR. Throughout, numbers indicate the percentage of cells in the indicated gate(s).

To further differentiate tetramer binding to KIR3DL1 and TCR, we probed PBMC from subject 12 with B57-RTR tetramer and AlexaFluor700-conjugated DX9 (Figure 2B,C). DX9, when used alone, showed about 0.7% of CD8 T cells and 3.4% of NK cells were KIR3DL1 positive (DX9 only plots, Figure 2C). Similarly, nearly 1% of CD8 T cells and 4.7% of NK cells bound B57-RTR when used alone (tetramer only plots, Figure 2C). Co-staining revealed that 0.13% of tetramer-positive CD8 T cells were DX9-negative, supporting an interaction with CD8 T cells that lack KIR3DL1, presumably via TCR (tetramer and DX9 plots, Figure 2C). These B57-RTR^{pos} DX9^{neg} CD8 T cells were CD56^{neg}CD16^{neg}, while most B57-RTR^{neg} DX9^{pos} CD8 T cells were CD56^{dim} and/or CD16^{dim}, suggesting KIR3DL1 expression occurs primarily on T cells with an NK-cell-like phenotype. Only a few NK cells were B57-RTR^{pos} in the presence of DX9, consistent with binding of B57-RTR to NK cells mediated by KIR3DL1. When used alone, DX9^{pos} and B57-RTR^{pos} NK cells shared the same phenotypes—predominantly CD56^{pos} (dim and bright) with varying levels of CD16—consistent with binding to the same population of NK cells.

To evaluate whether B57-RTR tetramer binding to enriched, expanded HSV-2-specific CD8 T-cell lines was mediated by KIR3DL1 or TCR interactions, we preincubated CD8 T-cell lines from subjects 5 and 12 with DX9 or isotype control antibody before staining with tetramer. The evaluation of CD56 and CD16 was included to assess the expression of these markers in cell lines originally sorted as T cells (CD3-positive). Little CD16 expression was observed, but some CD8 T cells expressed CD56 (Figure 2D), consistent with an activated phenotype [35]. Only a small decrease in tetramer staining was observed in the presence of DX9 compared to isotype control in CD8 T cells without NK markers (10–12% reduction in binding) or with NK markers (7–8% reduction), supporting the CD8 T-cell recognition of B57-RTR through TCR rather than KIR3DL1 (Figure 2E).

3.3. *KIR3DL1 Binding Is Independent from HLA-B*57:01 Genotype or HSV-2 Antigen Experience*

We next tested whether HSV-2 infection or *HLA-B*57:01* genotype were necessary for the interaction of B57-RTR with NK cells. NK cell binding was evaluated using PBMC of three subjects: 1) one HSV-seronegative subject without prior exposure to the HSV-2 RTR epitope, 2) one HSV-2 seropositive subject that did not possess the HLA-B*57:01 allele, and 3) HSV-2 seropositive subject 5 (Figure 1) from which the B57-RTR epitope was identified. Tetramer binding was assessed for cells pre-incubated with DX9, isotype control antibody, or no blocking antibody. NK cells were gated by flow cytometry as in Figure 2B. The B57-RTR tetramer bound >4% of NK cells from each subject, suggesting independence from HLA type or prior antigenic exposure (Figure 3A). DX9 blocked tetramer binding for all three subjects tested, confirming interaction with NK cell KIR3DL1. The phenotype of tetramer-positive NK cells from samples stained with no blocking antibody was compared to the tetramer negative cells. We observed that B57-RTR tetramer-positive NK cells had a lower proportion CD16^{neg}CD56^{bright} cells than did the tetramer-negative cells (Figure 3B; Chi test, $p < 0.0001$ for all subjects). This is consistent with CD56^{bright} NK cells lacking KIR expression, since CD56^{dim} (KIR-suppressed) NK cells upregulate CD56 upon activation [17,35].

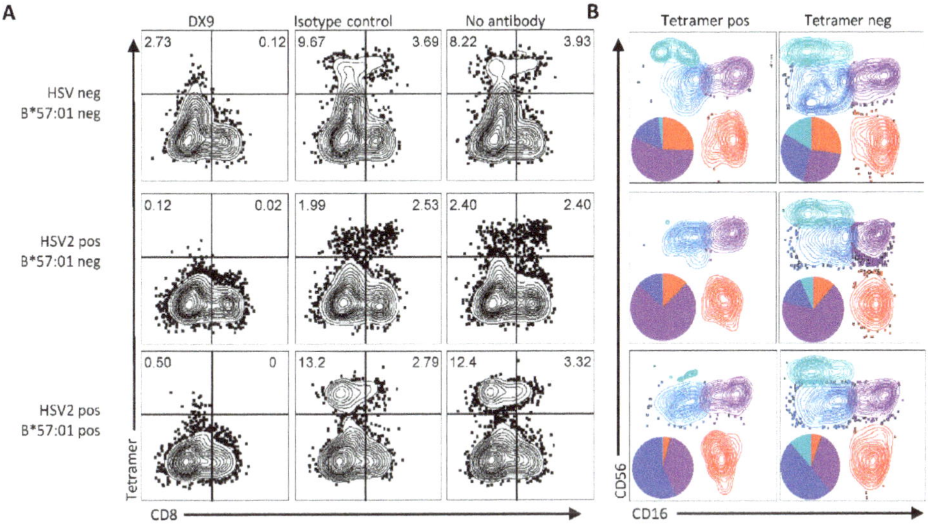

Figure 3. (**A**) B57-RTR tetramer binds gated NK cells in PBMC regardless of HLA type and HSV infection status in a KIR3DL1-dependent manner. Subject characteristics are displayed at left. Gating of NK cells as in Figure 2B. (**B**) Comparison of CD16 and CD56 expression between B57-RTR tetramer-positive and tetramer-negative NK cells from each donor. Contour plots show NK subpopulations of $CD56^{bright}CD16^{neg}$ (light blue), $CD56^{dim}CD16^{neg/dim}$ (dark blue), $CD56^{dim}CD16^{bright}$ (purple), and $CD56^{neg}CD16^{bright}$ (red) populations. Piecharts show relative proportions of each subpopulation within B57-RTR tetramer-positive or tetramer-negative NK cells in matching colors. Tetramer-positive NK cells had fewer $CD56^{bright}CD16^{neg}$ NK cells (light blue) relative to tetramer-negative NK cells.

4. Discussion

The HLA-Bw4 allele HLA-B*57:01 can be advantageous in the control of viral disease, as evidenced in detail for HIV-1, but is detrimental for persons receiving abacavir. Understanding the mechanism(s) by which HLA-B*57:01 benefits control of viral infections, whether through TCR-mediated activation of T cells or by tuning of HLA-KIR interactions, is of medical importance. In this paper, we show that HLA-B*57:01 complexed to a peptide epitope from the HSV-2 ICP22 protein (encoded by ORF *US1*) can activate CD8 T cells, and also bind with NK cells in a KIR3DL1-dependent manner. This indicates that the sensing of the same viral peptide may independently influence two cell populations to modulate HSV-2 infection.

This is the first known example of KIR3DL1 binding by an HLA-bound human herpesvirus peptide, but HLA-B*57:01 interactions are known for other viral systems. Several epitopes from HIV-1, including the TW10 epitope from HIV-1 Gag, when complexed with HLA-B*57:01 can bind KIR3DL1 [34]. A similar finding was described for an HLA-B*57:01-restricted CD8 T-cell epitope in dengue virus NS1 [36]. Circulating NK cells from persons with severe dengue that bound the B57-NS1 tetramer were activated ex vivo, consistent with a functional role in vivo. Similar to our present report, NK cells binding the B57-NS1 tetramer had an intermediate CD56 ($CD56^{dim}$) expression [36], consistent with the expected phenotype of a mature NK cell that expresses KIR [17,37]. A study examining NK phenotypes in subjects with recurrent HSV-2 showed no skewing of NK cells to a highly differentiated or terminal effector phenotype, suggesting HSV-2 peptides have no influence on KIR engagement, albeit the HLA-types of the study subjects were unknown [37]. As KIR3DL1 is expressed by a subset of both NK and T cells, multiple levels of immune crosstalk are plausible during HSV infection. For example, KIR3DL1 positive NK cells could be modulated by HSV-2 infection as a manifestation of trained immunity [38]. The study of large cohorts of HLA-B*57:01-positive persons with and without HSV-2 infection will be required to evaluate this. Dynamic changes in KIR3DL1-expressing NK or T cells during primary infection or reactivation of HSV-2 are also possible and could be studied using serial sampling. Our findings could permit deeper scrutiny at the HLA-KIR genotype level of the impact of HSV-2 infection on NK function.

At this time, the potential roles for viral peptide-HLA interactions with KIR3DL1-expressing lymphocytes are speculative. KIR3DL1 ligation is thought to inhibit NK cell activation. Therefore, the interaction of HLA-B*57:01 complexed to the HSV-2 RTR peptide on the surface of HSV-2 infected cells could prevent NK cell activation, and thus facilitate immune escape from NK cell host defense. Counterbalancing this, profound reduction in cell surface HLA class I expression via the inhibition of a transporter associated with antigen processing (TAP) by HSV [39] decreases HLA interactions with inhibitory KIRs and thus increases the susceptibility of infected cells to NK cell killing [40]. HLA alleles (and their peptides) differ in their dependence on TAP for transport to the cell surface [41,42] and in the degree they are impacted by virally encoded HLA-reducing genes [43]. It is, therefore, of interest to determine how strongly HSV-1 and HSV-2 inhibit HLA-B*57:01 on the cell surface. Adding complexity, CD8 T cells as well as NK cells can express KIR3DL1 (Figure 2C). Thus, B57-RTR could provide a general inhibitory signal to CD8 T cells with diverse TCRs, contributing to broad immune escape. In addition, the extracellular domains of KIR3DL1 and KIR3DS1 are largely identical. Prior work shows that viral peptide bound to HLA-B*57:01 can bind KIR3DS1 [16], albeit other data suggest KIR3DS1 binds poorly to HLA-B*57:01 [15], whereby the binding ability of either receptor is influenced by the peptide composition. As KIR3DS1 contributes to lymphocyte activation, future work on the potential interaction of HLA-B*57:01 with KIR3DS1 could reveal active NK- or T-cell sensing of HSV-infected cells.

T cells are thought to mediate abacavir toxicity in persons with an *HLA-B*57:01* genotype [7]. Abacavir can occupy a pocket in the peptide-binding groove of HLA-B*57:01 to displace the C-terminal amino acid of bound peptide. This could lead to the formation of neo-epitopes that had not participated in negative selection in the thymus and, thus, auto-reactivity. Abacavir binding can also cause the loss of CD8 T-cell responses to pathogens, as recently observed for a HLA-B*57:01-restricted epitope from the gamma-herpesvirus Epstein–Barr virus (EBV) [8]. A similar observation has been made for the antibiotic flucloxacillin, which also generates hypersensitivity reactions related to CD8 T cells, albeit by a mechanism that appears to increase self protein-derived neo-epitopes by modifying lysine residues within peptide sequences [44]. Arguably, altered conformations of self peptides could also impact interactions between HLA-B*57:01 and KIR3DL1, as proposed for other KIR types [45]. For instance, if an altered epitope prevents the KIR3DL1-mediated inhibition of NK cells or KIR3DL1-positive T cells, an undesired auto-reactive response could ensue. Arguing against this, there are no reports of abacavir toxicity correlating with KIR alleles or haplotypes. The incomplete penetrance of abacavir toxicity in HLA-B*57:01-positive persons remains unexplained and could be related to a combination of KIR variation and viral infection. In our study, we included a peptide from HSV-1 orthologous to the HSV-2 RTR epitope and sequence-related peptides in human IFNL2 and IFNL3. CD8 T-cell lines cross-reacted with the HSV-1 peptide, but not the self-peptides. No change in reactivity was observed to any peptide in the presence of abacavir, suggesting that these particular HSV-2-specific CD8 T cells may not participate in abacavir-induced hypersensitivity. Since epitope-specific CD8 T cells occur as a polyclonal swarm within-person [46], studies of RTR-specific clonal T cells with defined TCRs could uncover modulation by abacavir. We have not yet examined the effect of abacavir on the interaction between HLA-B*57:01-RTR and KIR3DL1 to examine the possibility that the drug influences KIR3DL1 function.

In summary, both TCR on CD8 T cells, and the KIR3DL1 molecule, can bind to complexes of HLA-B*57:01 and a peptide epitope that we newly identified in the HSV-2 ICP22 protein encoded by gene *US1*. The specificity of binding was confirmed by the heterotopic expression of KIR3DL1 in human KIR-negative non-human primate COS-7 cells, and by blocking with the KIR3DL1-specific DX9 mAb. NK cells are known to play an important role in host defense against HSV, but further work will be required to determine if the molecular interaction reported in this paper has functional consequences for recognition or escape from cells including NK cells and KIR-expressing T cells in persons with diverse KIR haplotypes.

Author Contributions: The study was devised, planned and managed by D.M.K. and K.J.L.; K.J.L. analyzed experimental data and wrote the manuscript; K.J.L., V.L.C. and L.D. conducted experimental procedures. All authors have read and agreed to the published version of the manuscript.

Funding: Funding was provided by NIH contract HHSN75N93019C00063 (DMK).

Institutional Review Board Statement: The study was conducted in accordance with the Declaration of Helsinki, and approved by the Institutional Review Board (or Ethics Committee) of the University of Washington (protocol code STUDY00004400).

Informed Consent Statement: Informed consent was obtained from all subjects involved in the study.

Data Availability Statement: The data presented in this study are available upon reasonable request from the corresponding author.

Acknowledgments: The authors wish to thank the clinical and administrative teams at the Virology Research clinic, led by Anna Wald, for subject recruitment and sample acquisition efforts.

Conflicts of Interest: D.M.K. and K.J.L. are co-inventor of patents owned by their institution on HSV-2 vaccines. D.M.K. has received research funding from Sanofi Pasteur for the immunogenicity testing of HSV-2 vaccines. The funders of this study had no role in the design of the study; in the collection, analyses, or interpretation of data; in the writing of the manuscript, or in the decision to publish the results.

References

1. International HIV Controllers Study. The major genetic determinants of HIV-1 control affect HLA class I peptide presentation. *Science* **2010**, *330*, 1551–1557. [CrossRef] [PubMed]
2. Dinges, W.L.; Richardt, J.; Friedrich, D.; Jalbert, E.; Liu, Y.; Stevens, C.E.; Maenza, J.; Collier, A.C.; Geraghty, D.E.; Smith, J.; et al. Virus-specific CD8+ T-cell responses better define HIV disease progression than HLA genotype. *J. Virol.* **2010**, *84*, 4461–4468. [CrossRef] [PubMed]
3. Chen, H.; Ndhlovu, Z.M.; Liu, D.; Porter, L.C.; Fang, J.W.; Darko, S.; Brockman, M.A.; Miura, T.; Brumme, Z.L.; Schneidewind, A.; et al. TCR clonotypes modulate the protective effect of HLA class I molecules in HIV-1 infection. *Nat. Immunol.* **2012**, *13*, 691–700. [CrossRef] [PubMed]
4. Genovese, L.; Nebuloni, M.; Alfano, M. Cell-Mediated Immunity in Elite Controllers Naturally Controlling HIV Viral Load. *Front. Immunol.* **2013**, *4*, 86. [CrossRef] [PubMed]
5. Boulet, S.; Song, R.; Kamya, P.; Bruneau, J.; Shoukry, N.H.; Tsoukas, C.M.; Bernard, N.F. HIV protective KIR3DL1 and HLA-B genotypes influence NK cell function following stimulation with HLA-devoid cells. *J. Immunol.* **2010**, *184*, 2057–2064. [CrossRef]
6. Mallal, S.; Nolan, D.; Witt, C.; Masel, G.; Martin, A.M.; Moore, C.; Sayer, D.; Castley, A.; Mamotte, C.; Maxwell, D.; et al. Association between presence of HLA-B*5701, HLA-DR7, and HLA-DQ3 and hypersensitivity to HIV-1 reverse-transcriptase inhibitor abacavir. *Lancet* **2002**, *359*, 727–732. [CrossRef]
7. Ostrov, D.A.; Grant, B.J.; Pompeu, Y.A.; Sidney, J.; Harndahl, M.; Southwood, S.; Oseroff, C.; Lu, S.; Jakoncic, J.; de Oliveira, C.A.; et al. Drug hypersensitivity caused by alteration of the MHC-presented self-peptide repertoire. *Proc. Natl. Acad. Sci. USA* **2012**, *109*, 9959–9964. [CrossRef]
8. Sooda, A.; Rwandamuriye, F.; Wanjalla, C.N.; Jing, L.; Koelle, D.M.; Peters, B.; Leary, S.; Chopra, A.; Calderwood, M.A.; Mallal, S.A.; et al. Abacavir inhibits but does not cause self-reactivity to HLA-B*57:01-restricted EBV specific T cell receptors. *Commun. Biol.* **2022**, *5*, 133. [CrossRef]
9. Downing, J.; D'Orsogna, L. High-resolution human KIR genotyping. *Immunogenetics* **2022**, 1–11. [CrossRef]
10. Campbell, K.S.; Purdy, A.K. Structure/function of human killer cell immunoglobulin-like receptors: Lessons from polymorphisms, evolution, crystal structures and mutations. *Immunology* **2011**, *132*, 315–325. [CrossRef]
11. Petrushkin, H.; Norman, P.J.; Lougee, E.; Parham, P.; Wallace, G.R.; Stanford, M.R.; Fortune, F. KIR3DL1/S1 Allotypes Contribute Differentially to the Development of Behcet Disease. *J. Immunol.* **2019**, *203*, 1629–1635. [CrossRef] [PubMed]
12. Harrison, G.F.; Leaton, L.A.; Harrison, E.A.; Kichula, K.M.; Viken, M.K.; Shortt, J.; Gignoux, C.R.; Lie, B.A.; Vukcevic, D.; Leslie, S.; et al. Allele imputation for the killer cell immunoglobulin-like receptor KIR3DL1/S1. *PLoS Comput. Biol.* **2022**, *18*, e1009059. [CrossRef]

13. Gardiner, C.M.; Guethlein, L.A.; Shilling, H.G.; Pando, M.; Carr, W.H.; Rajalingam, R.; Vilches, C.; Parham, P. Different NK cell surface phenotypes defined by the DX9 antibody are due to KIR3DL1 gene polymorphism. *J. Immunol.* **2001**, *166*, 2992–3001. [CrossRef] [PubMed]
14. O'Connor, G.M.; Yamada, E.; Rampersaud, A.; Thomas, R.; Carrington, M.; McVicar, D.W. Analysis of binding of KIR3DS1*014 to HLA suggests distinct evolutionary history of KIR3DS1. *J. Immunol.* **2011**, *187*, 2162–2171. [CrossRef] [PubMed]
15. Vivian, J.P.; Duncan, R.C.; Berry, R.; O'Connor, G.M.; Reid, H.H.; Beddoe, T.; Gras, S.; Saunders, P.M.; Olshina, M.A.; Widjaja, J.M.; et al. Killer cell immunoglobulin-like receptor 3DL1-mediated recognition of human leukocyte antigen B. *Nature* **2011**, *479*, 401–405. [CrossRef]
16. O'Connor, G.M.; Vivian, J.P.; Gostick, E.; Pymm, P.; Lafont, B.A.; Price, D.A.; Rossjohn, J.; Brooks, A.G.; McVicar, D.W. Peptide-Dependent Recognition of HLA-B*57:01 by KIR3DS1. *J. Virol.* **2015**, *89*, 5213–5221. [CrossRef]
17. Jacobs, R.; Hintzen, G.; Kemper, A.; Beul, K.; Kempf, S.; Behrens, G.; Sykora, K.W.; Schmidt, R.E. CD56bright cells differ in their KIR repertoire and cytotoxic features from CD56dim NK cells. *Eur. J. Immunol.* **2001**, *31*, 3121–3127. [CrossRef]
18. Biron, C.A.; Byron, K.S.; Sullivan, J.L. Severe herpesvirus infections in an adolescent without natural killer cells. *N. Engl. J. Med.* **1989**, *320*, 1731–1735. [CrossRef]
19. Mace, E.M.; Hsu, A.P.; Monaco-Shawver, L.; Makedonas, G.; Rosen, J.B.; Dropulic, L.; Cohen, J.I.; Frenkel, E.P.; Bagwell, J.C.; Sullivan, J.L.; et al. Mutations in GATA2 cause human NK cell deficiency with specific loss of the CD56(bright) subset. *Blood* **2013**, *121*, 2669–2677. [CrossRef]
20. Mace, E.M.; Orange, J.S. Emerging insights into human health and NK cell biology from the study of NK cell deficiencies. *Immunol. Rev.* **2019**, *287*, 202–225. [CrossRef]
21. Lenart, M.; Kluczewska, A.; Szaflarska, A.; Rutkowska-Zapala, M.; Wasik, M.; Ziemianska-Pieta, A.; Kobylarz, K.; Pituch-Noworolska, A.; Siedlar, M. Selective downregulation of natural killer activating receptors on NK cells and upregulation of PD-1 expression on T cells in children with severe and/or recurrent Herpes simplex virus infections. *Immunobiology* **2021**, *226*, 152097. [CrossRef] [PubMed]
22. John, M.; Heckerman, D.; James, I.; Park, L.P.; Carlson, J.M.; Chopra, A.; Gaudieri, S.; Nolan, D.; Haas, D.W.; Riddler, S.A.; et al. Adaptive interactions between HLA and HIV-1: Highly divergent selection imposed by HLA class I molecules with common supertype motifs. *J. Immunol.* **2010**, *184*, 4368–4377. [CrossRef] [PubMed]
23. Tigges, M.A.; Koelle, D.; Hartog, K.; Sekulovich, R.E.; Corey, L.; Burke, R.L. Human CD8+ herpes simplex virus-specific cytotoxic T-lymphocyte clones recognize diverse virion protein antigens. *J. Virol.* **1992**, *66*, 1622–1634. [CrossRef] [PubMed]
24. Ashley, R.L.; Militoni, J.; Lee, F.; Nahmias, A.; Corey, L. Comparison of Western blot (immunoblot) and glycoprotein G-specific immunodot enzyme assay for detecting antibodies to herpes simplex virus types 1 and 2 in human sera. *J. Clin. Microbiol.* **1988**, *26*, 662–667. [CrossRef] [PubMed]
25. Peng, T.; Phasouk, K.; Bossard, E.; Klock, A.; Jin, L.; Laing, K.J.; Johnston, C.; Williams, N.A.; Czartoski, J.L.; Varon, D.; et al. Distinct populations of antigen-specific tissue-resident CD8+ T cells in human cervix mucosa. *JCI Insight* **2021**, *6*, 1771. [CrossRef] [PubMed]
26. Magaret, A.S.; Wald, A.; Huang, M.L.; Selke, S.; Corey, L. Optimizing PCR positivity criterion for detection of herpes simplex virus DNA on skin and mucosa. *J. Clin. Microbiol.* **2007**, *45*, 1618–1620. [CrossRef] [PubMed]
27. Koelle, D.M. Expression cloning for the discovery of viral antigens and epitopes recognized by T-cells. *Methods* **2003**, *29*, 213–226. [CrossRef]
28. Johnston, C.; Zhu, J.; Jing, L.; Laing, K.J.; McClurkan, C.M.; Klock, A.; Diem, K.; Jin, L.; Stanaway, J.; Tronstein, E.; et al. Virologic and immunologic evidence of multifocal genital herpes simplex virus 2 infection. *J. Virol.* **2014**, *88*, 4921–4931. [CrossRef]
29. Jing, L.; Laing, K.J.; Dong, L.; Russell, R.M.; Barlow, R.S.; Haas, J.G.; Ramchandani, M.S.; Johnston, C.; Buus, S.; Redwood, A.J.; et al. Extensive CD4 and CD8 T Cell Cross-Reactivity between Alphaherpesviruses. *J. Immunol.* **2016**, *196*, 2205–2218. [CrossRef]
30. Jing, L.; Haas, J.; Chong, T.M.; Bruckner, J.J.; Dann, G.C.; Dong, L.; Marshak, J.O.; McClurkan, C.L.; Yamamoto, T.N.; Bailer, S.M.; et al. Herpes simplex virus type 1 T-cells antigens in humans revealed by cross-presentation and genome-wide screening. *J. Clin. Investig.* **2012**, *122*, 654–673. [CrossRef]
31. Vita, R.; Overton, J.A.; Greenbaum, J.A.; Ponomarenko, J.; Clark, J.D.; Cantrell, J.R.; Wheeler, D.K.; Gabbard, J.L.; Hix, D.; Sette, A.; et al. The immune epitope database (IEDB) 3.0. *Nucleic Acids Res.* **2015**, *43*, D405–D412. [CrossRef] [PubMed]
32. Lucas, A.; Lucas, M.; Strhyn, A.; Keane, N.M.; McKinnon, E.; Pavlos, R.; Moran, E.M.; Meyer-Pannwitt, V.; Gaudieri, S.; D'Orsogna, L.; et al. Abacavir-reactive memory T cells are present in drug naive individuals. *PLoS ONE* **2015**, *10*, e0117160. [CrossRef] [PubMed]
33. Koelle, D.M.; Chen, H.B.; Gavin, M.A.; Wald, A.; Kwok, W.W.; Corey, L. CD8 CTL from genital herpes simplex lesions: Recognition of viral tegument and immediate early proteins and lysis of infected cutaneous cells. *J. Immunol.* **2001**, *166*, 4049–4058. [CrossRef] [PubMed]
34. Fadda, L.; O'Connor, G.M.; Kumar, S.; Piechocka-Trocha, A.; Gardiner, C.M.; Carrington, M.; McVicar, D.W.; Altfeld, M. Common HIV-1 peptide variants mediate differential binding of KIR3DL1 to HLA-Bw4 molecules. *J. Virol.* **2011**, *85*, 5970–5974. [CrossRef] [PubMed]
35. Van Acker, H.H.; Capsomidis, A.; Smits, E.L.; Van Tendeloo, V.F. CD56 in the Immune System: More Than a Marker for Cytotoxicity? *Front. Immunol.* **2017**, *8*, 892. [CrossRef] [PubMed]

36. Townsley, E.; O'Connor, G.; Cosgrove, C.; Woda, M.; Co, M.; Thomas, S.J.; Kalayanarooj, S.; Yoon, I.K.; Nisalak, A.; Srikiatkhachorn, A.; et al. Interaction of a dengue virus NS1-derived peptide with the inhibitory receptor KIR3DL1 on natural killer cells. *Clin. Exp. Immunol.* **2016**, *183*, 419–430. [CrossRef]
37. Bjorkstrom, N.K.; Svensson, A.; Malmberg, K.J.; Eriksson, K.; Ljunggren, H.G. Characterization of natural killer cell phenotype and function during recurrent human HSV-2 infection. *PLoS ONE* **2011**, *6*, e27664. [CrossRef]
38. Palgen, J.L.; Tchitchek, N.; Huot, N.; Elhmouzi-Younes, J.; Lefebvre, C.; Rosenbaum, P.; Dereuddre-Bosquet, N.; Martinon, F.; Hocini, H.; Cosma, A.; et al. NK cell immune responses differ after prime and boost vaccination. *J. Leukoc. Biol.* **2019**, *105*, 1055–1073. [CrossRef]
39. Hill, A.; Jugovic, P.; York, I.; Russ, G.; Bennink, J.; Yewdell, J.; Ploegh, H.; Johnson, D. Herpes simplex virus turns off the TAP to evade host immunity. *Nature* **1995**, *375*, 411–415. [CrossRef]
40. Huard, B.; Fruh, K. A role for MHC class I down-regulation in NK cell lysis of herpes virus-infected cells. *Eur. J. Immunol.* **2000**, *30*, 509–515. [CrossRef]
41. Geng, J.; Zaitouna, A.J.; Raghavan, M. Selected HLA-B allotypes are resistant to inhibition or deficiency of the transporter associated with antigen processing (TAP). *PLoS Pathog.* **2018**, *14*, e1007171. [CrossRef] [PubMed]
42. Tey, S.K.; Khanna, R. Autophagy mediates transporter associated with antigen processing-independent presentation of viral epitopes through MHC class I pathway. *Blood* **2012**, *120*, 994–1004. [CrossRef] [PubMed]
43. Mwimanzi, F.; Toyoda, M.; Mahiti, M.; Mann, J.K.; Martin, J.N.; Bangsberg, D.; Brockman, M.A.; Goulder, P.; Kirchhoff, F.; Brumme, Z.L.; et al. Resistance of Major Histocompatibility Complex Class B (MHC-B) to Nef-Mediated Downregulation Relative to that of MHC-A Is Conserved among Primate Lentiviruses and Influences Antiviral T Cell Responses in HIV-1-Infected Individuals. *J. Virol.* **2018**, *92*, e01409–e01417. [CrossRef] [PubMed]
44. Puig, M.; Ananthula, S.; Venna, R.; Kumar Polumuri, S.; Mattson, E.; Walker, L.M.; Cardone, M.; Takahashi, M.; Su, S.; Boyd, L.F.; et al. Alterations in the HLA B*57:01 Immunopeptidome by Flucloxacillin and Immunogenicity of Drug-Haptenated Peptides. *Front. Immunol.* **2020**, *11*, 629399. [CrossRef]
45. Fadda, L.; Borhis, G.; Ahmed, P.; Cheent, K.; Pageon, S.V.; Cazaly, A.; Stathopoulos, S.; Middleton, D.; Mulder, A.; Claas, F.H.; et al. Peptide antagonism as a mechanism for NK cell activation. *Proc. Natl. Acad. Sci. USA* **2010**, *107*, 10160–10165. [CrossRef]
46. Dong, L.; Li, P.; Oenema, T.; McClurkan, C.L.; Koelle, D.M. Public TCR use by herpes simplex virus-2-specific human CD8 CTLs. *J. Immunol.* **2010**, *184*, 3063–3071. [CrossRef]

viruses

Article

Studies of Infection and Experimental Reactivation by Recombinant VZV with Mutations in Virally-Encoded Small Non-Coding RNA

Punam Bisht [1], Biswajit Das [1], Tatiana Borodianskiy-Shteinberg [1], Paul R. Kinchington [2] and Ronald S. Goldstein [1,*]

[1] Mina and Everard Goodman Faculty of Life Sciences, Bar-Ilan University, Ramat-Gan 52900, Israel; punam.bisht.pb@gmail.com (P.B.); biswamicrobio@gmail.com (B.D.); shteinbergt@gmail.com (T.B.-S.)
[2] Departments of Ophthalmology and of Microbiology and Molecular Genetics, University of Pittsburgh, Pittsburgh, PA 15213-2588, USA; kinchingtonp@upmc.edu
* Correspondence: ron.goldstein@biu.ac.il

Abstract: Locked-nucleotide analog antagonists (LNAA) to four varicella zoster virus small non-coding RNA (VZVsncRNA 10–13) derived from the mRNA of the open reading frame (ORF) 61 gene individually reduce VZV replication in epithelial cells and fibroblasts. To study the potential roles VZVsncRNA 10–13 have in neuronal infection we generated two recombinant VZV; one in which 8 nucleotides were changed in VZVsncRNA10 without altering the encoded residues of ORF61 (VZVsnc10MUT) and a second containing a 12-nucleotide deletion of the sequence common to VZVsncRNA12 and 13, located in the ORF61 mRNA leader sequence (VZVsnc12-13DEL). Both were developed from a VZV BAC with a green fluorescent protein (GFP) reporter fused to the N terminal of the capsid protein encoded by ORF23. The growth of both mutant VZV in epithelial cells and fibroblasts was similar to that of the parental recombinant virus. Both mutants established productive infections and experimental latency in neurons derived from human embryonic stem cells (hESC). However, neurons that were latently infected with both VZV mutant viruses showed impaired ability to reactivate when given stimuli that successfully reactivated the parental virus. These results suggest that these VZVsncRNA may have a role in VZV latency maintenance and/or reactivation. The extension of these studies and confirmation of such roles could potentially inform the development of a non-reactivating, live VZV vaccine.

Keywords: varicella zoster virus; latency; reactivation; human neuron culture; non-coding RNA

Citation: Bisht, P.; Das, B.; Borodianskiy-Shteinberg, T.; Kinchington, P.R.; Goldstein, R.S. Studies of Infection and Experimental Reactivation by Recombinant VZV with Mutations in Virally-Encoded Small Non-Coding RNA. *Viruses* **2022**, *14*, 1015. https://doi.org/10.3390/v14051015

Academic Editors: Charles Grose, Ravi Mahalingam and Joel Rovnak

Received: 31 March 2022
Accepted: 6 May 2022
Published: 10 May 2022

1. Introduction

Human Herpesvirus-3 or Varicella-Zoster virus (VZV) is a neurotropic alphaherpesvirus that causes varicella (chickenpox) after primary infection, and then establishes a life-long latent infection of peripheral ganglionic neurons [1]. Reactivation from latency in response to poorly defined stimuli and/or immune decline results in the development of reactivation disease, usually manifesting as painful Herpes Zoster (shingles; HZ). HZ is frequently complicated, most often by chronic pain states extending beyond the resolution of the infection termed postherpetic neuralgia or PHN [2]. Most non-human species and cells derived from them do not support full productive VZV infection, except the guinea pig whose cells were used to attenuate a clinical virus (Parent of Oka or POka) and generate the vOka vaccine strain [3]. The human-specificity of VZV infection and consequent lack of animal models has hampered the study of VZV latency and reactivation. This is in contrast to the ready in-vitro and in-vivo experimental infection of rodent and rabbit neurons by the related alphaherpesvirus herpes simplex that has allowed myriad studies of these processes for HSV1/2.

However, three recent experimental developments have permitted the advancement of our understanding of these clinically important processes. The first was the development of models of human neuronal infection, such as human ganglionic tissues in culture or grafted to severe compromised immunodeficient mice (SCID-hu DRG mice (reviewed in [4]). The second was the development of experimental models using human stem cell-derived neurons that recapitulate VZV latent infection in-vitro that can be experimentally reactivated [5]. The use of these models has revealed, for instance, that the vaccine strain vOka, infects neurons latently but is impaired for reactivation [6]. They have also been used to demonstrate that the Jun N- terminal kinase (JNK) pathway contributes to the lytic/latent decision process [7]. The third important advance was the discovery of a family of novel transcripts termed VLT in human ganglia obtained from VZV seropositive individuals at death. These are positionally and directionally similar to the herpes simplex virus (HSV) latency-associated transcript (LAT) and latency transcripts reported for other neurotropic alphaherpesviruses [8]. Multiple splice variants of VLT are found in cultured cells productively infected with VZV and evidence has been presented that unlike LAT of HSV, some spliced variants of VLT can encode proteins. It is still not clear how VZV VLT and the long-studied HSV LAT contribute to the infection of neurons, the latent state, and reactivation from it.

A focus of recent research has been the potential role of virally and host-encoded small non-coding RNAs (sncRNA), particularly microRNA (miR). Of the nine human herpesviruses, seven have been reported to express miRNA, with at least 29 miRNAs being reported to be encoded by HSV-1 [9]. The roles of several HSV-1 miRNAs have been evaluated for their potential contribution to productive and latent infection of neurons and the lytic/latent decision process (i.e., [10], reviewed in [11]). In addition to miRNA, other small RNAs such as sncRNA have been reported to be expressed in mammalian cells that can also regulate gene expression at the transcriptional level [12].

The encoding of many miR by the human herpesviruses, and especially the alphaherpesviruses HSV1 and HSV2, led us to search for sncRNA in VZV. We first predicted that 24 20–24 nt small non-coding sncRNA (VZVsncRNA to be encoded by the VZV genome [13] from bioinformatic analysis of NGS data of small (>200 nt) RNA in VZV-infected cells. Stem-loop reverse-transcriptase quantitative TaqMan PCR (SL-rt-qPCR) detected 23 of 24 of these sncRNA in multiple infected cell types in culture [14]. Transfection of locked nucleotide analog antagonists (LNAA) to several of the VZV-encoded small RNA sequences (VZVsncRNA) significantly reduced viral spread, suggesting that the predicted VZVsncRNA may have roles in VZV infection.

More recently, we found that LNAA directed to VZVsncRNA10–13, 4 VZVsncRNA that are encoded by the mRNA for the ORF61 gene and opposite to the putative VZV latency-associated transcript VLT, consistently reduced both the viral spread and plaque number [15]. LNAA-mediated inhibition of VZV sncRNA12 not only decreased viral replication but also resulted in a significant increase in VLT levels expressed during lytic infection. ORF61 encodes a protein that is the VZV homolog of HSV ICP0, a pro-lytic non-specific transactivator that functions in part to counteract intrinsic and innate responses to HSV infection through an E3 ubiquitin ligase activity [16]. Much stronger anti-viral effects were observed when combinations of antagonists to these four VZVsncRNA were found in the epithelial cell-line ARPE-19 and in primary human fibroblasts [17].

In view of the proposed roles of miR encoded by LAT in HSV1 reactivation, we here experimentally address the possibility that VZVsncRNA10–13 may influence VZV latency and reactivation. Two VZV mutant BACs were generated in which the sequences encoding for VZVsncRNA10 and VZVsncRNA12/13 were altered without changing the coding of ORF61 using BAC recombineering methods. The two VZV mutant viruses derived appeared not to result in significant differences in productive infection of either ARPE19 cells or in cultured neurons when compared to their parent VZV. Both mutant viruses also established experimental latency in hESC-derived neurons including the silencing of lytic gene expression, the expression of VLT RNAs, and the presence of viral genomes detected by PCR. When neurons latently infected by both VZVsncRNA mutant viruses were challenged with stimuli that experimentally reactivate VZV, their reactivation appeared impaired compared to cultures of the neurons latently infected with the parental VZV. These data are consistent with the possibility that these VZVsncRNA, which are encoded opposite to VLT, may have a role in the VZV latency/reactivation switch.

2. Materials and Methods

2.1. Cells

MeWo (human melanoma cell line ATCC HTB-65) were cultured in DMEM, containing 10% FBS, 2 mM glutamine, 100 U/mL penicillin, 100 µg/mL streptomycin, and 0.25 µg/mL amphotericin B. ARPE19 (human retinal pigmented epithelium (ATCC CL4000) were maintained in DMEM F12 with the glutamine and antibiotics.

2.2. Neuronal Differentiation from hESC

The human embryonic stem cell (hESC) line H9 (WA09 (WiCell Madison WI, USA)) was maintained on murine embryo fibroblast line (ATCC CRL-1503) in Nutristem (Biological Industries, Israel) medium and differentiated into neurons using a modification of the agarose microwell method of Birenboim et al. [18]. Briefly, hESC were dissociated with Accutase (Sigma-Aldrich, St. Louis, MO, USA), and 1×10^6 cells were seeded into 256-well agarose microwell dishes made from silicone molds (Sigma-Aldrich). The cells aggregated for 4 days in the molds in a medium consisting of GMEM (Gibco/Life Technologies, Carlsbad, CA, USA) 1% penicillin/streptomycin (Biological Industries (BI), 1% L-glutamate (BI) 1% pyruvate (BI) 10% KSR—knockout serum replacement (Gibco) 1% non-essential amino acids (BI) 0.1 um mercaptoethanol (Sigma-Aldrich), containing bone morphogenetic protein (BMP)4 inhibitors SB431542 (10 µM) and dorsomorphin (2 µM) (Tocris Bioscience). The aggregates were then fed for an additional 10 days in a medium lacking the BMP4 inhibitors. The aggregates were then plated on coverslips or wells of 24 well plates coated with polylysine/laminin or tissue culture plates in a differentiation medium consisting of DMEM/F12 with neural growth and survival factors NGF, BDNF, NT3, and GDNF (Alomone Labs) and B27 supplement (Gibco). Approximately 10–20 aggregates were plated on each coverslip/well. Dividing cells were eliminated from the cultures 2 days after seeding using mitotic inhibitors 24 µM 5-fluorodeoxyuridine (F0503, Sigma-Aldrich) and 0.6 µM cytosine arabinoside (C6645, Sigma-Aldrich) added 4 days after plating. Twenty-four µM Uridine (U3750, Sigma-Aldrich) was added to offset the toxicity of the fluordeoxyuridine ([19,20]). Cultures were maintained for a total of 21 days after removal from the molds. Immunofluorescent staining for neurofilament proteins was performed with antibody 2H3 (NF-M) deposited to the Developmental Studies Hybridoma Bank by Jessell, T.M. and Dodd, J., and Sigma-Aldrich (Merck) N4124 (NF-H).

2.3. Generation of VZV Mutant in VZVsncRNA

The two VZV mutant viruses were generated using a bacterial artificial chromosome (VZV BAC) based on the parent Oka (POka strain [21], that contained EGFP fused to the N terminus of ORF23 (VZVORF23GFP) detailed previously [12] and housed in the *E. coli* strain GS1783 (kind gift of G Smith Northwestern University, Chicago, IL, USA). GS1783 contains heat-inducible expression of the λ red recombination genes and an arabinose inducible expression of the homing enzyme IsceI. Briefly, primers developed to contain the sncRNA10 mutant or sncRNA 12/13 deletion were used to amplify a kanamycin resistance cassette from the plasmid pEGFP KAN-in [13] using the primers listed in Table 1.

Table 1. Oligonucleotides used for making mutant viruses.

S.N.	Primer Name	Primer Sequence (5′→3′)
1	VZVsncRNA10MUTFW	CTTTACTCGGATGGCTTGATGATCAACTTGCGGACTGTACCAGCCTGAGATAGTCG AACCAACAAAAATGTTGATATAGGATGACGACGATAAGTAGC
2	VZVsncRNA10MUTREV	GCCCGTAAATACCTATATAGTTTAATATCAAACATTTTTGTTGGTTCGACTATCTC AGGCTGGTACAGTTCCGCAAGTCAGGGTAATGCCAGTGTTAC
3	SNC1213DelFW	GCTACCGCCCGCTAATATGGTATCCATGGTAACAACTGGCTGTATTCTACCAAACA CGTAGCAGAACTGAGGATGACGACGATAAGTAGG
4	SNC1213delRev	TCACAATTTAGAACGCATGGCAGTTCTGCTACGTGTTTGGTAGAATACAGCCAG TTGTTACAGGGTAATGCCAGTGTTAC

The amplification products were PCR amplified and then recombined into VZV BAC GFP23 after electroporation as detailed previously [21] and modified by Erazo et al. [22]. Colonies selected on LB-agar plates containing kanamycin + chloramphenicol were then screened for correct insertion by RFLP analyses and then subjected to a second recombination event concurrent with arabinose induction of the expression of IScel to counterselect for loss of the kanamycin resistance cassette. Clones were again screened for removal of the cassette by RFLP analyses and sequencing across the junctions. Selected colonies were stored as 50% glycerol stocks.

Mutant VZV (VZVsnc10MUT and VZVsnc12-13DEL) were derived from the BACs following transfection of BAC DNA into MeWo cells using Lipofectamine 3000 (Thermo-Fischer Scientific, Waltham, MA, USA), as detailed in [22]. The viruses were verified for homogeneity of green fluorescence, which was found to be stable over extensive passaging so that all plaques remained GFP positive. Cell-free and debris VZV for infection were generated as detailed previously [5].

2.4. Stem-Loop Taqman qPCR for VZVsncRNA Quantification and Detection

The expression of the VZVsncRNA was carried out by Stem-loop TaqMan qPCR as detailed previously [13]). In brief, small (<200 nt) RNA obtained from ARPE19 cells infected with cell-free VZV were extracted using the Hybrid-R kit (Geneall, Seoul, Korea). Primers were designed using published algorithms [23] and the probe used was LGC Biosearch Technologies # DLO-RFB-5 (Middlesex, UK)) (Table 2). cDNA for VZVsncRNA10mut or VZVsncRNA12-13 as well as hsa-mir26 were prepared in separate reactions using MMLV reverse transcriptase (Promega, # M1701). RNA was treated with DNase (AMPD1, Sigma-Aldrich). All qPCR reactions were performed in triplicate and results were averaged to compensate for pipetting errors. Human hsa-mir26 shows little variation in expression between cells infected and uninfected by multiple viruses [4] and was used to normalize the expression of small RNA.

Table 2. Primers used for detecting transcripts in VZV-infected cells by qRT-PCR.

ORF	FW Primer Sequence (5′→3′)	Rev Primer Sequence (5′→3′)
ORF31	CCGTGGGATTATTGGTTTTG	CGACGGTTCAGTGTTTTGTG
ORF61	AAAGCCTGACTTTTTGGGGT	CAAACCTGGACCTGGAAAGA
ORF63	ATTGAGGCGCCGAATGTTC	CTTCACCACCATCATCAGATACG
GAPDH	CACATGGCCTCCAAGGAGTAA	TGAGGGTCTCTCTCTTCCTCTTG
VLT exon 3–4	TGGACGATCACGGTAGTCCT	CGGAAAAACCATGCCGTGTT

2.5. qRT-PCR for Viral Transcripts

Total RNA was extracted using the Hybrid-R kit for large RNA/small RNA (Geneall, Seoul, Korea) from hESC-derived neurons that were mock-infected or infected with 100 PFU of cell-free VZV23GFP, VZVsnc10MUT or VZVsnc12-13DEL Infections were monitored using GFP expression indicating productive infection for 7 days, or from neurons infected with 100 PFU of cell-free VZV23GFP, VZVsnc10MUT or VZVsnc12-13DEL ACV for 10 days and without ACV for 15 days that did not express GFP (latently infected). RNA was treated with DNase (AMPD1, Merck, Darmstadt, Germany). For the cDNA synthesis, 3 µg of total RNA was reverse transcribed using an oligo dT primer and Moloney murine leukaemia (M-MMLV) reverse transcriptase (# M1701; Promega, Madison, WI, USA). Negative control samples were obtained by performing the same cDNA synthesis reaction in the absence of reverse transcriptase. Three independent biological experiments were performed, and qPCR was performed in triplicates for each target. Primer- probe- probe sets for the VZV genes ORF 61, ORF 63, ORF 31, and VLT3F-4R sequence are listed in Table 1. Results are shown as averages with error bars representing the SEM.

2.6. Latency/Reactivation Experiments

After a minimum of 21 days of terminal differentiation of neurons into cultures that contained extensive axonal outgrowth, cells were pre-treated with acyclovir (ACV, 400 uM) for 24 Hrs, and then incubated with cell-free VZV (VZVsncRNA10 or VZVsncRNA12-13DEL and VZV23GFP). Cultures were exposed to 100 PFU of the virus in the presence of 400 uM of ACV and centrifuged at 180 g for 10 min, followed by incubating for a period of 2 Hr. at 37 °C. After the removal of the virus, the neurons were maintained in the presence of ACV for 10 days with daily media changes. Media without the ACV was then used to maintain the cultures for up to 2 weeks post-infection, with media changes every 3 days. Cultures were examined regularly over the incubation period for GFP expression microscopically, but no GFP fluorescence was observed. After two weeks, latently infected wells were subjected to a reactivation stimulus involving treating the neurons with LY294002 hydrochloride (10 µM, Tocris, CAT. #1130) and phorbol ester phorbol 12-myristate 13-acetate (PMA) (40 ng/mL) and were incubated at 33 °C for 5d. Reactivation of virus in live cultures was detected as GFP fluorescence representing the expression of ORF23. Three biological repeats of latency experiments each with 3 latently infected wells were carried out for each mutant with parallel VZVGFP23 controls. To confirm that GFP expression corresponded to reactivation including production of infective virions, 7d after application of the chemical stimuli the neurons were harvested and seeded on naïve ARPE19 cells, and infectious foci were observed using GFP fluorescence.

2.7. Multistep Growth-Kinetics

The multistep growth kinetics of VZVsncRNA10 or VZVsncRNA12-13DEL and VZV23GFP viruses were determined by inoculating ARPE19 cells with 1000 PFU of VZV debris-fraction. Infected cells were subsequently trypsinized at 2 h ("time 0") or 1, 2, 3, 4, and 6 dpi, serially diluted, and titrated by seeding 10-fold dilutions of harvested cells onto fresh ARPE19 monolayers. Seven days after seeding, viral plaques were fixed with 4% (w/v) paraformaldehyde containing crystal violet for 30 min and plaques were counted.

3. Results

3.1. Construction of VZV with Synonymous Mutation of VZVsncRNA10 and Deletion in the Overlap Region between VZVsncRNA 12&13

Transfection of LNAA antagonists to VZVsncRNA10, 12, and 13, singly or in combinations, significantly reduces VZV productive infection. This suggests that the VZVsncRNA themselves modulate lytic infection by VZV [15]. In order to further study the roles of these VZVsncRNA in VZV infections, two recombinant VZV mutants' viruses were derived in the background of a VZV BAC that contains a GFP reporter gene fused to ORF23, previously shown capable of establishing lytic infections in hESC-derived human neurons and model VZV latency and experimental reactivation ([6,24]). The BACs were generated by λred-mediated recombineering methods used in our laboratories [25]. The VZVsncRNA10 was mutated by changing 8 nt without altering the ORF61-encoded amino acid sequence, yielding a virus termed VZVsnc10MUT. The second recombinant VZV contained a deletion of 12 of the nucleotides common to VZVsncRNA12 and 13 that are located immediately upstream of the ORF61 coding sequence and presumably lies in the ORF61 mRNA non-coding leader. VZV derived from this BAC was termed VZVsnc12-13DEL (Figure 1A,B). All mutated sequences were confirmed in the BACs by sequencing across the junctions.

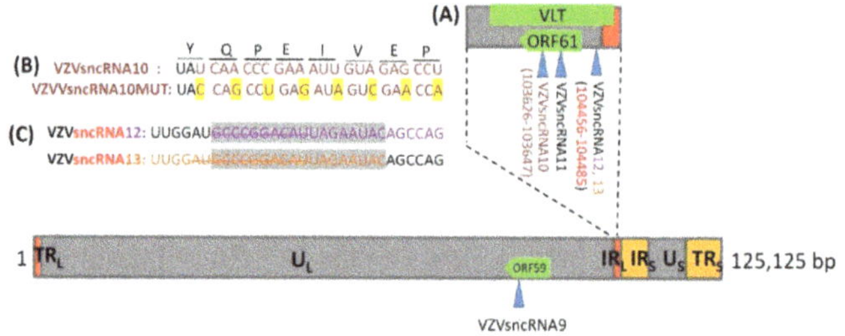

Figure 1. Genomic localization of VZVsncRNA10, VZVsncRNA12 VZVsncRNA13 expression and mutagenesis (**A**) depicts the region of the VZV genome containing the VZVsncRNA studied. Arrowheads indicate the VZVsncRNA coded by the lower viral DNA strand and deriving from a right to left primary transcript that is predicted to encode ORF61. (**B**) Mutations introduced into VZVsnc10MUT: the upper sequence shows the original sequence while the lower sequence shows the synonymous mutations introduced by mutagenesis highlighted in yellow. (**C**) Sequences of VZVsncRNA12 (purple letters) and VZVsncRNA13 (orange letters) deleted in VZVsnc12-13DEL l are indicated with a strikeout. The portion of VZVsncRNA12 and VZVsncRNA13 that overlap is highlighted in gray. The sncRNA is shown in the correct direction.

Both mutation strategies resulted in BACs that yielded viable VZV. Initial characterization of the mutant VZV was performed by infecting ARPE19 cells with each virus, extracting RNA at 5 dpi, and subjecting it to Taqman stem-loop quantitative reverse transcriptase PCR (SL-qRT-PCR) for the sncRNA, using primers matching the native and mutated sequences [13]. As expected, VZV sncRNAmut10 infected cells did not show any expression of VZVsncRNA10, while cells infected with VZV derived from the parental BAC showed strong expression (Figure 2A). In addition, stem-loop primers directed to the mutated sequence did not result in PCR amplification from RNA of cells infected with the mutant virus (or the parent virus) suggesting that either the mutated sequence was not expressed as an alternative sncRNA or that it was rapidly degraded in the host cells. Similar assays of RNA from cells infected with VZVsnc12-13DEL indicated complete abrogation of expression of both VZVsncRNA12 and VZVsncRNA 13 compared to parent control virus when sncRNA12 and 13 specific primers were used (Figure 2B), and these also detected abundant levels in another pOka-derived virus, VZV66GFP (Figure 2 and Refs. [13,14]).

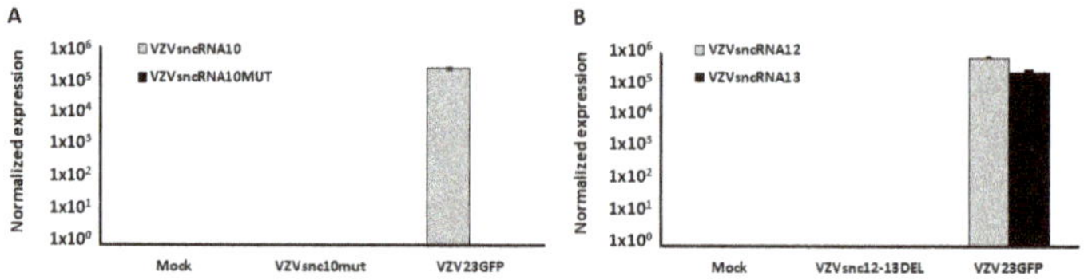

Figure 2. VZVsncRNA10MUT and VZVsncRNA12-13DEL do not express the corresponding sncRNA in VZV-infected cells. ARPE19 cells were infected with a cell-free virus from VZV23GFP (parental virus), VZVsnc12-13DEL, or VZVsncRNA10MUT. RNA size-selected for <200 nt was extracted at 5 dpi and VZVsncRNA10, VZVsncRNA10MUT, VZVsncRNA12, and VZVsncRNA13 expression was assayed by SL-qRT-PCR. (**A**) shows the expression of VZVsncRNA10 for infections with VZVsncRNA10mut and parental virus and (**B**) shows the expression data for VZVsncRNA12 and VZVsncRNA13 in cultures infected with VZVsncRNA12–13DEL and parent viruses. Expression is normalized to cellular miR26, whose expression has been shown not to be affected by viral infection [4].

3.2. Growth Rates of VZVsnc10MUT and VZVsnc12-13DEL Are Similar to Those of Their Parental Virus in ARPE-19 Cells

As mentioned above, antagonizing VZVsncRNA10,12 and 13 with LNAA reduced viral spread and generation of infectious viruses [15]. This raised the possibility that mutating their sequences could affect viral replication. We therefore conducted growth curve analyses for the mutants after a cell-free infection with a cell-free viral "debris" fraction (cellular debris derived from VZV-infected cells containing high-titers of infectious virus but no living cells [5]). The growth curves were generated from standard plaque assays and compared to those obtained from the parental VZVGFP23 virus. Both VZVsnc10MUT (Figure 3A) and VZVsnc12-13DEL (Figure 3B) displayed similar replication kinetics over a 6d period yielding a slight but not significant increase in infectious virus production in comparison to those of the parental VZV23GFP.

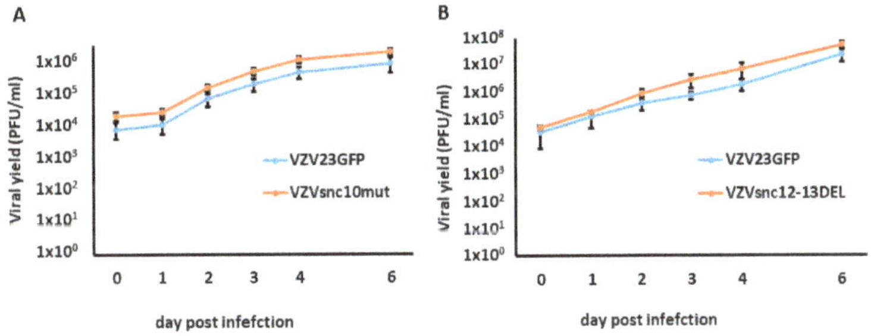

Figure 3. VZVsnc10MUT or VZVsnc12-13DEL growth curves compared to VZVGFP23 in lytically infected ARPE-19 cells. ARPE19 cells were infected with VZVsncRNA10mut or VZVsnc12-13DEL debris [5] and subsequently harvested from one well by trypsinization. Cells were then serially diluted and titrated for time 0 (0 h) or subsequent times into fresh monolayers of ARPE-19 cells. At 7 dpi, the cells were fixed and stained for 30 min, and the plaques were counted on dried and stained plates to give values of viral yield in PFU. (**A**) Represents the growth of VZV23GFP compared to VZVsncRNA10mut and (**B**) represents the growth of VZV23GFP compared to VZVsnc12-13DEL. $n = 3$ independent experiments for each mutant virus.

3.3. VZVsnc10MUT and VZVsnc12-13DEL Can Productively Infect Human Embryonic Stem Cell-Derived Neurons

We next addressed the question of whether neurons, the reservoir of latent VZV after varicella, supported productive infection by recombinant viruses with mutations in ORF61-derived VZVsncRNA. Human embryonic stem cell (hESC)-derived neurons were made using agarose micromolds (as described previously [18]) and then matured for 1 month on coverslips in 24-well plates. Neuronal cultures were then exposed to 100 PFU of cell-free VZVsnc10MUT, VZVsnc12-13DEL or VZV23GFP and neuronal infection was monitored by observing GFP fluorescence microscopically indicating the synthesis of VZVORF23 capsid protein. An extensive, spreading productive infection was observed for all three viruses, shown at 7 dpi in Figure 4A–C. As expected, no fluorescence was observed in mock-infected neurons (Figure 4D). This result indicates that the mutation/deletion of these VZVsncRNA does not prevent infection leading to viral replication of VZV in human neurons. Unfortunately, the heterogenous nature of our hESC-derived neuron cultures in terms of both number and neuronal phenotypes precludes performing growth curve analyses for the mutant viruses in neurons as we did in ARPE19 cells and fibroblasts.

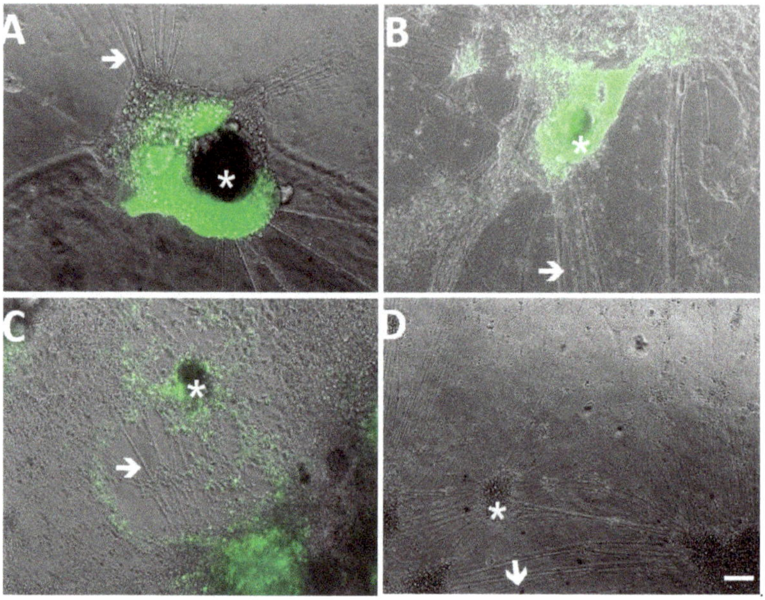

Figure 4. hESC-derived neurons are productively infected by recombinant VZV lacking expression of VZVsncRNA10 and VZVsncRNA12 and 13. Panel A-D show merged images of GFP fluorescence and phase contrast. Neurons differentiated for 21d were infected with cell-free VZV23GFP, VZVsnc10MUT or VZVsnc12-13DEL virus. Extensive productive infection of neurons was observed at 5 dpi for VZVsnc10MUT (**A**), VZVsnc12-13DEL (**B**) and VZV23GFP (**C**) as shown by expression of ORF23GFP. (**D**) Shows neurons in a mock-infected well. Arrowheads indicate neurites and asterisks indicate clusters of neuronal cell bodies. Scale Bar = 100 mm.

3.4. VZV Mutated for VZVsncRNA Derived from the mRNA for ORF61 Can Establish Experimental Latency in hESC-Derived Neurons

VZVsncRNA10-12 are derived from the mRNA for ORF61 and would be complementary to the recently identified VLT transcript expressed in cadaveric dorsal root ganglia of VZV- seropositive individuals. VLT has positional and antisense activities similar to other herpesvirus latency-associated transcripts and has been implicated to be involved in neuronal latency and/or reactivation [26]. In order to determine if a latency-like infection could be established by the mutant VZV in neurons, we used a model previously used by us [6] and others [27] in which acyclovir (ACV) treatment is used to inhibit lytic replication after infection of neuronal cultures with a cell-free virus.

Neurons were infected in the presence of ACV and cultured a further 10d in its presence. ACV was then withdrawn, and the cultures maintained an additional 15d. As a positive control for neuronal infection, some wells of neurons were infected without ACV treatment. At the end of the incubation period RNA was extracted from the neurons and the expression of transcripts for ORF61, ORF31, ORF63, and VLT exons 3–4 were measured by qRT-PCR (Figure 5).

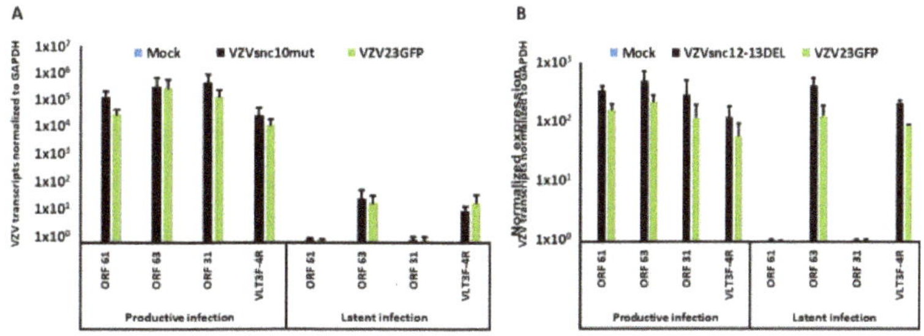

Figure 5. VZV mutant in VZVsncRNAs establishes experimental latency in hESC-derived neurons. (**A**) hESC-derived neurons were infected with 100 PFU of cell-free VZV. Latent infections were obtained by treatment with ACV (see methods for exact protocol). In the absence of ACV, all three viruses infected neurons productively (see Figure 4). RNA was harvested at 5 dpi from the wells not treated with ACV and at 25 dpi from presumably latently infected neurons in wells treated with ACV and lacking GFP expression. RT-qPCR analysis for IE transcripts ORF61 and ORF63, late transcript ORF31, and transcripts of VLT exons 3 and 4 of VLT VLTexon3–4 (4) was then performed. (**A**) VZVsnc10mut productively infected neurons (left 4 sets of bars) when not treated with ACV. ACV-treated latently infected wells (right 4 sets of bars) expressed very low levels of ORF61 and late transcript ORF31 consistent with our previous report [6]. The only transcripts detected at higher levels were those of VLT exons 3 and 4 of VLT (VLTexon3–4) (4) and a portion of ORF63 that is co-expressed with VLT [7] consistent with the establishment of latency. $N = 3$ independent experiments. (**B**) RT-qPCR analysis of ACV-untreated, productively infected (left 4 sets of bars) and ACV-treated, latently infected (right 4 sets of bars) neurons infected with VZVsnc12-13DEL revealed the same expression patterns of transcripts as observed for infections with VZVsnc10mut. That is, neurons productive infected with VZVsnc12-13DEL expressed all 4 tested transcripts at high levels, while the latently infected neurons only expressed transcripts expected to be present in latency. Results show the average expression levels from 3 independent experiments normalized to the expression of GAPDH.

Neurons that did not receive ACV treatment were productively infected by all three viruses and expressed all 4 of the transcripts assayed for. In contrast, neurons that were infected in the presence of ACV showed greatly reduced expression of lytic RNAs from ORF31 and ORF61 while the RNAs for latency associated ORF63 and VLT were expressed at relatively high levels. The primers for ORF63 identified the portion of the transcript known to be also expressed as a fusion with VLT [8]. This data indicates that the mutant viruses were able to establish experimental latency in a manner indistinguishable from the parental virus and other recombinant VZV used in our laboratory.

3.5. VZV Mutated in VZVsncRNA10 or Loss of VZVsncRNA12/13 Display Impaired Reactivation from Latency

After determining that the mutant viruses established latency in our model, latently infected neurons were treated with stimuli that we previously found to elicit reactivation and the generation of infectious viruses [6]. Neurons latently infected with VZVsnc10MUT, VZVsnc12-13DEL and VZV23GFP were maintained for 15d after ACV withdrawal and did not show GFP fluorescence. These cultures were then treated with LY294002 (10 μM) and phorbol ester phorbol 12-myristate 13-acetate (PMA) (40 ng/mL) and were incubated at 33 °C for a further 5 days. Reactivation of the virus in live cultures was followed by observation of GFP fluorescence indicating the expression of ORF23.

In three independent experiments, neurons infected with parental virus VZV23GFP expressed GFP in 1/3 of wells (3 wells GFP+ of 9 latently infected wells) of latently infected neurons (Figure 6A), consistent with results using a VZV expressing GFP fused to ORF66 [6]. In contrast, expression of GFP was only observed in one well in one of three experiments (that is, in 1 well GFP+ of 9 latently infected wells) with neurons latently infected with

VZVsnc10MUT (Figure 6B). No GFP expression indicating reactivation was observed in 9 wells containing neurons latently infected with VZVsnc12-13DEL and given reactivation stimuli in the three experiments (Figure 6C).

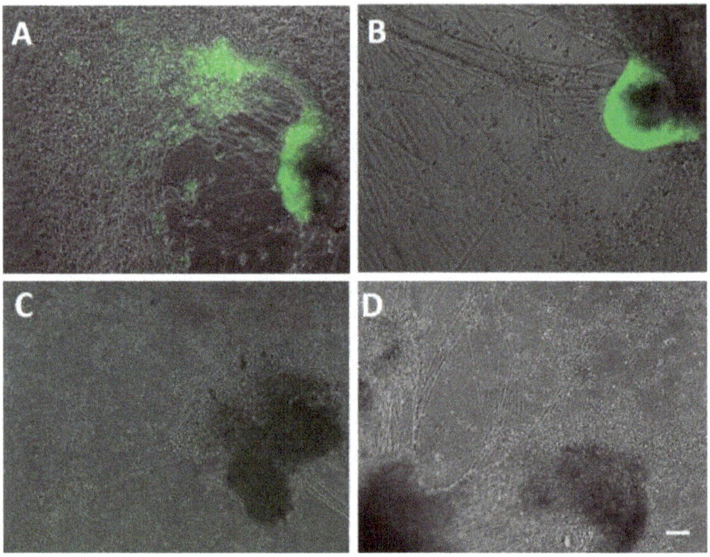

Figure 6. VZVsnc10MUT and VZVsnc12-13DEL are impaired for induced reactivation in hESC-derived neurons. Experimental latency was established in hESC-derived neurons using ACV treatment as described in [6] and the Methods. Ten to 15 days after ACV withdrawal there was no spontaneous reactivation as observed by lack of GFP fluorescence indicating production of the ORF23GFP fusion protein. The neurons then received a reactivating stimulus of P3K-inhibitor and PMA as detailed in the methods. Panels show a site of reactivated infection for VZV23GFP (**A**) which were detected in 3 wells of 9 infections performed in parallel to infection with the viruses mutant for VZVsncRNA. (**B**) The only well-containing neurons infected latently with VZVsnc10MUT that reactivated in three experiments including a total of 9 latently infected wells. No reactivation was observed in any of the 9 wells infected with the VZVsnc12-13DEL virus (**C**). (**D**) shows neurons in a mock-infected well receiving the reactivation stimulus. Three independent experiments for each virus VZV23GFP, MZVsnc10MUT, or VZVsnc12-13DEL were performed each including 3 wells infected with each virus. Size bar = 100 μm.

In order to confirm that the observed GFP expression from capsid protein ORF23 corresponded to complete reactivation including generation of the infective virus, we harvested neurons 5d after application of the chemical stimuli to the neurons and seeded them on naïve ARPE19 cells and monitored for the appearance of GFP-fluorescent foci of infection and cytopathic effect indicating a productive infection. Only re-seeded neurons that expressed GFP after receiving a reactivation stimulus were able to establish loci of infection, these eventually developed into plaques evident after crystal violet staining (not shown). Taken together, these results are consistent with the hypothesis that VZVsncRNA are involved in the reactivation of latent VZV.

4. Discussion

It has become clear that many biological processes are modulated by non-protein-coding RNA, particularly short RNAs. The short recognition sequence of these RNAs results in their ability to target multiple genes for regulation. miR is the most studied of these types of molecules, and it has been long known that viruses of the herpesvirus family encode many miR. The discovery that the HSV-1 LAT RNA encoded several miRNAs resulted in numerous studies to examine how they may regulate the expression of viral and host genes and the processes involved in lytic/latent decisions. miRNA have now been identified from multiple regions of HSV-1, and miRNA is encoded not only by LAT but also by transcripts of ICP0. Despite intense study, it is not yet known how these miR contribute to HSV-1 latency/reactivation [28]. Several miRNA loci have been deleted with minor or no effects, while mutation/deletion of other HSV-1 miRNA has influenced reactivation in rodent neurons infected with HSV-1 [29]. Most of the small non-coding RNA in VZV-infected cells and that encoded opposite to the VZV putative latency-associated transcript, VLT, are not predicted to fold into miR with current software tools. However, two sncRNA without classical miR structure have recently been reported to also be encoded by the HSV LAT, and these may affect the expression of the HSV receptor HVEM [30].

We have previously reported that chemically antagonizing these VZVsncRNA has significant effects on VZV replication and that combining several antagonists with them has stronger, additive effects. It was therefore somewhat surprising to not find significant effects on VZV replication in two cell lines infected with the VZV deletion or synonymous mutations viruses. It must be noted that the deletion of many VZV genes results in little to no effect on viral replication in cells in conventional monolayer cell culture, but when analyzed in the context of infection of human tissues in the SCID-Hu model, or in 3D in-vitro models such as intact skin these deletions have significant and dramatic effects on viral replication. Experiments are underway examining these and other VZV mutated in additional VZVsncRNA in such models.

The location of VZVsncRNA10-13, along with the observation that HSV LAT-derived miR may play a role in latent infection/reactivation, stimulated our analyses into their possible roles in VZV latency and reactivation, using our hESC-derived neuron model. We reproducibly found that the two sncRNA mutant viruses were able to initiate infection and replicate productively in neurons and establish a latency-like state using ACV. When recombinant pOka-based virus expressing GFP is used for modeling latency, 1/3 of latently infected cultures (shown by in-situ hybridization for VZV DNA or VZV transcripts at long time periods after removal of virus/ACV) reactivate when the appropriate stimulus is provided [6]. This reactivation was productive, as evidenced by the generation of GFP-viral fusion protein and its spread over time to additional neurons, but also by the and the ability of the reactivated neurons to infect naïve ARPE19 cells. In the present study, we found that synonymous mutations in VZVsncRNA10 reduced the ability of VZV to reactivate in response to a stimulus, while deleting 12 nt common to VZVsncRNA12–13 eliminated reactivation altogether. Ideally, the specificity of these effects should be tested by restoring the VZVsncRNA by transfection of agonists. Unfortunately, transfection to neurons is not experimentally feasible. An alternative method of introduction of the VZVsncRNA sequences would be the development of AAV expressing them, and current experiments our laboratory are working out details of providing VZVsncRNA sequences with AAV serotypes that infect neurons (see [26] in this volume).

We still do not know the mechanism of action of the VZVsncRNA that do not seem to be miR, and additional studies will be necessary to elucidate how they exert their effects on reactivation and other aspects of VZV infection. We have observed that ORF61 and one of the VLT transcripts appear to be present at approximately the same levels in productively and latently infected neurons infected with the mutant VZV as in the parental virus (Figure 5). However, because the mutations introduced are in ORF61 and its 5′UTR, it is important to determine the effects of the mutations on ORF61 protein expression and localization as well as other VLT transcripts, and experiments addressing these issues now ongoing.

The current varicella-zoster vaccine given to children results in latent infection of peripheral neurons that can reactivate, albeit at a lower frequency than that of wild-type virus strains. Recent experiments in vitro [6] and the clinical experience show that vOka is impaired for reactivation, and usually results in milder disease when it occurs. However, our findings that deletion of 12 nt that are not required for viral replication and establishment of an immune response, but which can apparently abrogate reactivation, suggest the deletions generated could form the basis of a safer vaccine strain virus that is able to replicate in order to generate an immune response but unable to reactivate. The strategy of developing a VZV vaccine by making mutations modulating neuronal infection was recently published [31], utilizing a VZV mutant for ORF7, previously shown to impact VZV replication in skin and neurons [32]. This recent study demonstrates the interest and importance of developing safer vaccines for HZ and varicella disease.

Author Contributions: Conceptualization, R.S.G. and P.R.K.; experimental design, P.B., P.R.K. and R.S.G., methodology, P.B., B.D. and T.B.-S.; formal analysis, P.B., B.D. and T.B.-S.; resources, P.R.K.; writing—original draft preparation, P.B. and B.D.; writing—review and editing, R.S.G., P.R.K. and T.B.-S.; project administration, T.B.-S.; funding acquisition, R.S.G. and P.R.K. All authors have read and agreed to the published version of the manuscript.

Funding: This work was supported by an Israel-US Binational Science Foundation award (#2017259) and awards from the National Institutes of Health from NIAID (AI122640 & AI156527) to P.R.K and R.S.G); and an Israel Science Foundation award (#613/20) to R.S.G.

Institutional Review Board Statement: Not applicable.

Informed Consent Statement: Not applicable.

Data Availability Statement: The data in this study are available in this article, as well as upon reasonable request from the corresponding author.

Acknowledgments: P.R.K. wishes to acknowledge NEI CORE grant P30 EY08098 and unrestricted support from the Eye & Ear Foundation of Pittsburgh and the Research to Prevent Blindness Inc, NY. P.R.K. and R.S.G. wish to dedicate this manuscript to the memory of our friend and central figure in the study of the molecular virology of VZV, Randall Cohrs. We both owe Randy a great deal for his insightful discussions and important studies that greatly advanced our field.

Conflicts of Interest: The authors declare no conflict of interest.

References

1. Zerboni, L.; Sen, N.; Oliver, S.L.; Arvin, A.M. Molecular mechanisms of varicella zoster virus pathogenesis. *Nat. Rev. Microbiol.* **2014**, *12*, 197–210. [CrossRef] [PubMed]
2. Zerboni, L.; Arvin, A.M. The pathogenesis of varicella-zoster virus neurotropism and infection. In *Neurotropic Viral Infections*; Cambridge University Press: Cambridge, UK, 2008; pp. 225–250.
3. Takahashi, M. Development of a live varicella vaccine–past and future. *Jpn. J. Infect. Dis.* **2000**, *53*, 47–55. [PubMed]
4. Watson, S.; Mercier, S.; Bye, C.; Wilkinson, J.; Cunningham, A.L.; Harman, A.N. Determination of suitable housekeeping genes for normalisation of quantitative real time PCR analysis of cells infected with human immunodeficiency virus and herpes viruses. *Virol. J.* **2007**, *4*, 130. [CrossRef] [PubMed]
5. Sloutskin, A.; Goldstein, R.S. Laboratory preparation of Varicella-Zoster Virus: Concentration of virus-containing supernatant, use of a debris fraction and magnetofection for consistent cell-free VZV infections. *J. Virol. Methods* **2014**, *206*, 128–132. [CrossRef]
6. Markus, A.; Lebenthal-Loinger, I.; Yang, I.H.; Kinchington, P.R.; Goldstein, R.S. An in vitro model of latency and reactivation of varicella zoster virus in human stem cell-derived neurons. *PLoS Pathog.* **2015**, *11*, e1004885. [CrossRef]
7. Depledge, D.P.; Ouwendijk, W.J.D.; Sadaoka, T.; Braspenning, S.E.; Mori, Y.; Cohrs, R.J.; Verjans, G.M.G.M.; Breuer, J. A spliced latency-associated VZV transcript maps antisense to the viral transactivator gene 61. *Nat. Commun.* **2018**, *9*, 1167. [CrossRef]
8. Ouwendijk, W.J.D.; Depledge, D.P.; Rajbhandari, L.; Lenac Rovis, T.; Jonjic, S.; Breuer, J.; Venkatesan, A.; Verjans, G.M.G.M.; Sadaoka, T. Varicella-zoster virus VLT-ORF63 fusion transcript induces broad viral gene expression during reactivation from neuronal latency. *Nat. Commun.* **2020**, *11*, 6324. [CrossRef]
9. Han, Z.; Liu, X.; Chen, X.; Zhou, X.; Du, T.; Roizman, B.; Zhou, G. miR-H28 and miR-H29 expressed late in productive infection are exported and restrict HSV-1 replication and spread in recipient cells. *Proc. Natl. Acad. Sci. USA* **2016**, *113*, E894–E901. [CrossRef]
10. Umbach, J.L.; Kramer, M.F.; Jurak, I.; Karnowski, H.W.; Coen, D.M.; Cullen, B.R. MicroRNAs expressed by herpes simplex virus 1 during latent infection regulate viral mRNAs. *Nature* **2008**, *454*, 780–783. [CrossRef]
11. Piedade, D.; Azevedo-Pereira, J.M. The Role of microRNAs in the Pathogenesis of Herpesvirus Infection. *Viruses* **2016**, *8*, 156. [CrossRef]
12. Weinberg, M.S.; Morris, K.V. Transcriptional gene silencing in humans. *Nucleic Acids Res.* **2016**, *44*, 6505–6517. [CrossRef] [PubMed]
13. Markus, A.; Golani, L.; Ojha, N.K.; Borodianskiy-Shteinberg, T.; Kinchington, P.R.; Goldstein, R.S. Varicella-Zoster Virus Expresses Multiple Small Noncoding RNAs. *J. Virol.* **2017**, *91*, e01710-17. [CrossRef] [PubMed]

14. Golani-Zaidie, L.; Borodianskiy-Shteinberg, T.; Bisht, P.; Das, B.; Kinchington, P.R.; Goldstein, R.S. Bioinformatically-predicted varicella zoster virus small non-coding RNAs are expressed in lytically-infected epithelial cells and neurons. *Virus Res.* **2019**, *274*, 197773. [CrossRef] [PubMed]
15. Bisht, P.; Das, B.; Kinchington, P.R.; Goldstein, R.S. Varicella-Zoster Virus (VZV) Small Noncoding RNAs Antisense to the VZV Latency-Encoded Transcript VLT Enhance Viral Replication. *J. Virol.* **2020**, *94*, e00123-20. [CrossRef] [PubMed]
16. Perusina Lanfranca, M.; Mostafa, H.H.; Davido, D.J. HSV-1 ICP0: An E3 Ubiquitin Ligase That Counteracts Host Intrinsic and Innate Immunity. *Cells* **2014**, *3*, 438–454. [CrossRef]
17. Das, B.; Bisht, P.; Kinchington, P.R.; Goldstein, R.S. Locked-nucleotide antagonists to varicella zoster virus small non-coding RNA block viral growth and have potential as an anti-viral therapy. *Antivir. Res.* **2021**, *193*, 105144. [CrossRef] [PubMed]
18. Birenboim, R.; Markus, A.; Goldstein, R.S. Simple generation of neurons from human embryonic stem cells using agarose multiwell dishes. *J. Neurosci. Methods* **2013**, *214*, 9–14. [CrossRef]
19. Martin, D.P.; Wallace, T.L.; Johnson, E.M. Cytosine arabinoside kills postmitotic neurons in a fashion resembling trophic factor deprivation: Evidence that a deoxycytidine-dependent process may be required for nerve growth factor signal transduction. *J. Neurosci.* **1990**, *10*, 184–193. [CrossRef]
20. Morris, E.J.; Geller, H.M. Induction of neuronal apoptosis by camptothecin, an inhibitor of DNA topoisomerase-I: Evidence for cell cycle-independent toxicity. *J. Cell Biol.* **1996**, *134*, 757–770. [CrossRef]
21. Tischer, B.K.; Kaufer, B.B.; Sommer, M.; Wussow, F.; Arvin, A.M.; Osterrieder, N. A self-excisable infectious bacterial artificial chromosome clone of varicella-zoster virus allows analysis of the essential tegument protein encoded by ORF9. *J. Virol.* **2007**, *81*, 13200–13208. [CrossRef]
22. Erazo, A.; Yee, M.B.; Osterrieder, N.; Kinchington, P.R. Varicella-zoster virus open reading frame 66 protein kinase is required for efficient viral growth in primary human corneal stromal fibroblast cells. *J. Virol.* **2008**, *82*, 7653–7665. [CrossRef] [PubMed]
23. Mohammadi-Yeganeh, S.; Paryan, M.; Mirab Samiee, S.; Soleimani, M.; Arefian, E.; Azadmanesh, K.; Mostafavi, E.; Mahdian, R.; Karimipoor, M. Development of a robust, low cost stem-loop real-time quantification PCR technique for miRNA expression analysis. *Mol. Biol. Rep.* **2013**, *40*, 3665–3674. [CrossRef] [PubMed]
24. Markus, A.; Grigoryan, S.; Sloutskin, A.; Yee, M.B.; Zhu, H.; Yang, I.H.; Thakor, N.V.; Sarid, R.; Kinchington, P.R.; Goldstein, R.S. Varicella-zoster virus (VZV) infection of neurons derived from human embryonic stem cells: Direct demonstration of axonal infection, transport of VZV, and productive neuronal infection. *J. Virol.* **2011**, *85*, 6220–6233. [CrossRef] [PubMed]
25. Tischer, B.K.; Von Einem, J.; Kaufer, B.; Osterrieder, N. Two-step red-mediated recombination for versatile high-efficiency markerless DNA manipulation in Escherichia coli. *BioTechniques* **2006**, *40*, 191–197.
26. Wu, B.W.; Yee, M.B.; Goldstein, R.S.; Kinchington, P.R. Antiviral Targeting of Varicella Zoster Virus Replication and Neuronal Reactivation Using CRISPR/Cas9 Cleavage of the Duplicated Open Reading Frames 62/71. *Viruses* **2022**, *14*, 378. [CrossRef]
27. Sadaoka, T.; Depledge, D.P.; Rajbhandari, L.; Venkatesan, A.; Breuer, J.; Cohen, J.I. In vitro system using human neurons demonstrates that varicella-zoster vaccine virus is impaired for reactivation, but not latency. *Proc. Natl. Acad. Sci. USA* **2016**, *113*, E2403–E24012. [CrossRef]
28. Bhela, S.; Rouse, B.T. Are miRNAs critical determinants in herpes simplex virus pathogenesis? *Microbes Infect.* **2018**, *20*, 461–465. [CrossRef]
29. Barrozo, E.R.; Nakayama, S.; Singh, P.; Vanni, E.A.H.; Arvin, A.M.; Neumann, D.M.; Bloom, D.C. Deletion of Herpes Simplex Virus 1 MicroRNAs miR-H1 and miR-H6 Impairs Reactivation. *J. Virol.* **2020**, *94*, e00639-20. [CrossRef]
30. Tormanen, K.; Wang, S.; Matundan, H.H.; Yu, J.; Jaggi, U.; Ghiasi, H. Herpes Simplex Virus 1 Small Noncoding RNAs 1 and 2 Activate the Herpesvirus Entry Mediator Promoter. *J. Virol.* **2022**, *96*, e0198521. [CrossRef]
31. Wang, W.; Pan, D.; Fu, W.; Ye, X.; Han, J.; Yang, L.; Jia, J.; Liu, J.; Zhu, R.; Zhang, Y.; et al. Development of a skin- and neuro-attenuated live vaccine for varicella. *Nat. Commun.* **2022**, *13*, 824. [CrossRef]
32. Selariu, A.; Cheng, T.; Tang, Q.; Silver, B.; Yang, L.; Liu, C.; Ye, X.; Markus, A.; Goldstein, R.S.; Cruz Cosme, R.S.; et al. ORF7 of Varicella Zoster Virus is a Neurotropic Factor. *J. Virol.* **2012**, *86*, 8614–8624. [CrossRef] [PubMed]

viruses

MDPI

Article

Two Consecutive Prolines in the Fusion Peptide of Murine β-Coronavirus Spike Protein Predominantly Determine Fusogenicity and May Be Essential but Not Sufficient to Cause Demyelination

Abass Alao Safiriyu [1], Manmeet Singh [1], Abhinoy Kishore [1], Vaishali Mulchandani [1], Dibyajyoti Maity [2], Amrutamaya Behera [1], Bidisha Sinha [1], Debnath Pal [2] and Jayasri Das Sarma [1,*]

[1] Department of Biological Sciences, Indian Institute of Science Education and Research Kolkata, Mohanpur 741246, India; saa19rs001@iiserkol.ac.in (A.A.S.); manmeetsinghbiotech@gmail.com (M.S.); abhinoy@gmail.com (A.K.); vm17ip003@iiserkol.ac.in (V.M.); ab16m081@iiserkol.ac.in (A.B.); bidisha.sinha@iiserkol.ac.in (B.S.)

[2] Department of Computational and Data Sciences, Indian Institute of Science, Bengaluru 560012, India; djmaity@gmail.com (D.M.); dpal@iisc.ac.in (D.P.)

* Correspondence: dassarmaj@iiserkol.ac.in

Abstract: Combined in silico, in vitro, and in vivo comparative studies between isogenic-recombinant Mouse-Hepatitis-Virus-RSA59 and its proline deletion mutant, revealed a remarkable contribution of centrally located two consecutive prolines (PP) from Spike protein fusion peptide (FP) in enhancing virus fusogenic and hepato-neuropathogenic potential. To deepen our understanding of the underlying factors, we extend our studies to a non-fusogenic parental virus strain RSMHV2 (P) with a single proline in the FP and its proline inserted mutant, RSMHV2 (PP). Comparative in vitro and in vivo studies between virus strains RSA59(PP), RSMHV2 (P), and RSMHV2 (PP) in the FP demonstrate that the insertion of one proline significantly resulted in enhancing the virus fusogenicity, spread, and consecutive neuropathogenesis. Computational studies suggest that the central PP in Spike FP induces a locally ordered, compact, and rigid structure of the Spike protein in RSMHV2 (PP) compared to RSMHV2 (P), but globally the Spike S2-domain is akin to the parental strain RSA59(PP), the latter being the most flexible showing two potential wells in the energy landscape as observed from the molecular dynamics studies. The critical location of two central prolines of the FP is essential for fusogenicity and pathogenesis making it a potential site for designing antiviral.

Keywords: β-Coronavirus; mouse hepatitis virus-A59/MHV-A59; mouse hepatitis virus-2/MHV2 spike protein; fusion peptide/FP; cell-to-cell fusion (fusogenicity); neuropathogenesis; hepatitis; demyelination; structural rigidity

Citation: Safiriyu, A.A.; Singh, M.; Kishore, A.; Mulchandani, V.; Maity, D.; Behera, A.; Sinha, B.; Pal, D.; Das Sarma, J. Two Consecutive Prolines in the Fusion Peptide of Murine β-Coronavirus Spike Protein Predominantly Determine Fusogenicity and May Be Essential but Not Sufficient to Cause Demyelination. *Viruses* **2022**, *14*, 834. https://doi.org/10.3390/v14040834

Academic Editor: Charles Grose

Received: 17 March 2022
Accepted: 14 April 2022
Published: 17 April 2022

1. Introduction

Fusion peptide (FP) is one of the functional segments of virus-host attachment spike protein that is important for viral fusion to either host cell plasma membrane or endosomal membrane [1,2]. During the initial stage of the fusion cascade, spike protein unfolds and extends to expose FP to the target membrane for anchoring via the fusion core domain [3,4]. Mounting evidence from a large body of literature highlights that FP is a valuable target for the development of pan-CoV therapeutics owing to its conserved nature and contribution to mediating the fusion of the viral and host cell membrane, and driving its fusion mechanism across the coronavirus (CoV) family [5–12]. The ongoing COVID-19 pandemic highlights the immediate requirement to develop effective therapeutics against the existing human-β-coronavirus, SARS-CoV-2, and other infectious coronavirus strains, or those emerging in the near future [13–15]. While a lot of work has been done on SARS-CoV-2, its highly infectious nature limits its regular laboratory use. Limited experimental evidence, therefore, exists for human β-coronavirus, but alternate evidence from prototype murine-β-coronavirus may shed some light on the understanding of the intricate mechanism of the FP-mediated fusion process and its associated pathogenesis [16,17].

Two very closely related strains of murine-β-coronavirus MHV-A59 and MHV2 differ in their fusogenic properties and hepato-neuropathogenesis [18–22]. MHV-A59 is highly fusogenic and causes acute hepatitis, meningoencephalomyelitis with chronic progressive demyelination concurrent with axonal loss and is denoted as a neurotrophic strain [22,23]. In contrast, non-neurotropic strain MHV-2 sharing 91% genome identity with MHV-A59, and 83% pairwise spike sequence identity, causes only meningitis and is unable to invade the brain parenchyma. A series of detailed comparative in vitro and in vivo studies previously demonstrated that spike protein is one of the major determinants of cell-to-cell fusion, viral infectivity, viral antigen spread, and its consequent neuropathology, demyelination, and axonal loss [18,19,21,24]. Target RNA recombination using a reverse genetic system engineering two isogenic spike gene recombinant strains of MHV, RSA59, and RSMHV2 have given interesting insight. The spike gene recombinant strain of MHV, RSA59 where the spike gene was replaced from parental neurotropic and demyelinating strain MHV-A59, and RSMHV2 where the spike gene was derived from parent hepatotropic non-demyelinating strain MHV2 share the same genetic background except for the spike gene [21,25]. Both RSA59 and RSMHV2 can efficiently infect neurons but RSA59 can spread from neuron to neuron more specifically from gray matter to white matter following axonal transport and can release at the nerve end to directly infect oligodendrocytes in the white matter through cell-to-cell fusion [26,27]. Thus, RSA59 can evade the immune system, silently infect white matter oligodendrocytes, and cause moderate to severe demyelination. In contrast, RSMHV2 as mentioned earlier can infect the neuron but is impaired in axonal transport and cannot reach the white matter oligodendrocytes. Impaired axonal transport and lack of cell-to-cell fusion properties contribute to impaired demyelination in RSMHV2. Previous studies have demonstrated that RSA59 can reach the optic nerve via retrograde axonal transport and cause optic neuritis including neuron inflammation, demyelination, and axonal loss [24,28]. In contrast, RSMHV2 was unable to follow retrograde transport and thus was unable to induce optic neuritis. Retrograde axonal transport of RSA59 is also known to damage retinal ganglionic cells as a consequence of cell-to-cell fusion, whereas RSMHV2 is impaired in causing retinal ganglionic cell loss due to lack of retrograde axonal transport and infection to retinal ganglionic cells [24,28]. Another seminal study combining exogenous spike protein trafficking as well as in vivo and in vitro viral spread and dissemination demonstrated that irrespective of the presence of known murine-coronavirus viral entry receptor CEACAM1, spike protein by itself can initiate the fusion process [17,29]. Different strains of CoV induce differential cell-to-cell fusion and are responsible for the differential pathogenicity and disease severity. In summary, the spike protein plays a vital role in CoV-induced cell-to-cell fusion and pathogenesis. To delineate the minimum essential motif required for fusogenicity, studies were targeted to the FP of the S2 domain of the spike protein owing to its key role in early events in cell-to-cell fusion necessary for intercellular viral spread. Two consecutive central proline residues in the FP domain were identified to play a crucial role in the event of the fusion process. In one of the previous studies, it has been demonstrated by generating a proline deletion mutant that a proline deletion or insertion in the FP may significantly alter murine CoV(m-CoV) induced fusogenicity, viral antigen spread, infectivity, and its consequent neuropathological event, demyelination, and axonal loss [17].

We had previously demonstrated that the deletion of single proline from the centrally located double prolines in the FP of fusogenic and demyelinating strain RSA59 (PP) led to the loss of fusogenicity due to the loss of structural rigidity around the FP neighborhood from the lack of double prolines [16,17]. In this study, we add one proline at the center of the FP of fusion impaired non-demyelinating strain RSMHV2 (P) and investigate whether only two consecutive prolines with the parental neighboring amino acids within a membrane environment are sufficient to provide the rigidity required for syncytia formation. We, therefore, generated a mutant RSMHV2 (PP) in which one proline was added to RSMHV2 (P) and compared with RSA59 (PP) and RSMHV2 (P). Comparative studies reveal that RSMHV2 (PP) can induce cell-to-cell fusion, spread through the neuron, and can occasionally reach the white matter and thus can induce discrete myelin loss, but is unable to form a concentrated demyelinating plaque. An increased ability to cause cell-to-cell fusion, viral antigen spread, and consecutive myelin loss could be due to the increased rigid structure induced by one additional proline in the FP neighborhood but it was restricted as compared to RSA59 (PP). Previous in silico studies had identified the cause to be the transformation of the FP structure from helix-turn-helix-turn-helix in RSA59 (PP) to a helix-loop-helix in RSA59 (P). We find the same in this study from the comparison of RSMHV2 (PP) versus RSMHV2 (P). With the local torsional flexibility, the secondary structure of RSMHV2 (PP) is identical to RSA59 (PP). Two prolines are required for efficient viral spread from the brain to the spinal cord and from the brain to the retina via the optic nerve to the eye by retrograde axonal transport. Altered rigidity of the FP may contribute significantly to cell-to-cell fusion and viral spread which may help in designing the therapeutic strategies for anti-pan-CoV.

2. Materials and Methods

2.1. Viruses

A recombinant isogenic demyelinating strain of RSA59 (PP) and non-demyelinating strain RSMHV2 (P) expressing enhanced-Green Fluorescent Protein (EGFP) have been used for understanding spike mediated fusogenicity and hepato-neuropathogenesis in our series of previous studies [25,26,30]. Briefly, RSMHV2 (PP) expressing EGFP was engineered by inserting one additional proline through quick-change site-directed mutagenesis combined with targeted RNA recombination next to the existing central proline of the parental RSMHV2 (P) FP of spike protein [21,31]. The plasmids were isolated and linearized. Synthetic RNA was obtained using an in vitro transcription kit. The synthetic capped RNA as a donor and fMHV as a recipient virus were subjected to targeted RNA recombination methods [25]. The resultant recombinant strains were selected and sequenced for verification. Overall, the viruses are isogenic recombinant strains of MHV-A59 expressing EGFP but differ in spike gene, where RSA59 (PP) possesses spike gene of the parental strain of MHV-A59, RSMHV2 (P) possesses spike gene of the fusion impaired non-demyelinating parental strain of MHV2 and mutant RSMHV2 (PP) also possesses the spike gene of the fusion impaired non-demyelinating parental strain of MHV2 with one proline insertion in the FP.

2.2. Isolation and Enrichment of Primary Neuron from Neonatal Day 0 Mouse Brain

Whole brains were harvested from day 0 pups and meninges were carefully removed from brain tissue. Homogenized brain tissues were incubated in a rocking water bath set at 37 °C for 30 min in Hanks' balanced salt solution (Gibco), containing 300 µg/mL DNase I (Sigma, Tokyo, Japan) and 0.25% trypsin (Sigma). Enzyme-dissociated cells were triturated in the presence of 0.25% FBS, followed by a wash and centrifugation (300× g for 10 min). The pellet was again resuspended in Hanks' balanced salt solution and passed through a 70-µm nylon mesh. A second wash and centrifugation (300× g for 10 min) were performed, and finally, the cell pellet was diluted to 10^6 cells/mL with (DMEM) containing 1% HEPES, 1% penicillin, 1% streptomycin, 1% non-essential amino acid, 1% L-glutamine, 10% FBS. Cells were plated on PolyDiLysin and Laminin (PDL/Lam) coated culture plates and allowed to adhere for 1 day in a humidified CO_2 incubator at 37 °C. After 24 h, all non-adherent cells were removed, and cells were switched to a serum-free, growth medium (Neurobasal medium containing B27). The purity of isolated neuronal cells was determined by double-label immunofluorescence with neuronal marker anti-MAP2 and astrocyte marker anti-GFAP (glial fibrillary acidic protein).

2.3. Maintenance of Secondary Cells in Culture

L2 is a murine lung epithelial cell line (CCL-149) and Neuro2a is a mouse neuroblastoma cell line (CCL-131) that was obtained from the American Tissue Culture Collection (ATCC). DBT is a mouse astrocytic delayed brain tumor cell line (JCRB1580) that was obtained from the Japanese Collection of Research Bioresources Cell Bank. Neuro2A cells were cultured in Minimum Essential Medium (MEM) containing 10% FBS and DMEM containing 7.5% $NaHCO_3$, 1M HEPES, 1% penicillin, 1% streptomycin, 1% non-essential amino acid, 1% L-glutamine, 10% FBS was used to maintain for L2 and DBT cells. All the cells were maintained in continuous culture up to 5 passages for experimentation. Cells were cultured on etched coverslips as per requirement.

2.4. Viral Infection in Primary Neuronal Cells, Lung Epithelial Cells L2, Neuroblastoma Cells Neuro2A, and Astrocytoma Cells DBT

Confluent monolayers of primary neurons were infected with 2 MOI and, L2, Neuro2A cells, and DBT cells were infected with 0.5 MOI of RSA59 (PP), RSMHV2 (P), and RSMHV2 (PP) inoculum prepared in respective media containing 2% FBS. Infected cells were incubated with the virus at 37 °C in a humidified incubator with 5% CO_2 for 1 h 15 min with intermittent rocking at 15 min intervals for efficient viral adsorption. The viral inoculum was discarded and the infected cells were cultured with their respective growth medium in presence of 5% CO_2 at 37 °C. After 8 h, 12 h, 16 h, and 24 h post-infection infected cells were either processed for studying viral antigen spread and syncytia formation by EGFP fluorescence or immunofluorescence colocalizing viral antigen with different neuroglial cell-specific markers.

2.5. Immunofluorescence on Cultured Cells

Infected cells were washed with 1X PBS containing Ca^{2+} and Mg^{2+}, fixed in 4% PFA, washed with 1X PBS, and then mounted on glass slides using Mowiol 4–88 mounting media with DAPI. For neuron enrich primary culture characterization, cells were permeabilized with Triton X-100 in PBS. The cells were incubated with a blocking solution of PBS containing Triton X-100 and goat serum (GS). The cells were then incubated with primary antibodies rabbit anti-MAP2 (neuron marker), mouse anti-NFM (neuron marker), mouse anti-GFAP (astrocyte marker), at dilution 1:200 and mouse anti-H8H9 (same as anti-GalC, matured oligodendrocyte marker) at dilution 1:50 prepared in PBS/GS/Triton X-100 for 1 h. The antibodies used with their source and dilutions are tabulated in Table 1. The primary antibody labeled cells were then washed thrice with 1X PBS for 5 min and then incubated with respective secondary antibodies prepared in PBS/GS/Triton X-100 for 1 h, as specified in Table 1. The fluorescence-labeled cells were carefully washed with PBS carefully avoiding exposure to light and then mounted. The slides were then observed for EGFP fluorescence. Images were acquired with a Nikon eclipse Ti2 epifluorescence microscope with Nikon DS-Qi2 coupled camera and analyzed using ImageJ software.

2.6. Quantification of RSA59 (PP), RSMHV2 (P), and RSMHV2 (PP) Induced Cell-to-Cell Fusion In Vitro

The merged images of EGFP and DAPI channels were used for the quantification. The mean nuclei per syncytium formed in Neuro2A cells were quantified by counting the number of DAPI-stained nuclei inside the EGFP expressing syncytia. A total of 30 frames of 20× images were taken from three independent experiments for every time point for each strain of the virus. Randomly selected 25 syncytia were considered for precise quantification of nuclei stained with DAPI within a syncytium and graphically presented for L2 and Neuro2A cells. As the syncytia formed by RSA59 (PP) in DBT were very big, the size of each nucleus was determined and used to calculate the number of nuclei in each syncytium by using software algorithms. The syncytia formed by RSMHV2 (P) and RSMHV2 (PP) were not distinct, thus clusters of infected cells were considered as syncytia for DBT cells.

Like in Neuro2A, each cluster was cropped out and the nuclei from the DAPI channel was counted. ImageJ 1.52 g (Fiji) software was used for quantification.

2.7. Inoculation of Mice

Four-week-old MHV-free C57BL/6J mice were intracranially inoculated with a 50% LD50 dose of 20,000 PFU/mL RSA59 (PP), RSMHV2 (PP), or 100 PFU/mL RSMHV2 (P) as previously described [16,30]. Mice were monitored for the signs and symptoms of disease and possible mortality. Mock-infected groups received PBS with 0.75% BSA and were housed in the same conditions as infected groups. On days 3, 5, and 6 (acute stage) and day 30 (chronic stage) p.i., mice were anesthetized, transcardially perfused with PBS followed by 4% PFA, and successively the brains, spinal cords, and livers were harvested. Six mice (N = 6 in two experiments) were inoculated in each infection group for histopathological analysis on days 3, 6, and 30. Three mice (N = 3) were used for each infection group alongside mock-infected mice for immunofluorescence analysis on day 5 p.i. spinal cord cryosections.

2.8. Histopathological Analyses

Mice were sacrificed at days 3, 6, and 30 p.i. and perfused transcardially with PBS followed by 4% paraformaldehyde (PFA) in PBS. Liver, brain, and spinal cord tissues were harvested, postfixed in 4% PFA overnight, tissues were transferred to 70% ethanol and processed for tissue routine paraffin sectioning. Five μm thick longitudinal liver sections, sagittal brain sections, and cross-sections of spinal cord tissues were prepared for further histological and immunohistochemical analysis. Liver and spinal cord (cervical, thoracic and lumbar) tissue sections were processed for H&E and Luxol Fast Blue (LFB) staining, respectively. Liver and spinal cord pathology were blindly analyzed by two other investigators.

2.9. Immunohistochemical Analysis and Quantification of Viral Antigen

Serial sagittal sections of the brain and cross-sections of spinal cord tissues were stained using the avidin-biotin-immuno-peroxidase technique (Vector Laboratories, Burlingame, CA, USA) with 3, 3′ diaminobenzidine (DAB) as substrate and anti-N antibody as primary antibody, which was a kind gift from Dr. Julian Leibowitz of Texas A&M, College Station, Texas. The degree of viral antigen staining in different neuroanatomic regions of infected mice brain tissues was evaluated based on the scoring scale; score 0: no apparent viral antigen staining; 1: very small foci of viral antigen-positive cells; 2: widespread but small foci of viral antigen-positive cells; 3: widespread large foci of viral antigen-positive cells. To quantify anti-viral staining in the spinal cord, Fiji (ImageJ 1.52 g) software was used [32]. Image analysis was performed using the basic densitometric thresholding features of Fiji. The image was first captured at 10× magnification which allowed the entire section of the spinal cord to be visualized within a single image frame. The RGB image was color-deconvoluted into three different colors to separate the DAB-specific staining. The background labeling was also subtracted from all images, and then the contrast was slightly enhanced to improve the resolution. The perimeter of the spinal cord was digitally outlined, and the area was calculated in μm^2. A threshold value was defined for each image. The magnitude of viral staining was defined as the percentage area of staining (ratio of target-stained area to total selected area multiplied by 100), as previously described [17]. Two other investigators blindly analyzed all the sections following the above-described method.

2.10. Immunofluorescence on Spinal Cord Cryosections

Spinal cord tissues harvested on day 5 p.i.were fixed in PBS containing 4% PFA for 8 h and then placed in PBS containing 10% sucrose at 4 °C overnight and subsequently transferred to PBS containing 30% at 4 °C for 24 h. Tissues were embedded in OCT medium (cryomatrix) and coronally sectioned at 8μm with the aid of a cryotome (Thermo Scientific,

Waltham, MA, USA). Sectioned tissues were placed on the charged slides and kept at −80 °C till the time of immunofluorescence. Frozen sections were immunofluorescently labeled as previously described [27,30,33]. Briefly, sections were postfixed with ice-cold 95% ethanol for 20 min and washed with PBS at room temperature for 10 min followed by incubation with 1M glycine in PBS for 1 h to reduce nonspecific cross-linking, and 1 mg/mL NaBH4 for 10 min to decrease autofluorescence at room temperature in a humidified chamber. The sections were again PBS washed thrice for 5 min each before being incubated with a blocking solution containing 2.5% goat serum in PBS (GS) and 0.5% triton-X-100 in PBS for 1 h at room temperature. Incubation of sections with primary antibodies; virus-specific anti-nucleocapsid (anti-N) and anti-MAP2 prepared in GS at 1:50 and 1:200 dilutions respectively were done overnight at 4 °C. The section was then PBS washed thrice for 5 min before incubation with secondary antibodies as mentioned in Table 1. All incubations were carried out in a humidified chamber. Tissue sections were then washed with PBS thrice for 5 min and subsequently mounted with Mowiol 4–88 mounting medium containing DAPI. A green channel was used to visualize the viral antigen. MAP2 staining was viewed in the red channel. Visualization and images were acquired with a Nikon eclipse Ti2 epifluorescence microscope with Nikon DS-Qi2 coupled camera and processed with Image J (Fiji) software.

Table 1. Antibodies and respective dilutions used for immunofluorescence.

Primary Antibody	Dilution	Secondary Antibody	Dilution
Rabbit polyclonal Anti-MAP2 (Sigma)	1:200	Alexa fluor 568 Donkey Anti Rabbit (Invitrogen)	1:800
Mouse monoclonal Anti-NFM (Sigma)	1:200	FITC Goat Anti Mouse IgG (Jackson immunoresearch, West Grove, PA, USA)	1:250
Mouse monoclonal Anti-GFAP (Sigma)	1:200	FITC Goat Anti Mouse IgG (Jackson immunoresearch, West Grove, PA, USA)	1:250
Anti-H8H9 (same as anti-Gal C) (mouse monoclonal antibody against mature Oligodendrocytes) [34]	1:50	FITC Goat Anti Mouse IgG (Jackson immunoresearch, West Grove, PA, USA)	1:250
Anti-N (anti-nucleocapsid of MHV) (Gift from Dr. Julian Leibowitz of Texas A&M, College Station, TX)	1:25	FITC Goat Anti Mouse IgG (Jackson immunoresearch, West Grove, PA, USA)	1:250

2.11. Detection of Demyelination and Quantification

Spinal cord (cervical, thoracic and lumbar) tissue sections were processed for Luxol Fast Blue (LFB) staining to detect myelin loss. Total white matter area and areas with myelin loss (marked by a plaque of no LFB stain within white matter) were determined on day 30 RSA59 (PP), RSMHV2 (P), and RSMHV2 (PP) p.i. mice. Four-six LFB-stained spinal cord cross-sections from each mouse were randomly selected and analyzed using Fiji software (ImageJ 1.52 g) [17]. The total number of mice in each group was 5–6. The total perimeter of the white matter regions in each cross-section was outlined and calculated by adding up the dorsal and ventrolateral white matter areas in each section. The total perimeter of the demyelinated regions was also outlined and added for each section separately. The percentage of spinal cord demyelination per section per mouse was obtained by dividing the total area of the demyelinating plaque over the total area of the calculated white matter and then multiplied by 100, as previously described [17].

2.12. Structure Modeling and Molecular Dynamics Simulation of RSMHV2 (PP)

The structure of the trimeric fusion domain of RSMHV2 (PP) was obtained through homology modeling using MODELLER with RSA59 (PP) as the template [35]. All-atom MD simulation of this modeled structure was performed with GROMACS 5.1.4 using the CHARMM force field with cmap following the protocol previously described [17]. The simulated region of the RSMHV2 (PP) fusion domain, residues 909–1154 of each chain, corresponds to the same region of RSMHV2 (P) previously studied. The structure was centered in a cubic box with a dimension chosen by GROMACS to ensure a distance of 1 nm between the protein edge and box face. The system was solvated with water. Na^+ and Cl^- ions were added to neutralize the charge on the protein and ensure a final concentration of 0.1 M. Periodic boundary condition in all directions was applied. The system was energy minimized with the steepest descent algorithm, and the minimization was stopped when the maximum force in the system dropped below 1000 kj mol^{-1} nm^{-1}. This was followed by equilibration under NVT and NPT ensembles for 2 ns each. Berendsen thermostat at a temperature of 300 K and Parrinello-Rahman barostat at 1 atm was used in the simulations [36,37]. The production MD was run for 500 ns with a time step of 2 fs with frames saved every 10 ps yielding 50,000 frames for analysis.

The comparative analysis of this MD trajectory with the previous simulations [17] was done with MD DaVis, an analysis tool developed in-house (https://github.com/djmaity/md-davis, accessed on 18 January 2022). First, the free energy landscapes were calculated using root mean square deviation and radius of gyration following the protocol of Tavernelli et al., 2003 [38]. The free energy landscapes of the full trajectory showed that the systems were still evolving for the first 100 ns. Therefore, the trajectory from 100–500 ns was used for replotting the free energy landscapes and subsequent analysis. Next, the secondary structure was calculated with the GROMACS do_dssp tool. The percentage of frames for which a specific secondary structure was observed was calculated at each residue location. The torsional flexibility was also calculated, defined as the circular standard deviation of the φ and ψ backbone torsional angles. Finally, the hydrogen bonds with the central prolines in the FP were calculated with the GROMACS hbond tool and the percentage of frames with the hydrogen bonds was tabulated.

2.13. Statistical Analysis

The data from syncytia formation in the cultured cells was computed and analyzed as described above. Unpaired Student's *t*-test was used to compare two strains of virus at each time point p.i. The level of significance for viral antigen and LFB staining in different neuroanatomic regions and spinal cord sections were calculated using one-way ANOVA followed by Tukey's multiple comparisons test. All data were plotted and analyzed using GraphPad Prism 6.01 software. The level of significance and the means are presented in a scatter-bar diagram. A p-value < 0.05 was considered statistically significant and presented as * p. The standard error of the mean (SEM) was presented with each mean as an error bar in the plots.

3. Results

3.1. Sequence Comparison between RSA59 (PP), RSMHV2 (P), RSMHV2 (PP) Fusion Peptide

The mutant RSMHV2 (PP) was generated by adding one proline to FP by site-directed mutagenesis and targeted RNA recombination as discussed in the "Material and Methods". The entire FP of RSA59 (PP), RSMHV2 (P), and RSMHV2 (PP) were sequenced and compared with the known predicted MHV-A59 FP sequence (gene bank accession number: 9629812) and MHV2 (gene bank accession number: AF201929) FP within the spike gene fusion domain (Figure 1A). A consensus sequencing was carried out by RT-PCR of viral mRNA amplified from RSA59 (PP), RSMHV2 (P) and RSMHV2 (PP) infected L2 cells [17]. The FP sequence of RSA59 (PP) and RSMHV2 (P) was identical to the published sequence of our previous work [17]. RSMHV2 (PP) FP sequence was identical to RSMHV2 (P) with an addition proline in the FP (Figure 1B).

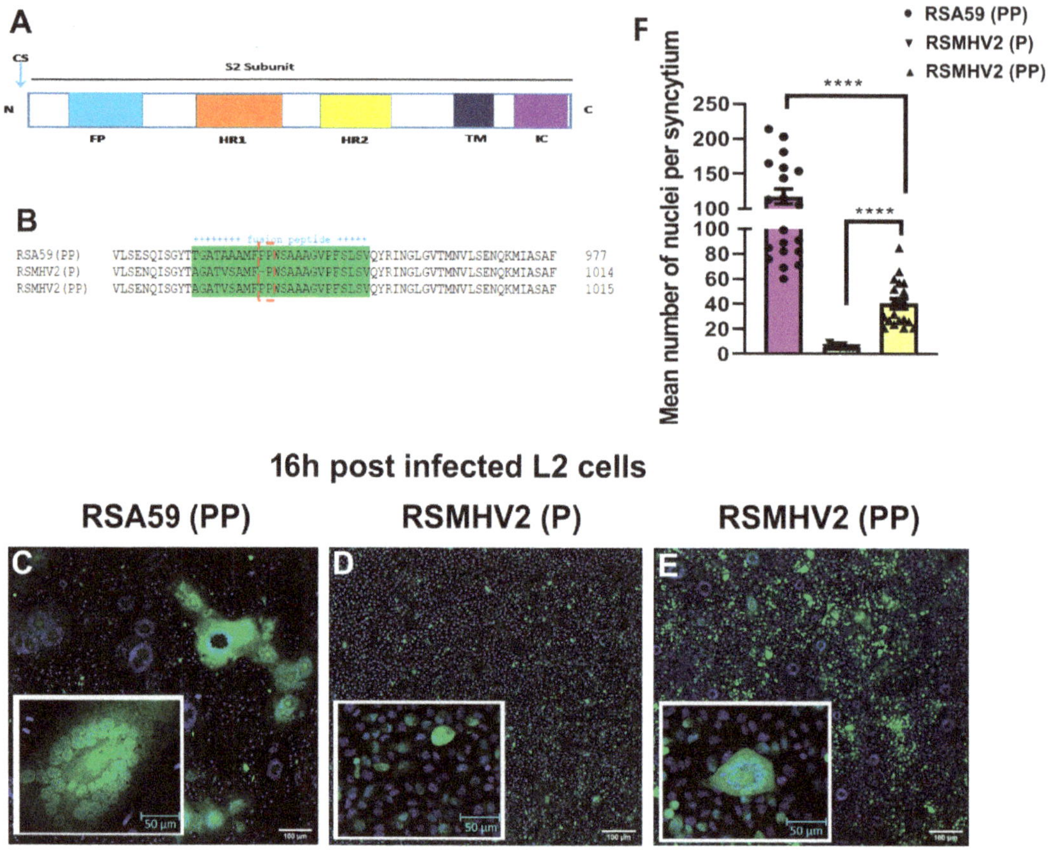

Figure 1. Sequence comparison of fusion peptide and fusogenic properties of RSA59 (PP), RSMHV2 (P), and RSMHV2 (PP) in L2 cells. Schematic of the S2 subunit of coronavirus Spike gene, showing cleavage site (CS), FP, heptad repeat 1 (HR1), heptad repeat 2 (HR2), transmembrane domain (TM), and intracellular tail (IC) (**A**). FP region of RSA59 (PP), RSMHV2 (P), and RSMHV2 (PP) sequences are aligned using ClustalW tool, dashed lines show central double proline (PP) or single proline (P) in RSA59 (PP), RSMHV2 (PP), and RSMHV2 (P), respectively (**B**). A monolayer of L2 cells was infected with RSA59 (PP), RSMHV2 (P), and RSMHV2 (PP) at MOI 0.5. The cells were incubated for 16 h at 37 °C with 5% CO_2, fixed with 4% paraformaldehyde, and mounted in DAPI containing mounting media. Epifluorescence microscopy was performed, and images were acquired and further processed using Image J (Fiji) software. The green fluorescence is the intrinsic viral EGFP that is incorporated into the genome. Images of DAPI (blue) and EGFP (green) were merged to construct the images presented. RSA59 (PP) and RSMHV2 (PP) infection cause syncytia formation (**C**,**E**) but RSMHV2 (P) infection rarely formed syncytia (**D**). The mean nuclei per syncytium were quantified and scatter-bar plotted (**F**). Experiments were repeated five times. Significance level was taken at $p < 0.05$ following unpaired t-test analysis. **** $p < 0.00001$.

3.2. *Divergent Fusogenicity of RSA59 (PP), RSMHV2 (P), and RSMHV2 (PP) in L2 Cells*

Cell-to-cell fusion property (fusogenicity) of RSA59 (PP), RSMHV2 (P), and RSMHV2 (PP) was examined in the confluent culture of L2 cells at an MOI 0.5. RSA59 (PP), RSMHV2 (P), and RSMHV2 (PP) significantly differ in their fusogenic and cytopathic properties. Briefly, upon RSA59 (PP) infection, L2 cells started to fuse as early as 8 h and formed profuse syncytia which started to increase in size with time until 16 h post-infection (p.i.)

(Figure 1C). Beyond16 h p.i., most of the giant syncytia started to dissolve as all of the infected cells had lysed. In contrast, RSMHV2 (P) infected individual cells rarely formed syncytia even 24 h after p.i., only a few infected cells (2–3) formed a cluster as observed previously (Figure 1D) [17]. Interestingly, infection of RSMHV2 (PP) with two consecutive prolines in the FP significantly increased the size of the multinucleated cluster of cells denoted as syncytia formation, as compared to RSMHV2 (P) (Figure 1E). The growth curve performed for RSMHV2 (P) and RSMHV2 (PP) until 24 h starting from the point of adherence (0 h) is identical hence depicting that no significant alteration in viral replication and growth can be observed after the addition of one proline (Figure S1). The experiments were repeated 5 times under the same conditions, and in most of the cases, distinctive multinucleated syncytia were observed to be formed by RSMHV2 (PP), which is different from RSMHV2 (P), but on the other hand, the number of multinucleated cells in the syncytia are much less compared to RSA59 (PP). The quantification of the fusion index as discussed in "Material and Methods" was plotted in a scatter-bar diagram (Figure 1F).

3.3. *RSA59 (PP), RSMHV2 (P), and RSMHV2 (PP) Differ in Their Neuronal Tropism, Spread through Neuron and Syncytia Formation in Primary Neuronal Culture, Neuroblastoma Cell Line, Neuro2a, and Delayed Brain Tumor (DBT), Astrocytoma Cell Lines*

Mixed neuroglial cultures enriched in primary neurons as discussed in "Material and Methods" were immunolabeled with anti-MAP2 antibody (neuron marker), anti-NFM antibody (neuron marker), anti-GFAP antibody (astrocyte marker), anti-H8H9 antibody (matured oligodendrocyte marker). Cells were counterstained with DAPI (nuclear stain). Intrinsic EGFP fluorescence denotes viral antigen. Visual manual counting of immunostained cells revealed that 65–70% of the cells in the culture are MAP2 positive indicating neurons (Figure 2D). Neurofilament neuronal staining frequently showed colocalization with MAP2 (Figure 2A). Immunostaining of the culture with astrocyte marker GFAP and differentiated oligodendrocyte marker H8H9 revealed that some cells are astrocytes and oligodendrocytes, respectively (Figure 2B,C). Visual manual counting of immunostained cells showed that 15–20% cells are GFAP positive astrocytes and 8–10% cells are H8H9 positive oligodendrocytes. Scatter-bar diagram in (Figure 2D), depicts the percentage of different neuroglial cells in the mixed neuroglial culture. These primary neuron enriched cultures were infected with RSA59 (PP), RSMHV2 (P), and RSMHV2 (PP) at an MOI of 2, and after 24 h p.i. cells were stained with neuronal marker MAP2 and counterstained with DAPI. Colocalization of intrinsic viral-EGFP fluorescence with neuronal markers revealed that a large number of MAP2+ neurons are infected, forming profuse syncytia. RSMHV2 (P) can infect neurons at a much lower efficiency. In contrast, the efficiency to infect neurons is significantly increased in RSMHV2 (PP). Moreover, colocalization studies revealed that RSA59 (PP) can infect neurons and also can spread through neurons to cause cell-to-cell fusion (Figure 2E–I). RSMHV2 (P) has limited ability to infect neurons and even if it infects the viral antigen spread is significantly restricted in the culture (Figure 2J–N). The infectivity and the viral spread increased in RSMHV2 (PP) compared to RSMHV2 (P) but remains much less for RSA59 (PP) as also observed in L2 cell infection (Figure 2O–S). Detailed quantification of differential neuronal tropism has been depicted in a scatter-bar diagram (Figure 2T). In summary, the results demonstrate that the addition of one proline significantly increases the neuronal tropism and viral antigen spread in RSMHV2 (PP) compared to its parental strain RSMHV2 (P), still, the addition of one proline in the backbone of RSMHV2 (P) was not able to match the ability to infect neuron, viral antigen spread and, cell-to-cell fusion as observed in RSA59 (PP).

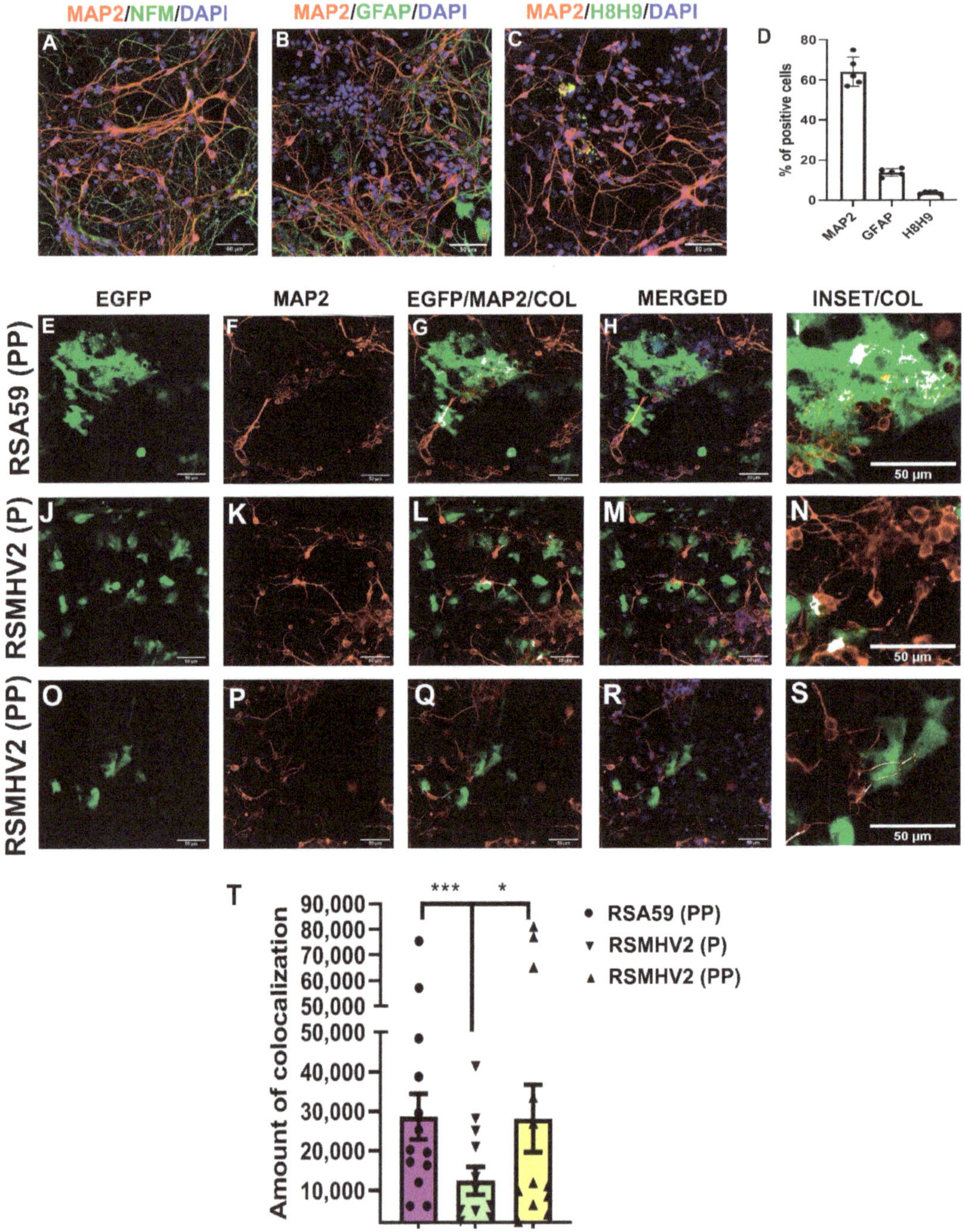

Figure 2. Characterization and infection of primary neuronal culture with RSA59 (PP), RSMHV2 (P), and the one proline added mutant RSMHV2 (PP). Mixed neuroglial culture enriched in primary neuron immunolabeled with anti-MAP2 antibody (Red; neuronal marker), anti-NFM antibody (Green;

neuron marker) (**A**), anti-GFAP antibody (Green; astrocytic marker) (**B**), anti-H8H9 antibody (Green; matured oligodendrocyte marker) (**C**). Cells were counterstained with DAPI (Blue; nuclear stain). (**D**), Visual manual counting of immunostained cells revealed that culture consists of 65% MAP2 positive neurons. Primary neuronal cultures were infected with RSA59 (PP) (**E–I**), RSMHV2 (P) (**J–N**), and RSMHV2 (PP) (**O–S**) at an MOI of 2, and after 24 h p.i. cells were stained with MAP2 and counterstained with DAPI. Representative images of the cells infected with the virus are shown where the EGFP panel shows the virus-infected cells, MAP2 channel denotes neurons in the culture and colocalization shows that the neurons were infected with the virus. The merged panel depicted the neurons are EGFP positive hence infected with the virus and DAPI represents the nucleus. Higher magnification images of the colocalization are highlighted in insets. (**T**) Colocalization of MAP2 and EGFP was represented in a scatter-bar plot. Experiments were repeated two times (N = 2) with three replicates per experiment per virus (n = 6). Significance level was taken at $p < 0.05$ following unpaired t-tests analysis. * $p < 0.05$, *** $p < 0.001$.

The differential neuronal tropism of RSA59 (PP), RSMHV2 (P), and RSMHV2 (PP) were further confirmed in Neuro2a, neuroblastoma cells in continuous culture. Detailed immunofluorescence characterization revealed that 95% of Neuro2a cells in culture are NFM/MAP2 positive (data not shown), demonstrating homogeneity of neuronal cells in culture. Briefly, 85–90% of confluent Neuro2a cells were infected with RSA59 (PP), RSMHV2 (P), and RSMHV2 (PP) at 0.5 MOI. Time kinetic studies of cell-to-cell fusion from 12–24 h revealed that RSA59 (PP) profusely infected Neuro2a cells in culture and started to form syncytia as early as 12 h p.i. which increased with time to form giant syncytia as shown in Figure 3A–D until 24 h p.i. RSMHV2 (P) and RSMHV2 (PP) showed individually infected cells at 12 h but not many obvious multinucleated cells were observed in any of the infected cultures (Figure 3E,I). Whereas, after 16 h p.i. RSMHV2 (P) showed 2–4 cells in the cluster, no obvious syncytia were observed (Figure 3F,G). The rate of syncytia formation did not alter much even after 24 h p.i. (Figure 3H). In contrast RSMHV2 (PP) at 16 hp.i. formed multinucleated moderate size syncytia comprising 10–15 nuclei (Figure 3J,K). The size and number of the syncytia did not increase at 24 h p.i. in infected cells (Figure 3L). A scatter-bar diagram of the mean number of nuclei or cells per cluster clearly shows the significant difference in the size of the syncytia between RSA59 (PP) with RSMHV2 (P) or RSMHV2 (PP) at 12 h,16 h, and 24 h p.i. (Figure 3M). But the differences between RSMHV2 (P) and RSMHV2 (PP) were maximum and significant only at 16 h p.i. (Figure 3M). Time kinetics studies in Neuro2a further confirmed that RSA59 (PP) with RSMHV2 (P) and RSMHV2 (PP) differ in their neuronal tropism, viral spread, and syncytia formation. Further, these studies confirm that the addition of one proline in RSMHV2 (PP) significantly increases the efficiency of syncytia formation compared to parental strain RSMHV2 (P), but is significantly less able to form syncytia compared to RSA59 (PP).

Our observations in the neuron-enrich primary culture as well as neuronal cell line, Neuro2a culture where the addition of one proline in RSMHV2 (P) significantly increased syncytia formation though smaller in size and fewer in number than the syncytia observed in RSA59 (PP) prompted us to investigate the fusogenicity of RSMHV2 (PP) in astrocytoma Delayed Brain Tumor (DBT) cells. The DBT cells are known to be frequently used for the productive replication of different MHV strains. Confluent monolayers of DBT cells were infected with RSA59 (PP), RSMHV2 (P), or RSMHV2 (PP) at MOI 0.5. Similar to Neuro2a cells, RSA59 (PP) started forming giant syncytia as early as 8 h p.i. in the DBT cell line (Figure 4A), and that significantly increased in size and number by 12 h p.i. (Figure 4B), and right after that started to dissolve due to profuse viral replication. Like neuronal cell infection, RSMHV2 (P) and RSMHV2 (PP) can efficiently infect DBT cells in culture, which increases with time between 8 h–12 h p.i. At 8 h p.i. no multinucleated cell clustering was observed either in RSMHV2 (P) or RSMHV2 (PP) infected culture (Figure 4C,E). Clustering of multinucleated cells was significantly observed in the RSMHV2 (PP) cluster at 12 h p.i. (Figure 4F); whereas, RSMHV2 (P) infected cells still failed to form clusters (Figure 4D). The size and number of RSMHV2 (PP) clustering remained low as compared to RSA59 (PP) clustering. Quantification of the mean number of cells per syncytium at 8 h and 12 h p.i.is depicted in a scatter-bar diagram (Figure 4G).

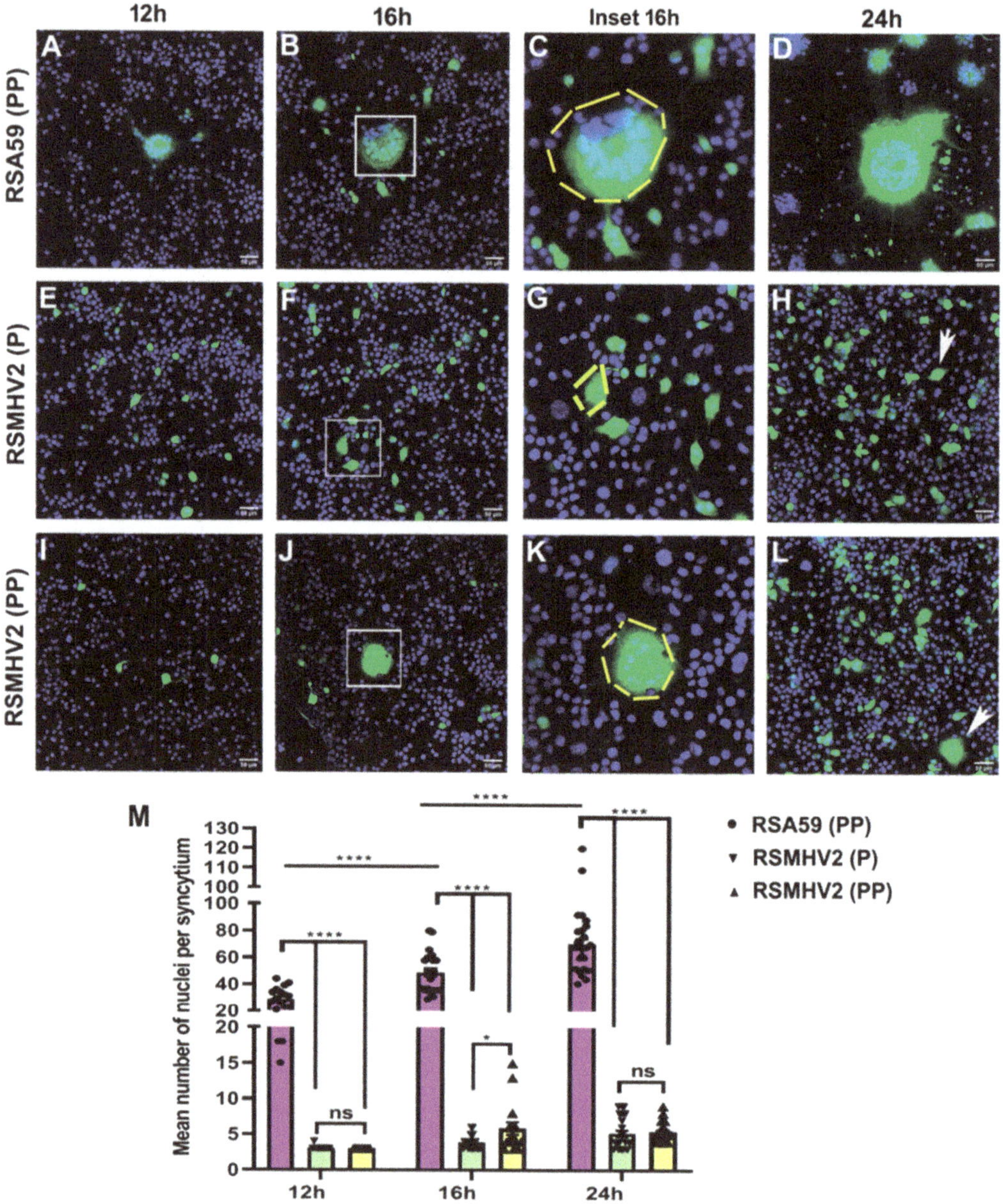

Figure 3. Comparison of syncytia formation in Neuro2a cell infected by RSA59 (PP), RSMHV2 (P), and the one proline added mutant RSMHV2 (PP). Neuro2a cells were infected with RSA59 (PP), RSMHV2 (P), and RSMHV2 (PP) at 0.5 MOI. The green fluorescence is due to EGFP that is integrated into the viral genome. Images of DAPI (Blue) and EGFP (Green) were merged to construct the final

images presented here; (**A–D**) panels are from RSA59 (PP) infected Neuro2a cells at 12 h (**A**) 16 h, (**B**,**C**) is a magnified image of 16 h, (**D**) at 24 h, respectively. Panel E-H shows Neuro2a cells infected by RSMHV2 (P), (**E**) at 12 h, (**F**) 16 h, (**G**), is a magnified image of 16 h, (**H**) at 24 h, respectively. Arrowheads indicate 1–2 cells in a cluster. (**I–L**) show Neuro2a cells infected by RSMHV2 (PP) where (**I**) is 12 h p.i., (**J**) at 16 h, (**K**) magnified 16 h, and (**L**) 24 h p.i, respectively. The mean nuclei per syncytia were counted and plotted in a scatter bar (**M**). Experiments were repeated three times (N = 3) with three replicates per experiment per virus per time point p.i (n = 9). Significance level was taken at p < 0.05 following unpaired *t*-test analysis. * $p < 0.05$, **** $p < 0.0001$; ns means not significant.

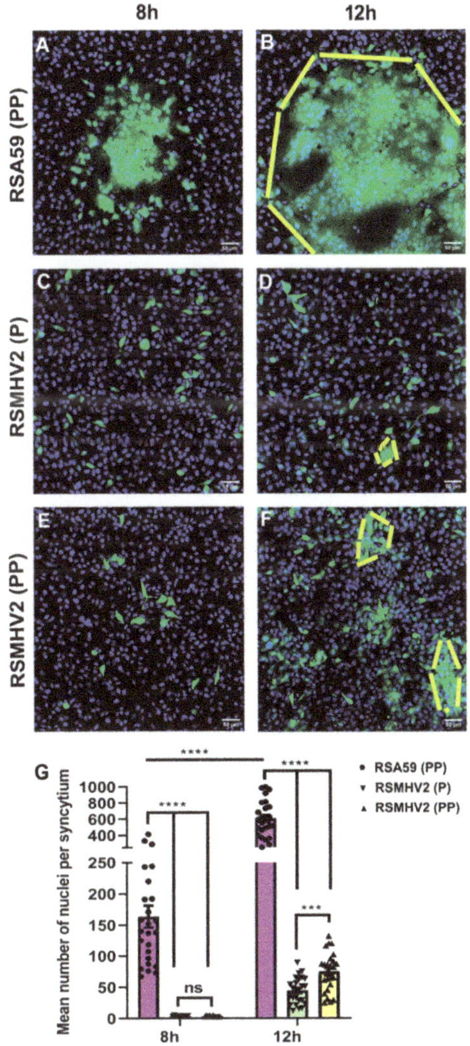

Figure 4. Effect of double proline on the fusogenic ability of RSA59 (PP), RSMHV2 (P), and RSMHV2 (PP) in DBT cell line. DBT cells were infected with RSA59 (PP), RSMHV2 (P), and RSMHV2 (PP) at an MOI of 0.5. Viral Antigen is denoted by EGFP fluorescence. Images are composite

of DAPI and FITC channels; Panel (**A**,**B**) are from RSA59 (PP) infected DBT cells at 8 h and 12 h p.i, respectively. Large syncytia are indicated within the dashed lines. Panel (**C**,**D**) and (**E**,**F**) depicts DBT cells infected by RSMHV2 (P) and RSMHV2 (PP), respectively at 8 h and 12 h p.i. Dashed lines indicate clusters of infected cells in panels (**D**,**F**). The mean nuclei per syncytium in panels (**A**,**B**), and clusters of infected cells in panels (**C–F**) were counted and plotted in the scatter-bar diagram (**G**). Experiments were repeated three times (N = 3) with three replicates per experiment per virus per time point p.i (n = 9). Significance level was taken at $p < 0.05$ following unpaired t-tests analysis. *** $p < 0.001$; **** $p < 0.0001$.

3.4. RSMHV2 (P) and RSMHV2 (PP) Differ in Their Ability to Induce Necrotizing Hepatitis at Day 3 and Day 6 p.i.

Four-week-old C57BL/6J mice were infected intracranially with a 50% of LD50 dose of RSMHV2 (P) at 100 pfu and RSMHV2 (PP) at 20,000 pfu. Upon infection, on day 3 and day 6 p.i. mice were euthanized, and liver tissues were harvested for pathological studies. Briefly, harvested liver tissues from RSMHV2 (P), and RSMHV2 (PP) infected mice were processed, 5 µm thin longitudinal sections were stained with Hematoxylin and Eosin (H &E). Bright-field microscopic observations revealed that both RSMHV2 (P) and RSMHV2 (PP) induce large foci of hepatic lesions throughout the liver sections. Detailed quantification of the area of the hepatic lesions showed that the size of the hepatic lesions was much bigger in RSMHV2 (PP), at day 3 p.i. (Figure 5B), but the differences in the size of the hepatic lesions are much more evident at day 6 p.i. (Figure 5A) than RSMHV2 (P) (Figure 5A,C). The quantification data has been plotted in a scatter-bar diagram (Figure 4E). RSMHV2 (P), and RSMHV2 (PP) induce moderate to severe hepatitis like their parental stain but the insertion of one proline significantly increased the rate of hepatitis as denoted by the increase in the size of the hepatic lesion both at day 3 and day 6 p.i.

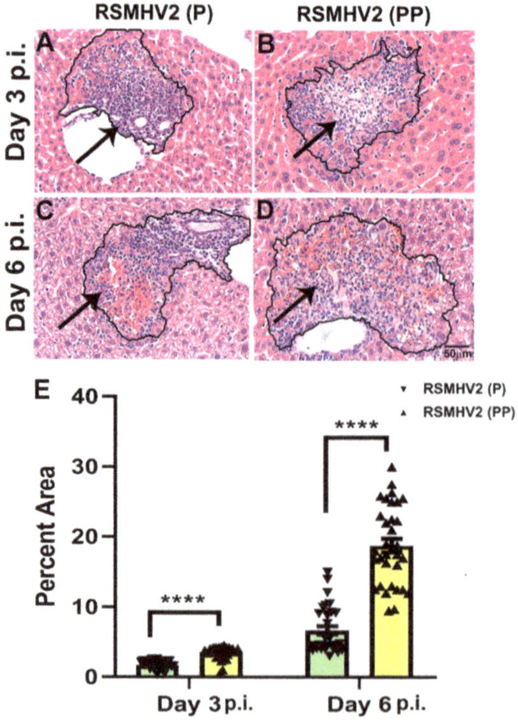

Figure 5. Liver pathology consisting of moderate to severe hepatitis following RSMHV2 (P) and RSMHV2 (PP) infection. Five-micron thin liver sections of day 3 and day 6 post-infected mice were

stained with Hematoxylin & Eosin. Liver pathology consists of moderate to severe necrotizing and non-necrotizing hepatitis in both RSMHV2 (P) (**A**,**C**) and RSMHV2 (PP) (**B**,**D**) at day 3 and day 6 p.i., respectively. The size of hepatic lesions was quantified on day 3 and day 6 p.i., and the percent area of hepatic lesions was plotted in the scatter-bar diagram (**E**). The level of significance was calculated by unpaired Student's *t*-test with Welch's correction. n = 6 mice each day per virus infection and **** $p < 0.0001$.

3.5. RSMHV2 (P), and RSMHV2 (PP) Differ in Their Ability to Distribute Viral Antigen in Different Neuroanatomic Regions on Day 3 and Day 6 p.i.

Four-week-old C57BL/6J mice were infected intracranially with a 50% of LD50 dose of RSMHV2 (P) at 100 pfu and RSMHV2 (PP) at 20,000 pfu. Day 3 and day 6 p.i. mice were euthanized, and brain tissues were harvested for pathological studies. Briefly, harvested brain tissues from RSMHV2 (P), and RSMHV2 (PP) infected mice were processed, 5–10 µm thin mid-sagittal sections were examined for the viral antigen distribution in different neuroanatomical regions. Brain sections were immunohistochemically stained with an anti-nucleocapsid antibody, Anti-N. Immunostaining data revealed that on day 3 RSMHV2 (P) infected brain samples viral antigen distribution is restricted mainly in the olfactory bulb, meninges, near the lateral-ventricle/ subventricular zone at the lining of the subependymal layer of the 4th ventricle (Figure 6A,C,E,G,I,K). As compared to RSMHV2 (P), on day 3 RSMHV2 (PP) infected brain section show more viral antigen that are invading the different neuroanatomical regions including the site of inoculation (lateral geniculate nuclei), olfactory bulb, meninges, and within the brain parenchyma, in the basal forebrain, ventral striatum, lateral ventricular nuclei and the subependymal layer of 4th ventricle (Figure 6B,D,F,H,J,L). Though no significant differences in the neuroanatomic spread of RSMHV2 (P) and RSMHV2 (PP) were observed, it seems that RSMHV2 (PP) invades the brain parenchyma with higher efficiency and replicates profusely in the olfactory bulb, basal forebrain, ventricular striatum, and cerebral cortex as compared to RSMHV2 (P).

On day 6 p.i.of mice, the diffused viral antigen is mainly present in the olfactory bulb and posterior region of the brain, brain stem in RSMHV2 (P) (Figure 6M,O,Q,S,U,W). In contrast, in RSMHV2 (PP) viral antigen was still present in the olfactory bulb, cerebral cortex, ventral striatum/basal forebrain, lateral ventricle, and brainstem region (Figure 6N,P,R,T,V,X). Much widespread dissemination of virus antigen in RSMHV2 (PP) infected mouse brain section compared with RSMHV2 (P) both at day 3 and day 6 p.i. indicated that the addition of one proline resulted in increased viral antigen spreading and proliferation. The viral antigen distribution of RSMHV2 (PP) and RSMHV2 (P) was compared in different neuroanatomical regions of the brain as mentioned in the "Material and Methods" section. RSMHV2 (PP) versus RSMHV2 (P) significantly differed in the amount of viral antigen in the olfactory bulb and cerebral cortex on day 3 p.i. On day 6 p.i. in addition to the olfactory bulb and cerebral cortex there are significant changes in the ventral striatum and brain stem region. No significant differences were observed at the lateral ventricle/subventricular zone on day 3 and day 6 p.i. The mean differences and standard error of the distribution of viral antigen in different anatomical regions are plotted as a scatter-bar diagram (Figure 6Y,Z).

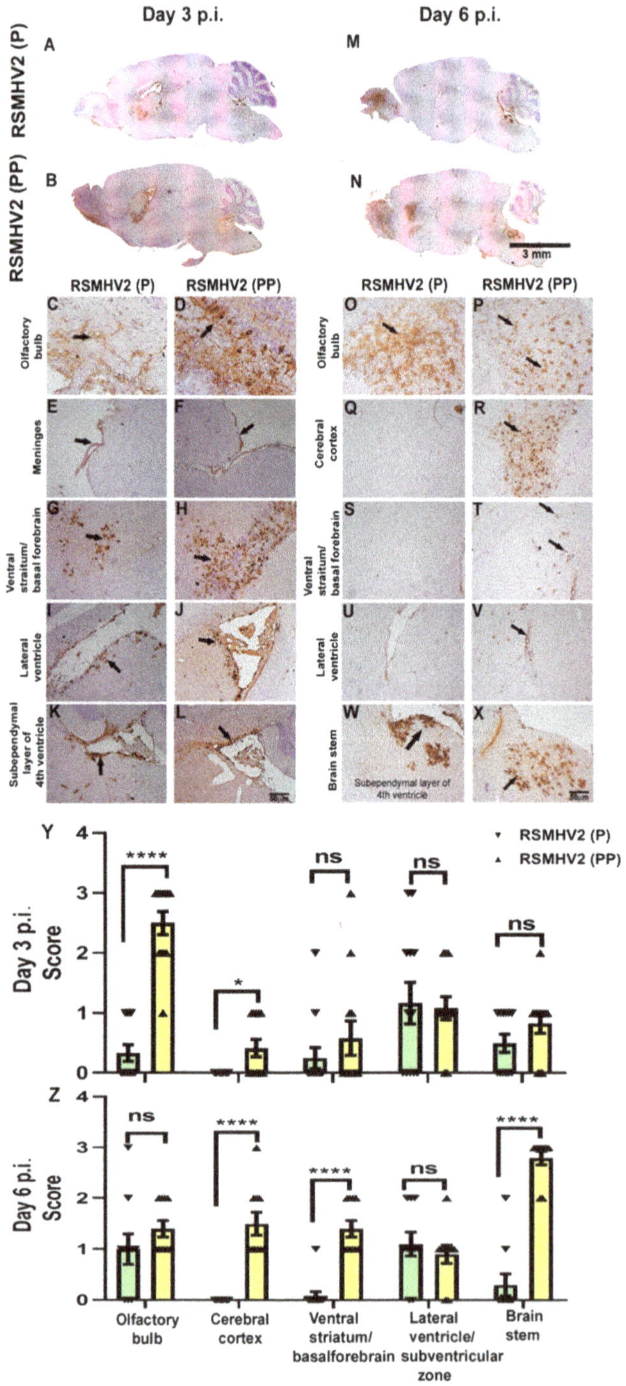

Figure 6. Differential distribution of viral antigen in different neuroanatomical regions from RSMHV2 (P) and RSMHV2 (PP) at days 3 and 6 p.i. brains. Five-micron thick mid-sagittal brain

sections from RSMHV2 (P) and RSMHV2 (PP) infected mice were immunohistochemically stained with anti-N antibody (viral antigen). Scanned images of a whole mid-sagittal section of the brain from RSMHV2 (P) (**A**) and RSMHV2 (PP) (**B**) on day 3 and (**M,N**) at day 6, respectively. On day 3, both RSMHV2 (P) and RSMHV2 (PP) infect and spread to similar neuroanatomic regions but the amount of viral antigen was significantly more in RSMHV2 (PP) compared to RSMHV2 (P). In RSMHV2 (P), viral antigen distribution was observed in the olfactory bulb (**C**), the subpial layer of meninges (**E**), ventral striatum/basal forebrain (**G**), lateral ventricle (**I**), and the subependymal layer of the 4th ventricle (**K**). Similarly, in RSMHV2 (PP) infected brain sections, viral antigen distribution was observed in different regions: olfactory bulb (**D**), meninges (**F**), ventral striatum/basal forebrain (**H**), lateral ventricle (**J**), and also to the subependymal layer of 4th ventricle (**L**). On day 6 p.i, both RSMHV2 (P) and RSMHV2 (PP), viral antigen distribution was observed in the olfactory bulb (**O,P**), and the subependymal layer of the 4th ventricle (**W,X**). Meanwhile, RSMHV2 (PP) was able to invade deep in parenchyma compared to RSMHV2 (P) in the cerebral cortex (**Q,R**), ventral striatum/basal forebrain (**S,T**), and lateral ventricle (**U,V**), and brainstem. Arrows indicate viral antigen staining. Quantification was done based on the scores 0: no infection; 1: very small foci of infection; 2: widespread small foci of infection; 3: widespread with large foci of infection. Day 3 p.i (**Y**) showed a significant difference in the olfactory bulb and cerebral cortex. Day 6 p.i. (**Z**) revealed a significant difference in the cerebral cortex, ventral striatum/basal forebrain, and brainstem (N = 6 mice per virus infection).The level of significance was determined by unpaired Student's *t*-test with Welch's correction * $p < 0.05$, **** $p < 0.0001$.

3.6. RSMHV2 (P) and RSMHV2 (PP) Differ in Their Ability to Neuronal Spread from Gray Matter to White Matter and Invade into the White Matter in Spinal Cord

Four-week-old C57BL/6J mice were either mock-infected or infected intracranially with a 50% of LD50 dose of RSMHV2 (P) at 100 pfu and RSMHV2 (PP) at 20,000 pfu. Day 3 and day 6 p.i. mice were euthanized, and spinal cord tissues were harvested for histopathological studies. Briefly, harvested spinal cord tissues from Mock-infected, RSMHV2 (P), and RSMHV2 (PP) infected mice were processed, 5–10 μm thin cross-sections were examined for the viral antigen distribution in gray and white matter neurons (Figure 7A–F). Spinal cord cross-sections were immunostained with Anti-N indicating viral antigen. Mock infected spinal cord sections were negative for viral antigen staining (Figure 7A,D). The viral spread of RSMHV2 (P) as previously shown was restricted mainly to the gray matter and occasionally to the gray-white matter junction at day 3 and day 6 p.i.(Figure 7B,E). In contrast, in the RSMHV2 (PP) infected spinal cord viral antigen is present in the gray matter, gray-white matter junction, and also in the white matter of the spinal cord. RSMHV2 (PP) is significantly replicating and spreading to gray-white matter junctions and white matter as compared to RSMHV2 (P), but the differences are much more pronounced at day 6 p.i. than on day 3 p.i. (Figure 7C,F). The quantification data has been presented in a scatter-bar diagram (Figure 7G).

To confirm the viral antigen spread, spinal cord cryosections were prepared as described in the "Material and Methods" section. Intrinsic EGFP fluorescence was used to detect viral antigen spread in gray matter, and from gray to white matter. The neuroanatomical distribution of the gray-white matter was corroborated by its corresponding LFB stained sections which clearly demarcated gray matter from white matter in the spinal cord (Figure 7H). RSA59 (PP) is a known fusogenic demyelinating strain of MHV and is known to spread from gray matter to white matter by day 5 p.i. Previous studies demonstrated that the successful axonal transport of virus from gray matter to white matter p.i. is key to early-stage myelitis and chronic stage demyelination– given this reason it was used as a control for the rest of the studies [26,27]. In this study, we compared the viral antigen distribution of RSMHV2 (PP) and RSMHV2 (P) with RSA59 (PP) in cryosections. Data revealed that while in RSMHV2 (P) infected spinal cord sections at day 5 viral antigen was restricted to the gray matter (Figure 7J), in RSMHV2 (PP) viral antigen was also primarily present in the gray matter with occasional distribution from gray matter to white matter like RSA59 (PP) (Figure 7I). Insertion of proline in RSMHV2 (PP) increased the ability of RSMHV2 (PP) to traffic to the white matter but could not attain the ability of RSA59 (PP) to invade the white matter (Figure 7K).

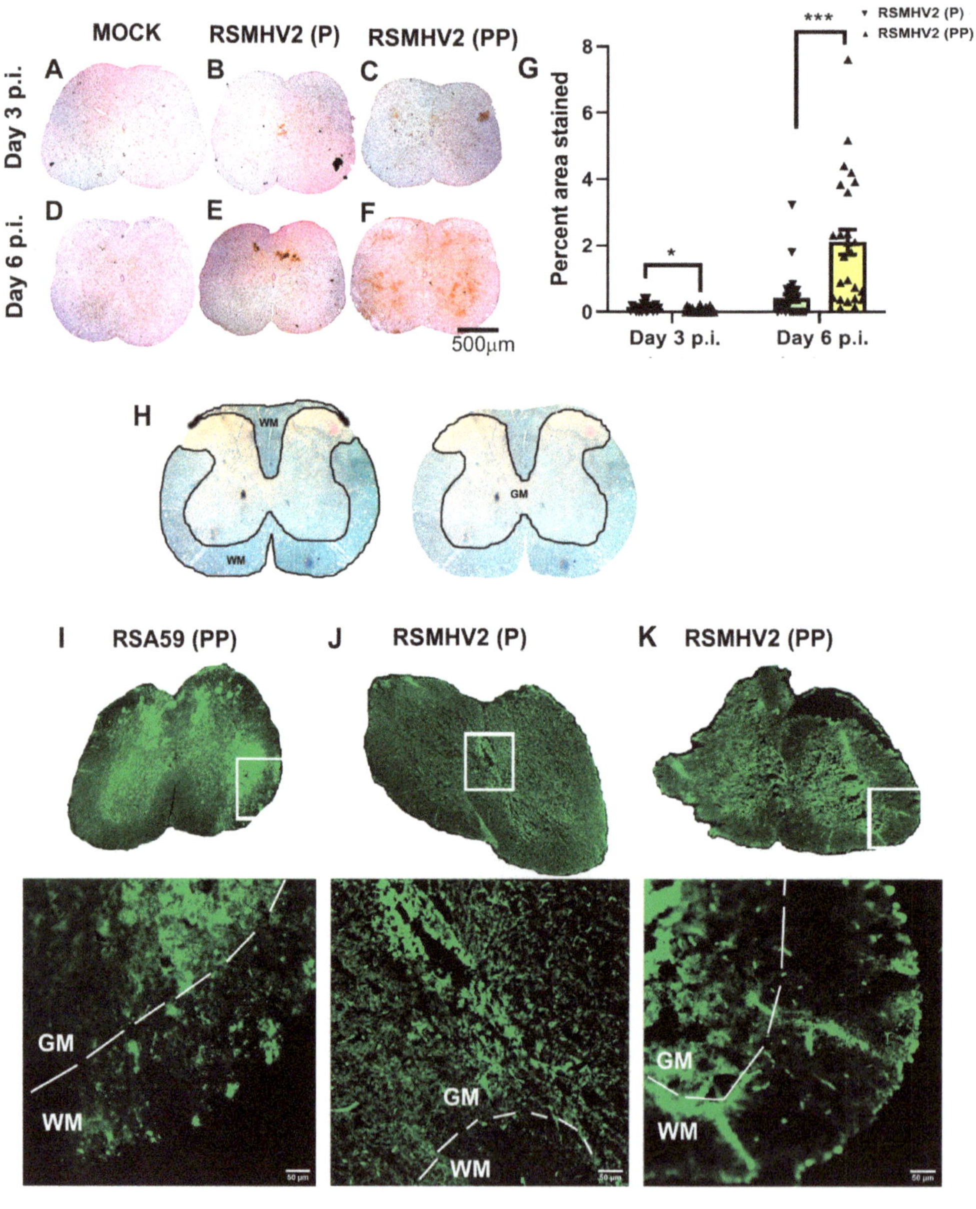

Figure 7. Differential distribution of viral antigen in RSMHV2 (P) and RSMHV2 (PP) infected mice spinal cord. Day 3 and Day 6 p.i. mock and viral infected spinal cord cross-sections were stained immunohistochemically (IHC) with anti-N antibody (**A–F**). No obvious staining was observed in any

mock-infected sections (**A**,**D**). On day 3 p.i. in RSMHV2 (P) infected spinal cord viral antigen distribution was observed restricted around the central canal (**B**), whereas in RSMHV2 (PP) viral antigen distribution was apparent throughout the spinal cord including gray matter, gray-white matter junction with occasional distribution to the white matter (**C**). On day 6 p.i. RSMHV2 (PP) replicated profusely in the gray matter also with frequent distribution to the white matter of spinal cord (**F**) but in RSMHV2 (P) infected spinal cord, viral antigen was restricted mostly in the gray matter (**E**). Cross-section of spinal cord showing white and gray matters mapped out in an LFB staining (**H**). Day 5 p.i. infected spinal cord cryosections were prepared for fluorescent microscopy. Dashed lines demarcate the grey and white matter in cryosections denoted by GM and WM, respectively. RSA59 (PP) infected spinal cord showed widespread viral antigen indicated by EGFP in both grey and white matter (**I**). Similar to IHC, RSMHV2 (PP) infected spinal cord showed viral antigen present in gray and white matter (**K**) whereas RSMHV2 (P) infected spinal cord showed a restricted spread of viral antigen around the central canal in gray matter with no viral antigen spread to the white matter (**J**). The viral antigen staining was quantified and plotted as a scatter-bar diagram (**G**). The level of significance was determined by unpaired Student's *t*-test with Welch's correction. N = 6 mice per day per virus infection. * $p < 0.05$; *** $p < 0.001$.

3.7. RSMHV2 (PP) and RSMHV2 (P) Differ in Their Axonal Transport Contributing to Viral Antigen Spread from Gray Matter to White Matter

RSA59 (PP) is known to follow intraneuronal transport from gray to white matter and release at the nerve ending to infect white matter oligodendrocytes [26,27]. Whereas RSMHV2 (P) is known to infect neurons but is impaired in axonal transport and is not released at the nerve end. To examine whether the addition of one proline significantly altered the axonal transport of RSMHV2 (PP), spinal cord cryosection was immunostained with MAP2 and compared with RSA59 (PP) and RSMHV2 (P) infected spinal cord cryosections. Intrinsic EGFP fluorescence denotes viral antigen. Most of the colocalized images were taken from gray-white matter junctions to follow the axonal transport either from the dorsal or the ventral horn. Colocalization studies demonstrated that in RSA59 (PP) viral antigen colocalizes with MAP2 denoting the axonal translocation (Figure 8A–C,S,D,F,T). In RSMHV2 (P) not much colocalization was obvious either in the dorsal horn or the ventral horn (Figure 8G–L,U,V). Interestingly, RSMHV2 (PP) occasionally colocalized with MAP2 which demonstrates that RSMHV2 (PP) can conduit through the axon to reach the white matter (Figure 8M–R,W,X). Colocalization studies demonstrate that the addition of one proline significantly increased the ability of axonal transport of RSMHV2 (PP), and thus may allow it to invade the white matter.

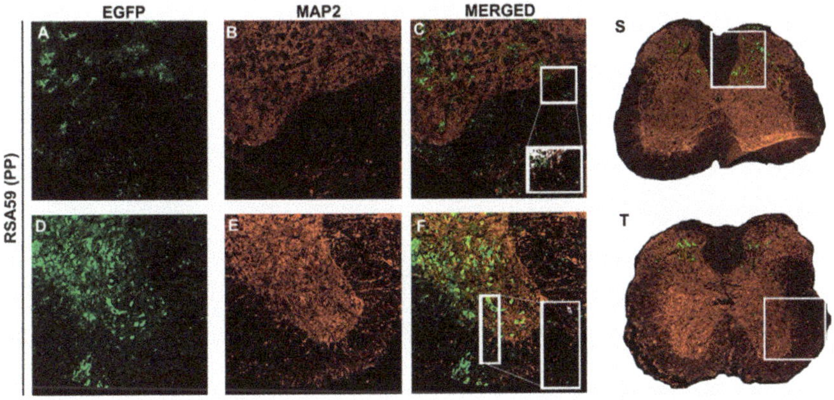

Figure 8. *Cont.*

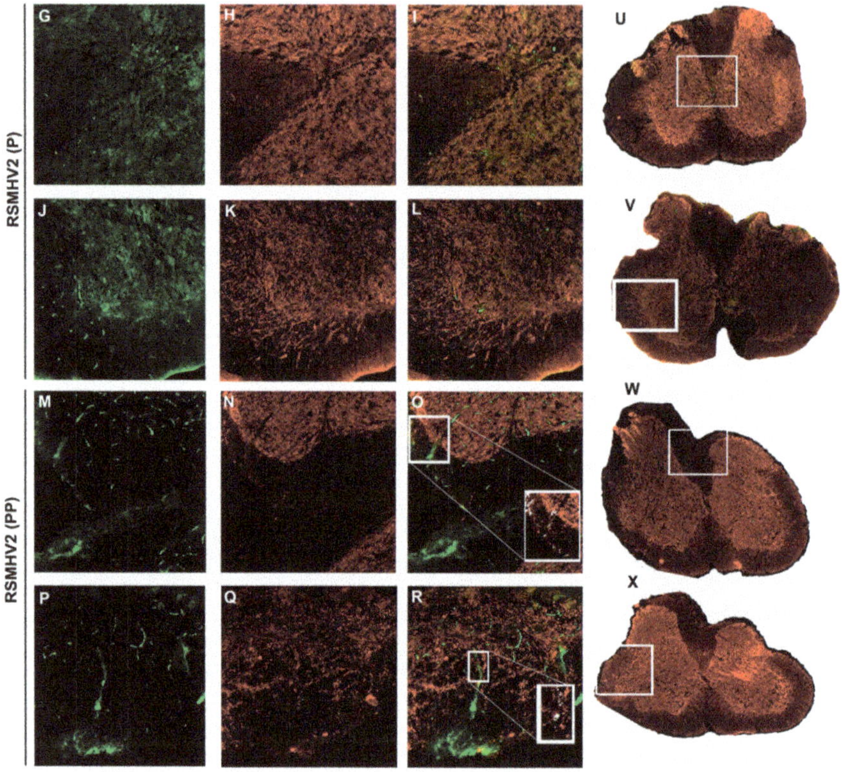

Figure 8. Colocalization of viral antigen with neuronal microtubule marker (MAP2) of RSA59 (PP), RSMHV2 (P), and RSMHV2 (PP) infected spinal cord cryosections. Day 5 post-infected spinal cord cryosections immuno-fluorescently labeled with MAP2 (Red) (**B**,**E**,**H**,**K**,**N**,**Q**). Intrinsic EGFP fluorescence denotes viral antigen (**A**,**D**,**G**,**J**,**M**,**P**). Merged images of EGFP and MAP2 (**C**,**F**,**I**,**L**,**O**,**R**). RSA59 (PP) infected spinal cord sections are (**A–F**,**S**,**T**), RSMHV2 (P) are from (**G–L**,**U**,**V**), RSMHV2 (PP) are (**M–R**,**W**,**X**). Dorsal column gray-white matter junctions are in panels (**A–C**) (RSA59 (PP)), (**G–I**) (RSMHV2 (P)), and P-R (RSMHV2 (PP)). The ventral white matter and adjacent ventral horns from the infected section are shown as RSA59 (PP) (**D**,**F**), RSMHV2 (P) (**J–L**), RSMHV2 (PP) (**P–R**). The insets in the merged images show magnified views of the co-localization of virus-infected cells with the Anti-MAP2 and dendrites at the gray-white matter junction. In the insets, co-localization is seen as white with infected axons.

3.8. RSMHV2 (PP) Induced Mild to Moderate Myelin Damage Compared to No Demyelination of RSMHV2 (P), but the Intensity of the Demyelination Was Much Less Compared to Parental Demyelinating Strain RSA59 (PP) as Studied on Day 30 p.i.

Four-week-old C57BL/6J mice were either mock-infected or infected intracranially with 50% of LD50 dose of RSA59 (PP) at 20,000 pfu, RSMHV2 (P) at 100 pfu, and RSMHV2 (PP) at 20,000 pfu. At the chronic stage, day 30 p.i. mice were euthanized and spinal cord tissues were harvested for histopathological studies. Briefly, harvested spinal cord tissues from mock-infected, RSA59 (PP), RSMHV2 (P), and RSMHV2 (PP) infected mice were processed, 5–10 μm thin cross-sections were stained with Luxol Fast Blue (LFB) to examine for demyelination. No evident demyelination was observed in mock-infected spinal cord sections (Figure 9A). RSA59 (PP), as known in previous studies produces discrete large demyelinating plaques to profuse confluent demyelination both in the dorsal column, the anterior and ventral horn of white matter at multiple levels of spinal cord tissue [26,30]. Representative images are shown in Figure 9B. No obvious myelin loss was observed in RSMHV2 (P) infected spinal cord sections (Figure 9C), in contrast, RSMHV2 (PP) infection-induced sparse myelin loss in dorsal column, ventral/anterior horn but no large discrete plaques or confluent plaques were observed throughout the spinal cord section (Figure 9D). The quantification of demyelination was performed as discussed in the "Material and Methods" section and the scatter-bar diagram of the quantified data demonstrates that RSMHV2 (PP) induced significant demyelination compared to RSMHV2 (P), but the intensity of the myelin loss is significantly low compared to RSA59 (PP) (Figure 9E). The addition of one proline significantly increased the demyelinating properties of RSMHV2 (PP) compared to no demyelination by one proline containing parental strain RSMHV2 (P), but not sufficient enough to form demyelinating plaques observed in the case of RSA59 (PP).

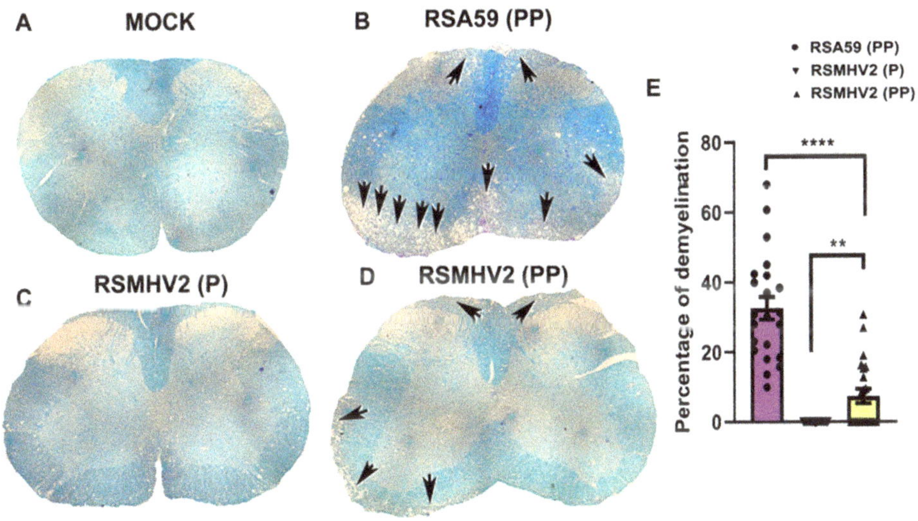

Figure 9. Presence of double proline in the FP of RSMHV2 (PP) induces sparse myelin loss during day 30 p.i. (chronic stage of infection). Five to ten-micron thick spinal cord sections from either mock-infected (**A**) or RSA59 (PP) (**B**), RSMHV2 (P) (**C**), RSMHV2 (PP) (**D**) infected mice were stained with LFB to detect myelin loss. No myelin loss was observed in mock-infected or RSMHV2 (P) infected spinal cord sections (**A**,**C**), respectively. In contrast, large demyelinating plaques were observed at multiple levels of the spinal cord from RSA59 (PP) infected mice (**B**). RSMHV2 (PP) infected mice showed discrete yet fewer demyelinating axons without any obvious demyelinating plaques in sections of the spinal cord (**D**) (Black arrows indicate demyelinated area). Quantification of demyelination was plotted in the scatter-bar diagram (**E**), and the level of significance was calculated by unpaired Student's *t*-test. N = 5 mice per virus infection. ** $p < 0.001$; **** $p < 0.0001$.

3.9. Rigidity in the Fusion Peptide from the Addition of Proline to Spike Protein from RSMHV2 (P)

Molecular dynamics was used to the structural properties of the spike protein variants. The free energy landscape of spike from RSMHV2 (PP) has a single potential well like the spike from RSMHV2 (P), albeit shallower and in contrast to RSA59 (PP) spike which has two deep potential wells (Figure 10A). The torsional flexibility shows that the local dynamics of the FP from spike in RSMHV2 (PP) are almost identical to the variant from RSA59 (PP) (Figure 10B), also corroborated by the root-mean-square fluctuation. The secondary structure of the FP in RSMHV2 (PP) is also similar to the helix-loop-helix-loop helix arrangement in spike from RSA59 (PP). In contrast, the spike from RSMHV2 (P) is missing the central helix and has a helix-loop-helix arrangement (Figure 10B). The disorder is caused by the lack of H-bonds in the region due to the absence of double proline in the FP of spike from RSMHV2 (P). Despite the local flexibility of the FP in spike from RSMHV2 (P), the deep free energy well indicates a relatively rigid structure. This is explained by the fact that the free energy landscapes in Figure 10A encompass both the tertiary and quaternary dynamics of the spike protein fusion domain. Therefore, the local tertiary dynamics of spike from RSMHV2 (PP) resemble the spike from RSA59 (PP), but the global quaternary dynamics resemble spike from RSMHV2 (P).

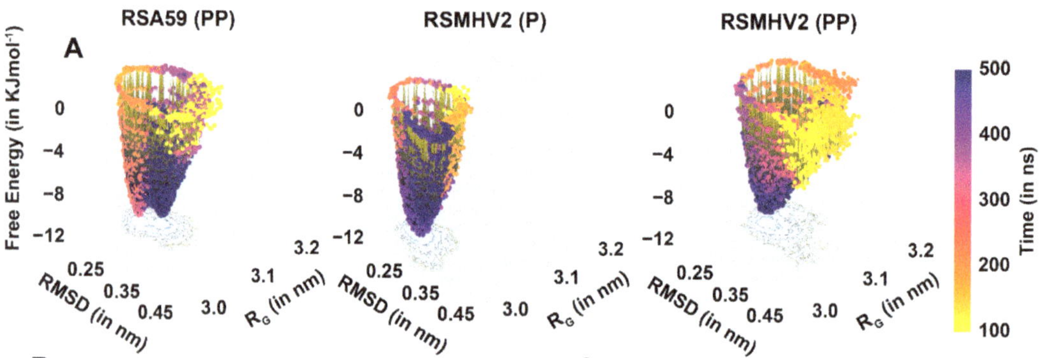

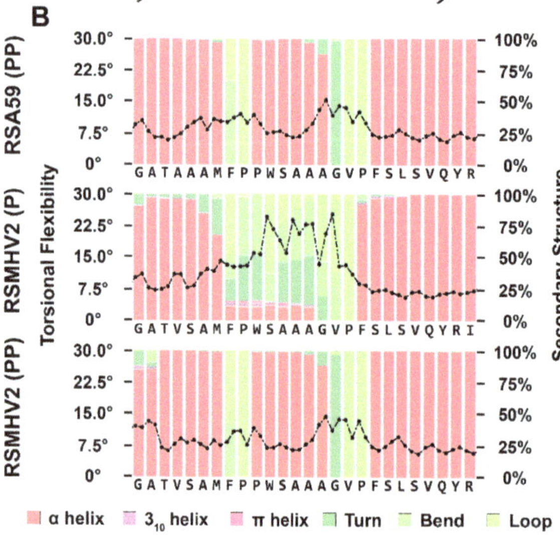

C

Donor	Acceptor	H-bond Occurrence
RSA59 (PP)		
A-SER-941 OG-HG	A-PRO-938 O	90.13 %
A-ALA-942 N-H	A-PRO-939 O	91.31 %
B-TRP-940 N-H	B-PRO-938 O	9.16 %
B-SER-941 N-H	B-PRO-938 O	85.64 %
B-SER-941 OG-HG	B-PRO-938 O	91.22 %
B-ALA-943 N-H	B-PRO-939 O	10.71 %
C-SER-941 OG-HG	C-PRO-938 O	91.48 %
C-ALA-942 N-H	C-PRO-939 O	98.51 %
C-SER-951 OG-HG	C-PRO-938 O	4.53 %
RSMHV2 (P)		
C-GLN-1123 NE2-HE21	C-PRO-976 O	23.62 %
RSMHV2 (PP)		
A-SER-979 N-H	A-PRO-976 O	88.05 %
A-ALA-980 N-H	A-PRO-976 O	94.37 %
A-ALA-981 N-H	A-PRO-977 O	8.38 %
A-GLN-1124 NE2-HE21	A-PRO-977 N	89.70 %
B-SER-979 OG-HG	B-PRO-976 N	96.36 %
B-ALA-980 N-H	B-PRO-977 O	7.18 %
C-SER-979 N-H	C-PRO-976 O	90.24 %
C-ALA-980 N-H	C-PRO-976 O	95.68 %
C-ALA-981 N-H	C-PRO-977 O	5.25 %

Figure 10. Results from the analysis of MD simulations of Spike from RSA59 (PP), RSMHV2 (P), and RSMHV2 (PP) in water. (**A**) Free energy landscapes of spike proteins from RSA59 (PP), RSMHV2 (P), and RSMHV2 (PP) from root-mean-square deviation and radius of gyration of the trajectory between 100–500 ns. The trajectory points are plotted on the surface and colored according to time as given by the color bar. (**B**) Torsional flexibility and secondary structure of the fusion domain in the three simulations. The two markers for each residue correspond to φ and ψ backbone torsion angle. The stacked bar for each residue shows the percentage of frames for which the secondary structure was observed. Since the three chains of the spike protein are identical and their results were similar, the results for the three chains were averaged. (**C**) Hydrogen bonds with the central proline/prolines in the FP. The donor and acceptors are given in the following format 'Chain-Residue-Residue number Atom name'; the donor is also suffixed with '-Hydrogen atom name'.

4. Discussion

The presence of two consecutive central prolines (PP) within the FP contributes to the efficiency of the fusion process. The data from the current study clearly emphasizes this, given a non-fusogenic non-demyelinating strain RSMHV2 (P) is mutated to a fusogenic and weakly demyelinating strain RSMHV2 (PP) on insertion of an additional proline adjacent to the central proline in the FP. The extent of this difference can be gauged from the use of a 20,000 pfu inoculation dose for experimentation with RSA59 (PP) and RSMHV2 (PP), compared to only 100 pfu for RSMHV2 (P) in the in vivo study (Figure S2). This difference in inoculation volume is based on the feasibility of experimental design else the readout of syncytia formation becomes erroneous when a large pfu of the virus is used without commensurate virus fusion ability. In vitro comparison of RSMHV2 (PP) with its parental strain RSMHV2 (P) and RSA59 (PP) show different rates and intensity of infection with primary neonatal neuronal cells, Neuro2A cells, L2 cells, and Delayed Brain Tumor astrocytoma cells; but the neuroglial cell infection was significantly different. Overall, the RSMHV2 (PP) showed moderate but significant cell-to-cell fusion and syncytia formation compared to RSMHV2 (P), but significantly low fusogenicity compared to RSA59 (PP). In vivo studies also show consistent results, with RSMHV2 (PP) producing a severe necrotic lesion in the liver compared to the RSMHV2 (P)—the caveat being the 50% of the LD50 dose of 100 pfu used for RSMHV2 (P), compared to 20,000 pfu for RSMHV2 (PP) and RSA59 (PP). The insertion of an adjacent proline appears to boost 50% of the LD50 dose from 100 pfu to 20,000 pfu in mice.

The decreased susceptibility of RSMHV2 (PP) compared to RSMHV2 (P) may also occur considering that the cell-to-cell spread helped the mutant strain to evade the immune responses. Though this hypothesis is beyond the scope of this study, it would be interesting to understand whether the addition of one proline significantly altered the immune modulation in mounting the host immunity. Our studies demonstrate that though both RSMHV2 (PP) and RSMHV2 (P) differ in their susceptibility, they successfully invade brain parenchyma but significantly differ in viral antigen distribution in some of the neuroanatomic regions, namely, ventral striatum, basal forebrain, cerebral cortex, and brain stem. Some significant differences in the viral antigen distribution are observed in the olfactory bulb during day 3 p.i., but none on day 6 p.i. None are observed in the leptomeninges either, nor in the ventricular and subventricular regions of the brain indicating that RSMHV2 (PP) and RSMHV2 (P) do not differ in their ability to cause meningitis. While striking differences were observed in the viral antigen distribution of both the strains in the spinal cord, in RSMHV2 (P), the viral antigen is mainly restricted to the gray matter both at day 3 and 5 p.i, but RSMHV2 (PP) did not restrict to the gray-white matter demarcation and was almost equally distributed throughout the spinal cord. As RSMHV2 (P) was restricted to gray matter in the acute stages of infection, there was no myelin loss in the chronic phase. RSMHV2 (PP) showed moderate demyelination pertaining to the spread of the virus along the axon through the white matter in the acute stage of infection. The intensity of the demyelination of RSMHV2 (PP) is significantly less compared to the parental demyelinating strain RSA59 (PP).

The comparative role of PP in the aggressive virulence of RSA59 (PP) was previously investigated by deleting one proline in the PP of the FP, yielding strain RSA59 (P) [17]. RSA59 (P) significantly lowered neural cell syncytia formation and viral titers post-infection in vitro. Transcranial inoculation of C57Bl/6 mice with RSA59 (PP) or RSA59 (P) yielded similar degrees of necrotizing hepatitis and meningitis, but only RSA59 (PP) can produce widespread encephalitis extending deeply into the brain parenchyma. Both the virus variants are mostly cleared from the brain by day 6 post-infection. In addition, RSA59 (PP) is also known to cause optic neuritis and demyelination in the optic nerve as well as retinal ganglionic cells (RGC) loss. As a mechanism, it was evident that upon intracranial inoculation into the lateral geniculate nuclei of the brain, RSA59 (PP) in addition to anterograde axonal transport also follow retrograde axonal transport through the optic nerve to reach the retinal ganglionic cells and in due course causes optic neuritis, optic nerve demyelination, and RGC loss. Thus, an intact dyad of proline residues in the FP of the targeted-recombinant meningio-encephalomyelitis and demyelinating MHV strain may regulate the translocation of the virus-antigen along the axons and subsequent neurodegeneration.

Which FP feature is likely critical in establishing the virus phenotype—the location of the proline residue in the FP, or the physicochemical property of the amino acid at that position? Previous studies on MHV-A59 FP have shown that the substitution of the proline residue at position 938 with lysine (P938K) partially impaired fusion while replacing the same proline residue with a leucine residue did not have any effect on fusion [31]. The mutagenesis/substitution of the methionine residue at position 936 with lysine (M936K) or leucine (M936L) did not affect fusion [32]. Thus, the data suggest that both the location and the physicochemical property of the amino acid are important. The importance of location can be hypothesized from a recently proposed membrane-protein contact initiation model in SARS-CoV-2 Spike [33]. The FP in the Spike trimeric quaternary structure is juxtaposed in a location that facilitates its early contact with the host membrane after attachment with the host cell receptor. The trimeric architecture of Spike ensures three copies of the FP are present and the alignment of each of the FPs is midway between a pair of receptor binding domains that can guide the initial contact with the host membrane. The precise geometric/positional requirement of the proline appears to coincide with the positional requirement of FP for facile early membrane-protein contact. The presence of proline at the required position also highlights the importance of its physicochemical property. Proline is distinct due to its imino acid stereochemistry compared to an amino acid. Consequently, it cannot donate hydrogen bonds in the tertiary structure, and its side-chain ring is fused to the protein backbone thereby restricting the polypeptide chain flexibility around its location. Among all protein residues, proline restricts the protein backbone flexibility the most which can be seen from its smallest area occupancy in the Ramachandran map [34].

The presence of PP is expected to sterically create a larger segment of local backbone restriction compared to a single-proline in the FP. It is expected to induce rigidity also in its immediate spatial neighborhood. This is corroborated by the MD simulation of MHV2 spike protein where the PP in the FP region is significantly restricted in RSMHV2 (PP) compared to RSMHV2 (P). The local fluctuation graph around FP matches closely with RSA59 (PP). The local rigidity is also supplanted by more local hydrogen bonds in PP containing FP and is driven by the stereochemically-imposed rigidity as the single P containing FP in RSMHV (P) has limited hydrogen bonds. The local rigidity in the PP neighborhood is, however, quite contrasting to the global flexibility of the S2 domain. The S2 domain of RSMHV2 (PP) appears to have a shallower energy well compared to RSMHV2 (P), although both have a single potential well. This differs from the two energy wells from the S2 domain of RSA59 (PP). A greater number of wells in the energy landscape indicates more interconvertible conformational states of the protein and also reflects a low transition barrier when the lowest energy states in these wells are conformationally close to each other. This is favorable for triggering conformational transition needed for change to post-fusion structure in comparison to a deep single energy well as in RSMHV2 (P) where the chances of these transitions are dampened. This may explain the order of fusogenic potential seen in the three viruses considered in this study.

The critical location of the FP in the spike protein and its demonstrated ability to alter fusogenicity and pathogenesis make it an apt site for a potential antiviral. A series of studies exist that attempt to identify the minimum essential region required for the fusogenic property for designing the therapeutic mimetic peptides [34,35]. These premised that the internal FP owing to its hydrophobicity and location adjacent to the heptad repeat segments within the S2 spike domain of MHV-A59 may be responsible for the fusogenic properties associated with its hepato-neuropathogenicity [16,17,31,32]. Dissecting the minimal essential motif of the Spike protein FP may enable us to design a mimetic peptide to set the stage for competition to reduce virus-induced neuroinflammation. This strategy is advantageous because the S2 domain of spike is much more evolutionarily conserved compared to the receptor binding or the N-terminal domain. As such, this will also allow us to build a therapeutic approach that applies to pan-CoV situations.

Altogether, combining proline insertion deletion studies highlight insights that fusion-dependent viral spread could be important in general for virus-host cell interaction. Interestingly, many fusion proteins do not have a central proline in their FP; however, residues therein that improve FP rigidity and cause the neighborhood structure to become more ordered in the membrane environment improve kinetic efficiency of the fusion process, making them suitable targets as per insights of our study. Overall, our work is useful in understanding the mechanisms employed by the virus fusion apparatus and may help in guiding the development of therapeutic interventions to restrict cell-to-cell viral antigen spread and its consecutive viral infection.

5. Conclusions

The presence of two consecutive central prolines (PP) within the FP of spike protein of β-coronavirus contributes to the efficiency of the fusion process. Comparative in vitro and in vivo studies between virus strains RSA59(PP), RSMHV2 (P), and RSMHV2 (PP) in the FP demonstrate that the insertion of one proline significantly resulted in enhancing the virus fusogenicity, spread, and consecutive neuropathogenesis. The intensity of the demyelination as a read-out of the neuropathogenesis of RSMHV2 (PP) is significantly less compared to the parental demyelinating strain RSA59 (PP), but significantly more compared to RSMHV2 (P). The critical location of two central prolines of the FP predominantly determines fusogenicity and may be essential but not sufficient to cause demyelination.

Supplementary Materials: The following supporting information can be downloaded at: https://www.mdpi.com/article/10.3390/v14040834/s1, Figure S1: Growth kinetics of RSMHV2 (P) and RSMHV2 (PP). Figure S2: Viral antigen spread in brain parenchyma of RSA59 (PP), RSMHV2 (P), and RSMHV2 (PP) infected at 200 PFU at day 5 p.i.

Author Contributions: Conceptualization, J.D.S.; methodology, A.A.S., M.S., A.K., D.M., D.P. and J.D.S. software, A.A.S., M.S., V.M., D.P. and J.D.S.; validation, D.P. and J.D.S.; formal analysis, A.A.S., M.S., A.K., V.M., D.M., A.B. and B.S.; investigation, A.A.S., M.S., A.K., D.M., D.P. and J.D.S.; resources, J.D.S.; data curation, A.A.S., M.S., A.K., D.M., D.P. and J.D.S.; writing—original draft preparation, A.A.S., V.M., D.P. and J.D.S.; writing—review and editing, V.M., D.P. and J.D.S.; visualization, A.A.S., M.S. and J.D.S.; supervision, J.D.S.; project administration, J.D.S.; funding acquisition, J.D.S. All authors have read and agreed to the published version of the manuscript.

Funding: This research received no external funding.

Institutional Review Board Statement: The use of animals and all experimental procedures were reviewed and approved by the institutional animal care and use committee at the Indian Institute of Science Education and Research Kolkata. The animal protocols adhered to the guidelines of the Committee for the Purpose of Control and Supervision of Experiments on Animals (CPCSEA), India. The name of the approved protocol is Development of unique animal model to understand the viral-induced etiology of human central nervous system auto-immune disease Multiple Sclerosis (MS), protocol No.: IISERK/IAEC/AP/2015/02.03, originally approved in the year of 2015, recently extended 13 July 2020 and valid through 12 July 2022.

Informed Consent Statement: Not applicable.

Data Availability Statement: Not applicable.

Acknowledgments: This manuscript is dedicated in the memory of Randall J. Corhs for his constant inspiration and encouragement. We also thank and acknowledge the IISER Kolkata Microscopy facility. We thank the Department of Biotechnology (DBT), New Delhi, for supporting the computational facilities. We thank IISER Kolkata for providing fellowship to A.A.S., A.K., CSIR for providing fellowship to M.S., and PMRF for providing fellowship to V.M.. We also thank respective Institutional funding of D.M., A.B. for their fellowship.

Conflicts of Interest: The authors declare no conflict of interest.

References

1. Earp, L.J.; Delos, S.E.; Park, H.E.; White, J.M. The many mechanisms of viral membrane fusion proteins. *Curr. Top. Microbiol. Immunol.* **2005**, *285*, 25–66. [CrossRef] [PubMed]
2. Durell, S.R.; Martin, I.; Ruysschaert, J.M.; Shai, Y.; Blumenthal, R. What studies of fusion peptides tell us about viral envelope glycoprotein-mediated membrane fusion (review). *Mol. Membr. Biol.* **1997**, *14*, 97–112. [CrossRef] [PubMed]
3. Lamb, R.A.; Jardetzky, T.S. Structural basis of viral invasion: Lessons from paramyxovirus F. *Curr. Opin. Struct. Biol.* **2007**, *17*, 427–436. [CrossRef] [PubMed]
4. White, J.M.; Whittaker, G.R. Fusion of Enveloped Viruses in Endosomes. *Traffic* **2016**, *17*, 593–614. [CrossRef]
5. Gallagher, T.M.; Buchmeier, M.J. Coronavirus spike proteins in viral entry and pathogenesis. *Virology* **2001**, *279*, 371–374. [CrossRef]
6. Heald-Sargent, T.; Gallagher, T. Ready, set, fuse! The coronavirus spike protein and acquisition of fusion competence. *Viruses* **2012**, *4*, 557–580. [CrossRef]
7. Xia, S.; Yan, L.; Xu, W.; Agrawal, A.S.; Algaissi, A.; Tseng, C.-T.K.; Wang, Q.; Du, L.; Tan, W.; Wilson, I.A.; et al. A pan-coronavirus fusion inhibitor targeting the HR1 domain of human coronavirus spike. *Sci. Adv.* **2019**, *5*, eaav4580. [CrossRef]
8. Bosch, B.J.; van der Zee, R.; de Haan, C.A.; Rottier, P.J. The coronavirus spike protein is a class I virus fusion protein: Structural and functional characterization of the fusion core complex. *J. Virol.* **2003**, *77*, 8801–8811. [CrossRef]
9. Kirchdoerfer, R.N.; Cottrell, C.A.; Wang, N.; Pallesen, J.; Yassine, H.M.; Turner, H.L.; Corbett, K.S.; Graham, B.S.; McLellan, J.S.; Ward, A.B. Pre-fusion structure of a human coronavirus spike protein. *Nature* **2016**, *531*, 118–121. [CrossRef]
10. Li, F. Structure, Function, and Evolution of Coronavirus Spike Proteins. *Annu. Rev. Virol.* **2016**, *3*, 237–261. [CrossRef]
11. Sainz, B., Jr.; Rausch, J.M.; Gallaher, W.R.; Garry, R.F.; Wimley, W.C. Identification and characterization of the putative fusion peptide of the severe acute respiratory syndrome-associated coronavirus spike protein. *J. Virol.* **2005**, *79*, 7195–7206. [CrossRef] [PubMed]
12. Qiu, Z.; Hingley, S.T.; Simmons, G.; Yu, C.; Das Sarma, J.; Bates, P.; Weiss, S.R. Endosomal proteolysis by cathepsins is necessary for murine coronavirus mouse hepatitis virus type 2 spike-mediated entry. *J. Virol.* **2006**, *80*, 5768–5776. [CrossRef] [PubMed]
13. Deng, S.Q.; Peng, H.J. Characteristics of and Public Health Responses to the Coronavirus Disease 2019 Outbreak in China. *J. Clin. Med.* **2020**, *9*, 575. [CrossRef] [PubMed]
14. Hui, D.S.; Azhar, E.I.; Madani, T.A.; Ntoumi, F.; Kock, R.; Dar, O.; Ippolito, G.; McHugh, T.D.; Memish, Z.A.; Drosten, C.; et al. The continuing 2019-nCoV epidemic threat of novel coronaviruses to global health—The latest 2019 novel coronavirus outbreak in Wuhan, China. *Int. J. Infect. Dis.* **2020**, *91*, 264–266. [CrossRef]
15. Tang, D.; Comish, P.; Kang, R. The hallmarks of COVID-19 disease. *PLoS Pathog.* **2020**, *16*, e1008536. [CrossRef]
16. Rout, S.S.; Singh, M.; Shindler, K.S.; Das Sarma, J. One proline deletion in the fusion peptide of neurotropic mouse hepatitis virus (MHV) restricts retrograde axonal transport and neurodegeneration. *J. Biol. Chem.* **2020**, *295*, 6926–6935. [CrossRef]
17. Singh, M.; Kishore, A.; Maity, D.; Sunanda, P.; Krishnarjuna, B.; Vappala, S.; Raghothama, S.; Kenyon, L.C.; Pal, D.; Das Sarma, J. A proline insertion-deletion in the spike glycoprotein fusion peptide of mouse hepatitis virus strongly alters neuropathology. *J. Biol. Chem.* **2019**, *294*, 8064–8087. [CrossRef]
18. Das Sarma, J.; Fu, L.; Hingley, S.T.; Lai, M.M.; Lavi, E. Sequence analysis of the S gene of recombinant MHV-2/A59 coronaviruses reveals three candidate mutations associated with demyelination and hepatitis. *J. Neurovirol.* **2001**, *7*, 432–436. [CrossRef]
19. Das Sarma, J.; Fu, L.; Hingley, S.T.; Lavi, E. Mouse hepatitis virus type-2 infection in mice: An experimental model system of acute meningitis and hepatitis. *Exp. Mol. Pathol.* **2001**, *71*, 1–12. [CrossRef]
20. Navas, S.; Seo, S.H.; Chua, M.M.; Das Sarma, J.; Lavi, E.; Hingley, S.T.; Weiss, S.R. Murine coronavirus spike protein determines the ability of the virus to replicate in the liver and cause hepatitis. *J. Virol.* **2001**, *75*, 2452–2457. [CrossRef]
21. Das Sarma, J.; Fu, L.; Tsai, J.C.; Weiss, S.R.; Lavi, E. Demyelination determinants map to the spike glycoprotein gene of coronavirus mouse hepatitis virus. *J. Virol.* **2000**, *74*, 9206–9213. [CrossRef] [PubMed]
22. Lavi, E.; Gilden, D.H.; Wroblewska, Z.; Rorke, L.B.; Weiss, S.R. Experimental demyelination produced by the A59 strain of mouse hepatitis virus. *Neurology* **1984**, *34*, 597. [CrossRef] [PubMed]
23. Perlman, S.; Jacobsen, G.; Olson, A.L.; Afifi, A. Identification of the spinal cord as a major site of persistence during chronic infection with a murine coronavirus. *Virology* **1990**, *175*, 418–426. [CrossRef]
24. Shindler, K.S.; Kenyon, L.C.; Dutt, M.; Hingley, S.T.; Das Sarma, J. Experimental optic neuritis induced by a demyelinating strain of mouse hepatitis virus. *J. Virol.* **2008**, *82*, 8882–8886. [CrossRef] [PubMed]
25. Das Sarma, J.; Scheen, E.; Seo, S.H.; Koval, M.; Weiss, S.R. Enhanced green fluorescent protein expression may be used to monitor murine coronavirus spread in vitro and in the mouse central nervous system. *J. Neurovirol.* **2002**, *8*, 381–391. [CrossRef] [PubMed]
26. Das Sarma, J.; Kenyon, L.C.; Hingley, S.T.; Shindler, K.S. Mechanisms of primary axonal damage in a viral model of multiple sclerosis. *J. Neurosci.* **2009**, *29*, 10272–10280. [CrossRef]
27. Kenyon, L.C.; Biswas, K.; Shindler, K.S.; Nabar, M.; Stout, M.; Hingley, S.T.; Grinspan, J.B.; Das Sarma, J. Gliopathy of Demyelinating and Non-Demyelinating Strains of Mouse Hepatitis Virus. *Front. Cell Neurosci.* **2015**, *9*, 488. [CrossRef]
28. Shindler, K.S.; Chatterjee, D.; Biswas, K.; Goyal, A.; Dutt, M.; Nassrallah, M.; Khan, R.S.; Das Sarma, J. Macrophage-mediated optic neuritis induced by retrograde axonal transport of spike gene recombinant mouse hepatitis virus. *J. Neuropathol. Exp. Neurol.* **2011**, *70*, 470–480. [CrossRef]

29. Sadasivan, J.; Singh, M.; Sarma, J.D. Cytoplasmic tail of coronavirus spike protein has intracellular targeting signals. *J. Biosci.* **2017**, *42*, 231–244. [CrossRef]
30. Das Sarma, J.; Iacono, K.; Gard, L.; Marek, R.; Kenyon, L.C.; Koval, M.; Weiss, S.R. Demyelinating and nondemyelinating strains of mouse hepatitis virus differ in their neural cell tropism. *J. Virol.* **2008**, *82*, 5519–5526. [CrossRef]
31. Kuo, L.; Godeke, G.J.; Raamsman, M.J.; Masters, P.S.; Rottier, P.J. Retargeting of coronavirus by substitution of the spike glycoprotein ectodomain: Crossing the host cell species barrier. *J. Virol.* **2000**, *74*, 1393–1406. [CrossRef] [PubMed]
32. Donnelly, D.J.; Gensel, J.C.; Ankeny, D.P.; van Rooijen, N.; Popovich, P.G. An efficient and reproducible method for quantifying macrophages in different experimental models of central nervous system pathology. *J. Neurosci. Methods* **2009**, *181*, 36–44. [CrossRef] [PubMed]
33. Bose, A.; Basu, R.; Maulik, M.; Das Sarma, J. Loss of Cx43-Mediated Functional Gap Junction Communication in Meningeal Fibroblasts Following Mouse Hepatitis Virus Infection. *Mol. Neurobiol.* **2018**, *55*, 6558–6571. [CrossRef] [PubMed]
34. Chakrabarti, P.; Pal, D. The interrelationships of side-chain and main-chain conformations in proteins. *Prog. Biophys. Mol. Biol.* **2001**, *76*, 1–102. [CrossRef]
35. Sali, A.; Blundell, T.L. Comparative protein modelling by satisfaction of spatial restraints. *J. Mol. Biol.* **1993**, *234*, 779–815. [CrossRef] [PubMed]
36. Parrinello, M.; Rahman, A. Polymorphic transitions in single crystals: A new molecular dynamics method. *J. Appl. Phys.* **1981**, *52*, 7182–7190. [CrossRef]
37. Berendsen, H.J.C.; Postma, J.P.M.; Van Gunsteren, W.F.; DiNola, A.; Haak, J.R. Molecular dynamics with coupling to an external bath. *J. Chem. Phys.* **1984**, *81*, 3684–3690. [CrossRef]
38. Tavernelli, I.; Cotesta, S.; Di Iorio, E.E. Protein Dynamics, Thermal Stability, and Free-Energy Landscapes: A Molecular Dynamics Investigation. *Biophys. J.* **2003**, *85*, 2641–2649. [CrossRef]

viruses

MDPI

Article

Development of Robust Varicella Zoster Virus Luciferase Reporter Viruses for In Vivo Monitoring of Virus Growth and Its Antiviral Inhibition in Culture, Skin, and Humanized Mice

Megan G. Lloyd [1], Michael B. Yee [2], Joseph S. Flot [2], Dongmei Liu [1], Brittany W. Geiler [1], Paul R. Kinchington [2,*] and Jennifer F. Moffat [1,*]

1 Department of Microbiology and Immunology, SUNY Upstate Medical University, Syracuse, NY 13210, USA; gribblem@upstate.edu (M.G.L.); liud@upstate.edu (D.L.); geilerb@upstate.edu (B.W.G.)

2 Department of Ophthalmology, School of Medicine, University of Pittsburgh, Pittsburgh, PA 15213, USA; myee@krystalbio.com (M.B.Y.); jsf@pitt.edu (J.S.F.)

* Correspondence: kinchingtonp@upmc.edu (P.R.K.); moffatj@upstate.edu (J.F.M.); Tel.: +1-412-647-6319 (P.R.K.); +1-315-464-5464 (J.F.M.)

Abstract: There is a continued need to understand varicella-zoster virus (VZV) pathogenesis and to develop more effective antivirals, as it causes chickenpox and zoster. As a human-restricted alphaherpesvirus, the use of human skin in culture and mice is critical in order to reveal the important VZV genes that are required for pathogenesis but that are not necessarily observed in the cell culture. We previously used VZV-expressing firefly luciferase (fLuc), under the control of the constitutively active SV40 promoter (VZV-BAC-Luc), to measure the VZV spread in the same sample. However, the fLuc expression was independent of viral gene expression and viral DNA replication programs. Here, we developed robust reporter VZV viruses by using bacterial artificial chromosome (BAC) technology, expressing luciferase from VZV-specific promoters. We also identified two spurious mutations in VZV-BAC that were corrected for maximum pathogenesis. VZV with fLuc driven by ORF57 showed superior growth in cells, human skin explants, and skin xenografts in mice. The ORF57-driven luciferase activity had a short half-life in the presence of foscarnet. This background was then used to investigate the roles for ORF36 (thymidine kinase (TK)) and ORF13 (thymidylate synthase (TS)) in skin. The studies reveal that VZV-ΔTS had increased sensitivity to brivudine and was highly impaired for skin replication. This is the first report of a phenotype that is associated with the loss of TS.

Keywords: varicella-zoster virus; antiviral drugs; bacterial artificial chromosome; luciferase; bioluminescence imaging; skin organ culture; humanized mice

Citation: Lloyd, M.G.; Yee, M.B.; Flot, J.S.; Liu, D.; Geiler, B.W.; Kinchington, P.R.; Moffat, J.F. Development of Robust Varicella Zoster Virus Luciferase Reporter Viruses for In Vivo Monitoring of Virus Growth and Its Antiviral Inhibition in Culture, Skin, and Humanized Mice. *Viruses* **2022**, *14*, 826. https://doi.org/10.3390/v14040826

Academic Editors: Charles Grose, Ravi Mahalingam and Joel Rovnak

Received: 28 March 2022
Accepted: 12 April 2022
Published: 15 April 2022

1. Introduction

Varicella-zoster virus (VZV) is the human neurotrophic alphaherpesvirus that causes varicella (chickenpox) during primary infection, and herpes zoster (shingles) upon the reactivation from latency in the sensory nerve ganglia. Zoster is particularly morbid, as it is frequently complicated by acute and chronic pain. While vaccines for both varicella (targeted to children) and zoster (targeted to adults 50+ years old) have decreased the incidence of the disease in the countries where they are used [1,2], VZV disease prevalence remains high worldwide. The vaccines are far from global use, and zoster vaccines still have low uptake in the countries where it can be obtained [3–5]. Several antiviral drugs are available to treat VZV infections, including acyclovir, foscarnet, and brivudine [6–8]. Much remains to be learned about VZV pathogenesis, and there is still a need for more effective and improved antiviral therapies.

The human-restricted nature of VZV greatly complicates its study, particularly as there are no effective animal models for the latency, reactivation, or reactivated disease [9,10].

Furthermore, the genetic manipulation of VZV has proven problematic, as the classical approaches that have been used for the related neurotropic virus, herpes simplex virus type 1 (HSV-1), do not work for VZV without strong selection [11]. The VZV infectivity in culture is highly cell-associated and yields low titers of cell-free virus, which are often composed of many infectious particles per plaque forming unit. This makes the purification of rare recombinants from a wild-type background difficult without a strong selection process [12,13]. The development of isogenic cloned DNA systems has provided the means to develop recombinant viruses more easily, starting with the overlapping cosmids for the vaccine virus strain (vOka) [12], and then for the parent strain of Oka (pOka) (wild type; [14,15]). Cosmid systems use 4–6 cloned DNAs that contain the complete genome as large overlapping fragments [14,15]. VZV genes are manipulated by DNA fragment subcloning, plasmid-based mutagenesis, and the rebuilding back into the cosmid. The co-transfection into permissive cells with other overlapping cosmid DNAs covering the entire VZV genome results in recombination to yield full genomes and virus, which eliminates the need for plaque purification [12].

The cosmid systems were improved with the development of VZV bacterial artificial chromosomes (BACs), where the entire VZV genome is present as a single DNA replicon [16,17]. The replication in *E. coli* required the insertion of a ~10 Kbp replicon into the genome. Two pOka BACs were developed with the insertion of the replicon into the same pOka cosmid set, which was followed by virus derivation and the retransformation of *E. coli* with circular DNA from virus-infected cells [18,19]. The VZV pOka BAC that was developed by Tischer et al. contained the BAC replicon in an internal repeat region of the genome (BAC-T1) [18]. BAC-T1 was used to generate several viruses, including VZV where VZV ORF23 and ORF62 were fluorescently tagged [20]. BAC-T1 was further refined by adding repeat sequences that flank the replicon elements to allow the self-excision of the F1 replicon (BAC DX) [18,21]. A second VZV BAC was developed by Zhu and colleagues, where the replicon was placed in the large intergenic noncoding region between ORF60 and 61 of the pOka cosmids [19,22]. However, recent studies suggest that this VZV region contains a promoter-driving mRNA that is expressed during lytic infection and viral latency [23], and it is not clear how the insertion affects the VZV replication. BAC systems have permitted the development of many recombinant viruses, which have helped inform researchers which genes are required for growth in culture, tissue, and the human host. For instance, specific gene deletions across the VZV genome suggest that the 17 to 20 ORFs are not required for growth in cell culture [24]. Many genes that are not required for growth in culture have been found to be important for growth in specific cell types or in organized tissues, such as skin and T cells [24].

The goal here was to refine the use of VZV reporter viruses for use in cultured cells, tissue explants, and humanized mouse models. Tissue explant and humanized mouse models are often limited in supply or are prohibitively costly. Thus, the use of bioluminescence to track the virus spread helps overcome the need for large numbers of mice. Firefly luciferase (fLuc) can be measured in the same sample or mouse at multiple times by using bioluminescence imaging [25]. Previously, we used a VZV that is termed VZV-BAC-Luc, which expresses the firefly luciferase from the constitutively active SV40 promoter [19,26]. Unfortunately, the luciferase expression from VZV-BAC-Luc does not necessarily reflect the VZV replication rates in tissues, as it is expressed independently of viral gene expression. Thus, we sought to identify a kinetically late promoter to drive the expression of luciferase that more accurately reflects viral replication. During this work, we also identified and corrected two unexpected mutations in the BAC-T1 and BAC-DX systems that map to ORF40, which encodes the major capsid protein; and to ORF50, which encodes glycoprotein gM. We then used a corrected BAC containing luciferase that was driven by the ORF57 gene to construct VZV lacking thymidine kinase (TK) and thymidylate synthase (TS). An analysis of the VZV ΔTK and ΔTS reporter viruses reveals insights that are important for antiviral drug development, and an unexpected phenotype for VZV TS in human skin organ culture.

2. Materials and Methods

2.1. Cells and Viruses

Human foreskin fibroblasts (HFFs) (CCD-1137Sk; American Type Culture Collection (ATCC), Manassas, VA, USA) were used up to passage 20. The human retinal pigment epithelial cell line, ARPE-19 (CRL-2302; ATCC), was grown to passage 36–40. Human malignant melanoma cells (MeWos; HTB-65; ATCC) were grown to passage 40–50. Human Tert-immortalized retinal epithelial cells (htRPE) (CRL4000; ATCC) were passaged likewise to 40. All cells were grown in Dulbecco's Modified Eagle Medium (DMEM) with 4.5 g/L glucose, L-glutamine, and sodium pyruvate (DMEM 1X, Corning, Manassas, VA, USA), supplemented with up to 10% heat-inactivated fetal bovine serum (Benchmark FBS; Gemini Bio Products, West Sacramento, CA, USA), penicillin–streptomycin (5000 IU/mL), and amphotericin B (250 µg/mL). For the MeWo cell cultivation, the medium was supplemented with nonessential amino acids (Mediatech, Herndon, VA, USA). The viruses used in this study include: VZV pOka (parental Oka) (accession number: AB097933) [14]; VZV-BAC-Luc (from H. Zhu [19]) and VZV-ORF57-Luc [27]; VZV-ORF9-Luc; VZV-ORF14-Luc; VZV-ORF17-Luc; VZV-ORF57-ΔTK; VZV-ORF57-ΔTS; and VZV-ORF57-ΔTKΔTS. The construct development and the viruses are shown in Figure 1. All recombinant VZV, except for VZV-BAC-Luc, were developed from BAC constructs by using the pOka BACs detailed previously [18,21].

2.1.1. Generation of Luciferase Reporter Viruses

VZV luciferase reporter viruses were generated by placing the coding region of the luciferase downstream of the selected ORFs as the in-frame C protein fusions separated by a sequence encoding the T2A-directed "ribosome skipping" motif. For this purpose, a plasmid construct was developed in which the ORF for the firefly luciferase was amplified by PCR from the plasmid pGL3basic (Promega Corp) by using the primers, fLucF and fLucR (all primer sequences are presented in Table 1). The resulting DNA fragment was digested with EcoRI and BamHI (sites underlined) and was cloned into the vector, pmCherryC1 (Clontech Inc., San Jose, CA, USA), and was cut with EcoRI and BglII in order to place it in-frame with pmCherry. The resulting plasmid, pmCherryT2ALuc, contained mCherry fused in-frame to luciferase, separated by the 22 amino acid T2A motif, and it expressed functional luciferase activity from the luciferase with a single proline residue on the amino terminal end in plasmid-transfected HEK293T cells (data not shown). This plasmid was modified by cutting with EcoRI and BglII (sites downstream of the mCherry-T2A-luciferase ORF) to insert a bacterial promoter-driven zeomycin resistance gene cassette, which was amplified by PCR with primers to add BglII and terminal EcoRI flanking sites (primers: *ZeoF* and *ZeoR*; Table 1). The resulting plasmid (pmCherry-T2A-Luc-Zeo) was used as a template for the PCR amplification of the entire T2A-Luc-Zeo cassette, using primers with long 5′ extensions that added 38–40 bp flanking homology arms to the PCR product to direct the recombination into the VZV BAC-mediated l-red site-specific recombination. The methods of recombineering have been detailed by us and others previously [28,29].

The in-frame fusion to gC (encoded by ORF14), with the alteration of its stop codon to a coding residue, used the primers, gClucF2 and gClucR3. The PCR product was amplified, and the gel was purified and transformed for recombineering into pOka-DX BAC after 20 min of 42 °C heat induction by using the pGS1783 bacterial host (the kind gift of Gregory Smith, Northwestern University, Evanston, IL, USA), as detailed previously [29]. Bacterial colonies resistant to chloramphenicol and zeomycin were then validated for the BAC DNA integrity and for the correct in-frame insertion into ORF14 by restriction fragment length polymorphism (RFLP) analyses and Sanger sequencing across the fusion insert join. Virus was derived by the co-transfection of the BAC-purified DNA in MeWo or hTRPE cells, as recently described [28,29]. A similar strategy was used to generate VZV containing the T2A-luciferase-zeo^r that is: fused to the C terminus of ORF9, using the primers, ORF9-lucF2 and ORF9lucR3; fused to the C terminus of ORF17, using the primers, ORF17lucF2 and ORF17lucR2; and fused to the C terminus of ORF57, using the primers, 57lucF and 57lucR.

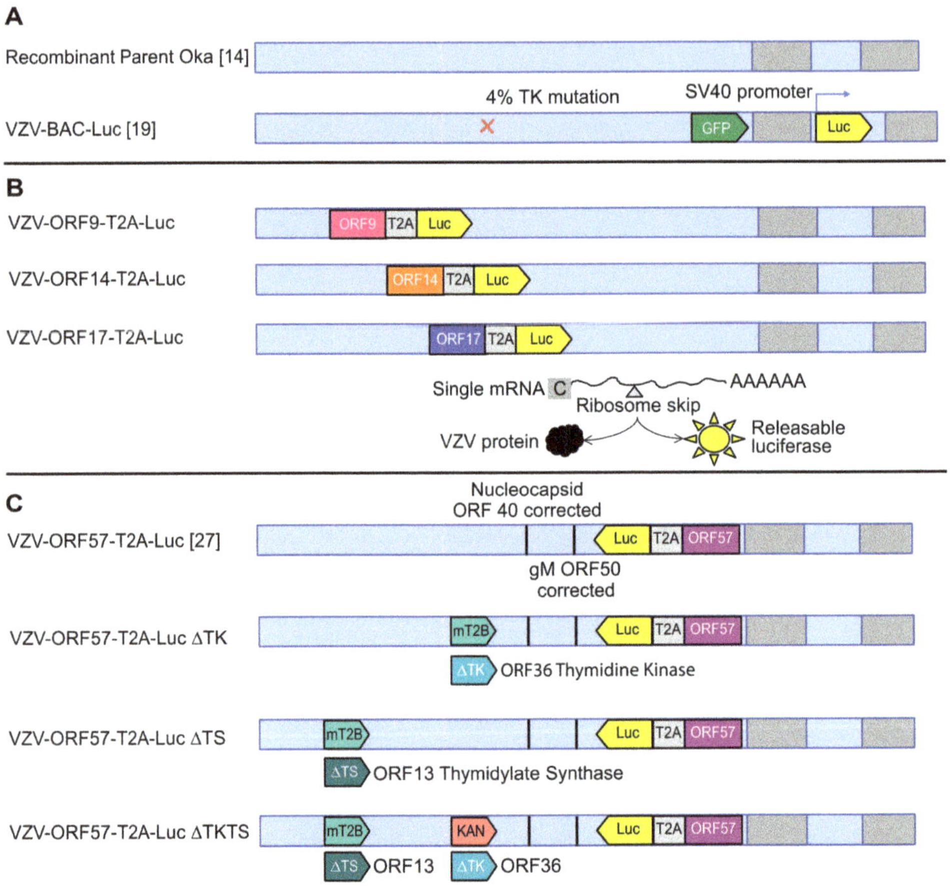

Figure 1. Constructs for VZV reporter viruses. The diagram depicts the presence of the recombinant sequences used to alter pOka [14] with BAC recombineering, as detailed in the text. Viruses shown in (**A**) include parent Oka and VZV-BAC-Luc, derived from the BAC-Zhu pOka system [14,19]. VZV-BAC-Luc was generated from pOka (passage 9), as previously described [19], and included a small population of TK-virus, which arose through unknown mechanisms. (**B**)—upper section: viruses derived from an uncorrected BAC system, as detailed by the Osterreider group, which include VZV-ORF9-T2A-Luc; VZV-ORF14-T2A-Luc; and VZV-ORF17-T2A-Luc. (**B**)—lower section: diagram of the viral protein and luciferase expression mediated by the T2A ribosome skipping motif, which is used to generate two proteins from one mRNA. (**C**) VZV-ORF57-T2A-Luc [27], VZV-ORF57-ΔTK, VZV-ORF57-ΔTS, and VZV-ORF57-ΔTKΔTS were generated in a BAC with corrected spurious mutations identified in ORF40 and ORF50 (Tischer VZV-BAC [18]), as detailed in the text. In VZV-ORF57-Luc, ORF13 (TS), ORF36 (TK), or both, were deleted through site-directed recombination events to replace the partial (ORF36) or entire (ORF13) ORFs with either the mturq2blue (mT2B) fluorescent reporter gene or a kanamycin resistance cassette. The colors used for each virus are carried through the rest of the figures for continuity.

Table 1. Primers used to generate VZV viruses.

Gene Target	Forward Primer (5′-3′)	Reverse Primer (5′-3′)
fLuc Firefly (Photinus pyralis) Luciferase	GAG<u>GGATCC</u>GGTTCCGGAGAGGGCAGAG- GAAGTCTGCTAACATGCGGTGACGTCGAG- GAGAATCCTGGCCCAATGGAAGACGCC AAAAACATA	AATTC<u>GAATTC</u>GCGCGCAGATCTTTACACGGCGATCTTTC- CGCCCTTCTTGGC
Zeo (Zeomycin resistance)	AGATCT<u>AGATCTC</u>GAGTAATGGAACGGACCGTGTTGAC	GCTGACGTCGAC<u>GAATTC</u>TGATCACTCAAGTTTCGAG- GTCGAGGTG
gCLuc (F2 and R3; ORF14)	CTTATCGCAGTTATCGCAACCCTATG- CATCCGTTGCTGTTCAATGGACGAGCTGTACAAG	AATAAAATGATATACACAGACGCGT- TTGGTTGGTTTCTGTCAGTCCTGCTCCTCAGCC
ORF9Luc (F2 and R3)	AGTAGGGCCCGTTCGGCATCAAGAACTG- ATGCGCGAAAATCAATGGACGAGCTGTACAAG	ACGTTTATTTATTATACATAATACCG- GGTAAACCGTTACTTCAGTCCTGCTCCTCAGCC
ORF17Luc (F2 and R2)	CTCCACTCCCACTAAACACTGTATTAA- CAAAATATTGGAATTCAATGGACGAGCTGTACAAG	AGCAAAATAAAACAATGAACCATTAAG- TCGCTCTTATGTGTGTCAGTCCTGCTCCTCAGCC
57Luc (ORF57)	ACGTTGAGGAGCCTTGCAGGTT- GGGTGCCGCGCTTCACCGTCAATGGACGAGCTGTACAAG	TTTATATTTAACGGCTTTTAATTTGAAGAC- ACCTATCCTCTCAGTCCTGCTCCTC
40REP (ORF40)	ATATAGATATTACGTTTATCATGCCAATG- GGAGTGTTTCAGGCGAATTCCATG- GACAGATATACACGACAGGATGACGACGATAAGTAGG	CAGTTGAAAAATCGCCGGCGTGTCGTGTA- TATCTGTCCATGGAATTCGCCTG- AAACACTCCCATTGGCACAGGGTAATGCCAGTGTTAC
50REP (ORF50)	TGAATTATCCAAATTCGCCAATTAAGC- GTATCCATTTGATGATCTAAAGCTT- CCACCTCGGGTGTCGTGGTGTCGTAC- GGGGATGACGACGATAAGTAGG	TCTGAAAAAGTCTCACCGTACGACAC- CACGACACCCGAGGTGGAAGCTTTAGATCA- TCAAATGGATACGCTTAATAGGGTAATGCCAGTGTTAC
TS (TS T2B-For and endR2)	TATCAAGTGGTCGTTTGTATTTAACG- ATTATTACCGGTACCGGTACCATG- GTGAGCAAGGGCGAGGAG	ACATCTACTGTCTTGACAACATTTAA- AAATCCATTAAAGATTATCTGGATCCTACCTTTC
TK (TK T2BFor and endR3)	CGCCGAAGAATTTTTACACCACTT- TGCAATAACACCAAACGGTACCATGGTGAGC- AAGGGCGAGGAG	TGTGTATCATCTTTTTACTGGTACATAC- GTAAATACTAGGTTATCTGGATCCTACCTTTC
TKkan-del (For and Rev)	CGCCGAAGAATTTTTACACCACTTTGCA- ATAACACCAAACGGATGACGACGATAAGTAGG	TGTGTATCATCTTTTTACTGGTACATAC- GTAAATACTAGGAGGGTAATGCCAGTGTTAC

The restriction sites engineered to sequences to enable cloning, as detailed in the text, are <u>underlined</u>.

2.1.2. Correction of Spurious Mutations

ORF14, 9 and 17- luciferase reporter viruses were derived from the pOkaDX BAC background and were detailed previously [18,21]. The ORF57 luciferase reporter virus was derived from a corrected version of this BAC (named, pOka DX-RR), which had been corrected for two newly identified spurious mutations that were revealed by pOka BAC DX sequencing (Table 2). One mutation in pOka DX was in ORF40, which encodes the major capsid protein and causes A428T nonsynonymous change. The ORF40 SNP was also present in the nonexcisable version of the BAC [18] but was not found in the pOka cosmids that were used to derive the Tisher BAC [14], nor in the independently derived Zhu BAC that was used to generate VZV-BAC-Luc [19]. It was not found in any other VZV sequence (of 93 genome sequences analyzed).

A second nonsynonymous mutation in ORF50, which encodes glycoprotein gM, caused V26A. The ORF50 mutation may reflect a naturally arising variant, since it was found in 4 VZV-reported sequences of the 93 scanned. The correction of the two mutations used primers that were designed to add silent restriction sites near the SNPs for the rapid identification of positive constructs, specifically a novel noncoding EcoRI site for ORF40 and a novel noncoding HindIII site for that in ORF50. The primers were used to amplify the kanamycin resistance cassette from pEPSkan2, and the gel-purified PCR products were recombined into their respective sites and were then reversed [29]. The BACs were verified for the inserted restriction sites introduced at each step and were verified by DNA sequencing across the mutations. ORF40 was corrected first, followed by the correction of ORF50, using the primers, 40REPF + 40REPR, and 50REPF + 50REPR, respectively.

Table 2. Spurious mutations detected in pOka-DX BAC and their corrections.

Location of Mutation	Identification of Mutation and the Correction
ORF40 mutation in pOka DX	72661 atg cca atg gga gtg ttt cag **A**ca aac tcc atg gac aga tat aca cga cac ... M P M G V F Q $\mathbf{T_{428}}$ N S M E R Y T R H
ORF40 Correction in pOka DX-RR	72661 atg cca atg gga gtg ttt cag **G**<u>cG aaT tcc</u> atg gac aga tat aca cga cac ... M P M G V F Q $\mathbf{A_{428}}$ N S M E R Y T R H
ORF50 mutation in pOka DX	gaa aaa gtc tca ccg tac gac acc acg aca ccc gag g**C**g gaa gcg tta gat cat caa atg E_{11} K V S P Y D T T T P E **A** E A L D H Q M
ORF50 correction in pOka DX-RR	gaa aaa gtc tca ccg tac gac acc acg aca ccc gag g**T**g <u>gaa gcT</u> tta gat cat caa atg E_{11} K V S P Y D T T T P E **V** E A L D H Q M

Note: Lower case letters denote the sequence identical to the parent, while capital letters represent the spurious mutation in pOka DX (**bolded**), or the changed residues made to accompany the correction to introduce a novel restriction site (<u>underlined</u>). The single letter amino acid code of the region of the protein affected is shown below each sequence, with the key mutant residue **bolded**.

2.1.3. Deletion of ORF13 and ORF36 in the BAC DXRR 57luczeo Background

The partial deletion of the VZV thymidine kinase (ORF36), and the full deletion of thymidylate synthase (ORF13), used a strategy in which the genes were replaced with the mturq2blue (mT2B) fluorescent reporter gene. We first developed the plasmid, p-mTurq2Blue-KAN-in, which contained a recombination-reversable kanamycin resistance cassette derived from the plasmid, pEPS-kan2 [18]. The entire Turq2Blue-KANin cassette was then PCR amplified with the primers, TS-T2B-For and TSendR2 (to delete TS), and the primers, TKT2BFor and TKendR3 (to delete part of TK). The gel-purified PCR fragments were independently recombined into pOka BAC DX-RR57luczeo and were reversed to remove the kan^r cassette and to restore the mT2B ORF. This strategy placed the mT2B reporter under the control of the TS and TK promoters in each virus, respectively, when the VZV was derived as detailed above. As expected, the plaques were blue-fluorescent when illuminated with 434 nm. The partial ORF36 replacement recombined the mT2B gene in-frame with the 43rd codon of the TK ORF because the start of TK is very close to ORF35 and was suspected to contain regulatory sequences that regulate the ORF 35 expression. The partial gene replacement strategy preserved 220 bp of the sequence upstream of ORF35. To delete both the TK and TS, a kanamycin resistance cassette from pEPSkan2 was PCR amplified with the primers, TKkan-delF and TKkan-delR, and was then recombined into the BAC containing the replacement of TS with mT2B and selected for the gain of kanamycin resistance. The viruses from the BACs were derived from at least two independently derived constructs and they all showed the same phenotype in subsequent studies.

2.2. Compounds

The compounds used include commercially purchased cidofovir (CDV) (BEI Resources, Manassas, VA, USA), foscarnet (Millipore Sigma, Burlington, MA, USA), acyclovir (Millipore Sigma, Burlington, MA, USA), and brivudine (BVdU) (Fisher Scientific, Hampton, NH, USA). For the cell-based assays, dry compounds were dissolved in water or dimethyl sulfoxide (DMSO) and were stored at −20 °C. Stocks were diluted in complete tissue culture medium to working concentrations (CDV = 0.078–5 µM; BVdU = 0.000027–0.02 µM; ACV = 0.0026–40 µM; foscarnet = 1 mM). For the mouse experiments, CDV was prepared in water (2.5 mg/mL) and stored at 4 °C.

2.3. Virus Growth Kinetics in Cells

The virus growth kinetics were evaluated in HFFs or ARPE-19 cells, as previously described [26,30]. Briefly, cells were grown to confluence and were then infected with cell-free VZV at a ratio of 1:50 or 1:100, and were incubated at 37 °C. At 2 h post-infection, additional media were added to each well. The virus spread was measured with daily bioluminescence imaging. Prior to scanning, the media were removed and replaced with D-luciferin (300 μg/mL in PBS; XenoLight™ D-Luciferin Potassium Salt, Perkin Elmer, Waltham, MA, USA) for 40 min. The medium was replaced daily after imaging. The bioluminescent signal has been significantly correlated to the viral spread (pfu) in previous studies [19] and is confirmed here. VZV was added to HFF monolayers and imaged daily for bioluminescence. Cells were then harvested after imaging to determine the virus yield, measured by standard infectious center assays [19].

2.4. Total Flux Half-Life in Cells Infected with VZV-ORFx-Luc Viruses

Each VZV-ORFx-Luc virus and the parental VZV-BAC-Luc virus were evaluated for decay in the luciferase activity, measured as the Total Flux, in the presence of 1 mM of viral DNA replication inhibitor, foscarnet. Trypsinized and counted HFFs were resuspended in tissue culture medium and mixed with VZV-infected HFFs at a 1:50 ratio. Approximately 5×10^5 HFFs with VZV in 2 mL of medium were added to 42 individual 35 mm tissue culture dishes (Corning, Corning, NY, USA) and were incubated at 37 °C in a humidified CO_2 incubator for 36–40 h. Three dishes were scanned to measure the Total Flux at time zero (0 h). Then, 1 mM of foscarnet was added to the media in all dishes to inhibit the viral DNA polymerase and the dishes were returned to the incubator. At 2–12 h intervals, three dishes were selected at random and individually scanned by IVIS over 12–14 timepoints, which varied for each VZV reporter virus, with many clustered around the expected time when the Total Flux signals neared 50%. At each timepoint, the media were removed and replaced with D-luciferin (300 μg/mL in PBS) for 40 min at 37 °C, and were then scanned in the IVIS™ 50 instrument to record the bioluminescence, measured as the Total Flux. The triplicate values at each timepoint were divided by the average Total Flux at 0 h to calculate the percent change in the Total Flux. The time when the Total Flux reached 50% was determined by using nonlinear regression analysis to fit a dose vs. response curve (F constrained to 50).

2.5. Efficacy in Cultured Cells

Antiviral activity against VZV-ORF57-Luc, VZV-ORF57-ΔTK, and VZV-ORF57-ΔTS was evaluated by previously established dose response assays [31,32]. Briefly, ARPE-19 cells were seeded in 24-well tissue-culture-treated plates for 3 d prior to infection. Cells were infected with up to 500 pfu per well and were incubated at 37 °C with 5% CO_2. Cidofovir and brivudine were added at varying concentrations 2 h post infection and were returned to 37 °C. Infected and treated cells were incubated for 3 d post infection. Prior to scanning, the medium was replaced with D-luciferin (300 μg/mL in PBS) for 40 min at 37 °C. The bioluminescence was measured using the IVIS™ 50 instrument (Caliper Life Sciences/Xenogen, Hopkinton, MA, USA). The virus yield, reported as the fold change, was calculated as the Total Flux (photons/s/cm^2/steradian) at each concentration divided by the Total Flux for the untreated wells.

2.6. Preparation of Skin

Human skin from fetal and adult sources was used. Fetal skin (18–20 weeks gestational age) (Advanced Bioscience Resources, Alameda, CA, USA) was obtained according to all local, state, and federal guidelines. Adult human skin was obtained with informed consent from healthy adults who were undergoing reduction mammoplasty surgeries at SUNY Upstate Medical University in Syracuse, New York. The adult human skin collection was managed under approved Institutional Review Board protocols and procedures (SUNY Upstate, Institutional Review Board #1140572), as well as all local, state, and federal guidelines. Fetal and adult skin preparation has been described previously [27,33]. Briefly, skin was cleaned with povidone iodine and ethanol and washed in sterile media. Adult skin was thinned using a Weck knife and Goulian guard (0.028") to remove any underlying adipose and excess dermal tissue. Prepared skin was then cut into approximately 1 cm^2 pieces and cultured on NetWells for the skin organ culture (Corning, NY, USA), or was implanted into mice, as previously described [26,34,35].

2.7. Skin Organ Culture

Fetal or adult human skin was cultured on NetWells (Corning, NY, USA), which hold tissue at the air–liquid interface. The skin was infected with VZV by scarification and was cultured as previously described [26,27,33]. The infected skin was incubated at 37 °C for fetal tissue, and at 35 °C for adult tissue. Virus spread was measured by bioluminescence imaging every day, or every other day, by scanning the skin in the IVIS™ 50. Prior to scanning, skin tissue was submerged in D-luciferin (300 μg/mL in PBS) for 40 min. The medium was replaced every other day after imaging.

2.8. Animal Procedures

Animal procedures were performed as previously described [26,35], with approved protocols, and were monitored by the Institutional Animal Care and Use Committee (IACUC) at SUNY Upstate Medical University. All studies were performed in accordance with all state and federal laws and regulations. Briefly, fetal human skin xenografts were introduced subcutaneously above the left flank of 5–6-week-old male *scid-beige* (CB.17; CB17/Icr-*Prkdc^{scid}/IcrIcoCrl*) mice (Charles River, Wilmington, MA, USA). Three to four weeks postimplantation, xenografts were exposed through an incision on the mouse back and were inoculated by intradermal injection with cell-associated VZV (1×10^4–10^5 pfu/mL; 60 μL injection; grown in HFFs). Prior to imaging, mice were injected with D-luciferin (15 mg/mL in PBS) for 10 min. Virus spread was monitored by daily bioluminescence imaging with the IVIS™ 200 for 9 d. Mice were monitored daily for weight loss and signs of distress.

2.9. Bioluminescence Imaging

The imaging procedure was previously described in [26]. Briefly, for skin organ culture, tissue was scanned with the IVIS™ 50 instrument and images acquired for 30 s–1 min (Caliper Life Sciences/Xenogen, Hopkinton, MA, USA). For mouse studies, animals were scanned with the IVIS™ 200 instrument and images acquired for 5 min, maximum. The VZV infection was measured as the Total Flux (photons/s/cm^2/steradian) in a region-of-interest (ROI) encircling each skin piece or drawn over the xenograft(s). The fold change was calculated as the daily Total Flux divided by the lowest and/or initial Total Flux value.

2.10. Statistical Analysis

All calculations and graphs were created using GraphPad Prism (Graph-Pad Software, San Diego, CA, USA). The dose response curves were analyzed using nonlinear regression analysis. All other data was analyzed using one-way ANOVA and various post hoc tests or a Student's *t*-test. A $p \leq 0.05$ was considered statistically significant.

3. Results

3.1. Construction of Robust Reporter VZV Viruses

VZV-BAC-Luc was the "gold standard" reporter virus in our previous work assessing antivirals against VZV [26,35]. The luciferase reporter allows the virus growth to be measured over time in the same sample by using IVIS, which is useful in cases where considerable sample-to-sample variability may exist, or when the availability of skin pieces or mice is limited or prohibitively costly. Unfortunately, the luciferase gene in VZV-BAC-Luc is under the control of the constitutively active SV40 promoter [19], which was suspected to be rapidly transcribed upon the genome insertion into the nucleus so that the luciferase expression was not reliant on viral gene expression, and likely occurred before viral DNA replication. Here, we aimed to create improved reporter VZVs, in which the luciferase was expressed with late kinetics and, thus, was sensitive to the inhibition of viral DNA synthesis. The construction of the reporter viruses is described in detail in the methods (see Methods 2.1; Figure 1). Briefly, the luciferase gene was inserted in-frame with the C terminal residue of the VZV open reading frames (ORFs), 9, 14, 17, and 57, by using an in-frame T2A ribosome skipping motif between the VZV ORF and the luciferase gene. As luciferase is not active as a fusion protein, the T2A motif releases active luciferase with a single N terminal proline that does not affect the activity. The first-generation virus set (VZV-ORF9-Luc; VZV-ORF14-Luc; and VZV-ORF17-Luc) was found to contain mutations in ORF40 and ORF50, which were also found in the pOkaDX BAC [18]. The second-generation of VZV had the two spurious mutations of corrected and contained luciferase inserted after ORF57 (VZV-ORF57-Luc, previously described in [27]). VZV-ORF57-Luc was eventually selected as the optimal reporter virus on the basis of the experiments to follow. The third generation of VZV reporter viruses were developed from the VZV-ORF57-Luc BAC and with either VZV ORF36 (thymidine kinase (TK)), ORF13 (thymidylate synthase (TS)), or both genes, deleted to be used in drug discovery/activation mechanism studies. All viruses were viable and were assessed for the growth kinetics in cells, skin, and in SCIDhu mice to identify those with high levels of luciferase activity that would extinguish upon the cessation of viral DNA replication (such as from treatment with antiviral compounds).

3.2. Comparison of VZV-BAC-Luc to VZV-ORFx-Luc Reporter Viruses in Cells and Skin

It is well documented that mutations or changes to VZV ORFs can affect virulence in skin that is not necessarily observed in cell culture [15,34,36–38]. Thus, we conducted a comprehensive study of the VZV reporter viruses in cell culture and organized skin in order to identify virulence differences that might be a consequence of the insertion of *luc*. First, HFF cells were inoculated with cell-free VZV reporter viruses, at an approximate MOI of 0.01. In a second experiment, human fetal skin organ culture (SOC) explants were inoculated by scarification with a suspension of VZV infected HFFs for each virus (cell-associated virus; 60 μL; 1×10^4–10^5 pfu/mL) and placed on NetWells at the air–media interface. The VZV spread in the HFFs and SOC was measured daily by bioluminescence imaging and was reported as the fold change of the Total Flux. The Total Flux (photons/s/cm^2/steradian) is a measure of the radiance that reflects the number of VZV-infected cells and the level of luciferase activity per cell. As expected, VZV-BAC-Luc spread exponentially during the first 4 d post infection in HFFs (Figure 2A), and the reporter activity increased through approximately 5–7 d post infection in the SOC (Figure 2B). VZV-ORF17-Luc showed a similar growth pattern that was equivalent to VZV-BAC-Luc in both the cell culture and the SOC, while both VZV-ORF14-Luc and VZV-ORF57-Luc reached a much higher Total Flux than VZV-BAC-Luc in the HFFs (Figure 2A; $p < 0.0001$; one-way ANOVA with Dunnett's post hoc test). In the SOC, there was no apparent growth difference, with the exception that, while VZV-ORF9-Luc reached a significantly higher Total Flux in the HFFs (Figure 2A; $p < 0.001$; one-way ANOVA with Dunnett's post hoc test), the activity was considerably lower in the SOC and declined over time (Figure 2B; not significant; $p = 0.068$; one-way ANOVA with Kruskal–Wallis post hoc test). This result suggests that the VZV ORF9-Luc virus may have growth impairments in organized skin tissue, but the other late promoter luciferase viruses were considerably improved in the signal reporter activity over the previously used VZV BAC-Luc.

We previously reported that VZV-BAC-Luc bioluminescence significantly correlated with VZV pfu [19]. We considered it necessary to confirm that the bioluminescence from the cells infected with each of the new VZV-ORF reporter viruses similarly reflected the virus spread. As such, the correlation between the Total Flux and the VZV infectivity (infectious foci or pfu) was analyzed in the HFFs daily for 4 days (Figure 2C). After the bioluminescence measurements were collected, the cells were harvested with trypsin, and the ability to form infectious centers was determined by an infectious center assay [19]. For all the VZV reporter viruses, the bioluminescence was highly correlated with the VZV spread (Figure 2C; nonlinear regression analysis with log–log line; $R^2 > 0.90$). However, the pattern of luciferase activity varied for each virus. For VZV-ORF9-Luc, VZV-ORF14-

Luc, and VZV-ORF17-Luc, the bioluminescence increased sharply with a marginal level of virus spread. However, for VZV-ORF57-Luc, the bioluminescence and virus spread increased gradually and in tandem, similar to VZV-BAC-Luc. This suggests that ORF57 had the optimal correlation to infectivity, where luciferase was driven by a VZV-specific late viral promoter.

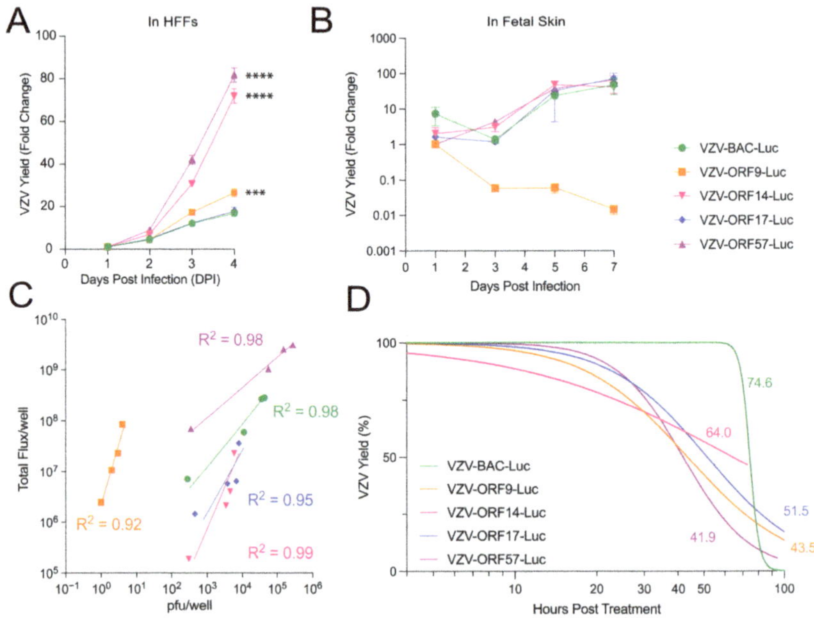

Figure 2. Evaluation of VZV-BAC-Luc and VZV-ORFx-Luc growth kinetics and bioluminescence in tissue culture and human fetal skin. Cells and skin were infected with cell-free or cell-associated virus, respectively, and grown at 37 °C. Cell culture experiments (**A**,**C**,**D**) were performed in HFFs, while skin organ culture (**B**) was performed in fetal skin. (**A**,**B**) VZV yield was measured by bioluminescence imaging and calculated as the fold change from the average Total Flux (photons/s/cm^2/steradian), divided by the lowest Total Flux value (DPI 1 for cells, or DPI 1-3 for SOC). (**C**) Correlation coefficients of luciferase and virus plaque number were determined for each VZV reporter virus on the basis of the relationship of pfu/well to average Total Flux per well. (**D**) HFFs and VZV were co-cultured for approximately 40 h prior to foscarnet treatment (1 mM) to block viral DNA replication. Values next to each curve represent the time (in h) for bioluminescence to decrease by 50% after treatment started and are shown in the corresponding color for each virus (individual points omitted for clarity of graph). Each point represents the mean ± SEM. Statistical analyses included one-way ANOVA with Dunnett's post hoc test (**A**,**B**, *** $p < 0.001$; **** $p < 0.0001$). Nonlinear regression analysis with (**C**) log–log line or (**D**) dose response—inhibition was used for best-fit lines. n = 3 biological replicates for cell-based assays; n = 6 biological replicates for skin organ culture.

It was then deemed necessary to determine the effects of halting the viral DNA synthesis on the bioluminescence from each reporter virus. The *luc* gene in VZV-BAC-Luc is under the control of the constitutively active SV40 promoter [19], and we have observed that bioluminescence persists for an extended period after the DNA synthesis is halted by antiviral compounds (data not shown). If the expression of a VZV late gene was dependent on viral DNA synthesis, then we would predict that its promoter activity might stop or decline when the replication was inhibited. To determine the Total Flux half-life after blocking the viral DNA synthesis, the HFFs were infected with VZV reporter viruses for 36–40 h, and then 1 mM of foscarnet (phosphonoformate) was added to inhibit the viral

DNA polymerase during the exponential growth phase. The cultures were then scanned over an additional 96 h to measure the decrease in the Total Flux as an indicator of the luciferase activity decay. Intriguingly, the Total Flux half-life varied, depending on the ORF promoter that was regulating the luciferase expression (Figure 2D). In line with previous observations, VZV-BAC-Luc had the longest half-life, at 74.6 h, and the VZV-ORF14-Luc half-life was nearly as long, at 64 h. However, the Total Flux half-life for VZV-ORF17-Luc was intermediate, at 51 h, while VZV-ORF9-Luc and VZV-ORF57-Luc had the shortest half-lives, at 43.5 and 41.9 h, respectively. On the basis of these results and the growth patterns in the HFFs and the SOC, VZV-ORF57-Luc was selected as the optimal VZV reporter virus.

3.3. Comparison of VZV-BAC-Luc and VZV-ORF57-Luc in a SCIDhu Mouse Model

While VZV-ORF57-Luc performed as well as VZV-BAC-Luc in skin organ culture (Figure 2B), it was not known whether the viruses replicated equally in skin xenografts, nor if it would be responsive to in vivo antiviral treatment. We previously showed that VZV-ORF57-Luc can infect skin xenografts in SCIDhu mice [27], but the relative growth kinetics were not assessed. SCIDhu mice with human fetal skin xenografts were inoculated by intra-xenograft injection with equivalent levels of VZV-BAC-Luc or VZV-ORF57-Luc, and the spread was measured daily by bioluminescence imaging with the IVIS™ 200 instrument. The groups were vehicle alone or were treated with cidofovir (10 mg/kg), which is a broad-spectrum antiviral drug that inhibits VZV DNA polymerase [39]. The treatment was administered by intraperitoneal (i.p.) injection from DPI 3-9. The fold change in the VZV yield was calculated as the daily Total Flux divided by the average Total Flux on the lowest day for each individual mouse (typically, DPI 2 or 3). VZV-BAC-Luc grew efficiently in vivo, reaching its highest point on DPI 8, with a final fold change of 21.8 ± 8 (Figure 3A; mean ± SEM). The VZV-ORF57-Luc reporter activity was still increasing, which suggests that the virus was still growing at DPI 9, and that it had a significantly higher fold change, reaching 164 ± 63 (Figure 3B) (mean ± SEM; $p = 0.0374$; Student's t-test). As expected, cidofovir significantly reduced the yield for both viruses (Figure 3; $p < 0.05$; Student's t-test). VZV-ORF57-Luc's reporter activity indicated that it replicated better in SCIDhu mice and that it was responsive to antiviral treatment, which suggests that it is robust and an optimal reporter VZV virus.

3.4. VZV Thymidylate Synthase (TS), but Not Thymidine Kinase (TK), Is Required for Virulence

We then used the VZV-ORF57-Luc background to construct a set of reporter viruses lacking the TK (from ORF36) and/or TS (from ORF13) genes, to evaluate the role of these nucleotide-modifying enzymes in the activation of novel antiviral compounds, and to understand their mechanism of action (Figure 1). VZV ORF36 encodes the thymidine kinase, which is required to phosphorylate many nucleoside analogs [40], while ORF13 encodes thymidylate synthase, which is involved in increasing the pool of thymidine in quiescent cells and may be involved in some nucleoside activation mechanisms [41]. Both enzymes play key roles in pathways that are affected by antiviral compounds. ORF13, ORF36, or both, were deleted and replaced with marker genes, which resulted in VZV-ORF57-ΔTK, VZV-ORF57-ΔTS, and VZV-ORF57-ΔTKΔTS. After the genotypes were confirmed by RFLP and regional sequencing, the phenotypes were evaluated in cells and adult human skin organ culture. We previously showed that VZV grows similarly in fetal versus adult skin (data not shown) [27]. The ARPE-19 cell monolayers were inoculated with cell-free VZV at an approximate MOI of 0.01, whereas the skin explants were infected by scarification through cell-associated VZV that was placed on the skin surface (30 μL inoculum; 1×10^4–10^5 pfu/mL). The skin explants were then placed on NetWells at the air–media interface. As expected, VZV-ORF57-Luc grew well in ARPE-19 cells over 4 d, and in adult skin explants over 7 d (Figure 4). VZV-ORF57-ΔTK had equivalent growth kinetics in cells and skin (Figure 4A,D). VZV-ORF57-ΔTS was slightly impaired in ARPE-19 cells, but was significantly impaired in skin compared to VZV-ORF57-Luc (Figure 4B,D; $p < 0.01$; one-way ANOVA with Dunnett's post hoc test). When both the VZV TK and TS were deleted, the growth in the cell culture was slightly less but did not reach significance (Figure 4C). However, similar to VZV-ORF57-ΔTS, VZV-ORF57-ΔTKΔTS showed minimal spread in skin organ culture (Figure 4D; $p < 0.05$; one-way ANOVA with Dunnett's post hoc test). Together, these results indicate that, while TK was dispensable for skin growth under these conditions, the expression of TS may facilitate the growth and spread of VZV in confluent ARPE-19 cells and it is required for spread in skin. To the best of our knowledge, this is the first demonstration of a growth phenotype for a VZV TS-deletion mutant.

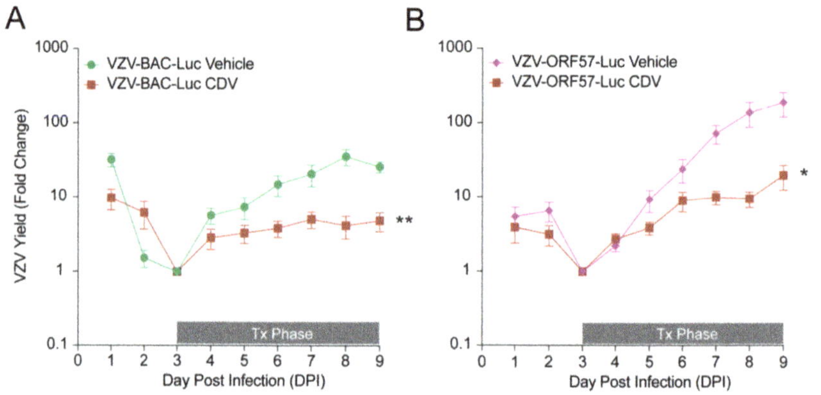

Figure 3. Evaluation of VZV-BAC-Luc and VZV-ORF57-Luc growth kinetics and responses to antiviral treatment in a SCIDhu mouse model. SCID *beige* mice were implanted with a single subcutaneous xenograft of fetal human skin. Xenografts were inoculated 3–4 weeks later with (**A**) VZV-BAC-Luc or (**B**) VZV-ORF57-Luc by intra-xenograft injection (1×10^4–10^5 pfu/mL; 60 μL injection; grown in HFFs). Mice were treated with vehicle (water) or 10 mg/kg/day cidofovir (CDV) by intraperitoneal injection from DPI 3 to 9 (Tx Phase). VZV yield was measured daily by bioluminescence imaging and the fold change was calculated as Total Flux each day divided by the lowest Total Flux value per mouse (usually taken on DPI 2 or 3). Virus growth kinetics were evaluated for statistical significance on DPI 9. Symbols represent mean ± SEM. Student's *t*-test; asterisks indicate significance between vehicle and cidofovir groups (* $p < 0.05$; ** $p < 0.01$). n = 6–10 mice per group.

3.5. VZV TS Deletion Increases Sensitivity to Antiviral Drugs

Mutations in VZV TK have long been associated with antiviral resistance to acyclovir and brivudine (BVdU), which are antivirals that require an initial phosphorylation by VZV TK for activation [42]. Cidofovir does not require TK phosphorylation because of its phosphonate moiety. It is also known that blocking viral or cellular thymidylate synthase can improve the potency of nucleoside analogs, such as acyclovir [43]. However, it was not known how deletions in TK and TS would affect the antiviral sensitivity of the VZV-ORF57-Luc virus. Thus, the potency (50% effective concentration (EC_{50})) of acyclovir, brivudine, and cidofovir against VZV-ORF57-Luc, VZV-ORF57-ΔTK, and VZV-ORF57-ΔTS was evaluated in confluent ARPE-19 cells. The antiviral activity was assessed by bioluminescence imaging at 3 d post infection (DPI), which was previously shown to be highly correlated to the VZV pfu [19] (Figure 2C). The EC_{50} for each antiviral compound was calculated as the average Total Flux at each concentration divided by the average Total Flux of the untreated wells [26]. Here, VZV-ORF57-Luc was sensitive to both brivudine and cidofovir, with EC_{50} values of 3.0 nM and 1.7 μM, respectively (Figure 5A). As expected, VZV-ORF57-ΔTK was resistant to brivudine but sensitive to cidofovir, with an EC_{50} value of 0.81 μM. Notably, VZV-ORF57-ΔTS showed considerably enhanced sensitivity to both brivudine and cidofovir, with EC_{50} values of 1.9 nM and 0.81 μM, respectively. Similar to brivudine, VZV-ORF57-Luc and VZV-ORF57-ΔTS were sensitive to acyclovir, with EC_{50} values of 7.6 μM and 0.37 μM, respectively (dose response curves not shown).

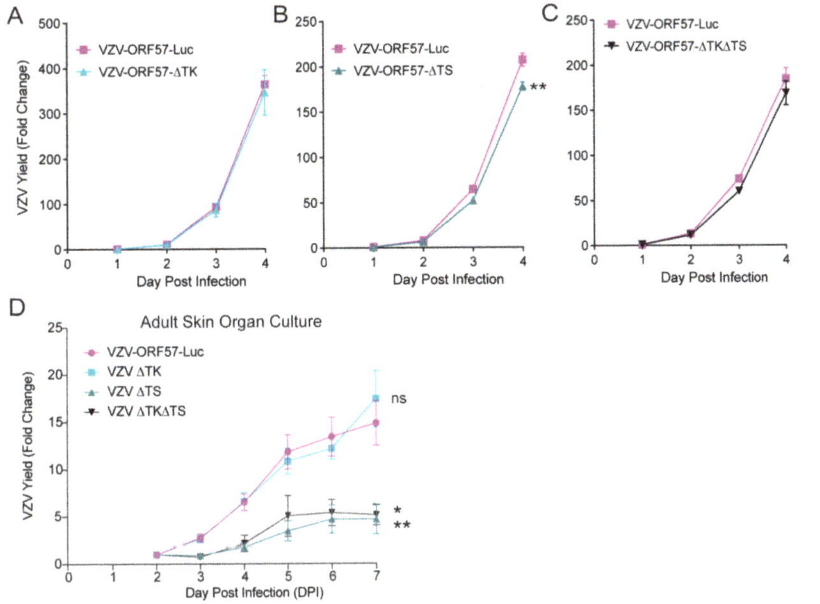

Figure 4. VZV TK is dispensable while VZV TS is required for optimal growth in cells and is critical for human skin organ culture. Cell-free VZV was added to ARPE-19 monolayers at 37 °C (**A–C**; 1.50 ratio) or introduced to adult human skin (**D**; 1 × 10^4–10^5 pfu/mL; 30 µL inoculum) by scarification, placed on NetWells, and incubated at 35 °C. Virus spread was monitored daily by bioluminescence imaging. Cell cultures were infected with up to 500 pfu/well. Skin was infected with 300–3000 pfu/piece of skin, depending on the viral titer that could be attained in ARPE-19 cells. VZV-ORF57-Luc was grown independently in ARPE-19 for each assay in (**A–C**), and directly compared to (**A**) VZV-ORF57-ΔTK, (**B**) VZV-ORF57-ΔTS, or (**C**) VZV-ORF57-ΔTKΔTS under the same conditions. (**D**) Growth kinetics of ΔTK, ΔTS, and ΔTKΔTS in adult human skin explants were evaluated for significance compared to parental VZV-ORF57-Luc on DPI 7. VZV yield was calculated as the average Total Flux each day divided by the Total Flux on (**A–C**) DPI 1 or (**D**) DPI 2. Each point and line represent the mean ± SEM. (**A–C**) Student's *t*-test or (**D**) one-way ANOVA with Dunnett's post hoc test; * $p < 0.05$; ** $p < 0.01$, ns = not significant. $n = 6$ biological replicates.

To further explore this phenomenon, we extended this study and evaluated the sensitivity of VZV-ORF57-ΔTS to brivudine. The dose response curves indicate that the enhanced sensitivity was greatest around the EC_{50} value. ARPE-19 monolayers were inoculated with cell-free VZV, treated with 1.25, 2.5, or 5 nM of brivudine for 3 d, and the VZV yield was measured as the reporter activity. VZV-ORF57-Luc was increasingly sensitive to higher concentrations of brivudine, while VZV-ORF57-ΔTK was resistant at all concentrations (Figure 5B). VZV-ORF57-ΔTS was significantly more sensitive to brivudine than the parent virus at 1.25 and 2.5 nM but did not show significance at 5 nM (Figure 5B; $p < 0.001$; one-way ANOVA; Tukey post hoc test). These data further demonstrate that the absence of VZV TS enhances the antiviral sensitivity to brivudine. Given that the TK and TS deletion viruses are isogenic, and that VZV-ORF57-Luc has a robust phenotype in all the models tested, we are now poised to study the mechanism of action of the novel antivirals as they develop. These viruses should prove to be highly useful to the field.

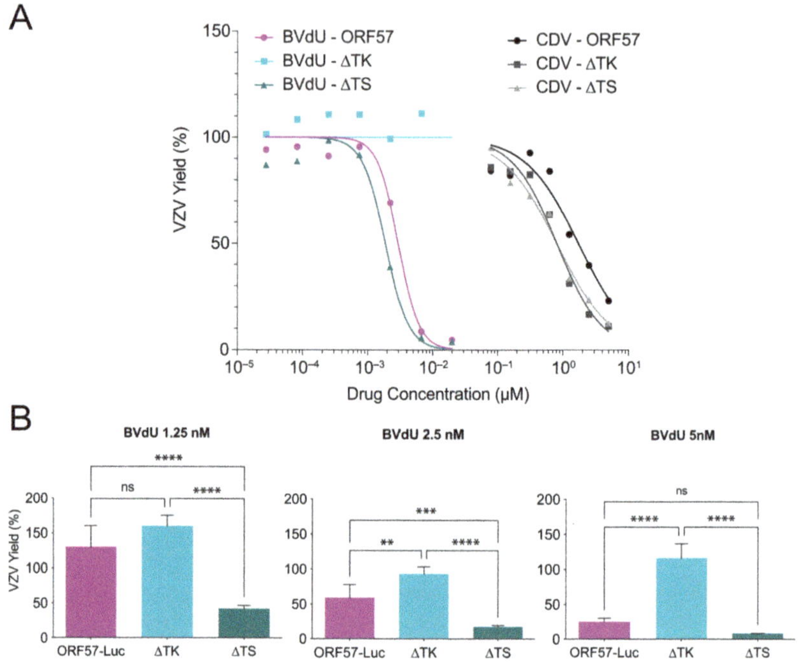

Figure 5. Loss of VZV TK and TS affect virus sensitivity to select antivirals in cell culture. Cell-free VZV was added to ARPE-19 monolayers at 37 °C and compounds were added at 2 h postinoculation (HPI). Antiviral compounds remained in culture medium until VZV yield was measured on DPI 3. Infected cells were treated with cidofovir (CDV) and brivudine (BVdU) (**A**) in a range of concentrations, or (**B**) at specific concentrations of BVdU (1.25, 2.5, or 5 nM). VZV yield was calculated as the average Total Flux of treated wells divided by the average Total Flux of untreated wells. (**A**) Each point represents the mean Total Flux with the best fit line. (**B**) Each bar represents the mean + SD. Asterisks indicate significance between VZV viruses for each treatment condition (ns = no significance; ** $p < 0.01$; *** $p < 0.001$; **** $p < 0.0001$). $p < 0.05$; one-way ANOVA; Tukey post hoc test. $n = 3$ (**A**) to 6 (**B**) biological replicates.

4. Discussion

Here, we constructed a series of robust VZV reporter viruses by using a BAC recombineering system for use in cell culture, skin organ culture, and in SCIDhu mice, and we validated their robustness in each system. We identified a specific strong viral promoter (a VZV-unique ORF57 gene of unknown function), where the luciferase activity was linked to viral replication and had a relatively short half-life. This virus was then used to evaluate the contribution of the VZV TK and TS genes to the viral replication in the cell and skin organ culture, which identified an unrecognized role for the VZV TS gene in human skin. This work sets the stage for the further analyses of the contribution of viral genes to the VZV growth in various model systems, and for the development and testing of novel antivirals to VZV.

Several luciferase reporter VZVs have been detailed previously. These reporters are beneficial tools, as the expression of luciferase can be used with IVIS systems for the imaging of the same samples over time. Firefly luciferase is particularly useful when the tissue or host platform are heterogenous, limited, prohibitively costly, or difficult to generate, such as human skin cultures and in xenotransplants harbored in SCIDhu mice. Oliver et al. detailed the development of VZV that reported growth in SCIDhu mice by expressing luciferase fused to the ORF68 (gE) and IE63 genes [44]. However, VZV with luciferase fused to ORF68 rapidly lost its reporter activity, which is likely because of the C terminal fusion of the gE lay in the reiterated elements of the genome. The stable reporter fusion to ORF63 required the addition to both ORF63 and its duplicated gene, ORF70 [44]. When only added to ORF63, the reporter was lost. In addition, IE63 is strongly suspected to be an immediate early gene, so it may be expressed from the genome rapidly and without the need for viral replication [45]. We previously reported an alternative strategy to express luciferase, in which viral-promoter-luciferase cassettes were placed at an ectopic locus, and specifically at the ORF65-ORF66 intergenic region, in recombinant viruses [46]. While this approach was successful, we could not eliminate the possibility that the ectopic placement of the promoter region did not contain a sufficient promoter sequence to confer full regulation in the context of the virus. We also could not eliminate the possibility that the ectopic site may have influenced the regulation and expression kinetics. An advantage of the system that was developed and used here is that the native promoter at its natural locus is employed to drive the luciferase expression, which minimizes the concerns of the positional contributions to the activity. In addition, the separation of the luciferase from the viral protein with a T2A motif avoids the intolerance of the luciferase-to-protein fusion, and it ensures minimal interference with the viral protein. When fully active, the T2A ribosome skipping motif results in a 21–23 residue that is a remnant of the motif that is left attached to the viral protein, and a single proline attached to the downstream luciferase gene.

There are, however, a few potential disadvantages to this strategy, but we have yet to encounter them. Specifically, the BAC system may develop spurious mutations, or regions of the genome may be unstable in *E. coli*. Indeed, we identified two spurious mutations in the pOka DX BAC that were not present in the pOka cosmids that were used to derive the BAC, but that were present in the earlier nonself-excising version of the BAC [18,21]. We expect that this arose during the derivation of the BACs from circular viral DNA. While the deletion of the sequences has not been identified specifically for the VZV, it is recognized that part of the UL origin of the replication in the HSV genome is deleted in most *E. coli* strains [47,48], and that it may affect HSV pathogenesis in a mouse model [49]. Additionally, the potential for sporadic SNP mutations to arise is always a concern, since each recombineering step requires a clonal step, and such changes are unlikely to be detected by restriction analyses. Such mutations could accumulate with multiple recombination events, which is most often conducted by using "scarless" sequential recombineering methods, as detailed by Tischer et al. [18,21], and as employed here. Here, we acknowledge that the robust ORF57 luciferase virus underwent five sequential recombination steps (1 = the luciferase insertion; 2 and 3 = the correction of ORF40 and its reversal; 4 and 5 = the correction of ORF50 and its reversal), but we did not identify any spurious or unexpected mutations in the genome.

An important consideration when designing VZV reporter strains is the location of the *luc* insertion and the gene to which it is fused. We have found that the insertion of fluorescent proteins within VZV ORFs may alter the infectivity and virulence in skin (authors' observations). This phenomenon has been described for other viruses, where the insertion of fluorescent reporter proteins interfered with the growth kinetics [50,51]. Thus, we specifically chose certain ORFs for the virus development. We found that *luc* insertion after the VZV ORFs, 14, 17, or 57, did not alter the growth kinetics in cells or human skin. By contrast, the *luc* insertion after the ORF9 reduced the VZV growth in skin, despite the T2A ribosome skipping motif that releases fLuc from the ORF9 protein. ORF9 interacts with IE62 to form complexes that are involved in tegument formation [52], and ORF9 is required for VZV replication [38]. It is possible that the *luc* insertion altered the ORF9 protein, thereby preventing it from interacting with the IE62 protein and performing its function during replication. Another possibility is that the *luc* sequences interrupted the mRNA and the noncoding RNA transcription through the ORF9 region [53]. While it is outside of the scope of this study, it would be interesting to create a library of VZV reporter strains with the T2A-*luc* (or other reporter genes) inserted after other candidate VZV ORFs. Such a library could potentially reveal phenotypes not previously described for different

VZV ORFs. For this, the recent detailed transcriptional map of the VZV genome will be important to take into consideration [53].

A goal of this project was to construct a reporter virus with *luc* expression under the transcriptional control of a late viral promoter. The annotation of the VZV transcriptome and the kinetics of the expression, as predicted from HSV, prompted the selection of the ORF17; the true late genes, ORF9 and ORF14; and the leaky late gene, ORF57, as potential regulators of the *luc* expression [53]. Transcript abundance was also considered, since high levels would produce a brighter bioluminescence signal. ORF9 is one of the most abundant VZV proteins, it is packaged into the virion tegument, and it is detected in the cytoplasm as soon as 2 h post infection [54]. We [29] and others [55] have reported the successful tagging of ORF9 at its C terminal end, and, thus, we expected it to be a good candidate for driving luciferase. The VZV transcription kinetics of ORF14, which encodes glycoprotein C, is atypical and unpredictable in cultured cells and it may be very delayed, with little protein detected until 3 DPI, after 4–6 rounds of VZV replication [56]. ORF17, a putative host shut-off protein, is required for VZV replication at 37 °C, but not at 33 °C [24,57], although this functional role is contested [58]. ORF17 can also be detected as early as 6–12 h post infection in cells [58]. ORF57 is a very short ORF that is unique to VZV and has no known function, but it is detected in the Golgi at 48 h post infection [59]. In this study, infected cells or skin explants were first scanned at 24 h post infection, at which time there was a strong bioluminescent signal. This indicates that the fLuc protein was synthesized from the viral mRNAs that were driven by these ORFs within the first 24 h.

VZV-BAC-Luc has long been the gold standard reporter virus. Strong luciferase activity enables the bioluminescence imaging of the VZV spread, which has been instrumental in understanding VZV pathogenesis and in evaluating the antiviral compounds in cells, skin, and in vivo [19,26,27,44]. Unfortunately, *luc* is controlled by the SV40 promoter [19], which is normally considered to be constitutively expressed and may not respond to the antiviral effects of novel compounds in the context of the virus. Here, we showed that the bioluminescence half-life for VZV-BAC-Luc in the presence of the DNA polymerase inhibitor, foscarnet, is 74.6 h, or just over 3 days (Figure 2C). This is problematic during the standard antiviral dose response studies that are conducted over 72 h because the bioluminescent signal may only decrease by half in the treated infected cells. This could potentially mask the effects of an antiviral compound that is evaluated by using the VZV-BAC-Luc strain. In skin organ culture and SCIDhu mice with skin xenografts, bioluminescence imaging usually coincides with the initiation of antiviral treatment. If the bioluminescent signal is sustained for 3 days, it may conceal how quickly a novel antiviral prevents the virus spread. Thus, a new VZV reporter strain, with a shorter bioluminescent signal half-life, was needed. Here, VZV-ORF57-Luc appears to be a superior and optimal reporter strain. It had the shortest bioluminescence half-life of 41.9 h, or less than 2 days, it produced a strong bioluminescent signal, and it was virulent in skin and SCIDhu mice. The characterization of VZV-ORF57-Luc is a promising step forward and will improve the evaluation of novel antivirals.

Acyclovir and its derivatives are the first-line antiviral therapy for VZV infections, and they are activated by TK phosphorylation. Mutations in the TK gene are the most common cause of resistance to acyclovir and other nucleoside analogs that lack a phosphonate group [6,60,61]. Nucleoside analogs remain the focus of most antiviral drug development, and so we constructed VZV-ORF57-Luc-ΔTK as a tool to study how nucleoside analogs are processed in VZV-infected cells. Here, we show that VZV-ORF57-ΔTK was resistant to acyclovir and brivudine, which must be phosphorylated in the cell to their active form, but not to cidofovir, which has a phosphonate group. The deletion of TK had no discernable phenotype on this strain. This reporter virus is isogenic with VZV-ORF57-Luc, and it will be invaluable for screening nucleoside analogs for activity against clinical strains that are resistant to acyclovir.

VZV encodes several proteins for nucleic acid metabolism that may impact the mechanism and activity of antiviral nucleoside analogs. VZV ORFs 18 and 19 encode a ribonucleotide reductase that generates deoxyribonucleotides, and the deletion of these genes potentiates the activity of acyclovir [62]. VZV ORF8 encodes a dUTPase that converts dUTP to dUMP [63], which is a substrate for ORF13 thymidylate synthase. The TS enzyme converts dUMP to dTMP in quiescent cells [41], which is then phosphorylated by VZV TK and cellular kinases to increase the dTTP pool for viral DNA synthesis. Nucleoside analogs are also substrates of VZV TK and compete with deoxythymidine and dTMP for the enzyme. It has long been noted that inhibiting TS increases the activity of acyclovir by reducing the dTMP levels. In cells that are infected with HSV-1, blocking the viral and cellular TS activity with 5-fluorouracil increases the potency of acyclovir [43]. We found a similar effect in this study, in that the potency of brivudine increased against VZV-ORF57-ΔTS. This enhancing effect was also observed for CDV and ACV, although they were not analyzed in depth. We also found that VZV-ORF57-ΔTS was impaired for virus spread in skin. The cells in adult skin are mostly quiescent, with little to no cellular TS. In this environment, VZV TS may be important for generating dTMP and adequate dTTP. To the best of our knowledge, this is the first time a growth phenotype has been described for a VZV TS mutant. It would be interesting to test the VZV-ORF57-ΔTS growth in replicating and quiescent cells, to measure the dTTP pools, and to discern the impact on the VZV replication and antiviral activity of nucleoside analogs. Such studies could fully resolve the VZV TS phenotype uncovered here.

In summary, we detail a strategy to generate VZV reporter viruses to assess the gene contributions to virulence in cells and skin, as well as to evaluate novel antiviral compounds. The reporter viruses were constructed from a recombinant pOka strain with a T2A ribosome skipping motif between the VZV ORF and the *luc* gene. VZV-ORF57-Luc appears to be an ideal reporter virus for its maintenance of the virulence in a human skin model and for the responsiveness to inhibition of DNA replication. We used this background to develop isogenic reporter viruses with deletions in ORF36 and ORF13 as additional tools for antiviral research. In cells and skin, we show that TK was dispensable, while TS was found to have an unprecedented requirement for virulence. Furthermore, the loss of TS enhanced the antiviral sensitivity to certain nucleoside analogues. Taken together, this work lays the foundation for the probing of the viral gene contributions to pathogenesis in human skin and for the evaluation of novel antiviral compounds.

Author Contributions: Conceptualization, M.G.L., M.B.Y., P.R.K. and J.F.M.; data curation, M.G.L., D.L., B.W.G. and J.F.M.; formal analysis, M.G.L., M.B.Y., P.R.K. and J.F.M.; funding acquisition, P.R.K. and J.F.M.; investigation, M.G.L., D.L. and J.F.M.; methodology, M.G.L., M.B.Y., D.L., P.R.K. and J.F.M.; project administration, P.R.K. and J.F.M.; resources, P.R.K. and J.F.M.; software, M.G.L., M.B.Y. and J.S.F.; supervision, P.R.K. and J.F.M.; validation, M.G.L., M.B.Y., J.S.F., D.L., B.W.G., P.R.K. and J.F.M.; visualization, M.G.L. and M.B.Y.; writing—original draft, M.G.L.; writing—review and editing, P.R.K. and J.F.M. All authors have read and agreed to the published version of the manuscript.

Funding: This research was funded, in part, by the National Institute of Allergy and Infectious Diseases (Division of Microbiology and Immunology), contract HHSN272201700030I (J.F.M.). PRK was supported by the NIH awards, AI122640, NS064022, an NEI CORE grant (P30 EY08098) and unrestricted support from Research to Prevent Blindness Inc NY and The Eye & Ear Foundation of Pittsburgh.

Institutional Review Board Statement: The study was conducted according to the guidelines of the Declaration of Helsinki and was approved by the Institutional Review Board of SUNY Upstate Medical University (Protocol number: 1140572; original date of approval: 17 November 2017). The animal study protocol was approved by the Institutional Animal Care and Use Committee at SUNY Upstate Medical University and was performed conforming to all local, state, and national guidelines (IACUC #282; protocol reapproved on 23 September 2020).

Informed Consent Statement: Informed written consent to donate tissue for research purposes was obtained from each study participant. The approved protocol conforms to all local, state, and national guidelines, as well as to the ethical guidelines of the 1975 Declaration of Helsinki.

Data Availability Statement: The data in this study are fully presented in this article. These data are also available upon reasonable request from the corresponding authors.

Acknowledgments: The authors wish to acknowledge the help and support of Prashant Upadhyaya in the Department of Surgery at SUNY Upstate Medical University for his help in obtaining patient consent and procuring skin from reduction mammoplasty surgeries. The authors also wish to thank Wanda Coombs in the Department of Microbiology and Immunology at SUNY Upstate Medical University for the excellent technical assistance with the humanized mouse assays.

Conflicts of Interest: The authors declare no conflict of interest.

References

1. Lopez, A.S.; Zhang, J.; Marin, M. Epidemiology of Varicella During the 2-Dose Varicella Vaccination Program—United States, 2005–2014. *MMWR. Morb. Mortal. Wkly. Rep.* **2016**, *65*, 902–905. [CrossRef] [PubMed]
2. Gershon, A.A.; Gershon, M.D. Pathogenesis and Current Approaches to Control of Varicella-Zoster Virus Infections. *Clin. Microbiol. Rev.* **2013**, *26*, 728–743. [CrossRef] [PubMed]
3. Draper, M.; Stergiopoulos, S. Shingles vaccination uptake in Massachusetts adults aged 50 years and older. *Vaccine* **2021**, *39*, 6781–6786. [CrossRef] [PubMed]
4. Lu, P.-J.; Hung, M.-C.; Srivastav, A.; Grohskopf, L.A.; Kobayashi, M.; Harris, A.M.; Dooling, K.L.; Markowitz, L.E.; Rodriguez-Lainz, A.; Williams, W.W. Surveillance of Vaccination Coverage Among Adult Populations—United States. *MMWR. Surveill. Summ.* **2021**, *70*, 1–26. [CrossRef]
5. Patterson, B.J.; Chen, C.-C.; McGuiness, C.B.; Glasser, L.I.; Sun, K.; Buck, P.O. Early examination of real-world uptake and second-dose completion of recombinant zoster vaccine in the United States from October 2017 to September. *Hum. Vaccines Immunother.* **2021**, *17*, 2482–2487. [CrossRef]
6. De Clercq, E.; Li, G. Approved Antiviral Drugs over the Past 50 Years. *Clin. Microbiol. Rev.* **2016**, *29*, 695–747. [CrossRef]
7. Poole, C.L.; James, S.H. Antiviral Therapies for Herpesviruses: Current Agents and New Directions. *Clin. Ther.* **2018**, *40*, 1282–1298. [CrossRef]
8. De Clercq, E. Fifty Years in Search of Selective Antiviral Drugs. *J. Med. Chem.* **2019**, *62*, 7322–7339. [CrossRef]
9. Mahalingam, R.; Gershon, A.; Gershon, M.; Cohen, J.I.; Arvin, A.; Zerboni, L.; Zhu, H.; Gray, W.; Messaoudi, I.; Traina-Dorge, V. Current In Vivo Models of Varicella-Zoster Virus Neurotropism. *Viruses* **2019**, *11*, 502. [CrossRef]
10. Haberthur, K.; Messaoudi, I. Animal Models of Varicella Zoster Virus Infection. *Pathogens* **2013**, *2*, 364–382. [CrossRef]
11. Roizman, B.; Knipe, D.M.; Whitley, R.J. (Eds.) *Fields Virology, Chapter 67: Herpes Simplex Viruses*, 5th ed.; Lippincott Williams & Wilkin: Philadelphia, PA, USA, 2007; pp. 2502–2601.
12. Cohen, J.I.; Seidel, K.E. Generation of varicella-zoster virus (VZV) and viral mutants from cosmid DNAs: VZV thymidylate synthetase is not essential for replication in vitro. *Proc. Natl. Acad. Sci. USA* **1993**, *90*, 7376–7380. [CrossRef]
13. Sawyer, M.H.; Inchauspe, G.; Biron, K.K.; Waters, D.J.; Straus, S.E.; Ostrove, J.M. Molecular analysis of the pyrimidine deoxyribonucleoside kinase gene of wild-type and acyclovir-resistant strains of varicella-zoster virus. *J. Gen. Virol.* **1988**, *69*, 2585–2593. [CrossRef]
14. Niizuma, T.; Zerboni, L.; Sommer, M.H.; Ito, H.; Hinchliffe, S.; Arvin, A.M. Construction of varicella-zoster virus recombinants from parent Oka cosmids and demonstration that ORF65 protein is dispensable for infection of human skin and T cells in the SCID-hu mouse model. *J. Virol.* **2003**, *77*, 6062–6065. [CrossRef]
15. Zerboni, L.; Hinchliffe, S.; Sommer, M.H.; Ito, H.; Besser, J.; Stamatis, S.; Cheng, J.; DiStefano, D.; Kraiouchkine, N.; Shaw, A.; et al. Analysis of varicella zoster virus attenuation by evaluation of chimeric parent Oka/vaccine Oka recombinant viruses in skin xenografts in the SCIDhu mouse model. *Virology* **2005**, *332*, 337–346. [CrossRef]
16. Nagaike, K.; Mori, Y.; Gomi, Y.; Yoshii, H.; Takahashi, M.; Wagner, M.; Koszinowski, U.; Yamanishi, K. Cloning of the varicella-zoster virus genome as an infectious bacterial artificial chromosome in Escherichia coli. *Vaccine* **2004**, *22*, 4069–4074. [CrossRef]
17. Yoshii, H.; Somboonthum, P.; Takahashi, M.; Yamanishi, K.; Mori, Y. Cloning of full length genome of varicella-zoster virus vaccine strain into a bacterial artificial chromosome and reconstitution of infectious virus. *Vaccine* **2007**, *25*, 5006–5012. [CrossRef]
18. Tischer, B.K.; Kaufer, B.B.; Sommer, M.; Wussow, F.; Arvin, A.M.; Osterrieder, N. A Self-Excisable Infectious Bacterial Artificial Chromosome Clone of Varicella-Zoster Virus Allows Analysis of the Essential Tegument Protein Encoded by *ORF9*. *J. Virol.* **2007**, *81*, 13200–13208. [CrossRef]
19. Zhang, Z.; Rowe, J.; Wang, W.; Sommer, M.; Arvin, A.; Moffat, J.; Zhu, H. Genetic Analysis of Varicella-Zoster Virus ORF0 to ORF4 by Use of a Novel Luciferase Bacterial Artificial Chromosome System. *J. Virol.* **2007**, *81*, 9024–9033. [CrossRef]
20. Markus, A.; Grigoryan, S.; Sloutskin, A.; Yee, M.B.; Zhu, H.; Yang, I.H.; Thakor, N.V.; Sarid, R.; Kinchington, P.R.; Goldstein, R.S. Varicella-Zoster Virus (VZV) Infection of Neurons Derived from Human Embryonic Stem Cells: Direct Demonstration of Axonal Infection, Transport of VZV, and Productive Neuronal Infection. *J. Virol.* **2011**, *85*, 6220–6233. [CrossRef]
21. Tischer, B.K.; von Einem, J.; Kaufer, B.; Osterrieder, N. Two-step red-mediated recombination for versatile high-efficiency markerless DNA manipulation in Escherichia coli. *Biotechniques* **2006**, *40*, 191–197.
22. Zhang, Z.; Huang, Y.; Zhu, H. An Efficient Protocol for VZV BAC-Based Mutagenesis. *Adv. Struct. Saf. Stud.* **2010**, *634*, 75–86. [CrossRef]

23. Depledge, D.P.; Sadaoka, T.; Ouwendijk, W.J.D. Molecular Aspects of Varicella-Zoster Virus Latency. *Viruses* **2018**, *10*, 349. [CrossRef] [PubMed]
24. Zhang, Z.; Selariu, A.; Warden, C.; Huang, G.; Huang, Y.; Zaccheus, O.; Cheng, T.; Xia, N.; Zhu, H. Genome-wide mutagenesis reveals that ORF7 is a novel VZV skin-tropic factor. *PLoS Pathog.* **2010**, *6*, e1000971. [CrossRef] [PubMed]
25. Luker, G.D.; Prior, J.L.; Song, J.; Pica, C.M.; Leib, D.A. Bioluminescence Imaging Reveals Systemic Dissemination of Herpes Simplex Virus Type 1 in the Absence of Interferon Receptors. *J. Virol.* **2003**, *77*, 11082–11093. [CrossRef]
26. Rowe, J.; Greenblatt, R.J.; Liu, D.; Moffat, J.F. Compounds that target host cell proteins prevent varicella-zoster virus replication in culture, ex vivo, and in SCID-Hu mice. *Antivir. Res.* **2010**, *86*, 276–285. [CrossRef]
27. Lloyd, M.G.; Smith, N.A.; Tighe, M.; Travis, K.L.; Liu, D.; Upadhyaya, P.K.; Kinchington, P.R.; Chan, G.C.; Moffat, J.F. A Novel Human Skin Tissue Model to Study Varicella-Zoster Virus and Human Cytomegalovirus. *J. Virol.* **2020**, *94*, 22. [CrossRef]
28. Erazo, A.; Yee, M.B.; Osterrieder, N.; Kinchington, P.R. Varicella-Zoster Virus Open Reading Frame 66 Protein Kinase Is Required for Efficient Viral Growth in Primary Human Corneal Stromal Fibroblast Cells. *J. Virol.* **2008**, *82*, 7653–7665. [CrossRef]
29. Warner, B.E.; Yee, M.B.; Zhang, M.; Hornung, R.S.; Kaufer, B.B.; Visalli, R.J.; Kramer, P.R.; Goins, W.F.; Kinchington, P.R. Varicella-zoster virus early infection but not complete replication is required for the induction of chronic hypersen-sitivity in rat models of postherpetic neuralgia. *PLoS Pathog.* **2021**, *17*, e1009689. [CrossRef]
30. Leisenfelder, S.A.; Moffat, J.F. Varicella-Zoster Virus Infection of Human Foreskin Fibroblast Cells Results in Atypical Cyclin Expression and Cyclin-Dependent Kinase Activity. *J. Virol.* **2006**, *80*, 5577–5587. [CrossRef]
31. De, C.; Liu, D.; Zheng, B.; Singh, U.S.; Chavre, S.; White, C.; Arnold, R.D.; Hagen, F.K.; Chu, C.K.; Moffat, J.F. β-L-1-[5-(E-2-bromovinyl)-2-(hydroxymethyl)-1,3-(dioxolan-4-yl)] uracil (l-BHDU) prevents varicella-zoster virus replication in a SCID-Hu mouse model and does not interfere with 5-fluorouracil catabolism. *Antivir. Res.* **2014**, *110*, 10–19. [CrossRef]
32. Lloyd, M.G.; Liu, D.; Legendre, M.; Markovitz, D.M.; Moffat, J.F. H84T BanLec has broad spectrum antiviral activity against human herpesviruses in cells, skin, and mice. *Sci. Rep.* **2022**, *12*, 1641. [CrossRef]
33. Taylor, S.L.; Moffat, J.F. Replication of Varicella-Zoster Virus in Human Skin Organ Culture. *J. Virol.* **2005**, *79*, 11501–11506. [CrossRef]
34. Moffat, J.F.; Zerboni, L.; Kinchington, P.R.; Grose, C.; Kaneshima, H.; Arvin, A.M. Attenuation of the vaccine Oka strain of varicella-zoster virus and role of glycoprotein C in alphaherpesvirus virulence demonstrated in the SCID-hu mouse. *J. Virol.* **1998**, *72*, 965–974. [CrossRef]
35. Lloyd, M.; Liu, D.; Lyu, J.; Fan, J.; Overhulse, J.; Kashemirov, B.; Prichard, M.; McKenna, C.; Moffat, J. An acyclic phosphonate prodrug of HPMPC is effective against VZV in skin organ culture and mice. *Antivir. Res.* **2022**, *199*, 105275. [CrossRef]
36. Moffat, J.F.; Zerboni, L.; Sommer, M.H.; Heineman, T.C.; Cohen, J.I.; Kaneshima, H.; Arvin, A.M. The ORF47 and ORF66 putative protein kinases of varicella-zoster virus determine tropism for human T cells and skin in the SCID-hu mouse. *Proc. Natl. Acad. Sci. USA* **1998**, *95*, 11969–11974. [CrossRef]
37. Besser, J.; Sommer, M.H.; Zerboni, L.; Bagowski, C.P.; Ito, H.; Moffat, J.; Ku, C.C.; Arvin, A.M. Differentiation of vari-cella-zoster virus ORF47 protein kinase and IE62 protein binding domains and their contributions to replication in human skin xenografts in the SCID-hu mouse. *J. Virol.* **2003**, *77*, 5964–5974. [CrossRef]
38. Che, X.; Reichelt, M.; Sommer, M.H.; Rajamani, J.; Zerboni, L.; Arvin, A.M. Functions of the ORF9-to-ORF12 gene clus-ter in varicella-zoster virus replication and in the pathogenesis of skin infection. *J. Virol.* **2008**, *82*, 5825–5834. [CrossRef]
39. De Clercq, E. Antiviral drugs in current clinical use. *J. Clin. Virol.* **2004**, *30*, 115–133. [CrossRef]
40. Balzarini, J.; McGuigan, C. Chemotherapy of varicella-zoster virus by a novel class of highly specific anti-VZV bicyclic pyrimidine nucleosides. *Biochim. Biophys. Acta (BBA) Mol. Basis Dis.* **2002**, *1587*, 287–295. [CrossRef]
41. Hew, K.; Dahlroth, S.-L.; Veerappan, S.; Pan, L.X.; Cornvik, T.; Nordlund, P. Structure of the Varicella Zoster Virus Thymidylate Synthase Establishes Functional and Structural Similarities as the Human Enzyme and Potentiates Itself as a Target of Brivudine. *PLoS ONE* **2015**, *10*, e0143947. [CrossRef]
42. Andrei, G.; Topalis, D.; Fiten, P.; McGuigan, C.; Balzarini, J.; Opdenakker, G.; Snoeck, R. In vitro-selected drug-resistant varicella-zoster virus mutants in the thymidine kinase and DNA polymerase genes yield novel phenotype-genotype asso-ciations and highlight differences between antiherpesvirus drugs. *J. Virol.* **2012**, *86*, 2641–2652. [CrossRef] [PubMed]
43. Prichard, M.; Prichard, L.E.; Shipman, C. Inhibitors of thymidylate synthase and dihydrofolate reductase potentiate the antiviral effect of acyclovir. *Antivir. Res.* **1993**, *20*, 249–259. [CrossRef]
44. Oliver, S.L.; Zerboni, L.; Sommer, M.; Rajamani, J.; Arvin, A.M. Development of recombinant varicella-zoster viruses expressing luciferase fusion proteins for live in vivo imaging in human skin and dorsal root ganglia xenografts. *J. Virol. Methods* **2008**, *154*, 182–193. [CrossRef] [PubMed]
45. Debrus, S.; Sadzot-Delvaux, C.; Nikkels, A.F.; Piette, J.; Rentier, B. Varicella-zoster virus gene 63 encodes an immedi-ate-early protein that is abundantly expressed during latency. *J. Virol.* **1995**, *69*, 3240–3245. [CrossRef]
46. Jones, M.; Dry, I.R.; Frampton, D.; Singh, M.; Kanda, R.; Yee, M.; Kellam, P.; Hollinshead, M.; Kinchington, P.R.; O'Toole, E.; et al. RNA-seq Analysis of Host and Viral Gene Expression Highlights Interaction between Varicella Zoster Virus and Keratinocyte Differentiation. *PLoS Pathog.* **2014**, *10*, e1003896. [CrossRef]
47. Balliet, J.W.; Min, J.C.; Cabatingan, M.S.; Schaffer, P.A. Site-directed mutagenesis of large DNA palindromes: Construction and in vitro characterization of herpes simplex virus type 1 mutants containing point mutations that eliminate the oriL or oriS initiation function. *J. Virol.* **2005**, *79*, 12783–12797. [CrossRef]

48. Weller, S.K.; Spadaro, A.; Schaffer, J.E.; Murray, A.W.; Maxam, A.M.; Schaffer, P.A. Cloning, sequencing, and func-tional analysis of oriL, a herpes simplex virus type 1 origin of DNA synthesis. *Mol. Cell Biol.* **1985**, *5*, 930–942.
49. Polvino-Bodnar, M.; Orberg, P.K.; Schaffer, P.A. Herpes simplex virus type 1 oriL is not required for virus replication or for the establishment and reactivation of latent infection in mice. *J. Virol.* **1987**, *61*, 3528–3535. [CrossRef]
50. Rostad, C.A.; Currier, M.C.; Moore, M.L. Fluorescent and Bioluminescent Reporter Myxoviruses. *Viruses* **2016**, *8*, 214. [CrossRef]
51. Fulton, B.O.; Palese, P.; Heaton, N.S. Replication-Competent Influenza B Reporter Viruses as Tools for Screening Anti-virals and Antibodies. *J. Virol.* **2015**, *89*, 12226–12231. [CrossRef]
52. Cilloniz, C.; Jackson, W.; Grose, C.; Czechowski, D.; Hay, J.; Ruyechan, W.T. The Varicella-Zoster Virus (VZV) ORF9 Protein Interacts with the IE62 Major VZV Transactivator. *J. Virol.* **2007**, *81*, 761–774. [CrossRef]
53. Braspenning, S.E.; Sadaoka, T.; Breuer, J.; Verjans, G.M.G.M.; Ouwendijk, W.J.D.; Depledge, D.P. Decoding the Architecture of the Varicella-Zoster Virus Transcriptome. *mBio* **2020**, *11*, e01568-20. [CrossRef]
54. Cai, M.; Wang, S.; Xing, J.; Zheng, C. Characterization of the nuclear import and export signals, and subcellular transport mechanism of varicella-zoster virus ORF. *J. Gen. Virol.* **2010**, *92*, 621–626. [CrossRef]
55. Lebrun, M.; Thelen, N.; Thiry, M.; Riva, L.; Ote, I.; Condé, C.; Vandevenne, P.; Di Valentin, E.; Bontems, S.; Sadzot-Delvaux, C. Varicella-zoster virus induces the formation of dynamic nuclear capsid aggregates. *Virology* **2014**, *454–455*, 311–327. [CrossRef]
56. Storlie, J.; Jackson, W.; Hutchinson, J.; Grose, C. Delayed biosynthesis of varicella-zoster virus glycoprotein C: Upregulation by hexamethylene bisacetamide and retinoic acid treatment of infected cells. *J. Virol.* **2006**, *80*, 9544–9556. [CrossRef]
57. Sato, H.; Callanan, L.D.; Pesnicak, L.; Krogmann, T.; Cohen, J.I. Varicella-zoster virus (VZV) ORF17 protein induces RNA cleavage and is critical for replication of VZV at 37 degrees C but not 33 degrees C. *J. Virol.* **2002**, *76*, 11012–11023. [CrossRef]
58. Desloges, N.; Rahaus, M.; Wolff, M.H. The varicella–zoster virus–mediated delayed host shutoff: Open reading frame 17 has no major function, whereas immediate–early 63 protein represses heterologous gene expression. *Microbes Infect.* **2005**, *7*, 1519–1529. [CrossRef]
59. Roviš, T.L.; Bailer, S.M.; Pothineni, V.R.; Ouwendijk, W.J.; Šimić, H.; Babić, M.; Miklić, K.; Malić, S.; Verweij, M.C.; Baiker, A.; et al. Comprehensive Analysis of Varicella-Zoster Virus Proteins Using a New Monoclonal Antibody Collection. *J. Virol.* **2013**, *87*, 6943–6954. [CrossRef]
60. Strasfeld, L.; Chou, S. Antiviral drug resistance: Mechanisms and clinical implications. *Infect. Dis. Clin.* **2010**, *24*, 809–833. [CrossRef]
61. Morfin, F.; Thouvenot, D.; De Turenne-Tessier, M.; Lina, B.; Aymard, M.; Ooka, T. Phenotypic and genetic characteriza-tion of thymidine kinase from clinical strains of varicella-zoster virus resistant to acyclovir. *Antimicrob. Agents Chemother.* **1999**, *43*, 2412–2416. [CrossRef]
62. Heineman, T.C.; Cohen, J.I. Deletion of the varicella-zoster virus large subunit of ribonucleotide reductase impairs growth of virus in vitro. *J. Virol.* **1994**, *68*, 3317–3323. [CrossRef] [PubMed]
63. Ross, J.; Williams, M.; Cohen, J.I. Disruption of the varicella-zoster virus dUTPase and the adjacent ORF9A gene results in impaired growth and reduced syncytia formation in vitro. *Virology* **1997**, *234*, 186–195. [CrossRef] [PubMed]

viruses

MDPI

Article

Dermatitis during Spaceflight Associated with HSV-1 Reactivation

Satish K. Mehta [1,*], Moriah L. Szpara [2], Bridgette V. Rooney [3], Douglass M. Diak [4], Mackenzie M. Shipley [2], Daniel W. Renner [2], Stephanie S. Krieger [5], Mayra A. Nelman-Gonzalez [5], Sara R. Zwart [6], Scott M. Smith [7] and Brian E. Crucian [7]

1 JES Tech, Human Health and Performance Directorate, Houston, TX 77058, USA
2 Center for Infectious Disease Dynamics, Departments of Biology, Biochemistry and Molecular Biology, Huck Institute for the Life Sciences, Pennsylvania State University, University Park, PA 16802, USA; moriah@psu.edu (M.L.S.); shipleym8@gmail.com (M.M.S.); dwr19@psu.edu (D.W.R.)
3 GeoControl Systems, Human Health and Performance Directorate, Houston, TX 77054, USA; breedge23@yahoo.com
4 Aegis Aerospace, Human Health and Performance Directorate, Houston, TX 77058, USA; douglass.m.diak@nasa.gov
5 KBR, Human Health and Performance Directorate, Houston, TX 77058, USA; stephanie.s.krieger@nasa.gov (S.S.K.); mayra.a.nelman@nasa.gov (M.A.N.-G.)
6 University of Texas Medical Branch, Preventive Medicine and Population Health, Galveston, TX 77555, USA; sara.zwart-1@nasa.gov
7 National Aeronautics and Space Administration (NASA) Johnson Space Center, Human Health and Performance Directorate, Houston, TX 77058, USA; scott.m.smith@nasa.gov (S.M.S.); brian.crucian-1@nasa.gov (B.E.C.)
* Correspondence: satish.k.mehta@nasa.gov

Abstract: Human alpha herpesviruses herpes simplex virus (HSV-1) and varicella zoster virus (VZV) establish latency in various cranial nerve ganglia and often reactivate in response to stress-associated immune system dysregulation. Reactivation of Epstein Barr virus (EBV), VZV, HSV-1, and cytomegalovirus (CMV) is typically asymptomatic during spaceflight, though live/infectious virus has been recovered and the shedding rate increases with mission duration. The risk of clinical disease, therefore, may increase for astronauts assigned to extended missions (>180 days). Here, we report, for the first time, a case of HSV-1 skin rash (dermatitis) occurring during long-duration spaceflight. The astronaut reported persistent dermatitis during flight, which was treated onboard with oral antihistamines and topical/oral steroids. No HSV-1 DNA was detected in 6-month pre-mission saliva samples, but on flight day 82, a saliva and rash swab both yielded 4.8 copies/ng DNA and 5.3×10^4 copies/ng DNA, respectively. Post-mission saliva samples continued to have a high infectious HSV-1 load (1.67×10^7 copies/ng DNA). HSV-1 from both rash and saliva samples had 99.9% genotype homology. Additional physiological monitoring, including stress biomarkers (cortisol, dehydroepiandrosterone (DHEA), and salivary amylase), immune markers (adaptive regulatory and inflammatory plasma cytokines), and biochemical profile markers, including vitamin/mineral status and bone metabolism, are also presented for this case. These data highlight an atypical presentation of HSV-1 during spaceflight and underscore the importance of viral screening during clinical evaluations of in-flight dermatitis to determine viral etiology and guide treatment.

Keywords: herpes; viral reactivation; spaceflight; dermatitis; stress; immune depression

Citation: Mehta, S.K.; Szpara, M.L.; Rooney, B.V.; Diak, D.M.; Shipley, M.M.; Renner, D.W.; Krieger, S.S.; Nelman-Gonzalez, M.A.; Zwart, S.R.; Smith, S.M.; et al. Dermatitis during Spaceflight Associated with HSV-1 Reactivation. *Viruses* **2022**, *14*, 789. https://doi.org/10.3390/v14040789

Academic Editor: Oliver Schildgen

Received: 2 March 2022
Accepted: 25 March 2022
Published: 11 April 2022

1. Introduction

Over the last two decades, our studies have shown that astronauts exhibit persistent immune system dysregulation due to stress and other unique features associated with spaceflight [1–3]. Further, we have illustrated that multiple herpesviruses persistently reactivate in astronauts during space missions. This is evidenced by the shedding of viral

DNA in body fluids, namely saliva, before, during, and after both short (up to 16 days) and long (≥180 days) duration space missions [4,5]. About 50% of astronauts reactivate and shed viral DNA for one or more of the nine known human herpesviruses during and after spaceflight [6]. Four common herpes viruses that have been detected during space flight include Epstein Barr virus (EBV), Varicella Zoster virus (VZV), herpes simplex virus 1 (HSV-1), and cytomegalovirus (CMV). Though the typical shedding of viral DNA is asymptomatic in most astronauts regardless of mission duration, live/infectious virus has been recovered in some cases. Whether crew develop symptoms or not, virus reactivation rates increase with spaceflight duration. This represents a significant health risk to crew assigned to extended missions to Mars and beyond.

Stress depresses immunity, which contributes to latent herpesvirus reactivation. This correlation has been demonstrated in astronauts during space missions of variable durations over the last 20 years. Activation of the hypothalamic-pituitary-adrenal (HPA) axis and the sympathetic-adrenal-medullary (SAM) axis during spaceflight results in increased levels of stress hormones, including cortisol, dehydroepiandrosterone (DHEA), epinephrine, and norepinephrine [7–9]. Increases in these stress hormones, along with decreased cell-mediated immunity (as evidenced by decrements in cytotoxic T-cell function [10] and Th1 > Th2 cytokine shifts [11]), are contributing factors in the reactivation of latent herpes viruses in astronauts [12].

Here, we report for the first-time detection of HSV-1 in saliva and in skin lesion samples. These samples were taken from an astronaut suffering persistent dermatitis lasting approximately 200 days during a long-duration spaceflight (≥180 days) aboard the International Space Station (ISS). This astronaut reported rash onset at/around flight day 24 (FD24), with red, bumpy lesions appearing on the hands, arms, chest, back, and neck. The astronaut started antihistamines at that time, along with topical and oral hydrocortisone (fluocinonide 0.05% applied twice daily and fexofenadine HCl 180 mg daily). Twenty-nine days later, the astronaut developed a cold sore at FD53 that was subsequently treated with Valtrex (2000 mg PO q12h for 2 days). The astronaut continued to have rash along hands, arms, chest, back, and neck from which a lesion swab sample was collected at FD82. Finally, the astronaut was treated again for a cold sore with Valtrex on FD168. Interestingly, this astronaut had a previously documented history of atopic dermatitis [13]. While no lesion swab samples were collected during the previous mission, saliva samples taken in flight were positive for EBV and VZV. Immunologic, endocrine, biochemical, and viral sequencing data from this long-duration ISS spaceflight suggest a viral association to the symptoms observed in this astronaut.

2. Materials and Methods

2.1. Study Subject

The subject is an astronaut who participated in several concurrent NASA experiments, which were approved by the Johnson Space Center Institutional Review Board. Written informed consent was provided prior to data acquisition. The astronaut provided verbal and written approval to publish these data. The NASA Lifetime Surveillance of Astronaut Health Project personnel reviewed and approved the sharing of data among the experiments and approved the publication of these data. The astronaut completed over 9 months onboard the ISS. Biological samples were collected before, during, and after the space mission, as described in Figure 1.

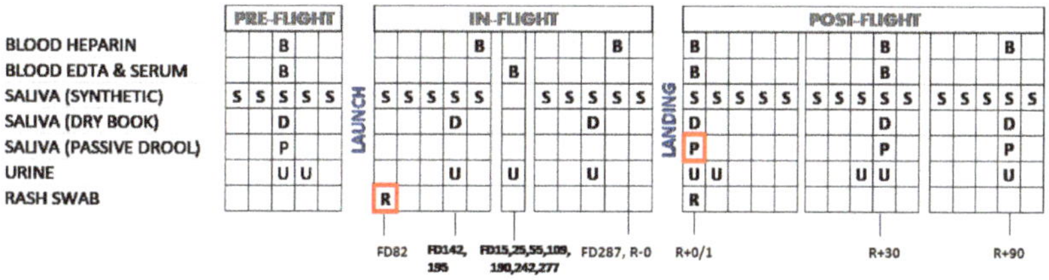

Figure 1. Schedule for collection of biological samples aboard ISS. Samples sent for sequencing are boxed in red. B—blood; S—liquid saliva; D—dry saliva; P—passive drool; U—urine; R—return after flight; FD—flight day. The subject was HSV-1 seropositive.

2.2. Saliva Sample Collection

Saliva was collected in three different ways (rolled, passive, and dry) to evaluate very specific outcome variables. Rolled saliva was collected for five consecutive days at each of the eight time points to evaluate viral load, alpha amylase, cortisol, and DHEA. Passive drool was collected to evaluate the viability of infectious virus therein. Dry saliva was collected to determine the cortisol and DHEA concentration changes occurring during the day from wake to sleep.

Saliva was collected for eight mission time points: two pre-flight (180 and 45 days before flight), three in-flight (early, FD120; mid, FD180; and late, FD240), and three post-flight (1, 30, and 90 days after return) using a synthetic salivette (SalivaBio Oral Swab (SOS), Salimetrics, State College, PA, USA) each time immediately after their sleep cycle, before eating and brushing their teeth [5]. These samples were stored frozen until processed for viral DNA analysis by real-time PCR. Samples collected in flight were stored frozen until returned to Earth by SpaceX vehicles and then transported to the Immunology Laboratory of Johnson Space Center for processing.

Diurnal dry saliva samples (5 per sampling day) were collected at awakening, wake +30 min, +6 h, +10 h, and then upon retiring, using a unique filter paper collection method. The subject wet the filter paper with saliva, which was then air-dried and stored at room temperature until return to Earth. All of a subject's samples were assayed in batch on the same plate. Filters were processed for cortisol and DHEA measurements as previously described [14]. Intra and inter-assay coefficients of variation for cortisol and DHEA were less than 5% and 10%, respectively, using this procedure.

To culture live virus, fresh saliva was collected by passive drool method one day after landing and processed immediately as follows: the sample was centrifuged, and the cell pellet was re-suspended in one milliliter of DMEM (Dulbecco's Modified Eagle Medium, Gibco, Invitrogen, Carlsbad, CA, USA) media. The cells were then plated atop human lung fibroblasts (HFL), then the culture vessel was centrifuged at 500 rpm for 5 min and then incubated at 37 °C with 5% humidified CO_2. The cultures were observed daily for the formation of viral plaques, which were in turn confirmed by HSV-1-specific antibody staining.

2.3. DNA Extraction from Biological Samples

A single peripheral blood sample using an 8.5 mL heparin anti-coagulated tube and a 24 h urine pool were also collected, each mission time point (Figure 1). Only one blood sample was taken at each in-flight time point. DNA was extracted from all the frozen and fresh saliva samples, as well as blood samples using a QIAamp DNA Blood Mini Kit (Qiagen, Hilden, Germany). DNA was extracted from urine samples using QIAamp Viral RNA mini kit (Qiagen, Hilden, Germany). The viral loads (viral copies/ng DNA recovered) for EBV, HSV-1, VZV, and CMV were determined by real-time PCR using TaqMan 7900 [4,5]. The primers and probes used for EBV, VZV, and CMV have been published previously [5].

2.4. Blood Samples

To correlate the reactivation of latent herpesviruses with immune system dysregulation, an immune assessment was conducted in conjunction with saliva collection. A blood sample was collected twice during flight; each collection was near the time of hatch closure for a returning Soyuz spacecraft. The blood samples were returned in ambient conditions to support live cellular functional analyses.

2.5. Immunology and Biochemical Assays

Peripheral leukocyte distribution, plasma cytokine analysis, T-cell function, and mitogen-stimulated cytokine profiles were determined as previously described [1,10]. Additional blood samples (using serum and EDTA plasma separator tubes) were collected at seven distinct time points as part of a separate 'Biochemical Profile' study activity (FD15, 25, 55, 109, 190, 242, and 277). Samples were centrifuged and immediately frozen in a −80 °C freezer onboard ISS for return to Earth as previously described [15]. Once the samples returned to Earth, the serum and plasma were aliquoted and frozen until batch analysis. Assays included general chemistry, vitamin and mineral status, hormone, bone metabolism, and renal stone assessments as described [16,17].

2.6. Skin Swab

A skin lesion swab was collected on FD82 using an EnviroTrans™ swab rinse kit containing 5 mL 0.85% saline with a swab (Hardy Diagnostics, Santa Maria, CA, USA). The sample was stored frozen at −80 °C until processed in the laboratory as follows: The sample was homogenized, and the cell pellet was separated by centrifuging at 14,000 rpm for 15 min. The DNA extraction for PCR was performed on one part of the sample while the other part was plated onto HFL cells (ATCC CCL-153) for viral culture. Viral load was measured in triplicate, and the average was normalized by the DNA concentration. DNA concentration was determined using a Qubit 2.0 Fluorometer and Invitrogen™ Quant-iT™ Qubit™ dsDNA HS Assay Kits (Invitrogen, Carlsbad, CA, USA). HSV-1 primers and probe sequences used for qPCR were as follows: forward primer (TGG TAT TGC CCA ACA CTT TCC), reverse primer (GCG CCA GGC ACA CAC AT), and probe (CGT GTC GCG TGT GGT).

2.7. Clinical Specimen DNA and HSV Genome Quantitation

Isolated DNA samples were sent to Pennsylvania State University for genome quantification. Samples included DNA isolated from both the cell pellet and supernatant of each virus-positive sample: one in-flight rash swab (FD 82) and one post-flight (R + 0) passive saliva sample. Following acquisition of the clinical specimens, 20 µg of linear polyacrylamide was added to each sample to serve as an inert co-precipitant. Total DNA was quantified using a Qubit® 2.0 fluorimeter with Qubit® compatible High-Sensitivity assay reagents (Invitrogen #Q32854). Viral DNA was quantified using qPCR to detect the type-common region of HSV glycoprotein B (gB) gene U_L27 as previously described [18,19]. This value was then used to infer the number of gB copies and, by inference, the total number of HSV copies in each sample.

2.8. Library Prep, Oligonucleotide Enrichment, and Illumina Deep Sequencing

Total sample DNA was sheared into ~800 base-pair (bp) fragments using a Covaris sonicator with settings as follows: 10% duty, power 60, 200 cycles/burst for 60 s at 4 °C. Sheared DNA was processed using the KAPA Hyperprep Kit (KAPA Biosystems #KR0961) compatible with Illumina® platform reagents. Following overnight ligation with Illumina indices at 4 °C, a post-ligation cleanup was performed before library DNA was amplified by PCR (10–14 cycles). The oligonucleotide bait library was a custom Arbor Biosciences target enrichment platform designed and validated in house [20]. The bait library included custom biotinylated DNA fragments (MyBaits®, Arbor Biosciences, Ann Arbor, MI, USA) that bind specifically to the DNA of HSV-1 strain 17 (GenBank JN555585.1). Each sample library was hybridized for ~40 h at 65 °C with the oligonucleotide probes, after which the baits and attached captured viral DNA were isolated using streptavidin-bound magnetic beads. Following enrichment, a post-capture PCR step (14 cycles) was performed before sequencing each bar-coded library on an Illumina MiSeq to obtain 300 bp paired-end sequence reads using v.3 chemistry.

2.9. De Novo Viral Genome Assembly

Following paired-end Illumina sequencing, the sequence reads were passed through a series of quality control filters to remove Illumina primers and adapters [21,22]. A BLAST database consisting of all known HSV-1 and HSV-2 genomes was constructed, and sequence reads with an e-value less than 10^2 were used to build consensus genomes. Consensus HSV-1 genomes were assembled de novo using a previously published viral genome assembly pipeline (VirGA) [23]. Briefly, eight SSAKE de novo assemblies were generated and combined into a draft genome for each sample using Celera and GapFiller [24–26]. The consensus HSV-1 genomes were annotated based on the HSV-1 reference genome (strain 17; GenBank ID JN555585) using sequence homology. The supernatant- and pellet-derived viral DNA was sequenced separately for each sample (in-flight swab and post-flight saliva), and the FASTQ data files were combined after Illumina sequencing. GenBank IDs and sequencing statistics for the two clinical genomes generated in this work are listed in Table 1.

Table 1. Sequencing statistics and GenBank IDs for two HSV-1 genomes from NASA astronaut.

Virus	GenBank ID	Input DNA ^ HSV Copy #	Enriched Library ^ HSV Copy #	Total # Reads *	% HSV	# Reads Used for Assembly *	Average Coverage
In-flight rash (FD82)	ON152715	1.7×10^3	5.7×10^5	7.8 million	80%	6.2 million	11,166X
Post-flight (R + 0 passive) saliva	ON152716	2.0×10^7	2.5×10^7	8.6 million	79%	6.8 million	11,828X

^ Values determined by qPCR on each sample after DNA isolation (see Methods for details). * All read counts refer to the number of forward reads only. "Total # reads" includes all forward reads, including unpaired and host-aligning reads. "# reads used for assembly" includes quality-trimmed, properly paired, HSV-specific forward reads.

2.10. Consensus Genome Comparison and Phylogenetic Analysis

Consensus genome comparisons between the two HSV-1 genomes were performed using trimmed versions of viral genomes (lacking the terminal repeats) so as to avoid over-representation of the internal and terminal repeats [27]. MAFFT was used to construct pairwise global nucleotide alignments between whole genome sequences [28]. ClustalW2 was used for pairwise amino acid alignments between open reading frames [29]. Custom Python scripts were used to calculate protein-coding differences and DNA variation between samples. The phylogenetic network was constructed with SplitsTree 4.14.5, using the uncorrected P distance and excluding all gaps [30]. See Supplementary Table S1 for GenBank IDs and references for HSV-1 isolates used to construct the phylogenetic network.

2.11. Minor Variant Detection and Validation

Each consensus genome was analyzed for the presence of minor variant (MV) loci, which are defined as alternative alleles that exist at a low frequency ($\leq$50%) in the viral population. VarScan v2.2.11 was used to detect MVs in each population of viral genomes [31]. Multiple parameters were applied to differentiate true MVs from potential sequencing artifacts [32]: minimum variant allele frequency $\geq$ 2% (0.02); base call quality $\geq$ 20; read depth at the position $\geq$ 100; independent reads supporting minor allele $\geq$ 5. Polymorphisms with directional strand bias $\geq$ 90% were excluded. SnpEff and SnpSift were used to annotate the MVs and identify their distribution and potential impact [33,34]. All MVs were visually inspected using the Integrative Genomics Viewer (IGV) v2.3.97 to verify raw sequence read support for the VarScan output.

2.12. Statistical Analyses

Pearson correlation constants were determined for relationships between angiotensin II and cytokines and markers of oxidative damage. One-way ANOVA with Sidak's multiple comparison tests was used to determine the significance of the averages of the 5 saliva samples collected at each time point compared to baseline (L-180) for DHEA, cortisol, and alpha amylase. Additionally, one-way ANOVA with Sidak's multiple comparison tests was used to calculate the significance of the averages between each phase of the mission (pre-, during-, and post-mission) for DHEA, cortisol, and alpha amylase.

3. Results

3.1. Stress Hormones and Biochemistry

The mean of salivary amylase, cortisol, and DHEA measured before, during, and after the flight are given in Figure 2. Salivary cortisol, a marker of activation of the HPA axis, revealed higher concentrations during flight, FD120, FD180, and FD240 as compared to before and after flight (Figure 2). A statistically significant increase was observed at FD240 (0.301 ± 0.042 µg/DL) when compared to baseline at L-180 (0.172 ± 0.040 µg/DL; $p = 0.023$). Circadian rhythm of cortisol done based upon five samples collected during the day (explained in Methods section for dry saliva) did not show any significant changes at any of the time points measured during the mission. Salivary DHEA also showed an increasing trend during early and mid-flight but did not show any significant changes between the phases. No significant increase in alpha amylase, a biomarker for assessing acute psychological stress in humans due to activation of the sympathetic nervous system, was observed during flight as compared to pre-flight at any of the time points tested. However, when combining the pre-flight, during-flight, and post-flight sessions, post-flight alpha amylase levels (mean ± SEM 20.1 ± 1.89 U/mL) were significantly decreased compared to both pre-flight (mean ± SEM 38.4 ± 7.48 U/mL; $p = 0.046$) and during flight (mean ± SEM 46.6 ± 6.95 U/mL; $p = 0.004$).

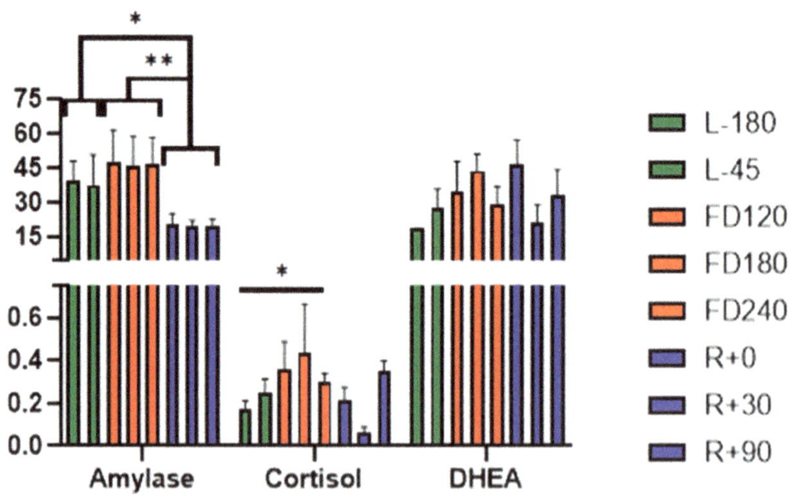

Figure 2. Mean distribution of salivary alpha amylase, cortisol, and DHEA before (green), during (red), and after (blue) the mission. Pre-mission samples are indicated by "L-" referencing the number of days before launch. In-flight mission samples are indicated as FD, flight day. Post-mission samples are similarly indicated by "R+" for the days post-return. Statistical significance is indicated by * $p < 0.05$ and ** $p < 0.01$.

Several analytes increased during flight are consistent with an inflammatory response and/or oxidative stress, including ferritin, and serum and urinary cortisol (Supplementary Table S2). Folate status was at the low end of the clinical normal range during the mission (in-flight concentrations ranged from 21 early during flight to 13 nmol/L by the end of the mission, normal range 21–64 nmol/L). Vitamin B6 (pyridoxal 5-phosphate) decreased over the course of the mission from 48 nmol/L before flight to 16 on landing day (normal range 11.3–302 nmol/L). Angiotensin II increased as much as 80% early during flight, and this mirrored many of the inflammatory cytokines (see raw data in Supplementary Table S2—Biochemical profile). Angiotensin II was correlated with TNFα (Pearson $r = 0.60$, $p = 0.04$) and markers of oxidative damage (oxidized LDL, Pearson $r = 0.72$, $p = 0.07$; 8-hydroxy 2′-deoxyguanosine, Pearson $r = 0.84$, $p = 0.02$).

3.2. Immune Status

The astronaut subject displayed a fairly typical pattern of immune changes previously characterized as being associated with spaceflight [4,10,11]. There was little alteration in the distribution of peripheral leukocytes and a mild reduction in T-cell function as determined by the induction of cellular activation antigens following mitogenic stimulation via Staphylococcal enterotoxins or antibodies to CD3 and CD28 as previously described [1,10] (data not shown). Mitogen-stimulated cytokine profiles did not display the characteristic reduction in cytokine induction observed in previous astronaut studies (data not shown). An assessment of plasma cytokine concentrations revealed an elevation during mission in several cytokines, which is not atypical for astronauts during spaceflight (Figure 3). Plasma cytokine concentrations across three pre-flight samples were very consistent. Several inflammatory cytokines, including IL-1, IL-1ra, and IL-12, as well as cytokines associated with adaptive immune responses including IFNα, IFNγ, and IL-4, were elevated throughout the mission, Figure 3.

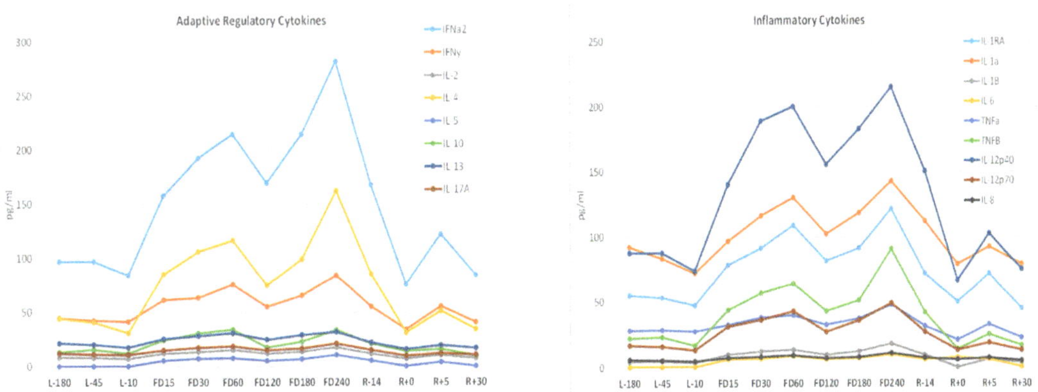

(a) (b)

Figure 3. Adaptive regulatory (**a**) and inflammatory (**b**) plasma cytokines concentrations (pg/mL) measured via multiplex array before, during, and after the spaceflight. Pre-flight samples are indicated by "L-" referencing the number of days before launch. In-flight (flight day) mission samples are indicated by "FD" followed by the number of days post launch. Post-flight samples are similarly indicated by "R+" for the days post-return.

3.3. Viral PCR Analysis

The levels of three herpesviruses, HSV-1, EBV, and VZV, were quantified in 40 saliva samples collected at different times before, during, and after the spaceflight by real-time PCR. No EBV or VZV was detected by real-time PCR in any of the samples collected before, during, and after spaceflight. None of the eight 24 h urine samples collected during this mission was positive for CMV. Even though no HSV-1 was detected before the flight, the saliva sample collected on the 82nd day of the spaceflight was positive for HSV-1 (4.8 copies/ng DNA). The rash swab collected from a skin lesion from the neck region on the same flight day was also positive for HSV-1 (5.3×10^4 copies/ng DNA). Saliva collected at landing (R + 0) showed 1.67×10^7 copies/ng of total DNA. The post-flight passive saliva yielded infectious HSV-1, as observed by culturing on HFL cells. Post-recovery saliva and skin swab samples collected 30 and 90 days after spaceflight did not test positive for viral DNA, which coincided with negative virus recovery by culture as well as a lack of symptoms.

3.4. Viral Genome Sequencing and Comparison

In order to gain insight on whether the in-flight rash associated with HSV-1 shedding was identical to the viral DNA detected in saliva at landing post-flight (R + 0), we subjected DNA isolated from these samples to library preparation and oligonucleotide-bait-based enrichment for HSV-1 DNA. We then used this virus-enriched library for high-throughput deep viral genome sequencing (see Table 1 for DNA isolation yield, enrichment, and genome metrics). The resulting sequence data were de novo assembled into full-length HSV-1 genomes using our previously published methods [23]. Comparison of these genomes to each other revealed them to be 99.9% identical at the consensus genome level, i.e., the most frequent base detected at each position in the viral genome. We compared the astronaut's two viral consensus genomes to a set of 51 previously described and sequenced HSV-1 genomes (Supplementary Table S1), using a network graph (Figure 4). This analysis revealed the astronaut's HSV-1 genomes to be most similar to HSV-1 genomes previously collected in China (CR38, HM585508 [27]) and Russia (L2, KT780616 [35]) (Figure 4). While the Russian sample is extant, the CR-38 sample was collected in the 1980s [27]. Relative to this collection of known viral genomes, there were only four novel and unique amino acid (AA) coding variations in these astronaut HSV-1s (Table 2).

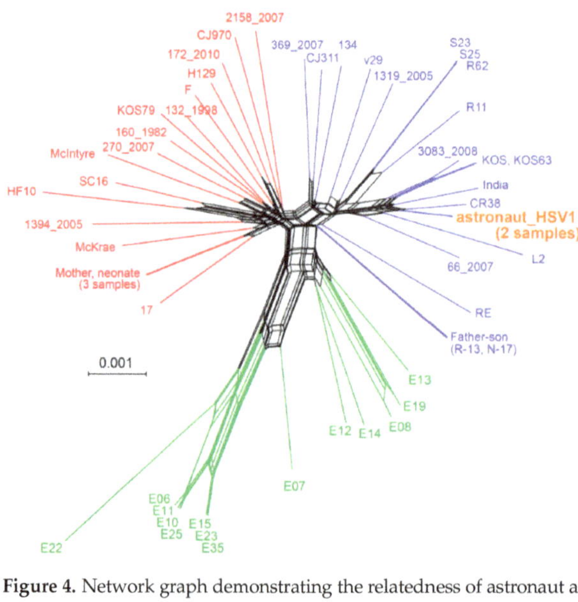

Figure 4. Network graph demonstrating the relatedness of astronaut and other HSV-1 genomes. The two astronaut-derived HSV-1 genomes (orange) clustered nearest to HSV-1 genomes from China (CR-38)

and Russia (L2). This graph-based SplitsTree network was generated from an alignment of the astronaut-derived HSV-1 genomes with 51 HSV-1 genomes that encompass the global genetic diversity of this virus. The Φ statistical test in SplitsTree4 found statistically significant evidence for recombination ($p = 0.0$), as expected based on prior analyses of HSV-1 phylogenies. The overall geographical origins of the prior HSV-1 genomes are European and North American (red); European, North American, Asian (blue), and African (green). The scale bar represents 0.1% nucleotide divergence. Supplementary Table S1 includes the names, accessions, and geographic origins of each strain.

Table 2. Unique amino acid variants found in astronaut's HSV-1 genomes.

Gene	Gene Product	AA Alignment Position	Astronaut HSV-1 AA	Wildtype AA
UL8	DNA helicase/primase	595	T	A
RL1	ICP34.5, neurovirulence factor	155	T	A
US2	Unknown function	166	Q	P
US5	Glycoprotein J	52	A	V

3.5. Minor Variants Indicate Higher Viral Genetic Diversity for In-Flight Sample

Since prior studies of direct-from-patient viral shedding have revealed within-host variation in the viral genome population, we next investigated whether the deep sequencing coverage of the astronaut's HSV-1 genomes would reveal any variation within the virus population of each sample (average coverage > 11,000-fold; see Table 2). We used a conservative limit of detection to identify only those minor variants present in ≥2% of the population (see Methods for full criteria). This analysis identified minor variants (MV) at 366 sites in the in-flight rash HSV-1 population, but only 24 sites in the post-flight saliva HSV-1 population (Figure 5). These minor variants were distributed throughout the viral genome (Figure 5). With the exception of four synonymous MVs that occur in a tandem repeat (the "PQ" repeats) in the gene UL36, encoding the tegument protein VP1/2, there were no MVs in common between these two samples.

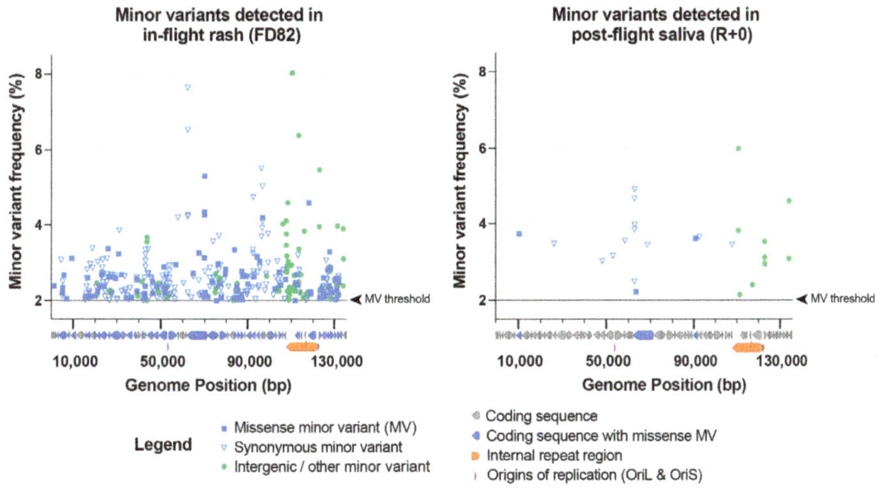

(a) (b)

Figure 5. The in-flight rash (FD82) HSV-1 sample contained far more minor variants in the viral genome population than the post-flight saliva (R + 0) HSV-1 sample. This included (**a**) 366 MVs in the in-flight rash vs. (**b**) 24 in the post-flight saliva HSV-1 samples. Minor variants (MVs) are plotted based on their location within the HSV-1 genome (x-axis) and the frequency of the detected MV allele (y-axis) at each position. Minor variants are color-coded according to their classification as missense (genic), synonymous (genic), or intergenic/other. A black line and arrowhead (right of each plot) indicate the MV detection threshold of 2% (0.02). A diagram of HSV-1 coding sequences is located below each x-axis to highlight the coding sequences that contain missense minor variants. Supplementary Table S3 includes the precise location, coverage depth (i.e., forward and reverse reads supporting the major vs. the minor allele), and a list of specific MV impacts within each gene.

4. Discussion

Herpes simplex virus-1 (HSV-1) is a highly prevalent and communicable neurotropic alpha herpesvirus that establishes latency in cranial nerve ganglia. This virus persists lifelong as a latent virus and is commonly referred to as oral herpes. Reactivation of HSV-1 can be asymptomatic or can lead to lesions or skin rashes at sites corresponding to the infected innervating ganglion. The most typical presentation is characterized by an oral "cold sore" lesion. Reactivation of latent herpesviruses has been reported in previous spaceflight studies, and it largely is an asymptomatic phenomenon. However, in some astronauts, herpesvirus reactivation may potentially lead to clinical diseases such as zoster and dermatitis. Here, we report for the first time, detection of HSV-1 both in saliva and in skin lesions of an astronaut during long-duration spaceflight. The astronaut evaluated had persistent dermatitis during flight, with atypical presentation but also cold sores, which were treated with a combination of antihistamines, topical and oral steroids, and Valtrex. No HSV-1 DNA was detected in the astronaut's samples collected before flight at 180 and 45 days before launch. Samples collected during spaceflight were found to have a viral HSV-1 load by real-time PCR, with 4.8 copies/ng in a saliva sample and 5.3×10^4 copies/ng in a lesion swab sample. The post-landing saliva samples yielded infectious HSV-1 with extremely high copy numbers (1.67×10^7 copy/ng DNA). This was confirmed by specific antibody staining as well as by PCR. HSV-1 genomes recovered from the in-flight skin lesion and post-orbit saliva were genotyped and found to have 99.9% homology.

It is generally established that reductions in immunocompetence, including T-cell function, can result in loss of "control" of latent herpesviruses and thus allow their reactivation. Such immune suppression is also typically associated terrestrially with either stress or aging. However, astronauts who are typically very fit individuals in an extreme environment also experience latent herpesvirus reactivation associated with immune suppression. This phenomenon likely results from a synergy of stress, microgravity, radiation, and/or prolonged isolation. As observed in space missions with previous astronauts, this astronaut did show increases in several salivary markers of stress (Figure 2) during flight compared to pre- and post-flight, confirming physiological stress changes within this extreme situation. Surprisingly, the current subject's stress levels did not manifest the broad suppression in T-cell function observed in previous studies by the reduction in mitogen-stimulated cytokine production. However, a more atypical immune dysregulation did occur, as evidenced by the elevation in many plasma cytokines spanning the entire mission duration (Figure 3). We speculate that this astronaut subject retained immune cellular functions and stress hormone stability more so than the astronaut norm, yet the dysregulation and stress imbalance was still sufficient to manifest HSV-1 reactivation and shedding. Alternatively, the reactivation may be associated with loss of virus control by an alternate and unmonitored cellular mechanism such as angiotensin II discussed below.

Our comparison of in-flight vs. post-flight HSV-1 genomes revealed that these viral genomes were nearly identical to one another (99.94%) at the consensus level. However, the in-flight HSV-1 genome had strikingly more minor variants (366) than the post-flight sample (24; see Figure 5). This numeric difference alone suggests that there are distinct environmental or clinical impacts on these viral populations during spaceflight. From an environmental perspective, the in-flight sample contains viral genomes that replicated in the presence of ionizing radiation, which may have impacted the rate of double-strand

DNA breaks, recombination, and DNA repair, as well as impacts on the host DNA damage response [36–39]. However, we cannot untangle that possibility from other clinical distinctions between these samples, such as their different status of viral reactivation. For instance, an active lesion and ongoing rash during space flight may entail different viral and host responses than asymptomatic viral shedding in saliva after landing. There is also a potential difference in immune surveillance, such as inflammatory or immune cell-killing states, at the site of a skin lesion versus at salivary sites of asymptomatic viral shedding. Distinguishing between these possibilities will necessitate analysis of further in-orbit shedding samples, both asymptomatic and symptomatic, as well as additional comparisons of pre- and post-flight samples from future space travel.

Nutritional status decreased during flight for some nutrients, particularly folate, vitamin B6, niacin, and vitamin C (see Supplementary Table S2). Total and LDL cholesterol increased as well as triglycerides. The increase in angiotensin II could have contributed to an inflammatory process that contributes to viral reactivation. Angiotensin II exerts an inflammatory effect on leukocytes and endothelial cells, contributing to reactive oxygen species formation, adhesion molecule, and chemokine release [40]. Angiotensin II also modulates T-cell activation through autocrine actions from endogenous production. T cells can produce angiotensin II, which stimulates the production of superoxide, which promotes T cells to produce TNFα [41]. In the current subject, angiotensin II concentration was positively correlated with both TNFα and markers of oxidative damage (oxLDL and 8-hydroxy 2′-deoxyguanosine). Supportive of this, there is a higher prevalence of HSV 1 activation in patients with chronic kidney disease as an inflammatory stimulus where angiotensin II is high [42].

Regardless of the cause, this subject showed symptomatic in-flight HSV-1 viral reactivation and asymptomatic, continued shedding of infectious HSV-1 virus upon landing, which potentially could have transferred and caused clinical disease in crew contacts, including uninfected or immunocompromised individuals, as well as newborn infants. Specifically, infants exposed to HSV-1 are at increased risk for developing severe and life-threatening complications such as fatal organ damage (liver, lungs, and heart), as well as viral meningitis and sores on the face and eyes. Thus, it is essential to develop spaceflight countermeasures to prevent HSV-1 reactivations and ensure the health of the crew, as well as the health of their contacts upon return. One countermeasure that has proven efficacious is viral-specific therapeutic vaccines such as Zostavax against VZV reactivation. Our previous work on the reactivation of VZV in astronauts led to the implementation of Zostavax as a prophylactic countermeasure to herpes zoster. Since its implementation in 2013, very little VZV reactivation has been observed in crew samples. Interestingly, with the reduction in VZV reactivation, we have observed an increasing incidence of subclinical HSV-1 reactivation in crew samples. Historically, HSV-1 reactivation rates have been very low, but they now seem to be on the rise, suggesting a potential inverse correlational relationship with VZV reactivation. That may be true for the subject of this case study. In a previous long-duration mission, this astronaut had dermatitis and VZV and EBV shedding, although EBV would be an unlikely source of skin lesions. In the current study, this astronaut had dermatitis that has been directly linked to HSV-1 reactivation, and there was no evidence, in any of the biological samples, for VZV reactivation. No VZV vaccination was given between the missions. Unfortunately, at this time, no HSV-1 vaccine is available, but several therapeutic live-attenuated and subunit vaccines are in clinical or pre-clinical trials [43]. Another potential countermeasure against both HSV-1 and VZV, especially for individuals for whom vaccines are either ineffective or contraindicated, could be prophylactic administration of an antiviral drug (valacyclovir or Valtrex). Interestingly, the astronaut from this study was treated with Valtrex (2000 mg PO q12h for 2 days) during this mission (flight days 53 and 168) upon outbreak of cold sores, but it did not seem to be efficacious. Perhaps prophylactic pre-dosing and/or using a higher dose during outbreak could have reduced or eliminated the viral and symptom burden. To that end, we are currently studying the effectiveness of this prophylactic antiviral countermeasure to potentially prevent VZV and HSV-1 (as well as EBV) shedding in Antarctic expeditioners, who have similar patterns of viral DNA shedding in saliva as astronauts. These and other countermeasures (i.e., specific resistive and aerobic exercise, nutritional supplementation, and stress relieving exercise) must continually be evaluated and updated to ensure the health and safety of our astronauts in space and their contacts upon return.

5. Conclusions

These data confirm that in some cases, the viral reactivation, stress imbalance, and immune dysregulation experienced by astronauts during spaceflight is not entirely a subclinical phenomenon and may indeed be associated with clinical processes requiring treatment. For this case, dermatological lesions present during space flight and immediately after returning to Earth coincided with elevated stress markers, circulating inflammatory cytokines, and HSV-1 DNA shedding in saliva and lesion swab. HSV-1 genomes recovered from in-flight skin lesions and post-orbit saliva were deep-sequenced and found to be nearly identical at the consensus genome level. However, the in-flight rash sample contained a far more diverse population of viral genomes than the post-orbit saliva sample. The reasons for this are not known but may include differences in immune activation state, exposure to ionizing radiation in space, or lesion vs. asymptomatic shedding. Astronaut saliva contains increasingly significant viral DNA, which can be infectious during and after spaceflight. For that reason, and in response to the data from the current astronaut subject indicating that the persistent skin rash may have a viral etiology, we recommend prophylactic (vaccine and/or antiviral) treatment, where available, to the astronauts before they go into space as a countermeasure. Further, a broader package of countermeasures to reduce viral reactivation and restore immune function should be considered for the upcoming "Artemis" deep space missions.

Supplementary Materials: The following supporting information can be downloaded at: https://www.mdpi.com/article/10.3390/v14040789/s1, Table S1: GenBank accession numbers and references; Table S2: Biochemical profile; Table S3: Minor variant details.

Author Contributions: Conceptualization, S.K.M. and B.E.C.; methodology, S.K.M. and B.E.C.; formal analysis, S.S.K., M.A.N.-G., B.V.R., M.L.S., M.M.S., D.W.R., S.R.Z. and S.M.S.; investigation, S.S.K., M.A.N.-G., B.V.R., M.M.S., D.W.R., S.R.Z. and S.M.S.; resources, S.S.K., M.A.N.-G., B.V.R., M.M.S. and D.W.R.; writing—original draft preparation, S.K.M., B.V.R. and M.L.S.; writing—review and editing, S.K.M., M.L.S., B.V.R., D.M.D., S.R.Z., S.M.S. and B.E.C.; visualization, S.K.M., M.L.S., B.V.R., D.M.D., D.W.R., S.R.Z., S.M.S. and B.E.C.; supervision, S.K.M.; funding acquisition, S.K.M., M.L.S. and B.E.C. All authors have read and agreed to the published version of the manuscript.

Funding: This research was funded by the NASA Human Research Program, Human Health and Countermeasures Element, via a directed study to Brian E. Crucian, and by National Institutes of Health R01 AI132692 (to M.L.S.).

Institutional Review Board Statement: The study was conducted in accordance with the Declaration of Helsinki and approved by the Institutional Review Board of NASA (PRO1611; last approved on 20 May 2021).

Informed Consent Statement: Informed consent was obtained from the subject involved in the study. Verbal and written informed consent were obtained from the patient to publish this paper.

Data Availability Statement: The data presented in this study are available on request from the corresponding author and approval through NASA. The data are not publicly available, with the exception of the two viral genomes deposited in GenBank.

Acknowledgments: The authors acknowledge the participation and cooperation of the crewmember in this study. Stress hormones were analyzed by the late Mark Laudenslager, Ph.D., from the University of Colorado, Denver, CO. We also acknowledge the late Randall Cohrs, Ph.D., for many valuable discussions with him on the study.

Conflicts of Interest: The authors declare no conflict of interest. The funders had no role in the design of the study, in the collection, analyses, or interpretation of data, in the writing of the manuscript, or in the decision to publish the results.

Legal: Trade names and trademarks are used in this report for identification only. Their usage does not constitute an official endorsement, either expressed or implied, by the National Aeronautics and Space Administration.

References

1. Crucian, B.; Stowe, R.; Mehta, S.; Uchakin, P.; Quiriarte, H.; Pierson, D.; Sams, C. Immune system dysregulation occurs during short duration spaceflight on board the space shuttle. *J. Clin. Immunol.* **2013**, *33*, 456–465. [CrossRef] [PubMed]
2. Crucian, B.E.; Stowe, R.P.; Pierson, D.L.; Sams, C.F. Immune System Dysregulation Following Short- vs Long-Duration Spaceflight. *Aviat. Space Environ. Med.* **2008**, *79*, 835–843. [CrossRef] [PubMed]
3. Crucian, B.; Sams, C. Immune system dysregulation during spaceflight: Clinical risk for exploration-class missions. *J. Leukoc. Biol.* **2009**, *86*, 1017–1018. [CrossRef] [PubMed]
4. Mehta, S.K.; Laudenslager, M.L.; Stowe, R.P.; Crucian, B.E.; Feiveson, A.H.; Sams, C.F.; Pierson, D.L. Latent virus reactivation in astronauts on the international space station. *NPJ Microgravity* **2017**, *3*, 11. [CrossRef]
5. Mehta, S.K.; Laudenslager, M.L.; Stowe, R.P.; Crucian, B.E.; Sams, C.F.; Pierson, D.L. Multiple latent viruses reactivate in astronauts during Space Shuttle missions. *Brain Behav. Immun.* **2014**, *41*, 210–217. [CrossRef]
6. Rooney, B.V.; Crucian, B.E.; Pierson, D.L.; Laudenslager, M.L.; Mehta, S.K. Herpes Virus Reactivation in Astronauts During Spaceflight and Its Application on Earth. *Front. Microbiol.* **2019**, *10*, 16. [CrossRef]
7. Stowe, R.P.; Pierson, D.L.; Barrett, A.D.T. Elevated Stress Hormone Levels Relate to Epstein-Barr Virus Reactivation in Astronauts. *Psychosom. Med.* **2001**, *63*, 891–895. [CrossRef]
8. Stowe, R.P.; Pierson, D.L.; Mehta, S.K. *Stress Challenges and Immunity in Space*; Springer: Berlin/Heidelberg, Germany, 2012. [CrossRef]
9. Stowe, R.P.; Sams, C.F.; Pierson, D.L. Adrenocortical and immune responses following short- and long-duration spaceflight. *Aviat. Space Environ. Med.* **2011**, *82*, 627–634. [CrossRef]
10. Crucian, B.; Stowe, R.P.; Mehta, S.; Quiriarte, H.; Pierson, D.; Sams, C. Alterations in adaptive immunity persist during long-duration spaceflight. *NPJ Microgravity* **2015**, *1*, 15013. [CrossRef]
11. Crucian, B.E.; Zwart, S.R.; Mehta, S.K.; Uchakin, P.; Quiriarte, H.D.; Pierson, D.L.; Sams, C.F.; Smith, S.M. Plasma Cytokine Concentrations Indicate That In Vivo Hormonal Regulation of Immunity Is Altered During Long-Duration Spaceflight. *J. Interf. Cytokine Res.* **2014**, *34*, 778–786. [CrossRef]
12. Mehta, S.; Crucian, B.; Stowe, R.; Simpson, R.; Ott, C.; Sams, C.; Pierson, D. Reactivation of latent viruses is associated with increased plasma cytokines in astronauts. *Cytokine* **2013**, *61*, 205–209. [CrossRef] [PubMed]
13. Crucian, B.; Johnston, S.; Mehta, S.; Stowe, R.; Uchakin, P.; Quiriarte, H.; Pierson, D.; Laudenslager, M.L.; Sams, C. A case of persistent skin rash and rhinitis with immune system dysregulation onboard the International Space Station. *J. Allergy Clin. Immunol. Pr.* **2016**, *4*, 759–762.e8. [CrossRef] [PubMed]
14. Laudenslager, M.L.; Calderone, J.; Philips, S.; Natvig, C.; Carlson, N.E. Diurnal patterns of salivary cortisol and DHEA using a novel collection device: Electronic monitoring confirms accurate recording of collection time using this device. *Psychoneuroendocrinology* **2013**, *38*, 1596–1606. [CrossRef] [PubMed]
15. Zwart, S.R.; Booth, S.L.; Peterson, J.W.; Wang, Z.; Smith, S.M. Vitamin K status in spaceflight and ground-based models of spaceflight. *J. Bone Miner. Res.* **2011**, *26*, 948–954. [CrossRef] [PubMed]
16. Morgan, J.L.L.; Zwart, S.R.; Heer, M.; Ploutz-Snyder, R.; Ericson, K.; Smith, S.M. Bone metabolism and nutritional status during 30-day head-down-tilt bed rest. *J. Appl. Physiol.* **2012**, *113*, 1519–1529. [CrossRef] [PubMed]
17. Smith, S.M.; Heer, M.; Wang, Z.; Huntoon, C.L.; Zwart, S.R. Long-Duration Space Flight and Bed Rest Effects on Testosterone and Other Steroids. *J. Clin. Endocrinol. Metab.* **2012**, *97*, 270–278. [CrossRef] [PubMed]
18. Johnston, C.; Zhu, J.; Jing, L.; Laing, K.J.; McClurkan, C.M.; Klock, A.; Diem, K.; Jin, L.; Stanaway, J.; Tronstein, E.; et al. Virologic and Immunologic Evidence of Multifocal Genital Herpes Simplex Virus 2 Infection. *J. Virol.* **2014**, *88*, 4921–4931. [CrossRef]
19. Ryncarz, A.J.; Goddard, J.; Wald, A.; Huang, M.-L.; Roizman, B.; Corey, L. Development of a High-Throughput Quantitative Assay for Detecting Herpes Simplex Virus DNA in Clinical Samples. *J. Clin. Microbiol.* **1999**, *37*, 1941–1947. [CrossRef]
20. Shipley, M.; Rathbun, M.M.; Szpara, M.L. Oligonucleotide Enrichment of HSV-1 Genomic DNA from Clinical Specimens for Use in High-Throughput Sequencing. *Metab. Cancer* **2020**, *2060*, 199–217. [CrossRef]
21. Szpara, M.L.; Tafuri, Y.R.; Parsons, L.; Shamim, S.R.; Verstrepen, K.J.; Legendre, M.; Enquist, L.W. A Wide Extent of Inter-Strain Diversity in Virulent and Vaccine Strains of Alphaherpesviruses. *PLoS Pathog.* **2011**, *7*, e1002282. [CrossRef]
22. Szpara, M.L.; Parsons, L.; Enquist, L.W. Sequence Variability in Clinical and Laboratory Isolates of Herpes Simplex Virus 1 Reveals New Mutations. *J. Virol.* **2010**, *84*, 5303–5313. [CrossRef] [PubMed]
23. Parsons, L.R.; Tafuri, Y.R.; Shreve, J.T.; Bowen, C.D.; Shipley, M.M.; Enquist, L.W.; Szpara, M.L. Rapid genome assembly and comparison decode intrastrain variation in human alphaherpesviruses. *mBio* **2015**, *6*, e02213-14. [CrossRef] [PubMed]

24. Myers, E.W.; Sutton, G.G.; Delcher, A.L.; Dew, I.M.; Fasulo, D.P.; Flanigan, M.J.; Kravitz, S.A.; Mobarry, C.M.; Reinert, K.H.J.; Remington, K.A.; et al. A Whole-Genome Assembly of Drosophila. *Science* **2000**, *287*, 2196–2204. [CrossRef] [PubMed]
25. Boetzer, M.; Pirovano, W. Toward almost closed genomes with GapFiller. *Genome Biol.* **2012**, *13*, R56. [CrossRef] [PubMed]
26. Warren, R.L.; Sutton, G.G.; Jones, S.J.M.; Holt, R.A.; Bateman, A. Assembling millions of short DNA sequences using SSAKE. *Bioinformatics* **2007**, *23*, 500–501. [CrossRef]
27. Szpara, M.L.; Gatherer, D.; Ochoa, A.; Greenbaum, B.; Dolan, A.; Bowden, R.J.; Enquist, L.W.; Legendre, M.; Davison, A.J. Evolution and Diversity in Human Herpes Simplex Virus Genomes. *J. Virol.* **2014**, *88*, 1209–1227. [CrossRef]
28. Kazutaka, K.; Misakwa, K.; Kei-ichi, K.; Miyata, T. MAFFT: A novel method for rapid multiple sequence alignment based on fast Fourier transform. *Nucleic Acids Res.* **2002**, *30*, 3059–3066. [CrossRef]
29. Larkin, M.A.; Blackshields, G.; Brown, N.P.; Chenna, R.; McGettigan, P.A.; McWilliam, H.; Valentin, F.; Wallace, I.M.; Wilm, A.; Lopez, R.; et al. Clustal W and Clustal X version 2.0. *Bioinformatics* **2007**, *23*, 2947–2948. [CrossRef]
30. Huson, D.H. SplitsTree: Analyzing and visualizing evolutionary data. *Bioinformatics* **1998**, *14*, 68–73. [CrossRef]
31. Koboldt, D.C.; Zhang, Q.; Larson, D.E.; Shen, D.; McLellan, M.D.; Lin, L.; Miller, C.A.; Mardis, E.R.; Ding, L.; Wilson, R.K. VarScan 2: Somatic mutation and copy number alteration discovery in cancer by exome sequencing. *Genome Res.* **2012**, *22*, 568–576. [CrossRef]
32. Depledge, D.P.; Kundu, S.; Jensen, N.J.; Gray, E.R.; Jones, M.; Steinberg, S.; Gershon, A.; Kinchington, P.R.; Schmid, D.S.; Balloux, F.; et al. Deep Sequencing of Viral Genomes Provides Insight into the Evolution and Pathogenesis of Varicella Zoster Virus and Its Vaccine in Humans. *Mol. Biol. Evol.* **2013**, *31*, 397–409. [CrossRef] [PubMed]
33. Cingolani, P.; Platts, A.; Wang, L.L.; Coon, M.; Nguyen, T.; Wang, L.; Land, S.J.; Lu, X.; Ruden, D.M. A program for annotating and predicting the effects of single nucleotide polymorphisms. *Fly* **2012**, *6*, 80–92. [CrossRef] [PubMed]
34. Cingolani, P.; Patel, V.M.; Coon, M.; Nguyen, T.; Land, S.J.; Ruden, D.M.; Lu, X. Using Drosophila melanogaster as a Model for Genotoxic Chemical Mutational Studies with a New Program, SnpSift. *Front. Genet.* **2012**, *3*, 35. [CrossRef] [PubMed]
35. Skoblov, M.Y.; Lavrov, A.V.; Bragin, A.G.; Zubtsov, D.A.; Andronova, V.L.; Galegov, G.A.; Skoblov, Y.S. The genome nucleotide sequence of herpes simplex virus 1 strain L2. *Russ. J. Bioorganic Chem.* **2017**, *43*, 140–142. [CrossRef]
36. Moreno-Villanueva, M.; Wong, M.; Lu, T.; Zhang, Y.; Wu, H. Interplay of space radiation and microgravity in DNA damage and DNA damage response. *NPJ Microgravity* **2017**, *3*, 14. [CrossRef]
37. Horneck, G.; Rettberg, P.; Kozubek, S.; Baumstark-Khan, C.; Schmitz, C.; Rink, H.; Schmitz, M.S. The Influence of Microgravity on Repair of Radiation-Induced DNA Damage in Bacteria and Human Fibroblasts. *Radiat. Res.* **1997**, *147*, 376. [CrossRef]
38. Horneck, G. Impact of microgravity on radiobiological processes and efficiency of DNA repair. *Mutat. Res. Mol. Mech. Mutagen.* **1999**, *430*, 221–228. [CrossRef]
39. Horneck, G.; Klaus, D.M.; Mancinelli, R.L. Space Microbiology. *Microbiol. Mol. Biol. Rev.* **2010**, *74*, 121–156. [CrossRef]
40. Dandona, P.; Dhindsa, S.; Ghanim, H.; Chaudhuri, A. Angiotensin II and inflammation: The effect of angiotensin-converting enzyme inhibition and angiotensin II receptor blockade. *J. Hum. Hypertens.* **2006**, *21*, 20–27. [CrossRef]
41. Hoch, N.E.; Guzik, T.J.; Chen, W.; Deans, T.; Maalouf, S.A.; Gratze, P.; Weyand, C.; Harrison, D.G. Regulation of T-cell function by endogenously produced angiotensin II. *Am. J. Physiol. Integr. Comp. Physiol.* **2009**, *296*, R208–R216. [CrossRef]
42. Pallos, D.; Ruivo, G.F.; Ferrari-Junior, S.H.; Pannuti, C.S.; Perozini, C.; Sarmento, D.J.S.; Palmieri, M.; Souza, A.C.M.F.; Tozetto-Mendoza, T.R.; Doglio, A.; et al. Periodontal disease and detection of human herpesviruses in saliva and gingival crevicular fluid of chronic kidney disease patients. *J. Periodontol.* **2020**, *91*, 1139–1147. [CrossRef] [PubMed]
43. Krishnan, R.; Stuart, P.M. Developments in Vaccination for Herpes Simplex Virus. *Front. Microbiol.* **2021**, *12*, 798927. [CrossRef] [PubMed]

viruses

MDPI

Article

Human Cytomegalovirus Infection Elicits Global Changes in Host Transcription by RNA Polymerases I, II, and III

Christopher B. Ball [1], Mrutyunjaya Parida [1], Ming Li [2], Benjamin M. Spector [1], Gustavo A. Suarez [1], Jeffery L. Meier [2] and David H. Price [1,*]

[1] Department of Biochemistry and Molecular Biology, University of Iowa, Iowa City, IA 52242, USA; christopher-ball-1@uiowa.edu (C.B.B.); mrutyunjaya-parida@uiowa.edu (M.P.); benjamin-m-spector@uiowa.edu (B.M.S.); gustavo-suarez@uiowa.edu (G.A.S.)

[2] Departments of Internal Medicine and Epidemiology, University of Iowa and Iowa City Veterans Affairs Health Care System, Iowa City, IA 52242, USA; ming-li@uiowa.edu (M.L.); jeffery-meier@uiowa.edu (J.L.M.)

* Correspondence: david-price@uiowa.edu

Abstract: How human cytomegalovirus (HCMV) infection impacts the transcription of the host genome remains incompletely understood. Here, we examine the global consequences of infection of primary human foreskin fibroblasts (HFFs) on transcription by RNA polymerase I, II, and III over the course of a lytic infection using PRO-Seq. The expected rapid induction of innate immune response genes is observed with specific subsets of genes exhibiting dissimilar expression kinetics. We find minimal effects on Pol II initiation, but increased rates of the release of paused Pol II into productive elongation are detected by 24 h postinfection and pronounced at late times postinfection. Pol I transcription increases during infection and we provide evidence for a potential Pol I elongation control mechanism. Pol III transcription of tRNA genes is dramatically altered, with many induced and some repressed. All effects are partially dependent on viral genome replication, suggesting a link to viral mRNA levels and/or a viral early–late or late gene product. Changes in tRNA transcription are connected to distinct alterations in the chromatin state around tRNA genes, which were probed with high-resolution DFF-ChIP. Additionally, evidence is provided that the Pol III PIC stably contacts an upstream −1 nucleosome. Finally, we compared and contrasted our HCMV data with results from published experiments with HSV-1, EBV, KSHV, and MHV68. We report disparate effects on Pol II transcription and potentially similar effects on Pol III transcription.

Keywords: HSV; EBV; KSHV; MHV68; PIC; transcription; productive elongation; mRNA; rRNA; tRNA

Citation: Ball, C.B.; Parida, M.; Li, M.; Spector, B.M.; Suarez, G.A.; Meier, J.L.; Price, D.H. Human Cytomegalovirus Infection Elicits Global Changes in Host Transcription by RNA Polymerases I, II, and III. *Viruses* **2022**, *14*, 779. https://doi.org/10.3390/v14040779

Academic Editors: Charles Grose, Ravi Mahalingam and Joel Rovnak

Received: 7 March 2022
Accepted: 30 March 2022
Published: 9 April 2022

Publisher's Note: MDPI stays neutral with regard to jurisdictional claims in published maps and institutional affiliations.

1. Introduction

Eukaryotic nuclear transcription is primarily executed by three distinct RNA polymerases, Pol I, Pol II, and Pol III. Pol I synthesizes abundant ribosomal RNA (rRNA) from clusters of rDNA repeats, accounting for approximately 30–40% of ongoing transcription in the nucleus [1]. Pol II is the sole enzyme responsible for the synthesis of protein-coding messenger RNA (mRNA) and is highly regulated to produce the unique transcriptional profiles of diverse metazoan cell types and to execute timely responses to environmental stimuli [2–4]. Pol II also transcribes long noncoding RNA, microRNA, and most small nuclear RNAs (snRNA). Pol III is dedicated to the transcription of short noncoding RNAs, including all transfer RNAs (tRNA), the U6 snRNA, and the 5S rRNA, three classes of Pol III genes that differ in their promoter architecture and the involvement of general transcription factors during initiation [5–7]. Interactions with distinct complements of basal machinery guide these polymerases to initiate at their target core promoters. The initiation and elongation steps of transcription are regulated and rate-limiting. Pol II transcriptional regulation has been intensely investigated for decades, the efforts of which have uncovered numerous control mechanisms operating at these stages [2–4]. Several strategies for Pol I and Pol III transcriptional control have been reported and evidence of regulatory crosstalk between Pol I, II, and III transcription also exists [1,8,9].

Mammalian dsDNA viruses that replicate in the nucleus rely on host RNA polymerases to execute their gene expression programs. Here, most attention has been directed towards the mechanisms by which viruses co-opt host Pol II to direct the transcription of their protein-coding and long noncoding RNA genes. The human herpesviruses are dsDNA viruses that replicate largely within the host cell nucleus and purpose Pol II for the transcription of the viral genome. Decades of herpesvirus research have revealed principles of regulated Pol II initiation at viral gene promoters, which are frequently similar in structure and composition to host gene promoters, and are engaged by host and viral factors in a coordinated manner to drive a temporal cascade of viral gene expression [10–13]. More recent studies have demonstrated that paradigms of Pol II elongation control, including pausing and P-TEFb-dependent release into productive elongation [14], that are reflected at host genes also pertain to transcription on the viral genome [12,15–17]. Interestingly, however, several reports have shown that the viral genome differs markedly from the host genome during lytic infection in the extent to which the genome is packaged as chromatin [18–22]. The host genome is largely packaged in nucleosomes, restricting transcription initiation to accessible nucleosome-free regions, while the viral genome is sparsely populated by nucleosomes, a property that enables rampant transcription initiation.

Infection by herpesviruses can lead to large-scale alterations of host transcription. Alpha and gamma herpesviruses employ host shut-off mechanisms that favor productive viral gene expression. In the case of herpes simplex virus I (HSV-1), a model member of the alphaherpesvirus subfamily, the major lytic transactivator ICP4 drives a competition between the host and viral promoters for Pol II and its initiation machinery that indiscriminately diminishes transcription initiation at host promoters by late times postinfection [19]. In addition, the HSV-1 tegument protein VHS mediates the global decay of the host and viral mRNAs during early infection, a function that is essential for the full-scale production of viral progeny [23,24]. In the case of the gammaherpesviruses Kapsosi's sarcoma-associated herpesvirus (KSHV) and murine gammaherpesvirus 68 (MHV68), a virally encoded RNA endonuclease drives widespread mRNA decay, which leads to the downmodulation of Pol II recruitment to host genes [25,26]. Here, mechanisms of anti-repression that permit robust viral gene expression are enacted. Very recent studies have shown that infection by HSV-1 and MHV68 each dramatically impact the cellular tRNA pool, actually driving an increase in the abundance of dozens of tRNAs [27,28]. Interestingly, infection was associated with a much larger increase in the abundance of immature and presumably nonfunctional pre-tRNAs than mature species, which exhibited a modest change or remained static. In one study, it was noted that the increase in pre-tRNA abundance corresponded to an increase in Pol III occupancy at tRNA genes, suggesting that Pol III transcription is induced in infected cells [28]. These studies indicated that herpesviruses may commonly regulate the tRNA pool, potentially to the benefit of productive infection, and that Pol III transcription is not subject to shut-off mechanisms akin to those that are thought to globally downregulate Pol II transcription of host genes. Notably, several other DNA viruses, including SV40 polyomavirus, Epstein–Barr virus, and adenovirus, as examples, have also been shown to increase tRNA expression, through mechanisms that involve, in part, increasing the abundance of limiting Pol III basal machinery [29–33].

Human cytomegalovirus (HCMV) is the prototypical member of the betaherpesvirus subfamily. HCMV is a nearly ubiquitous pathogen that infects more than half of the world's population and persists lifelong in its hosts. Although infection by HCMV is typically subclinical, immunocompromised populations are at risk of unchecked HCMV replication, which can lead to systemic infection and life-threatening disease [34]. Moreover, intrauterine HCMV infection is a common cause of morbidity in newborns and is the leading infectious cause of birth defects in the United States [35]. Dissimilar to HSV-1, KSHV, and MHV68, HCMV is not thought to elicit global host shutoff through transcriptional or post-transcriptional mechanisms. HCMV infection does induce innate and adaptive immune responses and profoundly impacts the levels of hundreds of mRNAs encoding factors that are involved in cell cycle progression, DNA replication, formation of the extracellular matrix, vesicular trafficking, and metabolism [36–38]. The effects of HCMV infection of Pol I and Pol III transcription have not, to our knowledge, been directly examined. Here, we further explore the transcriptional and epigenetic changes driven by lytic HCMV infection at host genes using PRO-Seq, a method that globally profiles Pol I, Pol II, and Pol III nascent transcripts [39,40] and DFF-ChIP, which enables complexes such as the Pol I, II, and III preinitiation complexes, in addition to their local chromatin environment, to be probed with high resolution [18].

2. Materials and Methods

2.1. Viruses, Cells, and Conditions of Infection

Primary human foreskin fibroblasts (HFFs) were derived from deidentified, discarded newborn infant foreskins. HFFs were maintained in Minimum Essential Medium (Gibco, 11095080) that was supplemented with 5% fetal bovine serum (Gibco, Waltham, MA, USA, 26140079) and 1% penicillin–streptomycin (Gibco, Waltham, MA, USA, 15140122). All infections were performed with contact-inhibited HFF and TB40/E BAC4 virus at an MOI of 3 as previously described [41]. Cells were treated with 1 μM Flavopiridol (Flavo) or DMSO vehicle control during the last hour of infection, and phosphonoformic acid (PFA, Sigma Aldrich, St. Louis, MO, USA) was added to the media at 400 μg/mL for the indicated conditions as described previously [41].

2.2. PRO-Seq and PRO-Cap

Spike-in quantitative PRO-Seq data analyzed in this study from TB40/E infected primary HFF were previously published and were prepared exactly as described in [18]. PRO-Cap data from HFF infected with TB40/E for 72 h were utilized to define transcription start sites at Pol II genes for truQuant analysis of pause region and gene body counts. The PRO-Cap dataset was previously published and prepared as described in [18]. All data are available at NCBI GEO GSE185763.

2.3. DFF-ChIP

DFF-ChIP data analyzed in this study from TB40/E infected primary HFFs were previously published. Sample processing and library construction were carried out exactly as described by Spector et al. [18]. Data are available at NCBI GEO GSE185763.

2.4. Generation of Tracks

Stranded PRO-Seq data tracks viewable with the UCSC genome browser (human genome assembly hg38) were generated as previously described [41], in a process that first involved trimming of adapter sequences with trim_galore v0.6.0 (available online: https://github.com/FelixKrueger/TrimGalore/releases/tag/0.6.6 (accessed on 3 March 2019)), strand-specific alignment of paired-end sequences with bowtie v1.2.3 (available online: http://bowtie-bio.sourceforge.net/index.shtml (accessed on 5 July 2019)), UMI-based removal of PCR duplicates with dedup (available online: https://github.com/P-TEFb/dedup (accessed on 14 February 2019)), and subsequent conversion of the deduplicated, aligned data into bedGraphs and bigwigs. Read coverage at each position was normalized by multiplication of spike-in correction factors. Spike-in correction factors were computed based on total reads in the library and reads mapping to the spike-in genome, as previously described [42], all details of which are provided in Supplementary Data File, PRO-Seq Stats. Tracks for DFF-ChIP data were generated as previously described.

2.5. TruQuant Analysis, Gene Body Clustering, FragMaps, and Metaplots

The pause region and gene body intervals of 11,593 genes transcribed in the human genome after 72 hpi HCMV infection were generated from the HFF PRO-Cap dataset using the truQuant program with default settings (available online: https://github.com/meierjl/

truQuant (accessed on 31 August 2020)). The bedtools coverage program v2.27.1 [43] (available online: https://bedtools.readthedocs.io/en/latest/index.html (accessed on 14 December 2017)) was used to determine the total number of 5′ or 3′ ends in pause region or gene body intervals, respectively. The gene body counts for all infected time points were divided by the count from the uninfected data to calculate the fold change. Genes with their associated fold changes at various infected time points were hierarchically clustered using the pheatmap library in R. Pearson correlation was chosen as the clustering distance metric and the number of clusters was retained at 10. The number of clusters was chosen using the within-cluster sum of squares (wcss) method in Python's sklearn.cluster package and KMeans library. All genes and their associated fold change profiles were scaled independently using the z-score formula and colored using the pheatmap function in R. Average gene body changes for all clusters were calculated using a Python script by averaging fold changes across time points and plotted as line graphs using MS Excel. Gene sets for specific clusters were plotted independently using the pheatmap function in R.

Median fold changes of 11,593 genes across all infected time points were determined using a Python script from the list of fold changes calculated by first dividing the total number of 5′ end counts in their pause regions and the total number of 3′ end counts in their gene body regions with associated counts of pause region and gene body regions from the uninfected datasets. Median fold changes were plotted using MS Excel. The pause ratio was computed for each gene in the DMSO-treated samples by dividing the number of 5′ ends in the pause region by the pause region length (150 bp), and then dividing this number by ratio of the number of 3′ ends in the gene body over the gene body length for each gene (based downstream of the pause region to the annotated CPS). Pause ratios for all genes in each sample were plotted as boxplot in MS excel.

The sum of H3K4me3 read densities across genomic intervals between −500 and +3000 bp from the MaxTSS of all genes in clusters 1, 3, and the repressed set were generated using bedtools coverage program. Read densities were plotted using MS Excel. FragMaps were generated from the HFF uninfected and 48 hpi H3K4me3 fragments of size between 18 bp and 400 bp that were present in the genomic intervals between −/+500 bp from the MaxTSSs of genes in the repressed set using the fragMap program (available online: https://github.com/P-TEFb/fragMap (accessed on 13 August 2021)).

2.6. *Bioinformatics Analysis of tRNA Transcription*

An annotation of 429 tRNA genes was downloaded from the genomic tRNA database (available online: http://gtrnadb.ucsc.edu/ (accessed on 1 December 2021)). The mature 5′ ends of tRNAs were used for further analysis. In total, 235 tRNA genes out of 429 were used to compare tRNA transcription between uninfected and infected time points. These genes exhibited most initiation from the position upstream of their mature 5′ end. Strand specific transcription of tRNA genes was measured at their 1 bp upstream position using bedtools coverage program [43]. Scatter plots of uninfected HFF PRO-Seq data were plotted against the 4 hpi, 12 hpi, 24 hpi, 48 hpi, 72 hpi, and 72 hpi PFA-treated samples using MS Excel. Read coverage of the 235 tRNA genes at the position 1 bp downstream from their mature 3′ ends was calculated using bedtools coverage program. The program was used to determine the total number of strand-specific fragments at this position for 48 hpi HCMV infection and uninfected DMSO-treated datasets. Fold change calculation was performed by dividing the fragment counts of 48 hpi HCMV infection over uninfected DMSO-treated datasets. The scatter plot of the fold change was plotted using MS Excel.

Heatmaps of −50 bp and +200 bp genomic intervals centered on the mature 5′ ends of 235 tRNA genes were created using the heatmap program (available online: https://github.com/P-TEFb/Heatmap (accessed on 4 March 2022)). The 5′ and 3′ ends of HFF PRO-Seq uninfected and 48 hpi DMSO-treated datasets were generated following the steps mentioned in our metaplots tutorial (available online: https://github.com/P-TEFb/Metaplot-and-Counting-fragments (accessed on 3 March 2022)). Parameters for these heatmaps were strand = yes, order = DESC, avgrows = 1, width = 2, height = 4, black = user-specified (1/10th of the maximum value in the heatmap table), and gamma = 0.5.

2.7. Motif Enrichment Analysis

Motif enrichment was performed using a previously documented motif analysis software [44] (available online: https://regulatory-genomics.org/motif-analysis/method/ (accessed on 17 February 2022)). Input regions were bed files containing 400 bp promoter regions padded around the 150 bp pause region (−200 bp upstream, +50 bp downstream) for genes in clusters 1, 3, 4, and 8. The background file consisted of such promoter regions from all 11,594 active gene promoters. Transcription factor motifs enriched in experimental regions versus the background were determined using the rgt-motifanalysis matching function with script –input-files input/experimental.bed input/Background.bed –organism hg38 –remove-strand-duplicates. Next, the statistical significance of enrichment over background was measured using the rgt-motifanalysis enrichment function, which performed a Fisher's exact test for each enriched transcription factor binding motif.

2.8. Statistics

Statistics for motif enrichment analysis were reported by the motif analysis software. For correlations of tRNA read coverage between DMSO and Flavo PRO-Seq datasets, a Pearson's correlation coefficient was computed in Microsoft Excel.

3. Results

3.1. Global Changes in Pol II Transcription during HCMV Infection

To investigate the effects of HCMV infection on transcription of the host genome, we carried out a time course of HCMV infection (TB40/E strain, MOI 3) in contact-inhibited HFF and performed spike-in quantitative PRO-Seq. Libraries were prepared from uninfected cells and cells infected for 4, 12, 24, 48, and 72 hpi, which sample the early-stage of lytic infection (4, 12 hpi), a mid-stage of infection close to the onset of viral genome replication (24 hpi), and late-stage (48, 72 hpi) that follows the onset of genome replication. Additional samples in which the uninfected or infected cells were treated for the final hour prior to harvest with Flavopiridol, a P-TEFb inhibitor that blocks Pol II pause-release [45], were prepared. In addition, a sample in which cells were infected for 72 h and treated from the onset of infection with phosphonoformic acid (PFA), an HCMV replication inhibitor, was prepared, and this condition was additionally combined with the Flavo treatment during the final hour of infection (Figure 1A). PRO-Seq involves the rapid isolation of native nuclei followed by a nuclear-run on reaction in the presence of biotinylated nucleotides [39,40]. Total RNA was afterwards isolated and biotinylated nascent transcripts were enriched with streptavidin beads under stringent conditions for the subsequent library construction. The efficiency of adding consecutive biotinylated nucleotides to the nascent RNA chain was extremely low, at least for Pol II [39]; this affords single-nucleotide resolution not offered by other run-on and metabolic labeling approaches. PRO-Seq does not distinguish between transcription by Pol I, II, or III, such that a single experiment theoretically enables one to query transcription by any of these polymerases.

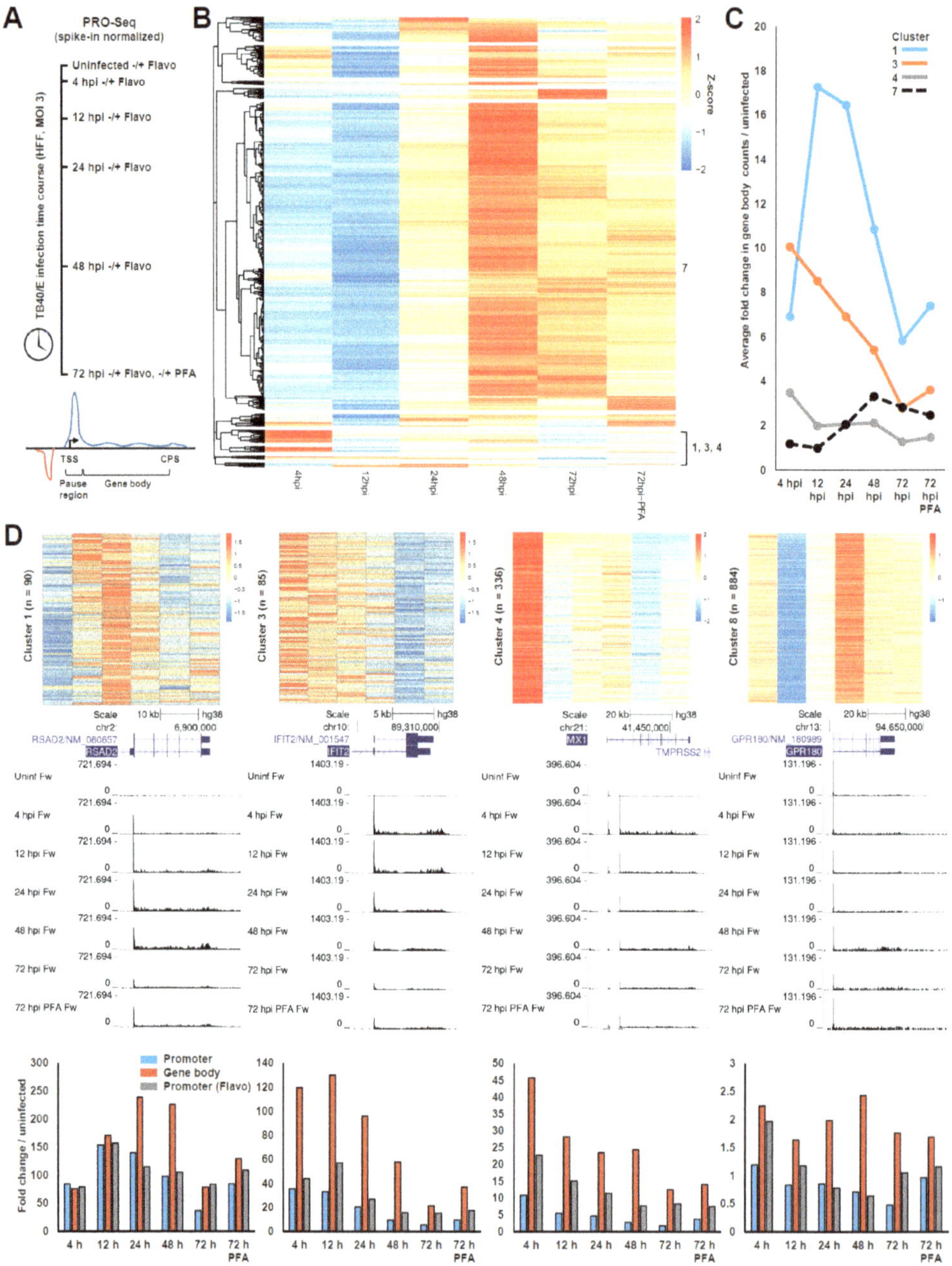

Figure 1. Effects of HCMV infection on Pol II transcription. (**A**) Schematic depicting experimental design. Spike-in quantitative PRO-Seq was performed with samples derived from contact-inhibited HFFs that were mock-infected or infected for 4, 12, 24, 48, or 72 h. For all time points, an additional sample in which the cells were treated with Flavo for the final hour of infection to block Pol II pause-release was collected. For the 72 h time point, the effect of blocking HCMV replication by treatment with PFA from the onset of infection was also tested −/+ Flavo during the final hour of infection. At the bottom, a diagram depicting the pause and gene body regions quantified for analysis

is shown; (**B**) Heatmap displaying hierarchical clustering of ratios in PRO-Seq gene body counts, which were computed for each gene at each time point over the uninfected control and are represented as Z-scores, which were computed for each gene. Certain clusters are identified at the right of the heatmap; (**C**) plot of average fold change in gene body counts for all genes within clusters 1, 3, 4, and 7 along the infection time course; (**D**) top: blown up views of heatmaps representing genes in clusters 1, 3, 4, and 8. Legends represent the Z-score. Middle: UCSC Genome Browser snapshots showing forward strand (Fw) PRO-Seq data from DMSO-treated samples, for example, genes representing the above clusters. Bottom: graphs displaying fold changes in pause region and gene body counts from DMSO-treated samples and fold change in pause region counts from Flavo-treated samples along the infection time course for the above genes.

The total number of sequenced reads in each sample, numbers of reads mapping to the host, viral, and spike-in (Spodoptera) genomes, percentage of reads mapping to the viral genome, and spike-in normalization factors for each sample are shown in Supplementary Data File, PRO-Seq Stats. All subsequent analyses utilized spike-in normalized data. Initial focus was directed towards the effects of HCMV infection on Pol II transcription. PRO-Cap data for HFFs infected for 72 h with TB40/E were utilized to annotate a set of transcriptionally active genes (n = 11,594) with our previously described tsrFinder and truQuant algorithms [12,41]. For every active gene, the number of 5′ ends in the Pol II pause region, defined by the truQuant algorithm [41], and the number of 3′ ends in the gene body region, beginning at the base after the pause region end and extending to the annotated cleavage and polyadenylation site (CPS), were quantified for each PRO-Seq dataset (Figure 1A, Supplementary Data File). As gene body counts were more highly correlated with transcriptional output than pause region counts, we took the ratio of gene body counts for all genes at every time point over the uninfected control and utilized these ratios to perform hierarchical clustering and identify groups of genes that exhibited similar patterns of change along the infection time course (Figure 1B). Ten clusters were selected using the elbow method. Clusters five, six, and seven, containing a total of 9605 (82%) active genes, reflected a major, unexpected trend in the data, which was a gradual increase in gene body counts along the time course, peaking at 48 or 72 h. Genes within clusters one (n = 90), three (n = 85), and four (n = 336) exhibited a sharp increase in gene body counts at 4 hpi over the uninfected control, but differed with respect to change in gene body counts at later time points (Figure 1C,D). These clusters were, as expected, highly enriched for genes involved in innate immunity, as revealed by the pathway and ontology analysis performed with Enrichr [46] (Supplementary Data File, GO-Terms and BioPlanet Pathways). For example, the cluster one gene RSAD2, an interferon-stimulated gene (ISG) [47], exhibited a great induction at 4 hpi that further increased at 12 hpi and peaked at 24–48 hpi (Figure 1D). The cluster three gene IFIT2, an ISG [48], was greatly induced at 4 hpi and gradually decreased as infection progressed. The well-characterized ISG MX1 [48] fell within cluster four, and was sharply induced at 4 hpi and exhibited a reduced but relatively stable level of gene body transcription between 12 and 48 hpi. Cluster eight genes (n = 884) contained genes such as the putative interferon-responsive gene GPR180 [49], which was modestly induced at 4 hpi, decreased at 12 hpi, and then in association with most other active genes, exhibited an upward trend in transcription during the later stages of infection (Figure 1D).

Intriguingly, we found that the trends in the pause region and gene body counts, plotted as a fold change over the uninfected control for each example gene (Figure 1D, lower panels), were frequently inconsistent, which indicated an uncoupling between initiation and release into productive elongation. Following initiation, paused Pol II either terminates or undergoes a P-TEFb-dependent release into productive elongation, a step that is known to be regulated and govern transcriptional output [50]. The amount of paused Pol II observed in the presence of Flavo is in part dependent on termination. However, due to the inhibition of P-TEFb, this amount is not influenced by release into productive elongation, and, therefore, more accurately represents the levels of initiation. Accordingly, treatment with Flavo partially corrected discrepant trends in the pause region and gene body counts in the presence of DMSO along the course of infection, indicating that a substantial fraction of paused Pol II at these genes entered into productive elongation (Figure 1D).

These patterns were representative of dozens of other genes induced by infection, many of which are known ISGs. The transcriptional mechanisms that control the induction of ISGs have been intensely investigated, but remain incompletely understood and are complex, involving numerous transcription factors (TFs) operating concertedly and dynamically across time [51–53]. This transcriptional response is wired to be transient, at least in part to limit the deleterious effects of long-term proinflammatory gene expression. In addition, HCMV encodes a large arsenal of immunomodulatory factors that counter innate and adaptive immune responses. HCMV is known to subvert innate immune responses through transcriptional [54], post-transcriptional [55], and post-translational mechanisms [56]. Our PRO-Seq data, which highlighted the dynamic nature of ISG induction, measured nascent transcription, and were not significantly impacted by post-transcriptional mechanisms regulating RNA abundance, represent a minable source of information that may be useful in unraveling the direct transcriptional response to HCMV infection and innate immune signaling more broadly. In this regard, we investigated the promoter regions of genes in clusters one, three, four, and eight for enriched TF motifs using a motif analysis software [44] (Figure S1). The region queried was 400 bp, padded around the pause region, which captured most of the sequence contained within a typical nucleosome-free region (NFR), and enrichment was tested relative to a background consisting of equivalent regions from all 11,594 active host gene promoters. Interestingly, genes in clusters three and four, which were enriched for ISGs that were most induced shortly after infection at 4 hpi, exhibited a highly significant enrichment of IRF and STAT-family TF motifs in the proximal promoter region. Cluster one genes, which exhibited a delayed but a robust induction, differed in that their proximal promoters were only weakly enriched for IRF-family TF motifs, and no enrichment of STAT-family TF motifs was detected. By contrast, cluster eight genes, which were mildly induced at 4 hpi, exhibited a weak enrichment for STAT and IRF-family TF motifs.

3.2. Late HCMV Infection Is Associated with Increased Rates of Release into Productive Elongation at Host Pol II Genes

The observation that gene body transcription trended upwards during late HCMV infection was unexpected and further explored. The median fold change in the pause region and gene body counts along the infection time course, as well as the median fold change in the number of pause region counts in the presence of Flavo which blocked the release into productive elongation, was plotted (Figure 2A). Clearly, the median number of gene body counts increased substantially along the course of infection, while the number of pause region counts modestly decreased. Blocking the release into productive elongation led to an increase in counts in the pause region at late times postinfection. Extending this analysis, the pause ratio for each gene, which measures the proportion of Pol II engaged in pause regions versus gene bodies (after normalization to the size of the regions), was calculated (Figure 2B). In uninfected cells and cells infected for 4 or 12 hpi, most genes exhibited a high pause ratio, but a near-global downward shift was detected at 24 hpi and the pause ratio dramatically reduced at 48 and 72 hpi, reflecting the widespread increase in gene body counts. Interestingly, blocking HCMV replication with PFA in cells infected for 72 h resulted in increased pause ratios compared to the 72 h time point, mirroring the result acquired at 24 hpi (pre-replication), and suggesting that the substantial decrease in pause ratios observed at late times may have required viral genome replication and/or increasing levels of viral transcription. The above described trends were well represented by the PRO-Seq signal over the EXOC4 gene, which, for clarity, was broken into the pause and gene body regions (Figure 2C). This result was also captured in cells treated with Flavo, as there was a notable increase in the 'receding wave' of Pol II elongating down the long EXOC4 gene at late times postinfection, following the acute inhibition of pause release.

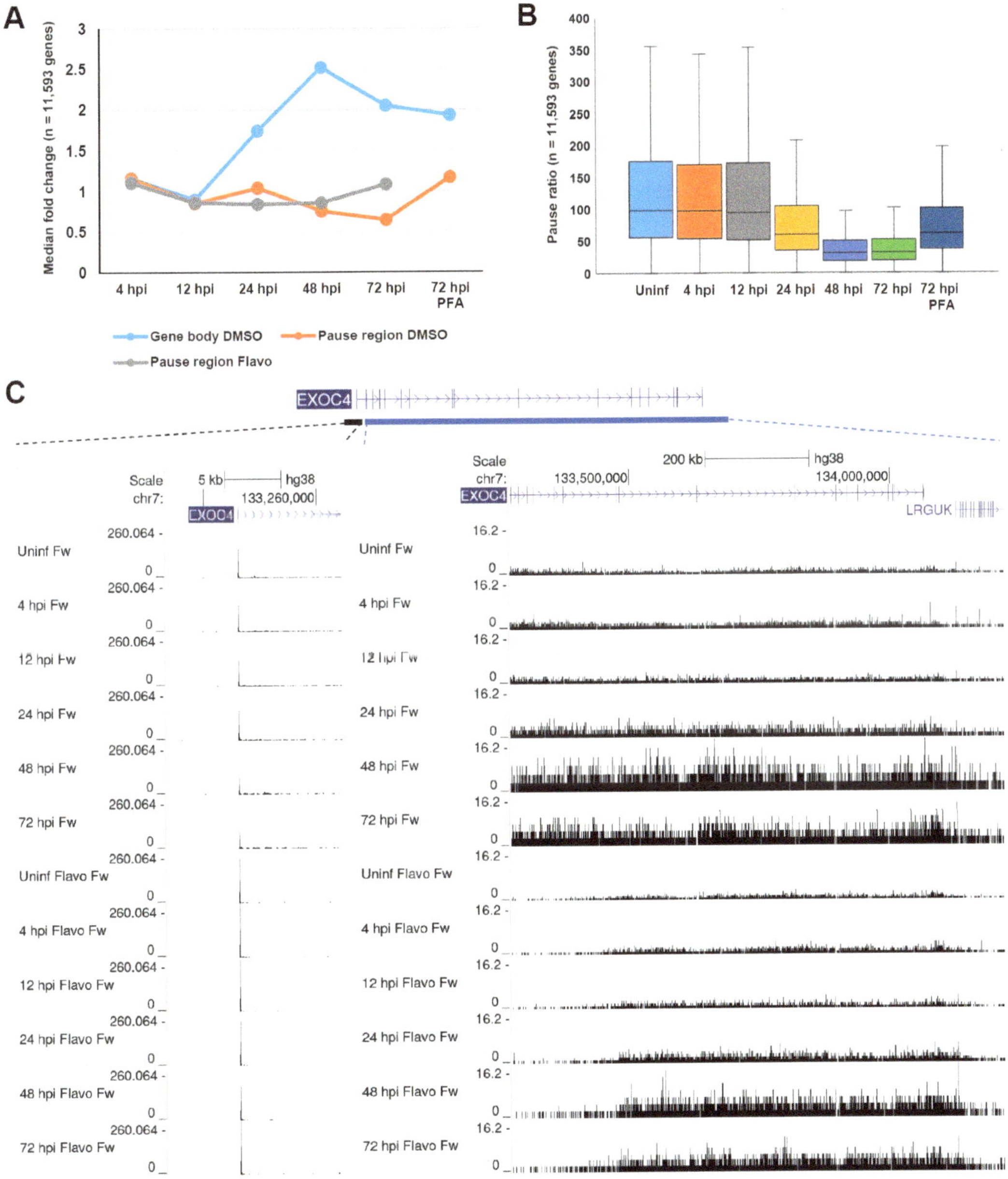

Figure 2. Late HCMV infection is associated with increased rates of release into productive elongation. (**A**) Graph displaying median fold change in pause region and gene body counts from DMSO-treated samples and fold change in pause region counts from Flavo-treated samples along the infection time course; (**B**) boxplot displaying calculated pause ratios for each gene across all samples; (**C**) genome browser snapshots of the pause region and gene body of the EXOC4 gene, showing a modest decrease in pausing and a substantial increase in amounts of productively elongating Pol II at late times postinfection.

3.3. Chromatin Changes at Pol II Genes Activated and Repressed by HCMV Infection

Next, we sought to address whether HCMV infection drove changes in the chromatin state of differentially transcribed genes. We recently reported DFF-ChIP for H3K4me3, an epigenetic modification that was enriched on the positioned nucleosomes flanking active Pol II promoters, in uninfected HFFs and HFFs infected for 48 h with HCMV TB40/E. DFF-ChIP utilizes the dsDNA-specific endonuclease DFF to digest chromatin prior to IP [18]. Importantly, DFF exhibits very little propensity to digest within nucleosomes, a property that enables sites of H3K4me3 nucleosome occupancy to be defined with high resolution [57]. ISGs are transcribed at very low levels under normal conditions, but their promoters are, nevertheless, reported to be open and primed for induction [53,58]. Congruent with these descriptions, we observed that the ISG OAS1 contained a well-defined nucleosome-free region (NFR) in uninfected cells flanked by H3K4me3-modified nucleosomes (Figure 3A). At 48 hpi, the NFR was unaltered, but the profile of H3K4me3 extended deeply into the gene body, which was consistent with the current model that Pol II elongation is associated with H3K4me3 installment [59,60]. H3K4me3 was recognized by the TAF3 subunit of TFIID [61] and the chromatin remodeler CHD1 [62], which is thought to also tether components of spliceosome. Thus, the increased load of H3K4me3 may function to facilitate the robust initiation and elongation at ISGs.

By contrast, 86 genes were observed to undergo a significant repression of gene body transcription across consecutive time points of the infection time course. For example, the transcription of the normally active COL1A2 gene was repressed more than five-fold at 48 hpi compared to the uninfected control (Figure 3B). Repression was initiated rapidly after infection, at 4 hpi, and deepened over time. H3K4me3 DFF-ChIP revealed a broad profile of the modification that extended far into the gene body in uninfected cells and was considerably retracted at 48 hpi. Strikingly, the COL1A2 NFR exhibited an encroachment of H3K4me3 nucleosomes at 48 hpi compared to the uninfected control, likely a consequence of the reduction in promoter activity. A gene ontology and pathway analysis revealed that this set of repressed genes was enriched for proteins involved in the organization of the extracellular matrix, including several collagens, lumican, and metallopeptidases (Supplementary Data File, GO-Terms, BioPlanet Pathways). The HCMV-induced repression of these genes has been reported [36,37,63], and may be linked to the well-documented observation that HCMV-infected cells exhibit cytopathic cell rounding and compromised substrate adhesion [64,65]. Our data added that the repression of these genes initiated rapidly following infection through a mechanism that ultimately resulted in the closure of the promoter NFRs. Metaplots were generated for H3K4me3 in uninfected cells and cells infected for 48 at all cluster one and cluster three genes, which were enriched for induced ISGs, in addition to the 86 defined repressed genes. These metaplots revealed an overall unchanged NFR for cluster one and three genes and increased H3K4me3 levels downstream in gene bodies. Repressed genes exhibited a retraction of the H3K4me3 signal from gene bodies and an increase in H3K4me3 nucleosomes within the NFR (Figure 3C). To view these data another way, we generated fragMaps of the H3K4me3 signal at repressed genes in uninfected cells and cells infected with HCMV for 48 h. A fragMap is a quantitative 2D plot of the level of fragments of various sizes versus the position [18], here centered on the MaxTSS for every gene represented. The fragMaps revealed a well-positioned +1 nucleosome downstream of the TSS, and in infected cells, fragments corresponding to a nucleosome in size (~150 bp) further extended into the NFR, which was consistent with promoter occlusion (Figure 3D).

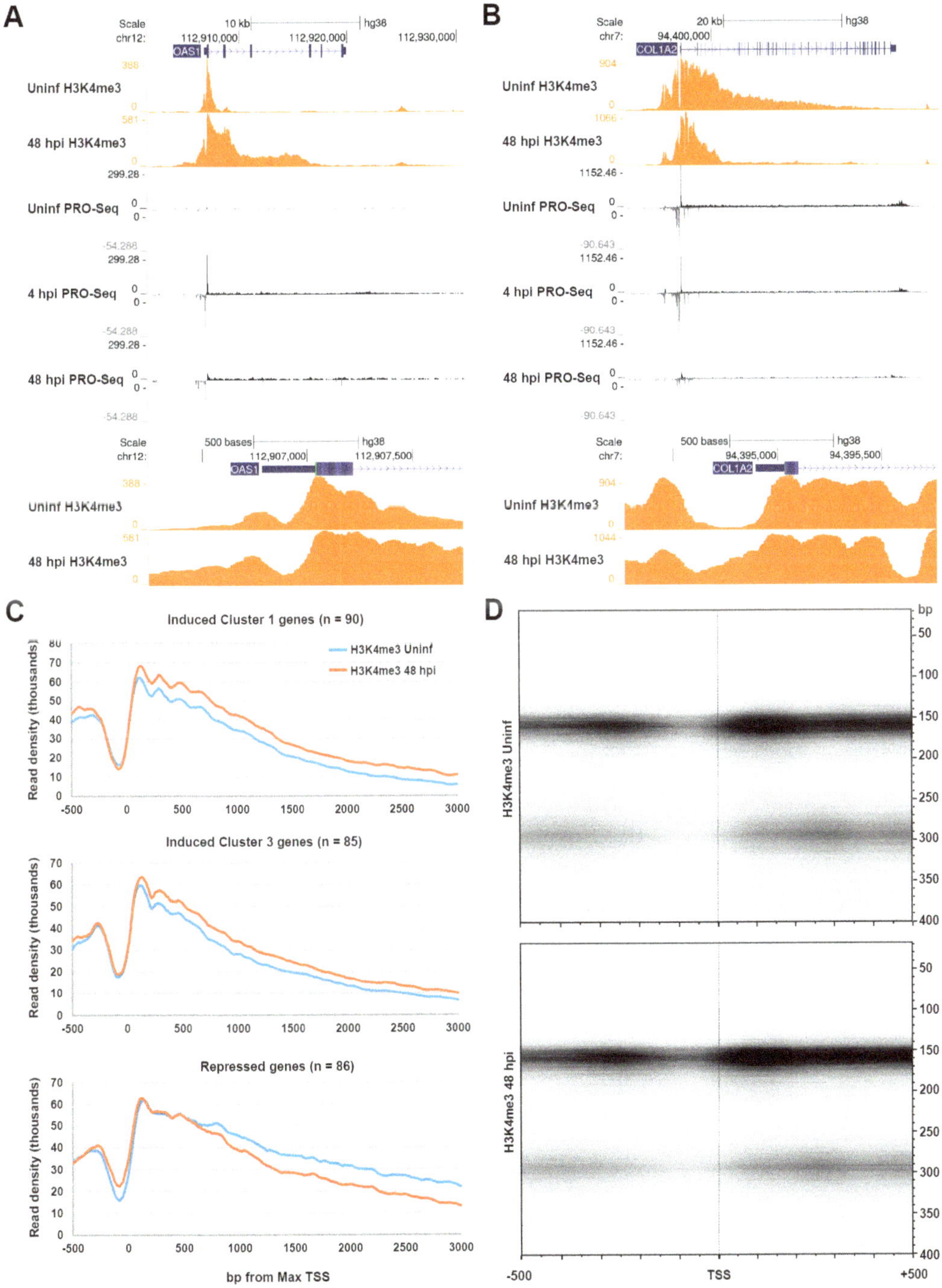

Figure 3. Changes in the chromatin status of genes induced and repressed by HCMV infection. (**A**) Top: genome browser snapshot of PRO-Seq data for the induced OAS1 gene in uninfected HFFs and HFFs infected for 4 or 48 h, and H3K4me3 DFF-ChIP data in uninfected HFFs and HFFs infected

for 48 h. Bottom: blown up view of H3K4me3 DFF-ChIP data at the OAS1 promoter region; (**B**) top: Genome browser snapshot of PRO-Seq data for the repressed COL1A2 gene in uninfected HFFs and HFFs infected for 4 or 48 h, and H3K4me3 DFF-ChIP data in uninfected HFFs and HFFs infected for 48 h. Bottom: blown up view of H3K4me3 DFF-ChIP data at the COL1A2 promoter region; (**C**) metaplots of H3K4me3 DFF-ChIP data in uninfected HFFs and HFFs infected for 48 h for cluster 1 genes and cluster 3 genes, which were enriched for induced ISGs and repressed genes. Summed H3K4me3 signals were plotted over a region spanning −500 to +3000 relative to the Max TSS; (**D**) FragMaps of H3K4me3 DFF-ChIP data at repressed genes in uninfected HFF and HFF infected for 48 h.

3.4. HCMV Infection Alters Transcription of the 45S rDNA Repeat by Pol I

Ribosomal RNA is synthesized by Pol I, which transcribes a single 45S precursor RNA that is, subsequently, processed to yield the 18S, 5.8S, and 28S ribosomal RNAs; 45S is encoded within an rDNA repeat that exists as a tandem array found on the short arms of chromosomes 13, 14, 15, 21, and 22. In the present hg38 assembly, eight rDNA repeats are annotated. Due to the remarkable abundance of ribosomal RNA in cells, mature 18S, 5.8S, and 28S rRNAs can contaminate PRO-Seq libraries despite extensive selection for biotinylated nascent transcripts, but we found that thorough washing eliminated most of this contaminating background signal over mature rRNA. To eliminate all background, the analysis of Pol I transcription was restricted to the 45S promoter region and 5′ portion of 45S that is not retained as a mature transcript. Initiation by Pol I occurred at a single sharp start site. Pol I is not thought to undergo promoter-proximal pausing like Pol II. However, unexpectedly, we observed an accumulation of transcripts all starting at the 5′ end of the 45S gene, extending up to about 200 bp downstream. This was reminiscent of Pol II pausing, though the transcripts associated with Pol I were about 2–3 times longer than typically seen for paused Pol II (Figure 4A). Very interestingly, as infection progressed, this signal corresponding to 'paused Pol I' was diminished, and a more elongating Pol I was detected downstream. At 72 hpi, this effect on Pol I transcription was partially reversed by treatment with PFA, suggesting that the apparent stimulation of Pol I elongation may depend on viral genome replication, increasing levels of viral mRNA and/or the involvement of a viral factor. These effects were also observed in the presence of Flavo, indicating that P-TEFb is not involved in regulating Pol I transcription and suggesting that the differences in Pol I transcription during late infection were not directly coupled to ongoing Pol II productive elongation (Figure S2).

We recently reported DFF-ChIP for TATA-binding protein (TBP) [18], a core general transcription factor that assembles into Pol I, Pol II, and Pol III preinitiation complexes (PICs), in HFFs infected for 48 h with HCMV TB40/E. Our DFF-ChIP approach was demonstrated to define the boundaries of Pol II PICs with an unprecedented resolution. Although not reported in the original publication, major TBP features were detected in the rDNA loci. Two were in the 45S promoter region and unexpectedly three other TBP-containing complexes were detected downstream of the region encoding the 45S precursor rRNA (Figure 4B,C). The two upstream features were present at all eight annotated rDNA repeats; they corresponded to PICs formed over the previously described spacer promoter and the rRNA promoter [66] (Figure 4C, left). Downstream of the 45S gene, in the five rDNA repeats that contained sequence information, three apparent PICs spaced approximately 700 bp apart were detected (Figure 4C, right). All of the PICs (both the upstream and downstream) drove transcription initiation, albeit at vastly different levels, starting 5–10 bp beyond the downstream edge of the PIC. Only initiation from the main PIC located immediately upstream of the 45S gene drove 45S transcription. Transcription from the downstream PICs was slightly increased during late HCMV infection, while transcription from the upstream spacer promoter was not impacted either at the level of initiation or elongation. A set of TBP fragMaps for the example shown demonstrated that these PICs all had similar 160 bp footprints (Figure 4C). Presumably, all of the PICs were occupied by the TBP-containing SL1 complex downstream of a region bound by UBF [67]. DFF digestion between these two subcomplexes was detected in the main PIC driving rRNA transcription (Figure 4C, left). The rRNA promoter was also unique among the regions directing PIC assembly, as it alone contained a TATAT sequence upstream of the TSS. It is not clear what role the downstream PICs possess, but it was noted that their upstream edges were intimately associated with TTF1 termination sites (Sal boxes, GGGTCGACCAG) [68] and, thus, could be involved in preventing transcription read-through into adjacent rDNA repeats (Figure 4D).

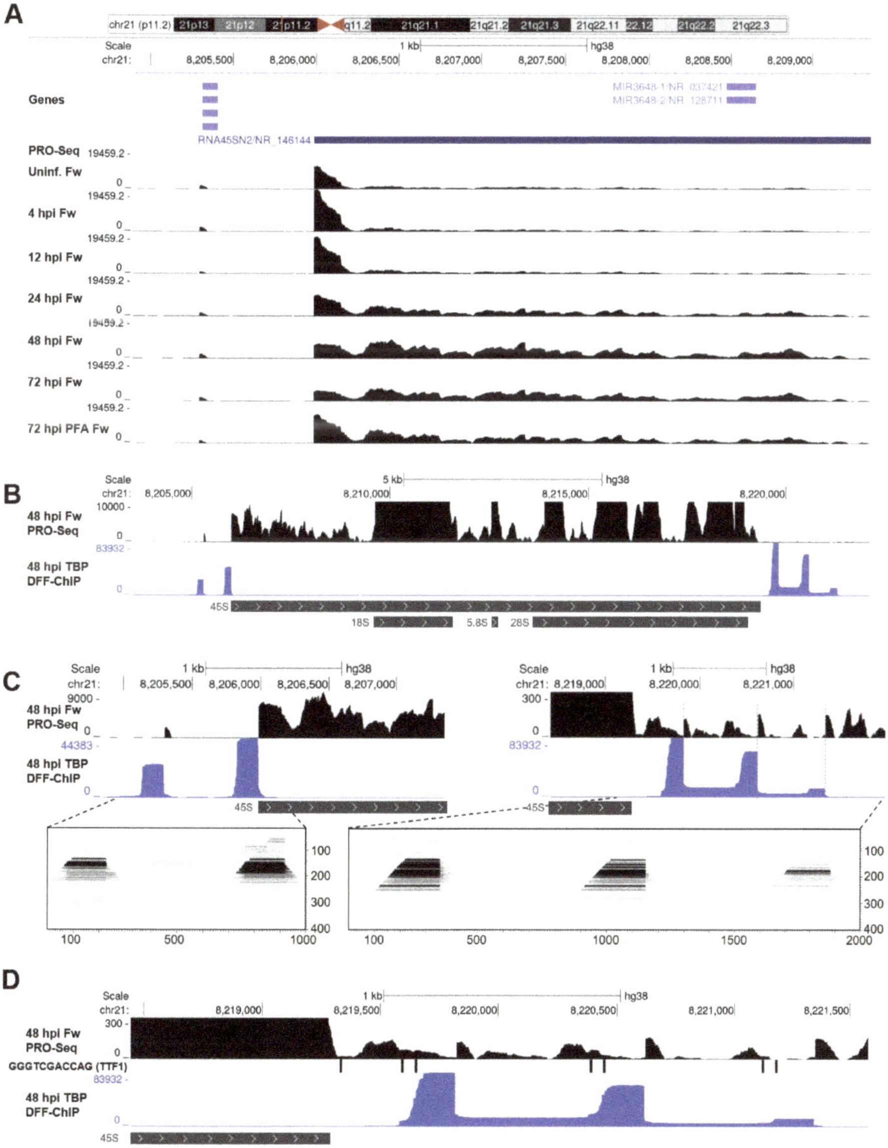

Figure 4. HCMV infection impacts Pol I transcription of rDNA. (**A**) UCSC Genome Browser snapshot of PRO-Seq data over the 45S promoter region and 5′ portion of 45S not retained in mature rRNA; (**B**) genome browser snapshot of TBP DFF-ChIP data at an rDNA repeat, with a schematic depiction

of 45S, 18S, 5.8S, and 28S below; (**C**) left: blown up view of TBP DFF-ChIP data at the 45S promoter region, indicating the positions of Pol I PICs, and a fragMap below corresponding the region indicated by the dashed lines. Right: blown up view of TBP DFF-ChIP data at the 45S downstream region, indicating the positions of Pol I PICs, and a fragMap below corresponding the region indicated by the dashed lines; (**D**) UCSC Genome Browser of TBP DFF-ChIP and PRO-Seq data at the 45S downstream region documenting evidence that the downstream PICs drive transcription initiation, and that the PICs are intimately associated at their upstream end with TTF1 termination sites.

3.5. HCMV Infection Dramatically Impacts tRNA Transcription by Pol III

In light of recent reports suggesting that infection by both alpha- and gammaherpesviruses induces tRNA expression [27,28,31], we next investigated whether HCMV, a betaherpesvirus, also affects tRNA transcription. The measurement of tRNAs by sequencing approaches has been limited, as mature tRNAs are refractory to reverse transcription due to their extensive base modifications and secondary structure. Several recently developed approaches have enabled better quantitative profiling of tRNAs [69–72]. As tRNAs are thought to be post-transcriptionally processed and modified, the measurement of nascent Pol III transcripts theoretically eliminates the restriction to sequencing posed by base modifications. In addition, secondary structure may be minimal on tRNAs that are not completely transcribed. Finally, mature tRNAs are extremely stable, with a half-life of days [72], and, thus, small changes measured in the mature tRNA population may belie major changes at the level of transcription. PRO-Seq could, therefore, be uniquely informative of changes in tRNA expression, with the caveat that post-transcriptional mechanisms also contribute to the abundance of functional tRNAs.

Dramatic changes in the transcription of most tRNAs were observed over the course of infection. For example, the transcription of tRNA-Met-CAT-5-1 was induced more than 400-fold at 48 hpi. Major increases in PRO-Seq signal were detected over the TSS and downstream of the mature tRNA 3′ end (Figure 5A, left). Interestingly, blocking HCMV replication with PFA was associated with only a partial induction of transcription, which suggests that accumulating levels of viral genomes and mRNA may drive tRNA induction. This result also suggests the involvement of a viral early-late or late factor in the effect on tRNA transcription, and is partly in keeping with a recent study, which showed that an HSV-1 mutant deficient for the viral DNA polymerase (UL30) did not induce tRNA expression to the same extent as the wild-type virus, nor did infected cells treated with DNA polymerase inhibitors [27]. By contrast, certain tRNAs, such as tRNA-Ser-TGA-1-1, were apparently repressed (Figure 5A, right). Interestingly, the PRO-Seq signal within tRNA-Ser-TGA-1-1 was characterized by sharp decreases that were not associated with termination signals and may represent Pol III pausing, which is further explored later.

We noticed that some tRNAs transcripts had 5′ end signals coinciding with the mature 5′ end of the tRNA (Figure S3A). Many transcripts contained 3′ ends upstream of the mature 3′ end, indicating that they may be nascent transcripts. However, this could be due to transcription actually starting at the mature 5′ end, 5′ end processing occurring co-transcriptionally, or some amount of processed tRNA contamination. Because of this uncertainty, we developed a method to separate those transcripts from those that started upstream of the mature 5′ end (Figure 5B). Using the PRO-Seq signal for the tRNA-Leu-CAG-2-1 gene as an example, we quantified the number of reads at the mature 5′ end seen in the annotation, the number of reads crossing the preceding base position from transcription that started upstream of the mature 5′ end (termed the 5′ upstream position), and the number of reads crossing the base following the mature 3′ end (termed the 3′ downstream position) (Figure 5B, Supplementary Data File, tRNA Analyses). The short segment of the pre-tRNA between the TSSs and the mature 5′ end is referred to as the leader sequence and is removed during 5′ end processing. It was noted for tRNA-Leu-CAG-2-1 that PRO-Seq fragments beginning at the mature 5′ end were associated with 3′ ends within the tRNA gene body or close to the mature 3′ end, whereas fragments containing the leader sequence were primarily associated with 3′ ends downstream of the mature 3′ end, reflecting Pol III elongation (Figure S3B). Pol III terminates while synthesizing polyuridine tracts, which destabilize the elongation complex [73]. Accordingly, sharp decreases in engaged Pol III were detected precisely in line with poly-T stretches on the nontemplate strand (Figure 5B). For subsequent analyses, we used tRNA genes that had a ratio of read coverage at the base preceding the mature 5′ end over the reads starting at the mature 5′ end of 0.8 or greater (269/429, or 63% of annotated tRNAs). A minimum read cutoff for the analysis was also imposed, resulting in a final list of 235 tRNAs. That the majority of transcribed tRNAs contained upstream leader sequences and many, by observation, exhibited termination over poly-T stretches verified that PRO-Seq captured nascent Pol III transcripts.

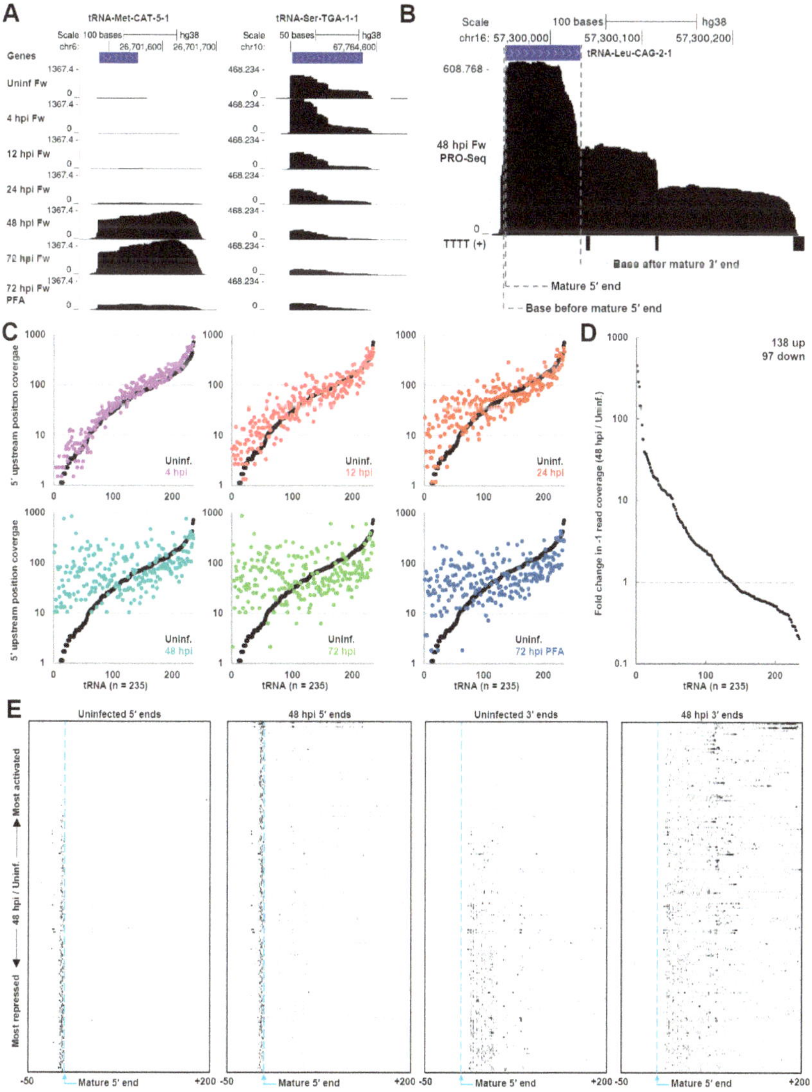

Figure 5. HCMV infection dramatically impacted Pol III transcription of tRNAs. (**A**) Genome browser snapshots of forward-strand (Fw) PRO-Seq data at tRNA genes that were significantly induced along the course of infection (left) repressed along the course of infection (right); (**B**) genome browser snapshot

of forward-strand PRO-Seq data of the tRNA-Leu-CAG-2-1 gene at 48 hpi. Sites where read coverage was quantified for analysis are indicated. Matches to the motif TTTT on the nontemplate strand detected with the UCSC genome browser short match function are indicated; (**C**) plots of read coverage at the 5′ upstream position in uninfected HFFs and HFFs infected for 4, 12, 24, 48, or 72 h, and 72 h + PFA. In each plot, the read coverage for each tRNA in the uninfected HFFs is displayed with the data for the infection time point, and the data are sorted by lowest to highest read coverage in uninfected cells. Data are plotted on a $\log_{10}$ scale; (**D**) Graphs depicting the fold-change in read coverage at the 5′ upstream position (48 hpi/uninfected) for all 235 analyzed tRNA genes, sorted from most induced to most repressed. Fold change was plotted on a $\log_{10}$ scale; (**E**) heatmaps of PRO-Seq 5′ ends and 3′ ends at tRNA genes in uninfected HFFs and HFFs infected for 48 h. Data are displayed for a region spanning −50 to +200 relative to the mature 5′ end and are sorted from top to bottom by most induced at 48 h to most repressed according to the ratio of read coverage at the 5′ upstream position.

To view the global effects of HCMV infection on tRNA transcription, scatterplots of read coverage at the 5′ upstream position, representing initiation, were generated comparing the uninfected control to each infection time point (Figure 5C). The data were sorted by an increasing read coverage in the uninfected control. At early times postinfection (4 hpi), differences in tRNA transcription were subtle, but were notable at 12 hpi and marked by 24 hpi. At 48 and 72 hpi, the global profile of tRNA transcription was remarkably transformed. The blockage of HCMV replication with PFA partially attenuated global effects on tRNA transcription. Interestingly, tRNAs that were virtually silent in uninfected cells were activated, many more than 100-fold, while repressed tRNAs tended to correlate with those that were already highly transcribed in uninfected cells. Of the 235 analyzed tRNAs, 138 (59%) exhibited an increase in transcription, while 97 (41%) exhibited a decrease (Figure 5D). Interestingly, the induced population of tRNAs was modestly enriched for a set containing AT-rich anticodons (45% GC, 55% AT) versus those that were repressed (50% GC, 50% GC). The effects on tRNA transcription were also reproduced in samples treated with Flavo, which indicated that the dramatically altered occupancy of Pol III was not affected by the loss of Pol II productive elongation for 1h (Figure S3C). Coverage at the 5′ upstream position correlated poorly between the samples uninfected at 48 hpi (r = 0.18), as expected, but each of the mock infected or time-matched samples −/+ Flavo correlated well (r = 0.78 to 0.97). Similar effects on tRNA transcription were also observed by plotting the coverage over the base downstream of the annotated 3′ end in uninfected and infected samples (Figure S3D).

To view the effects on tRNA transcription on a gene-by-gene basis, heatmaps of tRNA 5′ and 3′ ends were generated spanning a region of −50 to +200 bp relative to the mature 5′ ends (Figure 5E). The genes were sorted by most induced at 48 hpi to most repressed, top to bottom. The position of the mature 5′ end was indicated. Notably, for virtually all tRNAs, PRO-Seq 5′ ends were located upstream of the mature 5′ end, indicating that the data represented true Pol III initiation. The apparently inverse relationship between tRNA genes that were active in uninfected cells and those that were most induced at 48 hpi was highly evident. With infection, PRO-Seq 3′ ends extended well past the mature 3′ end at most tRNA genes, which are on average 74 nt downstream of the mature 5′ end. Interestingly, a recurrent pattern of 3′ ends in the heatmap for the uninfected sample was evident, and this pattern was diminished at 48 hpi, reflected instead by a shift in 3′ ends downstream. Barring the possibility that these apparent pauses in Pol III transcription arose from an undiscovered mapping artifact linked possibly to tRNA gene duplication, it seems that infection could alter tRNA transcription, in part, through the stimulation of Pol III elongation. However, a more extensive investigation would be needed to test this hypothesis.

3.6. HCMV Infection Alters the Chromatin State of Pol III-Transcribed Genes

Next, we investigated whether changes in Pol III transcription induced by infection were associated with changes in the chromatin structure around Pol III-transcribed genes, once more drawing upon our recently published DFF-ChIP data for H3K4me3 and TBP [18]. Similar to Pol II promoters, tRNA genes are couched within accessible regions of the genome [74]. H3K4me3 was also detected around Pol III promoters, though the abundance of this modification was often considerably less than that observed at Pol II promoters, and there was no consistent pattern of H3K4me3 flanking tRNA genes. In some cases, neighboring Pol II transcription impacts tRNA expression [9] and is likely to influence the epigenetic environment of tRNA genes. Three distinct effects on H3K4me3 were observed in association with tRNA induction by infection. Some tRNAs, such as tRNA-Val-TAC-4-1, tRNA-Met-CAT-5-1, and tRNA-Lys-TTT-7-1, which were nearly silent in uninfected cells, exhibited a marked gain in H3K4me3 in association with their induction (Figure 6A, left). At induced tRNAs, where H3K4me3-modified nucleosomes were detected downstream of the tRNA gene, such as tRNA-Met-CAT-4-2, tRNA-Gln-CGT-5-1, and tRNA-Pro-CGG-2-1, a downstream shift in the proximal H3K4me3 nucleosomes was often observed, which may be related to increased levels of Pol III elongation downstream of the tRNA gene (Figure 6A, middle). DFF-ChIP for TBP revealed prominent signals corresponding to Pol III PICs that, interestingly, were almost always associated with an upstream extension of approximately 150 bp that appeared to reflect association of the PIC with the -1 nucleosome (Figure 6). In yeast, the -1 nucleosome is more strongly positioned at tRNA genes than the downstream +1 nucleosome [74]. This is at contrast with Pol II promoters, where the +1 nucleosome was more strongly positioned, and interactions between the Pol II PIC and +1 nucleosome were detectable [18]. In view of this finding, at tRNA genes where H3K4me3 nucleosomes were detected primarily upstream, such as tRNA-Thr-CGT-5-1, tRNA-Ser-GCT-2-1, tRNA-Ser-AGA-2-4, and tRNA-Asp-GTC-2-7, stronger positioning of the −1 nucleosome was detected, and the signal profiles of H3K4me3 data suggested an increased contact between the −1 nucleosome and the PIC −/+ infection (Figure 6A, right). A fragMap for TBP at all tRNA genes revealed the size and position of the Pol III PIC relative to the major TSS and clearly showed an association between this PIC and the upstream −1 nucleosome (Figure 6B). The significance of this interaction was unclear, but it is possible that maintaining an interaction between the Pol III PIC and −1 nucleosome could block Pol II PIC assembly and transcriptional interference. Pol III also transcribes the 5S rRNA and a variety of snRNAs, including 7SK, which contain Class I and Class III promoters, respectively. An investigation of our PRO-Seq and DFF-ChIP data suggested that Pol III transcription of these genes was also induced by infection. Relating to this, it was previously noted that the Epstein–Barr virus induces the expression of 5S rRNA and 7SL [31]. Our TBP DFF-ChIP data revealed a unique struc-ture to PICs at Class I and Class III genes as compared to Class II tRNA genes, pre-sumably reflecting the involvement of different general transcription factors in the as-sembly of the complex. A diagram that depicts the size of the protected region detected by TBP DFF-ChIP in relation to the TSS for each promoter subtype, in addition to Pol II and Pol I promoters, is provided in Figure S4.

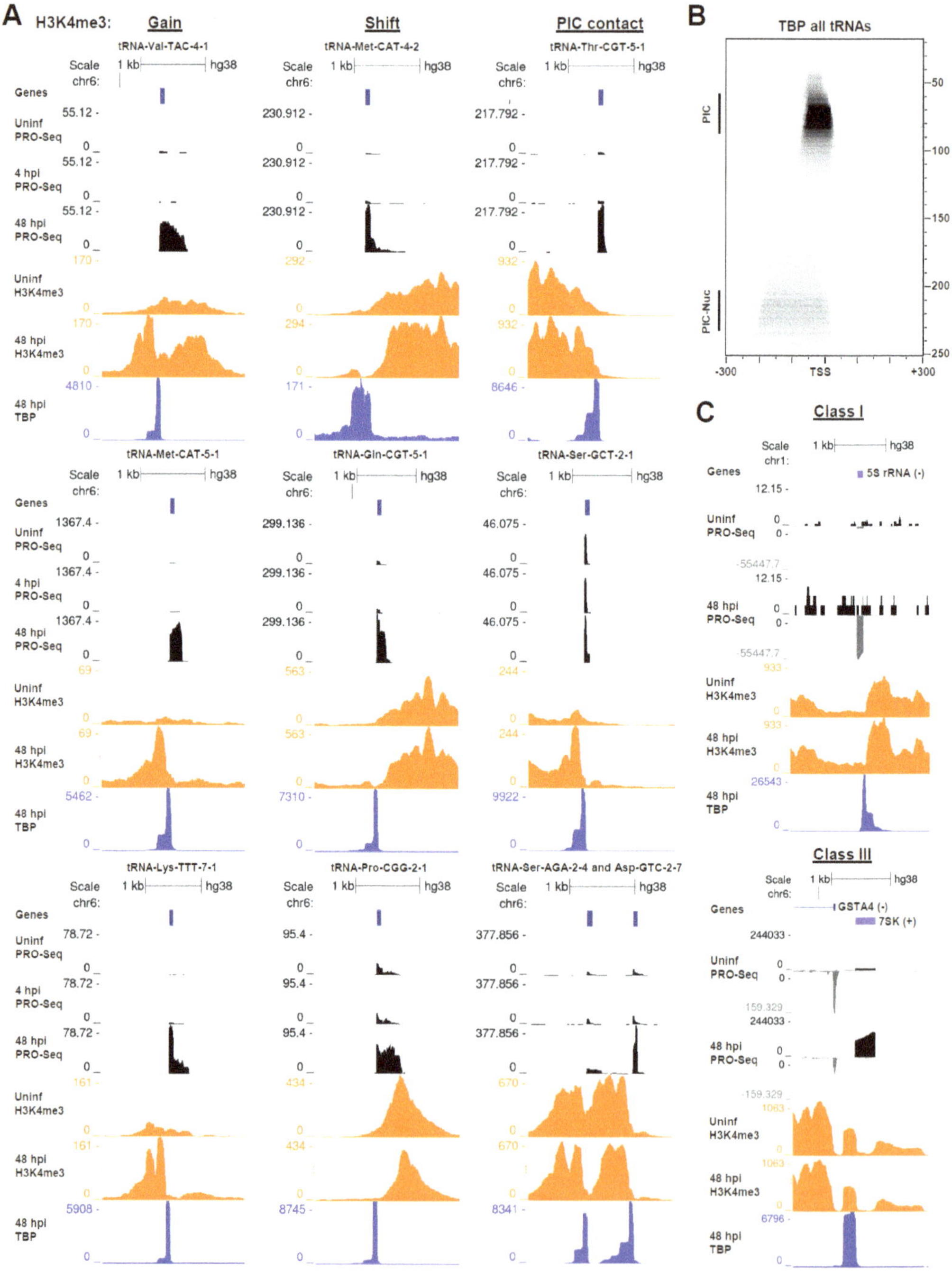

Figure 6. HCMV alters the chromatin state at induced tRNA genes. (**A**) Genome browser snapshots of forward strand PRO-Seq data in uninfected HFFs and HFFs infected for 4 or 48 h, DFF-ChIP for H3K4me3 in uninfected HFFs and HFFs 48 hpi, and DFF-ChIP for TBP in HFFs 48 hpi, at tRNA genes that represented three distinct changes in local chromatin observed in association with HCMV infection

(gain in H3K4me3, shift in H3K4me3 nucleosomes, and increased contact between the −1 nucleosome and the Pol III PIC). Three examples of each effect are shown; (**B**) FragMap of TBP DFF-ChIP data tRNA genes centered on the tRNA max TSS. The location of the Pol III PIC with respect to the TSS and association between the Pol III PIC and the upstream −1 nucleosome are visible; (**C**) genome browser snapshots of PRO-Seq data and DFF-ChIP data for H3K4me3 and TBP revealing an induction of the Class I 5S rRNA gene and Class III 7SK gene by HCMV infection, and features of TBP-containing PICs at these Pol III promoters.

4. Discussion

Here, we investigated the transcriptional effects of lytic HCMV infection in HFFs using PRO-Seq, which profiles nascent transcripts being generated by Pol I, II, and III. Our study provided surprising new insights, revealing, most significantly, that late HCMV infection was associated with a nearly global increase in the rate of the release of Pol II into productive elongation, that infection was seemingly linked to a change in Pol I transcription of the 45S rRNA at the level of elongation, and that, like HSV-1 and MHV68 [27,28], HCMV dramatically impacted tRNA expression at the level of transcription. These findings were connected to chromatin and epigenetic changes at affected genes, and together significantly extended our understanding of how HCMV manipulates the transcription of the host genome.

Our PRO-Seq data unexpectedly showed that that levels of Pol II productive elongation were substantially increased at late times postinfection (48, 72 h), while levels of initiation appeared to be largely unaffected. This observation is in alignment with the idea that HCMV does not shut off transcription of host genes as is thought to occur in the context of HSV-1, KSHV, and MHV68 infection. The trend was observed at most actively transcribed genes (Figures 1B and 2A,B) and appeared to be dependent on viral genome replication, as treatment with PFA from the onset of a 72 h infection restored pause ratios to a prereplication level (24 h). Pol II pause release is achieved by the action of P-TEFb, which is recruited to pause regions by BRD4 and other transcriptional coactivators, and critically phosphorylates the pausing and productive elongation factor DSIF. The phosphorylation of DSIF leads to the dissociation of the pausing factor NELF, after which the elongation complex is engaged by productive elongation factors to facilitate elongation through the gene body [14,75]. In cells, a portion of P-TEFb is free and able to mediate pause release, while the remainder exists within an inhibitory 7SK snRNP. P-TEFb is directly controlled by certain viruses, such as HIV, to drive productive viral gene expression [76], and P-TEFb release from the 7SK snRNP may be an important effector of HSV-1 reactivation from latency [77]. P-TEFb is generally required for productive transcription at the HCMV genome [12], reflective of its general role in Pol II transcription. At late times, following the onset of the HCMV genome replication, the transcription of the HCMV genome accounted for ~10–25% of the transcription in the cell (Supplementary Data File, PRO-Seq Stats), and, thus, a substantial fraction of free cellular P-TEFb must be needed to drive productive elongation on the HCMV genome. A previous study indicated that total P-TEFb levels are increased in HCMV-infected cells and that Cdk9 localization is altered during early and late times postinfection [78]. An increase in P-TEFb levels may explain our observed effects on pause release. Whether HCMV impacts the fraction of P-TEFb present in the snRNP or the total amount of P-TEFb is unknown and should be explored. In addition to uncovering this previously unreported broad effect of HCMV infection on pausing and productive elongation, our PRO-Seq data identified subsets of genes, consisting of many ISGs that were induced with differential kinetics in response to infection, and using DFF-ChIP for H3K4me3, we also showed that genes repressed by HCMV infection exhibited NFR closure.

Our PRO-Seq data unexpectedly suggested that Pol I transcription of the 45S rRNA is impacted by HCMV infection. Relatively few studies have investigated Pol I initiation and elongation in cells using next-generation sequencing approaches, which is in part related to limitations of existing annotations of rDNA repeats. Our data suggests that there is an enrichment of elongating Pol I at the 5′ end of the 45S gene, which is reminiscent of

Pol II pausing, and that late HCMV infection was associated with a decrease in this 5′ pileup and increased levels of Pol I elongation downstream. This effect also appeared to be dependent on viral replication, increased levels of viral mRNA, and/or a viral factor, as treatment with PFA led to its partial reversal. These data suggest that HCMV may drive productive rRNA transcription and perhaps downstream ribosome biogenesis to facilitate robust translation of viral mRNAs in a competitive environment where host mRNAs remain abundant. Additionally, our DFF-ChIP data for TBP sharply demarcated the boundaries of Pol I PICs in the 45S promoter region, and unexpectedly revealed PICs downstream of 45S that drive low levels of transcription initiation, are intimately associated with TTF1 termination sites, and could be involved in insulating tandem rDNA repeats from upstream Pol I transcription.

Finally, our PRO-Seq data revealed that HCMV dramatically induced the transcription of tRNAs. Our use of PRO-Seq to address effects on tRNA transcription is novel. Standard RNA-Seq approaches fail to accurately quantify the abundance of mature tRNAs due to extensive base modifications and secondary structure, which are both refractory to reverse transcription. Resultantly, methods that partially overcome these obstacles and enable a quantitative assessment of the levels of both mature tRNAs and pre-tRNAs have been developed [69–71,79,80]. PRO-Seq differs in that Pol III nascent transcripts are captured and sequenced. As tRNAs are post-transcriptionally processed, nascent Pol III transcripts lack base modifications. Also, the secondary structure within incompletely synthesized tRNAs may be minimal. Enrichment for nascent Pol III transcripts by PRO-Seq was shown through abundant evidence of unprocessed tRNA leader sequences, Pol III elongation downstream of the mature tRNA 3′ end, and termination over sites of poly-U incorporation. An advantage to measuring nascent transcripts is that mature tRNA half-lives are long, on the order of days, and, thus, small changes in the level of mature tRNAs could belie a major change at the level of transcription. In addition, effects on nascent transcription measured by PRO-Seq are likely to directly correlate with downstream changes in the levels of mature tRNA. We found that HCMV infection primarily resulted in an upregulation of tRNA transcription. Many tRNAs were induced more than 10-fold, and it was noted that induced tRNAs were modestly biased towards those containing AT-rich anticodons compared to repressed tRNAs. Interestingly, viral genome replication appeared to be important for the full induction of tRNA transcription, as treatment with PFA shifted the induction of tRNAs to prereplication levels observed at 24 h. This result suggests that tRNA induction may be linked to an increasing abundance of viral mRNAs, a result that is mirrored by a recent study of HSV-1 [27], and suggests the involvement of a viral factor in Pol III stimulation. Interestingly, and apparently unique to HCMV, the most induced tRNAs tended to be transcribed at low levels in uninfected cells, while repressed tRNAs tended to be very actively transcribed in uninfected cells. Thus, it appears that viruses from all major human herpesvirus subfamilies impact tRNA expression, possibly to the benefit of productive infection.

Interestingly, recent reports have suggested that HSV-1 and MHV68 generally lead to a greater increase in levels of pre-tRNAs than mature tRNAs [27,28]. It is possible that immature tRNAs may give rise to tRNA-derived fragments, which have recently emerged as important regulators of gene expression at the post-transcriptional level [81]. Whether tRNA transcripts that are induced by HCMV infection undergo complete maturation remains to be discerned. Regarding a mechanism for tRNA induction by HCMV, one possibility relates to the induction of Pol III initiation machinery, which has been reported to occur in the context of EBV infection [31]. Another interesting possibility relates to the function of Maf1, a Pol III transcriptional repressor that blocks the function of TFIIIB in Pol III PIC assembly [82,83]. Maf1 is rendered inactive by phosphorylation, which is thought to be mediated, at least in part, by the mTOR kinase [84,85]. The mTOR complex is a major player in the regulation of cellular metabolism and translation, and viral infection often triggers a cellular stress response that inactivates mTOR and, consequently, leads to a global downregulation of mRNA translation [86]. However, the HCMV UL38 protein, which is expressed with early kinetics, is thought to subvert this response by binding to and inhibiting TSC2, which with TSC1 indirectly represses mTOR, thereby promoting mTOR function [87,88]. Notably, some tRNAs began to be induced at 4 and 12 hpi, which was possibly consistent with early UL38 function, and UL38 protein levels apparently increase between 24 and 48 hpi [88]. This action may not only enable the translation of viral mRNAs, but could also lead to Maf1 inactivation by mTOR-mediated phosphorylation, with downstream consequences for tRNA transcription. Relating to this, HSV-1 is also a known mediate constitutive in mTOR activation, even in the context of nutrient deprivation, using similar mechanisms that impinge on upstream mTOR effectors [89,90]. However, possibly at contrast with these ideas, it was demonstrated that MHV68 infection induced tRNA upregulation in wild-type and Maf1-deficient cells [28].

Finally, our DFF-ChIP data provided several new insights regarding the chromatin status of tRNA genes and how it is impacted by infection. We observed that some induced tRNA genes exhibited gains in H3K4me3 at proximal nucleosomes, while tRNAs that were associated with H3K4me3 nucleosomes downstream exhibited a shift in the downstream nucleosomes to a more distal position in association with tRNA induction. The latter may be the result of increased levels of Pol III elongation downstream of the tRNA. Finally, our DFF-ChIP data resolved a striking connection between Pol III PICs and the upstream −1 nucleosome. At induced tRNAs, the profile of H3K4me3 signal appeared to suggest stronger positioning of the −1 nucleosome and increased contact with the Pol III PIC. A caveat to these interpretations is that H3K4me3 is likely substoichiometric with respect to nucleosomes flanking tRNA genes. As such, future studies should explore changes in the tRNA chromatin structure using approaches that are agnostic to the epigenetic status. Overall, our results positioned PRO-Seq and DFF-ChIP as useful tools for investigating the dynamics of tRNA transcription and chromatin.

5. Conclusions

In this study, we showed that HCMV infection elicited major changes in transcription by Pol I, II, and III using PRO-Seq. Our investigation of Pol II transcription unexpectedly revealed a substantial increase in the rates of Pol II pause release at late times postinfection that was reflected at most genes. This is in contrast with the effects of alpha and gamma-herpesviruses, which are thought to shut off host Pol II transcription. We showed that Pol I transcription was characterized by an apparent block to elongation at the 5′ end of the 45S gene that diminished during late infection, where increased levels of elongating Pol I were detected downstream. Finally, we reported on massive changes in Pol III transcription of tRNA genes, similar to recent reports on HSV-1 and MHV68 [27,28]. Hundreds of tRNAs were differentially transcribed during infection, and many were induced massively, up to nearly 500-fold. This effect largely required HCMV genome replication, indicating that tRNA induction may be coupled to increasing levels of viral mRNA or involve viral factors that remain to be identified.

Supplementary Materials: The following are available online at https://www.mdpi.com/article/10.3390/v14040779/s1, Supplementary Figure S1: Motif enrichment analysis in clustered regions, Figure S2: Reproducibility of Pol I transcription profile, Figure S3: Reproducibility of Pol III findings, Figure S4: Schematic depicting the regions protected by DFF-ChIP for TBP at Pol I, Pol II, and Pol III promoters, Supplementary Data File.

Author Contributions: Conceptualization, C.B.B., J.L.M. and D.H.P.; metho dology, C.B.B. and M.L.; software, M.P., C.B.B. and G.A.S.; validation, C.B.B. and D.H.P.; formal analysis, C.B.B., M.P. and D.H.P.; investigation, C.B.B. and D.H.P.; resources, M.L. and C.B.B.; data curation, C.B.B. and B.M.S.; writing—original draft preparation, C.B.B.; writing—review and editing, C.B.B., J.L.M. and D.H.P.; visualization, C.B.B.; supervision, D.H.P.; project administration, D.H.P.; funding acquisition, C.B.B., J.L.M. and D.H.P. All authors have read and agreed to the published version of the manuscript.

Funding: This research was funded by NIH, grant GM126908 to D.H.P., VA Merit Award I01 BX004434 to JLM, NIH grant AI130453 to JLM and DHP, and NIH T32-AI007533 to CBB.

Institutional Review Board Statement: The University of Iowa Institutional Review Board (IRB-01) determined on 02/08/17 that the use of de-identified, discarded human tissue for HCMV infection was not considered human subject research (IRB ID: 201702734).

Data Availability Statement: The raw and processed PRO-Seq and DFF-ChIP data analyzed in this study have been published in our previous study, and are available at NCBI GEO GSE185763.

Acknowledgments: We are delighted to contribute our study to this Special Issues of *Viruses*, honoring Randy Cohrs. Although some of us are relatively new to the herpesvirus field, we become acquainted with Randy at the International Herpesvirus Workshop. Clearly, he was one of the genuinely excellent scientists.

Conflicts of Interest: The authors declare no conflict of interest.

References

1. Russell, J.; Zomerdijk, J.C. The RNA polymerase I transcription machinery. *Biochem. Soc. Symp.* **2006**, *73*, 203–216. [CrossRef]
2. Chen, F.X.; Smith, E.R.; Shilatifard, A. Born to run: Control of transcription elongation by RNA polymerase II. *Nat. Rev. Mol. Cell Biol.* **2018**, *19*, 464–478. [CrossRef] [PubMed]
3. Jonkers, I.; Lis, J.T. Getting up to speed with transcription elongation by RNA polymerase II. *Nat. Rev. Mol. Cell Biol.* **2015**, *16*, 167–177. [CrossRef] [PubMed]
4. Field, A.; Adelman, K. Evaluating Enhancer Function and Transcription. *Annu. Rev. Biochem.* **2020**, *89*, 213–234. [CrossRef] [PubMed]
5. White, R.J. Transcription by RNA polymerase III: More complex than we thought. *Nat. Rev. Genet.* **2011**, *12*, 459–463. [CrossRef] [PubMed]
6. Turowski, T.W.; Tollervey, D. Transcription by RNA polymerase III: Insights into mechanism and regulation. *Biochem. Soc. Trans.* **2016**, *44*, 1367–1375. [CrossRef]
7. Dergai, O.; Hernandez, N. How to Recruit the Correct RNA Polymerase? Lessons from snRNA Genes. *Trends Genet.* **2019**, *35*, 457–469. [CrossRef]
8. Willis, I.M.; Moir, R.D. Signaling to and from the RNA Polymerase III Transcription and Processing Machinery. *Annu. Rev. Biochem.* **2018**, *87*, 75–100. [CrossRef]
9. Gerber, A.; Ito, K.; Chu, C.S.; Roeder, R.G. Gene-Specific Control of tRNA Expression by RNA Polymerase II. *Mol. Cell* **2020**, *78*, 765–778. [CrossRef]
10. Stinski, M.F.; Meier, J.L. Immediate-Early Viral Gene Regulation and Function. In *Human Herpesviruses: Biology, Therapy, and Immunoprophylaxis*; Arvin, A., Campadelli-Fiume, G., Mocarski, E., Moore, P.S., Roizman, B., Whitley, R., Yamanishi, K., Eds.; Cambridge University Press: Cambridge, UK, 2007.
11. Li, M.; Hu, Q.; Collins, G.; Parida, M.; Ball, C.B.; Price, D.H.; Meier, J.L. Cytomegalovirus late transcription factor target sequence diversity orchestrates viral early to late transcription. *PLoS Pathog.* **2021**, *17*, e1009796. [CrossRef]
12. Parida, M.; Nilson, K.A.; Li, M.; Ball, C.B.; Fuchs, H.A.; Lawson, C.K.; Luse, D.S.; Meier, J.L.; Price, D.H. Nucleotide Resolution Comparison of Transcription of Human Cytomegalovirus and Host Genomes Reveals Universal Use of RNA Polymerase II Elongation Control Driven by Dissimilar Core Promoter Elements. *mBio* **2019**, *10*, e02047-18. [CrossRef] [PubMed]
13. Isomura, H.; Stinski, M.F. Coordination of late gene transcription of human cytomegalovirus with viral DNA synthesis: Recombinant viruses as potential therapeutic vaccine candidates. *Expert. Opin. Ther. Targets* **2013**, *17*, 157–166. [CrossRef] [PubMed]
14. Peterlin, B.M.; Price, D.H. Controlling the elongation phase of transcription with P-TEFb. *Mol. Cell* **2006**, *23*, 297–305. [CrossRef] [PubMed]
15. Durand, L.O.; Roizman, B. Role of cdk9 in the optimization of expression of the genes regulated by ICP22 of herpes simplex virus 1. *J. Virol.* **2008**, *82*, 10591–10599. [CrossRef] [PubMed]
16. Zhao, Z.; Tang, K.W.; Muylaert, I.; Samuelsson, T.; Elias, P. CDK9 and SPT5 proteins are specifically required for expression of herpes simplex virus 1 replication-dependent late genes. *J. Biol. Chem.* **2017**, *292*, 15489–15500. [CrossRef]
17. Birkenheuer, C.H.; Baines, J.D. RNA Polymerase II Promoter-Proximal Pausing and Release to Elongation Are Key Steps Regulating Herpes Simplex Virus 1 Transcription. *J. Virol.* **2020**, *94*, e02035-19. [CrossRef]
18. Spector, B.M.; Parida, M.; Li, M.; Ball, C.B.; Meier, J.L.; Luse, D.S.; Price, D.H. Differences in RNA polymerase II complexes and their interactions with surrounding chromatin on human and cytomegalovirus genomes. *Nat. Commun.* **2022**. [CrossRef]
19. Dremel, S.E.; DeLuca, N.A. Herpes simplex viral nucleoprotein creates a competitive transcriptional environment facilitating robust viral transcription and host shut off. *eLife* **2019**, *8*, e51109. [CrossRef]
20. Dembowski, J.A.; DeLuca, N.A. Selective recruitment of nuclear factors to productively replicating herpes simplex virus genomes. *PLoS Pathog.* **2015**, *11*, e1004939. [CrossRef]
21. Kent, J.R.; Zeng, P.Y.; Atanasiu, D.; Gardner, J.; Fraser, N.W.; Berger, S.L. During lytic infection herpes simplex virus type 1 is associated with histones bearing modifications that correlate with active transcription. *J. Virol.* **2004**, *78*, 10178–10186. [CrossRef]

22. Zalckvar, E.; Paulus, C.; Tillo, D.; Asbach-Nitzsche, A.; Lubling, Y.; Winterling, C.; Strieder, N.; Mucke, K.; Goodrum, F.; Segal, E.; et al. Nucleosome maps of the human cytomegalovirus genome reveal a temporal switch in chromatin organization linked to a major IE protein. *Proc. Natl. Acad. Sci. USA* **2013**, *110*, 13126–13131. [CrossRef] [PubMed]
23. Shu, M.; Taddeo, B.; Zhang, W.; Roizman, B. Selective degradation of mRNAs by the HSV host shutoff RNase is regulated by the UL47 tegument protein. *Proc. Natl. Acad. Sci. USA* **2013**, *110*, E1669–E1675. [CrossRef] [PubMed]
24. Friedel, C.C.; Whisnant, A.W.; Djakovic, L.; Rutkowski, A.J.; Friedl, M.S.; Kluge, M.; Williamson, J.C.; Sai, S.; Vidal, R.O.; Sauer, S.; et al. Dissecting Herpes Simplex Virus 1-Induced Host Shutoff at the RNA Level. *J. Virol.* **2021**, *95*, e01399-20. [CrossRef] [PubMed]
25. Hartenian, E.; Gilbertson, S.; Federspiel, J.D.; Cristea, I.M.; Glaunsinger, B.A. RNA decay during gammaherpesvirus infection reduces RNA polymerase II occupancy of host promoters but spares viral promoters. *PLoS Pathog.* **2020**, *16*, e1008269. [CrossRef] [PubMed]
26. Pardamean, C.I.; Wu, T.T. Inhibition of Host Gene Expression by KSHV: Sabotaging mRNA Stability and Nuclear Export. *Front Cell Infect Microbiol.* **2021**, *11*, 648055. [CrossRef]
27. Dremel, S.E.; Sivrich, F.L.; Tucker, J.M.; Glaunsinger, B.A.; DeLuca, N.A. Manipulation of RNA polymerase III by Herpes Simplex Virus-1. *Nat. Commun.* **2022**, *13*, 623. [CrossRef] [PubMed]
28. Tucker, J.M.; Schaller, A.M.; Willis, I.; Glaunsinger, B.A. Alteration of the Premature tRNA Landscape by Gammaherpesvirus Infection. *mBio* **2020**, *11*, e02664-20. [CrossRef]
29. White, R.J.; Stott, D.; Rigby, P.W. Regulation of RNA polymerase III transcription in response to Simian virus 40 transformation. *EMBO J.* **1990**, *9*, 3713–3721. [CrossRef]
30. Felton-Edkins, Z.A.; White, R.J. Multiple mechanisms contribute to the activation of RNA polymerase III transcription in cells transformed by papovaviruses. *J. Biol. Chem.* **2002**, *277*, 48182–48191. [CrossRef]
31. Felton-Edkins, Z.A.; Kondrashov, A.; Karali, D.; Fairley, J.A.; Dawson, C.W.; Arrand, J.R.; Young, L.S.; White, R.J. Epstein-Barr virus induces cellular transcription factors to allow active expression of EBER genes by RNA polymerase III. *J. Biol. Chem.* **2006**, *281*, 33871–33880. [CrossRef]
32. Gaynor, R.B.; Feldman, L.T.; Berk, A.J. Transcription of class III genes activated by viral immediate early proteins. *Science* **1985**, *230*, 447–450. [CrossRef] [PubMed]
33. Hoeffler, W.K.; Roeder, R.G. Enhancement of RNA polymerase III transcription by the E1A gene product of adenovirus. *Cell* **1985**, *41*, 955–963. [CrossRef]
34. Wreghitt, T.G.; Abel, S.J.; McNeil, K.; Parameshwar, J.; Stewart, S.; Cary, N.; Sharples, L.; Large, S.; Wallwork, J. Intravenous ganciclovir prophylaxis for cytomegalovirus in heart, heart-lung, and lung transplant recipients. *Transpl. Int.* **1999**, *12*, 254–260. [CrossRef] [PubMed]
35. Dollard, S.C.; Grosse, S.D.; Ross, D.S. New estimates of the prevalence of neurological and sensory sequelae and mortality associated with congenital cytomegalovirus infection. *Rev. Med. Virol.* **2007**, *17*, 355–363. [CrossRef]
36. Hertel, L.; Mocarski, E.S. Global analysis of host cell gene expression late during cytomegalovirus infection reveals extensive dysregulation of cell cycle gene expression and induction of Pseudomitosis independent of US28 function. *J. Virol.* **2004**, *78*, 11988–12011. [CrossRef]
37. Zhu, H.; Cong, J.P.; Mamtora, G.; Gingeras, T.; Shenk, T. Cellular gene expression altered by human cytomegalovirus: Global monitoring with oligonucleotide arrays. *Proc. Natl. Acad. Sci. USA* **1998**, *95*, 14470–14475. [CrossRef]
38. Browne, E.P.; Wing, B.; Coleman, D.; Shenk, T. Altered cellular mRNA levels in human cytomegalovirus-infected fibroblasts: Viral block to the accumulation of antiviral mRNAs. *J. Virol.* **2001**, *75*, 12319–12330. [CrossRef]
39. Kwak, H.; Fuda, N.J.; Core, L.J.; Lis, J.T. Precise maps of RNA polymerase reveal how promoters direct initiation and pausing. *Science* **2013**, *339*, 950–953. [CrossRef]
40. Mahat, D.B.; Kwak, H.; Booth, G.T.; Jonkers, I.H.; Danko, C.G.; Patel, R.K.; Waters, C.T.; Munson, K.; Core, L.J.; Lis, J.T. Base-pair-resolution genome-wide mapping of active RNA polymerases using precision nuclear run-on (PRO-seq). *Nat. Protoc.* **2016**, *11*, 1455–1476. [CrossRef]
41. Li, M.; Ball, C.B.; Collins, G.; Hu, Q.; Luse, D.S.; Price, D.H.; Meier, J.L. Human cytomegalovirus IE2 drives transcription initiation from a select subset of late infection viral promoters by host RNA polymerase II. *PLoS Pathog.* **2020**, *16*, e1008402. [CrossRef]
42. Ball, C.B.; Nilson, K.A.; Price, D.H. Use of the nuclear walk-on methodology to determine sites of RNA polymerase II initiation and pausing and quantify nascent RNAs in cells. *Methods* **2019**, *159–160*, 165–176. [CrossRef] [PubMed]
43. Quinlan, A.R.; Hall, I.M. BEDTools: A flexible suite of utilities for comparing genomic features. *Bioinformatics* **2010**, *26*, 841–842. [CrossRef] [PubMed]
44. Gusmao, E.G.; Allhoff, M.; Zenke, M.; Costa, I.G. Analysis of computational footprinting methods for DNase sequencing experiments. *Nat. Methods* **2016**, *13*, 303–309. [CrossRef] [PubMed]
45. Chao, S.H.; Fujinaga, K.; Marion, J.E.; Taube, R.; Sausville, E.A.; Senderowicz, A.M.; Peterlin, B.M.; Price, D.H. Flavopiridol inhibits P-TEFb and blocks HIV-1 replication. *J. Biol. Chem.* **2000**, *275*, 28345–28348. [CrossRef] [PubMed]
46. Chen, E.Y.; Tan, C.M.; Kou, Y.; Duan, Q.; Wang, Z.; Meirelles, G.V.; Clark, N.R.; Ma'ayan, A. Enrichr: Interactive and collaborative HTML5 gene list enrichment analysis tool. *BMC Bioinform.* **2013**, *14*, 128. [CrossRef] [PubMed]
47. Kurokawa, C.; Iankov, I.D.; Galanis, E. A key anti-viral protein, RSAD2/VIPERIN, restricts the release of measles virus from infected cells. *Virus Res.* **2019**, *263*, 145–150. [CrossRef] [PubMed]

48. Schoggins, J.W. Interferon-Stimulated Genes: What Do They All Do? *Annu. Rev. Virol.* **2019**, *6*, 567–584. [CrossRef]
49. Rusinova, I.; Forster, S.; Yu, S.; Kannan, A.; Masse, M.; Cumming, H.; Chapman, R.; Hertzog, P.J. Interferome v2.0: An updated database of annotated interferon-regulated genes. *Nucleic Acids Res.* **2013**, *41*, D1040–D1046. [CrossRef]
50. Nilson, K.A.; Lawson, C.K.; Mullen, N.J.; Ball, C.B.; Spector, B.M.; Meier, J.L.; Price, D.H. Oxidative stress rapidly stabilizes promoter-proximal paused Pol II across the human genome. *Nucleic Acids Res.* **2017**, *45*, 11088–11105. [CrossRef]
51. Platanitis, E.; Decker, T. Regulatory Networks Involving STATs, IRFs, and NFkappaB in Inflammation. *Front. Immunol.* **2018**, *9*, 2542. [CrossRef]
52. DeFilippis, V.R.; Robinson, B.; Keck, T.M.; Hansen, S.G.; Nelson, J.A.; Fruh, K.J. Interferon regulatory factor 3 is necessary for induction of antiviral genes during human cytomegalovirus infection. *J. Virol.* **2006**, *80*, 1032–1037. [CrossRef] [PubMed]
53. Wang, W.; Xu, L.; Su, J.; Peppelenbosch, M.P.; Pan, Q. Transcriptional Regulation of Antiviral Interferon-Stimulated Genes. *Trends Microbiol.* **2017**, *25*, 573–584. [CrossRef] [PubMed]
54. Paulus, C.; Krauss, S.; Nevels, M. A human cytomegalovirus antagonist of type I IFN-dependent signal transducer and activator of transcription signaling. *Proc. Natl. Acad. Sci. USA* **2006**, *103*, 3840–3845. [CrossRef] [PubMed]
55. Song, J.; Lee, S.; Cho, D.Y.; Lee, S.; Kim, H.; Yu, N.; Lee, S.; Ahn, K. Human cytomegalovirus induces and exploits Roquin to counteract the IRF1-mediated antiviral state. *Proc. Natl. Acad. Sci. USA* **2019**, *116*, 18619–18628. [CrossRef]
56. Ziehr, B.; Vincent, H.A.; Moorman, N.J. Human Cytomegalovirus pTRS1 and pIRS1 Antagonize Protein Kinase R To Facilitate Virus Replication. *J. Virol.* **2016**, *90*, 3839–3848. [CrossRef]
57. Luse, D.S.; Parida, M.; Spector, B.M.; Nilson, K.A.; Price, D.H. A unified view of the sequence and functional organization of the human RNA polymerase II promoter. *Nucleic Acids Res.* **2020**, *48*, 7767–7785. [CrossRef]
58. Cui, K.; Tailor, P.; Liu, H.; Chen, X.; Ozato, K.; Zhao, K. The chromatin-remodeling BAF complex mediates cellular antiviral activities by promoter priming. *Mol. Cell Biol.* **2004**, *24*, 4476–4486. [CrossRef]
59. Soares, L.M.; He, P.C.; Chun, Y.; Suh, H.; Kim, T.; Buratowski, S. Determinants of Histone H3K4 Methylation Patterns. *Mol. Cell* **2017**, *68*, 773–785. [CrossRef]
60. Chen, K.; Chen, Z.; Wu, D.; Zhang, L.; Lin, X.; Su, J.; Rodriguez, B.; Xi, Y.; Xia, Z.; Chen, X.; et al. Broad H3K4me3 is associated with increased transcription elongation and enhancer activity at tumor-suppressor genes. *Nat. Genet.* **2015**, *47*, 1149–1157. [CrossRef]
61. Lauberth, S.M.; Nakayama, T.; Wu, X.; Ferris, A.L.; Tang, Z.; Hughes, S.H.; Roeder, R.G. H3K4me3 interactions with TAF3 regulate preinitiation complex assembly and selective gene activation. *Cell* **2013**, *152*, 1021–1036. [CrossRef]
62. Sims, R.J., 3rd; Millhouse, S.; Chen, C.F.; Lewis, B.A.; Erdjument-Bromage, H.; Tempst, P.; Manley, J.L.; Reinberg, D. Recognition of trimethylated histone H3 lysine 4 facilitates the recruitment of transcription postinitiation factors and pre-mRNA splicing. *Mol. Cell* **2007**, *28*, 665–676. [CrossRef] [PubMed]
63. Reinhardt, B.; Winkler, M.; Schaarschmidt, P.; Pretsch, R.; Zhou, S.; Vaida, B.; Schmid-Kotsas, A.; Michel, D.; Walther, P.; Bachem, M.; et al. Human cytomegalovirus-induced reduction of extracellular matrix proteins in vascular smooth muscle cell cultures: A pathomechanism in vasculopathies? *J. Gen. Virol.* **2006**, *87*, 2849–2858. [CrossRef] [PubMed]
64. Stanton, R.J.; McSharry, B.P.; Rickards, C.R.; Wang, E.C.; Tomasec, P.; Wilkinson, G.W. Cytomegalovirus destruction of focal adhesions revealed in a high-throughput Western blot analysis of cellular protein expression. *J. Virol.* **2007**, *81*, 7860–7872. [CrossRef] [PubMed]
65. Wroblewska, Z.; Wellish, M.C.; Wolinsky, J.S.; Gilden, D. Comparison of human cytomegalovirus growth in MRC-5 human fibroblasts, brain, and choroid plexus cells in vitro. *J. Med. Virol.* **1981**, *8*, 245–256. [CrossRef] [PubMed]
66. Moss, T.; Mars, J.C.; Tremblay, M.G.; Sabourin-Felix, M. The chromatin landscape of the ribosomal RNA genes in mouse and human. *Chromosome Res.* **2019**, *27*, 31–40. [CrossRef] [PubMed]
67. Mars, J.C.; Sabourin-Felix, M.; Tremblay, M.G.; Moss, T. A Deconvolution Protocol for ChIP-Seq Reveals Analogous Enhancer Structures on the Mouse and Human Ribosomal RNA Genes. *G3 (Bethesda)* **2018**, *8*, 303–314. [CrossRef]
68. Herdman, C.; Mars, J.C.; Stefanovsky, V.Y.; Tremblay, M.G.; Sabourin-Felix, M.; Lindsay, H.; Robinson, M.D.; Moss, T. A unique enhancer boundary complex on the mouse ribosomal RNA genes persists after loss of Rrn3 or UBF and the inactivation of RNA polymerase I transcription. *PLoS Genet.* **2017**, *13*, e1006899. [CrossRef]
69. Zheng, G.; Qin, Y.; Clark, W.C.; Dai, Q.; Yi, C.; He, C.; Lambowitz, A.M.; Pan, T. Efficient and quantitative high-throughput tRNA sequencing. *Nat. Methods* **2015**, *12*, 835–837. [CrossRef]
70. Gogakos, T.; Brown, M.; Garzia, A.; Meyer, C.; Hafner, M.; Tuschl, T. Characterizing Expression and Processing of Precursor and Mature Human tRNAs by Hydro-tRNAseq and PAR-CLIP. *Cell Rep.* **2017**, *20*, 1463–1475. [CrossRef]
71. Behrens, A.; Rodschinka, G.; Nedialkova, D.D. High-resolution quantitative profiling of tRNA abundance and modification status in eukaryotes by mim-tRNAseq. *Mol. Cell* **2021**, *81*, 1802–1815. [CrossRef]
72. Pan, T. Modifications and functional genomics of human transfer RNA. *Cell Res.* **2018**, *28*, 395–404. [CrossRef] [PubMed]
73. Arimbasseri, A.G.; Rijal, K.; Maraia, R.J. Transcription termination by the eukaryotic RNA polymerase III. *Biochim. Biophys. Acta* **2013**, *1829*, 318–330. [CrossRef] [PubMed]
74. Kumar, Y.; Bhargava, P. A unique nucleosome arrangement, maintained actively by chromatin remodelers facilitates transcription of yeast tRNA genes. *BMC Genom.* **2013**, *14*, 402. [CrossRef]
75. Zumer, K.; Maier, K.C.; Farnung, L.; Jaeger, M.G.; Rus, P.; Winter, G.; Cramer, P. Two distinct mechanisms of RNA polymerase II elongation stimulation in vivo. *Mol. Cell* **2021**, *81*, 3096–3109. [CrossRef] [PubMed]

76. Nilson, K.A.; Price, D.H. The Role of RNA Polymerase II Elongation Control in HIV-1 Gene Expression, Replication, and Latency. *Genet. Res. Int.* **2011**, *2011*, 726901. [CrossRef]
77. Alfonso-Dunn, R.; Turner, A.W.; Jean Beltran, P.M.; Arbuckle, J.H.; Budayeva, H.G.; Cristea, I.M.; Kristie, T.M. Transcriptional Elongation of HSV Immediate Early Genes by the Super Elongation Complex Drives Lytic Infection and Reactivation from Latency. *Cell Host Microbe* **2017**, *21*, 507–517. [CrossRef]
78. Tamrakar, S.; Kapasi, A.J.; Spector, D.H. Human cytomegalovirus infection induces specific hyperphosphorylation of the carboxyl-terminal domain of the large subunit of RNA polymerase II that is associated with changes in the abundance, activity, and localization of cdk9 and cdk7. *J. Virol.* **2005**, *79*, 15477–15493. [CrossRef]
79. Shigematsu, M.; Honda, S.; Loher, P.; Telonis, A.G.; Rigoutsos, I.; Kirino, Y. YAMAT-seq: An efficient method for high-throughput sequencing of mature transfer RNAs. *Nucleic Acids Res.* **2017**, *45*, e70. [CrossRef]
80. Pinkard, O.; McFarland, S.; Sweet, T.; Coller, J. Quantitative tRNA-sequencing uncovers metazoan tissue-specific tRNA regulation. *Nat. Commun.* **2020**, *11*, 4104. [CrossRef]
81. Krishna, S.; Raghavan, S.; DasGupta, R.; Palakodeti, D. tRNA-derived fragments (tRFs): Establishing their turf in post-transcriptional gene regulation. *Cell Mol. Life Sci.* **2021**, *78*, 2607–2619. [CrossRef]
82. Desai, N.; Lee, J.; Upadhya, R.; Chu, Y.; Moir, R.D.; Willis, I.M. Two steps in Maf1-dependent repression of transcription by RNA polymerase III. *J. Biol. Chem.* **2005**, *280*, 6455–6462. [CrossRef] [PubMed]
83. Goodfellow, S.J.; Graham, E.L.; Kantidakis, T.; Marshall, L.; Coppins, B.A.; Oficjalska-Pham, D.; Gerard, M.; Lefebvre, O.; White, R.J. Regulation of RNA polymerase III transcription by Maf1 in mammalian cells. *J. Mol. Biol.* **2008**, *378*, 481–491. [CrossRef] [PubMed]
84. Michels, A.A.; Robitaille, A.M.; Buczynski-Ruchonnet, D.; Hodroj, W.; Reina, J.H.; Hall, M.N.; Hernandez, N. mTORC1 directly phosphorylates and regulates human MAF1. *Mol. Cell Biol.* **2010**, *30*, 3749–3757. [CrossRef] [PubMed]
85. Kantidakis, T.; Ramsbottom, B.A.; Birch, J.L.; Dowding, S.N.; White, R.J. mTOR associates with TFIIIC, is found at tRNA and 5S rRNA genes, and targets their repressor Maf1. *Proc. Natl. Acad. Sci. USA* **2010**, *107*, 11823–11828. [CrossRef] [PubMed]
86. Martin, S.; Saha, B.; Riley, J.L. The battle over mTOR: An emerging theatre in host-pathogen immunity. *PLoS Pathog.* **2012**, *8*, e1002894. [CrossRef] [PubMed]
87. Bai, Y.; Xuan, B.; Liu, H.; Zhong, J.; Yu, D.; Qian, Z. Tuberous Sclerosis Complex Protein 2-Independent Activation of mTORC1 by Human Cytomegalovirus pUL38. *J. Virol.* **2015**, *89*, 7625–7635. [CrossRef]
88. Moorman, N.J.; Cristea, I.M.; Terhune, S.S.; Rout, M.P.; Chait, B.T.; Shenk, T. Human cytomegalovirus protein UL38 inhibits host cell stress responses by antagonizing the tuberous sclerosis protein complex. *Cell Host Microbe* **2008**, *3*, 253–262. [CrossRef]
89. Vink, E.I.; Lee, S.; Smiley, J.R.; Mohr, I. Remodeling mTORC1 Responsiveness to Amino Acids by the Herpes Simplex Virus UL46 and Us3 Gene Products Supports Replication during Nutrient Insufficiency. *J. Virol.* **2018**, *92*, e01377-18. [CrossRef]
90. Chuluunbaatar, U.; Roller, R.; Feldman, M.E.; Brown, S.; Shokat, K.M.; Mohr, I. Constitutive mTORC1 activation by a herpesvirus Akt surrogate stimulates mRNA translation and viral replication. *Genes Dev.* **2010**, *24*, 2627–2639. [CrossRef]

viruses

MDPI

Article

Tripartite-Motif 21 (TRIM21) Deficiency Results in a Modest Loss of Herpes Simplex Virus (HSV)-1 Surveillance in the Trigeminal Ganglia Following Cornea Infection

Amanda Berube [1], Grzegorz B. Gmyrek [1], Derek J. Royer [2] and Daniel J. J. Carr [1,3,*]

[1] Department of Ophthalmology, The University of Oklahoma Health Sciences Center, Oklahoma City, OK 73104, USA; amanda-berube@ouhsc.edu (A.B.); grzegorz-gmyrek@ouhsc.edu (G.B.G.)
[2] Department of Biology, East Texas Baptist University, Marshall, TX 75670, USA; droyer@etbu.edu
[3] Department of Microbiology and Immunology, The University of Oklahoma Health Sciences Center, Oklahoma City, OK 73104, USA
* Correspondence: dan-carr@ouhsc.edu

Abstract: Tripartite-motif 21 (TRIM21) is thought to regulate the type I interferon (IFN) response to virus pathogens and serve as a cytosolic Fc receptor for immunoglobulin. Since herpes simplex virus (HSV)-1 is sensitive to type I IFN and neutralizing antibody, we investigated the role of TRIM21 in response to ocular HSV-1 infection in mice. In comparison to wild type (WT) mice, TRIM21 deficient (TRIM21 KO) mice were found to be no more susceptible to ocular HSV-1 infection than WT animals, in terms of infectious virus recovered in the cornea. Similar pathology, in terms of neovascularization, opacity, and loss of peripheral vision function, was observed in both WT and TRIM21 KO mice. However, TRIM21 KO mice did possess a significant increase in infectious virus recovered in the trigeminal ganglia, in comparison to the WT animals. The increased susceptibility was not due to changes in HSV-1-specific CD4+ or CD8+ T cell numbers or functional capabilities, or in changes in type I IFN or IFN-inducible gene expression. In summary, the absence of TRIM21 results in a modest, but significant, increase in HSV-1 titers recovered from the TG of TRIM21 KO mice during acute infection, by a mechanism yet to be determined.

Keywords: HSV-1; cornea; trigeminal ganglia; T cells; interferon-γ; interferon stimulatory genes; TRIM21

Citation: Berube, A.; Gmyrek, G.B.; Royer, D.J.; Carr, D.J.J. Tripartite-Motif 21 (TRIM21) Deficiency Results in a Modest Loss of Herpes Simplex Virus (HSV)-1 Surveillance in the Trigeminal Ganglia Following Cornea Infection. *Viruses* **2022**, *14*, 589. https://doi.org/10.3390/v14030589

Academic Editors: Charles Grose, Ravi Mahalingam and Joel Rovnak

Received: 22 February 2022
Accepted: 10 March 2022
Published: 12 March 2022

1. Introduction

Herpes simplex virus (HSV)-1 is a double-stranded DNA virus that elicits a robust inflammatory response following ocular infection in the naive mouse, often resulting in significant tissue pathology [1–3]. While the incidence of HSV-1 infection in the human host overall is declining in industrialized countries, this pathogen continues to plague mankind and is now the leading cause of genital herpes infection in the United States and elsewhere [4,5]. Responsible for the success of the pathogen are the countermeasures elicited by virus-encoded proteins that facilitate immune evasion or suppression. For example, HSV-1-encoded glycoprotein (g)C, gE, and gI interfere with various aspects of antibody and complement action, including antibody-dependent cell cytotoxicity [6–8], whereas HSV-1-infected cell protein (ICP)47 blocks major histocompatibility complex (MHC) class I presentation of virus peptides to CD8+ cytotoxic T cells [9]. The virus also disrupts T cell receptor signaling at the linker for activation of T cell stage [10] and suppresses the invariant chain associated with MHC class II processing; thus, reducing MHC class II antigen presentation that is important in CD4+ T cell activation [11]. HSV-1, not only targets processes associated with the adaptive immune response, but also interferes with a number of anti-viral pathways associated with the innate immune response [12], including oligoadenylate synthetase/RNase L and STING/p204/IFI16 pathways, thought to be important in controlling HSV-1 replication in the cornea of mice [13,14]. HSV-1 also evades immune detection, establishing a 'latent' infection in neurons, where MHC class I molecules remain quiescent, even though the virus undergoes abortive reactivation with ongoing transcript expression in latently infected neurons [15,16]. The overall success of an HSV-1 infection depends upon the status of the naive host relative to the type I IFN response [3], which is influenced by the site of initial infection, as well as the likelihood of a sufficient (duration) and broad antigen exposure, to achieve significant coverage by the adaptive immune system against virus-encoded products.

Numerous cytosolic sensors detect viral RNA or DNA, which leads to the activation of type I IFNs [17–20]. Whereas type I IFNs are potent anti-viral compounds, chronic or excessive production of these cytokines has been associated with autoimmune disease and functional neuropathology [21–23]. Therefore, it is paramount to control the expression of IFN, once the resolution that prompts IFN expression occurs. The suppressor of cytokine signaling (SOCS) molecules, SOCS1 and SOCS3, have been found to inhibit Janus kinase signaling and, thus, dampen IFN expression [24]. Furthermore, SOCS1 and SOCS3 are induced by HSV-1, which would favor replication and spread of the virus in the host [25,26]. Another family of molecules that are upregulated by type I IFN and thought to influence downstream pathways activated by type I IFN are the tripartite-motif (TRIM) proteins [27]. The N-terminal domain of this family of proteins is known to possess E3 ligase activity, which targets ubiquitin and interferon stimulated protein of 15 kDa [28]. One specific TRIM protein, TRIM21, has been described as advantageous or deleterious to host resistance to virus infection based on targeting interferon regulator factor (IRF)3 [29,30]. In addition to the regulation of IFN downstream pathways, TRIM21 has also been found to be a cytosolic Fc receptor for IgG, IgM, and IgA molecules, which can then neutralize and degrade antibody-bound targets, including viruses [31,32].

In the present study, we investigated the role of TRIM21 in host resistance to ocular HSV-1 infection using TRIM21 deficient (TRMI21 KO), heterozygous (TRIM21+/−), and wild type (WT) mice, assessing virus replication and spread, the immune response, and cornea pathology parameters, to characterize and compare predicted changes between WT and TRIM21 KO mice. The results showed no difference in virus titer, inflammation, or pathology in the cornea of infected mice. However, there was a noticeable difference in virus replication, with 10-fold more infectious virus recovered from the trigeminal ganglia (TG) of TRIM21 KO mice compared to WT animals. This observation was consistent with an increase in lytic gene expression in the TG of TRIM21 KO mice compared to WT or TRIM21+/− animals. The increase in virus replication was not due to changes in the function of infiltrating cells, as CD4+ and CD8+ T cells isolated from the TG of infected mice all showed similar levels of IFN-γ expression post stimulation. Likewise, type I IFN and IFN-inducible gene expression levels within the TG were consistent between mouse genotypes. Therefore, the absence of TRIM21 expression specifically impacts the host control of HSV-1 replication in the TG by mechanisms yet to be determined

2. Materials and Methods

2.1. Mice and Infection

A breeder pair of heterozygous TRIM21 mice were obtained from Jackson Laboratory (Bar Harbor, ME, USA) and used to establish a colony. Offspring were genotyped to identify homozygous wild type (WT), heterozygous TRIM21 (TRIM21+/−), and deficient TRIM21 (TRIM21 KO) male and female mice, which were subsequently used in all experiments. Five- to six-week old mice were anesthetized with xylazine (5.0 mg/kg) and ketamine (100 mg/kg) administered intraperitoneally and then infected with HSV-1 (1000 plaque forming units (PFU/cornea) in 3 μL of phosphate buffered saline (PBS, pH = 7.4), following scarification of the cornea using a 25-gauge needle. All animals were housed in a specific pathogen-free vivarium at the Dean McGee Eye Institute on the University of Oklahoma Health Sciences Center campus, under the approved institutional animal use and care committee protocol, 19-060-ACHIX. At the indicated time, mice were exsanguinated, as previously described [33].

2.2. Virus and Plaque Assay

HSV-1 (strain McKrae) was propagated in the Vero African green monkey cell line (American Type Culture Collection, Manassas, VA, USA) with a stock (1.2×10^8 PFU/mL) stored at −80 °C until use. Virus titers in the cornea and TG were determined by plaque assay on day 3 and 7 post infection (pi), as previously described [34].

2.3. Real Time Reverse Transcriptase (RT)-Polymerase Chain Reaction (PCR)

The cornea and WT mice were collected at the indicated days pi or in uninfected WT mice. RNA was isolated using Trizol (Thermofisher, Waltham, MA, USA), and cDNA was generated using an iSCRIPT cDNA synthesis kit (Bio-Rad, Hercules, CA, USA). Forward and reverse oligonucleotide primers were used to amplify the targeted genes of interest by RT-PCR, as noted below Table 1:

Table 1. Primer Sequences for Targeted Genes.

	Forward	Reverse
IFNa4	5′-TTC TGC AAT GAC CTC CAT CA-3′	5′-GGC ACA GAG GCT GTG TTT CT-3′
mSTING	5′-CCT AGC CTC GCA CGA ACT TG-3′	5′-CGC ACA GCC TTC CAG TAG C-3′
DAI	5′-GGA AGA TCT ACC ACT CAC GTC-3′	5′-CCT TGT TGG CAG ATG ATG TTG-3′
Oasla	5′-CTT TGA TGT CCT GGG TCA TGT-3′	5′-GCT CCG TGA AGC AGG TAG AG-3′
ICP27	5′- GCA TCC TTC GTG TTT GTC AT-3′	5′- ACC AAG GGT CGC GTA GTC-3′
TK	5′-ATA CCG ACG ATC TGC GAC CT-3′	5′-TTA TTG CCG TCA TAG CGC GG-3′
gB	5′-TCT GCA CCA TGA CCA AGT G-3′	5′-TGG TGA AGG TGG TGG ATA TG-3′

A proprietary set of forward and reverse primers were obtained from a commercial vendor (BioRad) targeting GAPDH, TRIM21, IFNα1, IFNβ, and Bst2 to amplify targeted genes. The oligonucleotide primers are validated by the vendor for specificity and primer/dimer formation. Relative values of gene expression were determined via ΔΔCt method, using GAPDH to normalize samples [35]. Amplification of targeted gene using the gene amplicon format was as follows: 220 ng cDNA of sample was initially activated at 95 °C for 30 s, followed by a denaturation (95 °C for 5 s) and a annealing/extension step (60 °C for 30 s) repeated 40 times and a final hold step at 4 °C. A SYBER green ITaq Supermix (Bio-Rad) was used as part of the amplification, to detect amplified product. Melt curves (65–95 °C, 0.5 °C increments with 5 s/step) were used to establish a single product amplification. A CFX Connects thermal cycler (Bio-Rad) was used for all PCR experiments to monitor target amplification, with CFX Manager software to analyze the data. Threshold cycles were determined after subtracting the background fluorescence of each sample, where the relative light units achieved a level 10 standard deviations above the baseline relative light units.

2.4. Spectral Domain-Optical Coherence Tomography (SD-OCT)

Prior to and following infection, the corneas of mice were imaged using a Bioptigen SD-OCT system (Durham, NC, USA) for corneal thickness. Images were acquired using Bioptigen SD-OCT software (InVivoVue Clinic, Bioptigen, Morrisville, NC, USA).

2.5. Corneal Sensitivity

To measure the mechanosensory function of infected mice, a Cochet Bonnet esthesiometer (Luneau SAS, Rue Roger Bonnet, France) was employed. Briefly, infected mice, at the indicated time, were held firmly behind their neck and a monofilament varying in length from 0.5 to 6.0 cm was touched perpendicular to the cornea divided into four quadrants, and the length of the filament to elicit a blink response was recorded for each eye. The genotype of the infected mice was not apparent to the examiner until after the test. Uninfected mice served as the control to establish baseline readings.

2.6. Cornea Pathology

The opacity of the cornea was determined at 0, 7, and 30 days pi, following the removal of the cornea button from exsanguinated mice and assaying the tissue for absorbance of 500 nm using FLUOstar Omega plate reader (BMG Labtech, Offenburg, Germany), as described previously [33]. Following the measurement, the cornea was fixed using 4% paraformaldehyde (Sigma-Aldrich, St. Louis, MO, USA) for 30 min and washed in PBS containing 1% Triton X-100 (Sigma Aldrich). Following an incubation overnight with PBS containing 10% donkey serum (Abcam, Boston, MA, USA), the tissue was labeled for blood (CD31+) and lymphatic (LYVE1+) vessels, as previously described [36]. Alternatively, corneas were stained with rabbit anti-Trim21 (Abcam) and FITC-conjugated goat anti-HSV-1 antibody (Abcam) at a 1:1000 dilution. Following an incubation overnight with PBS containing 10% donkey serum (Abcam), the tissue was stained with donkey anti-rabbit Alexa Fluor 568 (Invitrogen, Eugene, OR, USA) at a 1:1000 dilution. Following an additional washing step, incisions were made into the cornea into quarters, such that the cornea could be placed on a slide for whole mount imaging in Dapi mounting media containing 50% glycerol. An Olympus FV1200 scanning confocal microscope (Center Valley, PA, USA) was used to sequentially scan labeled corneas and generate z-stacked images. The total area positive for vessels per field of view (4 quadrants/cornea) was analyzed and quantified using Metamorph software (Molecular Devices Inc., San Jose, CA, USA).

2.7. Flow Cytometry

TG from exsanguinated WT, TRIM21+/− and TRIM21 KO mice were collected at day 7 pi. Single-cell suspensions were generated using a 2 mL Wheatley Dounce homogenizer (Fisher Scientific, Waltham, MA, USA) in Dulbecco's modified eagle medium (DMEM) containing high glucose, L-glutamine, and pyruvate (Gibco Life Technologies, Grand Island, NY, USA) containing 10% fetal bovine serum (FBS, Gibco Life Technologies) and antibiotic/antimycotic solution (Thermofisher), and passed through a 70-µm cell strainer (Cell Treat Scientific Products, Pepperell, MA, USA). Cells were then washed in PBS (supplemented with 2% FBS; herein referred to as staining buffer, SB) and finally resuspending in the same buffer. Next, the cell suspension was stained with amine-reactive fluorescent dye Zombie Aqua (Biolegend, San Diego, CA, USA), used to assess live vs. dead status of TG cells. Non-specific binding sites of TG single-cell suspensions were blocked with 800 ng of CD16/CD32 (Catalog # 16-0161-82, clone 93, EBioscience, San Diego, CA, USA) for 15 min at 4 °C. The cells were then labeled with a combination of anti-CD45 (clone 30-F11) conjugated with Pacific Blue, (Catalog # 103126), -CD3 (clone 17A2) conjugated with PE-Cy7 (Catalog # 100220), -CD4 (clone GK1.5) conjugated with APC (Catalog #100412), and -CD8 (clone 53-6.7) conjugated with APC-Cy7 (Catalog #100714) (all from BioLegend, San Diego, CA, USA) at 800 ng/antibody/sample. To identify HSV-1-specific CD8+ T cells, cells were initially stained with 500 ng of one of the following PE-or Alexa Fluor 488-conjugated tetramers: H-2K(b)/SSIEFARL (glycoprotein B, gB), H-2K(b)/INNTFLHL (infected cell protein [ICP]8), or H-2K(b)/QTFDFGRL (ICP6) (all from NIH Tetramer Core Facility, Emory University, Atlanta, GA, USA), and incubated for 30 min on ice. Next, the cells were washed with PBS/2% FBS and fixed with 2% paraformaldehyde. To determine the number of HSV-1 -specific CD4 T cells, the cell suspension was stained with a combination of anti-CD45, -CD3, -CD4, and -CD8, along with BV421-conjugated tetramer I-A(b) IPPNWHIPSIQDA (glycoprotein D, gD, from NIH Tetramer Core Facility) for 1 h at 37 °C. Subsequently, the cells were washed with PBS/2% FBS and fixed with 2% paraformaldehyde until acquisition 24–48 h later using a spectral flow cytometer Aurora (Cytek Biosciences, Fremont, CA, USA) and analyzed using FlowJo software (FlowJo LLC, Medford, OR, USA).

For evaluation of interferon gamma (IFN-γ) producing T cells, the single TG cell suspension was placed into RPMI (supplemented with 10 mM D-glucose, 10% FBS, 1mM sodium pyruvate 2 mM L glutamine, and 1× antibiotic-antimycotic solution) and stimulated with 100 nM phorbol 12-myristate 13-acetate (PMA) and 1 μM ionomycin (both from MilliporeSigma, Burlington, MA, USA) for 6 h in the presence of brefeldin A (GolgiPlug from BD Biosciences/ThermoFisher Scientific). After the incubation period, the cells were washed with SB and stained with Zombie Aqua, followed by washing and staining with antibody cocktail (consisting of anti-CD45, CD3, CD4, or CD8) and PE-, AlexaFluor488, or BV421-conjugated HSV-1 tetramers, as described above, with the same incubation conditions. Finally, the cells were fixed and permeabilized with Foxp3/Transcription Factor Staining Buffer Set (Thermo Fischer Scientific) according to the manufacturer recommendations. At the permeabilization step, anti-mouse IFN-γ Ab conjugated with FITC (Thermo Fischer Scientific) was added to the cell suspension, followed by incubation at room temperature for 45 min. Subsequently, the cells were washed with permeabilization buffer, acquired using a spectral flow cytometer Aurora (Cytek Biosciences), and analyzed using FlowJo software (FlowJo LLC).

2.8. Statistical Analysis

Data were analyzed for significance using ($p < 0.05$) Prism 9.0 software (GraphPad, San Diego, CA, USA), with tests used to determine the significance indicated in each figure legend.

3. Results

3.1. HSV-1 Infection Induces TRIM21 Expression in the Cornea

TRIM21 expression has previously been reported to be upregulated in the cornea of mice in response to HSV-1 infection [37]. To confirm this observation in our model, we initially evaluated TRIM21 gene expression over time post HSV-1 infection. Whereas, there was little detection of TRIM21 mRNA expression in the uninfected cornea, TRIM21 levels dramatically rose 2 days pi and maintained that level through day 7 pi (Figure 1A). However, by day 14 pi TRIM21 mRNA expression had dropped back down to baseline levels, suggesting a dynamic but temporal TRIM21 gene expression response in the cornea following HSV-1 infection. Similar to the cornea, TRIM21 mRNA was not readily detected in the TG of uninfected mice (Figure 1B). However, by day 2 pi, TRIM21 expression significantly rose and was maintained throughout the period of surveillance (day 14 pi), following HSV-1 infection (Figure 1B). We also investigated TRIM21 expression at the protein level and found that, consistent with mRNA expression, TRIM21 protein (red) was not detected in the uninfected cornea (Figure 1C) but was upregulated in HSV-1-infected cornea, expressed by cells that were (yellow) and were not infected with HSV-1 (Figure 1D). It should be noted that a large number of HSV-1-infected cells (green) did not express TRIM21 (Figure 1D). Taken together, HSV-1 elicits a significant upregulation of TRIM21 expression in infected tissue, which is temporal or chronic dependent on the tissue.

3.2. The Absence of TRIM21 Is Reflected in a Loss of Virus Surveillance in the TG, but Not in Cornea of HSV-1-Infected Mice

TRIM21 has previously been reported to contribute to host defense against lethal mouse adenovirus-1 infection following intraperitoneal administration [38], likely through a cGAS-dependent mechanism [39]. However, in the case of HSV-1 infection, targeting TRIM21 expression using siRNA silencing, it was reported that a loss of TRIM21 expression resulted in a decrease in HSV-1 shedding in the tear film during acute infection [37]. Since we had found that virus shedding does not necessarily correlate with HSV-1 tissue levels [40], virus titers were obtained from the cornea and TG at times pi in WT, TRIM21+/−, and TRIM21 KO mice. At 3 days pi, there was no difference in the amount of infectious virus recovered from the cornea (Figure 2A) or TG (Figure 2B) of WT, TRIM21+/−, and TRIM21 KO mice. Likewise, infectious virus titers obtained from the cornea of WT, TRIM21+/−, and TRIM21 KO mice were all similar in quantity at day 7 pi (Figure 2C).

However, infectious HSV-1 levels were significantly higher in the TG of TRIM21 KO mice compared to the WT animals at day 7 pi (Figure 2D).

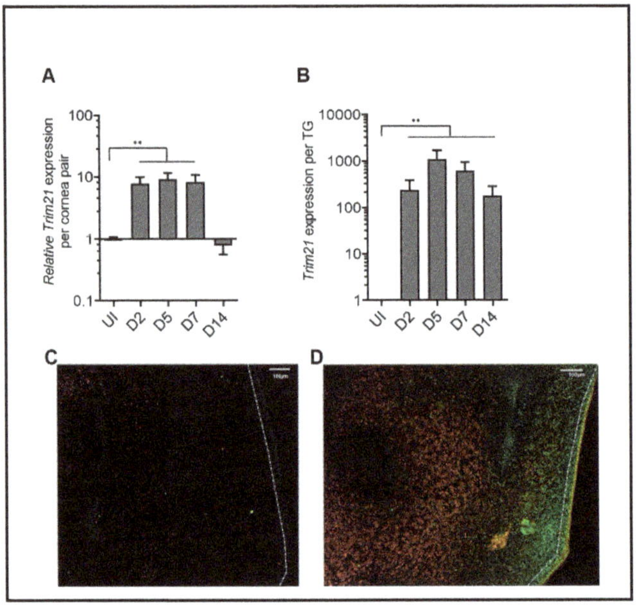

Figure 1. TRIM21 expression in the cornea and trigeminal ganglia. Male and female C57BL/6 (B6) mice were infected in the cornea with HSV-1 (10^3 plaque forming units (PFU) in 3 ul PBS. At the indicated time post infection (pi), the (**A**) corneas (5–7 samples/time point) or (**B**) trigeminal ganglia (TG) (3–5 samples/time point) of infected mice were removed and assessed for expression of TRIM21 by real-time RT-PCR. Bars represent the mean relative value + SEM, ** $p < 0.01$, comparing the indicated groups as determined by one-way ANOVA and Kruskal–Wallis multiple comparison test. Representative confocal images of (**C**) uninfected and (**D**) infected corneas stained for TRIM21 (red) and HSV-1 (green) antigen at 3 days post infection. Uninfected (UI) mice served as controls. The white scale bar = 100 µm.

3.3. The Loss of TRIM21 Does Not Modify Corneal Pathology Compared to WT Mice

Acute corneal infection of mice results in tissue pathology, including opacity and neovascularization [36,41–43]. Therefore, corneal opacity was quantified during acute infection (day 7 pi) into latency (day 30 pi). The results show that, similar to WT and TRIM21+/− mice, TRIM21 KO mice displayed the same degree of opacity at day 7 (Figure 3A) and day 30 (Figure 3B) pi. The opacity increased as the infection progressed from day 7 to day 30 pi, with significant differences comparing the TRIM21+/− and TRIM21 KO at day 7 to day 30 pi.

Consistent with these results, corneal neovascularization was also similar between genotypes, with WT, TRIM21+/− and TRIM21 KO showing the same degree of blood and lymphatic vessel genesis into the central cornea by day 30 pi (Figure 4). Representative corneal vascularization images for WT, TRIM21+/−, and TRIM21 KO show similar levels of blood and lymphatic vessel genesis into the central cornea (Figure 4A), summarized for area occupied by blood (Figure 4B) and lymphatic (Figure 4C) vessels at 30 days pi.

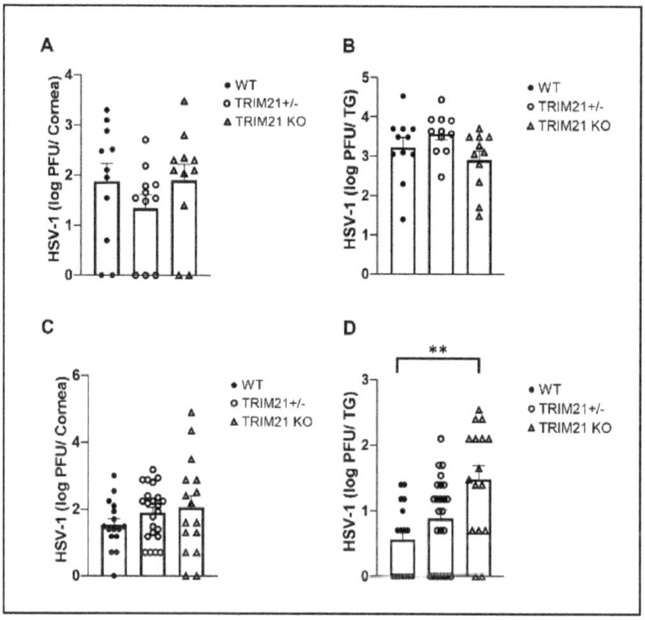

Figure 2. HSV-1 levels are elevated in the trigeminal ganglia of TRIM21 KO mice at day 7 post infection. Male and female C57BL/6 (WT), TRIM21+/−, and TRIM21 KO mice (n = 11–22/group) were infected with HSV-1 (10^3 PFU/cornea). At day 3 (**A**,**B**) or day 7 (**C**,**D**), the corneas (**A**,**C**) and trigeminal ganglia (TG) (**B**,**D**) of infected mice were removed and processed to determine viral load/tissue by standard plaque assay. Bars represent mean + SEM, ** $p < 0.01$ comparing the TRIM21 KO to WT mice day 7 pi, as determined by one-way ANOVA and Kruskal–Wallis multiple comparison test.

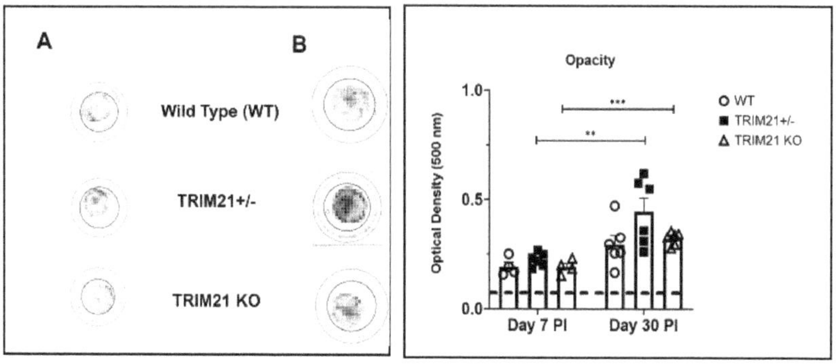

Figure 3. Corneal opacity is equivalent amongst WT, TRIM21+/−, and TRIM21 KO mice at different times post HSV-1 infection. Male and female C57BL/6 (WT), TRIM21+/−, and TRIM21 KO mice (n = 4–6/group) were infected with HSV-1 (10^3 PFU/cornea). At day 7 (**A**) or day 30 (**B**) post infection (PI), the mice were exsanguinated and the corneas were removed and assessed for opacity by optical density, using a plate reader, with tissue analyzed at 500-nm wavelength using a 30 × 30 matrix distributed over the cornea surface. Left panel is a representative cornea image from each mouse genotype captured at 500 nm. A summary of the results is shown in the right panel. Bars represent the mean + SEM. ** $p < 0.01$, *** $p < 0.001$ comparing the day 7 to day 30 pi timepoints, as determined by Kruskall–Wallis ANOVA and the Holm–Sidak post hoc t-test.

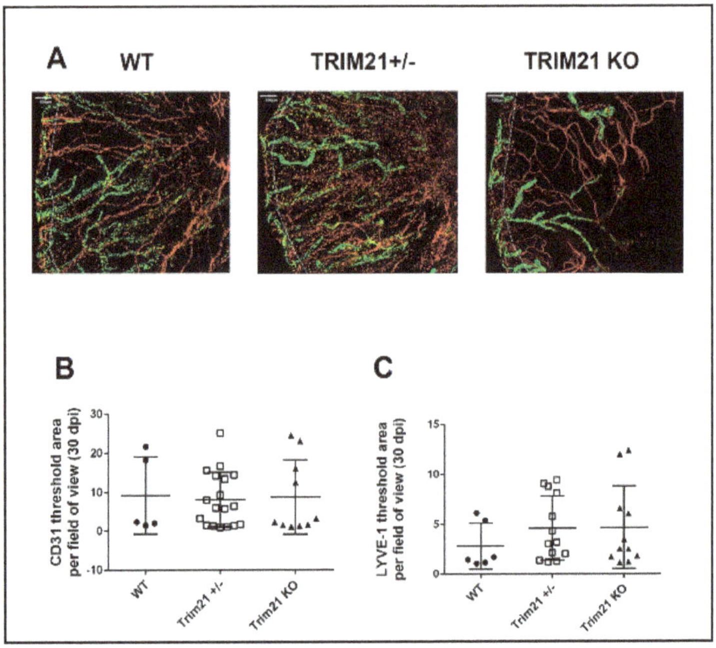

Figure 4. Corneal neovascularization is similar between WT, TRIM21+/− and TRIM21 KO mice in response to HSV-1 infection. Male and female C57BL/6 (WT), TRIM21+/−, and TRIM21 KO mice (n = 5–18/group) were infected with HSV-1 (10^3 PFU/cornea). At 30 days post infection, the mice were exsanguinated, and the corneas were removed, processed, and stained for neovascularization measuring of lymphatic (green) and blood (red) vessel genesis into the central cornea, captured by confocal microscopy. (**A**) Representative images of corneas from WT, TRIM21+/−, and TRIM21 KO infected mice at day 30 pi. Dotted line depicts the limbus margins. Summary of the area occupied by (**B**) CD31+ blood and (**C**) LYVE-1+ lymphatic vessels for each group of mice. Horizontal bars represent the mean + SEM. The white scale bar = 100 μm.

HSV-1 infection is also known to elicit cornea edema, likely due to the expression of pro-inflammatory cytokines and recruitment of leukocytes, including neutrophils, via chemokines to the site of infection [44–46]. We previously utilized SD-OCT to assess corneal edema following HSV-1 infection of mice [47]. Therefore, we assessed whether the absence of TRIM21 impacted the level of cornea edema, following HSV-1 infection. WT, TRIM21+/−, and TRIM21 KO mice were infected with HSV-1 and evaluated for cornea edema over time, until day 30 pi. Prior to infection, mice were imaged by SD-OCT, to establish a baseline. The results showed that, although HSV-1 induced a 30–40% increase in cornea edema, there was no difference in the amount of swelling between the genotypes of animals tested (Figure 5A,B). Taken together, the absence of TRIM21 within the cornea was not reflected by quantitative changes in ocular pathology, suggesting that TRIM21 plays little, if any, role in the host response to local virus infection and tissue pathology within the cornea.

3.4. Visual Axis Function Is Reduced in WT and TRIM21 KO Mice Following HSV-1 Infectionn

The consequences of ocular HSV-1 infection in mice often results in the disruption of functional parameters associated with vision or the integrity of the visual axis. Specifically, HSV-1 infection of the cornea often leads to a loss in mechanosensory function from denervation, including a loss of substance P-expressing sensory fibers, which are thought to contribute to wound healing [48–52]. One means of measuring mechanosensory function of the cornea is the blink reflex. Using a Cochet Bonnet esthesiometer, HSV-1-infected WT, TRIM21+/−, and TRIM21 KO mice were evaluated for cornea sensitivity. All groups of infected mice showed a significant loss in the blink response at day 7 pi (Figure 6A). WT mice tended to recover by day 30 pi, whereas TRIM21+/− and TRIM21 KO mice still showed a 30% loss in function. However, due to variability between measurements within the same group of TRIM21+/− and TRIM21 KO animals, the loss at 30 days pi was not significant.

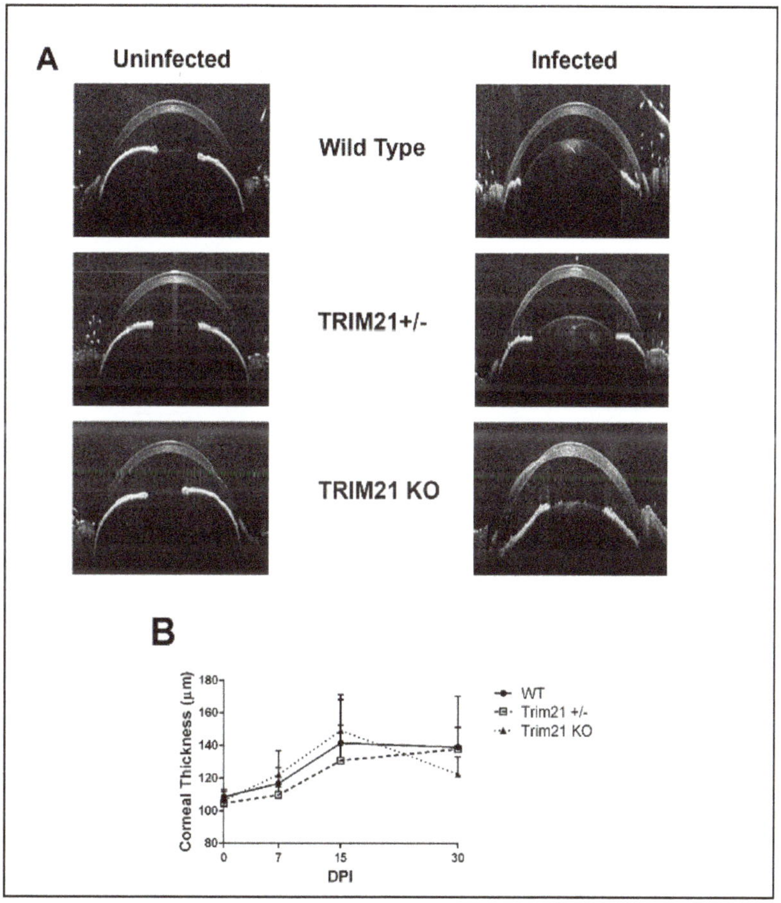

Figure 5. Cornea edema is similar between WT, TRIM21+/−, and TRIM21 KO mice in response to HSV-1 infection. Male and female C57BL/6 (WT), TRIM21+/−, and TRIM21 KO mice (n = 10–16/group/time point) were infected with HSV-1 (10^3 PFU/cornea). At the indicated time point before (0) or after infection, the mice were sedated and the corneas were measured for thickness by spectral-domain optical coherence tomography. (**A**) Representative figures for each mouse genotype prior to (uninfected) and at day 15 post infection (infected). (**B**) Summary of corneal thickness for each mouse genotype at each time point, taken prior to and after infection. The points represent the mean + SD.

The perception of optomotor movement by rodents, including mice, can be assessed using a virtual optokinetic tracking system [53]. To determine if changes in visual acuity occurred in WT, TRIM21+/− and TRIM21 KO mice following HSV-1 infection, we measured head movement under binocular viewing conditions and found that all groups of mice lost vision by day 15 pi, with a permanent loss of 25% measured at day 30 pi (Figure 6B). Thus, as in the tissue pathology results, the function of the visual axis appeared to be similarly compromised in all groups of mice evaluated post HSV-1 infection.

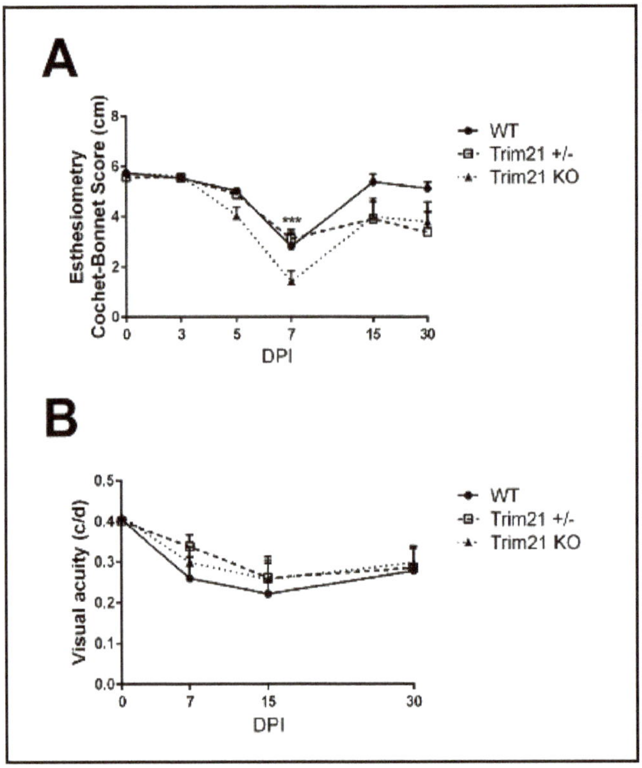

Figure 6. Performance of the visual axis drops in WT, TRIM21+/−, and TRIM21 KO mice following HSV-1 infection. Male and female C57BL/6 (WT), TRIM21+/− and TRIM21 KO mice (n = 5–15/group/ time point) were infected with HSV-1 (10^3 PFU/cornea). At the indicated time point, before (0) or after infection, the mice were evaluated for (**A**) mechanosensory function using a Cochet–Bonnet esthesiometer, to measure the blink reflex, or (**B**) assessed for visual acuity by optomotor kinetic tracking. All mice lost corneal sensation, with a maximum effect at 7 days post infection (DPI) and partial to complete recovery (in the case of WT mice) by 30 DPI (**A**). A 40% reduction in visual acuity was observed in all mouse groups, with a maximum effect at 15 DPI and partial recover by 30 DPI (**B**). Each point graphed represents the mean ± SEM. *** $p < 0.001$ comparing the 7 DPI to uninfected (0) time point, as determined by ANOVA and Tukey's multiple comparison test.

3.5. T Cell Infiltration and Function Are Similar between WT, TRIM21+/−, and TRIM21 KO Mice

Historically, T cells are thought to control HSV-1 replication in the TG during acute and latent infection following ocular HSV-1 challenge [54–57]. Since infectious virus detected in the TG of TRIM21 KO mice was equivalent to that found in the WT mice at day 3 pi, but elevated by day 7 pi, we interpreted these results as suggesting that the adaptive arm of the immune system likely plays a more significant role compared to the innate immune response within the TG of TRIM21 KO mice, relative to virus surveillance. Therefore, we initially investigated the T cell infiltrate and function in the TG at day 7 pi in WT, TRIM21+/−, and TRIM21 KO mice. The results showed no deficiency in the number of total (Figure 7A,C) or HSV-1-specific CD4+ or CD8+ T cells (Figure 7B,C) residing in the TG of HSV-1-infected TRIM21 KO mice compared to WT or TRIM21+/− animals.

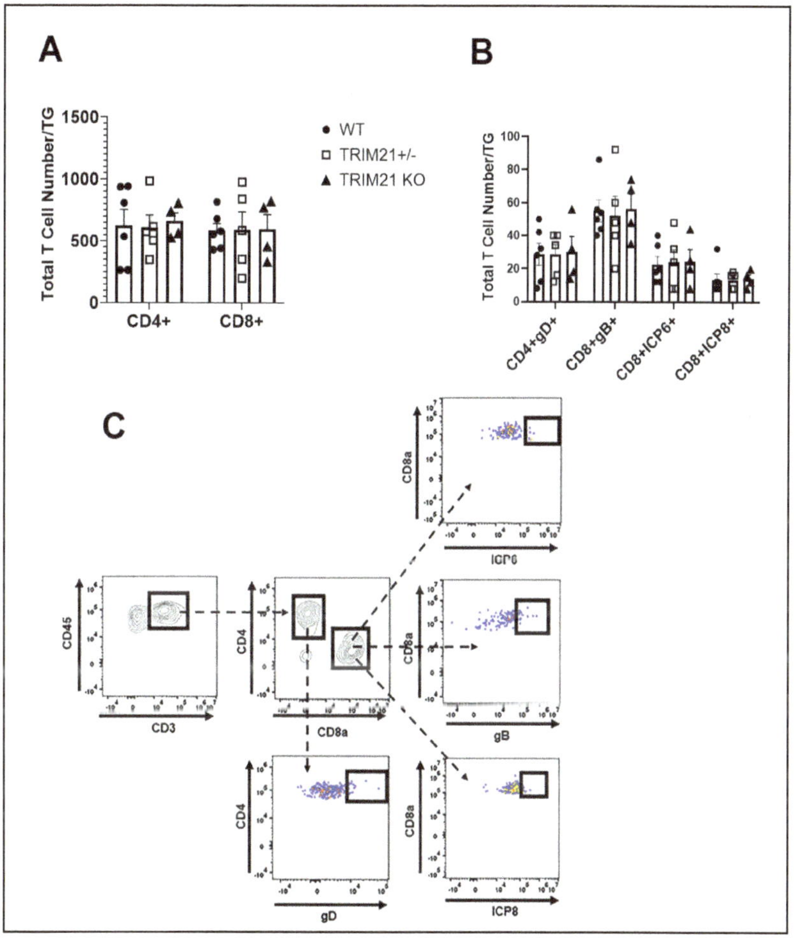

Figure 7. T cell infiltration in the TG of WT, TRIM21+/−, and TRIM21 KO mice at day 7 post infection is similar. Male and female C57BL/6 (WT), TRIM21+/−, and TRIM21 KO mice (n = 5–6/group) were infected with HSV-1 (10^3 PFU/cornea). At day 7 pi, the mice were exsanguinated, and the TG were processed to single cell suspensions and stained for (**A**) total CD4+ and CD8+ T cells and (**B**) HSV-1 gD-specific CD4+ T cells and HSV-1 ICP6-, ICP8-, and gB-specific CD8+ T cells. The results are summarized, with the bars representing the mean + SEM. Uninfected mice possessed less than 25 CD4+ or CD8+ T cells/TG, with no HSV-1 tetramer-positive cells detected. (**C**) Gating strategy in the identification of cell populations acquired for panels A and B.

Moreover, the function of HSV-1-specific CD4+ and CD8+ effector T cells, as measured by IFN-γ expression post-stimulation, did not reveal any significant differences in T cells from the WT, TRIM21+/−, and TRIM21 KO moue TG (Figure 8A,B). We interpreted these results as suggesting that T cell mobilization and function were not compromised in the TRIM21 KO mice, which would not explain the increase in virus titer captured in the TG of these animals at day 7 pi.

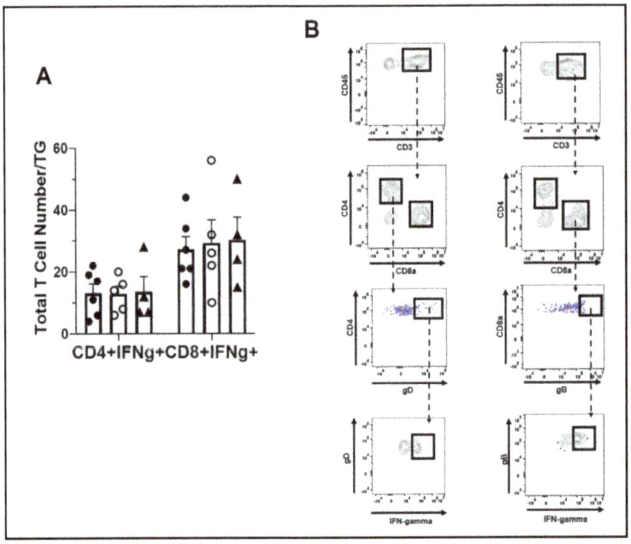

Figure 8. T cell function in the TG of WT, TRIM21+/− and TRIM21 KO mice, at day 7 post infection. Male and female C57BL/6 (WT), TRIM21+/−, and TRIM21 KO mice (n = 5–6/group) were infected with HSV-1 (10^3 PFU/cornea). (**A**) At day 7 pi, the mice were exsanguinated, and the TG were processed to single cell suspensions and cultured for 6 h in the presence of PMA and ionomycin in the presence of brefeldin. Following these culture conditions, the cells were washed and stained for IFN-γ expressing HSV-1 gD-specific CD4+ and gB-specific CD8+ T cells and analyzed by flow cytometry. The results are summarized, with the bars representing the mean + SEM. No IFN-γ expressing cells were noted in unstimulated cultures. (**B**) Gating strategy in the identification of cell populations acquired for panel A.

3.6. TG of TRIM21 KO Mice Display an Elevation in HSV-1 Lytic Gene Expression but Not Type I IFN or IFN-Inducible Genes Following Infection

Since greater susceptibility of TRIM21 KO mice to HSV-1 replication in the TG could not be explained by a deficiency in T cell infiltrate or function, other mechanisms were explored. TRIM21 is thought to be regulated by IFN regulatory elements, including IFN regulatory factor (IRF)1 and STING, as well as regulate type I IFN expression through IRF3 [37,58–60]. Moreover, type I IFNs and STING-mediated pathways contribute to the host defense against HSV-1 infection, by preventing replication of the virus [14,19,61–63]. Therefore, we, next evaluated levels of type I IFN and downstream effector pathways activated in response to type I IFN in the TG of HSV-1-infected mice. The results showed that the levels of type I IFNs, including IFN-α1, IFN-α4, and IFN-β, which have previously been found to be potent inhibitors of HSV-1 replication [35], were similar between TG from WT, TRIM21+/−, and TRIM21 KO mice (Figure 9A). Furthermore, downstream effector molecules of type I IFN-activated pathways that antagonize HSV-1, including tetherin (*Bst2*) [64] and oligoadenylate synthetases (OAS) [13,65] were found to be expressed at equivalent levels, as was STING and DAI in the TG of HSV-1-infected WT, TRIM21+/− and TRIM21 KO mice (Figure 9A), even though HSV-1 lytic gene expression (including ICP27, TK, and gB) was elevated in the TG of TRIM21 KO mice (Figure 9B). Therefore, the increased susceptibility of TRIM21 KO mice to HSV-1 replication in the TG, following ocular infection, is not due to changes in type I IFN expression or changes in the levels of IFN-inducible effector molecules OAS1a or tetherin found in the tissue.

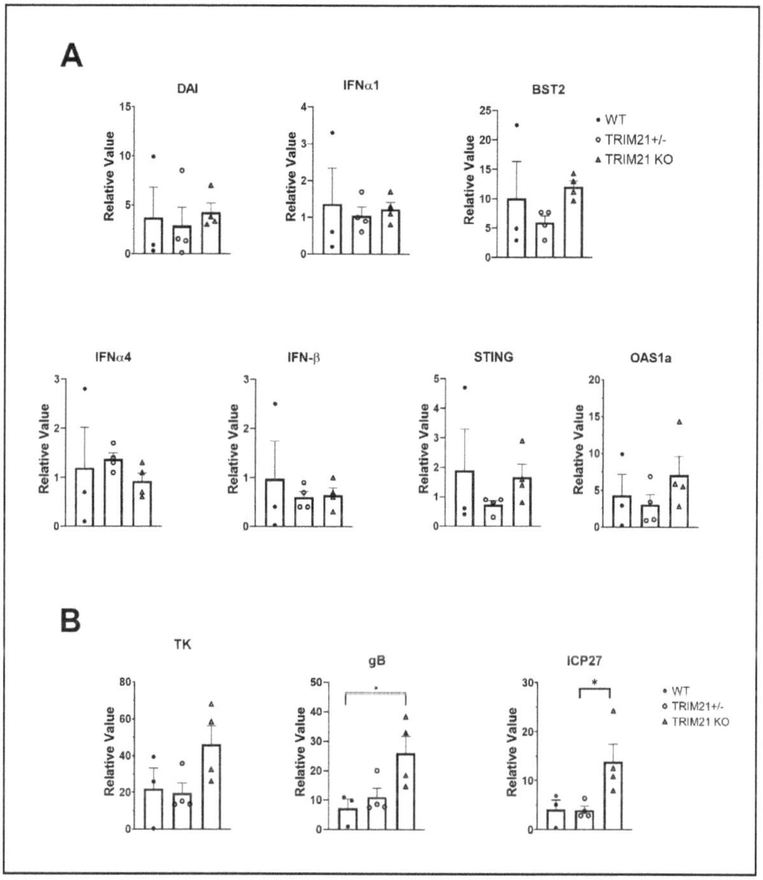

Figure 9. Elevation in HSV-1 lytic gene expression does not correlate with type I IFN gene expression. Female C57BL/6 (WT), TRIM21+/−, and TRIM21 KO mice (n = 4/group) were infected with HSV-1 (10^3 PFU/cornea). At day 7 pi, the mice were exsanguinated, and the TG were removed and processed for select gene expression by real-time RT-PCR. (**A**) Relative values for Bst2 (tetherin), DNA-dependent activator of interferon regulatory factors (DAI), IFN-α1, IFN-α4, IFN-β, STING, and oligoadenylate synthetase (OAS)1a and (**B**) HSV-1 lytic genes infected cell protein (ICP)27, thymidine kinase (TK), and glycoprotein B (gB) were determined using the ΔCt method, using uninfected genotypes as the controls to establish a baseline (relative value = 1). Bars represent the mean + SEM, * $p < 0.05$, comparing the indicated group as determined by Kruskall–Wallis ANOVA and the Holm–Sidak post hoc *t*-test.

4. Discussion

In the present study, we found local HSV-1-induced TRIM21 corneal expression during acute infection is temporal in nature returning to baseline levels by day 14 pi. However, TRIM21 does not appear to play a local role in host resistance against ocular HSV-1 challenge or virus-induced pathology, as there were no significant differences when comparing viral titer or virus-induced pathology in the cornea of WT to TRIM21 KO mice. In contrast to these results, the absence of TRIM21 was reflected by an elevation in the infectious virus content and virus lytic gene expression found in the TG by day 7 pi. As in the cornea, TRIM21 expression was significantly elevated in the TG following cornea infection. Unlike the cornea, TRIM21 expression did not dissipate in the TG following the clearance of infectious virus at day 14 pi. Since T cells are thought to contribute to virus surveillance in the TG during acute infection [56], we investigated the level and function of CD4+ and CD8+ T cells at the time when the virus load was elevated in the TRIM21 KO animals. We found there was no difference in the number of CD4+ or CD8+ T cells comparing WT, TRIM21+/−, and TRIM21 KO mice. Similarly to the incidence of T cell populations found in the HSV-1-infected TG, the number of IFN-γ-expressing CD4+ or CD8+ T cells was consistent amongst the WT, TRIM21+/−, and TRIM21 mice. While it is possible the location of effector T cells may not reside in close proximity to HSV-1-infected cells within the TG, to significantly influence virus replication, we surmise that it is more likely (an)other mechanism(s) may explain the loss in resistance to HSV-1 replication in the TG of TRIM21 KO mice.

TRIM21 is thought to regulate pathways associated with type I IFN response during virus infection. In human microglia, an increase in TRIM21 expression is thought to target and prevent IRF3 phosphorylation, resulting in the suppression IFN-β production in response to the single-stranded RNA virus, Japanese encephalitis virus [59]. Additional results were reported investigating the role of TRIM21 expression in mice, which showed TRIM21 KO mice yielded significantly less HSV-1 compared to WT controls, following intraperitoneal infection [66]. In this model, the negative effect of TRIM21 expression on control of HSV-1 infection was thought to be due to degradation of the cytosolic DNA sensor, DDX41, resulting in a loss of IFN-β production [66]. However, another study found TRIM21 stabilizes IRF3 expression resulting in an increase in resistance to virus infection; in this case, against Sendai virus [30]. As TRIM21 expression is tightly regulated by IRF pathways and the end-products, IFN-α and IFN-β, as well as IFN-γ [58], and as these cytokines are produced in response to HSV-1 infection [61,67,68], we investigated whether increased susceptibility to infection was reflected by changes in IFN levels or IFN-inducible gene expression. Even though there was an elevation in the HSV-1 lytic gene expression from TG of TRIM21 KO mice, we did not find any difference in IFN or IFN-inducible gene expression comparing the TRIM21 KO to WT or TRIM21+/− animals. Since the collection of tissue was a single time point, which coincided with when viral titers were assessed, it is possible earlier time points may have been more revealing. Alternatively, other antiviral mechanisms, including apoptosis [69,70] or necroptosis [71], may contribute to host defense against HSV-1 infection in the TG, as TRIM21 is associated with the tumor necrosis factor-related apoptosis-inducing ligand (TRAIL)-induced necrosome [72] and, thus, may prevent the use of elements within the cell that facilitate virus replication. At this point, we surmise that the microenvironment, which can include infiltrating myeloid-derived cells and NK cells, as well as the virus pathogen, greatly influences the positive or negative regulation of TRIM21 expression, which ultimately exerts influence on host resistance to virus infection, in this case, HSV-1.

It should be noted that the present study modeled acute infection into latency of naive mice. However, we did not evaluate latent virus content in our model and, therefore, cannot comment on whether the absence of TRIM21 influences the establishment of HSV-1 latency. It is possible that in the context of a prophylactic vaccine or upon repeat infection with a non-persistent pathogen, the Ig-Fc function of TRIM21 may be instrumental in the outcome of host resistance.

Author Contributions: Conceptualization, D.J.J.C. and D.J.R.; methodology, A.B., G.B.G. and D.J.J.C.; data analysis, A.B., G.B.G. and D.J.J.C.; writing—original draft, D.J.J.C.; writing—review and editing, D.J.J.C. and D.J.R.; supervision, D.J.J.C.; project administration, D.J.J.C.; funding acquisition, D.J.J.C. All authors have read and agreed to the published version of the manuscript.

Funding: This work was supported by NIH R01 AI053108, NEI core grant P30 EY021725, and an unrestricted grant from Research to Prevent Blindness. The funders had no role in data collection and interpretation, or the decision to submit the work for publication.

Institutional Review Board Statement: Not applicable.

Informed Consent Statement: Not applicable.

Data Availability Statement: The datasets generated and/or analyzed during the current study are available from the corresponding author upon reasonable request.

Conflicts of Interest: The authors declare no conflict of interest.

References

1. Aravalli, R.N.; Hu, S.; Rowen, T.N.; Palmquist, J.M.; Lokensgard, J.R. Cutting edge: TLR2-mediated proinflammatory cytokine and chemokine production by microglial cells in response to herpes simplex virus. *J. Immunol.* **2005**, *175*, 4189–4193. [CrossRef] [PubMed]
2. Gimenez, F.; Suryawanshi, A.; Rouse, B.T. Pathogenesis of herpes stromal keratitis—A focus on corneal neovascularization. *Prog Retin Eye Res.* **2013**, *33*, 1–9. [CrossRef] [PubMed]
3. Bryant-Hudson, K.; Conrady, C.D.; Carr, D.J. Type I interferon and lymphangiogenesis in the HSV-1 infected cornea—Are they beneficial to the host? *Prog. Retin. Eye Res.* **2013**, *36*, 281–291. [CrossRef] [PubMed]
4. Chemaitelly, H.; Nagelkerke, N.; Omori, R.; Abu-Raddad, L.J. Characterizing herpes simplex virus type 1 and type 2 seroprevalence declines and epidemiological association in the United States. *PLoS ONE* **2019**, *14*, e0214151. [CrossRef] [PubMed]
5. Ayoub, H.H.; Chemaitelly, H.; Abu-Raddad, L.J. Characterizing the transitioning epidemiology of herpes simplex virus type 1 in the USA: Model-based predictions. *BMC Med.* **2019**, *17*, 57. [CrossRef]
6. Fries, L.F.; Friedman, H.M.; Cohen, G.H.; Eisenberg, R.J.; Hammer, C.H.; Frank, M.M. Glycoprotein C of herpes simplex virus 1 is an inhibitor of the complement cascade. *J. Immunol.* **1986**, *137*, 1636–1641.
7. Lubinski, J.M.; Wang, L.; Soulika, A.M.; Burger, R.; Wetsel, R.A.; Colten, H.; Cohen, G.H.; Eisenberg, R.J.; Lambris, J.D.; Friedman, H.M. Herpes simplex virus type 1 glycoprotein gC mediates immune evasion in vivo. *J. Virol.* **1998**, *72*, 8257–8263. [CrossRef]
8. Dubin, G.; Socolof, E.; Frank, I.; Friedman, H.M. Herpes simplex virus type 1 Fc receptor protects infected cells from antibody-dependent cellular cytotoxicity. *J. Virol.* **1991**, *65*, 7046–7050. [CrossRef] [PubMed]
9. Hill, A.; Jugovic, P.; Russ, G.; Bennink, J.; Yewdell, J.; Ploegh, H.; Johnson, D. Herpes simplex virus turns off the TAP to evade host immunity. *Nature* **1995**, *375*, 411–415. [CrossRef] [PubMed]
10. Sloan, D.D.; Han, J.; Sandifer, T.K.; Stewart, M.; Hinz, A.J.; Yoon, M.; Johnson, D.C.; Spear, P.G.; Jerome, K.R. Inhibition of TCR signaling by herpes simplex virus. *J. Immunol.* **2006**, *176*, 1825–1833. [CrossRef] [PubMed]
11. Neumann, J.; Eis-Hübinger, A.M.; Koch, N. Herpes simplex virus type 1 targets the MHC class II processing pathway for immune evasion. *J. Immunol.* **2003**, *171*, 3075–3083. [CrossRef]
12. Zhu, H.; Zheng, C. The Race between Host Antiviral Innate Immunity and the Immune Evasion Strategies of Herpes Simplex Virus 1. *Microbiol. Mol. Biol. Rev.* **2020**, *84*, e00099-20. [CrossRef]
13. Austin, B.A.; James, C.; Silverman, R.H.; Carr, D.J. Critical role for the oligoadenylate synthetase/RNase L pathway in response to IFN-beta during acute ocular herpes simplex virus type 1 infection. *J. Immunol.* **2005**, *175*, 1100–1106. [CrossRef] [PubMed]
14. Conrady, C.D.; Zheng, M.; Fitzgerald, K.A.; Liu, C.; Carr, D.J. Resistance to HSV-1 infection in the epithelium resides with the novel innate sensor, IFI-16. *Mucosal Immunol.* **2012**, *5*, 173–183. [CrossRef] [PubMed]
15. Pereira, R.A.; Tscharke, D.C.; Simmons, A. Upregulation of class I major histocompatibility complex gene expression in primary sensory neurons, satellite cells, and Schwann cells of mice in response to acute but not latent herpes simplex virus infection in vivo. *J. Exp. Med.* **1994**, *180*, 841–850. [CrossRef]
16. Ma, J.Z.; Russell, T.A.; Spelman, T.; Carbone, F.R.; Tscharke, D.C. Lytic gene expression is frequent in HSV-1 latent infection and correlates with the engagement of a cell-intrinsic transcriptional response. *PLoS Pathog.* **2014**, *10*, e1004237. [CrossRef] [PubMed]
17. Yoneyama, M.; Kikuchi, M.; Natsukawa, T.; Shinobu, N.; Imaizumi, T.; Miyagishi, M.; Taira, K.; Akira, S.; Fujita, T. The RNA helicase RIG-I has an essential function in double-stranded RNA-induced innate antiviral responses. *Nat. Immunol.* **2004**, *5*, 730–737. [CrossRef] [PubMed]
18. Zhang, Z.; Kim, T.; Bao, M.; Facchinetti, V.; Jung, S.; Ghaffari, A.; Qin, J.; Cheng, G.; Liu, Y. DDX1, DDX21, and DHX36 helicases form a complex with the adaptor molecule TRIF to sense dsRNA in dendritic cells. *Immunity* **2011**, *34*, 866–878. [CrossRef] [PubMed]
19. Unterholzner, L.; Keating, S.E.; Baran, M.; Horan, K.A.; Jensen, S.B.; Sharma, S.; Sirois, C.M.; Jin, T.; Latz, E.; Xiao, T.S.; et al. IFI16 is an innate immune sensor for intracellular DNA. *Nat. Immunol.* **2010**, *11*, 997–1004. [CrossRef] [PubMed]
20. Ishikawa, H.; Barber, G.N. STING is an endoplasmic reticulum adaptor that facilitates innate immune signalling. *Nature* **2008**, *455*, 674–678. [CrossRef] [PubMed]
21. Rönnblom, L. The importance of the type I interferon system in autoimmunity. *Clin. Exp. Rheumatol.* **2016**, *34*, 21–24. [PubMed]
22. Kretschmer, S.; Lee-Kirsch, M.A. Type I interferon-mediated autoinflammation and autoimmunity. *Curr. Opin. Immunol.* **2017**, *49*, 96–102. [CrossRef] [PubMed]
23. Campbell, I.L.; Krucker, T.; Steffensen, S.; Akwa, Y.; Powell, H.C.; Lane, T.; Carr, D.J.; Gold, L.H.; Henriksen, S.J.; Siggins, G.R. Structural and functional neuropathology in transgenic mice with CNS expression of IFN-alpha. *Brain Res.* **1999**, *835*, 46–61. [CrossRef]
24. Naka, T.; Fujimoto, M.; Tsutsui, H.; Yoshimura, A. Negative regulation of cytokine and TLR signalings by SOCS and others. *Adv. Immunol.* **2005**, *87*, 61–122. [PubMed]

25. Frey, K.G.; Ahmed, C.M.I.; Dabelic, R.; Jager, L.D.; Noon-Song, E.N.; Haider, S.M.; Johnson, H.M.; Bigley, N.J. HSV-1-induced SOCS-1 expression in keratinocytes: Use of a SOCS-1 antagonist to block a novel mechanism of viral immune evasion. *J. Immunol.* **2009**, *183*, 1253–1262. [CrossRef]
26. Yokota, S.I.; Yokosawa, N.; Okabayashi, T.; Suzutani, T.; Miura, S.; Jimbow, K.; Fujii, N. Induction of suppressor of cytokine signaling-3 by herpes simplex virus type 1 contributes to inhibition of the interferon signaling pathway. *J. Virol.* **2004**, *78*, 6282–6286. [CrossRef] [PubMed]
27. McNab, F.W.; Rajsbaum, R.; Stoye, J.P.; O'Garra, A. Tripartite-motif proteins and innate immune regulation. *Curr. Opin. Immunol.* **2011**, *23*, 46–56. [CrossRef] [PubMed]
28. Meroni, G.; Diez-Roux, G. TRIM/RBCC, a novel class of 'single protein RING finger' E3 ubiquitin ligases. *Bioessays* **2005**, *27*, 1147–1157. [CrossRef] [PubMed]
29. Higgs, R.; Gabhann, J.N.; Ben Larbi, N.; Breen, E.P.; Fitzgerald, K.A.; Jefferies, C.A. The E3 ubiquitin ligase Ro52 negatively regulates IFN-beta production post-pathogen recognition by polyubiquitin-mediated degradation of IRF3. *J. Immunol.* **2008**, *181*, 1780–1786. [CrossRef]
30. Yang, K.; Shi, H.X.; Liu, X.Y.; Shan, Y.F.; Wei, B.; Chen, S.; Wang, C. TRIM21 is essential to sustain IFN regulatory factor 3 activation during antiviral response. *J. Immunol.* **2009**, *182*, 3782–3792. [CrossRef] [PubMed]
31. McEwan, W.A.; Tam, J.C.; Watkinson, R.E.; Bidgood, S.R.; Mallery, D.L.; James, L.C. Intracellular antibody-bound pathogens stimulate immune signaling via the Fc receptor TRIM21. *Nat. Immunol.* **2013**, *14*, 327–336. [CrossRef] [PubMed]
32. Mallery, D.L.; McEwan, W.A.; Bidgood, S.R.; Towers, G.J.; Johnson, C.M.; James, L.C. Antibodies mediate intracellular immunity through tripartite motif-containing 21 (TRIM21). *Proc. Natl. Acad. Sci. USA* **2010**, *107*, 19985–19990. [CrossRef] [PubMed]
33. Filiberti, A.; Gmyrek, G.B.; Montgomery, M.L.; Sallack, R.; Carr, D.J.J. Loss of Osteopontin Expression Reduces HSV-1-Induced Corneal Opacity. *Investig. Ophthalmol. Vis. Sci.* **2020**, *61*, 24. [CrossRef] [PubMed]
34. Halford, W.P.; Gebhardt, B.M.; Carr, D.J. Persistent cytokine expression in trigeminal ganglion latently infected with herpes simplex virus type 1. *J. Immunol.* **1996**, *157*, 3542–3549. [PubMed]
35. Harle, P.; Cull, V.; Agbaga, M.P.; Silverman, R.; Williams, B.R.; James, C.; Carr, D.J. Differential effect of murine alpha/beta interferon transgenes on antagonization of herpes simplex virus type 1 replication. *J. Virol.* **2002**, *76*, 6558–6567. [CrossRef]
36. Wuest, T.R.; Carr, D.J. VEGF-A expression by HSV-1-infected cells drives corneal lymphangiogenesis. *J. Exp. Med.* **2010**, *207*, 101–115. [CrossRef]
37. Tan, T.; Xia, L. TRIM21 Aggravates Herpes Simplex Virus Epithelial Keratitis by Attenuating STING-IRF3-Mediated Type I Interferon Signaling. *Front. Microbiol.* **2020**, *11*, 703. [CrossRef] [PubMed]
38. Vaysburd, M.; Watkinson, R.E.; Cooper, H.; Reed, M.; O'Connell, K.; Smith, J.; Cuickshanks, J.; James, L.C. Intracellular antibody receptor TRIM21 prevents fatal viral infection. *Proc. Natl. Acad. Sci. USA* **2013**, *110*, 12397–12401. [CrossRef] [PubMed]
39. Watkinson, R.E.; McEwan, W.A.; Tam, J.C.; Vaysburd, M.; James, L.C. TRIM21 Promotes cGAS and RIG-I Sensing of Viral Genomes during Infection by Antibody-Opsonized Virus. *PLoS Pathog.* **2015**, *11*, e1005253. [CrossRef] [PubMed]
40. Wuest, T.; Austin, B.A.; Uematsu, S.; Thapa, M.; Akira, S.; Carr, D.J. Intact TRL 9 and type I interferon signaling pathways are required to augment HSV-1 induced corneal CXCL9 and CXCL10. *J. Neuroimmunol.* **2006**, *179*, 46–52. [CrossRef]
41. Twardy, B.S.; Channappanavar, R.; Suvas, S. Substance P in the corneal stroma regulates the severity of herpetic stromal keratitis lesions. *Investig. Ophthalmol. Vis. Sci.* **2011**, *52*, 8604–8613. [CrossRef] [PubMed]
42. Zheng, M.; Deshpande, S.; Lee, S.; Ferrara, N.; Rouse, B.T. Contribution of vascular endothelial growth factor in the neovascularization process during the pathogenesis of herpetic stromal keratitis. *J. Virol.* **2001**, *75*, 9828–9835. [CrossRef]
43. Brissette-Storkus, C.S.; Reynolds, S.M.; Lepisto, A.J.; Hendricks, R.L. Identification of a novel macrophage population in the normal mouse corneal stroma. *Investig. Ophthalmol. Vis. Sci.* **2002**, *43*, 2264–2271.
44. Thomas, J.; Gangappa, S.; Kanangat, S.; Rouse, B.T. On the essential involvement of neutrophils in the immunopathologic disease: Herpetic stromal keratitis. *J. Immunol.* **1997**, *158*, 1383–1391. [PubMed]
45. O'Brien, W.J.; Guy, J.; Taylor, J.L. Pathogenesis of corneal oedema associated with herpetic eye disease. *Br. J. Ophthalmol.* **1990**, *74*, 723–730. [CrossRef] [PubMed]
46. Yan, X.T.; Tumpey, T.M.; Kunkel, S.L.; Oakes, J.E.; Lausch, R.N. Role of MIP-2 in neutrophil migration and tissue injury in the herpes simplex virus-1-infected cornea. *Investig. Ophthalmol. Vis. Sci.* **1998**, *39*, 1854–1862.
47. Royer, D.J.; Carr, M.M.; Gurung, H.R.; Halford, W.P.; Carr, D.J.J. The Neonatal Fc Receptor and Complement Fixation Facilitate Prophylactic Vaccine-Mediated Humoral Protection against Viral Infection in the Ocular Mucosa. *J. Immunol.* **2017**, *199*, 1898–1911. [CrossRef]
48. Hamrah, P.; Cruzat, A.; Dastjerdi, M.H.; Zheng, L.; Shahatit, B.M.; Bayhan, H.A.; Dana, R.; Pavan-Langston, D. Corneal sensation and subbasal nerve alterations in patients with herpes simplex keratitis: An in vivo confocal microscopy study. *Ophthalmology* **2010**, *117*, 1930–1936. [CrossRef] [PubMed]
49. Yun, H.; Rowe, A.M.; Lathrop, K.L.; Harvey, S.A.; Hendricks, R.L. Reversible nerve damage and corneal pathology in murine herpes simplex stromal keratitis. *J. Virol.* **2014**, *88*, 7870–7880. [CrossRef] [PubMed]
50. Chucair-Elliott, A.J.; Zheng, M.; Carr, D.J. Degeneration and regeneration of corneal nerves in response to HSV-1 infection. *Investig. Ophthalmol. Vis. Sci.* **2015**, *56*, 1097–1107. [CrossRef] [PubMed]
51. Chucair-Elliott, A.J.; Jinkins, J.; Carr, M.M.; Carr, D.J. IL-6 Contributes to Corneal Nerve Degeneration after Herpes Simplex Virus Type I Infection. *Am. J. Pathol.* **2016**, *186*, 2665–2678. [CrossRef] [PubMed]

52. Suvas, S. Role of Substance P Neuropeptide in Inflammation, Wound Healing, and Tissue Homeostasis. *J. Immunol.* **2017**, *199*, 1543–1552. [CrossRef]
53. Douglas, R.M.; Alam, N.M.; Silver, B.D.; McGill, T.J.; Tschetter, W.W.; Prusky, G.T. Independent visual threshold measurements in the two eyes of freely moving rats and mice using a virtual-reality optokinetic system. *Vis. Neurosci.* **2005**, *22*, 677–684. [CrossRef] [PubMed]
54. Liu, T.; Khanna, K.M.; Chen, X.; Fink, D.J.; Hendricks, R.L. CD8(+) T cells can block herpes simplex virus type 1 (HSV-1) reactivation from latency in sensory neurons. *J. Exp. Med.* **2000**, *191*, 1459–1466. [CrossRef] [PubMed]
55. Noisakran, S.; Carr, D.J. Lymphocytes delay kinetics of HSV-1 reactivation from in vitro explants of latent infected trigeminal ganglia. *J. Neuroimmunol.* **1999**, *95*, 126–135. [CrossRef]
56. Simmons, A.; Tscharke, D.; Speck, P. The role of immune mechanisms in control of herpes simplex virus infection of the peripheral nervous system. *Curr. Top. Microbiol. Immunol.* **1992**, *179*, 31–56. [PubMed]
57. Royer, D.J.; Conrady, C.D.; Carr, D.J. Herpesvirus-Associated Lymphadenitis Distorts Fibroblastic Reticular Cell Microarchitecture and Attenuates CD8 T Cell Responses to Neurotropic Infection in Mice Lacking the STING-IFNα/β Defense Pathways. *J. Immunol.* **2016**, *197*, 2338–2352. [CrossRef]
58. Sjöstrand, M.; Ambrosi, A.; Brauner, S.; Sullivan, J.; Malin, S.; Kuchroo, V.K.; Espinosa, A.; Wahren-Herlenius, M. Expression of the immune regulator tripartite-motif 21 is controlled by IFN regulatory factors. *J. Immunol.* **2013**, *191*, 3753–3763. [CrossRef] [PubMed]
59. Manocha, G.D.; Mishra, R.; Sharma, N.; Kumawat, K.L.; Basu, A.; Singh, S.K. Regulatory role of TRIM21 in the type-I interferon pathway in Japanese encephalitis virus-infected human microglial cells. *J. NeuroInflamm.* **2014**, *11*, 24. [CrossRef] [PubMed]
60. Li, D.; Wu, R.; Guo, W.; Xie, L.; Qiao, Z.; Chen, S.; Zhu, J.; Huang, C.; Huang, J.; Chen, B.; et al. STING-Mediated IFI16 Degradation Negatively Controls Type I Interferon Production. *Cell Rep.* **2019**, *29*, 1249–1260.e4. [CrossRef]
61. Hendricks, R.L.; Weber, P.C.; Taylor, J.L.; Koumbis, A.; Tumpey, T.M.; Glorioso, J.C. Endogenously produced interferon alpha protects mice from herpes simplex virus type 1 corneal disease. *J. Gen. Virol.* **1991**, *72 Pt 7*, 1601–1610. [CrossRef] [PubMed]
62. Halford, W.P.; Veress, L.A.; Gebhardt, B.M.; Carr, D.J. Innate and acquired immunity to herpes simplex virus type 1. *Virology* **1997**, *236*, 328–337. [CrossRef] [PubMed]
63. Reinert, L.S.; Lopušná, K.; Winther, H.; Sun, C.; Thomsen, M.K.; Nandakumar, R.; Mogensen, T.H.; Meyer, M.; Vægter, C.; Nyengaard, J.R.; et al. Sensing of HSV-1 by the cGAS-STING pathway in microglia orchestrates antiviral defence in the CNS. *Nat. Commun.* **2016**, *7*, 13348. [CrossRef] [PubMed]
64. Royer, D.J.; Carr, D.J. A STING-dependent innate-sensing pathway mediates resistance to corneal HSV-1 infection via upregulation of the antiviral effector tetherin. *Mucosal Immunol.* **2016**, *9*, 1065–1075. [CrossRef]
65. Al-khatib, K.; Williams, B.R.; Silverman, R.H.; Halford, W.; Carr, D.J. The murine double-stranded RNA-dependent protein kinase PKR and the murine 2′,5′-oligoadenylate synthetase-dependent RNase L are required for IFN-beta-mediated resistance against herpes simplex virus type 1 in primary trigeminal ganglion culture. *Virology* **2003**, *313*, 126–135. [CrossRef]
66. Zhang, Z.; Bao, M.; Lu, N.; Weng, L.; Yuan, B.; Liu, Y.J. The E3 ubiquitin ligase TRIM21 negatively regulates the innate immune response to intracellular double-stranded DNA. *Nat. Immunol.* **2013**, *14*, 172–178. [CrossRef] [PubMed]
67. Cantin, E.M.; Hinton, D.R.; Chen, J.; Openshaw, H. Gamma interferon expression during acute and latent nervous system infection by herpes simplex virus type 1. *J. Virol.* **1995**, *69*, 4898–4905. [CrossRef]
68. Thacore, H.R.; Mount, D.T.; Chadha, K.C. Interferon system of human cornea cells: Interferon production, characterization, and development of antiviral state. *J. Interferon Res.* **1982**, *2*, 401–408. [CrossRef] [PubMed]
69. Koyama, A.H.; Adachi, A. Induction of apoptosis by herpes simplex virus type 1. *J. Gen. Virol.* **1997**, *78 Pt 11*, 2909–2912. [CrossRef]
70. Krzyzowska, M.; Kowalczyk, A.; Skulska, K.; Thörn, K.; Eriksson, K. Fas/FasL Contributes to HSV-1 Brain Infection and Neuroinflammation. *Front Immunol.* **2021**, *12*, 714821. [CrossRef] [PubMed]
71. Huang, Z.; Wu, S.; Liang, Y.; Zhou, X.; Chen, W.; Li, L.; Wu, J.; Zhuang, Q.; Chen, C.; Li, J.; et al. RIP1/RIP3 binding to HSV-1 ICP6 initiates necroptosis to restrict virus propagation in mice. *Cell Host Microbe* **2015**, *17*, 229–242. [CrossRef] [PubMed]
72. Mélanie, S.E.; Florence, F.; Kara-Ali Ghania, H.; Mélanie, L.; Perrine, U.; Bonnet Marion, C.; Isabelle, G.; Emmanuelle, C.; Charles, P.; Michel, S.; et al. TRIM21, a New Component of the TRAIL-Induced Endogenous Necrosome Complex. *Front Mol. Biosci.* **2021**, *8*, 645134.

viruses

MDPI

Article

Antiviral Targeting of Varicella Zoster Virus Replication and Neuronal Reactivation Using CRISPR/Cas9 Cleavage of the Duplicated Open Reading Frames 62/71

Betty W. Wu [1,2], Michael B. Yee [2], Ronald S. Goldstein [3] and Paul R. Kinchington [2,*]

1 Graduate Program in Microbiology and Immunology, School of Medicine, University of Pittsburgh, Pittsburgh, PA 15213, USA; bww20@pitt.edu
2 Department of Ophthalmology, School of Medicine, University of Pittsburgh, Pittsburgh, PA 15213, USA; michaelyee80@gmail.com
3 Faculty of Life Sciences, Bar-Ilan University, Ramat Gan 5900002, Israel; ron.goldstein@biu.ac.il
* Correspondence: kinchingtonp@upmc.edu; Tel.: +1-412-647-6319

Abstract: Varicella Zoster Virus (VZV) causes Herpes Zoster (HZ), a common debilitating and complicated disease affecting up to a third of unvaccinated populations. Novel antiviral treatments for VZV reactivation and HZ are still in need. Here, we evaluated the potential of targeting the replicating and reactivating VZV genome using Clustered Regularly Interspaced Short Palindromic Repeat-Cas9 nucleases (CRISPR/Cas9) delivered by adeno-associated virus (AAV) vectors. After AAV serotype and guide RNA (gRNA) optimization, we report that a single treatment with AAV2-expressing *Staphylococcus aureus* CRISPR/Cas9 (saCas9) with gRNA to the duplicated and essential VZV genes ORF62/71 (AAV2-62gRsaCas9) greatly reduced VZV progeny yield and cell-to-cell spread in representative epithelial cells and in lytically infected human embryonic stem cell (hESC)-derived neurons. In contrast, AAV2-62gRsaCas9 did not reduce the replication of a recombinant virus mutated in the ORF62 targeted sequence, establishing that antiviral effects were a consequence of VZV-genome targeting. Delivery to latently infected and reactivation-induced neuron cultures also greatly reduced infectious-virus production. These results demonstrate the potential of AAV-delivered genome editors to limit VZV productive replication in epithelial cells, infected human neurons, and upon reactivation. The approach could be developed into a strategy for the treatment of VZV disease and virus spread in HZ.

Keywords: varicella zoster virus; latency; reactivation; genome cleavage; CRISPR/Cas9; AAV; antiviral therapies

Citation: Wu, B.W.; Yee, M.B.; Goldstein, R.S.; Kinchington, P.R. Antiviral Targeting of Varicella Zoster Virus Replication and Neuronal Reactivation Using CRISPR/Cas9 Cleavage of the Duplicated Open Reading Frames 62/71. *Viruses* **2022**, *14*, 378. https://doi.org/10.3390/v14020378

Academic Editors: Charles Grose, Ravi Mahalingam and Joel Rovnak

Received: 14 January 2022
Accepted: 10 February 2022
Published: 12 February 2022

1. Introduction

Varicella-zoster virus (VZV) is the human alphaherpesvirus that causes varicella (chickenpox) during primary infection and herpes zoster (HZ; commonly called "shingles") when the virus reactivates from the latent state, often decades after the initial infection [1]. Without immune boosting through the use of HZ vaccines, it is estimated that one-third of the population will develop HZ in their lifetimes, with incidence rising with age and declining immune status caused by natural senescence, disease, or iatrogenic causes. HZ remains a public health concern because it is often complicated by scarring, bacterial infections, and acute pain that can be debilitating. A significant fraction of HZ patients develop difficult-to-treat chronic pain states termed post-herpetic neuralgia (PHN), which can be so severe that they reduce quality-of-life. HZ may also be followed by neurological, gastrointestinal, and vascular diseases [2–4], as well as potentially blinding complications that develop after facial zoster [5].

HZ incidence and severity are reduced by boosting the existing VZV-specific immunity using vaccines. The first HZ licensed vaccine (used since 2005) was based on a higher dose version of the live-attenuated VZV strain used in the varicella vaccine, and it reduced HZ incidence by half and the disease burden by two-thirds [6]. However, it was contraindicated in immunocompromised patients who could develop vaccine virus-induced disease [7,8]. A more recent Federal Drug Administration (FDA)-approved vaccine is based on the novel AS01B adjuvant and purified VZV glycoprotein E (gE). This subunit vaccine has higher efficacy against HZ, but requires two doses, has frequent side effects or injection-site reactions and is not used worldwide [9–11]. Hence, the uptake of the HZ vaccines remains low. HZ disease can respond to antiviral treatment if it is initiated early, but current antivirals only target active viral replication. The latent state remains refractory to any known antiviral treatment [5,12,13]. Thus, VZV diseases still impact millions of people worldwide and there is a need for improved HZ treatments/prevention.

One of the main hurdles in developing novel VZV therapeutic strategies has been the difficulty in modeling VZV pathogenesis in animal models. VZV demonstrates high human specificity, and does not fully replicate or cause disease in rodents, compared to the modeling of infections and diseases caused by the related herpes simplex virus type 1 (HSV-1) [14]. Indeed, there is no in vivo immunocompetent model of human VZV-induced primary disease, a VZV latent state that is reactivatable, or HZ-like disease states resulting from reactivation [1,15]. Attempts to experimentally reactivate VZV from latently infected human cadaver ganglia have also not yet been successful [16,17]. VZV does replicate in fetal human tissues harbored in severe compromised immunodeficient (SCID) mice, which can be used to evaluate pathogenesis and antiviral studies [18–20], but their use can be cost-prohibitive and requires special animal-experimentation permissions. However, neuronal culture models have been developed that harbor latent VZV that can be experimentally reactivated. We previously reported neuronal cultures derived from human embryonic stem cells (hESC) that model VZV neuronal lytic replication, axonal transport, neuron-to-neuron spread, and a prolonged viral persistent state that could be experimentally reactivated by the interruption of the NGF signaling or alteration of histone/chromatin architecture [21–23]. These and similar neuronal models used by other groups [24–30] have now established the means to probe the VZV latent state and investigate the potential targeting of latent and lytic replicating genomes using gene editing. Targeted CRISPR/Cas9 and specific homing meganucleases have been explored to target HSV-infected neurons and in murine neuronal models of disease and latency in vivo [31–33], as well as human cytomegalovirus (hCMV) and Epstein–Barr virus in cultured immune-cell models [34–36]. However, gene-editing strategies have not yet been studied for antiviral targeting of VZV replication or its latent state, as far as we are aware.

In this study, the potential of gene editing to target VZV genomes as a means to prevent lytic, latent, and reactivated infections was investigated. The *Staphylococcus aureus* CRISPR/Cas9 (saCas9) system was exploited because it has a high specificity and small size, permitting efficient packaging into AAV together with guide RNA [37,38]. VZV essential and duplicated genes present in the internal and terminal genome repeat regions were targeted, allowing for the cleaving of genomes at more than one position using a single vector. The VZV genome contains three duplicated genes in the repeated sequences bounding the short unique region, ORFs 62/71, 63/70 and 64/69 [39,40]. Of these, ORFs 62/71 and 63/70 encode regulatory proteins [41,42] that have been shown to be essential [43,44]. Furthermore, ORF62 encodes IE62, the VZV ortholog of the well-characterized HSV-1 ICP4 transcriptional transactivator that is required for the expression of all HSV-1 early and late genes. The HSV-1 ICP4 activates transcription by recruiting the host-cell transcriptional machinery to the genome [45]. VZV IE62 enhances the infectivity of transfected VZV DNA [41,46] and has a sufficient functional-conserved structure to HSV ICP4 so that it can partly replace ICP4 in the HSV genome leading to the production of an infectious virus [47,48]. We selected AAV vectors for delivery because AAVs have been used both in animal models as well as clinical trials for neuronal delivery, with promising therapeutic delivery potential [49–53]. Here, we demonstrate that AAV2-packaged saCas9 targeting VZV ORFs 62/71 can greatly reduce VZV lytic replication in epithelial cells and lytic-infected human neurons, severely curtail VZV growth and damage the virus following reactivation from latency in neuron cultures.

2. Materials and Methods

2.1. Cells and Viruses

All cell lines except the NIH-registered human embryonic stem cell line Wa09 (H9) were purchased from ATCC. Cell-culture reagents were obtained from Thermo Fisher Scientific (Waltham, PA, USA) unless otherwise noted. The Wa09 (H9) was obtained from WiCell (Madison, WI, USA) and differentiated into neurons as detailed previously [22]. Retinal-pigmented epithelial (ARPE-19) and HEK 293 cells were grown in Dulbecco's Minimal Essential Media (DMEM, #10569-010) supplemented with 10% fetal bovine serum (FBS; #S11150, R&D Systems, Minneapolis, MN, USA), 100 units/mL penicillin + 100 mg/mL streptomycin + 0.25 mg/mL amphotericin B as an antibiotics/antimycotic solution (#ABL02, Caisson, Smithfield, UT, USA). Human melanoma (MeWo) cells were maintained in Minimum Essential Media (MEM) supplemented with 10% FBS and antibiotics/antimycotic.

All infection studies used virus or viral recombinants based on the Parent of Oka (pOka) strain, a wildtype clinical isolate that was the parent of the live attenuated varicella and zoster vaccines. Cell-associated VZV were prepared as previously described [54], as infected ARPE-19 cells that were frozen after mitotic inhibition with growth media containing 0.01 mM mitomycin C (#A4452, ApexBio, Houston, TX, USA) for 3–4 h. at 37 °C. Cell-associated VZV was slow-frozen at −80 °C overnight in media containing 10% DMSO before long-term liquid-nitrogen storage. Frozen aliquots were titrated in triplicate for subsequent infections. Cell-free VZV was prepared using a previously published protocol [55], stored aliquoted in liquid nitrogen, and titrated after freezing for use in subsequent infections.

2.2. CRISPR/Cas9 Plasmid and gRNA Design

Three optimal gRNA target sequences with PAM motifs and guide length that were predicted to target saCas9 to each of ORF62 and ORF63 were selected using an online protocol from the Zhang lab [37] and Software (Benchling Inc., San Francisco, CA, USA). Oligonucleotides (Table 1) were synthesized by Integrated DNA Technologies (Integrated DNA Technologies INC., Coralville, IA, USA). Oligonucleotide annealing and cloning into the backbone vector pX601 (pX601-AAV-CMV::NLS-SaCas9-NLS-3xHA-bGHpA;U6::BsaI-sgRNA was a gift from Feng Zhang; Addgene plasmid # 61591; http://n2t.net/addgene:61591; RRID:Addgene_61591 accessed on 10 January 2022) were performed as published online (https://media.addgene.org/cms/filer_public/6d/d8/6dd83407-3b07-47db-8adb-4fada30bde8a/zhang-lab-general-cloning-protocol-target-sequencing_1.pdf accessed on 10 January 2022). Briefly, the vector was digested with BsaI and a dsDNA formed by the hybridization of the two oligonucleotides was ligated into the pX601 plasmid, where pX601 expresses saCas9 and the inserted gRNA from the same vector. A vector expressing *Streptococcus Pyogenes* CRISPR/Cas9 targeting the intergenic region between UL3 and UL4 was developed using the pX330 vector (pX330-U6-Chimeric_BB-CBh-hSpCas9 was a gift from Feng Zhang (Addgene plasmid # 42230; http://n2t.net/addgene:42230; RRID:Addgene_42230 accessed on 10 January 2022), as previously described [56,57]. Primers used for making gRNA are listed in Table 1.

2.3. Preparation of AAV

AAV reporter vectors of different serotypes expressing the green fluorescent protein under the chicken β-actin promoter CAG were acquired from Virovek (Virovek, Hayward, CA, USA). High titer (10^{13} viral genome copies (GC) per mL) ORF62-targeting AAV2 and a pX601 vector-only control lacking gRNA, were prepared by Penn Vector Core's Gene Therapy Program (https://gtp.med.upenn.edu/, accessed on 10 January 2022). Lower titer (approximately 10^{10} GC/mL) control or ORF62-targeting AAV2 for use in preliminary experiments were made by transfecting HEK293 cells with the control pX601 (with no gRNA) or ORF62-targeting plasmid (pX601+62-1 gRNA), along with plasmids expressing AAV2 replication gene (pRC2-mi342 vector) and a helper plasmid that expresses adenoviral helper proteins (pHelper; Takara, Kusatsu, Shiga, Japan) using the Xfect reagent (#631318, Takara). 6 h after transfection, media was changed to fresh DMEM with 10% FBS and cells were further incubated at 37 °C 5% CO_2 for 72 h. Cells were then harvested into an AAV lysis buffer (50 mM Tris, 150 mM NaCl, 0.05% Tween-20, pH 8.5) and subjected to three freeze–thaw cycles. AAV were then purified using a published protocol in which a discontinuous 15%–25%–40%–60% iodixanol gradient was used. Virus at the interface between 40% and 60% iodixanol was harvested after centrifugation at 200,000× *g* for 2 h. at 18 °C [58,59]. Samples were then concentrated into 0.001% Pluronic F68 with 200 mM NaCl in D-PBS using Amicon Ultra-15 centrifugal filters (#C7715, Millipore Sigma, Burlington, MA, USA) and spun at 3700× *g* for 20 min at 4 °C. AAV stocks were then aliquoted and stored at −80 °C until use. An aliquot of each virus produced was titered by quantifying AAV genomes using qPCR as detailed below.

Table 1. gRNA Oligos.

Gene	Direction	Primer Sequence (5′ → 3′)
ORF62-1	F R	CACCGCTGGTTGAAGTCCCGATACGGA AAACTCCGTATCGGGACTTCAACCAGC
ORF62-2	F R	CACCGCCGGCTTTTTACCCGAGATGGA AAACTCCATCTCGGGTAAAAAGCCGGC
ORF62-3	F R	CACCGCAGCGCTCTACACCCCAACGCG AAACCGCGTTGGGGTGTAGAGCGCTGC
ORF63-1	F R	CACCGATACGCGGGTGCAGAAACCG AAACCGGTTTCTGCACCCGCGTATC
ORF63-2	F R	CACCGAAGACGGGTTCATTGAGGCG AAACCGCCTCAATGAACCCGTCTTC
ORF63-3	F R	CACCGTTGAATTTCGGGATTCCGACG AAACCGTCGGAATCCCGAAATTCAAC
UL3-4	F R	CACCGGTGACGAGCGCGATCCGGC AAACGCCGGATCGCGCTCGTCACC

2.4. *Quantification of Viral Genomes*

The pX601-based AAV vectors were quantified by SYBR qPCR as described previously [60], using PowerUp SYBR Green Master Mix (#25741, Thermo Fisher Scientific) and primers that amplified a 93 bp fragment of the saCas9 gene in all pX601 vectors (Table 2, "SaCas9"). Linearized plasmid DNA served as the standard, and all measurements were performed in triplicate. PCR conditions were 50 °C for 2 min; 95 °C hot start for 5 min; 45 cycles of: 95 °C for 15 s, 8 °C for 10 s and 72 °C for 45 s, followed by a dissociation curve analysis (95 °C for 15 s, 60 °C for 60 s, 95 °C for 15 s) to exclude the nonspecific amplification and formation of primer-dimers.

Table 2. Primers for Quantification of AAVs.

Gene	Direction	Primer Sequence (5′ → 3′)
SaCas9	L R	AGAAATACGTGGCCGAACTG TCACGTAGTCGCTGGTCTTG
ORF49 (VZV genome)	F R	CGGTCGAGGAGGAATCTGTG CCGTTGCACGTAACAAGCTC

For quantification of VZV genomes, the PowerUp SYBR Green Master Mix was again used with primers that amplified a 60 bp fragment of the VZV ORF49 gene (Table 2, "ORF49

(VZV genome)"). DNA isolated from cell-free VZV-pOka of known genome copy number content was used to generate a standard curve for genome number. The qPCR conditions were 50 °C for 2 min, 95 °C hot start for 5 min; 45 cycles of: 95 °C for 15 s, 58 °C for 10 s and 72 °C for 45 s. Dissociation curve analyses from 60 to 95 °C were performed as described above.

2.5. Generation of Recombinant VZV

VZV were generated using a modified version of an established bacterial artificial chromosome (BAC) system (pOka DX), based on the VZV parental Oka strain (pOka) containing a self-excisable mini-F replicon [61,62]. The parent BAC (pOka DXRR 57Luc-ZeoR) was partly detailed elsewhere [63], and is corrected for two nonsynonymous coding mutations in ORFs 40 and 50 that differed from the sequence of the parental pOka strain. It also contained a firefly luciferase reporter fused in-frame to the ORF57 gene using T2A ribosome skipping motif, and is expressed from the ORF57 late promoter (PRK and MBY, manuscript in preparation). The BAC was manipulated using two-step, markerless λ-Red-mediated recombination as previously described [61,62] in the E. coli strain GS1783 (a gift from Dr. Gregory Smith, Northwestern University, Chicago, IL, USA). The GS1783 contains a cassette of L-arabinose-inducible expression of the IsceI homing restriction enzyme and the 42 °C heat-inducible expression of the λ-Red recombination genes. All PCR amplifications for cloning and recombineering were performed using high fidelity PrimeSTAR GXL polymerase (#R050B, Takara). Final manipulated BACs were evaluated by restriction fragment length polymorphism (RFLP) analyses and BACs made into viruses were sequenced across the relevant sites of insertion and/or mutagenesis.

A VZV dual reporter BAC and virus were developed in which the fluorescent mCherry reporter gene was fused to the N-terminus and in-frame with ORF23, which encodes a minor capsid protein. This was derived by PCR amplification from the plasmid pmCherry kan, which contained a reversable IsceI site containing a kanamycin resistance cassette (kan^r) derived from the plasmid pEPS-kan2 (a gift from Dr Gregory Smith, Northwestern University, Chicago, IL, USA). Resolving the internal kan^r cassette by inducing a further recombination concurrent with the induction of IsceI expression resulted in BACs with restored mCherry-ORF23. The VZV derived from this BAC was termed VZV DR (dual reporter) and expressed mCherry from the ORF23 promoter and luciferase from the ORF57 promoter. Then, the ORF71 coding sequence was replaced with a PCR-amplified ampicillin resistance cassette (amp^r), selecting for gain of ampicillin resistance in addition to resistance to chloramphenicol (chm^R is directed by the replicon) and zeomycin resistance (directed by a cassette inserted downstream of the luciferase at ORF57). VZV derived from this new BAC (VZV DR-Δ71) contained two quantifiable reporters and one copy of the ORF62 gene. Evaluation of virus at passage 6 by Southern blotting confirmed that the virus contained a restored ORF71 derived from ORF62 gene (VZV-DR-Δ71: data not shown). The VZV BAC DR-Δ71 was then further subjected to a recombinatorial mutation of the ORF62 gRNA-targeted sequence region to derive VZV DR-62gmut mutant BACs and viruses. Two primers were used to amplify kan^r from pEPS-kan2 (Table 3, "gRNA region primers"). These changed eleven base pairs of the DNA sequence of ORF62 targeted by the 62-1 gRNA, without altering the encoded ORF62 amino acids (Table 3 "Wobble base pair changes"). The PCR amplified DNA was recombined into ORF62 in the BAC VZVΔ71, then selecting BACs for gain of kanamycin, chloramphenicol, and zeomycin resistance. RFLP was used to identify BACs and the kan^r was reversed out of the BAC by a second recombination and the induction of IsceI expression, as detailed elsewhere [61,62]. Sequencing across the region in the BAC and of the resulting virus after passage 6 confirmed the presence of the desired engineered mutations.

Table 3. Primer Design for VZV DR-62gmut BAC Recombineering.

Gene	Direction	
62 gRNA region primers (5′ → 3′) [a]	F	GTCATGGTGGGACGGGAACATGAGATCGTTTCAATTCCCagt GTcagtGGcCTgCAgCCtGAACCCAGAAGGATG*ACTACGATAAGTAGG*
	R	TTGTGTTAGCTCTTCGCCAACATCTTCCGTTCTGGGTTCaGGc TGcAGgCCactgACactGGGAATTGAAA*GGGTAATGCCAGTGTTAC*
Wobble base pair changes [b]	VZV DR	TCC GTA TCG GGA CTT CAA CCA G → S_{68} V_{69} S_{70} G_{71} L_{72} Q_{73} P_{74}
	VZV DR-62gmut	agt GTc agt GGc CTg CAg CCt G → S_{68} V_{69} S_{70} G_{71} L_{72} Q_{73} P_{74}

[a] = 62 gRNA region primers that were used to alter the gRNA region. Uppercase non-italic letters denote bases that match sequence to that of VZV ORF62 pOka wild-type sequence; bases in italics denote sequence that recognizes the kanamycin resistance cassette for PCR amplification. Lowercase letters represent altered bases designed to change the 62-1 gRNA recognition site. [b] = Wobble base pair changes that were engineered into the gRNA recognition sequence are shown for VZV DR-62mut, while the parental wild-type sequence is shown for VZV DR. The top line in each segment shows the DNA sequence and the bottom line shows the encoded amino acids as one-letter amino acids and their residue numbers in the ORF62 protein. Lower case letters indicate the mutated bases engineered by the primers.

2.6. Virus Derivation and Growth Curves

The virus was generated from the BACs that were just detailed after their transfection into MeWo cells, using Lipofectamine 3000 (#L3000015, Thermo Fisher Scientific) with 2.5 μg of purified BAC DNA and 100 ng each of plasmids expressing ORF61 and ORF62 DNA under the constitutive hCMV immediate-early promoter, as detailed previously [64,65]. Upon the appearance of fluorescent plaques, infected cultures were amplified by trypsinization and replated with uninfected ARPE-19 cells until they exhibited >80% cytopathic effect. Viruses were further amplified for a minimum of six passages to self-excise the BAC replicon elements and restore ORF71. Virus stocks were generated from ARPE-19-infected cultures showing >80% cytopathic effect, after treating cultures with 0.01 mM mitomycin C for 3–4 h. at 37 °C prior to harvesting for cryopreservation, as detailed previously [65].

Virus growth curves were determined as detailed previously [64,65], initiating infections in confluent 6-wells with approximately 400 pfu of virus (per 0.5×10^6 cells in a single well) from cryopreserved, pre-titered, mitomycin C-treated aliquots of infected ARPE-19 cells. All assays were performed in triplicate. At the desired times after infection, infectious virus-associated cells were quantified by trypsinizing the monolayers and then seeding serial dilutions onto ARPE-19 monolayers grown in 6 wells seeded 24 h. prior and at 80% confluency. At 4–5 dpi, the formed plaques were counted under a fluorescent microscope, and averaged from the triplicate values. Titers were normalized to the exact titer of virus at day 0 determined from the inoculate and then normalized to parental virus VZV DR.

2.7. Neuron Cultures

Neuron cultures were prepared using a previously published protocol [22] based on the differentiation of H9 human embryonic stem cells (hESC) cultured on feeder cells for 1–2 weeks. H9 cells were then further amplified on a feeder-free platform. This used StemFlex™ medium supplemented with StemFlex™ supplement (Thermo Fisher Scientific) and antibiotic/antimycotic. Cells were grown on 6-well dishes pre-coated with GelTrex Basement Membrane Mix (#A1413302, Thermo Fisher Scientific). Colonies were passaged either by manual dissection or using ReLeSR™ Enzyme-free cell selection and passaging reagent (Stemcell Tech., Vancouver, BC, Canada), per the manufacturer's instructions. Following either method, ROCK inhibitor Y-27632 (#S1049, Selleckchem, Houston, TX, USA) was added to the culture media for 24 h. to increase cell survival and deplete dividing cells. Neural precursor cells (NPCs) were generated from hESC by co-culture with the PA6 mouse stromal fibroblast cell line (RIKEN BioResource Center, Tsukuba, Ibaraki Province, Japan) as detailed previously [66,67]. For terminal differentiation, neurospheres were added to culture dishes coated sequentially with Poly-D Lysine (PDL) and GelTrex and cultured for 14 days in a Neuron Medium (Neurogenic Medium supplemented with CultureOne (#A3320201, Thermo Fisher Scientific).

For the fluorescent monitoring of VZV infections of neurons, neurospheres were seeded on coated 12-well glass bottom plates (P12-1.5P, Cellvis, Mountainview, CA, USA) and differentiated in neuron differentiation media for at least 14 days. Infections were initiated with PBS-washed, mitomycin-C treated VZV-infected ARPE-19 cells. Medium was changed the next day and every 2–3 days after. Latent infections were initiated with cell-free virus stocks of a previously detailed recombinant VZV (VZV66GFP) that expressed ORF66 as a GFP-ORF66 fusion [22,68] Latently infected neurons were then maintained in media containing 50 μM acyclovir (ACV, #A1915, Tokyo Chemical Industry, Tokyo, Japan) to block sporadic lytic initiation events, as previously described [23]. Latent infections were monitored for at least one week after infection in the absence of ACV by fluorescent microscopy screening to remove from consideration any cultures showing breakthrough GFP expression as a marker for productive infection. For reactivation, the media was then replaced with the same media lacking ACV and NGF, but including 50 μg/mL anti-NGF (Biolegend, 617904) and 50 ng/mL 12-O-tetradecanoylphorbol-13-acetate (TPA; #1201 Tocris Biochemicals, Bristol, UK). Reactivation was monitored for 1–2 weeks by microscopic screening for GFP expression. To quantify the reactivated virus, treated neurons were dissociated by manual dislodging and trituration, then by replating dilutions of the neurons onto confluent ARPE-19 monolayers. Foci of infection were counted 4 days later, using a fluorescent microscope.

2.8. Fluorescent Microscopy

Live neuronal cultures were monitored using a Nikon TI fluorescent microscope with a 10× air objective (N.A. 0.30). Imaging of fixed neuron cultures followed previous detailed procedures [22,23]. Neurons were immunocytochemically identified by chicken anti-beta tubulin III (Novus Biologicals, Moon Township, PA, USA) in 10% heat inactivated goat serum (HIGS) in PBS. Bound antibodies were detected using secondary goat anti chicken antibodies linked to AlexaFlor-594 (#A-11042, Thermo Fisher Scientific,). Cultures were then mounted in mounting media containing 4′,6-diamidino-2-phenylindole (DAPI) to stain nuclei. Images for all samples in each experiment were captured under identical acquisition settings and processed using Metamorph software (Version 7.7, Molecular Devices, San Jose, CA, USA). VZV fluorescent foci in infected epithelial cells were imaged at 4 dpi after growth at 34 °C after fixing in 4% paraformaldehyde and washed with 1× PBS. Multiple images containing entire individual foci that did not touch any borders or other plaques were acquired under identical acquisition settings with CellSens software (Olympus, Tokyo, Japan) on an Olympus IX83 microscope with a 10× (N.A.030) air objective. Data was exported and analyzed for size using Metamorph.

2.9. Flow Cytometry and Statistical Analyses

Flow cytometry was used to quantify GFP-fluorescent positive HEK293 cells in the initial evaluation of the efficiency of gRNAs to target genes and block protein expression. All flow cytometry samples were collected and analyzed using a BD FACSAria cytometer (Becton, Dickenson, and Co., Franklin Lakes, NJ, USA) and FlowJo software (FlowJo, Ashland, OR, USA). Statistical analyses were performed using GraphPad Prism software (GraphPad, La Jolla, CA, USA). Where indicated, error bars represent standard deviation (SD) or standard error of the mean (SEM) as specified in the figure legends.

3. Results

3.1. Characterization of VZV-Specific gRNAs for In Vitro Specificity

The goal of the studies was to assess the potential of using gene editing as an anti-viral strategy targeting VZV, not only in lytic-infected cells that are permissive for VZV, but also in reactivating neuron cultures. We recently detailed a hESC-derived neuron culture system that can host a VZV model latent state that can subsequently be reactivated [23]. Gene editing has potential to target the latent genome before reactivation. Gene editing strategies targeting HSV-1 have suggested that two sites of double stranded breaks (DSBs) are more efficient in reducing the HSV genome load and progeny virus production compared to single genome DSB sites, which can repair. The latter can still be disruptive as a consequence of errors in DNA repair that result in indels in a single-copy critical gene [33,69]. Our gRNA-based saCas9 antiviral strategy targeted duplicated essential VZV genes. Both ORF63/70 and ORF62/71 lie in the large reiterated genomic regions bounding the short unique region. This region also contains the ORF64/69 gene pair, but this pair has been shown to be dispensable for in vitro replication in culture, and in human T cells and human fetal skin that is harbored in SCID-hu mice [43]. VZV ORF 62/71 protein functions as the major transcriptional activator (transactivator) in VZV [41], while ORF63 protein has regulatory [70] and anti-apoptotic [71] activities, and ORF63 transcription is associated with latency [72]. One report has suggested that ORF63 is not essential for VZV culture growth [73], while others report ORF63 as absolutely required for VZV replication [43].

Three gRNA candidates for each gene were selected based on an optimal on/off-target specificity, the presence of the PAM motif, and length. The appropriate hybridized oligonucleotides were then cloned into the AAV-based pX601 backbone plasmid (Table 1) co-expressing saCas9. We then conducted a preliminary study to demonstrate that each gRNA construct was able to target the respective gene. Each pX601-based plasmid or control (pUC19 and pX330-UL 3/4) plasmid was co-transfected with corresponding CMV IE promoter-driven ORF-GFP fusion constructs into HEK293 cells. We reasoned that cleavage by the targeted saCas9 should prevent translation of the protein and reduce expression of the GFP reporter, measured by flow cytometry. Comparison of the three ORF62-specific gRNA constructs revealed that each greatly reduced the expression of GFP from the CMV-62GFP reporter as compared to control plasmid co-transfected cells (Figure 1a). A similar but slightly different protocol was then used to evaluate ORF63-targeting gRNAs, in which a single set amount of CMV-ORF63-GFP fusion (500 ng) was co-transfected with control plasmids pUC19 or pX601 containing the 62-1gRNA or the 3 different 63gRNA templates. All of the 63gRNA showed a reduced level of GFP expression from both ORF63 N-terminal (GFP ORF63)- or C-terminal (ORF63 GFP)-tagged expression plasmids, when compared to the non-targeting CRISPR/Cas9 constructs. Taken together, the data suggested that all gRNAs could efficiently target the respective gene. We then selected the "62-1" ORF62-targeting construct (62-1gR-saCas9) for further development.

3.2. *AAV Serotypes for Delivery to VZV Permissive Epithelial Cells and Human Stem Cell-Derived Neurons*

AAV capsid serotypes confer different tissue tropisms [74–77]. We considered it important to identify an AAV serotype that transduces epithelial cell lines that support VZV replication, in addition to being able to transduce hESC-derived neurons that can harbor model VZV reactivatable latent states. A preliminary study was conducted with four commercially acquired AAV neurotropic serotypes containing CAG-GFP reporters (serotypes 2, 5.2, 8.2, and 9), which have been previously exploited for murine-neuron delivery [78]. AAV 5.2 and 8.2 recombinant serotypes contain modifications of the corresponding WT serotypes that are potentially better able to escape cellular vesicles once endocytosed, avoiding lysosomal breakdown (Virovek, Hayward, CA, USA). AAV transduction with each serotype was performed at equivalent genome copy levels per cell and the transduction of MeWo and ARPE 19 cells was assessed by determining the fraction of cells showing GFP expression, using flow cytometry after 3 days incubation (Figure 2a). AAV2 appeared to be efficient in transducing both MeWo and ARPE19 cells, as GFP expression was detected in nearly 100% of the cells for both lines. Other serotypes did not result in GFP expression in the majority of cells of both lines. The ability of each serotype to transduce and express GFP in hESC-derived neurons was also confirmed. Fluorescent microscopy images (Figure 2b, enlarged in Supplemental Figure S1) indicate that all four serotypes resulted in GFP expression in the hESC-derived neuron culture platform. For AAV2, the GFP intensity appeared weaker than that seen after neuron transduction with AAV5.2, but AAV5.2 appeared to transduce a lower proportion of neurons compared to AAV2.

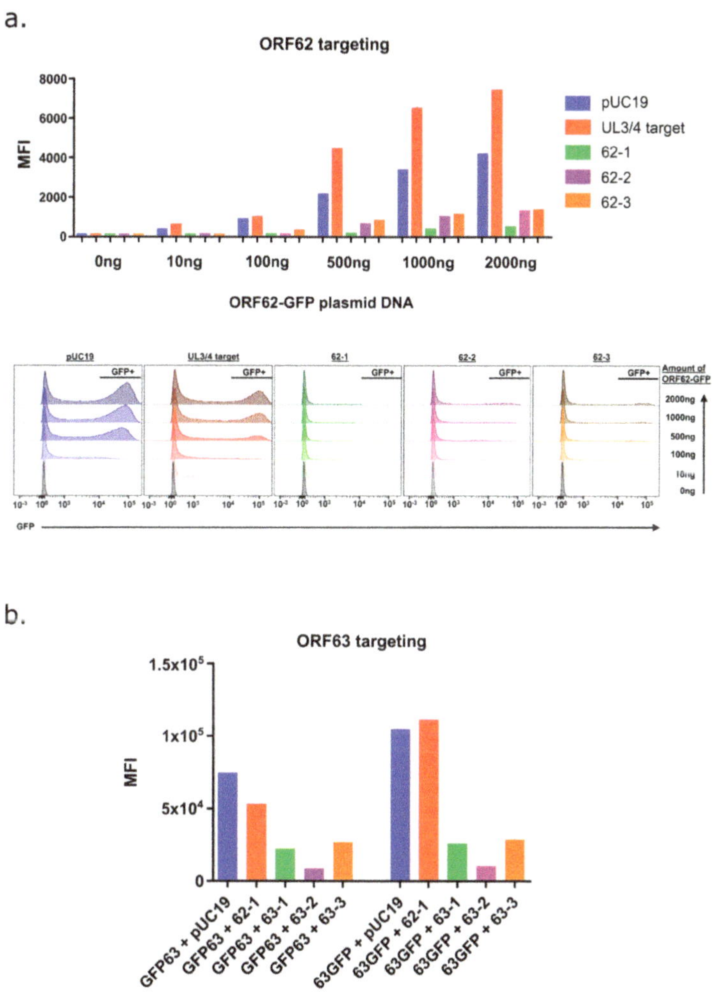

Figure 1. A preliminary study to evaluate ORF62- and ORF63-targeting gRNAs in pX601 to prevent gene expression. (**a**) HEK 293 cells in 6-well plates were transfected with a range (0–2000 ng) of CMV-ORF62-GFP plasmid, along with 2 µg of plasmid pUC19; a pX330-based plasmid with a gRNA template derived from HSV UL3/4; or pX601 AAV plasmids containing the templates for ORF62 gRNAs 1–3, constructed as detailed in the Methods. To maintain the same level of DNA in each transfection, pUC19 was added. At 36 h post transfection, cells were harvested by trypsinization and analyzed by flow cytometry to determine the mean fluorescence intensity (MFI) for each treatment. (**b**) To determine the optimal gRNAs for targeting ORF63 in pX601, HEK293 cells in 6-wells were co-transfected with 500 ng of CMV-promoter-driven ORF63-GFP plasmid expressing a C-terminal tagged GFP (63GFP) or a similar plasmid-driving expression of N-terminal tagged GFP (GFP63); each with either 1500 ng of pUC19 DNA, the px601 containing the 62-1gR template, or one of three selected pX601 plasmids containing the ORF63-gR templates as detailed in Table 1. After 72 h, the cells were subject to flow cytometry and the MFI was determined using FlowJo software. The data represent the MFI of single transfections performed in parallel and were not statistically evaluated.

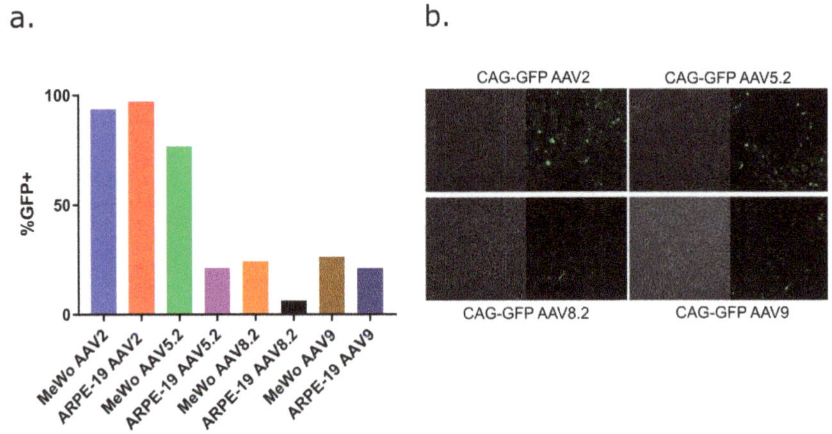

Figure 2. A preliminary study of AAV serotypes able to deliver to epithelial cells and neurons. (**a**) The fraction of two VZV permissive epithelial cell lines (MeWo and ARPE-19) that expressed GFP, as determined by flow cytometry, after infection with 5 µL of 10^{12} GC/mL AAV per well of a 6 well plate with AAV-GFP serotypes 2, 5.2, 8.2 or 9. AAV-mediated expression was measured by flow cytometry at 3 days post transduction and is represented as a fraction of the total cell population, using gates selected by analyses of untransduced cells. Of the four serotypes evaluated on MeWo and ARPE-19 cells, AAV-2 was the most efficient at transducing both cell types to express GFP. The study represents single transductions and was not statistically evaluated. (**b**) To qualitatively establish the transduction of hESC-derived neurons, neurons cultured in 12-well dishes were transduced with 1.25×10^{11} GC per 12-well of each GFP-expressing AAV serotype. The AAV-mediated expression of GFP was imaged at multiple (>5) non-overlapping random positions at 6 days. Representative images are shown that reflect GFP expression in neuron cultures. Further quantification could not be determined due to the heterogeneous and three-dimensional nature of the neuron cultures.

These observations are consistent with previous studies reporting the ability of AAV2 and AAV5 to transduce ARPE-19 cells [79] and murine neurons [80]. The three-dimensional nature of hESC-derived neuron cultures [22] did not permit more accurate quantitation of neuronal transduction, but given the ability of AAV2 to efficiently transduce VZV permissive epithelial cells close to 100%, we selected AAV2 for packaging of the pX601-based vectors containing saCas9, with or without the 62-1 guide RNA.

3.3. *AAV-62-1gR-saCas9 Reduces VZV Lytic Replication in Epithelial Cells*

High-titer preparations of AAV2-packaged saCas9 with (AAV-62-1gR-saCas9) or without (AAV-saCas9) the 62-1 gRNA sequence template were obtained at more than 10^{12} GC/mL and subsequently evaluated for their ability to prophylactically reduce lytic replication in VZV-infected ARPE-19 cells. ARPE-19 were mock treated or transduced with each AAV at 10^4 GC/cell and 4 days later, cultures were infected with a low dose (500 pfu per well of a 6-well plate) of mitomycin C-treated, cell-associated VZV DR or VZV DR-Δ71 (see Methods in Section 2). Southern blot analyses of DNA from VZV DR-Δ71 after 6 passages in culture established that ORF71 in the virus was restored by the reduplication of ORF62 (data not shown), as expected from similar previous mutagenic studies of ORF62 in the BAC from our group [64]. Parallel infected cultures were harvested multiple times after VZV infection and assessed for the expression of the kinetically late ORF57 promoter-expressed luciferase activity.

In ARPE-19 cells not pretreated with AAV before VZV infection, or in cells pretreated with AAV-saCas9 (no guide RNA), VZV luciferase activity reported an increase in growth of both VZV DR and VZV DR-Δ71 over time. Strikingly and in contrast, cells pretreated with 10^4 GC/cell of ORF62-targeting AAV showed a dramatic reduction in the VZV reporter activity over time as compared to untreated and AAV-saCas9 controls for both viruses. This indicated that AAV-62-1gR-saCas9 was highly effective at reducing progeny virus production (Figure 3a). The VZV derived from a BAC in which ORF71 was deleted gave very similar results to VZV DR, as expected. An assessment of viral growth visualized by mCherry expression in infected foci (Figure 3b) supported the luciferase studies, in that both focus size (Figure 3c) and the number of focus numbers at 4 dpi (Figure 3d) were vastly reduced in cells pretreated with AAV-62-1gR-saCas9, compared to controls, for both VZV DR and VZV DR-Δ71. These data indicate that the AAV-62-1gR-saCas9 blocks lytic replication of VZV in infected epithelial cells. Furthermore, the reduction in foci numbers developing in AAV-62-1gR-saCas9 pretreated cells strongly suggests that some infected cells failed completely to initiate the productive spreading foci of infection.

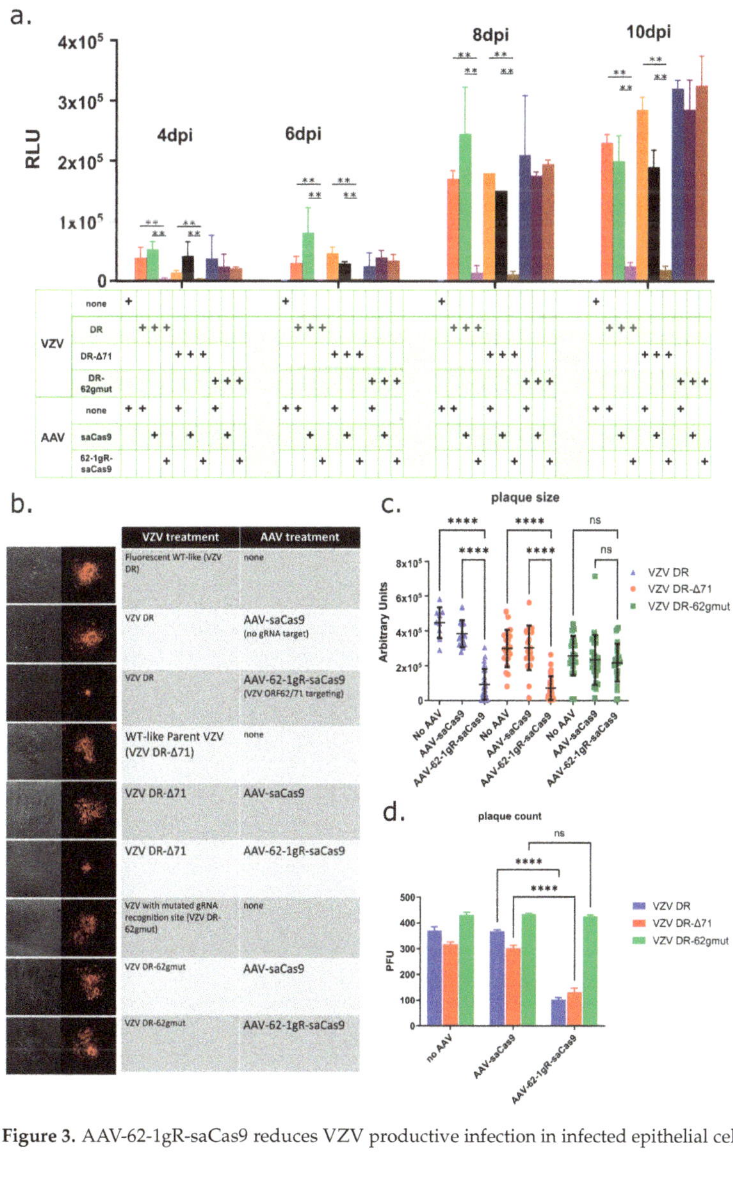

Figure 3. AAV-62-1gR-saCas9 reduces VZV productive infection in infected epithelial cells. (**a**) The

12-well plates of near confluent ARPE-19 cells were pre-treated with 10^4 GC/cell of the different AAV and at 4 days post transduction, cells were infected with 500 PFU/well of cell-associated VZV. VZV replication was measured using luciferase-reporter expression from the ORF57 gene at 4, 6, 8, and 10 days after VZV infection. Cell extracts were diluted 1000-fold prior to the assay. Significant differences between treatments, defined as $p < 0.01$ using a 2-way ANOVA with Geisser-Greenhouse's correction, are indicated by two asterisks (**). The box below the graph illustrates the components added to each condition and each condition was performed in triplicate. RLU = Relative Light Units (**b**) Representative images of infectious foci by phase-contrast imaging (first column, in gray) and by live cell fluorescence for the ORF23 promoter-driven mCherry (second column, in red) at 4 days after VZV infection in ARPE-19 cells. The boxes to the right illustrate the virus (third column) and the AAV (fourth column). (**c**) Quantitation of fluorescent focus size determined from at least 9 and less than 31 isolated, nonoverlapping foci measured for each sample condition (**** = $p < 0.0001$, 1-way ANOVA with Tukey's multiple comparisons test). The average size is indicated by the horizontal bar. (**d**) Average visible focus counts per 6 well that were observable under 10× objective. The number of foci/well (bars represent counts from triplicate wells) were clearly reduced in AAV-62-1gR-saCas9-pretreated samples infected with WT-like VZV, compared to cells infected with vectors expressing SaCas9 without guide RNA ("AAV-saCas9") (**** = $p < 0.0001$, 2-way ANOVA with Tukey's multiple comparisons test). All data shown is from one of two experiments with similar results. Error bars represent STD. ns = not significant.

The AAV-62-1gR-saCas9 was designed to specifically target VZV at the ORF62/71 gene, but a well-known issue with gene editing technologies is the possibility of off-target effects. It is conceivable that a cellular factor required for VZV replication was damaged by off-target activities that would result in similar observations to those reported above. Therefore, as an additional control, two VZV recombinants in the background of VZV DR-Δ71 were derived, in which the guide RNA target sequence was mutated at 11 bases in the codons of ORF62 (VZV DR-62gmut clones #18 and #36) that maintained the same encoded open reading frame residues for ORF62. The expectation was that the substitutions of 11 bases in the Cas9 recognition region of ORF62 would render the 62-1 gRNA no longer able to recognize and cleave at this location (Table 3). If the effect was acting though off-target activities, it would still impair the viral replication of such viruses. Growth curve analyses of the two VZV mutants after low-MOI infection of ARPE-19 monolayers indicated that the rate of luciferase expression over time was only slightly less than that of VZV DR, suggesting the silent mutation of codons might have a very minor influence on growth (Figure 4). However, testing the growth of one of these viruses (VZV DR-62gmut # 1–18) in the same conditions as those used in the experiments shown in Figure 3 revealed that in AAV-62-1gR-saCas9-pretreated cells, the mutant VZV grew at rates similar to that in cells that were not AAV-pretreated or pretreated with the AAV-saCas9 control. The mutant virus did not show the dramatic reduction in luciferase activity, plaque size and plaque number as seen for the parental VZV DR and VZV DR-Δ71 viruses. These experiments strongly suggest that the reduction of VZV replication by pretreatment with AAV-62-1gR-saCas9 was a consequence of the specific targeting of VZV ORF62/71 DNA.

3.4. AAV-62-1gR-saCas9 Reduces VZV Lytic Replication in hESC-Derived Human Neuron Cultures

The ability of AAV-62-1gR-saCas9 treatment to block lytic replication in hESC-derived neurons was then evaluated. We have previously shown that hESC-derived neuron cultures are able to support the spreading of productive VZV infections [22,23,81]. Neurons that were differentiated from neurospheres for 3 weeks were transduced with approximately 10^4 GC/cell of AAV or left untreated. Four days later, the cultures were infected with 500 PFU of mitotically inhibited VZV DR-infected ARPE-19 cells. Viral growth was monitored visually at the same position in the cultures by microscopy over a 10-day period, and representative live-cell images of the foci of fluorescence in the neuron cultures at day 10 post infection are shown in the first column of Figure 5a. In non-AAV treated, AAV-saCas9 pre-treated and AAV-62-1gR-saCas9 pretreated neurons infected by VZV DR-62gmut, foci of VZV infected neurons (indicated by red fluorescence) clearly developed over the 10-day incubation period (Figure 5a). However, neuron cultures treated with AAV-62-1gR-saCas9 that were infected with VZV DR or VZV DR-Δ71 developed only very small foci of red fluorescence that involved only a few neurons, far fewer than seen in the controls. This suggested the virus in this group was not able to spread and form foci as efficiently as infected neurons in wells receiving other treatments (Figure 5a). To obtain a more quantitative assessment, the infected neuron cultures at 10 dpi were dislodged, triturated, serial diluted and re-seeded onto monolayers of ARPE-19 cells. Infectious centers that formed from all non-AAV-treated or control AAV-saCas9-pretreated neurons after 5 days were approximately the same size, as were the infectious centers formed from AAV-62-1gR-saCas9 pretreated cultures infected with VZV DR-62gmut virus. However, plaques formed on ARPE-19 cells co-cultured with neurons that were infected with VZV DR or VZV DR-Δ71 after pretreatment with AAV-62-1gR-saCas9 were not only dramatically reduced in number, but at the highest concentration of neurons seeded onto ARPE-19 monolayers, only tiny foci of infection involving a few fluorescent cells were seen. No wildtype-sized plaques developed after neuron seeding on ARPE-19 under these conditions. Even taking such small foci of infection as positive, the number of infectious centers for VZV DR and VZV DR-Δ71 on neurons pretreated with AAV-62-1gR-saCas9 were more than a hundred-fold fewer compared to the non AAV-treated, mock AAV-treated, or VZV DR-62gmut-infected controls (Figure 5b). The strikingly reduced size of the plaques formed by progeny virus after AAV-62-1gR-saCas9 treatment suggested that virus produced from treated neurons was unable to replicate to wild type levels. These results strongly suggest that AAV-62-1gR-saCas9 pretreatment not only efficiently reduces the production of the progeny virus in infected neurons but also, when virus is produced, it is damaged and severely impaired for further replication.

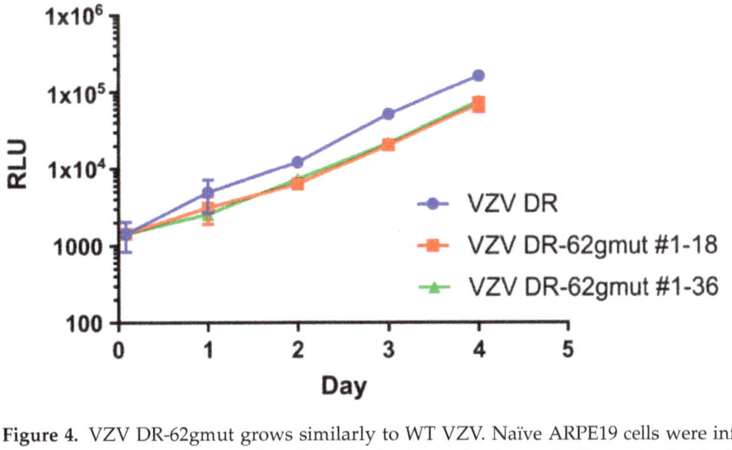

Figure 4. VZV DR-62gmut grows similarly to WT VZV. Naïve ARPE19 cells were infected with equivalent amounts of VZV-infected ARPE-19 mitomycin C treated cells, as detailed in the methods. Luciferase activity was measured daily. Graphs show the growth of two independently isolated clones ("#1–18" and "#1–36") of VZV DR-62gmut and VZV DR WT-like reporter virus. Each point represents the average of four replicates and the data shown is representative of two independent experiments. Error bars represent STD.

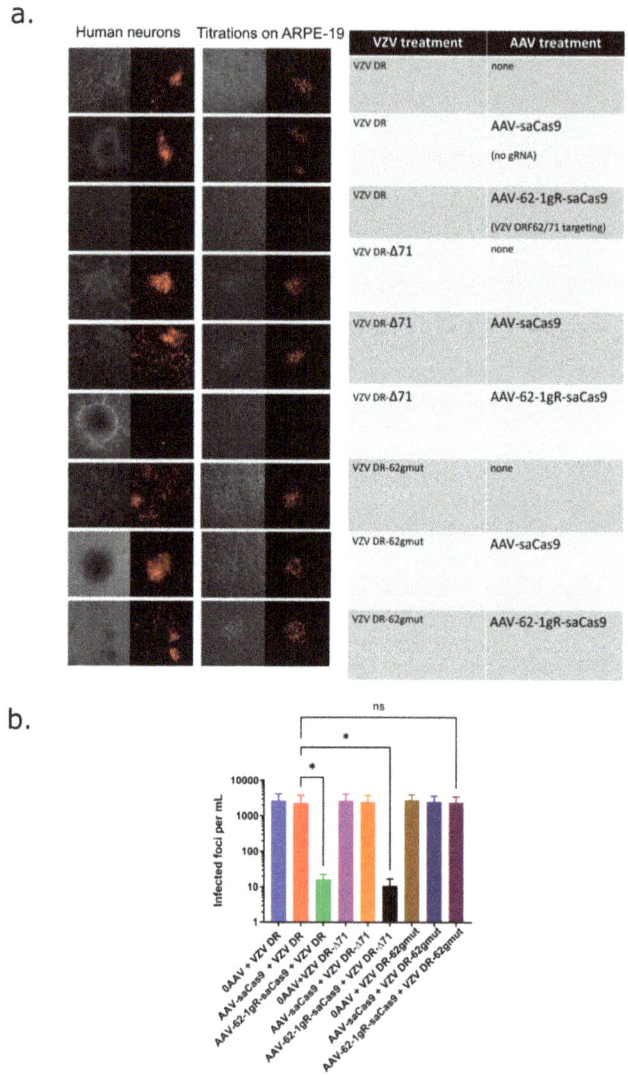

Figure 5. Pretreatment with AAV-62-1gR-saCas9 knockdown of VZV lytic replication in exogenously infected human neuron cultures. (**a**) The 12-well plates of hESC neuron cultures were transduced with 104 GC/cell of AAV-62-1gR-saCas9 constructs/well. Four days later, neurons were infected with 500 PFU of mitotically inhibited fluorescent (red)-cell associated VZV-infected cells to initiate lytic infections. Infections were subsequently monitored by fluorescence microscopy and representative images of mCherry reporter expression of infected cells and the same fields observed by phase contrast microscopy were acquired 10 days after VZV infection. (**b**) After imaging, duplicate infected neuron cultures were scraped, triturated, and then seeded onto confluent ARPE-19 cells to quantify the number of VZV-infected neurons by the number of infectious foci they generated. Foci were counted after 5 days. Representative images of foci formed on ARPE-19 at day 5 from virus obtained from the treated neuron cultures are shown in column 2 of Figure 5a. Data represents quadruplicate measurements and significant differences between treatments depicted by asterisks (* = $p < 0.01$, ordinary one-way ANOVA with multiple comparisons; ns = not significant). Similar results were obtained in two independent experiments. Error bars depict the STD.

3.5. *AAV-62-1gR-saCas9 Reduces VZV Replication Following Reactivation in Model Latently Infected Neuron Cultures*

A goal of VZV targeting by gene editing is to reduce the capacity of latent genomes to reactivate and induce zoster disease, not only by reducing spread at the periphery, but also in the ganglia, to reduce its intra-ganglionic spread after a reactivation event has initiated. Even without complete elimination of reactivation, a reduction of lytic replication in the ganglia could potentially limit sensory damage caused by VZV reactivation. The hESC neuron cultures used here have been shown to support a latent VZV infection (defined by the prolonged absence of any indicators of productive infection and lytic gene reporter expression), which can then be experimentally reactivated (to renew lytic reporter gene expression and produce virus that spreads to other neurons) [23]. We thus explored the potential of AAV-62-1gR-saCas9 as a treatment for preventing VZV reactivation and subsequent replication. Latent infections were established by exposing neuron cultures to cell-free virus of a previously characterized recombinant VZV that expresses GFP linked to ORF66 (VZV66GFP), and then incubated in the presence of 50 μM acyclovir (ACV) to inhibit lytic replication for 7 days. Cultures were then incubated for 7 days in the absence of ACV. None of the cultures contained GFP fluorescence indicating lytic infection. Cultures containing latently infected neurons were then transduced with approximately 10^4 GC/cell AAV or mock transduced, and then incubated in media without ACV. At 7 days post AAV transduction, GFP positive cells were again not observed, indicating that the AAV transduction itself did not induce VZV reactivation. A set of cultures was reactivated by treating with a combination of NGF withdrawal, antibody-mediated NGF depletion and the addition of 50 ng/mL 12-O-tetradecanoylphorbol-13-acetate (TPA), and cultures were monitored daily for GFP fluorescence indicating reactivation events for 7 days. After imaging, the neurons were dislodged, triturated, and seeded onto confluent ARPE-19 cells for infectious center assay (Figure 6b).

None of the neuron cultures latently infected with VZV66GFP and not subjected to reactivation stimuli expressed GFP. When such neurons were seeded on ARPE19 cells, infectious centers did not form. In contrast, the latently infected neurons receiving the reactivation stimulus developed numerous GFP-positive foci and formed multiple infectious centers when seeded onto ARPE-19 monolayers, indicating that productive reactivation had occurred (Figure 6a and Supplemental Figure S2). Latently infected neurons treated with AAV-saCas9 without guide RNA before receiving the reactivation stimulus also developed multiple GFP positive plaques in the neuronal cultures and generated infectious foci on ARPE-19 cells. Importantly, in latently infected cultures pretreated with AAV-62-1gR-saCas9 prior to receiving a reactivation stimulus, there were no visible GFP positive centers of infection forming, even at 7–10 days post stimulus. Subsequent seeding of these neurons onto ARPE-19 cells did result in the formation of a few small GFP positive foci, but these were significantly fewer and considerably smaller in size (Figure 6a, second column; enlarged in Supplemental Figure S2). These data indicate that reactivation events were not completely prevented and virus still formed after treatment with AAV-62-1gR-saCas9, the AAV treatment greatly reduced the number of infectious progenies after reactivation induction.

To investigate whether the reduction in the number of productively infected neurons in reactivated cultures pre-treated with AAV-62-1gR-saCas9 was due to a reduction in the number reactivating genomes or a result of reduction of viral spread to additional neurons, SYBR-based qPCR quantification of VZV genomes was performed. There was an increase of more than 10-fold in genome copies at 21 days-post VZV infection in reactivated samples with or without AAV pretreatment compared to neurons not receiving a reactivation stimulus. While AAV-62-1gR-saCas9 pretreatment resulted in a significantly reduced level of genomes as compared to controls, levels of genomes measured were still higher than those seen in latently infected cultures not receiving any reactivation stimulus (Figure 6b,c). These data indicate that AAV-62-1gR-saCas9 pretreatment effectively reduced the viral burden after reactivation, but did not indicate if the latent genome load was reduced by the treatment. Thus, AAV-CRISPR/Cas9 strategy for targeting duplicated VZV genes is clearly an effective antiviral approach, greatly reducing the production of progeny by damaging genomes.

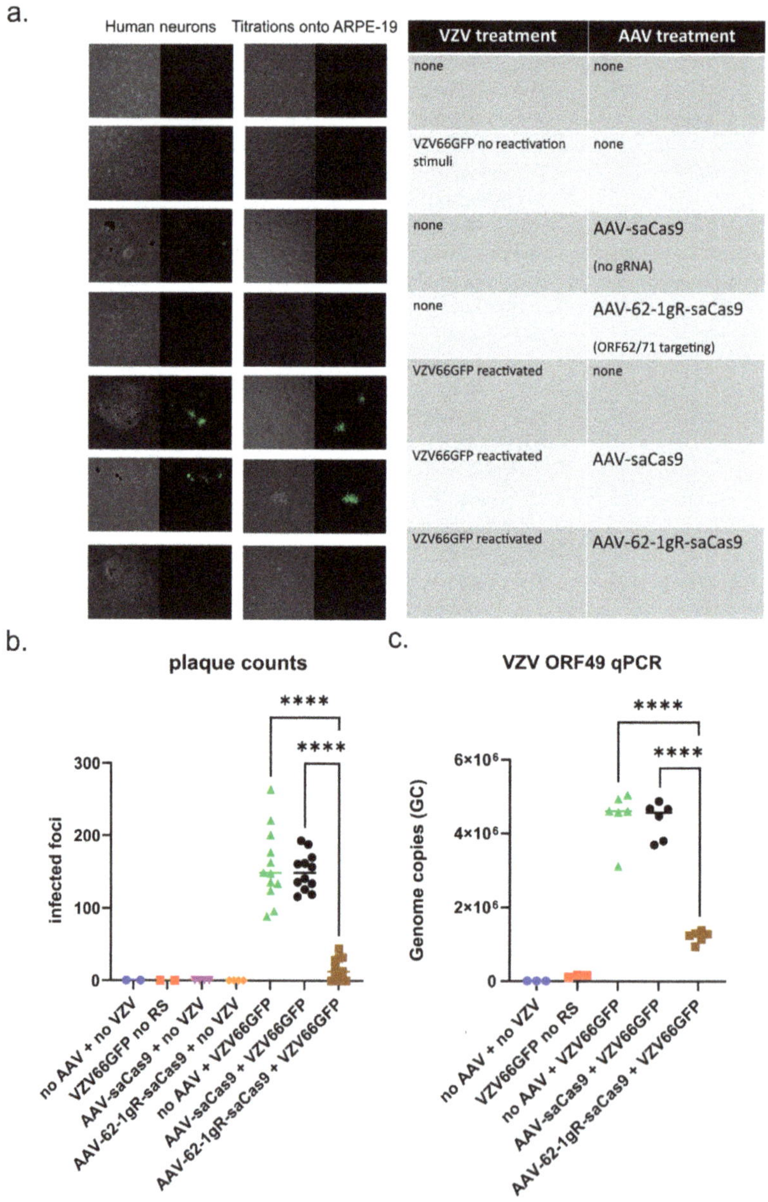

Figure 6. AAV-62-1gR-saCas9 treatment of latently infected neurons reduces VZV progeny and spread following reactivation induction. (**a**) hESC-derived neurons in 12-well plates were infected with 500 PFU of cell-free virus expressing GFP (VZV ORF66 GFP) in the presence of 50 μM acyclovir (ACV) to establish latent infections. After 7 days in the presence of ACV, approximately 10^4 GC/cell of different AAV (or mock transduction) was added to the appropriate wells. ACV was then removed,

and cultures incubated for a further 7 days. At 7-day post AAV transduction, GFP-negative neuron wells were left untreated or were stimulated to reactivate latent genomes by changing to neuron growth media lacking NGF and containing 50 µg/mL anti-NGF and 50 ng/mL TPA. Reactivation was then monitored by fluorescent microscopy and the representative images of the cultures were acquired at 7 days after the reactivation stimulus. Neurons were then scraped, triturated, and seeded onto ARPE-19 cells not treated with AAV, and infectious centers forming on ARPE-19 monolayers at 4 days from each of the experimental conditions were imaged (last two columns). (**b**) The number of foci forming on ARPE-19 monolayers in each 12-well plate were then counted and are shown as plaque-forming units. Each point represents the count from equal proportions of individual neuron cultures (**** = $p < 0.0001$, ordinary one-way ANOVA with multiple comparisons). (**c**) DNA was extracted from 1/3 of each neuron culture treated under the different conditions, and VZV DNA was quantified using a SYBR-based qPCR using primers against a region of VZV ORF49 (Table 2). Copy number values were determined by comparison to a standard curve determined using known concentrations of VZV genomes (with a range of 10^1–10^7 genome copies). Data is representative of results from two independent experiments with similar results. RS = reactivation stimuli.

4. Discussion

The goal of this study was to establish proof of principle that the CRISPR targeting of VZV genomes could be used as an antiviral strategy eventually leading to therapeutic applications. The results obtained here add to a growing body of knowledge in which several human DNA viruses (and RNA viruses with DNA genome intermediates) have been targeted by gene editing [31,33,57,82–84]. While gene editing can be performed with designer meganucleases and transcription factor-like endonucleases (TALENS), the complexity of their design and expense of their generation has made the more simple and widely available RNA-guided CRISPR/Cas technology more attractive. CRISPR/Cas also has the advantage that multiple sites on the viral genome can be targeted by increasing the number of gRNAs, rather than needing to express two or more enzymes. CRISPR-based genome editing been used to target viruses including hepatitis B virus (HBV) [85,86], human immunodeficiency virus (HIV) [87,88], and several herpesviruses [89] such as EBV [90], CMV [91], Kaposi's Sarcoma Virus [92], and HSV-1 [31–33]. In the case of EBV, antiviral gene editing has targeted both the viral genome [90] as well as key cellular components required for EBV replication [93]. To our knowledge, CRISPR/Cas-mediated antiviral targeting has not yet been applied to VZV, although it has been used as a research tool to generate recombinant VZV [94].

Here, we demonstrated efficient antiviral activity using AAV-delivered gRNA targeting a duplicated gene to reduce VZV replication, spread and virus production in both lytic-infected epithelial cells and hESC derived neurons. Importantly, we further showed that it dramatically reduced viral replication upon reactivation from latency in vitro. Duplicated genes were targeted because of the increase in frequency of cleavage events as well as the potential to result in the division of the genome into two segments, each incapable of replicating if both cuts occurred in the same genome. This strategy eliminates the need for two separate vectors targeting unique region genes in order to generate dual-cleaved genomes, as performed in the study of Aubert et al., who showed that cleavage of the related HSV genome at two sites was more effective at reducing reactivation frequency than a single cleavage event [33]. Aubert et al. used two designer meganucleases to target different sites in HSV-1 that required simultaneous delivery by two different vectors and targeted duplicated genes in the repeated regions of the HSV genome. However, even with single cleavage events, error-prone DNA repair mechanisms that mutate the VZV ORF62 gene appear to reduce progeny viral replication due to the critical roles of IE62 in expression [34].

An important aspect of developing therapies from CRISPR/Cas enzymes is achieving an efficient delivery of the required gRNA and enzymes to the appropriate cell types. This is especially challenging for VZV, where the reservoir of latent viral genomes resides in ganglionic neurons throughout the peripheral nervous system. Therefore, the first steps in this project were to optimize the delivery of these molecules with a focus on human neurons. While both lentiviruses and AAV have been exploited for efficient gene delivery, current AAV vectors have the advantage in that they show little or no integration into the host genome [95]. However, the gene-packaging limits of AAV necessitate the use of smaller gene-editing enzymes, since the widely used Streptococcus pyogenes CRISPR/Cas9 is above the normal AAV packaging limit. Many smaller alternatives have recently been overviewed [96] and the Staphylococcus aureus CRISPR/Cas9 was chosen for its reported higher specificity, lower off-target activity in mouse neuroblastoma and liver cell lines [37], and the fact that it is within the packaging limits of AAV vectors [38]. Surveying four different AAV serotypes that have been shown to be neurotropic revealed that AAV2 efficiently delivered GFP to both hESC-derived neurons and two VZV-permissive epithelial cell lines. While we only tested four serotypes, it is possible that other serotypes could be found to improve transduction or decrease the number of AAV required. We do note, however, that AAV2 is one of the most commonly used and tested serotypes used in human studies and has been applied for gene delivery to repair multiple genetic diseases, some of which are now being evaluated in clinical trials [97,98].

Our studies show that delivering AAV-62-1gR-saCas9 decreased VZV progeny infectivity dramatically in both epithelial cells and hESC-derived neurons, compared in no AAV and no gRNA ("AAV-saCas9") controls. This was shown by the reduction of both progeny-virus plaque size and numbers. Intriguingly, VZV that survived AAV-62-1gR-saCas9 treatment generated significantly smaller plaques when seeded onto naïve VZV-permissive cells, particularly VZV derived from AAV-targeted neurons. This strongly suggests that the virus produced in AAV-62-1gR-saCas9-treated cells is damaged, most likely as a result of genome cleavage and error-prone DNA repair, resulting in indels from the activation of the dsDNA damage response. Given that ORF62 is an essential gene whose expression is required for the expression of most other VZV genes, we postulate that some of the still-replicating but impaired viruses in these plaques are damaged but still have ORF62 in frame that permits expression of an ORF62 protein with some functional mutations. The site of the gRNA target in region 1 of ORF62/71 is that which the Cas9 acted upon, because VZV with mutations in that region ("VZV DR-62gmut") showed only a minor loss of replication efficiency compared to the parental strain, but were completely resistant to pretreatment with AAV-62-1gR-saCas9. This establishes that the Cas9 did not influence VZV replication because of off-targeting of the host genome, although we cannot exclude that off-targeting effects might have also occurred that did not affect VZV replication.

One of the exciting potential applications of gene editing as an antiviral strategy is that it cannot only target productive replication of the virus, but can also damage the latent genomes and potentially prevent reactivation [89]. Treatment of latently infected and reactivated cultures of hESC derived neurons [23] with AAV-62-1gR-saCas9 resulted in greatly decreased spread of reactivation foci compared to controls. The number and size of foci that developed from reactivated neurons seeded onto ARPE-19 from AAV-62-1gR-saCas9-treated reactivated cultures also significantly decreased. Results from quantitative PCR measuring genome copies in the reactivated neurons revealed that treatment with AAV-62-1gR-saCas9 reduced replication but did not indicate a reduction of the genome load. Taken together, these results suggest that AAV-62-1gR-saCas9, in addition to reducing the viral load in lytic/productively infected cells, is an effective strategy to reduce replication in reactivated neurons. Of note, we have not yet been able to demonstrate whether latent genomes were cleaved by the targeting AAV-62-1gR-saCas9. We did not observe a loss of genome numbers in latently infected neurons after AAV-62-1gR-saCas9 treatment, but this may be a technical issue due to the very low levels of latent VZV DNA in neuronal cultures, which were insufficient to detect a significant change resulting from CRISPR/Cas9 treatment. A recent report found that gene editing of lytic HSV-1 is efficient but the editing of the chromatin-silenced latent genome required the expression of ICP0. Furthermore, ICP0 is known to alter protective chromatin, which may block the latent genome from being accessed by gene editors [83]. Similar strategies to target HSV-latent and lytic replicating genomes in an in vivo murine model of latency showed statistically significant but relatively minor reductions in the latent genome load, and also suggested relatively poor activities of CRISPR/Cas9 compared to meganuclease-mediated targeting [31,33,57,69,83]. A future possibility would be to evaluate the targeting of ORF62 by a designer meganuclease and to determine if it is more effective than the CRISPR/Cas9 strategy. However, we feel that a better direction would be to seek improvement in targeting by expressing multiple gRNAs to additional viral targets simultaneously with the single CRISPR/Cas9 nuclease.

The results presented here demonstrate the potential of the CRISPR/Cas9 strategy to reduce both epithelial manifestations of the reactivated disease and potential damage in the ganglia that could result from ganglionic spread after a reactivation event has initiated. Of course, much remains to be worked out in translating this approach to prevent zoster and VZV disease as a therapy. A possible initial application of AAV-mediated gene editing as an antiviral strategy could be for VZV-induced retinal diseases. The retina has been a prime target for advancing AAV-mediated gene therapy strategies, and there are promising results from both animal models and human studies that AAV-mediated delivery can restore vision loss for specific inherited defects [99]. VZV replication in the retinal tissues is known to result in rare but devastating blinding diseases such as peripheral outer retinal necrosis (PORN), acute retinal necrosis (ARN) and chorioretinitis [100]. It is not uncommon for these patients to respond poorly to classic antiviral therapies, because of delayed diagnoses and/or antiviral resistance [101]. We speculate that AAV-mediated antiviral gene editing in the eye following intravitreal delivery could be an alternative treatment strategy to prevent VZV-induced retinal damage. Future work in the treatment of zoster and other VZV diseases will require additional optimization for in vivo delivery as well as the early detection of triggered VZV reactivation events in vitro, neither of which have been well-addressed for CRISPR/Cas9 therapy for other herpesviruses to date. In addition, there is a need to overcome the host antiviral immune response to AAV vectors, since clinical trials have demonstrated that this can prevent the long-term use and expression of AAV-delivered genes [102]. There remains a need for antivirals to zoster, because some studies suggest that zoster is showing increased incidence longitudinally in populations under 50 that are not yet eligible for immunization with the zoster vaccines [103]. Nevertheless, these results serve as an important proof-of-principle of the CRISPR/Cas9 approach for treatment of a widespread and painful human disease for which there are currently few therapeutic options.

Supplementary Materials: The following supporting information can be downloaded at: https://www.mdpi.com/article/10.3390/v14020378/s1, Figure S1: Enlarged images from Figure 2b to show the GFP expression from different AAV serotypes after transduction into neuron cultures; Figure S2: Enlarged microscopy images from Figure 6a, showing that AAV-62-1gR-saCas9 treatment of latently infected neurons reduces VZV progeny and spread following reactivation induction.

Author Contributions: Conceptualization, B.W.W., P.R.K. and R.S.G.; methodology, B.W.W. and M.B.Y.; software, B.W.W.; validation, B.W.W. and P.R.K.; formal analysis, B.W.W. and P.R.K.; investigation, B.W.W. and M.B.Y.; resources, P.R.K. and M.B.Y.; data curation, B.W.W.; writing—original draft preparation, B.W.W.; writing—review and editing, P.R.K. and R.S.G.; visualization, B.W.W.; supervision, P.R.K.; project administration, P.R.K.; funding acquisition, P.R.K. and R.S.G. All authors have read and agreed to the published version of the manuscript.

Funding: This work was supported by an Israel-US Binational Science Foundation award (#2017259; PRK and RSG), awards from the National Institutes of Health from NIAID (AI122640), and a P30 CORE grant from NEI (EY08098). BWW was also supported by a predoctoral award under T32 AI049820 (N.A. DeLuca, PI).

Institutional Review Board Statement: Not Applicable.

Informed Consent Statement: Not Applicable.

Data Availability Statement: The data in this study is available in this article, as well as upon reasonable request from the corresponding author.

Acknowledgments: The authors wish to acknowledge unrestricted support from the Eye & Ear Foundation of Pittsburgh and the Research to Prevent Blindness Inc., NY. The authors also wish to acknowledge advice and assistance from Leah Byrne, on the preparation of AAV; and Kira Lathrop, MAMS, for assisting with image processing.

Conflicts of Interest: The authors declare no conflict of interest.

References

1. Zerboni, L.; Sen, N.; Oliver, S.L.; Arvin, A.M. Molecular mechanisms of varicella zoster virus pathogenesis. *Nat. Rev. Microbiol.* **2014**, *12*, 197–210. [CrossRef]
2. Johnson, R.W. Herpes Zoster and Postherpetic Neuralgia. *Expert Rev. Vaccines* **2014**, *9*, 21–26. [CrossRef] [PubMed]
3. Gershon, A.A.; Gershon, M.D. Pathogenesis and current approaches to control of varicella-zoster virus infections. *Clin. Microbiol. Rev.* **2013**, *26*, 728–743. [CrossRef] [PubMed]
4. Edell, A.R.P.; Cohen, E.J. Herpes Simplex and Herpes Zoster Eye Disease. *Eye Contact Lens Sci. Clin. Pract.* **2013**, *39*, 311–314. [CrossRef] [PubMed]
5. Pavan-Langston, D. Herpes Zoster. Antivirals and Pain Management. *Ophthalmology* **2008**, *115*, 13–20. [CrossRef]
6. Oxman, M.N.; Levin, M.J.; Johnson, G.R.; Schmader, K.E.; Straus, S.E.; Gelb, L.D.; Arbeit, R.D.; Simberkoff, M.S.; Gershon, A.A.; Davis, L.E.; et al. A vaccine to prevent herpes zoster and postherpetic neuralgia in older adults. *N. Engl. J. Med.* **2005**, *352*, 2271–2284. [CrossRef]
7. Chun, C.; Weinmann, S.; Riedlinger, K.; Mullooly, J.P.; Houston, H.; Schmid, D.S.; Seward, J.F. Laboratory Characteristics of Suspected Herpes Zoster in Vaccinated Children. *Pediatr. Infect. Dis. J.* **2011**, *30*, 721–723. [CrossRef]
8. Tseng, H.F.; Schmid, D.S.; Harpaz, R.; LaRussa, P.; Jensen, N.J.; Rivailler, P.; Radford, K.; Folster, J.; Jacobsen, S.J. Herpes zoster caused by vaccine-strain varicella zoster virus in an immunocompetent recipient of zoster vaccine. *Clin. Infect. Dis.* **2014**, *58*, 1125–1128. [CrossRef]
9. Lal, H.; Cunningham, A.L.; Godeaux, O.; Chlibek, R.; Diez-Domingo, J.; Hwang, S.-J.; Levin, M.J.; McElhaney, J.E.; Poder, A.; Puig-Barberà, J.; et al. Efficacy of an Adjuvanted Herpes Zoster Subunit Vaccine in Older Adults. *N. Engl. J. Med.* **2015**, *372*, 2087–2096. [CrossRef]
10. Cunningham, A.L.; Lal, H.; Kovac, M.; Chlibek, R.; Hwang, S.-J.; Díez-Domingo, J.; Godeaux, O.; Levin, M.J.; McElhaney, J.E.; Puig-Barberà, J.; et al. Efficacy of the Herpes Zoster Subunit Vaccine in Adults 70 Years of Age or Older. *N. Engl. J. Med.* **2016**, *375*, 1019–1032. [CrossRef]
11. Lal, H.; Poder, A.; Campora, L.; Geeraerts, B.; Oostvogels, L.; Vanden Abeele, C.; Heineman, T.C. Immunogenicity, reactogenicity and safety of 2 doses of an adjuvanted herpes zoster subunit vaccine administered 2, 6 or 12 months apart in older adults: Results of a phase III, randomized, open-label, multicenter study. *Vaccine* **2018**, *36*, 148–154. [CrossRef] [PubMed]
12. Koshy, E.; Mengting, L.; Kumar, H.; Jianbo, W. Epidemiology, treatment and prevention of herpes zoster: A comprehensive review. *Indian J. Dermatol. Venereol. Leprol.* **2018**, *84*, 251–262. [CrossRef] [PubMed]
13. Sauerbrei, A. Diagnosis, antiviral therapy, and prophylaxis of varicella-zoster virus infections. *Eur. J. Clin. Microbiol. Infect. Dis.* **2016**, *35*, 723–734. [CrossRef] [PubMed]
14. Guedon, J.M.G.; Yee, M.B.; Zhang, M.; Harvey, S.A.K.; Goins, W.F.; Kinchington, P.R. Neuronal changes induced by Varicella Zoster Virus in a rat model of postherpetic neuralgia. *Virology* **2015**, *482*, 167–180. [CrossRef]
15. Sato, H.; Pesnicak, L.; Cohen, J.I. Varicella-zoster virus ORF47 protein kinase, which is required for replication in human T cells, and ORF66 protein kinase, which is expressed during latency, are dispensable for establishment of latency. *J. Virol.* **2003**, *77*, 11180–11185. [CrossRef]
16. Cohrs, R.J.; Badani, H.; Baird, N.L.; White, T.M.; Sanford, B.; Gilden, D. Induction of varicella zoster virus DNA replication in dissociated human trigeminal ganglia. *J. Neurovirol.* **2017**, *23*, 152. [CrossRef]
17. Ouwendijk, W.J.D.; Choe, A.; Nagel, M.A.; Gilden, D.; Osterhaus, A.D.M.E.; Cohrs, R.J.; Verjans, G.M.G.M. Restricted Varicella-Zoster Virus Transcription in Human Trigeminal Ganglia Obtained Soon after Death. *J. Virol.* **2012**, *86*, 10203–10206. [CrossRef]
18. Rowe, J.; Greenblatt, R.J.; Liu, D.; Moffat, J.F. Compounds that target host cell proteins prevent varicella-zoster virus replication in culture, ex vivo, and in SCID-Hu mice. *Antivir. Res.* **2010**, *86*, 276–285. [CrossRef]
19. Zerboni, L.; Arvin, A. Neuronal Subtype and Satellite Cell Tropism Are Determinants of Varicella-Zoster Virus Virulence in Human Dorsal Root Ganglia Xenografts In Vivo. *PLoS Pathog.* **2015**, *11*, e1004989. [CrossRef]
20. De, C.; Liu, D.; Zheng, B.; Singh, U.S.; Chavre, S.; White, C.; Arnold, R.D.; Hagen, F.K.; Chu, C.K.; Moffat, J.F. β-l-1-[5-(E-2-bromovinyl)-2-(hydroxymethyl)-1,3-(dioxolan-4-yl)] uracil (l-BHDU) prevents varicella-zoster virus replication in a SCID-Hu mouse model and does not interfere with 5-fluorouracil catabolism. *Antivir. Res.* **2014**, *110*, 10–19. [CrossRef]
21. Sloutskin, A.; Kinchington, P.R.; Goldstein, R.S. Productive vs. non-productive infection by cell-free varicella zoster virus of human neurons derived from embryonic stem cells is dependent upon infectious viral dose. *Virology* **2013**, *443*, 285–293. [CrossRef] [PubMed]
22. Markus, A.; Grigoryan, S.; Sloutskin, A.; Yee, M.B.; Zhu, H.; Yang, I.H.; Thakor, N.V.; Sarid, R.; Kinchington, P.R.; Goldstein, R.S. Varicella-zoster virus (VZV) infection of neurons derived from human embryonic stem cells: Direct demonstration of axonal infection, transport of VZV, and productive neuronal infection. *J. Virol.* **2011**, *85*, 6220–6233. [CrossRef] [PubMed]

23. Markus, A.; Lebenthal-Loinger, I.; Yang, I.H.; Kinchington, P.R.; Goldstein, R.S. An In Vitro Model of Latency and Reactivation of Varicella Zoster Virus in Human Stem Cell-Derived NeuronsA. *PLoS Pathog.* **2015**, *11*, e1004885. [CrossRef]
24. Sadaoka, T.; Depledge, D.P.; Rajbhandari, L.; Venkatesan, A.; Breuer, J.; Cohen, J.I. In vitro system using human neurons demonstrates that varicella-zoster vaccine virus is impaired for reactivation, but not latency. *Proc. Natl. Acad. Sci. USA* **2016**, *113*, E2403–E2412. [CrossRef] [PubMed]
25. Sadaoka, T.; Rajbhandari, L.; Shukla, P.; Jagdish, B.; Lee, H.; Lee, G.; Venkatesan, A. Human stem cell derived sensory neurons are positioned to support varicella zoster virus latency. *bioRxiv* **2020**. [CrossRef]
26. Christensen, J.; Steain, M.; Slobedman, B.; Abendroth, A. Differentiated neuroblastoma cells provide a highly efficient model for studies of productive varicella-zoster virus infection of neuronal cells. *J. Virol.* **2011**, *85*, 8436–8442. [CrossRef] [PubMed]
27. Baird, N.L.; Bowlin, J.L.; Cohrs, R.J.; Gilden, D.; Jones, K.L. Comparison of varicella-zoster virus RNA sequences in human neurons and fibroblasts. *J. Virol.* **2014**, *88*, 5877–5880. [CrossRef]
28. Depledge, D.P.; Ouwendijk, W.J.D.; Sadaoka, T.; Braspenning, S.E.; Mori, Y.; Cohrs, R.J.; Verjans, G.M.G.M.; Breuer, J. A spliced latency-associated VZV transcript maps antisense to the viral transactivator gene 61. *Nat. Commun.* **2018**, *9*, 1167. [CrossRef] [PubMed]
29. Baird, N.L.; Zhu, S.; Pearce, C.M.; Viejo-Borbolla, A. Current In Vitro Models to Study Varicella Zoster Virus Latency and Reactivation. *Viruses* **2019**, *11*, 103. [CrossRef] [PubMed]
30. Laemmle, L.; Goldstein, R.S.; Kinchington, P.R. Modeling Varicella Zoster Virus Persistence and Reactivation—Closer to Resolving a Perplexing Persistent State. *Front. Microbiol.* **2019**, *10*, 1634. [CrossRef]
31. Aubert, M.; Madden, E.A.; Loprieno, M.; Feelixge, H.S.D.; Stensland, L.; Huang, M.-L.; Greninger, A.L.; Roychoudhury, P.; Niyonzima, N.; Nguyen, T.; et al. In vivo disruption of latent HSV by designer endonuclease therapy. *JCI Insight* **2016**, *1*, e88468. [CrossRef] [PubMed]
32. Aubert, M.; Boyle, N.M.; Stone, D.; Stensland, L.; Huang, M.-L.; Magaret, A.S.; Galetto, R.; Rawlings, D.J.; Scharenberg, A.M.; Jerome, K.R. In vitro Inactivation of Latent HSV by Targeted Mutagenesis Using an HSV-specific Homing Endonuclease. *Mol. Ther. Nucleic Acids* **2014**, *3*, e146. [CrossRef]
33. Aubert, M.; Strongin, D.E.; Roychoudhury, P.; Loprieno, M.A.; Haick, A.K.; Klouser, L.M.; Stensland, L.; Huang, M.-L.; Makhsous, N.; Tait, A.; et al. Gene editing and elimination of latent herpes simplex virus in vivo. *Nat. Commun.* **2020**, *11*, 4148. [CrossRef] [PubMed]
34. Van Diemen, F.R.; Kruse, E.M.; Hooykaas, M.J.G.; Bruggeling, C.E.; Schurch, A.C.; van Ham, P.M.; Imhof, S.M.; Nijhuis, M.; Wiertz, E.J.H.J.; Lebbink, R.J. CRISPR/Cas9-Mediated Genome Editing of Herpesviruses Limits Productive and Latent Infections. *PLoS Pathog.* **2016**, *12*, e1005701. [CrossRef] [PubMed]
35. Wang, J.; Quake, S.R. RNA-guided endonuclease provides a therapeutic strategy to cure latent herpesviridae infection. *Proc. Natl. Acad. Sci. USA* **2014**, *111*, 13157–13162. [CrossRef]
36. Dang, C.H.; Aubert, M.; De Silva Feelixge, H.S.; Diem, K.; Loprieno, M.A.; Roychoudhury, P.; Stone, D.; Jerome, K.R. In vivo dynamics of AAV-mediated gene delivery to sensory neurons of the trigeminal ganglia. *Sci. Rep.* **2017**, *7*, 927. [CrossRef]
37. Ran, F.A.; Cong, L.; Yan, W.X.; Scott, D.A.; Gootenberg, J.S.; Kriz, A.J.; Zetsche, B.; Shalem, O.; Wu, X.; Makarova, K.S.; et al. In vivo genome editing using Staphylococcus aureus Cas9. *Nature* **2015**, *520*, 186–191. [CrossRef]
38. Wu, Z.; Yang, H.; Colosi, P. Effect of Genome Size on AAV Vector Packaging. *Mol. Ther.* **2010**, *18*, 80. [CrossRef]
39. Davison, A.J.; Scott, J.E. The complete DNA sequence of varicella-zoster virus. *J. Gen. Virol.* **1986**, *67 Pt 9*, 1759–1816. [CrossRef]
40. Davison, A.J.; Scott, J.E. DNA sequence of the major inverted repeat in the varicella-zoster virus genome. *J. Gen. Virol.* **1985**, *66 Pt 2*, 207–220. [CrossRef]
41. Moriuchi, M.; Moriuchi, H.; Straus, S.E.; Cohen, J.I. Varicella-Zoster Virus (VZV) Virion-Associated Transactivator Open Reading Frame 62 Protein Enhances the Infectivity of VZV DNA. *Virology* **1994**, *200*, 297–300. [CrossRef]
42. Kinchington, P.R.; Hougland, J.K.; Arvin, A.M.; Ruyechan, W.T.; Hay, J. The varicella-zoster virus immediate-early protein IE62 is a major component of virus particles. *J. Virol.* **1992**, *66*, 359–366. [CrossRef]
43. Sommer, M.H.; Zagha, E.; Serrano, O.K.; Ku, C.C.; Zerboni, L.; Baiker, A.; Santos, R.; Spengler, M.; Lynch, J.; Grose, C.; et al. Mutational analysis of the repeated open reading frames, ORFs 63 and 70 and ORFs 64 and 69, of varicella-zoster virus. *J. Virol.* **2001**, *75*, 8224–8239. [CrossRef]
44. Sato, B.; Ito, H.; Hinchliffe, S.; Sommer, M.H.; Zerboni, L.; Arvin, A.M. Mutational analysis of open reading frames 62 and 71, encoding the varicella-zoster virus immediate-early transactivating protein, IE62, and effects on replication in vitro and in skin xenografts in the SCID-hu mouse in vivo. *J. Virol.* **2003**, *77*, 5607–5620. [CrossRef] [PubMed]
45. Dremel, S.E.; Deluca, N.A. Herpes simplex viral nucleoprotein creates a competitive transcriptional environment facilitating robust viral transcription and host shut off. *Elife* **2019**, *8*, e51109. [CrossRef] [PubMed]
46. Cohen, J.I. The varicella-zoster virus genome. *Curr. Top. Microbiol. Immunol.* **2010**, *342*, 1–14. [CrossRef] [PubMed]
47. Felser, J.M.; Kinchington, P.R.; Inchauspe, G.; Straus, S.E.; Ostrove, J.M. Cell lines containing varicella-zoster virus open reading frame 62 and expressing the "IE" 175 protein complement ICP4 mutants of herpes simplex virus type 1. *J. Virol.* **1988**, *62*, 2076–2082. [CrossRef] [PubMed]
48. Disney, G.H.; Everett, R.D. A herpes simplex virus type 1 recombinant with both copies of the Vmw175 coding sequences replaced by the homologous varicella-zoster virus open reading frame. *J. Gen. Virol.* **1990**, *71 Pt 11*, 2681–2689. [CrossRef] [PubMed]

49. Niemeyer, G.P.; Herzog, R.W.; Mount, J.; Arruda, V.R.; Tillson, D.M.; Hathcock, J.; Van Ginkel, F.W.; High, K.A.; Lothrop, C.D., Jr. Long-term correction of inhibitor-prone hemophilia B dogs treated with liver-directed AAV2-mediated factor IX gene therapy. *Blood* **2009**, *113*, 797–806. [CrossRef]
50. Harding, T.; Koprivnikar, K.; Tu, G.; Zayek, N.; Lew, S.; Subramanian, A.; Sivakumaran, A.; Frey, D.; Ho, K.; VanRoey, M.; et al. Intravenous administration of an AAV-2 vector for the expression of factor IX in mice and a dog model of hemophilia B. *Gene Ther.* **2004**, *11*, 204–213. [CrossRef]
51. Mount, J.D.; Herzog, R.W.; Tillson, D.M.; Goodman, S.A.; Robinson, N.; McCleland, M.L.; Bellinger, D.; Nichols, T.C.; Arruda, V.R.; Lothrop, C.D., Jr.; et al. Sustained phenotypic correction of hemophilia B dogs with a factor IX null mutation by liver-directed gene therapy. *Blood* **2002**, *99*, 2670–2676. [CrossRef]
52. Nathwani, A.C.; Tuddenham, E.G.; Rangarajan, S.; Rosales, C.; Jenny, M.; Linch, D.C.; Chowdary, P.; Riddell, A.; Pie, A.J.; Harrington, C.; et al. Adenovirus-associated virus vector-mediated gene transfer in hemophilia B. *N. Engl. J. Med.* **2011**, *365*, 2357–2365. [CrossRef]
53. Manno, C.S.; Pierce, G.F.; Arruda, V.R.; Glader, B.; Ragni, M.; Rasko, J.J.; Ozelo, M.C.; Hoots, K.; Blatt, P.; Konkle, B.; et al. Successful transduction of liver in hemophilia by AAV-Factor IX and limitations imposed by the host immune response. *Nat. Med.* **2006**, *12*, 342–347. [CrossRef] [PubMed]
54. Guedon, J.-M.; Zhang, M.; Glorioso, J.; Goins, W.; Kinchington, P. Relief of pain induced by varicella-zoster virus in a rat model of post-herpetic neuralgia using a herpes simplex virus vector expressing enkephalin. *Gene Ther.* **2014**, *21*, 694. [CrossRef] [PubMed]
55. Sloutskin, A.; Goldstein, R.S. Laboratory preparation of Varicella-Zoster Virus: Concentration of virus-containing supernatant, use of a debris fraction and magnetofection for consistent cell-free VZV infections. *J. Virol. Methods* **2014**, *206*, 128–132. [CrossRef] [PubMed]
56. Cong, L.; Ran, F.A.; Cox, D.; Lin, S.; Barretto, R.; Habib, N.; Hsu, P.D.; Wu, X.; Jiang, W.; Marraffini, L.A.; et al. Multiplex genome engineering using CRISPR/Cas systems. *Science* **2013**, *339*, 819–823. [CrossRef] [PubMed]
57. Russell, T.A.; Stefanovic, T.; Tscharke, D.C. Engineering herpes simplex viruses by infection-transfection methods including recombination site targeting by CRISPR/Cas9 nucleases. *J. Virol. Methods* **2015**, *213*, 18–25. [CrossRef] [PubMed]
58. Choi, V.W.; Asokan, A.; Haberman, R.A.; Samulski, R.J. Production of recombinant adeno-associated viral vectors for in vitro and in vivo use. *Curr. Protoc. Mol. Biol.* **2007**. *Chapter 16*, Unit 16.25. [CrossRef]
59. Zolotukhin, S.; Byrne, B.; Mason, E.; Zolotukhin, I.; Potter, M.; Chesnut, K.; Summerford, C.; Samulski, R.; Muzyczka, N. Recombinant adeno-associated virus purification using novel methods improves infectious titer and yield. *Gene Ther.* **1999**, *6*, 973–985. [CrossRef]
60. Aurnhammer, C.; Haase, M.; Muether, N.; Hausl, M.; Rauschhuber, C.; Huber, I.; Nitschko, H.; Busch, U.; Sing, A.; Ehrhardt, A.; et al. Universal real-time PCR for the detection and quantification of adeno-associated virus serotype 2-derived inverted terminal repeat sequences. *Hum. Gene Ther. Methods* **2012**, *23*, 18–28. [CrossRef]
61. Tischer, B.K.; Von Einem, J.; Kaufer, B.B.; Osterrieder, N. Two-step Red-mediated recombination for versatile high-efficiency markerless DNA manipulation in Escherichia coli. *Biotechniques* **2006**, *40*, 191–197. [CrossRef] [PubMed]
62. Tischer, B.K.; Kaufer, B.B.; Sommer, M.H.; Wussow, F.; Arvin, A.M.; Osterrieder, N. A self-excisable infectious bacterial artificial chromosome clone of varicella-zoster virus allows analysis of the essential tegument protein encoded by ORF9. *J. Virol.* **2007**, *81*, 13200–13208. [CrossRef] [PubMed]
63. Lloyd, M.G.; Smith, N.A.; Tighe, M.; Travis, K.L.; Liu, D.; Upadhyaya, P.K.; Kinchington, P.R.; Chan, G.C.; Moffat, J.F. A Novel Human Skin Tissue Model To Study Varicella-Zoster Virus and Human Cytomegalovirus. *J. Virol.* **2020**, *94*, 1082–1102. [CrossRef]
64. Erazo, A.; Yee, M.B.; Osterrieder, N.; Kinchington, P.R. Varicella-zoster virus open reading frame 66 protein kinase is required for efficient viral growth in primary human corneal stromal fibroblast cells. *J. Virol.* **2008**, *82*, 7653–7665. [CrossRef] [PubMed]
65. Warner, B.E.; Yee, M.B.; Zhang, M.; Hornung, R.S.; Kaufer, B.B.; Visalli, R.J.; Kramer, P.R.; Goins, W.F.; Kinchington, P.R. Varicella-zoster virus early infection but not complete replication is required for the induction of chronic hypersensitivity in rat models of postherpetic neuralgia. *PLoS Pathog.* **2021**, *17*, e1009689. [CrossRef]
66. Pomp, O.; Brokhman, I.; Ben-Dor, I.; Reubinoff, B.; Goldstein, R.S. Generation of peripheral sensory and sympathetic neurons and neural crest cells from human embryonic stem cells. *Stem Cells* **2005**, *23*, 923–930. [CrossRef]
67. Pomp, O.; Brokhman, I.; Ziegler, L.; Almog, M.; Korngreen, A.; Tavian, M.; Goldstein, R.S. PA6-induced human embryonic stem cell-derived neurospheres: A new source of human peripheral sensory neurons and neural crest cells. *Brain Res.* **2008**, *1230*, 50–60. [CrossRef]
68. Eisfeld, A.J.; Turse, S.E.; Jackson, S.A.; Lerner, E.C.; Kinchington, P.R. Phosphorylation of the varicella-zoster virus (VZV) major transcriptional regulatory protein IE62 by the VZV open reading frame 66 protein kinase. *J. Virol.* **2006**, *80*, 1710–1723. [CrossRef]
69. Velusamy, T.; Gowripalan, A.; Tscharke, D.C. CRISPR/Cas9-Based Genome Editing of HSV. *Methods Mol. Biol.* **2020**, *2060*, 169–183. [CrossRef]
70. Jones, J.O.; Sommer, M.; Stamatis, S.; Arvin, A.M. Mutational analysis of the varicella-zoster virus ORF62/63 intergenic region. *J. Virol.* **2006**, *80*, 3116–3121. [CrossRef]
71. Hood, C.; Cunningham, A.L.; Slobedman, B.; Arvin, A.M.; Sommer, M.H.; Kinchington, P.R.; Abendroth, A. Varicella-zoster virus ORF63 inhibits apoptosis of primary human neurons. *J. Virol.* **2006**, *80*, 1025–1031. [CrossRef]

72. Ouwendijk, W.J.D.; Depledge, D.P.; Rajbhandari, L.; Lenac Rovis, T.; Jonjic, S.; Breuer, J.; Venkatesan, A.; Verjans, G.M.G.M.; Sadaoka, T. Varicella-zoster virus VLT-ORF63 fusion transcript induces broad viral gene expression during reactivation from neuronal latency. *Nat. Commun.* **2020**, *11*, 6324. [CrossRef]
73. Ambagala, A.P.N.; Krogmann, T.; Qin, J.; Pesnicak, L.; Cohen, J.I. A varicella-zoster virus mutant impaired for latency in rodents, but not impaired for replication in cell culture. *Virology* **2010**, *399*, 194. [CrossRef]
74. Taymans, J.M.; Vandenberghe, L.H.; Van Den Haute, C.; Thiry, I.; Deroose, C.M.; Mortelmans, L.; Wilson, J.M.; Debyser, Z.; Baekelandt, V. Comparative analysis of adeno-associated viral vector serotypes 1, 2, 5, 7, and 8 in mouse brain. *Hum. Gene Ther.* **2007**, *18*, 195–206. [CrossRef]
75. Gao, G.; Vandenberghe, L.; Wilson, J. New recombinant serotypes of AAV vectors. *Curr. Gene Ther.* **2005**, *5*, 285–297. [CrossRef]
76. Cearley, C.N.; Wolfe, J.H. Transduction characteristics of adeno-associated virus vectors expressing cap serotypes 7, 8, 9, and Rh10 in the mouse brain. *Mol. Ther.* **2006**, *13*, 528–537. [CrossRef]
77. Cearley, C.N.; Wolfe, J.H. A single injection of an adeno-associated virus vector into nuclei with divergent connections results in widespread vector distribution in the brain and global correction of a neurogenetic disease. *J. Neurosci.* **2007**, *27*, 9928–9940. [CrossRef] [PubMed]
78. Haery, L.; Deverman, B.E.; Matho, K.S.; Cetin, A.; Woodard, K.; Cepko, C.; Guerin, K.I.; Rego, M.A.; Ersing, I.; Bachle, S.M.; et al. Adeno-Associated Virus Technologies and Methods for Targeted Neuronal Manipulation. *Front. Neuroanat.* **2019**, *13*, 93. [CrossRef] [PubMed]
79. Auricchio, A.; Kobinger, G.; Anand, V.; Hildinger, M.; O'Connor, E.; Maguire, A.M.; Wilson, J.M.; Bennett, J. Exchange of surface proteins impacts on viral vector cellular specificity and transduction characteristics: The retina as a model. *Hum. Mol. Genet.* **2001**, *10*, 3075–3081. [CrossRef] [PubMed]
80. Davidson, B.L.; Stein, C.S.; Heth, J.A.; Martins, I.; Kotin, R.M.; Derksen, T.A.; Zabner, J.; Ghodsi, A.; Chiorini, J.A. Recombinant adeno-associated virus type 2, 4, and 5 vectors: Transduction of variant cell types and regions in the mammalian central nervous system. *Proc. Natl. Acad. Sci. USA* **2000**, *97*, 3428–3432. [CrossRef] [PubMed]
81. Dukhovny, A.; Sloutskin, A.; Markus, A.; Yee, M.B.; Kinchington, P.R.; Goldstein, R.S. Varicella-zoster virus infects human embryonic stem cell-derived neurons and neurospheres but not pluripotent embryonic stem cells or early progenitors. *J. Virol.* **2012**, *86*, 3211–3218. [CrossRef]
82. Lin, H.; Li, G.; Peng, X.; Deng, A.; Ye, L.; Shi, L.; Wang, T.; He, J. The Use of CRISPR/Cas9 as a Tool to Study Human Infectious Viruses. *Front. Cell. Infect. Microbiol.* **2021**, *11*, 774. [CrossRef] [PubMed]
83. Oh, H.S.; Neuhausser, W.M.; Eggan, P.; Angelova, M.; Kirchner, R.; Eggan, K.C.; Knipe, D.M. Herpesviral lytic gene functions render the viral genome susceptible to novel editing by CRISPR/Cas9. *Elife* **2019**, *8*, e51662. [CrossRef] [PubMed]
84. Neuhausser, W.; Oh, H.; Eggan, P.; Angelova, M.; Kirchner, R.; Eggan, K.; Knipe, D. Screening Method for CRISPR/Cas9 Inhibition of a Human DNA Virus: Herpes Simplex Virus. *Bio-Protoc.* **2020**, *10*, e3748. [CrossRef] [PubMed]
85. Stone, D.; Long, K.R.; Loprieno, M.A.; Feelixge, H.S.D.S.; Kenkel, E.J.; Liley, R.M.; Rapp, S.; Roychoudhury, P.; Nguyen, T.; Stensland, L.; et al. CRISPR-Cas9 gene editing of hepatitis B virus in chronically infected humanized mice. *Mol. Ther. Methods Clin. Dev.* **2021**, *20*, 258. [CrossRef]
86. Yan, K.; Feng, J.; Liu, X.; Wang, H.; Li, Q.; Li, J.; Xu, T.; Sajid, M.; Ullah, H.; Zhou, L.; et al. Inhibition of Hepatitis B Virus by AAV8-Derived CRISPR/SaCas9 Expressed From Liver-Specific Promoters. *Front. Microbiol.* **2021**, *12*, 665184. [CrossRef]
87. Kunze, C.; Börner, K.; Kienle, E.; Orschmann, T.; Rusha, E.; Schneider, M.; Radivojkov-Blagojevic, M.; Drukker, M.; Desbordes, S.; Grimm, D.; et al. Synthetic AAV/CRISPR vectors for blocking HIV-1 expression in persistently infected astrocytes. *Glia* **2018**, *66*, 413–427. [CrossRef]
88. Yin, C.; Zhang, T.; Qu, X.; Zhang, Y.; Putatunda, R.; Xiao, X.; Li, F.; Xiao, W.; Zhao, H.; Dai, S.; et al. In Vivo Excision of HIV-1 Provirus by saCas9 and Multiplex Single-Guide RNAs in Animal Models. *Mol. Ther.* **2017**, *25*, 1168–1186. [CrossRef]
89. Chen, Y.C.; Sheng, J.; Trang, P.; Liu, F. Potential Application of the CRISPR/Cas9 System against Herpesvirus Infections. *Viruses* **2018**, *10*, 291. [CrossRef]
90. Yuen, K.S.; Chan, C.P.; Wong, N.H.M.; Ho, C.H.; Ho, T.H.; Lei, T.; Deng, W.; Tsao, S.W.; Chen, H.; Kok, K.H.; et al. CRISPR/Cas9-mediated genome editing of Epstein-Barr virus in human cells. *J. Gen. Virol.* **2015**, *96*, 626–636. [CrossRef]
91. Adamson, C.S.; Nevels, M.M. Bright and Early: Inhibiting Human Cytomegalovirus by Targeting Major Immediate-Early Gene Expression or Protein Function. *Viruses* **2020**, *12*, 110. [CrossRef]
92. Brackett, K.; Mungale, A.; Lopez-Isidro, M.; Proctor, D.A.; Najarro, G.; Arias, C. CRISPR Interference Efficiently Silences Latent and Lytic Viral Genes in Kaposi's Sarcoma-Associated Herpesvirus-Infected Cells. *Viruses* **2021**, *13*, 783. [CrossRef]
93. Akidil, E.; Albanese, M.; Buschle, A.; Ruhle, A.; Pich, D.; Keppler, O.T.; Hammerschmidt, W. Highly efficient CRISPR-Cas9-mediated gene knockout in primary human B cells for functional genetic studies of Epstein-Barr virus infection. *PLoS Pathog.* **2021**, *17*, e1009117. [CrossRef] [PubMed]
94. Braspenning, S.E.; Lebbink, R.J.; Depledge, D.P.; Schapendonk, C.M.E.; Anderson, L.A.; Verjans, G.M.G.M.; Sadaoka, T.; Ouwendijk, W.J.D. Mutagenesis of the varicella-zoster virus genome demonstrates that vlt and vlt-orf63 proteins are dispensable for lytic infection. *Viruses* **2021**, *13*, 2289. [CrossRef] [PubMed]
95. Mendell, J.R.; Al-Zaidy, S.A.; Rodino-Klapac, L.R.; Goodspeed, K.; Gray, S.J.; Kay, C.N.; Boye, S.L.; Boye, S.E.; George, L.A.; Salabarria, S.; et al. Current Clinical Applications of In Vivo Gene Therapy with AAVs. *Mol. Ther.* **2021**, *29*, 464–488. [CrossRef] [PubMed]

96. Broeders, M.; Herrero-Hernandez, P.; Ernst, M.P.T.; van der Ploeg, A.T.; Pijnappel, W.W.M.P. Sharpening the Molecular Scissors: Advances in Gene-Editing Technology. *iScience* **2020**, *23*, 100789. [CrossRef] [PubMed]
97. McCullough, K.T.; Boye, S.L.; Fajardo, D.; Calabro, K.; Peterson, J.J.; Strang, C.E.; Chakraborty, D.; Gloskowski, S.; Haskett, S.; Samuelsson, S.; et al. Somatic Gene Editing of GUCY2D by AAV-CRISPR/Cas9 Alters Retinal Structure and Function in Mouse and Macaque. *Hum. Gene Ther.* **2019**, *30*, 571–589. [CrossRef]
98. Ramlogan-Steel, C.A.; Murali, A.; Andrzejewski, S.; Dhungel, B.; Steel, J.C.; Layton, C.J. Gene therapy and the adeno-associated virus in the treatment of genetic and acquired ophthalmic diseases in humans: Trials, future directions and safety considerations. *Clin. Experiment. Ophthalmol.* **2019**, *47*, 521–536. [CrossRef]
99. Da Costa, B.L.; Levi, S.R.; Eulau, E.; Tsai, Y.-T.; Quinn, P.M.J. Prime Editing for Inherited Retinal Diseases. *Front. Genome Ed.* **2021**, *3*, 775330. [CrossRef]
100. Kedar, S.; Jayagopal, L.N.; Berger, J.R. Neurological and Ophthalmological Manifestations of Varicella Zoster Virus. *J. Neuroophthalmol.* **2019**, *39*, 220–231. [CrossRef]
101. Schoenberger, S.D.; Kim, S.J.; Thorne, J.E.; Mruthyunjaya, P.; Yeh, S.; Bakri, S.J.; Ehlers, J.P. Diagnosis and Treatment of Acute Retinal Necrosis: A Report by the American Academy of Ophthalmology. *Ophthalmology* **2017**, *124*, 382–392. [CrossRef] [PubMed]
102. Nidetz, N.F.; McGee, M.C.; Tse, L.V.; Li, C.; Cong, L.; Li, Y.; Huang, W. Adeno-associated viral vector-mediated immune responses: Understanding barriers to gene delivery. *Pharmacol. Ther.* **2020**, *207*, 107453. [CrossRef] [PubMed]
103. Goldman, G.S.; King, P.G. Review of the United States universal varicella vaccination program: Herpes zoster incidence rates, cost-effectiveness, and vaccine efficacy based primarily on the Antelope Valley Varicella Active Surveillance Project data. *Vaccine* **2013**, *31*, 1680–1694. [CrossRef] [PubMed]

viruses

Article

Visualization of Marek's Disease Virus Genomes in Living Cells during Lytic Replication and Latency

Tereza Vychodil [1], Darren J. Wight [1], Mariana Nascimento [1], Fabian Jolmes [2], Thomas Korte [2], Andreas Herrmann [2,3] and Benedikt B. Kaufer [1,4,*]

[1] Institut für Virologie, Freie Universität Berlin, Robert von Ostertag-Straße 7-13, 14163 Berlin, Germany; tereza.vychodil@fu-berlin.de (T.V.); d.wight@fu-berlin.de (D.J.W.); m.nascimento@fu-berlin.de (M.N.)
[2] Department of Biology, Molecular Biophysics, Humboldt-Universität zu Berlin, Invalidenstraße 42, 10115 Berlin, Germany; jolmes@picoquant.com (F.J.); thomas.korte@rz.hu-berlin.de (T.K.); andreas.herrmann@rz.hu-berlin.de (A.H.)
[3] Institut für Chemie und Biochemie, Freie Universität Berlin, Altensteinstr. 23a, 14195 Berlin, Germany
[4] Veterinary Centre for Resistance Research (TZR), Freie Universität Berlin, 14163 Berlin, Germany
* Correspondence: b.kaufer@fu-berlin.de; Tel.: +49-30-838-51936

Abstract: Visualization of the herpesvirus genomes during lytic replication and latency is mainly achieved by fluorescence in situ hybridization (FISH). Unfortunately, this technique cannot be used for the real-time detection of viral genome in living cells. To facilitate the visualization of the Marek's disease virus (MDV) genome during all stages of the virus lifecycle, we took advantage of the well-established tetracycline operator/repressor (TetO/TetR) system. This system consists of a fluorescently labeled TetR (TetR-GFP) that specifically binds to an array of *tetO* sequences. This *tetO* repeat array was first inserted into the MDV genome (vTetO). Subsequently, we fused TetR-GFP via a P2a self-cleaving peptide to the C-terminus of the viral interleukin 8 (vIL8), which is expressed during lytic replication and latency. Upon reconstitution of this vTetO-TetR virus, fluorescently labeled replication compartments were detected in the nucleus during lytic replication. After validating the specificity of the observed signal, we used the system to visualize the genesis and mobility of the viral replication compartments. In addition, we assessed the infection of nuclei in syncytia as well as lytic replication and latency in T cells. Taken together, we established a system allowing us to track the MDV genome in living cells that can be applied to many other DNA viruses.

Keywords: Marek's disease virus; live-cell genome visualization; lytic replication; T cells; latency; genome integration; TetO/TetR system

Citation: Vychodil, T.; Wight, D.J.; Nascimento, M.; Jolmes, F.; Korte, T.; Herrmann, A.; Kaufer, B.B. Visualization of Marek's Disease Virus Genomes in Living Cells during Lytic Replication and Latency. *Viruses* **2022**, *14*, 287. https://doi.org/10.3390/v14020287

Academic Editors: Charles Grose, Ravi Mahalingam and Joel Rovnak

Received: 30 November 2021
Accepted: 26 January 2022
Published: 29 January 2022

1. Introduction

Marek's disease virus (MDV), also known as Gallid alphaherpesvirus 2 (GaHV-2), is a highly oncogenic herpesvirus that belongs to the genus Mardivirus. MDV infects chickens and causes neurological disorders, immunosuppression, paralysis, and deadly T cell lymphomas in various organs [1]. The virus enters the host through the respiratory tract where it infects macrophages and dendritic cells [2,3] that transport the virus to lymphoid organs. Here, this cell-associated virus is passed on to B and T cells in which it can replicate lytically [4–6]. In addition, MDV establishes latency predominantly in CD4+ T cells [7,8] and integrates its genome into the host telomeres [9,10]. During latency, only a few genes are expressed, including the major oncogene Meq (MDV005 and MDV076), splice variants of the viral chemokine vCXCL13 (aka. vIL-8; MDV003 and MDV078), [11,12] and the viral telomerase RNA [13,14]. Latently infected cells can also be transformed, resulting in the rapid formation of T cell lymphomas [15]. In addition, these cells can transport the virus to the feather follicle epithelia, where the virus replicates, is shed into the environment, and, thereby, spreads to naïve chickens [9].

Despite many years of research, many molecular processes involved in MDV replication and integration remain poorly understood. This is mostly due to the cell associated nature of the virus, its slow replication cycle, and the limited availability of tools. In recent years, viruses harboring fluorescent proteins have drastically expanded our knowledge on MDV replication and other processes during infection [16–22]. While virus proteins can be easily visualized in living infected cells by fusing them to, for example, green fluorescent protein (GFP), it remained impossible to visualize the virus genome in living cells. This genome visualization would provide valuable tools to assess the molecular processes including replication, integration, and latency.

In this study, we used the well-established tetracycline operator/repressor (*tetO*/TetR) system to visualize MDV genomes in living cells. The *tetO*/TetR system consists of a fluorescently labeled TetR protein (TetR-GFP) that specifically binds *tetO* repeat sequences as a dimer and, thereby, provides an increased fluorescent signal at the *tetO* insertion site [23]. This tool is commonly used in cell biology where cellular chromosome loci harboring *tetO* repeats are visualized to assess chromosome dynamics [24,25]. To monitor the MDV genome during infection, we inserted 112× *tetO* repeats into the virus genome using two-step Red-mediated mutagenesis [26]. TetR-GFP or TetR-mCherry were either expressed from the virus genome or stably expressed by the host cells. This system allowed us to visualize the MDV genome during lytic replication and in latently infected cells, providing important insights into the formation of replication compartments (RC), the structures in the nucleus in which the viral genome is replicated, and the infection of T cells.

2. Materials and Methods

2.1. Cells

Chicken embryo cells (CECs) were prepared from Valo specific-pathogen free (SPF) 11-day-old embryonated chicken eggs (Valo BioMedia, Osterholz-Scharmbeck, Germany), as described previously [27]. CECs were maintained in minimum essential medium (PAN Biotech, Aidenbach, Germany) supplemented with 10% fetal bovine serum (FBS, PAN Biotech) and 1% penicillin/streptomycin at 37 °C in a 5% CO_2 environment. ESCDL-1, cell line derived from chicken embryonic stem cells [28] were maintained in DMEM Ham's F12 (PAN Biotech) and supplemented with 10% FBS and 1% penicillin/streptomycin at 37 °C under a 5% CO_2 atmosphere. The reticuloendotheliosis virus-transformed chicken T-cell line 855-19 was kindly provided by Prof. Thomas Göbel (Ludwig Maximilian University of Munich, Germany). Cells were maintained in RPMI (PAN Biotech) and supplemented with 10% FBS, 1% sodium pyruvate (PAN Biotech), 1% non-essential amino acids (Biochrom; Berlin, Germany), and 1% penicillin/streptomycin at 41 °C under a 5% CO_2 atmosphere. All stable cell lines were confirmed by PCR to be mycoplasma-free.

2.2. Generation of Recombinant Viruses

The recombinant viruses harboring the *tetO*/TetR system components were generated based on a previously generated bacterial artificial chromosome (BAC) clone of the very virulent RB1B strain [29], in which most of the internal repeat short and long regions were deleted ($\Delta IR_L\Delta IR_S$) [15]. This deletion is rapidly restored upon reconstitution and facilitates a rapid manipulation of the repeat regions using two-step Red-mediated mutagenesis, as described previously [26,30]. First, transfer plasmids were generated that allowed the insertion of the components into the virus genome. To obtain the TetR-GFP transfer plasmid, the TetR-GFP cassette containing a nuclear localization signal (NLS) was amplified from the a p128tetR-GFP plasmid, kindly provided by Susan Gasser (Friedrich Miescher Institute, Switzerland) [25], and cloned into pcDNA3.1 using BamHI and EcoRI. The kanamycin selection cassette (kana_I-SceI) was amplified from pEPkan-S1, homologue sequences for its removal inserted via the primer overhangs and cloned into pcDNA3.1_TetR-GFP. This mutagenesis cassette was subsequently used to insert TetR-GFP (i) after the strong HSV-1 thymidine kinase (TK) promoter within the mini-F cassette (vTetR) or (ii) fused to the C-terminus of vCXCL13 via a P2A self-cleaving peptide (vTetO/TetR), resulting in expression of TetR-GFP and vCXCL13 as separate proteins, using two-step Red-mediated mutagenesis, as described previously [26,30].

To obtain the TetO transfer plasmid, the 112× *tetO* repeats array (4.5 kbp) of pRS306tet02x112 (Susan Gasser, Friedrich Miescher Institute, Switzerland) was cloned into a vector containing long homologue sequences of UL45 and UL46 (MDV058 and MDV059, respectively). The *tetO* sequence was inserted between UL45 and UL46 using BamHI and BglII. The kana_I-SceI cassette with homologue sequences for its removal were inserted into the UL45 homologue arm. This mutagenesis cassette was subsequently used to insert *tetO* between UL45 and UL46 of the virus genome using two-step Red-mediated mutagenesis (Figure 1A, vTetO and vTetO-TetR). This UL45/UL46 was chosen as it previously allowed the insertion of foreign genes without affecting the MDV replication [16]. In addition, either GFP or E2-Crimson, driven by the HSV-1 TK promoter, was inserted in the mini-F vector to facilitate the detection of infected cells (for vTetO and vTetO-TetR, respectively).

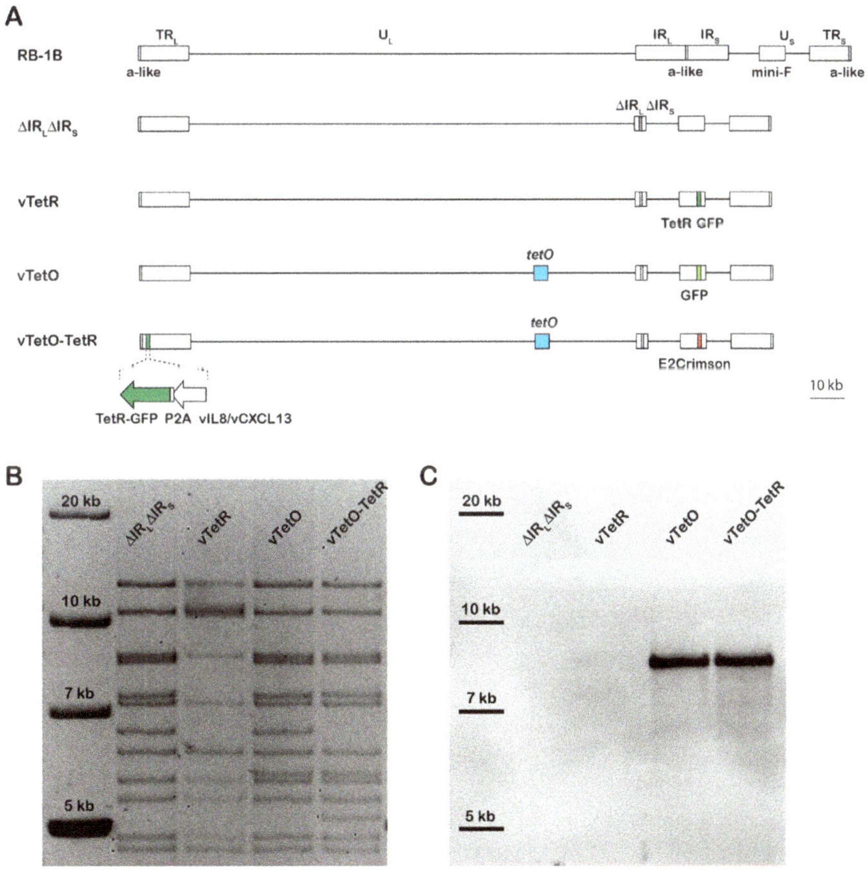

Figure 1. Generation of recombinant viruses. (**A**) Overview of the MDV genome and the recombinant BAC clones. The MDV genome consists of a unique long and short region (U_L and U_S) flanked by the terminal repeat long and short (TR_L, TR_S) and the internal repeat long and short (IR_L, IR_S). The a–like sequences harboring the cleavage and packaging signals are indicated. The recombinant viruses were generated based in the $\Delta IR_L \Delta IR_S$ BAC lacking most of the IR_L and IR_S region, which is rapidly restored upon reconstitution. TetR-GFP is depicted in dark green, *tetO* sequence in blue, GFP in light green, and E2-Crimson in red; (**B**) RFLP of indicated recombinant BAC genomes using NdeI; (**C**) Southern blot of the same RFLP gel using a specific TetO probe to confirm the presence and length of the *tetO* sequences in the virus genome.

All recombinant virus genomes were confirmed by restriction fragment length polymorphism (RFLP) analyses, Southern blotting, Sanger sequencing, and next-generation sequencing (NGS; Illumina MiSeq) of the entire virus genome. All primers used for cloning and mutagenesis can be found in Table 1.

Table 1. Oligonucleotide sequences used in this study.

Construct/Steps	Direction	Sequence (5′–3′)
pCDNA3.1 TetR-GFP	for	TAGATGAGCTCGGATCCATGCCAAAGAAGAAGCGTAAG
	rev	GATGGATATCTGCAGAATTCTCATCCCATGCCATTGGT
TetR-GFP-kana transfer	for	TACAAGACACGTGCTGAAGTCAAGTTTGAAGGTAGGGATAACAGGGTAATCGATTT
	rev	ACTTCAGCACGTGTCTTGTAGTTCCCGTCATCGCCAGTGTTACAACCAATTAACC
TetR-GFP in mini-F (vTetR)	ep for	TTAAGGTGACACGCGCGGCCTCGAACACAGCTGCAGGCCGATGTACGGGCCAGATATACG
	ep rev	CGTCGACCCGGGTACCTCTAGATCCGCTAGCGCTTTATGTCTTCCCAATCCTCCCC
P2A-TetR-GFP into vIL-8/vCXCL13	ep for	ATTGAGCCCACACCTCCTACTATTGGTTCCCATATCTGTCTTGGTTCCGGAGCCACGAACTTCTCT CTGTTAAAGCAAGCAGGAGACGTGGAAGAAAACCCCGGTCCTATGCCAAAGAAGAAGCGTAAG
	ep rev	AAAGTGCCTTCTTTTAATTACAGGAGGTAGCAATTAATCATCCCATGCCATTGGTAATCC
TetO-kana transfer	for	GACAGTAGATCTACCTGTATACTACCCACCATTGTAGGGATAACAGGGTAATCGATTT
	rev	ACAGGTAGATCTACTGTCCCGTAGTCTAAATATGCCAGTGTTACAACCAATTAACC
eGFP in mini-F	ep for	GGTGACACGCGCGGCCTCGAACACAGCTGCAGGCCATGGTGAGCAAGGGCGAGG
	ep rev	CGTCGACCCGGGTACCTCTAGATCCGCTAGCGCTTTACTTGTACAGCTCGTCCATGCC
E2-Crimson into mini-F	ep for	TGCCCTTGCTAGGGTTCTTCACACGAGCCTCGCCTTATTAAATGGGCTCCGGTGCCCGTC
	ep rev	CCCGAGGCCTCGTGGGGCACCTATTTGCGCGGAGGAAGGCCCATAGAGCCCGGGCCATC
TetO DIG-probe		DIG-TCCCTATCAGTCATAGAGAAAAGTGAAAGTCGAGTTTACCAC
iNOS (qPCR)	for	GAGTGGTTTAAGGAGTTGGATCTGA
	rev	TTCCAGACCTCCCACCTCAA
	probe	FAM-CTCTGCCTGCTGTTGCCAACATGC-TAMRA
UL30 (qPCR)	for	AAGCGGAATCGGTTTACAAG
	rev	GGAGTTGCTGTTAGAATACGGA
	probe	FAM-TCGACGAGTTTCTTCCTCCTCGTTG-TAMRA
Mycoplasma test	for	GGGAGCAAACAGGATTAGATACCCT
	rev	TGCACCATCTGTCACTCTGTTAACCTC

ep, en passant mutagenesis primer; for, forward primer; rev, reverse primer; FAM, 6-carboxyfluorescein; TAMRA, 6-carboxytetramethylrhodamine; DIG, Digoxigenin.

2.3. Southern Blotting

To confirm the insertion of the *tetO* repeats, DNA of BAC clones were digested with indicated enzymes and separated on a 0.8% agarose gel. DNA in the agarose gel was denatured (0.5 M NaOH, 1.5 M NaCl) and transferred to a positively charged nylon membrane (Immobilon-NY+, Merck Millipore, Darmstadt, Germany) and incubated with TetO DIG-labeled probe (Table 1), as described previously [9]. *TetO* repeats were detected using an anti-DIG alkaline phosphatase-labeled antibody (Roche GmbH, Mannheim, Germany).

2.4. Illumina MiSeq Sequencing

BAC DNA of the vTetO and vTetO-TetR clones were sequenced using Illumina MiSeq (Illumina, San Diego, CA, USA). Sequencing libraries were prepared using the NEBNext® Ultra™ II DNA Library Prep Kit for Illumina® (New England Biolabs, Ipswich, MA, USA). The generated Illumina reads were processed with Trimmomatic v.0.39 [31] and mapped against the RB-1B TetO (GenBank accession no. OM350391) and RB-1B TetO TetR-GFP (GenBank accession no. OM350392) GenBank references, respectively, using the Burrows-Wheeler aligner v.0.7.17 [32]. Single nucleotide polymorphisms (SNPs), insertions, and deletions were assessed with FreeBayes v.1.1.0-333 [33]. Data were merged by position and mutation using R v.3.2.3; the coverage was additionally assessed and generated using Geneious R11 software.

2.5. Plaque Size Assay and Growth Kinetics

Recombinant viruses were reconstituted using calcium-phosphate transfection of CECs and ESCDL-1 with respective BAC clones, as described previously [34]. Spread of recombinant viruses and replication properties in vitro were determined by plaque size assay and multi-step growth kinetics. For plaque size assays, one million CECs were infected with 100 plaque forming units (PFU) of each virus (passage 5 and 7) and the area of 50 randomly selected plaques were measured using ImageJ (https://imagej.nih.gov/ij/, accessed on 4 June 2020) and normalized against $\Delta IR_L \Delta IR_S$. In addition, plaques were also measured 6 days after BAC DNA transfection. For multi-step growth kinetics, one million CECs were infected with 100 PFU in 6-well plates per virus and cultured for 6 days. Every day, one well per plate was harvested and stored at −80 °C. After 6 days, DNA of all samples was extracted using Zymo Quick DNA Viral kit (Zymo Research Europe GmbH, Freiburg, Germany), according to the manufacturer's instructions. MDV genome copy numbers were determined by qPCR using specific primers and probe for the MDV polymerase (MDV043, UL30). UL30 copy numbers were normalized against the chicken inducible nitric oxide synthase (iNOS), as described previously [35]. Primers and probes used for qPCR are listed in Table 1.

2.6. Assessment of TetO Stability by Nanopore Sequencing

To determine if the array of *tetO* repeats is stably maintained in the virus, extrachromosomal DNA of CECs infected with higher passages of vTetO and vTetO-TetR (passage 9) was extracted using Hirt extraction, as described previously [36,37]. Briefly, infected cells from 150 mm dishes were trypsinized and pelleted by centrifugation at 4 °C and 800× *g* for 5 min. The pellet was washed with ice-cold PBS, centrifuged again, resuspended in 400 µL of Hirt lysis buffer (10 mM Tris-HCl, 20 mM EDTA, 1.2% SDS, pH 8.0) and incubated for 20 min at room temperature. Next, 200 µL of 5 M NaCl was added and incubated at 4 °C for ≥16 h followed by centrifugation at 4 °C and 15.000× *g* for 30 min to pellet proteins and chromosomes. The extrachromosomal DNA in the supernatants was then purified by phenol-chloroform extraction, precipitated with isopropanol, and washed with ethanol. The obtained viral genomes were used for nanopore sequencing. Nanopore libraries were prepared using the SQK-LSK110 Ligation Sequencing Kit (Oxford Nanopore Technologies, Oxford, UK) and sequenced using a Flongle flow cell (FLO-FLG001, Oxford Nanopore Technologies, Oxford, UK) on a MinION sequencer (MK-1B, Oxford Nanopore Technologies, Oxford, UK). The resulting Nanopore reads were mapped against the RB-1B TetO and RB-1B TetO TetR-GFP GenBank references using Minimap2 [38] embedded on Nanopore's MinKNOW GUI. Alignments were viewed using IGV Web App [39,40].

2.7. Generation of T Cell Line Stably Expressing TetR-mCherry

To generate a cell line stably expressing TetR-mCherry, the 855-19 chicken T-cell line was transduced with the pQCXIN-TetR-mCherry retroviral system (kindly provided by Tom Misteli from National Cancer Institute, Addgene plasmid #59417) [41] and selected in the presence of 1200 µg/mL Geneticin (Roth, Karlsruhe, Germany). The newly generated cell line was confirmed to be mycoplasma-free by PCR.

2.8. Wide-Field Microscopy

To investigate if the recombinant viruses efficiently express TetR-GFP, we infected CECs and ESCDL-1 cells with vTetR and/or vTetO-TetR. At 4 dpi, we stained the nuclei with Hoechst 33342 (Invitrogen, Carlsbad, CA, USA) for 30 min and fixed the cells with 4% paraformaldehyde (PFA). Wide-field images of infected cells were taken using Axio Imager M1 (Zeiss, Oberkochen, Germany) equipped with Axio Cam MRm camera (Zeiss) with a 100×/1.4 Oil Plan-Apochromat objective (Zeiss). Images were further processed in ImageJ.

2.9. Confocal Microscopy and Live-Cell Imaging

For live imaging, cells were grown in pre-coated µ-Slide ibiTreat plates with polymer coverslip (Ibidi, Gräfelfing, Germany). Cells were maintained in medium supplemented with

0.2 M HEPES (Roth). 855-19 cells were also immobilized in 0.25% low melting-point agarose (Gibco BRL, Carlsbad, CA, USA). Nuclei were stained with Hoechst 33342 for 30 min. To confirm TetR binding specificity, cells were incubated with 2 µg/mL tetracycline (Roth) for 1 h.

Live microscopy was performed using (i) a VisiScope spinning disk confocal system (Visitron Systems, Puchheim, Germany, CSU-W1; Yokogawa, Tokyo, Japan) built on a Nikon Eclipse Ti inverted microscope equipped with an iXon Ultra 888 EMCCD camera and an OkoLab gas and temperature controller (OkoLab, Ottaviano, Italy) to maintain a 5% CO_2 atmosphere at 37 °C or (ii) a FluoView1000 inverted confocal microscope (Olympus, Tokyo, Japan). If not indicated otherwise, the VisiScope spinning disc confocal microscope was used. Images were captured with (i) 60×/1.4 and 100×/1.45 Oil Plan-Apo objectives (Nikon) using VisiView software (v.4.3.0.6; Visitron Systems, Puchheim, Germany) or (ii) 60× water immersion objective (UPlanSApo) using Olympus FluoView software (v 4.02; Olympus, Tokyo, Japan). Z-step size between focal planes was 0.5 µm and final 2D images are visualized as maximum intensity projection. All images and videos were processed with ImageJ software.

2.10. Lymphocyte Infection

To infect 855-19 T-cells expressing TetR-mCherry, one million CECs were infected with 10,000 PFU of vTetO. After 6 days, CECs were overlaid with one million T-cells and incubated together for 24 h in RPMI (PAN Biotech) supplemented with 10% FBS, 1% sodium pyruvate (PAN Biotech), 1% non-essential amino acids (Biochrom, Berlin, Germany), and 1% penicillin/streptomycin at 41 °C under a 5% CO_2 atmosphere. The next day, T-cells were carefully harvested and the infection rate was determined by flow cytometry detecting eGFP expressed in infected cells using a CytoFLEX S system (Beckman Coulter, Krefeld, Germany). Infected T-cells were maintained in culture for 14 days to allow the virus to establish latency and integrate its genome into host chromosomes. Images were taken 1, 3, and 14 days post infection (dpi) using the VisiScope spinning disc confocal microscope.

2.11. Fluorescence In Situ Hybridization (FISH)

Interphase nuclei were prepared from infected 855-19 T-cells 3 dpi (lytic) and 14 dpi (latent), as described previously [42]. Briefly, MDV genomes were detected using a set of PCR-based MDV probes and visualized using Cy3 Streptavidin (1:1000 dilution; GE Healthcare, Munich, Germany) [43]. Images of interphases were taken using Axio Imager M1 (Zeiss) and analyzed with the ImageJ software.

2.12. Statistical Analysis

Statistical analyses were performed using GraphPad Prism software v.8.0.2. Data were considered significantly different for p values of $\leq$0.05. Description of all applied statistical tests can be found in the respective figure legends. If not noted differently, all experiments were repeated at least three independent times.

3. Results

3.1. Generation of Recombinant Viruses

To visualize MDV genomes in living cells, we generated recombinant viruses that harbor the *tetO*-repeats using the very virulent RB-1B strain [35,44]. The *tetO* repeat cassette (4.5 kbp) was inserted in between UL45 and UL46, a locus well established for the insertion of foreign genes without affecting virus replication [16]. The TetR fused to a fluorescence protein and a nuclear localization signal (NLS) and was then either expressed by the virus or the cells depending on the application. To exclude that the TetR protein alone forms unspecific aggregates in the nucleus of chicken cells, we generated a virus expressing high levels of TetR-GFP driven by the strong TK promoter in the mini-F in the absence of *tetO* (vTetR). Since we observed that high expression levels of TetR-GFP driven by the TK promoter reduced the signal to noise ratio, we tested various loci for the insertion of the protein. The optimal expression was achieved by fusing TetR-GFP via a P2A self-cleaving peptide to the C-terminus of vCXCL13 (aka. vIL-8). This facilitated TetR-GFP expression during both lytic replication and latency at optimal levels. To identify infected cells, we inserted the GFP into the virus only containing *tetO* (vTetO) and the far-red protein E2-Crimson into the vTetO-TetR double insertion virus (vTetO-TetR) into the mini-F (Figure 1A). The resulting clones were analyzed by RFLP (Figure 1B) and Sanger sequencing. The presence and length of the *tetO* repeats in the viral genome were confirmed by Southern blotting (Figure 1C). The specific TetO probe detected the fragment with the expected size in both vTetO and vTetO-TetR, indicating that the full length *tetO* cassette was inserted during mutagenesis. Illumina MiSeq next generation sequencing was performed and confirmed that no additional mutations are present in the recombinant virus genomes.

3.2. Characterization of Replication Properties

To examine if the insertion of *tetO* and TetR-GFP affects virus replication, we performed plaque size assays and multi-step growth kinetics. No significant difference was observed in plaque size assays performed after transfection compared to the parental virus (Figure 2A). Similarly, no significant difference was detected in plaque size assays upon serial passaging of the viruses (passage 7; Figure 2B). These results were confirmed by multi-step growth kinetic analyses (Figure 2C), highlighting that insertion of *tetO* and TetR-GFP has no significant impact in MDV replication. To further investigate the stability of the *tetO* repeats in the viral genome, we isolated extrachromosomal DNA from CECs infected with high-passaged virus (passage 9) and performed nanopore sequencing on the MDV genomes. The results from nanopore sequencing demonstrated that *tetO* is stably maintained in the UL45-UL46 locus in both vTetO and vTetO-TetR, highlighting that the system could even be used with high passage stocks.

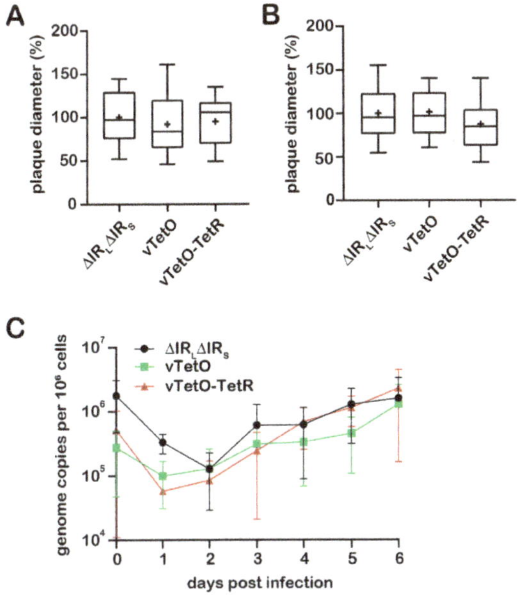

Figure 2. In vitro characterization of the recombinant viruses. Plaque size assays after (**A**) transfection and (**B**) infection with passaged viruses, as indicated. Data are shown as means of three independent experiments with means (+), medians (line within the bar), and standard deviations ($p > 0.05$, one-way analysis of variance (ANOVA) Dunnett's test). (**C**) Representative multi-step growth kinetics of indicated viruses with standard deviations ($p > 0.05$, one-way analysis of variance (ANOVA) Kruskal–Wallis test).

3.3. Visualization of the Virus Genome during Lytic Replication and Specificity of TetR Binding

To visualize MDV genomes during lytic replication, we infected primary CECs and the chicken ESCDL-1 cell line with 100 pfu of vTetO-TetR, counterstained with Hoechst 33342 at four dpi imaged the cells. In both CECs and ESCDL-1, we consistently observed one to two replication compartments (RCs) per nucleus that were visually separated from each other (Figure 3A). To ensure that TetR alone does not establish unspecific aggregates, we infected CECs with vTetR and vTetO-TetR. In cells infected with vTetR we observed a uniform TetR-GFP signal in the nucleus. In contrast, in cells infected with vTetO-TetR we detected specific signal for the virus genome, forming RCs within the nucleus (Figure 3B). To further validate the specificity of the TetO/TetR signal, we used tetracycline to induce a conformational change of the TetR DNA-binding domain, resulting in the dissociation of TetR-GFP from the *tetO* repeats [45]. CECs were infected with vTetO-TetR for 4 days and imaged before and after the addition of tetracycline. Neighboring infected cells harbored RCs in their nucleus before adding tetracycline. Upon addition of the drug, TetR-GFP dissociates and the specific signal pattern was lost, highlighting that the observed genome staining is highly specific (Figure 3C).

3.4. Genesis and Mobility of Replication Compartments

Next, we investigated the formation of the RCs in infected cells. We infected CECs with vTetO-TetR for 4 days, counterstained the nuclei with Hoechst 33342, and monitored the development of RCs in newly infected cells via live cell imaging for 21 h. Separated RCs in the nuclei were observed early in infection (Figure 4A). During the 21 h period, the RCs gradually increased in size (RC 1 from 6.6 μm^2 to 20.5 μm^2; RC 2 from 1.5 μm^2 to 5.7 μm^2).

Based on the previous findings that RCs have properties of phase-separated condensates (liquid-liquid phase separation) and proteins move freely within RCs [46], we set out to investigate the mobility of the viral DNA in these structures. At 4 dpi, we tracked infected nuclei for 1 h and captured stacks of 15 focal planes and analyzed the maximum-intensity projection (MIP) (Figure 4B) and rendered 3D images (Supplementary Video S1) of a representative nucleus. Although the nuclei themselves were moving during the imaging period, we observed that the shape of the RCs did not change and many of the dotted structures stayed in the same place suggesting that the viral DNA is rather immotile.

3.5. Detection of Both Infected and Uninfected Nuclei inside A Syncytium

MDV has been previously shown to induce syncytia in duck embryo fibroblasts (DEF) [47,48]. Using our vTetO-TetR expressing E2-Crimson in the cytoplasm, we frequently observed that MDV can also induce syncytia in infected primary CECs. We, therefore, infected CECs with vTetO-TetR to assess if all nuclei were infected and showed replication compartments. Nuclei of the cells were stained with Hoechst 33342 and we imaged the cells using a spinning-disk confocal microscope at 4 dpi. Interestingly, we observed that some nuclei harbor RCs while others did not. In some nuclei, single dots were detected (Figure 5A), suggesting that only one or few viral genomes were present. In other syncytia, no virus signal was detected in some of the nuclei (Figure 5B), indicating that MDV does not replicate its genome in all nuclei of a syncytium.

3.6. Visualization of Viral Genomes during Lytic Replication and Latency in T Cells

To assess lytic replication and latency in T cells, we co-seeded CECs highly infected with vTetO with uninfected 855-19 T cells stably expressing TetR-mCherry. Imaging started after the T cells settled on the infected CEC monolayer. Cells were imaged using the spinning-disk confocal microscope every ten minutes for 21 h. DNA replication was detectable at 8 hpi and gradually increased until 22 hpi (Figure 6A). Intriguingly, we observed smaller and sparse RCs in T cells when compared with CECs and ESCDL-1 cells, indicating that MDV replication in T cells may differ.

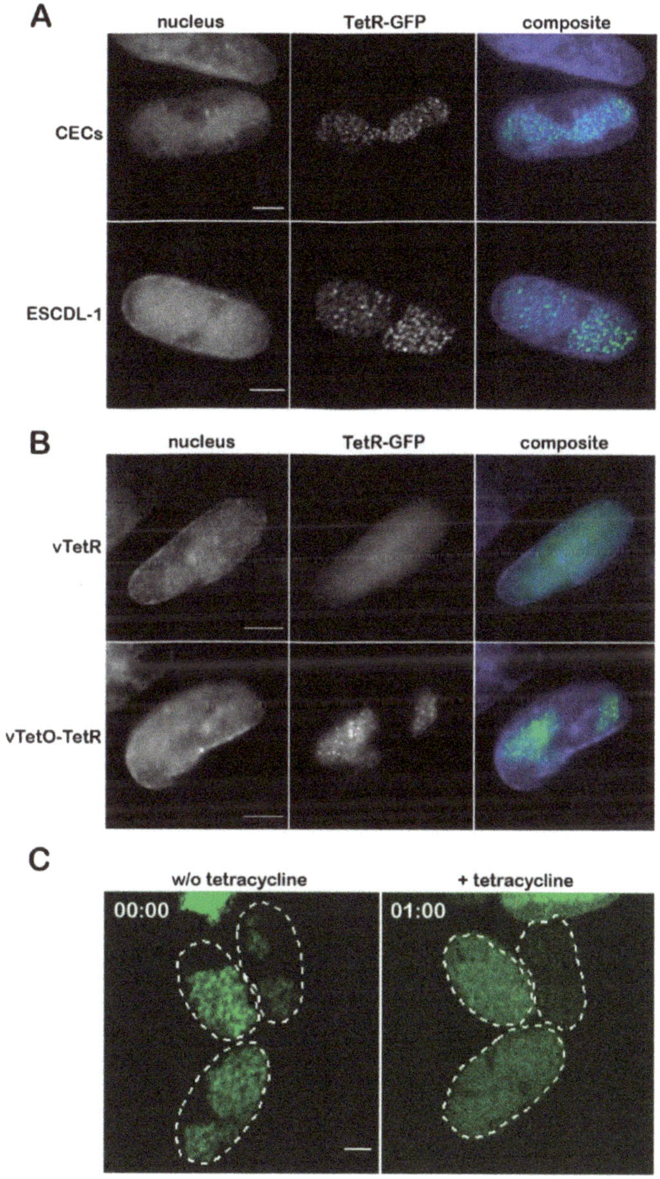

Figure 3. Detection of virus genomes in infected cells. (**A**) CECs and ESCDL-1 cells were infected with vTetO-TetR, fixed with 4% PFA at 4 dpi, and imaged with an Axio Imager M1 using a 100× oil objective. Scale bar corresponds to 5 µm; (**B**) CECs were infected with either vTetR or vTetO-TetR, fixed at 4 dpi using VisiScope spinning-disk confocal microscope with a 100× oil objective. Scale bar corresponds to 5 µm; (**C**) CECs infected with vTetO-TetR were treated with tetracycline. Images are shown before (**left** panel) and 1 h after the addition of tetracycline (**right** panel). Image stacks of nine focal planes were captured with a z-step size of 0.5 µm and displayed as maximum-intensity projection (MIP). The perimeter of the nuclei are depicted as dotted lines. Time (in hh:mm) and scale bar (5 µm) are indicated.

A

B

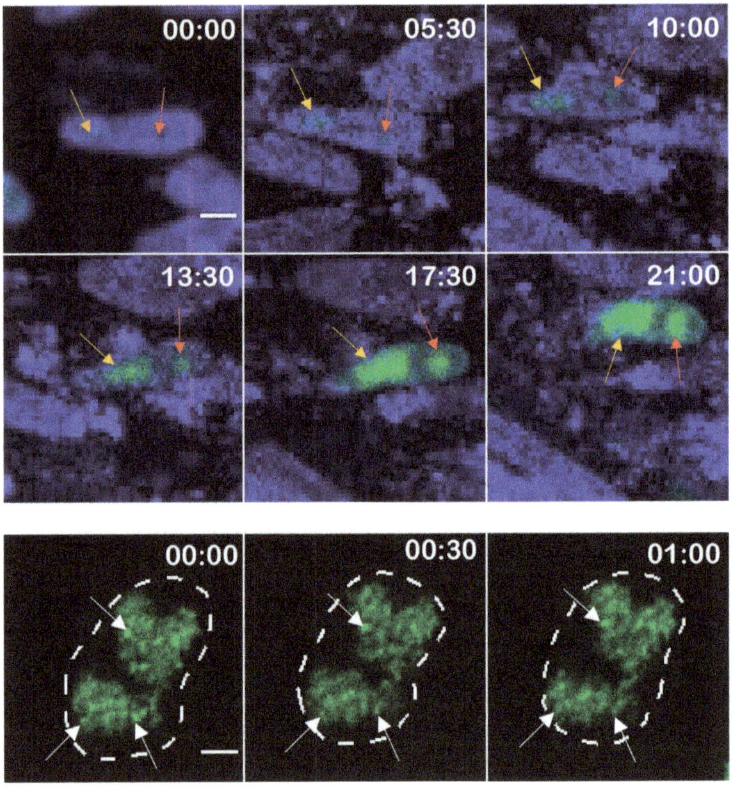

Figure 4. Genesis and mobility of replication compartments (RC). (**A**) The development of RCs in infected cells were monitored for 21 h. Sequential images were taken with spinning-disk confocal microscope using a 60× water immersion objective every ten minutes. Multiple Z-stacks were processed as maximum-intensity projection (MIP). RC 1 and RC 2 are indicated with an orange arrow and red arrow, respectively. Nuclei are shown in blue, TetR-GFP in green. Time in hh:mm, scale bar 5 µm; (**B**) the mobility of the RCs in infected cells was monitored for 1 h. Stacks of 15 focal planes with a z-step size of 0.5 µm were captured at 5 min intervals using a VisiScope microscope. Images represent the MIP of each time point. Nucleus contour is depicted as dotted line; white arrows indicate non-moving dotted structures. Time in hh:mm, scale bar 3 µm.

To further investigate infection and the establishment of latency in T cells, vTetO infected lymphocytes were harvested at 24 hpi and maintained in cell culture for 14 days. At 3 dpi, few lytically infected cells were still observed (Figure 6B). Intriguingly, we also detected cells with a single dot in the nucleus (white arrow) after 3 dpi (Figure 6B). These pictures were confirmed with FISH preparations of infected cells and are consistent with recently published work done on 885-19 T cells [43]. Moreover, to ascertain that this single bright staining represents a latent virus genome, we analyzed the infected T cells also 14 dpi when no replication was observed. Single specific spots were observed during the latent phase upon infection of these T cells (Figure 6C, white arrow), highlighting that our system is sensitive enough to detect these latent virus genomes.

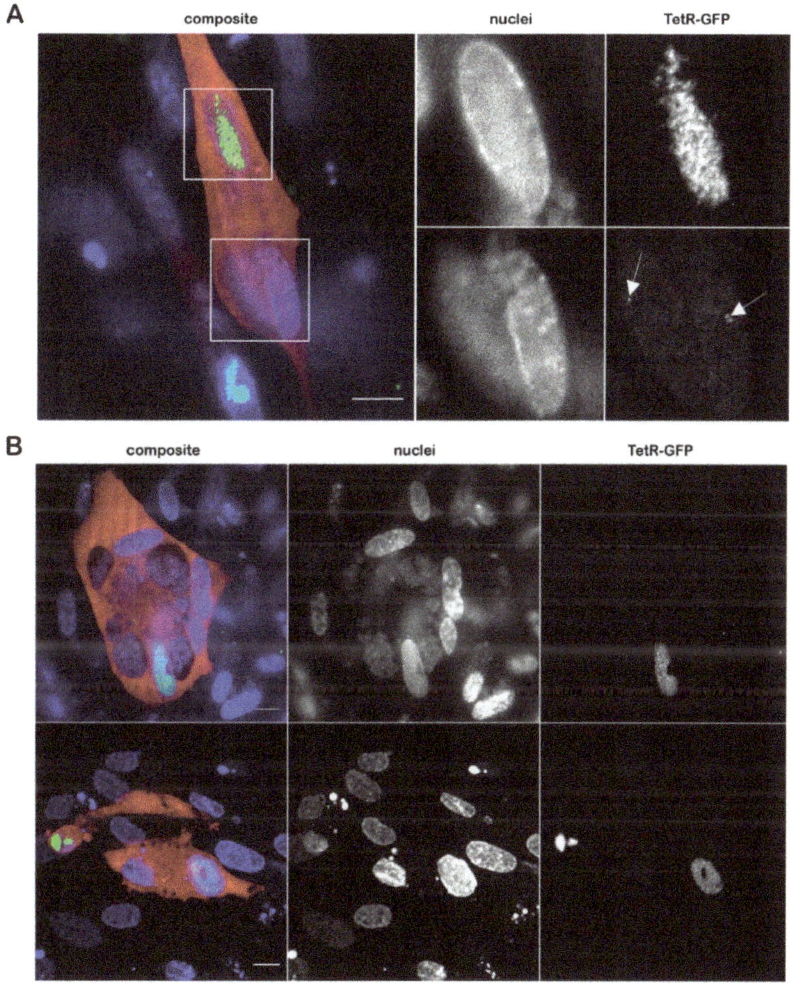

Figure 5. MDV can induce syncytia in CECs but replication is only detected in some of the nuclei. (**A**) Nuclei within a syncytium with varying TetR-GFP intensities. Image stacks of 21 focal planes with a z-step size of 0.5 µm are displayed as MIP. Nuclei were stained with Hoechst 33342 for 30 min and captured using WF-DAPI. Single or few MDV genomes are indicated by white arrows; (**B**) representative images of MDV syncytia. E2-Crimson in cytoplasm in red, nuclei in blue, and TetR-GFP in green. Scale bar 10 µm.

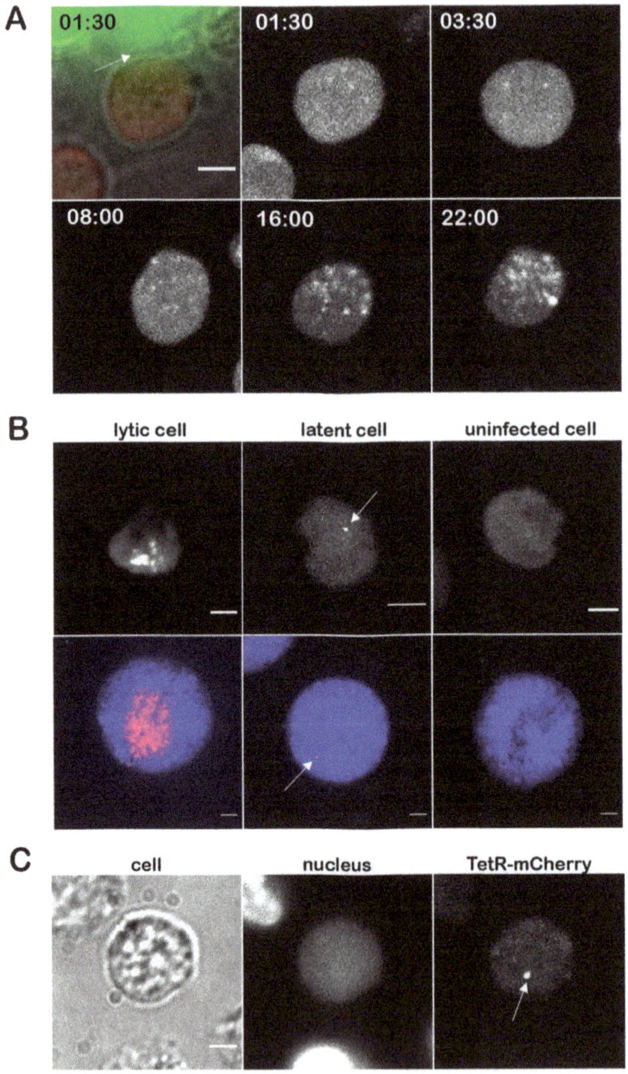

Figure 6. Lytic replication and latency in T cells. (**A**) 855-19 T cells stably expressing TetR-mCherry (multi-color image) were seeded on an infected CECs monolayer. The site of contact between infected CEC (green) and uninfected T cell (red) is shown with a white arrow. Time course images are shown as maximum intensity projections of the red channel (TetR-mCherry). Images were taken using a 60× Oil objective. 13 focal planes with a z-stack size of 1.7 μm were imaged every ten minutes for 21 h. Time in hh:mm, scale bar 3 μm (**B**) representative images of 855-19 T cells expressing TetR-mCherry three days after virus infection (upper line) in comparison with FISH images (lower line; DAPI in blue, viral DNA in red). Live cell images are shown as maximum intensity projection of z-stacks in the red channel. Arrows highlight single dots in the nucleus of latently infected cells. Scale bar 3 μm; (**C**) representative image of infected T cells at 14 dpi. One specific dot (arrow) was detected in maximum-intensity projection of multiple focal planes in TetR-mCherry.

4. Discussion

The visualization of viral genomes in living cells has an immense potential in research on a broad range of processes including antiviral responses to the genome, viral replication, and genome maintenance in latently infected cells. Most processes involving the virus genome during MDV infection remain poorly understood. Therefore, we set out to develop a system that facilitates visualization of the MDV genome in living cells.

To visualize MDV genomes, we inserted the *tetO* repeats into the MDV genome between UL45 and UL46 genes (vTetO). Plaque size assays and multiple step growth kinetics revealed that insertion of the *tetO* repeats into this site does not affect MDV replication and cell-to-cell spread in culture. This was not surprising as this locus has been previously successfully used (e.g., for the insertion of a GFP expression cassette) [16]. Southern blotting confirmed the expected size of the *tetO* repeats and Illumina MiSeq NGS ensured that no additional SNPs were present in the recombinant viruses. More importantly, nanopore sequencing confirmed that the *tetO* repeats in the MDV genome are stable for at least nine passages, highlighting that the system could even be used with high passage viruses and for long-term analyses.

Next, we expressed the fluorescently tagged TetR either from the virus genome or from a plasmid stably maintained in the target cell. To ensure the reliability of the *tetO*/TetR system, we validated the specificity of the TetR staining by two independent approaches. First, we confirmed that overexpression of TetR-GFP driven by the strong TK promoter does not cause unspecific staining in the absence of *tetO*, for example, due to the formation of aggregates in the nucleus. In addition, we validated the TetR specificity via the addition of tetracycline, resulting in a conformational change of the TetR DNA-binding domain [45]. This resulted in the expected release of the TetR-GFP from the *tetO* repeats present in the virus and resulted in a diffuse staining observed in the absence of *tetO* (Figure 3C). Although we detected even distribution of TetR-GFP (Figure 3B) in the entire nucleus, the amount of TetR expressed from the TK-promoter was too high to be able to detect single viral genomes. In addition, most of the viral genes (including the TK promoter in the mini-F) are silenced during latency. Therefore, we decided to fuse TetR-GFP via a P2a self-cleaving peptide to the C-terminus of vCXCL13 in vTetO (Figure 1A), as splice variants containing the last exon (exon 3) are expressed during both lytic replication and latency at moderate levels [12,49].

Using the optimized *tetO*/TetR system, we could efficiently visualize RCs in lytically infected primary cells and cell lines (Figures 3A and 5A). We could also observe the establishment and growth of the RCs using live-cell microscopy. In previous studies on HSV-1, it was shown that the vast majority of RCs are initiated from a single incoming viral genome at distinct sites in the nucleus [50,51]. Due to the strictly cell-associated nature of MDV, we could not perfectly time the infection, but could observe a similar phenotype. Our results indicate that it takes MDV RCs about 24 h to reach their full size upon entry into the nucleus. In addition, the observed one to two RCs within a nucleus possessed a different size and shape, although they were gradually growing (Figure 4A). In comparison, HSV-1 infected cells with a multiplicity of infection of 20 commonly harbor three to five RCs per nucleus depending on the cell line [52]. In T cells, MDV establishes usually just a diffuse RC that differs from the ones seen in fibroblasts. This could be due to the smaller size of T cell nuclei, their previously reported rigidity, and compact nature [53–55]. Our system, thereby, revealed a cell type dependent difference between the RCs.

Based on previous studies with HSV-1 [46,56], it is thought that viral genomes are phase-separated from chromosomal DNA and have the properties of liquid-liquid phase-separated condensates. It was shown that proteins can freely diffuse through RC, whereas the viral genomes appear to be much slower. However, these studies were done using HSV-1 amplicon plasmids that do not reflect the entire complexity of replicating herpesviruses [46,57]. Here, we observed that, although the infected cells and nuclei moved during the recording period, we detected only minimal changes in the shape of RC and position of the dotted structures within the RC (Figure 4B). This could be explained by the cohesion of concatemeric viral genomes during branched rolling circle replication, whereas proteins can freely diffuse through the RCs. We will assess the interaction of proteins with the MDV genome in this context using our MDV system in future studies.

Using the vTetO-TetR system, we observed that MDV can induce fusion of cultured CECs, resulting in the formation of syncytia. We examined this CPE, which was previously only reported in DEFs, and observed that some nuclei within a syncytium do not possess RC (Figure 5A). We noticed a weak but diffuse TetR-GFP signal in those nuclei, indicating that the fusion protein was, to some extent, transported into the uninfected nuclei within the syncytium due to its NLS.

To shed more light on the course of infection in T cells and the establishment of latency, we set out to explore the behavior of viral genomes early in infection. Since almost all MDV genes are silenced during latency, we generated T cell line stably expressing TetR-mCherry independent of the virus life cycle (Figure 6B). Replication foci were detected around 8 hpi that increased over the course of time (Figure 6A), indicating that lytic replication progresses in some of the T cells, as shown previously [4–6]. After 14 dpi we were able to detect single bright dots in the nucleus, as detected by FISH. This is consistent with previous studies that found only one or few integration sites in primary T cells and T cell lines latently infected in vitro [6,43], while multiple integration sites were detected in MDV-induced tumor cells [9,10]. Since no replication was detected at this time point, these most likely correspond to latent virus genomes (Figure 6B,C). Comparable signals were found in infected T cells already at 3 dpi, suggesting that MDV can enter in the quiescent stage early after infection. We will expand on these observations, especially in the context of MDV integration in the future.

Taken together, we established a system that facilitates the visualization of MDV genomes in living cells. This system provided exciting insights into the virus live cycle, including the number of RCs in different cell types, the establishment and expansion of RCs, the formation of syncytia, and the infection of T cells resulting in either lytic replication or latency. This *tetO*-TetR-based system, established in this study, will contribute to our understanding of MDV, replication, genome integration, and the establishment of latency in future studies.

Supplementary Materials: The following supporting information can be downloaded at: https://www.mdpi.com/article/10.3390/v14020287/s1, Video S1: 3D image of RC in nucleus.

Author Contributions: Conceptualization, B.B.K. and T.V.; methodology, B.B.K., T.V. and D.J.W.; formal analysis, T.V. and M.N.; investigation, T.V., D.J.W., M.N. and F.J.; resources, B.B.K., T.K. and A.H.; writing—original draft preparation, T.V. and B.B.K.; writing—review and editing, T.V., B.B.K., M.N., D.J.W., F.J., A.H. and T.K.; visualization, T.V. and M.N.; supervision, B.B.K.; project administration, B.B.K.; funding acquisition, B.B.K. All authors have read and agreed to the published version of the manuscript.

Funding: This research was funded by European Research Council, grant number Stg 677673 awarded to B.B.K. The publication of this article was funded by Freie Universität Berlin.

Institutional Review Board Statement: Not applicable.

Informed Consent Statement: Not applicable.

Data Availability Statement: Not applicable.

Acknowledgments: We are grateful to Ann Reum, Andelé Conradie, and Yu You for their technical assistance. We also thank Thomas Göbel for providing the chicken T cell line 855-19 used in this study.

Conflicts of Interest: The authors declare no conflict of interest. The funders had no role in the design of the study; in the collection, analyses, or interpretation of data; in the writing of the manuscript; or in the decision to publish the results.

References

1. Schat, K.A.; Nair, V. Neoplastic diseases. In *Diseases of Poultry*, 13th ed.; Swayne, D.E., Ed.; John Wiley & Sons, Inc.: Hoboken, NJ, USA, 2013; pp. 513–673.
2. Barrow, A.; Burgess, S.C.; Baigent, S.J.; Howes, K.; Nair, V. Infection of macrophages by a lymphotropic herpesvirus: A new tropism for Marek's disease virus. *J. Gen. Virol.* **2003**, *84*, 2635–2645. [CrossRef] [PubMed]
3. Chakraborty, P.; Vervelde, L.; Dalziel, R.G.; Wasson, P.S.; Nair, V.; Dutia, B.M.; Kaiser, P. Marek's disease virus infection of phagocytes: A de novo in vitro infection model. *J. Gen. Virol.* **2017**, *98*, 1080–1088. [CrossRef] [PubMed]
4. Bertzbach, L.D.; Laparidou, M.; Härtle, S.; Etches, R.J.; Kaspers, B.; Schusser, B.; Kaufer, B.B. Unraveling the role of B cells in the pathogenesis of an oncogenic avian herpesvirus. *Proc. Natl. Acad. Sci. USA* **2018**, *115*, 11603–11607. [CrossRef] [PubMed]
5. Bertzbach, L.D.; Van Haarlem, D.A.; Härtle, S.; Kaufer, B.B.; Jansen, C.A.; Haarlem, V. Marek's Disease Virus Infection of Natural Killer Cells. *Microorganisms* **2019**, *7*, 588. [CrossRef] [PubMed]
6. Schermuly, J.; Greco, A.; Härtle, S.; Osterrieder, N.; Kaufer, B.B.; Kaspers, B. In vitro model for lytic replication, latency, and transformation of an oncogenic alphaherpesvirus. *Proc. Natl. Acad. Sci. USA* **2015**, *112*, 7279–7284. [CrossRef] [PubMed]
7. Jarosinski, K.W.; Tischer, B.K.; Trapp, S.; Osterrieder, N. Marek's disease virus: Lytic replication, oncogenesis and control. *Expert Rev. Vaccines* **2006**, *5*, 761–772. [CrossRef] [PubMed]
8. Schat, K.A.; Chen, C.L.; Calnek, B.W.; Char, D. Transformation of T-lymphocyte subsets by Marek's disease herpesvirus. *J. Virol.* **1991**, *65*, 1408–1413. [CrossRef] [PubMed]
9. Kaufer, B.B.; Jarosinski, K.W.; Osterrieder, N. Herpesvirus telomeric repeats facilitate genomic integration into host telomeres and mobilization of viral DNA during reactivation. *J. Exp. Med.* **2011**, *208*, 605–615. [CrossRef] [PubMed]
10. Greco, A.; Fester, N.; Engel, A.T.; Kaufer, B.B. Role of the Short Telomeric Repeat Region in Marek's Disease Virus Replication, Genomic Integration, and Lymphomagenesis. *J. Virol.* **2014**, *88*, 14138–14147. [CrossRef] [PubMed]
11. Parcells, M.S.; Arumugaswami, V.; Prigge, J.T.; Pandya, K.; Dienglewicz, R.L. Marek's disease virus reactivation from latency: Changes in gene expression at the origin of replication. *Poult. Sci.* **2003**, *82*, 893–898. [CrossRef] [PubMed]
12. Parcells, M.S.; Lin, S.-F.; Dienglewicz, R.L.; Majerciak, V.; Robinson, D.R.; Chen, H.-C.; Wu, Z.; Dubyak, G.R.; Brunovskis, P.; Hunt, H.D.; et al. Marek's Disease Virus (MDV) Encodes an Interleukin-8 Homolog (vIL-8): Characterization of the vIL-8 Protein and a vIL-8 Deletion Mutant MDV. *J. Virol.* **2001**, *75*, 5159–5173. [CrossRef] [PubMed]
13. Kaufer, B.B.; Trapp, S.; Jarosinski, K.W.; Osterrieder, N. Herpesvirus Telomerase RNA(vTR)-Dependent Lymphoma Formation Does Not Require Interaction of vTR with Telomerase Reverse Transcriptase (TERT). *PLoS Pathog.* **2010**, *6*, e1001073. [CrossRef] [PubMed]
14. Kheimar, A.; Previdelli, R.L.; Wight, D.J.; Kaufer, B.B. Telomeres and Telomerase: Role in Marek's Disease Virus Pathogenesis, Integration and Tumorigenesis. *Viruses* **2017**, *9*, 173. [CrossRef] [PubMed]
15. Mwangi, W.N.; Smith, L.P.; Baigent, S.J.; Beal, R.K.; Nair, V.; Smith, A.L. Clonal Structure of Rapid-Onset MDV-Driven CD4+ Lymphomas and Responding CD8+ T Cells. *PLoS Pathog.* **2011**, *7*, e1001337. [CrossRef]
16. Liu, Y.; Li, K.; Cui, H.; Gao, L.; Liu, C.; Zhang, Y.; Gao, Y.; Wang, X. Comparison of different sites in recombinant Marek's disease virus for the expression of green fluorescent protein. *Virus Res.* **2017**, *235*, 82–85. [CrossRef]
17. Denesvre, C.; Rémy, S.; Trapp-Fragnet, L.; Smith, L.P.; Georgeault, S.; Vautherot, J.-F.; Nair, V. Marek's disease virus undergoes complete morphogenesis after reactivation in a T-lymphoblastoid cell line transformed by recombinant fluorescent marker virus. *J. Gen. Virol.* **2016**, *97*, 480–486. [CrossRef]
18. Jarosinski, K.W.; Donovan, K.M.; Du, G. Expression of fluorescent proteins within the repeat long region of the Marek's disease virus genome allows direct identification of infected cells while retaining full pathogenicity. *Virus Res.* **2015**, *201*, 50–60. [CrossRef]
19. Rémy, S.; Blondeau, C.; Le Vern, Y.; Lemesle, M.; Vautherot, J.-F.; Denesvre, C. Fluorescent tagging of VP22 in N-terminus reveals that VP22 favors Marek's disease virus (MDV) virulence in chickens and allows morphogenesis study in MD tumor cells. *Vet. Res.* **2013**, *44*, 125. [CrossRef]
20. Mao, W.; Kim, T.; Cheng, H.H. Visualization of Marek's disease virus in vitro using enhanced green fluorescent protein fused with US10. *Virus Genes* **2013**, *47*, 181–183. [CrossRef]
21. Jarosinski, K.W.; Arndt, S.; Kaufer, B.B.; Osterrieder, N. Fluorescently Tagged pUL47 of Marek's Disease Virus Reveals Differential Tissue Expression of the Tegument Protein In Vivo. *J. Virol.* **2011**, *86*, 2428–2436. [CrossRef]
22. Prigge, J.T.; Majerciak, V.; Hunt, H.D.; Dienglewicz, R.L.; Parcells, M.S. Construction and Characterization of Marek's Disease Viruses Having Green Fluorescent Protein Expression Tied Directly or Indirectly to Phosphoprotein 38 Expression. *Avian Dis.* **2004**, *48*, 471–487. [CrossRef] [PubMed]
23. Fuchs, J.; Lorenz, A.; Loidl, J. Chromosome associations in budding yeast caused by integrated tandemly repeated transgenes. *J. Cell Sci.* **2002**, *115*, 1213–1220. [CrossRef] [PubMed]
24. Mirkin, E.V.; Chang, F.S.; Kleckner, N. Protein-Mediated Chromosome Pairing of Repetitive Arrays. *J. Mol. Biol.* **2014**, *426*, 550–557. [CrossRef] [PubMed]
25. Michaelis, C.; Ciosk, R.; Nasmyth, K. Cohesins: Chromosomal Proteins that Prevent Premature Separation of Sister Chromatids. *Cell* **1997**, *91*, 35–45. [CrossRef]
26. Tischer, B.K.; von Einem, J.; Kaufer, B.; Osterrieder, N. Two-step red-mediated recombination for versatile high-efficiency markerless DNA manipulation in Escherichia coli. *Biotechniques* **2006**, *40*, 191–197. [PubMed]

27. Schat, K.A.; Purchase, H.G.; Amer Assn of Avian Pathologist. Cell-culture methods. In *A Laboratory Manual for the Isolation and Identification of Avian Pathogens*, 4th ed.; Swayne, D.E., Ed.; American Association of Avian Pathologists: Kennett Square, PA, USA, 1998.
28. Vautherot, J.-F.; Jean, C.; Fragnet-Trapp, L.; Rémy, S.; Chabanne-Vautherot, D.; Montillet, G.; Fuet, A.; Denesvre, C.; Pain, B. ESCDL-1, a new cell line derived from chicken embryonic stem cells, supports efficient replication of Mardiviruses. *PLoS ONE* **2017**, *12*, e0175259. [CrossRef]
29. Petherbridge, L.; Brown, A.C.; Baigent, S.J.; Howes, K.; Sacco, M.A.; Osterrieder, N.; Nair, V.K. Oncogenicity of Virulent Marek's Disease Virus Cloned as Bacterial Artificial Chromosomes. *J. Virol.* **2004**, *78*, 13376–13380. [CrossRef] [PubMed]
30. Tischer, B.K.; Kaufer, B.B. Viral Bacterial Artificial Chromosomes: Generation, Mutagenesis, and Removal of Mini-F Sequences. *J. Biomed. Biotechnol.* **2012**, *2012*, 472537. [CrossRef]
31. Bolger, A.M.; Lohse, M.; Usadel, B. Trimmomatic: A flexible trimmer for Illumina sequence data. *Bioinformatics* **2014**, *30*, 2114–2120. [CrossRef]
32. Li, H.; Durbin, R. Fast and accurate short read alignment with Burrows–Wheeler transform. *Bioinformatics* **2009**, *25*, 1754–1760. [CrossRef]
33. Garrison, E.; Marth, G. Haplotype-based variant detection from short-read sequencing. *arXiv* **2012**, arXiv:1207.3907.
34. Schumacher, D.; Tischer, B.K.; Fuchs, W.; Osterrieder, N. Reconstitution of Marek's Disease Virus Serotype 1 (MDV-1) from DNA Cloned as a Bacterial Artificial Chromosome and Characterization of a Glycoprotein B-Negative MDV-1 Mutant. *J. Virol.* **2000**, *74*, 11088–11098. [CrossRef] [PubMed]
35. Jarosinski, K.W.; Margulis, N.G.; Kamil, J.P.; Spatz, S.J.; Nair, V.K.; Osterrieder, N. Horizontal transmission of Marek's disease virus requires US2, the UL13 protein kinase, and gC. *J. Virol.* **2007**, *81*, 10575–10587. [CrossRef] [PubMed]
36. Trimpert, J.; Groenke, N.; Kunec, D.; Eschke, K.; He, S.; McMahon, D.; Osterrieder, N. A proofreading-impaired herpesvirus generates populations with quasispecies-like structure. *Nat. Microbiol.* **2019**, *4*, 2175–2183. [CrossRef] [PubMed]
37. Hirt, B. Selective extraction of polyoma DNA from infected mouse cell cultures. *J. Mol. Biol.* **1967**, *26*, 365–369. [CrossRef]
38. Li, H. Minimap2: Pairwise alignment for nucleotide sequences. *Bioinformatics* **2018**, *34*, 3094–3100. [CrossRef] [PubMed]
39. Robinson, J.T.; Thorvaldsdóttir, H.; Winckler, W.; Guttman, M.; Lander, E.S.; Getz, G.; Mesirov, J.P. Integrative genomics viewer. *Nat. Biotechnol.* **2011**, *29*, 24–26. [CrossRef]
40. Thorvaldsdóttir, H.; Robinson, J.T.; Mesirov, J.P. Integrative Genomics Viewer (IGV): High-performance genomics data visualization and exploration. *Brief. Bioinform.* **2013**, *14*, 178–192. [CrossRef]
41. Roukos, V.; Voss, T.C.; Schmidt, C.K.; Lee, S.; Wangsa, D.; Misteli, T. Spatial Dynamics of Chromosome Translocations in Living Cells. *Science* **2013**, *341*, 660–664. [CrossRef]
42. Kaufer, B.B. Detection of Integrated Herpesvirus Genomes by Fluorescence In Situ Hybridization (FISH). *Program. Necrosis* **2013**, *1064*, 141–152. [CrossRef]
43. You, Y.; Vychodil, T.; Aimola, G.; Previdelli, R.L.; Göbel, T.W.; Bertzbach, L.D.; Kaufer, B.B. A Cell Culture System to Investigate Marek's Disease Virus Integration into Host Chromosomes. *Microorganisms* **2021**, *9*, 2489. [CrossRef] [PubMed]
44. Gasser, S.M. Visualizing Chromatin Dynamics in Interphase Nuclei. *Science* **2002**, *296*, 1412–1416. [CrossRef] [PubMed]
45. Orth, P.; Cordes, F.; Schnappinger, D.; Hillen, W.; Saenger, W.; Hinrichs, W. Conformational changes of the Tet repressor induced by tetracycline trapping. *J. Mol. Biol.* **1998**, *279*, 439–447. [CrossRef] [PubMed]
46. Seyffert, M.; Georgi, F.; Tobler, K.; Bourqui, L.; Anfossi, M.; Michaelsen, K.; Vogt, B.; Greber, U.; Fraefel, C. The HSV-1 Transcription Factor ICP4 Confers Liquid-Like Properties to Viral Replication Compartments. *Int. J. Mol. Sci.* **2021**, *22*, 4447. [CrossRef]
47. Gupta, M.; Deka, D. Ramneek Sequence analysis of Meq oncogene among Indian isolates of Marek's disease herpesvirus. *Meta Gene* **2016**, *9*, 230–236. [CrossRef]
48. Boodhoo, N.; Gurung, A.; Sharif, S.; Behboudi, S. Marek's disease in chickens: A review with focus on immunology. *Vet. Res.* **2016**, *47*, 1–19. [CrossRef]
49. Jarosinski, K.W.; Schat, K.A. Multiple alternative splicing to exons II and III of viral interleukin-8 (vIL-8) in the Marek's disease virus genome: The importance of vIL-8 exon I. *Virus Genes* **2007**, *34*, 9–22. [CrossRef]
50. Sekine, E.; Schmidt, N.; Gaboriau, D.; O'Hare, P. Spatiotemporal dynamics of HSV genome nuclear entry and compaction state transitions using bioorthogonal chemistry and super-resolution microscopy. *PLoS Pathog.* **2017**, *13*, e1006721. [CrossRef]
51. Dembowski, J.A.; DeLuca, N.A. Temporal Viral Genome-Protein Interactions Define Distinct Stages of Productive Herpesviral Infection. *mBio* **2018**, *9*, e01182-18. [CrossRef]
52. Tomer, E.; Cohen, E.M.; Drayman, N.; Afriat, A.; Weitzman, M.D.; Zaritsky, A.; Kobiler, O. Coalescing replication compartments provide the opportunity for recombination between coinfecting herpesviruses. *FASEB J.* **2019**, *33*, 9388–9403. [CrossRef]
53. Jain, N.; Iyer, K.V.; Kumar, A.; Shivashankar, G.V. Cell geometric constraints induce modular gene-expression patterns via redistribution of HDAC3 regulated by actomyosin contractility. *Proc. Natl. Acad. Sci. USA* **2013**, *110*, 11349–11354. [CrossRef] [PubMed]
54. Borsos, M.; Torres-Padilla, M.-E. Building up the nucleus: Nuclear organization in the establishment of totipotency and pluripotency during mammalian development. *Genes Dev.* **2016**, *30*, 611–621. [CrossRef] [PubMed]
55. Skinner, B.M.; Johnson, E.E.P. Nuclear morphologies: Their diversity and functional relevance. *Chromosoma* **2017**, *126*, 195–212. [CrossRef] [PubMed]

56. McSwiggen, D.T.; Hansen, A.S.; Teves, S.S.; Marie-Nelly, H.; Hao, Y.; Heckert, A.B.; Umemoto, K.K.; Dugast-Darzacq, C.; Tjian, R.; Darzacq, X. Evidence for DNA-mediated nuclear compartmentalization distinct from phase separation. *eLife* **2019**, *8*, e47098. [CrossRef] [PubMed]
57. Sourvinos, G. Visualization of parental HSV-1 genomes and replication compartments in association with ND10 in live infected cells. *EMBO J.* **2002**, *21*, 4989–4997. [CrossRef]

viruses

MDPI

Article

Acyl-Coa Thioesterases: A Rheostat That Controls Activated Fatty Acids Modulates Dengue Virus Serotype 2 Replication

Laura A. St Clair †, Stephanie A. Mills †,‡, Elena Lian, Paul S. Soma, Aritra Nag §, Caroline Montgomery, Gabriela Ramirez, Nunya Chotiwan ‖, Rebekah C. Gullberg ¶ and Rushika Perera *

Center for Vector-Borne Infectious Diseases, Department of Microbiology, Immunology and Pathology, Colorado State University, Fort Collins, CO 80523, USA; stclairl@colostate.edu (L.A.S.C.); millsemail5@yahoo.com (S.A.M.); elian@rams.colostate.edu (E.L.); Paul.Soma@colosate.edu (P.S.S.); aritranag@yahoo.com (A.N.); carolinemontgomery70@gmail.com (C.M.); gabriela.ramirez@colostate.edu (G.R.); nunya.chotiwan@umu.se (N.C.); rgullber@stanford.edu (R.C.G.)

* Correspondence: Rushika.Perera@colostate.edu

† These authors contributed equally to this work.

‡ Current address: College of Medicine, Medical University of South Carolina, Charleston, SC 29425, USA.

§ Current address: College of Medicine, Case Western Reserve University, Cleveland, OH 44106, USA.

‖ Current address: Department of Clinical Microbiology, Section of Virology, Umeå University, 901 87 Umeå, Sweden.

¶ Current address: Department of Biology, Stanford University, Stanford, CA 94305, USA.

Abstract: During infection with dengue viruses (DENVs), the lipid landscape within host cells is significantly altered to assemble membrane platforms that support viral replication and particle assembly. Fatty acyl-CoAs are key intermediates in the biosynthesis of complex lipids that form these membranes. They also function as key signaling lipids in the cell. Here, we carried out loss of function studies on acyl-CoA thioesterases (ACOTs), a family of enzymes that hydrolyze fatty acyl-CoAs to free fatty acids and coenzyme A, to understand their influence on the lifecycle of DENVs. The loss of function of the type I ACOTs 1 (cytoplasmic) and 2 (mitochondrial) together significantly increased DENV serotype 2 (DENV2) viral replication and infectious particle release. However, isolated knockdown of mitochondrial ACOT2 significantly decreased DENV2 protein translation, genome replication, and infectious virus release. Furthermore, loss of ACOT7 function, a mitochondrial type II ACOT, similarly suppressed DENV2. As ACOT1 and ACOT2 are splice variants, these data suggest that functional differences and substrate specificities due to the location (cytosol and mitochondria, respectively) of these proteins may account for the differences in DENV2 infection phenotype. Additionally, loss of mitochondrial ACOT2 and ACOT7 expression also altered the expression of several ACOTs located in multiple organelle compartments within the cell, highlighting a complex relationship between ACOTs in the DENV2 virus lifecycle.

Keywords: lipids; dengue virus; acyl-CoA; acyl-CoA thioesterase; fatty acids; membranes; rheostat; fatty acyl-CoA

Citation: St Clair, L.A.; Mills, S.A.; Lian, E.; Soma, P.S.; Nag, A.; Montgomery, C.; Ramirez, G.; Chotiwan, N.; Gullberg, R.C.; Perera, R. Acyl-Coa Thioesterases: A Rheostat That Controls Activated Fatty Acids Modulates Dengue Virus Serotype 2 Replication. *Viruses* **2022**, *14*, 240. https://doi.org/10.3390/v14020240

Academic Editors: Charles Grose, Ravi Mahalingam and Joel Rovnak

Received: 13 December 2021
Accepted: 20 January 2022
Published: 25 January 2022

Publisher's Note: MDPI stays neutral with regard to jurisdictional claims in published maps and institutional affiliations.

1. Introduction

Dengue viruses (DENVs) are arthropod-borne viruses that are transmitted by the *Aedes aegypti* mosquito [1,2]. These viruses infect over 400 million people each year. They are obligate intercellular parasites that rely on host cell metabolic pathways to fulfill their energy requirements and access substrates necessary for generating progeny virions [3]. It is widely established that the lifecycle of DENVs is reliant upon host lipid metabolic pathways [3–6]. Specifically, these viruses alter endoplasmic reticulum membranes to form scaffolds for viral protein translation and assembly of viral replication complexes [6,7]. Moreover, during viral particle assembly, host cell membranes are co-opted and incorporated into the viral envelope as a structural component of the virus particle [3,6,7].

In our previous studies, we have shown that many lipid species are upregulated and are vital for the DENV serotype 2 (DENV2) lifecycle [8,9]. Specifically, host phospholipids and sphingolipids are increased in abundance, some to benefit viral replication and others as a host response to infection ([8,9], reviewed in [3]). Precursors of these molecules are composed of fatty acyl-CoAs, which are fatty acids that have been esterified to coenzyme A (CoA) [10,11]. These fatty acyl-CoAs (activated fatty acids) can then undergo further modifications to be integrated into more complex lipids or be shuttled towards β-oxidation for cellular energy production [11].

Acyl-CoA thioesterases (ACOTs) are a family of hydrolases that control the intercellular balance between fatty acyl-CoAs and free fatty acids (FFAs). Specifically, they cleave fatty acyl-CoA into FFA and coenzyme A [10,11]. There are 10 identified human ACOT enzymes, further categorized into two groups by their domain motifs: α-β hydrolase (type I) and "hot dog" domain (type II) (Figure 1A,B) [10,11]. These enzymes are distributed throughout the organelles of human host cells, including the cytoplasm, mitochondria and peroxisomes (Figure 1A,B) [10]. ACOTs have been implicated in the control of lipid metabolism by maintaining the ratios of fatty acyl-CoA and free fatty acids within each organelle [10,11]. Upon perturbation, they have been shown to cause increased proliferation of cancer cells, are implicated in neurodegenerative diseases, and play a protective role against diabetic cardiac damage [12–15].

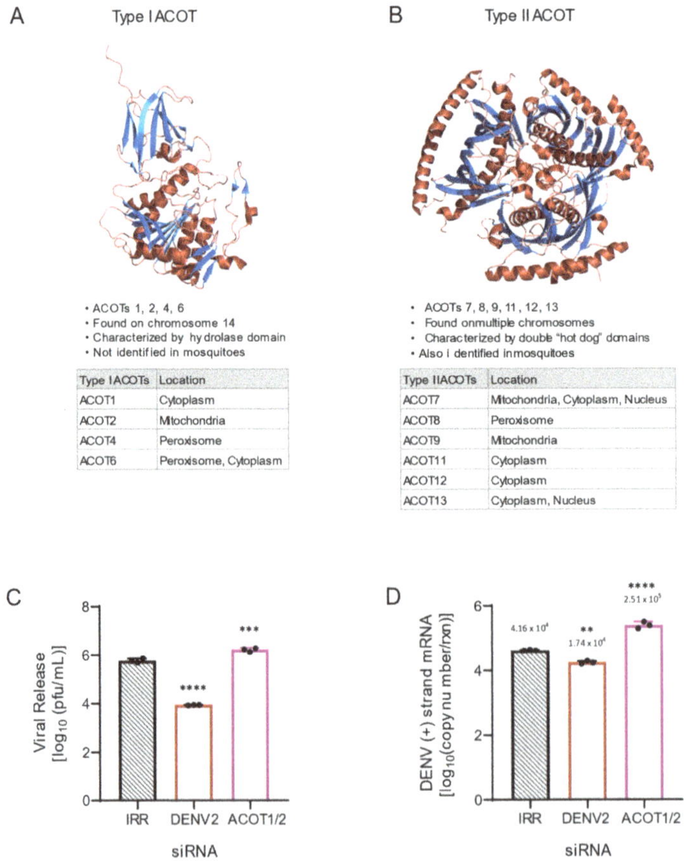

Figure 1. ACOT enzymes categorized by structural motifs and preliminary loss of function analysis of ACOT1/2. (**A**) Human ACOT2, Protein Data Bank identifier 3HLK is depicted. Type I ACOTs are characterized by a hydrolase domain that contains the active site. (**B**) Human ACOT7, Protein Data

Bank identifier 2QQ2 depicted. Type II ACOTs contain the "hot dog" domain composed of two alpha helixes surrounding a hydrophobic core. Although the composition of the active site is unknown within the "hot dog" domain, these enzymes are functionally analogous to type I ACOTs [10]. (**C**,**D**): Huh7 cells were transfected with an siRNA pool targeting both the ACOT1 and ACOT2 genes as well as indicated controls (IRR and DENV2) and then infected with DENV2 for 24 h (MOI = 3). (**C**) Infectious particle release was titrated via plaque assay. (**D**) Huh7 cells were collected, and relative copy number of viral RNA within cells was measured via qRT-PCR. qRT-PCR results were normalized to RPLP0. ACOT1/2: siRNA targeting acyl-CoA thioesterase 1 and 2, IRR: irrelevant siRNA control (no biological target for siRNA sequence), DENV2: siRNA targeting dengue virus, serotype 2. ((**A**,**B**): images were generated utilizing PyMOL Molecular Graphics System, Version 1.2r3pre, Schrödinger, LLC. Data base files from Protein Data Bank are available under CC0 1.0 Universal (CC0 1.0) Public Domain Dedication. C-D: one-way ANOVA with Dunnett's multiple comparisons test: ** $p \leq 0.01$, *** $p \leq 0.001$, **** $p \leq 0.0001$).

Given that fatty acyl-CoAs are integral to lipid metabolism and energy homeostasis, we investigated if perturbing fatty acyl-CoA homeostasis by altering ACOT enzyme expression modulated DENV2 infection. We also investigated if the ACOT enzyme location (cytoplasmic vs. mitochondrial) and/or specificity type (type I vs. type II) differentially influenced DENV2 replication. We examined three representative ACOT enzymes—ACOT1, ACOT2, and ACOT7—to understand their impact on the lifecycle of DENV2 in human hepatoma (liver) cells (Huh7s). siRNA-mediated loss of function studies of the type I ACOTs 1 and 2 together significantly increased DENV2 infectious particle release. However, isolated knockdown of ACOT2 significantly decreased DENV2 protein translation, genome replication, and infectious virus release. The loss of function of ACOT7, a mitochondrial type II ACOT, also similarly suppressed DENV2. Furthermore, our studies reveal a complex relationship between type I and type II ACOTs during DENV2 infection that suggests a functional interdependency of these enzymes.

2. Materials and Methods

2.1. Cell Lines and Viruses

The cell lines used in this study were as follows: Human hepatoma cells (Huh7) (unknown sex, From Dr. Charles Rice, Rockefeller University, New York, NY, USA) [16], Clone 15 (ATCC CCL-10) of the Baby Hamster Kidney Clone 21 cells (BHK-21), and C6/36 cells (ATCC CRL-1660, larva, unknown sex). Huh7 cells were maintained in Dulbecco's Modified Eagle Medium (DMEM) (Gibco, LifeTech, Carlsbad, CA, USA), while BHK-21 and C6/36 cells were maintained in Minimum Essential (MEM) (Gibco, LifeTech, Carlsbad, CA, USA). All culture media were supplemented with 2 mM L-glutamine (HyClone, Logan, UT, USA), 2 mM nonessential amino acids (HyClone, Logan, UT, USA), and 10% fetal bovine serum (FBS) (Atlas Biologicals, Fort Collins, CO, USA). C6/36 media was also supplemented with 25 mM HEPES buffer. Cells were incubated at 37 °C with 5% CO_2.

For this study, dengue virus serotype 2 was used (DENV2, strain 16681) [17,18]. The virus was passaged in C6/36 cells. Viral titer was determined via plaque assay on BHK-21 cells as previously described [19]. Virus infections were carried out at room temperature for 1 h, allowing for viral adherence. Subsequently, the virus was removed, and cells were washed with 1×PBS before the addition of media supplemented with 2 mM nonessential amino acids, 2 mM L-glutamine, and 2% FBS. Cells were incubated at 37 °C with 5% CO_2 for 24 h.

2.2. siRNA Transfection and Knockdown Confirmation

ACOT1/2, ACOT2, and ACOT7 loss of function was conducted by transfecting Huh7 cells with pooled siRNAs (Horizon Discovery/Dharmacon, Lafayette, CO, USA; Sigma-Aldrich, St. Louis, MO, USA—see Supplementary Table S1) as described previously [20] using RNAiMAX (Invitrogen, Carlsbad, CA, USA) and incubating for 48 h at 37 °C with

5% CO_2. At 48 h post transfection, cells were either collected for a cytotoxicity assay (described below), knockdown confirmation, or infected with DENV2 (as described above). At 24 hpi, cells and viral supernatant were collected for further analysis. Viral titration was completed via plaque assay. RNA was extracted from cells, and qRT-PCR analysis was used to confirm knockdown of ACOT2 and ACOT7 mRNA transcripts. Cellular transcripts were normalized to RPLP0, and the ACOT2 and ACOT7 siRNA-treated samples were then compared to an irrelevant control (IRR) using the delta-delta cq method described in [21]. The IRR control siRNA is a scrambled RNA with no homology to any human mRNA sequences (described in [20]). Cytotoxicity of siRNA treatment was determined using a 1:10 dilution of resazurin (ThermoFisher, Ward Hill, MA, USA) in cell culture media, and incubating cells for 1–2 h. Fluorescence was read on a Victor 1420 Multilabel plate reader (Perkin Elmer, Waltham, MA, USA) at 560 nm/590 nm (excitation/emission).

2.3. RNA Extraction and qRT-PCR

RNA was extracted from cells and viral supernatant using TRIzol or TRIzol LS (ThermoFisher), respectively, following standard TRIzol extraction methods. For qRT-PCR, the Brilliant III Ultra-Fast SYBR® Green one-step qRT-PCR kit (Agilent, Carlsbad, CA, USA) was used, and all reactions were set up according to manufacturer's protocols on a LightCycler 96-well real-time PCR machine (Roche, Indianapolis, IN, USA). The following cycling parameters were employed: 20 min at 50 °C for reverse transcription, 5 min at 95 °C, followed by 45 two-step cycles of 95 °C for 5 secs, and 60 °C for 60 secs. A melt curve followed each step, starting at 65 °C and ending at 97 °C. (Primer sequences are reported in Supplementary Table S1). A standard curve was generated using in vitro transcribed viral RNA from a DENV2 cDNA subclone (derived from strain 16681 full-length clone) in order to quantify DENV2 genome copies. All cellular RNA transcript copy numbers were normalized to Ribosomal Protein Lateral Stalk Subunit P0 (RPLP0) RNA using the delta-delta cq method as described previously [21,22].

2.4. Western Blotting

Huh7 cells were lysed in RIPA Buffer. Total protein in each sample was measured using the Qubit BR Protein Assay (ThermoFisher, Carlsbad, CA, USA) on a Qubit Flex Fluorometer. Equal total protein was loaded from each of the indicated cell lysates and run on a Criterion™ XT 4–12% Bis-Tris protein gel (Bio-Rad, Hercules, CA, USA). Protein was transferred at 4 °C to nitrocellulose membrane for 2 h at 50 V. Following transfer, blocking was performed overnight at 4 °C with a 5% milk solution in 1 × PBS supplemented with 0.1% Tween 20. The primary antibodies used were a 1:300 dilution of antibody against the DENV2 NS3 protein (mouse polyclonal antibody raised against the DENV2 NS3 protein as described in 20), and 1:100 dilution of β-actin (rabbit polyclonal antibody, Invitrogen). The secondary antibodies used were a 1:3000 dilution of goat-anti-mouse IRDye 800CW, and goat-anti-rabbit IRDye 680RD (Li-Cor Biosciences, Lincoln, NE, USA). Blots were imaged on a ChemiDoc MP Imaging System (Bio-Rad, Hercules, CA, USA), and quantified in ImageJ utilizing area under the curve analysis.

2.5. Statistical Analysis

In each figure and/or figure legend, statistical analysis details are noted. All results are expressed as mean values with standard deviation. Statistical significance was determined using a one-way Analysis of Variance (ANOVA) with Dunnett's multiple comparisons test, or an unpaired *t* test using Prism software version 9.0 (GraphPad Software, La Jolla, CA, USA).

3. Results

3.1. Combined Loss of ACOT1 and ACOT2 Function Increases DENV2 Genome Replication and Infectious Particle Release

As ACOTs act as a rheostat for intracellular levels of FFAs and fatty acyl-CoAs, we hypothesized that ACOT functionality, which purportedly limits the availability of fatty acyl-CoAs, may have an inhibitory effect on the DENV2 lifecycle. To determine if ACOT enzymes affect DENV2 replication, we used siRNA-mediated knockdown to decrease the expression of type I ACOTs 1 and 2 in Huh7 cells. ACOT1 is located in the cytoplasm and ACOT2 in the mitochondria. There is ~94% homology between the mRNAs of ACOT1 and ACOT2, with only an insertion of 90–110 nucleotides differentiating ACOT2 mRNA from ACOT1 mRNA [23–25]. We included a non-target, irrelevant siRNA (IRR) to control for off-target effects of siRNA treatment and a DENV2-specific siRNA as a positive control for viral inhibition. We found that loss of ACOT1 and ACOT2 together resulted in an increase (~178%) in infectious viral release (Figure 1C) and in viral genome replication (~470%, Figure 1D) compared to the IRR control. Due to the sequence similarity between ACOT1 and 2, we were unable to target siRNAs specific to ACOT1, but we were able to target specific siRNAs to ACOT2. Therefore, as shown below, we were able to parse out the influence of ACOT2 on this phenotype.

3.2. *Loss of ACOT2 Function Reduces DENV2 Replication, Infectious Particle Release and Infectivity*

Given the above observations from combined loss of ACOT1 and ACOT2 function, and that ACOT1 and ACOT2 share the same substrate specificity (18), we hypothesized that loss of ACOT2 function alone would also result in increased viral replication and release. Therefore, we repeated knockdown studies with siRNAs specifically targeting ACOT2 in Huh7 cells. Surprisingly, we found that loss of ACOT2 function resulted in a significant reduction (~76%) in infectious virus release (Figure 2A) compared to the IRR control. Additionally, DENV2 genome replication was reduced by ~50% following ACOT2 knockdown (Figure 2B). In our previous studies, we found that inhibition of specific enzymes involved in fatty acid metabolism resulted in the release of partially immature virions, thereby reducing particle infectivity [25]. To determine if loss of ACOT2 function also reduced particle infectivity, we compared the ratios of viral RNA copies to infectious virions (particle/pfu ratio) in the supernatant of each of our treatment groups (Figure 2C). The data showed that there was a significant increase in the DENV particle/pfu ratio following ACOT2 knockdown compared to the IRR control, thus suggesting a decrease in particle infectivity. We also found that viral protein translation was reduced by ~50% (Figure 2D and Supplementary Figure S1). We further confirmed that siRNA treatment was not cytotoxic to Huh7 cells (Figure 2E) and was also effective at reducing ACOT2 mRNA levels (Figure 2F). Overall, these data suggest that ACOT2 functionality is critical to the DENV2 lifecycle. This contrasts with the phenotype from the combined loss of function of ACOT1 and 2, suggesting that ACOT1 has a unique influence on viral replication.

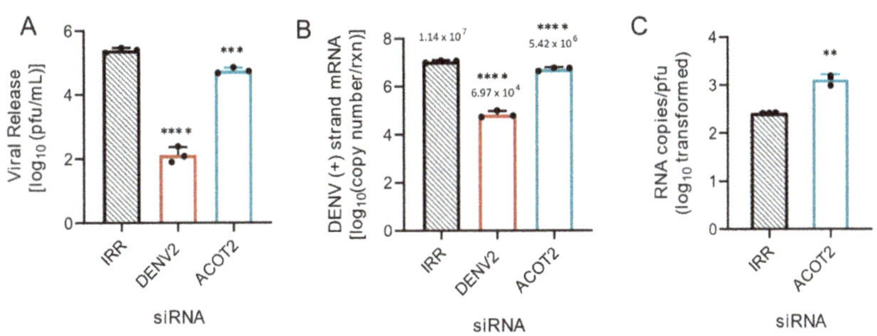

Figure 2. *Cont.*

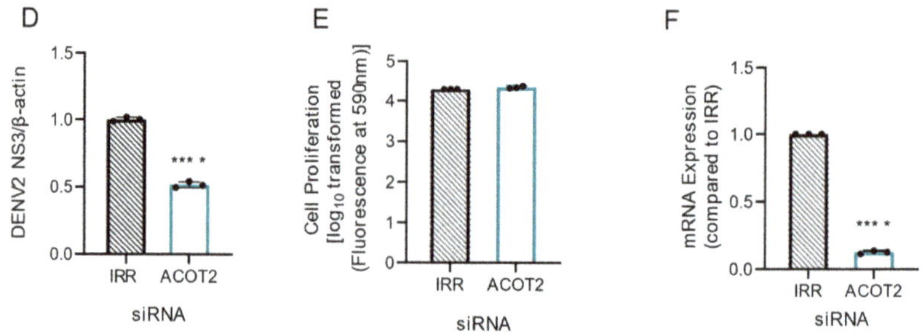

Figure 2. Loss of ACOT2 function reduces DENV2 genome replication and virus release. Huh7 cells were transfected with siRNA, targeting the ACOT2 gene or indicated controls (IRR and DENV2), and then infected with DENV2 for 24hr (MOI = 0.3). (**A**) Infectious virus release was titrated via plaque assay. (**B**) Huh7 cells were collected, and the relative copy number of viral RNA within cells was measured via qRT-PCR. Results were normalized to RPLP0. (**C**) Virus supernatant was collected at 24 hpi and split into two fractions. One fraction was titrated via plaque assay, while viral RNA from the other fraction was analyzed via qRT-PCR for total copy number of DENV2 positive-strand RNA. (**D**) Cell lysates were collected at 24 hpi, and analyzed via western blot. Samples were probed for DENV2 nonstructural protein 3, and normalized to β-actin. Li-cor IRDyes were used as secondary antibodies, and fluorescence intensity of each band was analyzed using area under the curve analysis in ImageJ. (**E**) A resazurin-reduction based cell viability assay was conducted to assess cytotoxicity of siRNA treatment. (**F**) Knockdown of ACOT2 mRNA was confirmed at 48 h post transfection via qRT-PCR. ACOT2: siRNA targeting acyl-CoA thioesterase 2, IRR: irrelevant siRNA control (no biological target for siRNA sequence), DENV2: siRNA targeting dengue virus, serotype 2. ((**A**,**B**): one-way ANOVA with Dunnett's multiple comparisons test, (**C**–**F**): unpaired *t* tests: ** $p \leq 0.01$, *** $p \leq 0.001$, **** $p \leq 0.0001$).

3.3. *Mitochondrial ACOTs Are Vital for Productive DENV2 Infection*

ACOT2 is suggested to be an important mediator of β-oxidation [26,27]. This is the process of gaining ATP from the breakdown of fatty acyl-CoA to 2-carbon acetyl-CoA molecules. As this process occurs in the mitochondria, mitochondrial ACOTs may be important in regulating the balance of FFAs and fatty acyl-CoAs through shuttling fatty acyl-CoAs into β-oxidation [26,27]. Previous studies indicate that β-oxidation is elevated during infection of host cells by DENVs (reviewed in [3]). Both type I and type II ACOTs exist in the mitochondria. ACOT7 is a type II ACOT, and is functionally homologous to ACOT2 (both hydrolyze long-chain FAs) [25]. Therefore, we investigated if loss of ACOT7 function also resulted in suppression of DENV2 infection. Similar to ACOT2 knockdown, we found that loss of ACOT7 function significantly reduced DENV2 infectious particle release (~77%) in Huh7 cells (Figure 3A). Interestingly, ACOT7 inhibition resulted in a greater reduction of viral protein translation (~75% reduction, Figure 3D and Supplementary Figure S1) and viral genome replication (~70% reduction, Figure 3B) compared to the effect of the loss of function of ACOT2 (Figure 2B). Similar to the ACOT2 study, loss of ACOT7 function significantly decreased particle infectivity (Figure 3C). We confirmed that these results were also not due to cytotoxicity of the siRNA treatment (Figure 3E), and that the ACOT7 siRNA effectively reduced ACOT7 mRNA levels (Figure 3F). Taken together, these studies suggest that two of the mitochondrial ACOTs have a vital role in the DENV2 lifecycle.

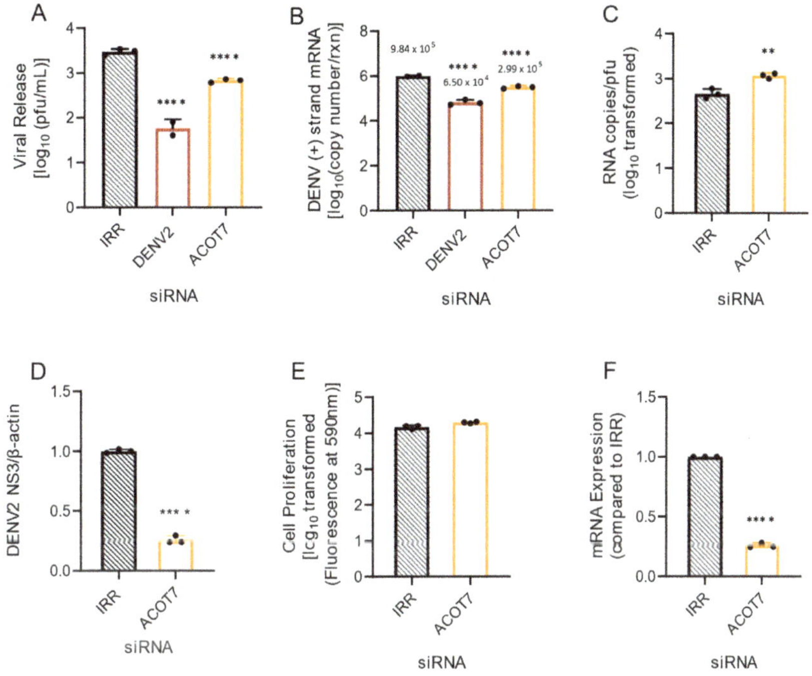

Figure 3. Loss of ACOT7 function suppresses DENV2 genome replication and infectious particle release. Huh7 cells were transfected with siRNA targeting the ACOT7 gene or the indicated controls (IRR and DENV2) and then subsequently infected with DENV2 for 24 h (MOI = 0.3). (**A**) Infectious virus release was titrated via plaque assay. (**B**) Huh7 cells were collected, and the relative copy number of viral RNA within cells was measured via qRT-PCR. Results were normalized to RPLP0. (**C**) Virus supernatant was collected at 24 hpi and split into two fractions. One fraction was titrated via plaque assay, while viral RNA from the other fraction was analyzed via qRT-PCR for total copy number of DENV2 positive-strand RNA. (**D**) Cell lysates were collected at 24 hpi and analyzed via western blot. Samples were probed for DENV2 nonstructural protein 3 and normalized to β-actin. Li-cor IRDyes were used as secondary antibodies, and the fluorescence intensity of each band was analyzed using area under the curve analysis in ImageJ. (**E**) A cell viability assay using resazurin was conducted to assess cytotoxicity of siRNA treatment. (**F**) Knockdown of ACOT7 mRNA was confirmed at 48 h post transfection via qRT-PCR. ACOT7: siRNA targeting acyl-CoA thioesterase 7, IRR: irrelevant siRNA control (no biological target for siRNA sequence), DENV2: siRNA targeting dengue virus, serotype 2. ((**A**,**B**): one-way ANOVA with Dunnett's multiple comparisons test, (**C**–**F**): unpaired *t* tests:, ** $p \leq 0.01$, **** $p \leq 0.0001$).

3.4. Both Type I and Type II ACOTs Are Differentially Expressed upon ACOT2 or ACOT7 Knockdown

Currently, there are ten human ACOTs known to exist in different subcellular compartments, and they have substrate specificity for a wide range of fatty acyl-CoAs [24,25]. However, the functional overlap and/or ability of each ACOT to compensate for the loss of function of the others is unknown. As mitochondrial ACOTs seem to play vital roles in fatty acid metabolism and energy production, we determined whether other ACOTs could compensate for loss of ACOT2 or ACOT7 function, both within and without the context of DENV2 infection. To investigate this, we carried out similar siRNA-mediated knockdown of ACOT2 and ACOT7 in both mock-infected and DENV2-infected Huh7 cells. An IRR siRNA control was also included. 24 h post infection (hpi), mock- and DENV2-infected cells were collected and cellular mRNA levels of each ACOT enzyme was determined via qRT-PCR.

In both mock-infected and DENV2-infected samples, we observed similar trends in the mRNA expression of other ACOTs upon knockdown of ACOT2 (Figure 4A,B) or ACOT7 (Figure 4C,D). Specifically, we observed that in both mock- and DENV2-infected Huh7 cells, knockdown of ACOT2 significantly reduced the mRNA expression of ACOTs 1, 6, 8, and 11 (Figure 4A,B). A decrease in ACOT9 and ACOT12 mRNA expression was only observed in DENV2-infected ACOT2 knockdown samples. We found loss of ACOT7 significantly reduced mRNA expression of ACOTs 2, 4, 8, 9, and 11 in both mock- and DENV2-infected cells, (Figure 4C,D). Additionally, in DENV2 infected cells, ACOT6 and ACOT13 mRNA expression was decreased in ACOT7 siRNA-treated cells. It should be noted that the siRNAs for ACOT2 and ACOT7 do not have any sequence similarity to the other ACOT mRNAs. Therefore, these data suggest that other ACOTs may be functionally dependent on the activity of ACOT2 and 7. Intriguingly, inhibiting ACOT7 also reduced expression of ACOT2 (Figure 4C,D); however, inhibition of ACOT2 did not impact the expression of ACOT7 (Figure 4A,B). Thus, while both mitochondrial ACOTs are functionally similar, neither compensates for the loss of the other at a transcriptional level, and ACOT2 functionality may rely upon ACOT7.

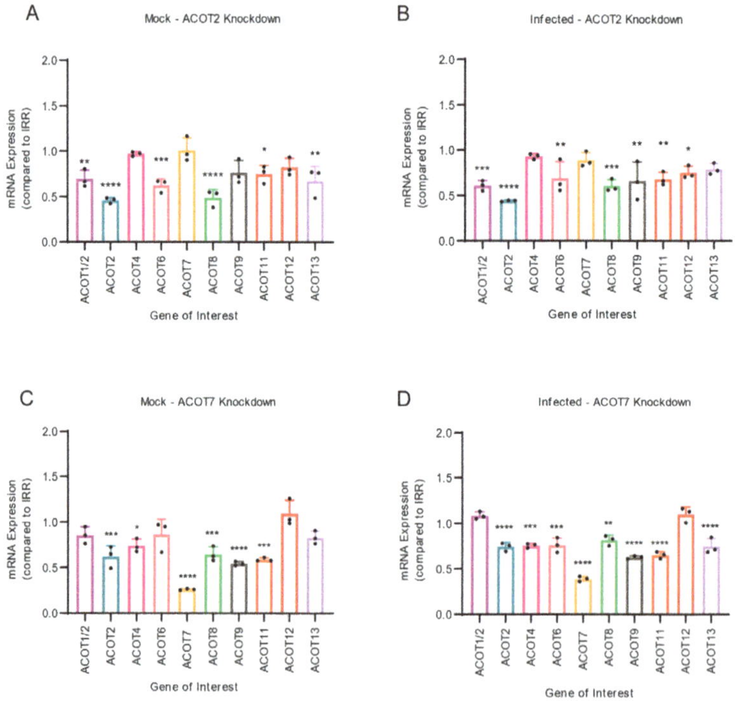

Figure 4. Inhibition of mitochondrial ACOTs underlines importance of mitochondrial ACOT functionality for expression of other ACOTs. Huh7 cells were treated with either ACOT2 (**A**,**B**) or ACOT7 (**C**,**D**) siRNA or an IRR control siRNA, and then (**A**,**C**) mock-infected or (**B**,**D**) DENV2-infected (MOI = 0.3). At 24 hpi, cells were collected and mRNA levels of all 10 human ACOTs was determined via qRT-PCR. ((**A**–**D**): one-way ANOVA with Dunnett's multiple comparisons tests, * $p \leq 0.05$, ** $p \leq 0.01$, *** $p \leq 0.001$, **** $p \leq 0.0001$).

3.5. Mitochondrial ACOTs Are Upregulated at Early Timepoints of Infection

In previous studies, it has been demonstrated that DENV2 infection results in both viral- and host-mediated modulation of enzymes involved in fatty acid metabolism [3–9,24].

As our loss of function studies of ACOT2 and ACOT7 indicated that these enzymes are vital for the DENV2 lifecycle, we analyzed whether ACOT2 and ACOT7 mRNA expression was altered over a time course of DENV2 infection. For this study, we collected both mock-infected and DENV2-infected cells at 0, 6, 24, and 48 hpi. These time points represent early, peak, and late viral replication. ACOT mRNA expression was analyzed using qRT-PCR (Figure 5). We observed that ACOT2 and ACOT7 mRNA expression was significantly upregulated at 6 hpi in DENV2-infected cells, but downregulated at 24 and 48 hpi (Figure 5A). A similar trend was noted when we compared mRNA expression between DENV2 and mock-infected samples at each timepoint, although the increased expression at 6 hpi was not statistically significant (Figure 5B). These results, combined with our results in Figures 2 and 3, suggest that ACOT2 and ACOT7 functionality is required for early stages of the DENV2 life cycle.

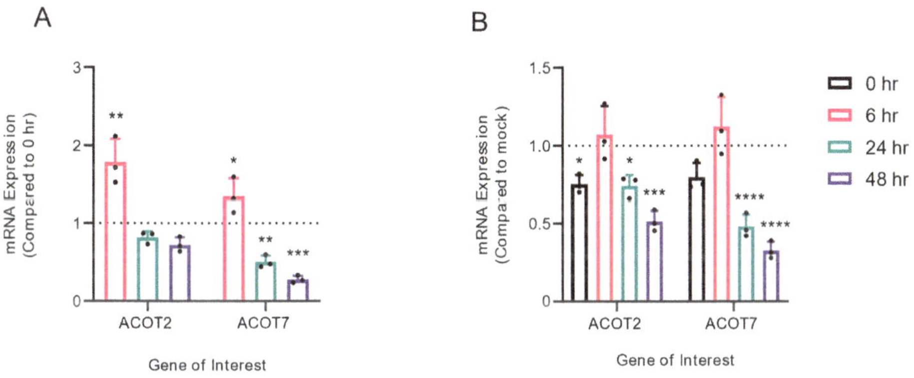

Figure 5. Mitochondrial ACOT mRNA expression is temporally altered during DENV2 infection. Huh7 cells were either mock-infected or DENV2-infected (MOI = 10). At 0, 6, 24, and 48 hpi, cells were collected and mRNA levels of ACOT2 and ACOT7 were assessed via qRT-PCR. DENV2-infected samples collected at 6, 24, and 48 hpi were compared either to (**A**) the 0 hpi DENV2-infected samples or to (**B**) mock infected samples at matched timepoints. Results are reported as an expression ratio (**A**) between each timepoint sample and the 0 hpi sample in DENV2-infected cells, or (**B**) between each timepoint sample and its respective mock-infected timepoint. ((**A**,**B**): one-way ANOVA with Dunnett's multiple comparisons test: * $p \leq 0.05$, ** $p \leq 0.01$, *** $p \leq 0.001$, **** $p \leq 0.0001$).

4. Discussion

The ACOT family of enzymes are suggested to be key regulators of the intercellular balance of activated fatty acids (fatty acyl-CoAs) and FFAs [10,25]. As previous studies have shown that DENVs are reliant upon activated fatty acids for completion of their lifecycle ([8,9,20,21,28,29], reviewed in [3]), our present study investigated whether ACOT enzymes were important modulators of DENV2 infection. The ACOT enzymes chosen represented both type I (ACOT1 and ACOT2) and type II ACOTs (ACOT7), as well as ACOTs found within the same organelle (ACOT2 and ACOT7 in the mitochondria). Interestingly, we found that the loss of function of these enzymes had a divergent effect on the DENV2 lifecycle that was independent of their substrate specificity, and, instead, highlighted a more nuanced relationship between DENV2 and the subcellular regulation of fatty acid metabolism. Specifically, we found that the loss of function of the cytoplasmic ACOT1 enzyme resulted in increased DENV2 genome replication and infectious virus release, indicating that its normal function may be antiviral. However, the loss of function of the mitochondrial ACOT2 and ACOT7 enzymes resulted in the suppression of multiple stages of the DENV2 lifecycle, indicating that the function of these enzymes is vital for effective DENV2 infection. Furthermore, our analyses revealed a functional dependency between enzymes within the ACOT family.

Because ACOTs act directly on activated fatty acids, it is predicted that they play a significant role in the modulation of fatty acid metabolism [10,11,21,25,30]. However, mechanistic insight as to the specific functions of ACOT enzymes is limited. One challenge is that the ester bond that links coenzyme A with fatty acids is labile, which limits the ability to obtain a true ratio of activated fatty acids and FFAs in cells with traditional methods [10,31]. Thus, much of what the field understands about the function of these enzymes is concluded from studies characterizing their biophysical and biochemical properties, or from mouse models studying the impact of ACOT gene deletion on cardiac, neurological, and metabolic disorders [10–15,23–27,31]. During the curation of this paper, we successfully developed an LC-MS assay that allowed us to distinguish between arachidonic acid, SH-CoA, and arachidonoyl-CoA (the primary substrate of ACOT7, discussed in 25 and 27). However, due to limitations in the scalability of siRNA treatment, the levels of these metabolites were below our limits of detection. As a future direction, use of knockdown cell lines may further characterize ACOT enzyme functionality. Moreover, an additional challenge was that ACOT1 and ACOT2 enzymes share 98% similarity at the protein level, and 94% nucleotide sequence homology [11,24]. This inhibited differentiation of these enzymes at the protein level. However, we were able to design specific siRNA and primer sequences that targeted the mitochondrial localization sequence of ACOT2, allowing us to characterize these enzymes based on their mRNA expression.

In an ACOT1 knockout mouse model, ACOT1 was shown to modulate liver fatty acid metabolism during fasting, and the loss of function of this enzyme led to increased triglyceride turnover and beta oxidation [23]. Importantly, infection with DENV2 mimics a fasting state in the cell, and previous studies have established that triglyceride hydrolysis and beta-oxidation are increased in DENV2-infected cells [20,28,29]. Thus, inhibition of ACOT1 may further enhance these processes, leading to a more favorable environment for DENV2 replication. Mitochondrial functional assays have suggested that ACOT2 and ACOT7 functions to decrease beta-oxidation overload, possibly by mediating efflux of fatty acids from the mitochondrial matrix [28,30,31]. Thus, loss of ACOT2 and ACOT7 function may have resulted in a buildup of fatty acyl-CoAs levels in the mitochondria, decreased turnover of CoA and fatty acid precursor molecules, and increased oxidative stress—all of which would antagonize the DENV2 lifecycle. Interestingly, we found that, following ACOT2 or ACOT7 knockdown, there was no compensation by any of the other mitochondrial ACOTs (ACOT2, 7 or 9). However, ACOT2 and ACOT7 knockdown decreased peroxisomal ACOT4, ACOT6, and ACOT8 expression. Peroxisomal ACOTs are important in the degradation of very long-chain fatty acids that cannot be directly shuttled to the mitochondria [10,11]. Peroxisomes are becoming recognized as important mediators for controlling or facilitating virus infection, including antiviral immune responses, interactions with viral capsid proteins, and influencing membrane fluidity [32,33]. Therefore, another possibility is that ACOT2 or ACOT7 knockdown is indirectly suppressing DENV2 replication through an altering of the expression of peroxisomal ACOTs, which may impact peroxisomal lipid degradation activity. The role of peroxisomes in these biochemical interactions between virus and host is as yet an uncharted territory.

In summary, we found there is a differential impact of ACOTs on DENV2 genome replication and infectious particle release that could be influenced by the subcellular location of these enzymes. We also observed that the loss of function of a single ACOT impacted the expression of multiple ACOTs in different cellular locations, highlighting the complexity of understanding how ACOTs influence viral replication. Overall, our current study underscores that DENV2 is reliant upon careful coordination of fatty acid metabolism to complete its lifecycle. Future studies aimed at characterizing the exact substrate specificities would increase our understanding of how the loss of function of specific enzymes regulates infection. Additionally, our data suggest that determining the functional relationship between ACOTs is warranted.

Supplementary Materials: The following are available online at https://www.mdpi.com/article/10.3390/v14020240/s1, Table S1: siRNA and oligonucleotide resources used in this study. Figure S1: Loss of function of mitochondrial ACOTs inhibits viral protein translation.

Author Contributions: Conceptualization, L.A.S.C., S.A.M., R.C.G. and R.P.; methodology, L.A.S.C., S.A.M., P.S.S., N.C., R.C.G. and R.P.; data curation, validation and formal analysis, L.A.S.C., S.A.M., E.L., C.M., A.N. and G.R.; writing—original draft preparation, L.A.S.C., S.A.M. and R.P.; writing—review and editing, L.A.S.C., S.A.M., E.L., G.R., N.C., R.C.G. and R.P.; visualization, L.A.S.C., S.A.M. and R.P.; supervision, R.P.; project administration, R.P.; funding acquisition, R.P. All authors have read and agreed to the published version of the manuscript.

Funding: This work was supported by the American Society for Microbiology Undergraduate Fellowship to S.A.M. and funding from the Department of Microbiology, Immunology and Pathology and the Office of the Vice President for Research, Colorado State University.

Conflicts of Interest: The authors declare no conflict of interest.

References

1. Halstead, S.B. Dengue antibody-dependent enhancement: Knowns and unknowns. *Microbiol. Spectr.* **2014**, *2*, 1–18. [CrossRef] [PubMed]
2. Messina, J.P.; Brady, O.J.; Scott, T.W.; Zou, C.; Pigott, D.M.; Duda, K.A.; Bhatt, S.; Katzelnick, L.; Howes, R.E.; Battle, K.E.; et al. Global spread of dengue virus types: Mapping the 70 year history. *Trends Microbiol.* **2014**, *22*, 138–146. [CrossRef] [PubMed]
3. Perera, R.; Kuhn, R.J. Chapter 4. "Host metabolism and its contribution in flavivirus biogenesis". In *Arboviruses: Molecular Biology, Evolution and Control*; Gubler, D., Vasilakis, N., Eds.; Caister Academic Press: Norfolk, UK, 2015; ISBN 978-191-019-021-0.
4. Zaitseva, E.; Yang, S.T.; Melikov, K.; Pourmal, S.; Chernomordik, L.V. Dengue virus ensures its fusion in late endosomes using compartment-specific lipids. *PLoS Pathog.* **2010**, *6*, e10001131. [CrossRef] [PubMed]
5. Neufeldt, C.J.; Cortese, M.; Acosta, E.G.; Bartenschlager, R. Rewiring cellular networks by members of the Flaviviridae family. *Nat. Rev. Microbiol.* **2018**, *16*, 125–142. [CrossRef] [PubMed]
6. Chatel-Chaix, L.; Bartenschlager, R. Dengue virus- and hepatitis C virus-induced replication and assembly compartments: The enemy inside—Caught in the web. *J. Virol.* **2014**, *88*, 5907–5911. [CrossRef] [PubMed]
7. Gillespie, L.K.; Hoenen, A.; Morgan, G.; Mackenzie, J.M. The endoplasmic reticulum provides the membrane platform for biogenesis of the flavivirus replication complex. *J. Virol.* **2010**, *84*, 10438–10447. [CrossRef] [PubMed]
8. Perera, R.; Riley, C.; Issac, G.; Hopf-Jannasch, A.S.; Moore, R.J.; Weitz, K.W.; Pasa-Tolic, L.; Metz, T.O.; Adamec, J.; Kuhn, R.J. Dengue virus infection perturbs lipid homeostasis in infected mosquito cells. *PLoS Pathog.* **2012**, *8*, e1002584. [CrossRef]
9. Chotiwan, N.; Andre, B.G.; Snachez-Vargus, I.; Grabowski, J.M.; Hopf-jannasch, A.; Hedrick, V.; Gough, E.; Nakayasu, E.; Blair, C.D.; Hill, C.A.; et al. Dynamic remodeling of lipids coincides with dengue virus replication in the midgut of *Aedes aegypti* mosquitoes. *PLoS Pathog.* **2018**, *14*, e1006853. [CrossRef]
10. Tillander, V.; Alexson, S.H.E.; Cohen, D.E. Deactivating fatty acids: Acyl-CoA thioesterase-mediated control of lipid metabolism. *Trends Endocrinol. Metab.* **2017**, *28*, 473–484. [CrossRef] [PubMed]
11. Hunt, M.C.; Siponen, M.I.; Alexson, S.E. The emerging role of acyl-CoA thioesterases and acyltransferases in regulating peroxisomal lipid metabolism. *Biochim. Biophys. Acta* **2012**, *1822*, 1397–1410. [CrossRef] [PubMed]
12. Jung, S.H.; Lee, H.C.; Hwang, H.J.; Park, H.A.; Moon, Y.; Kim, B.C.; Lee, H.M.; Kim, K.P.; Kim, Y.; Lee, B.L.; et al. Acyl-CoA thioesterase 7 is involved in cell cycle progression via regulation of PKCζ-p53-p21 signaling pathway. *Cell Death Dis.* **2017**, *8*, e2793. [CrossRef] [PubMed]
13. Wang, F.; Wu, J.; Qiu, Z.; Ge, X.; Liu, X.; Zhang, C.; Xu, W.; Wang, F.; Hua, D.; Qi, X.; et al. ACOT1 expression is associated with poor prognosis in gastric adenocarcinoma. *Hum. Pathol.* **2018**, *77*, 35–44. [CrossRef] [PubMed]
14. Ellis, J.M.; Wong, G.W.; Wolfgang, M.J. Acyl coenzyme A thioesterase 7 regulates neuronal fatty acid metabolism to prevent neurotoxicity. *Mol. Cell Biol.* **2013**, *33*, 1869–1882. [CrossRef] [PubMed]
15. Yang, S.; Chen, C.; Wang, H.; Rao, X.; Wang, F.; Duan, Q.; Chen, F.; Long, G.; Gong, W.; Zou, M.; et al. Protective effects of acyl-CoA thioesterase 1 on diabetic heart via PPARα/PGC1α signaling. *PLoS ONE* **2012**, *7*, e50376. [CrossRef] [PubMed]
16. Blight, K.J.; McKeating, J.A.; Rice, C.M. Highly permissive cell lines for subgenomic and genomic hepatitis C virus RNA replication. *J. Virol.* **2002**, *76*, 13001–13014. [CrossRef] [PubMed]
17. Yoksan, S.; Bhamarapravati, N.; Halstead, S.B. Dengue virus vaccine development: Study on biological markers of uncloned dengue 1–4 viruses serially passaged in primary kidney cells. In *Arbovirus Research in Australia, Proceedings of the Fourth Symposium*; St George, T.D., Kay, B.H., Blok, J., Eds.; CSIRO/QIMR: Brisbane, Australia, 1986; pp. 35–38.
18. Kinney, R.M.; Butrapet, S.; Chang, G.J.; Tsuchiya, K.R.; Roehrig, J.T.; Bhamarapravati, N.; Gubler, D.J. Construction of infectious cDNA clones for dengue 2 virus: Strain 16681 and its attenuated vaccine derivative, strain PDK-53. *Virology* **1997**, *230*, 300–308. [CrossRef] [PubMed]
19. Dulbecco, R.; Vogt, M. Some problems of animal virology as studied by the plaque technique. *Cold Spring Harb. Symp. Quant. Biol.* **1953**, *18*, 273–279. [CrossRef] [PubMed]

20. Heaton, N.S.; Perera, R.; Berger, K.L.; Khadka, S.; Lacount, D.J.; Kuhn, R.J.; Randall, G. Dengue virus nonstructural protein 3 redistributes fatty acid synthase to sites of viral replication and increases cellular fatty acid synthesis. *Proc. Natl. Acad. Sci. USA* **2010**, *107*, 17345–17350. [CrossRef] [PubMed]
21. Gullberg, R.C.; Steel, J.J.; Pujari, V.; Rovnak, J.; Crick, D.C.; Perera, R. Stearoyl-CoA desaturase 1 differentiates early and advanced dengue virus infections and determines virus particle infectivity. *PLoS Pathog.* **2018**, *14*, e1007261. [CrossRef]
22. Livak, K.J.; Schmittgen, T.D. Analysis of relative gene expression data using real-time quantitative PCR and the 2(-Delta Celta C(t)) method. *Methods* **2001**, *25*, 402–408. [CrossRef] [PubMed]
23. Franklin, M.P.; Sathyanarayan, A.; Mashek, D.G. Acyl-CoA thioesterase 1 (ACOT1) regulates PPARα to couple fatty acid flux with oxidative capacity during fasting. *Diabetes* **2017**, *66*, 2112–2123. [CrossRef] [PubMed]
24. Hunt, M.C.; Rautanen, A.; Westin, M.A.; Svensson, L.T.; Alexson, S.E. Analysis of the mouse and human acyl-CoA thioesterase (ACOT) gene clusters shows that convergent, functional evolution results in a reduced number of human peroxisomal ACOTs. *FASEB J.* **2006**, *20*, 1855–1864. [CrossRef] [PubMed]
25. Brocker, C.; Carpenter, C.; Nebert, D.W.; Vasiliou, V. Evolutionary divergence and functions of the human acyl-CoA thioesterase gene (ACOT) family. *Hum. Genom.* **2010**, *4*, 411–420. [CrossRef] [PubMed]
26. Svensson, L.T.; Engberg, S.T.; Aoyama, T.; Usuda, N.; Alexson, S.E.; Hashimoto, T. Molecular cloning and characterization of a mitochondrial peroxisome proliferator-induced acyl-CoA thioesterase from rat liver. *Biochem. J.* **1998**, *329 Pt 3*, 601–608. [CrossRef] [PubMed]
27. Bekeova, C.; Anderson-Pullinger, L.; Boye, K.; Boos, F.; Sharpadskaya, Y.; Herrmann, J.M.; Seifert, E.L. Multiple mitochondrial thioesterases have distinct tissue and substrate specificity and CoA regulation, suggesting unique functional roles. *J. Biol. Chem.* **2019**, *294*, 19034–19047. [CrossRef] [PubMed]
28. Heaton, N.S.; Randall, G. Dengue virus-induced autophagy regulates lipid metabolism. *Cell Host Microbe* **2010**, *8*, 422–432. [CrossRef] [PubMed]
29. Samsa, M.M.; Mondotte, J.A.; Iglesias, N.G.; Assunção-Mirando, I.; Barbosa-Lima, G.; Da Poian, A.T.; Bozza, P.T.; Gamarnik, A.V. Dengue virus capsid protein usurps lipid droplets for viral particle formation. *PLoS Pathog.* **2009**, *5*, e1000632. [CrossRef]
30. Moffat, C.; Bhatia, L.; Nguyen, T.; Lynch, P.; Wang, M.; Wang, D.; Ilkayeva, O.R.; Han, X.; Hirschey, M.D.; Claypool, S.M.; et al. Acyl-CoA thioesterase-2 facilitates mitochondrial fatty acid oxidation in the liver. *J. Lipid Res.* **2014**, *55*, 2458–2470. [CrossRef] [PubMed]
31. Fujita, M.; Momose, A.; Ohtomo, T.; Nishinosono, A.; Tanonaka, K.; Toyoda, H.; Morikawa, M.; Yamada, J. Upregulation of fatty acyl-CoA thioesterases in the heart and skeletal muscles of rats fed a high-fat diet. *Bio. Pharm. Bull.* **2011**, *34*, 87–91. [CrossRef] [PubMed]
32. Cook, K.C.; Moreno, J.A.; Jean Beltran, P.M.; Cristea, I.M. Peroxisome plasticity at the virus-host interface. *Trends Microbiol.* **2019**, *27*, 906–914. [CrossRef] [PubMed]
33. You, J.; Hou, S.; Malik-Soni, N.; Xu, Z.; Kumar, A.; Rachubinski, R.A.; Frappier, L.; Hobman, T.C. Flavivirus infection impairs peroxisome biogenesis and early antiviral signaling. *J. Virol.* **2015**, *89*, 12349–12361. [CrossRef] [PubMed]

viruses

MDPI

Article

Mutagenesis of the Varicella-Zoster Virus Genome Demonstrates That VLT and VLT-ORF63 Proteins Are Dispensable for Lytic Infection

Shirley E. Braspenning [1], Robert Jan Lebbink [2], Daniel P. Depledge [3], Claudia M. E. Schapendonk [1], Laura A. Anderson [1], Georges M. G. M. Verjans [1], Tomohiko Sadaoka [4,*,†] and Werner J. D. Ouwendijk [1,*,†]

1 Department of Viroscience, Erasmus MC, 3015 GD Rotterdam, The Netherlands; s.braspenning@erasmusmc.nl (S.E.B.); c.schapendonk@erasmusmc.nl (C.M.E.S.); l.anderson@erasmusmc.nl (L.A.A.); g.verjans@erasmusmc.nl (G.M.G.M.V.)
2 Department of Medical Microbiology, University Medical Center Utrecht, 3508 GA Utrecht, The Netherlands; R.J.Lebbink-2@umcutrecht.nl
3 Institute of Virology, Hannover Medical School, 30625 Hannover, Germany; Depledge.Daniel@mh-hannover.de
4 Division of Clinical Virology, Center for Infectious Diseases, Kobe University Graduate School of Medicine, Kobe 650-0017, Japan
* Correspondence: tomsada@crystal.kobe-u.ac.jp (T.S.); w.ouwendijk@erasmusmc.nl (W.J.D.O.); Tel.: +81-78-382-6272 (T.S.); +31-10-7032134 (W.J.D.O.)
† These authors contributed equally to the work.

Citation: Braspenning, S.E.; Lebbink, R.J.; Depledge, D.P.; Schapendonk, C.M.E.; Anderson, L.A.; Verjans, G.M.G.M.; Sadaoka, T.; Ouwendijk, W.J.D. Mutagenesis of the Varicella-Zoster Virus Genome Demonstrates That VLT and VLT-ORF63 Proteins Are Dispensable for Lytic Infection. *Viruses* **2021**, *13*, 2289. https://doi.org/10.3390/v13112289

Academic Editors: Charles Grose, Ravi Mahalingam and Joel Rovnak

Received: 14 October 2021
Accepted: 12 November 2021
Published: 16 November 2021

Publisher's Note: MDPI stays neutral with regard to jurisdictional claims in published maps and institutional affiliations.

Abstract: Primary varicella-zoster virus (VZV) infection leads to varicella and the establishment of lifelong latency in sensory ganglion neurons. Reactivation of latent VZV causes herpes zoster, which is frequently associated with chronic pain. Latent viral gene expression is restricted to the VZV latency-associated transcript (VLT) and VLT-ORF63 (VLT63) fusion transcripts. Since VLT and VLT63 encode proteins that are expressed during lytic infection, we investigated whether pVLT and pVLT-ORF63 are essential for VZV replication by performing VZV genome mutagenesis using CRISPR/Cas9 and BAC technologies. We first established that CRISPR/Cas9 can efficiently mutate VZV genomes in lytically VZV-infected cells through targeting non-essential genes ORF8 and ORF11 and subsequently show recovery of viable mutant viruses. By contrast, the VLT region was markedly resistant to CRISPR/Cas9 editing. Whereas most mutants expressed wild-type or N-terminally altered versions of pVLT and pVLT-ORF63, only a minority of the resulting mutant viruses lacked pVLT and pVLT-ORF63 coding potential. Growth curve analysis showed that pVLT/pVLT-ORF63 negative viruses were viable, but impaired in growth in epithelial cells. We confirmed this phenotype independently using BAC-derived pVLT/pVLT-ORF63 negative and repaired viruses. Collectively, these data demonstrate that pVLT and/or pVLT-ORF63 are dispensable for lytic VZV replication but promote efficient VZV infection in epithelial cells.

Keywords: varicella-zoster virus; VLT; VLT-ORF63; CRISPR/Cas9; BAC mutagenesis

1. Introduction

Varicella-zoster virus (VZV) is a widespread human alphaherpesvirus and the causative agent of both varicella (chickenpox) and herpes zoster (HZ or shingles) [1]. During primary infection (varicella), VZV gains access to sensory ganglia where it establishes lifelong latency [2,3]. In one-third of infected individuals latent VZV will reactivate to cause HZ later in life, a condition often complicated by chronic pain (post-herpetic neuralgia) for which currently no effective treatment is available [4,5]. Although a potent subunit vaccine, Shingrix, has recently been licensed to prevent HZ in the elderly and immunocompromised [6], the latent virus itself is not targeted and thus reactivation events are not excluded. However, the development of therapeutics that eradicate the latent virus is hampered by a limited understanding of mechanisms governing latency and reactivation.

During lytic infection VZV expresses at least 136 distinct polyadenylated RNAs, both coding and non-coding, that are expressed in a coordinated cascade [7,8]. By contrast, VZV latency is characterized by expression of the VZV latency-associated transcript (VLT) and, to a lesser extent, low-level expression of two fusion transcripts of VLT and RNA 63 (VLT63-1 and VLT63-2) [9,10]. VLT and VLT63s are also expressed as *True-Late* transcripts during lytic VZV infection. However, unlike the single VLT isoform and two VLT63 isoforms expressed during latency, numerous lytic VLT ($_{lyt}$VLT) and VLT63 ($_{lyt}$VLT63) isoforms are produced by the usage of alternative transcription start-sites, read-through transcription and alternative splicing. Whereas $_{lyt}$VLT and $_{lyt}$VLT63-2 encode VLT protein (pVLT), $_{lyt}$VLT63-1 encodes for a pVLT-ORF63 fusion protein, in which canonical pORF63 is C-terminally linked to a N-terminal part of pVLT (118 of 136 aa) [10]. However, the function of pVLT and pVLT-ORF63 during lytic VZV infection remains elusive.

Similar to latency-associated transcripts of other alphaherpesviruses, the VLT locus is located partially antisense to ORF61 [the homologue of herpes simplex virus infected cell polypeptide 0 (*ICP0*)]. VLT RNA specifically downregulates the expression of RNA 61, which encodes the promiscuous transcriptional regulatory protein ORF61 (pORF61), in co-transfection experiments, suggesting that VLT may function in the establishment or maintenance of latency [9]. Ectopic in vitro expression of pVLT-ORF63 in latently VZV-infected human sensory neurons induces broad viral gene expression, suggesting that pVLT-ORF63 is a key regulator in the transition from latent to lytic infection [10]. However, the complexity of $_{lyt}$VLT transcription and the genomic position of VLT antisense to the essential gene ORF61 [11,12], hinders complete deletion of the VLT locus from the viral genome and complicates studies into VLT function.

Mutant herpesviruses can be generated using bacterial artificial chromosome (BAC)-based mutagenesis or by clustered regularly interspaced short palindromic repeats (CRISPR) CRISPR-associated protein 9 (Cas9) (CRISPR/Cas9) editing [13,14]. BAC-based mutagenesis remains the gold standard and allows for precise nucleotide insertions, deletions or substitutions to study viral protein functions, but are restricted to the genetic background of the parental BAC clone. By contrast, CRISPR/Cas9 editing results in the insertion or deletion of one to several nucleotides and can be used to edit viruses regardless of their origin, thereby providing a potential tool to combat herpesvirus infections in vivo. Furthermore, the application of CRISPR/Cas9 genome editing during lytic infection in vitro allows for direct competition between progeny viruses (natural selection), and hence the resulting viruses provide insight into the biological relevance of the targeted genomic region for viral replication.

In this study, we investigated the role of pVLT and pVLT-ORF63 during lytic VZV infection in epithelial cells using mutant viruses obtained through complementary CRISPR/Cas9 genome editing and BAC mutagenesis.

2. Materials and Methods

2.1. Cells and Viruses

Human retinal pigmented epithelial ARPE-19 cells were cultured in a 1:1 (v/v) mixture of DMEM (Lonza, Verviers, Belgium) and Ham's F12 (Gibco, Breda, The Netherlands) supplemented with 10% heat-inactivated FCS, and 5% penicillin-streptomycin with L-glutamine (Lonza, Verviers, Belgium). Human melanoma MeWo cells (ATCC HTB-65) and MeWo-Cre cells expressing Cre-recombinase [15] were cultured in DMEM + GlutaMax-I (Thermo Fisher Scientific, Tokyo, Japan) supplemented with 8% FCS. VZV strain EMC-1 is a low-passage clinical isolate and served as the parental virus for CRISPR/Cas9-based gene editing. BAC-derived VZV was based on the pOka strain and generated as described below. Cell-free VZV was obtained from ARPE-19 cells, as described [16,17]. All cells and viruses were maintained in a humidified CO_2 incubator at 37 °C.

2.2. DNA Isolation from Varicella-Zoster Virus (VZV)-Infected Cells

VZV-infected ARPE-19 cells were subjected to DNA isolation using the QIAamp DNA mini kit (Qiagen, Venlo, The Netherlands) according to manufacturer's instructions. DNA was eluted in 200 μL and used for PCR and Sanger sequencing.

2.3. RNA Isolation and cDNA Synthesis

VZV-infected cells were harvested in 1 mL Trizol (Thermo Fisher Scientific, Breda, The Netherlands), mixed with 200 μL chloroform and centrifuged for 15 min at 12,000× *g* at 4 °C. From the aqueous phase, RNA was isolated using the RNeasy Mini kit (Qiagen, Venlo, The Netherlands) according to manufacturer's instructions, including on-column DNAse I treatment. RNA concentration was determined using a Nanodrop spectrophotometer, and residual DNA was removed using the Turbo DNA-free kit (Ambion, Breda, The Netherlands). For cDNA synthesis, maximum 5 μg RNA was reverse transcribed with Superscript IV reverse transcriptase (RT+) and oligo-dT primers (Thermo Fisher Scientific, Breda, The Netherlands). Negative control samples were obtained by performing the same cDNA synthesis reaction in the absence of reverse transcriptase (RT−).

2.4. Quantitative Polymerase Chain Reaction (PCR) Analysis

Taqman quantitative PCR (qPCR) was performed in duplicate on RT- and RT+ cDNA on a 7500 Taqman PCR system using 4× Taqman Fast Advanced Master mix (Applied Biosystems, Breda, The Netherlands). Primer-probe sets for VZV genes RNA 29, RNA 61, RNA 63, VLT2_3 and VLT3_4 have been described previously [9] and the primer-probe set for ORF60 was described and validated in this study. Primer sequences are given in Table S2.

2.5. PCR and Sanger Sequencing

PCR was performed on extracted DNA or cDNA using AmpliTaq Gold DNA Polymerase (Thermo Fisher Scientific, Breda, The Netherlands) using primer pairs as indicated in Table S2 and the following PCR cycling parameters: initial denaturation at 95 °C for 10 min, followed by 40 cycles of 30 s at 95 °C, 30 s at 56 °C and 1 min at 72 °C. Extension was completed with a final step of 10 min at 72 °C. Resulting amplicons were either purified from gel using the Qiaquick Gel Extraction Kit (Qiagen, Venlo, The Netherlands) or used directly for sequence reaction, using the BigDye 3.1 Cycle Sequencing Kit (Applied Biosystems, Breda, The Netherlands) on the ABI PRISM 3130 XL Genetic Analyser.

2.6. Plasmid Construction and Generation of ARPE-19 Cells Stably Expressing pVLT

The pVLT coding sequence was amplified from cDNA of VZV EMC-1 infected ARPE-19 cells with NheI_VLT_Fw and VLT_Rv_BamHI primers (Table S2), cloned into the pcDNA3.1(+) backbone and the incorporated sequence was confirmed by Sanger sequencing. Stable cell lines ARPE-19 pVLT were then created by transfection with Genius transfection reagent (Westburg, Leusden, The Netherlands) of ARPE-19 cells at 70–80% confluency and subsequent selection with 1000 μg/mL geneticin (Gibco, Breda, The Netherlands) for at least 4 weeks. Expression of pVLT-encoding mRNA in the was confirmed by RT-qPCR as described above.

2.7. Generation of SpCas9-sgRNA Lentiviruses

crRNA target sequences (Table S1) were designed manually or using the CRISPOR website (http://crispor.tefor.net, accessed on 13 July 2017) and cloned as sgRNA downstream a human U6 promoter in the pSicoR-CRISPR-PuroR or pSicoR-CRISPR-BlastR vector [14]. All vectors were sequence verified by Sanger sequencing. For lentiviral transductions, viruses were produced in 293 T cells in 24-well plates using standard lentiviral production protocols and third-generation packaging vectors.

2.8. Transduction and Selection of ARPE-19 Cells

ARPE-19 cells were seeded one day prior to lentiviral transduction in a 48-well plate at 70% confluency. The next day, lentiviral stocks were diluted 1:5 in culture medium containing 4 μg/mL polybrene and spin-inoculated (1000× *g*, 15 min) on the ARPE-19 cells. Two days after transduction, medium was replaced for selection medium containing either 0.94 μg/mL puromycin (InvivoGen, Toulouse, France) or 20 μg/mL blasticidin (InvivoGen, Toulouse, France) and stable cell lines were obtained by continuous selection for 3 weeks. To generate ARPE-19 cells expressing two sgRNAs, cells stably expressing the first sgRNA were lentivirally transduced to express the second sgRNA, followed by a second round of antibiotic selection.

2.9. Generation of CRISPR/Cas9-Based Mutant Viruses

ARPE-19 cells stably expressing one or two sgRNA(s) and SpCas9 were inoculated with cell-free VZV EMC-1 (MOI = 0.01) in a 12-well plate. Cell-free virus was harvested at 2 days post infection (dpi) and a 2-fold serial dilution was used to infect fresh monolayers of ARPE-19 cells in a 96-well plate. In case of sgRNA VLT-9, the obtained cell-free VZV was also serially diluted on ARPE-19 cells stably expressing pVLT. After 5–6 days incubation, plates were screened and wells containing a single plaque were passaged on fresh ARPE-19 cells in a 12-well plate. Obtained virus cultures were subjected to another 2 (double sgRNAs) or 3 (sgRNA VLT-9) rounds of plaque purification on ARPE-19 cells. The resulting viruses were analyzed by Sanger sequencing of the VLT target region.

2.10. Illumina Sequencing of CRISPR/Cas9-Derived Mutant Viruses

ARPE-19 cells were infected with VZV EMC1-pVLTwt, EMC1-pVLTΔ1aa, EMC1-pVLTshift or EMC1-pVLTstop, and at 50–70% CPE infected cells were harvested. The cell pellet was resuspended in 200 μL PBS + 0.1% sodium deoxycholate, 50 μL 25 mM $MgCl_2$ and 2.5 μL Omnicleave (Lucigen, Middleton, WI, USA) and incubated for 1 h at 37 °C and DNA was subsequently isolated by the QIAamp DNA mini kit (Qiagen, Venlo, The Netherlands) as described. Paired-end Illumina sequencing libraries were prepared from 100 ng DNA using the KAPA HyperPlus library preparation kit (Roche, Almere, The Netherlands) according to the manufacturer's instructions with KAPA Unique Dual-Indexed Adapters Kit (Roche) for multiplexing. Libraries were pooled with unrelated samples and run on an Illumina MiSeq version 3 flowcell (2 × 300 cycles) (Illumina, San Diego, CA, USA). Resulting sequencing data were demultiplexed and trimmed to remove adapter sequences and low-quality 3′ nucleotides using TrimGalore(https://www.bioinformatics.babraham.ac.uk/projects/trim_galore/, v0.6.5). Trimmed sequence data were aligned against the VZV strain Dumas reference genome using bbmap (https://sourceforge.net/projects/bbmap/,) and parsed using SAMTools [18] and BEDTools [19]. Bam-readcount (https://github.com/genome/bam-readcount, v1.0.0), with the flag –d 1000000, was subsequently used to identify all reported SNPs and Indels prior to filtering using a custom script, variant_caller_v1.0.py, available at https://github.com/DepledgeLab/NAGATA.

2.11. Generation of Bacterial Artificial Chromosome (BAC) Mutant and Repaired Genomes by Redα/β-Mediated Linear Recombination

The pBpOka100798-104820 plasmid was generated by PCR amplification of the VLT locus (100,798–104,820; pOka [AB097933]) of pOka-BAC DNA [20] using primers pOka100798ecoF and pOka104820salR (Table S2) and subsequent cloning into pBlueScript II SK(-) (Agilent Technologies, Tokyo, Japan). The pBpOka100798-104820VLTM1I plasmid containing G->A mutation at position 102,824—changing the ATG (methionine; M) start codon of pVLT into ATA (isoleucine; I)—was generated from the pBpOka100798-104820 plasmid with primer VLTG3A (Table S2) using QuickChange Lightning Multi Site-Directed Mutagenesis Kit (Agilent Technologies) according to the manufacturer's instruction. Next, a DNA fragment containing an ampicillin-resistant gene and *sacB* gene flanked by 102,501–102,550 and 104,448–104,500 of pOka was PCR amplified from the pST76A-SR plasmid [21] using primers pOka102501ampsacF and pOka104500ampsacR (Table S2) and subsequently transformed into GS1783 *E. coli* harboring pOka-BAC genome [22], resulting in the replacement of 102,551–104,447 of pOka-BAC by an ampicillin resistant gene and *sacB* gene (pOka-BACampsac1st). To generate the pOka-BACVLTM1I genome a DNA fragment of 100,798–104,820 of pOka amplified from pBpOka100798-104820VLTM1I plasmid was transformed into GS1783 harboring pOka-BACampsac1st, and followed by negative selection with 5% sucrose. The repaired BAC genome pOka-BACVLTM1IR was similarly generated based on the pOka-BACVLTM1I. Briefly, the 102,551–104,447 region of pOka-BACVLTM1I was replaced by an ampicillin-resistant gene and *sacB* gene in GS1783 *E. coli*, resulting in pOka-BACampsac2nd and the pOka-BACVLTM1Irev genome was then generated by transformation of a DNA fragment of 100,798–104,820 of pOka amplified from pBpOka100798-104820 plasmid into GS1783 pOka-BACampsac2nd, followed by negative selection with 5% sucrose. Finally, pOka-BACVLTM1I and pOka-BACVLTM1IR were purified using the Genopure Plasmid Maxi Kit (Roche Diagnostics, Tokyo, Japan), subjected to restriction fragment length polymorphism analysis using BamHI or EcoRI and Sanger sequencing of the recombined region.

2.12. *Reconstitution of pOka-VLTM1I and pOka-VLTM1IR Viruses from BAC Genomes*

VZV strain pOka-based VLTM1I and its repaired VLTM1IR recombinant viruses were reconstituted in MeWo cells by transfection of pOka-BACVLTM1I and pOka-BACVLTM1IR genomes using Lipofectamine2000 (Thermo Fisher Scientific) and the BAC cassette within the reconstituted viruses was subsequently removed in MeWo-Cre cells [23,24].

2.13. *Immunofluorescence*

ARPE-19 cells on glass coverslips were inoculated with freshly harvested cell-free VZV EMC 1 edited by sgRNA8 2 and incubated for 3 days at 37 °C. Infected cells were fixed with 4% paraformaldehyde (PFA), permeabilized for 10 min with 0.1% Triton X-100 in PBS, blocked with 5% goat serum diluted in PBS + 0.05% Tween and incubated with monoclonal mouse IgG1 anti-ORF8 antibody (1:100, generously provided by Dr. S. Jonjic, University of Rijeka, Rijeka, Croatia [16]) and monoclonal mouse IgG2b anti-VZV glycoprotein E antibody (MAB8612, Millipore, Amsterdam, The Netherlands) diluted PBS + 0.05% Tween 20 (PBS-T) overnight at 4 °C. Cells were washed with PBS-T and incubated with AF488-conjugated goat anti-mouse IgG2b and AF555-conjugated goat anti-mouse IgG1 antibodies (Thermo Fisher Scientific, Breda, The Netherlands) diluted 1:500 in PBS-T. Finally, the coverslips were washed, incubated for 5 min in PBS with Hoechst 33342 (1:1000, Life Technologies, 20 mM), washed in PBS and mounted using Prolong Gold Antifade Mounting medium (Thermo Fisher Scientific, Breda, The Netherlands). Stained cells were analyzed using a Zeiss LSM 700 confocal laser scanning microscope (Zeiss, Oberkochen, Germany) with a magnification of 100× or 400×. Photoshop CC 2021 software (Adobe, San Jose, CA, USA) was used to adjust brightness and contrast.

2.14. *Infectious Focus Assay*

ARPE-19 cells were plated in a 24-well plate and the next day cell monolayers were infected with 100 PFU/well of cell-free VZV. At 6 dpi, plates were washed and fixed with 4% PFA in PBS. Subsequently, cells were permeabilized with 0.1% Triton X-100 in PBS for 10 min, blocked with 5% normal goat serum in PBS-T for 30 min, incubated with mouse anti-VZV glycoprotein E antibody (MAB8612, Millipore, Amsterdam, The Netherlands) diluted 1:3200 in PBS-T overnight at 4 °C. The next day, the plates were washed with PBS-T, incubated for 1 h at room temperature with 1:200 biotinylated rabbit anti-mouse Ig antibody (DAKO, Amstelveen, The Netherlands) in PBS-T, washed and incubated for 1 h at room temperature with 1:300 streptavidin-HRP (DAKO, Amstelveen, The Netherlands) in PBS-T, washed once in PBS-T and twice with PBS. Finally, the signal was visualized using 3-amino-9-ethylcarbazole (AEC) as a substrate. Infectious foci were measured using the Immunospot S6 Ultimate UV Image Analyzer and foci size was determined using the Immunospot software (Cellular Technology Limited, Cleveland, OH, USA).

2.15. Cell-Associated VZV Titer

Monolayers of ARPE-19 cells in a 24-well plate were infected with 1000 PFU/well VZV, harvested at 3 dpi and titrated on ARPE-19 cells in 96-well plates using five-fold serial dilutions. After 3 days, infectious virus titers were calculated using the Spearman–Karber formula and expressed as TCID50 per cm^2 cell culture.

2.16. Flow Cytometry

ARPE-19 cells were plated in a 96-well plate one day prior to infection. Cells were infected with 150 PFU/well cell-free virus and harvested at 24 h post infection (hpi), 48 hpi and 72 hpi. Subsequently, cells were fixed and permeabilized with BD Cytofix/Cytoperm, stained for VZV glycoprotein E (MAB8612, Millipore) in BD PermWash for 30 min at 4 °C, stained with secondary APC-conjugated goat anti-mouse Ig antibody (BD Biosciences, Vianen, The Netherlands) for 30 min at 4 °C and finally resuspended for analysis. Frequency of VZV-infected (APC-positive) cells was determined. Experiments were performed on a BD FACSLyric flow cytometer and the data was analyzed using FlowJo software (BD Biosciences).

2.17. Western Blotting

Mock- or VZV-infected ARPE-19 cells at 5 dpi were incubated in RIPA lysis buffer (0.01 M Tris-HCl [pH 7.4], 0.15 M NaCl, 1% sodium deoxycholate, 1% NP-40 and 0.1% SDS) on ice for 15 min, sonicated in a water bath for 10 min, and centrifuged at 20,000× *g* for 15 min. Proteins were separated on a 4–12% Nu-PAGE Gel (Thermo Fisher Scientific) and transferred onto PVDF membranes (0.2 µm). The membrane was blocked in 5% (*w*/*v*) skim milk in PBS + 0.1% Tween 20 at room temperature for 1 h and subsequently stained overnight at 4 °C with polyclonal rabbit anti-pVLT (1:2000) [9], polyclonal rabbit anti-pORF63 (1:30,000) [9] or monoclonal mouse anti-α-tubulin (1:30,000, clone B-5-1-2) antibodies diluted in blocking buffer. Next day, the membranes were stained for 1 h at room temperature with donkey anti-rabbit IgG HRP-linked or sheep anti-mouse IgG HRP-linked (GE Healthcare Bio-Sciences, Tokyo, Japan) antibodies diluted 1:3000 in blocking buffer. The signal was visualized using Chemi-Lumi One Super (Nacalai Tesque, Inc., Kyoto, Japan) and captured using LAS4000mini (GE Healthcare Bio-Sciences).

2.18. Statistical Analysis

Figures show individual data points or mean +/− SEM, all statistical analyses were performed with Graphpad Prism 9 software using the statistical test indicated in the figure legends, with * = $p < 0.05$, ** = $p < 0.01$, *** = $p < 0.001$. If no statistics are given, the difference was not significant.

3. Results

3.1. Successful Editing of Non-Essential VZV Genes by CRISPR/Cas9

To establish whether CRISPR/Cas9 can be used to edit VZV genomes during lytic infection, we initially targeted two VZV genes shown to be non-essential in cell culture: ORF8 and ORF11 [25,26]. To this end, we generated stable transduced human retina epithelial cells (ARPE-19) expressing both the *Streptococcus pyogenes* Cas9 (SpCas9) endonuclease and a single guide RNA (sgRNA) targeting the genomic region of interest (Figure 1A and Table S1). To counter the possibility of inefficient sgRNAs [27], we designed two sgRNAs for each ORF. We infected the stable cell lines with cell-free VZV (strain EMC-1), isolated DNA 2 days post-infection (dpi), and performed PCR and Sanger sequencing of the target regions (Figure 1A). Both ORF8 sgRNAs (ORF8-1 and ORF8-2), and one ORF11 sgRNA (ORF11-1) successfully induced editing of the VZV genome, as indicated by non-clonal Sanger traces around the sgRNA target site (Figure 1B,C). sgRNA ORF11-2 did not detectably induce VZV genome editing. To determine whether CRISPR/Cas9 edited VZV is viable and editing leads to functional disruption of viral protein expression, we isolated sgRNA ORF8-2 edited VZV, infected ARPE-19 cells and stained infectious foci for ORF8 protein (pORF8). While the majority of the foci completely lacked pORF8 expression, various levels of pORF8 were still detectable in some foci (Figure 1D). Sanger sequencing analysis on the mutant virus isolates ($n = 6$ per sgRNA), obtained after a single round of limited dilution and plaque purification, detected insertions or deletions (indels) in all analyzed isolates edited using sgRNA ORF8-2 or sgRNA ORF11-1 (Figure 1E). However, as CRISPR/Cas9 editing was applied during active lytic infection in the presence of a multitude of viral genomes, the edited populations after a single plaque purification still contained a mixture of genomes.

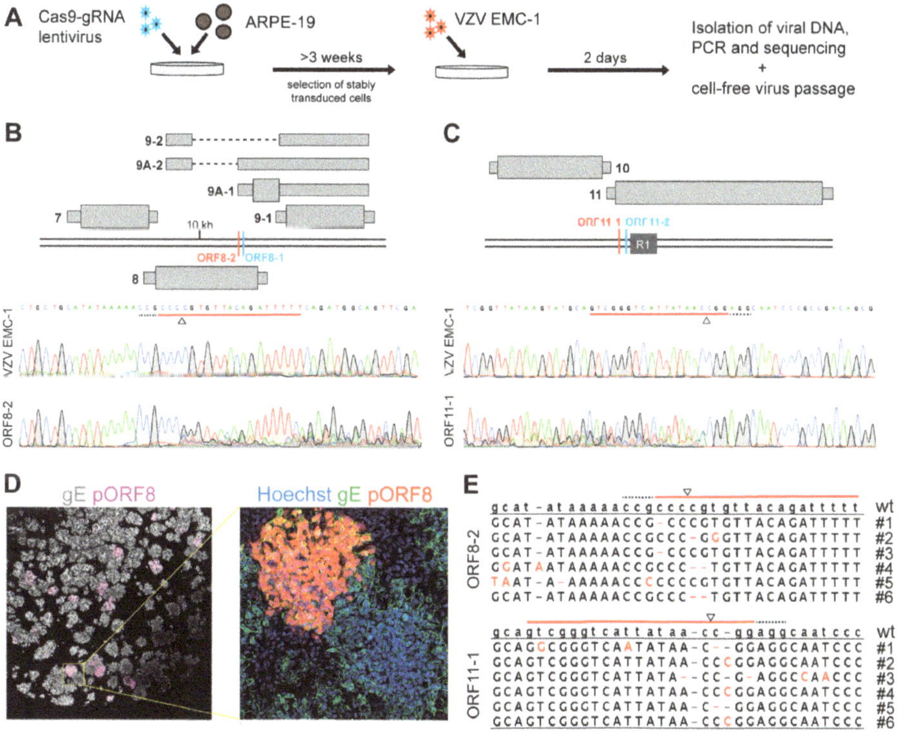

Figure 1. Editing of varicella-zoster virus (VZV) genomes using CRISPR/Cas9. (**A**) Experimental layout: ARPE-19 cells were transduced with a lentivirus encoding for both SpCas9 and a sgRNA, and selected for at least 3 weeks with appropriate antibiotics. Stably transduced cells were infected with cell-free VZV EMC-1 for two days, viral DNA was isolated and the target region was amplified by PCR and sequenced. (**B**,**C**) Top: Schematic illustration of the target region of ORF8 (**B**) and ORF11 (**C**): thick double black lines represent the VZV genome, currently annotated RNAs are depicted in grey—wide boxes are coding sequences (CDS), thin boxes are untranslated regions (UTRs) and dashed lines are intronic sequences—and DNA target sites of sgRNAs ORF8-1 and ORF8-2 (**B**) or sgRNAs ORF11-1 and ORF11-2 (**C**) are depicted. Bottom: Sanger sequencing traces showing the sgRNA target regions (red), PAM sequence (dashed black) and cleavage site with triangle for wild-type VZV EMC-1 DNA or from DNA extracted from VZV EMC-1 infected stably transduced ORF8-2 (**B**) and ORF11-1 cells (**C**). (**D**) Representative images of infected ARPE-19 cells with cell-free virus isolated from infected ORF8-2 cells stained by immunofluorescence for glycoprotein E (gray/green) and pORF8 (magenta/red). Nuclei were counterstained with Hoechst 33342 (blue), with 20× magnification (left) or 200× magnification (right). (**E**) Consensus Sanger sequences of 6 individual infectious foci after a single round of cell-free virus isolation and plaque purification following infection of stably transduced ORF8-2 (top) and ORF11-1 (bottom) cells. sgRNA target regions (red line), PAM sequence (dashed black line) and cleavage site with triangle are depicted.

To establish a workflow for the isolation of near-clonal mutant virus populations using CRISPR/Cas9 editing, we infected gRNA ORF8-2 expressing cells with VZV EMC-1, isolated the cell-free virus and performed three successive rounds of cell-free virus isolation, limiting dilution and plaque purification (Figure 2A). Sanger sequencing at each intermediate population showed that after 2–3 rounds of plaque purification, sequence traces were clonal, suggesting highly pure (or clonal) virus population (Figure 2B). In total, we recovered four mutant VZV isolates, of which three contained disruption of pORF8, as predicted by their nucleotide sequences and confirmed by immunofluorescence staining of ARPE-19 cells infected by these viruses. (Figure 2C). Thus, we established a CRISPR/Cas9 genome editing approach that can be applied to target VZV genomes during lytic infection and showed that mutant viruses can be purified by successive plaque purifications.

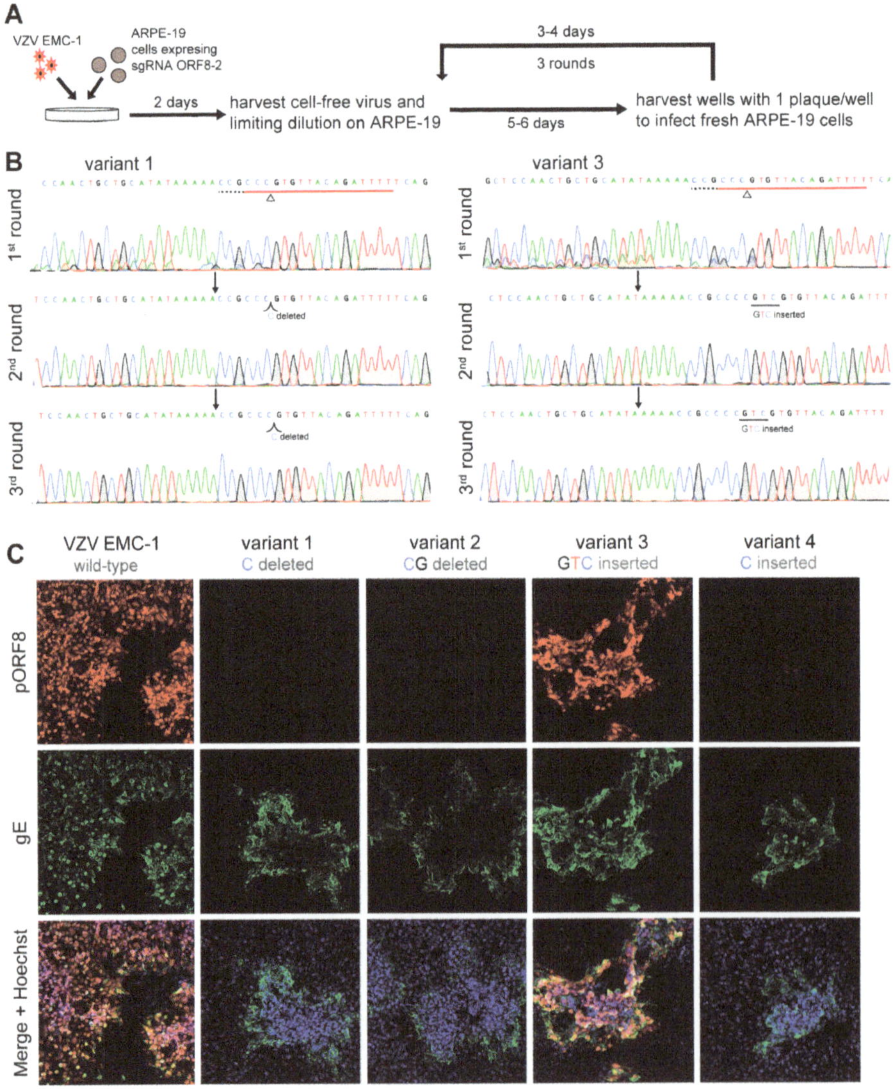

Figure 2. Plaque purification of VZV edited by gRNA ORF8-2. (**A**) Experimental design: Stably transduced ARPE-19 cells expressing sgRNA ORF8-2 were infected with VZV EMC-1. Two days post-infection, cell-free virus was harvested, and subjected

to three subsequent rounds of plaque purification. (**B**) Sanger sequencing after each round of plaque purification for variant 1 and 3 showing the sgRNA target regions (red), PAM sequence (dashed black), cleavage site with triangle and identified mutation. (**C**) Representative images of ARPE-19 cells infected with VZV EMC-1 or 4 ORF8 variant viruses derived from sgRNA ORF8-2 cells by plaque purifications stained by immunofluorescence for glycoprotein E (green) and pORF8 (red). Nuclei were counterstained with Hoechst 33342 (blue), with 10× magnification.

3.2. The VZV Latency-Associated Transcript (VLT) Locus Is Resilient to CRISPR/Cas9 Genome Editing

To target the VZV VLT locus and generate VLT mutant viruses, we designed six sgRNAs targeting three distinct regions of VLT that are present in all $_{\text{lyt}}$VLT isoforms, as well as in $_{\text{lyt}}$VLT63-1 and $_{\text{lyt}}$VLT63-2 (Figure 3A and Table S1). Two sgRNAs—VLT-1 and VLT-2—targeted VLT exon 1 and potentially disrupt splicing from upstream exons into VLT exon 1. Two sgRNAs—VLT-3 and VLT-4—were directed close to the only in-frame pVLT ATG start codon in located VLT exon 2 to interfere with pVLT/pVLT-ORF63 expression. Finally, two sgRNAs—VLT-5 and VLT-6—targeted the first intron of VLT, into which a large GFP-expression cassette was inserted previously by BAC-based mutagenesis [28], suggesting mutations in this region are tolerated. We generated stable cells lines for each sgRNA, infected these cells with cell-free VZV EMC-1 and performed Sanger sequencing on viral DNA extracted at 2 dpi. Strikingly, none of the sgRNAs induced detectable VZV genome editing (Figure 3B), independent of the multiplicity of infection or viral DNA replication (data not shown).

Conformational structures of the genome might render mammalian genes less accessible to endonucleases and thus difficult to target by CRISPR/Cas9 [29,30]. For some of those genes, combination of two low-efficient sgRNAs leads to efficient editing of the target region [31,32]. Hence, we generated four cell lines stably expressing a combination of two sgRNAs; one in exon 1, and the other in exon 2 (Figure 3A,C). VZV EMC-1 infection of all double sgRNA expressing cell lines lead to efficient deletion of VLT exon 1, intron 1 and part of exon 2 (±450 bp) from the VZV genome, as indicated by a smaller PCR product of the extracted viral DNA (Figure 3C). Next, we infected these double sgRNA cell lines to generate four groups of VLT mutant viruses (hereafter named A–D), followed by three successive rounds of cell-free virus isolation, limiting dilution and plaque purifications on parental ARPE-19 cells (Figure 3D). We obtained 12 virus isolates—three for each combination—on which we performed Sanger sequencing to determine the DNA sequence of the VLT region. Eleven out of 12 isolates contained deletions within the VLT region that correspond to the combination of sgRNAs used; within the four groups, 1–2 nucleotide differences were observed between mutant viruses (Figure 3E). In silico predictions on sequenced $_{\text{lyt}}$VLT/$_{\text{lyt}}$VLT63 transcripts showed that in 9 out of 11 isolates an upstream ATG start codon was positioned in-frame with the remainder of canonical pVLT/pVLT-ORF63 coding sequences, potentially resulting in N-terminally extended and altered proteins (Figure 3F,G). Importantly, two mutant isolates—C14 and D23—expressed $_{\text{lyt}}$VLT/$_{\text{lyt}}$VLT63 transcripts that were not predicted to encode for pVLT/pVLT-ORF63 (or variants thereof) (Figure 3E).

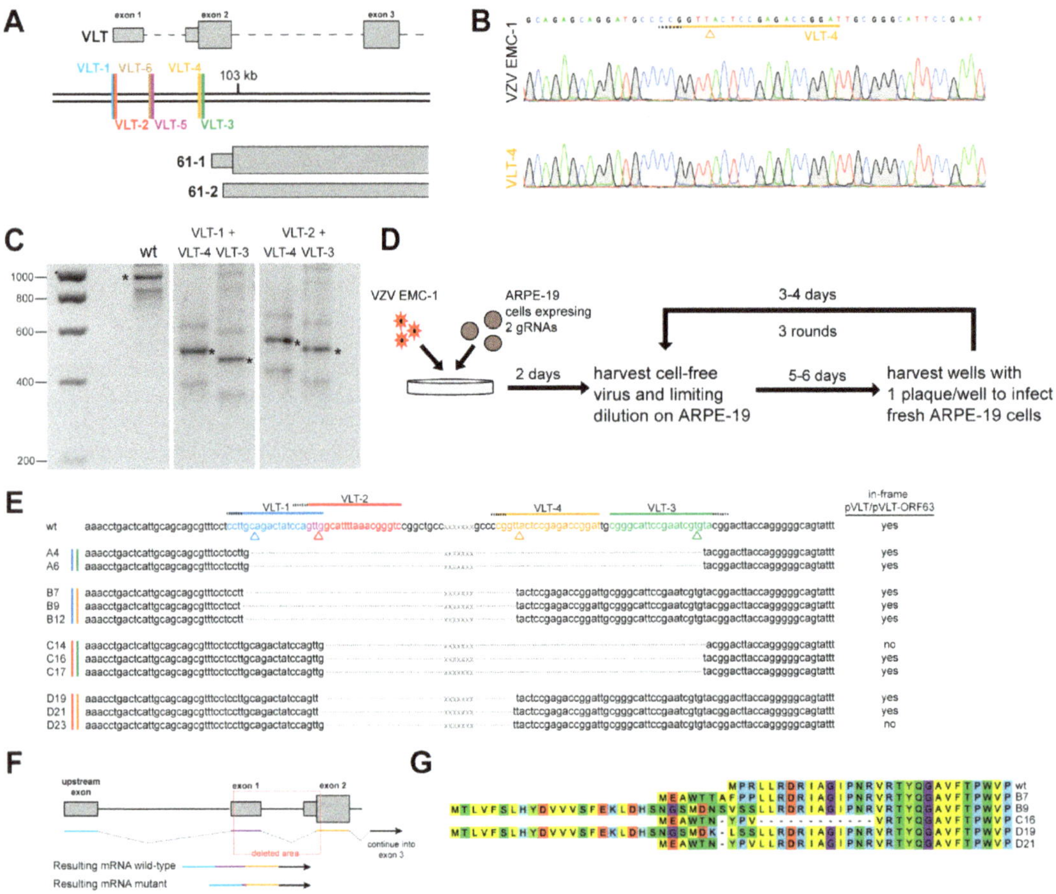

Figure 3. Generation of VZV latency-associated transcript (VLT) mutant viruses using combinations of two different sgRNAs. (**A**) Schematic illustration of the VLT target region: thick double black lines represent the VZV genome, annotated RNAs are depicted in grey—wide boxes are coding sequences (CDS), thin boxes are untranslated regions (UTRs) and dashed lines are intronic sequences—and the DNA target sites of sgRNAs VLT-1 to VLT-6 are depicted in differentially colored lines. (**B**) Sanger sequencing traces showing the sgRNA target region (yellow), PAM sequence (dashed black) and cleavage site (colored triangle) for wild-type VZV EMC-1 DNA (top) and DNA extracted from VZV EMC-1 infected stably transduced VLT-4 cells (bottom). (**C**) PCR products from VLT upstream exon C to VLT exon 2 on DNA isolated from parental ARPE-19 cells or from cells stably transduced with a combination of sgRNAs as indicated 2 days post-infection with VZV EMC-1. Correct band is indicated by an asterisk, additional bands result from duplication of primer binding site in R5. (**D**) Experimental design: Stably transduced ARPE-19 cells expressing two sgRNAs targeting the VLT region were infected with VZV EMC-1. Two days post-infection, cell-free virus was harvested, and subjected to three subsequent rounds of plaque purification. (**E**) VZV DNA sequence around sgRNA target sites of isolated plaque purified viruses divided into 4 groups: A—combination of sgRNA VLT-1 and sgRNA VLT-3, B—combination of sgRNA VLT-1 and sgRNA VLT-4, C—combination of sgRNA VLT-2 and VLT-3, D—combination of sgRNA VLT-2 and VLT-4. Dashed lines represent deleted nucleotides, whereas 'xxx' signifies the region in between the sgRNAs target sequences deleted in all mutants. PAM sequences (dashed black) and expected cleavage sites (colored triangles) are indicated. (**F**) Schematic illustration of VLT transcript generation from wild-type or mutant viruses from any upstream VLT exon (blue line), into exon 1 (purple line), exon 2 (orange line) and the remainder of VLT (black arrow). (**G**) Alignment of proteins generated in mutant viruses by in-frame splicing of an upstream ATG with the remainder of pVLT/pVLT-ORF63. Amino acids are colored to aid visual comparison.

3.3. Growth Characterization of VZV VLT Mutant Viruses

Next, we analyzed the expression levels of $_{\text{lyt}}$VLT and neighboring VZV transcripts in VZV VLT mutant virus isolates compared to VZV EMC-1 by RT-qPCR. Relative expression levels of $_{\text{lyt}}$VLT/$_{\text{lyt}}$VLT63 RNAs were reduced for all mutants compared to parental VZV EMC-1, especially in pVLT/pVLT-ORF63(-) isolates, C14 and D23 (Figure 4A). However, expression of RNA 60 and RNA 61 was also reduced in all mutant viruses, whereas expression of RNA 63 was not affected (Figure 4A). Together, these data show that large deletions in the VLT region impact relative expression of VLT and also its flanking genes, presumably by affecting regulatory sequences in the 5′UTR of RNA 60 and/or 3′UTR of RNA 61.

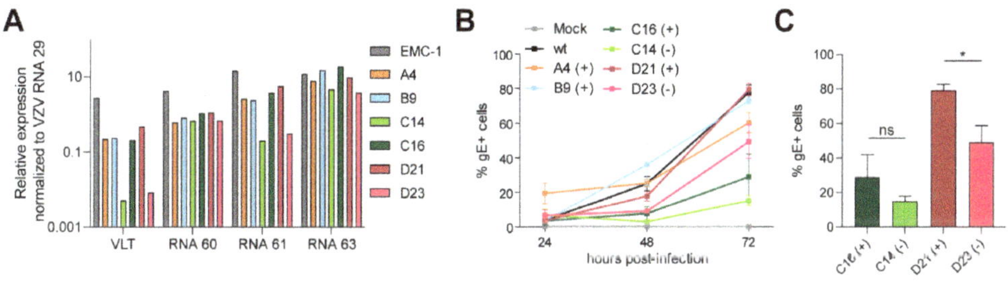

Figure 4. Characterization of VZV VLT mutant viruses generated using combinations of two sgRNAs. (**A**) Relative expression level of VLT RNAs, RNA 60, RNA 61 and RNA 63 in ARPE-19 cells asynchronously infected with either wild-type VZV EMC-1 or mutants A4, B9, C14, C16, D21 and D23. Expression levels were normalized to VZV RNA 29 to compensate for differences in infection. (**B**) Representative growth curves of wild-type virus, one mutant per group and two pVLT/pVLT-ORF63(-) viruses by flow cytometry of VZV infected cells (n = 2 independent experiments, n = 3 technical replicates). Percentages shown are VZV glycoprotein E (gE)—positive cells at 24, 48 and 72 hpi. None of the mutant viruses significantly differed from wild-type (one-way ANOVA). (**C**) Comparison between two pVLT/pVLT-ORF63(+) viruses (C16 and D21) and their pVLT/pVLT-ORF63(-) counterparts (C14 and D23) at 72 hpi. ns: not significant and * $p < 0.05$ by unpaired Student's *t*-test.

Subsequently, we compared viral growth of these VLT mutant isolates and VZV EMC-1 in ARPE-19 cells by flow cytometry. We observed substantial variability between biological replicates, which was independent of the virus used and most likely caused by the high particle to PFU ratio inherent to cell-free VZV stocks [33,34]. Although none of the mutant virus isolates significantly differed from VZV-EMC-1 (Figure 4B), the two pVLT/pVLT-ORF63(-) mutants (C14 and D23) appeared to be impaired in replication compared to pVLT/pVLT-ORF63(+) counterparts from the same group (C16 and D21 respectively) (Figure 4C). However, since these mutant viruses contain relatively large deletions in the VLT region that also affect expression of VLT, RNA 60, and RNA 61, we cannot exclude the possibility that the observed phenotype is partially unrelated to the absence of pVLT/pVLT-ORF63.

3.4. Generation of VZV pVLT/pVLT-ORF63 with Indel Mutations

To generate indel mutations within VLT, rather than larger deletions, we designed two additional sets of sgRNAs directed to the VLT locus. One set (n = 3 sgRNAs, designated VLT-7, -8 and -9) targets the 3′end of VLT exon 2, and could alter the pVLT/pVLT-ORF63 protein sequence (Figure 5A and Table S1). The other set (n = 3 sgRNAs, designated VLT-10, -11 and -12) targets one of the upstream exons of $_{\text{lyt}}$VLT/$_{\text{lyt}}$VLT63 [7]; presumably, deletion of only one of multiple possible alternative upstream exons would be compatible with VZV viability, as many other $_{\text{lyt}}$VLT/$_{\text{lyt}}$VLT63 variants remain available (Figure 5A and Table S1). Stable cell lines expressing these single sgRNAs and SpCas9 were infected with cell-free VZV EMC-1 and the viral DNA was sequenced at 2 dpi. None of the sgRNAs targeting the upstream $_{\text{lyt}}$VLT/$_{\text{lyt}}$VLT63 exon B induced any detectable mutations, nor did

2 out of 3 sgRNAs targeting VLT exon 2. Only a single sgRNA targeting exon 2 (VLT-9), induced efficient CRISPR/Cas9-based VZV genome editing (Figure 5B).

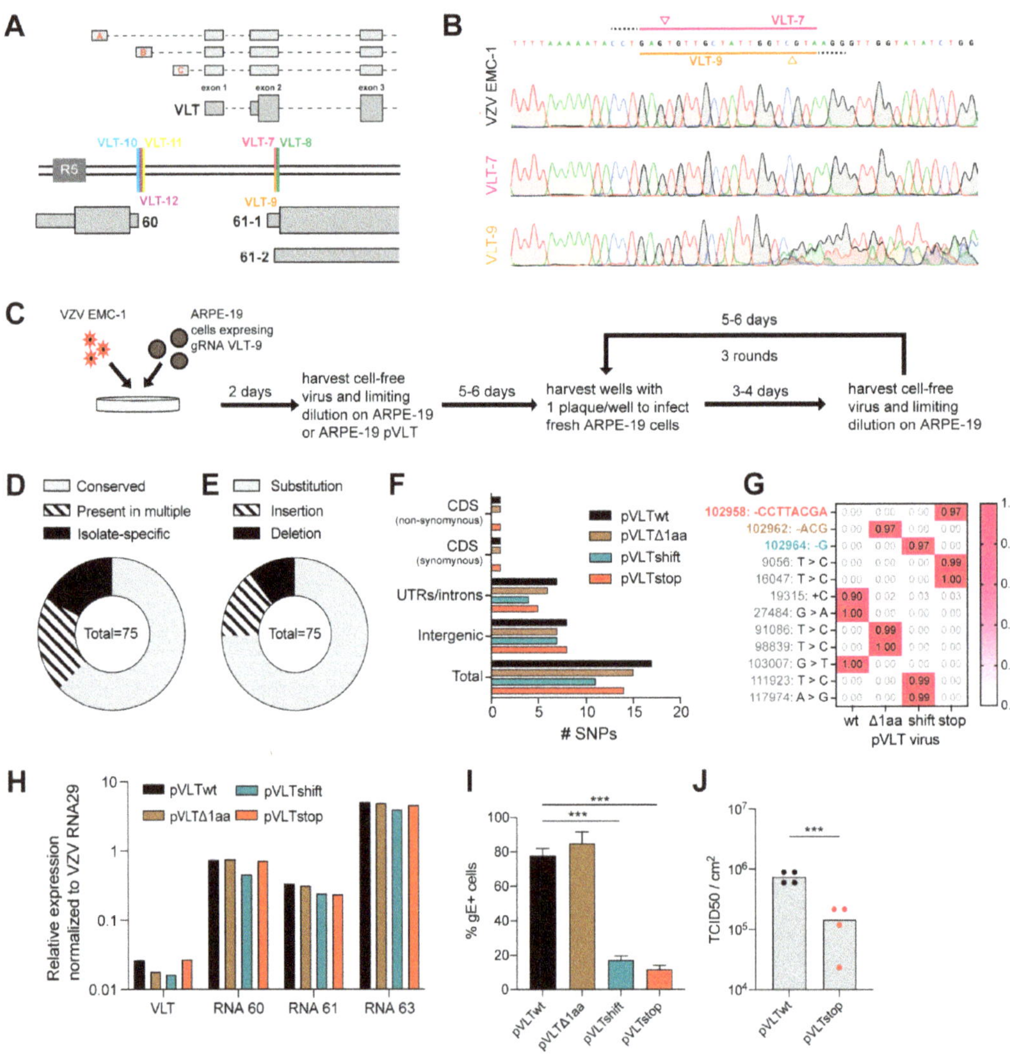

Figure 5. Generation of VZV pVLT/pVLT-ORF63 mutant viruses using a single sgRNA. (**A**) Schematic illustration showing the sgRNA target sites in the VLT region: thick double black lines represent the genome, annotated RNAs are depicted in grey—wide boxes are coding sequences (CDS), thin boxes are untranslated regions (UTRs) and dashed lines are intronic sequences—and the DNA target sites of sgRNAs VLT-7 to VLT-12 are depicted in differentially colored lines. Upper three transcripts represent most abundant alternative upstream exons used during lytic infection—named A-B-C and described in [9]. (**B**) Sanger sequence traces showing the sgRNA VLT-7/VLT-9 target region with PAM sequence (dashed black) and cleavage site (colored triangle) for wild-type VZV EMC-1 DNA (top) or DNA extracted from VZV EMC-1 infected stably transduced VLT-7 cells (middle) and VLT-9 cells (bottom). (**C**) Experimental layout: Stably transduced ARPE-19 cells expressing sgRNA VLT-9 were infected with VZV EMC-1. Two days post-infection, cell-free virus was harvested, and subjected to plaque purification on either ARPE-19 or ARPE-19 pVLT cells. Recovered viruses when then purified by three subsequent rounds of plaque purification. (**D–G**) Illumina sequencing of viral DNA of four selected mutant virus isolates identified 75 SNPs compared to reference. (**D**) Distribution of SNPs conserved between all four isolates ($n = 47$), SNPs present

in multiple isolates (n = 15) and isolate-specific SNPs (n = 13). (**E**) Type of SNPs identified in all isolates, substitutions (n = 56), insertions (n = 11) or deletions (n = 8). (**F**) Location of SNPs and number of SNPs outside the VLT coding sequence for each mutant viral isolate. (**G**) Proportion of reads of isolate-specific mutations for each mutant viral isolates, with VLT mutations highlighted in bold and colored: pVLTΔ1aa—brown, pVLTshift—blue and pVLTstop—red. (**H**) Relative expression level of VLT RNAs, RNA 60, RNA 61 and RNA 63 in ARPE-19 cells asynchronously infected with either pVLTwt, pVLTΔ1aa, pVLTshift or pVLTstop. Expression levels were normalized to VZV RNA 29 to compensate for differences in infection. (**I**) Percentage VZV glycoprotein E (gE)–positive cells by flow cytometry for pVLTwt, pVLTΔ1aa, pVLTshift or pVLTstop at 72 hpi. Representative data for n = 2 independent experiments, n = 3 technical replicates. *** $p < 0.001$ by unpaired Student's t-test. (**J**) Cell-associated VZV titers of monolayers infected with wt (black dot) or stop (red dot) virus at 72 hpi as measured by TCID50 per cm^2. *** $p < 0.001$ by unpaired Student's t-test.

Given that our pVLT/pVLT-ORF63(-) viruses with genomic deletions were impaired in replication compared to the corresponding pVLT/pVLT-ORF63(+) viruses (Figure 4), we performed an additional first round of limiting dilution of the cell-free virus obtained from VZV infection of the sgRNA VLT-9 cell line on either parental ARPE-19 cells or on ARPE-19 cells stably expressing pVLT (ARPE-19 pVLT), to complement for the potential loss of pVLT in trans (Figure 5C). Interestingly, all virus isolates recovered from parental ARPE-19 cells contained either no mutations in the VLT region or lacked three nucleotides, leading to an in-frame deletion of 1 amino acid (threonine 35). By contrast, virus isolates recovered from ARPE-19 pVLT cells contained other pVLT/pVLT-ORF63 mutations, such as frame-shift mutations or the formation of a premature stop codon, resulting in the loss of pVLT/pVLT-ORF63 coding potential (Table 1). We selected four viruses for further experiments and named them after their mutation; isolate A-VLT-5—pVLTΔ1aa, isolate P-VLT-5—pVLTshift, isolate P-VLT-8—pVLTstop, and the co-purified isolate A-VLT-7 with wild type VLT sequence—pVLTwt.

Table 1. Overview of viruses recovered from sgRNA VLT-9 expressing cells after plaque purification.

Virus Isolate [a]	Rescue Cells [b]	Nucleotide Sequence	Amino Acid Sequence	Mutation Type [c]	pVLT/pVLT-ORF63 [d]
VZV EMC-1	n.a.	gatataccaacccttacgaccaatagcaac	PDIPTLTTNSN	n.a.	yes
A-VLT-1	ARPE-19	gatataccaacccttaccaatagcaac	PDIPTLT NSN	Δ1aa	yes
A-VLT-5	ARPE-19	gatataccaacccttaccaatagcaac	PDIPTLT NSN	**Δ1aa**	yes
A-VLT-7	ARPE-19	gatataccaacccttacgaccaatagcaac	PDIPTLTTNSN	**wt**	yes
A-VLT-8	ARPE-19	gatataccaacccttaccaatagcaac	PDIPTLT NSN	Δ1aa	yes
A-VLT-9	ARPE-19	gatataccaacccttaccaatagcaac	PDIPTLT NSN	Δ1aa	yes
P-VLT-3	ARPE-19 pVLT	gatataccaacccttgcggccaatagcaac	PDIPTLADQ-	stop	no
P-VLT-4	ARPE-19 pVLT	gatataccaacccttaccaatagcaac	PDIPTLT NSN	Δ1aa	yes
P-VLT-5	ARPE-19 pVLT	gatataccaacccttacaccaatagcaac	PDIPTLPIATLR	**shift**	no
P-VLT-6	ARPE-19 pVLT	gatataccaacccttaccaatagcaac	PDIPTLT NSN	Δ1aa	yes
P-VLT-8	ARPE-19 pVLT	gatataccaacccaatagcaac	PDIPTQ-	**stop**	no

[a] Viruses name. [b] Cell type used in first plaque purification: parental ARPE-19 or pVLT-expressing ARPE-19. [c] Type of mutation: wild-type (no mutation), single amino acid deletion (Δ1aa), insertion of a premature stop codon (stop) and frame-shift mutation (shift). Viruses indicated in green and bold were studied in detail. [d] Predicted ability of the virus to encode pVLT/pVLT-ORF63 (variants). n.a.: not applicable.

Short-read Illumina sequencing of VZV pVLTwt, pVLTΔ1aa, pVLTshift and pVLTstop genomes identified 75 SNPs (Table S3) compared to the reference VZV Dumas DNA sequence (Genbank: NC_001348.1). The majority of these SNPs (n = 47) were conserved between all four VZV isolates and thus likely represent SNPs already present in the parental EMC-1 virus (Figure 5D). Another 15 SNPs were present in more than one VZV isolate but not shared between all, whereas 13 SNPs were specific to a single VZV isolate (Figure 5D). Most SNPs were substitutions, with insertions or deletions accounting for approximately one quarter of all SNPs (Figure 5E). The majority were located in intergenic regions, UTRs or introns, while only 23 non-VLT related SNPs affected the CDS of another VZV ORF (Figure 5F and Table S3). Most of these SNPs result in synonymous mutations and do not, therefore, impact the viral protein coding potential. One nonsynonymous SNP in the pVLTwt isolate resulted in a frameshift replacing the C-terminal 7AA with 11AA in pORF13, whereas pORF11 is extended by 3AA in the pVLTstop isolate (Table S3). Importantly, the high proportion of reads harboring the expected mutation for each VZV isolate in VLT exon 2 (Figure 5G), in the absence of any remaining contaminating wild-type VLT sequences in VZV pVLTΔ1aa, pVLTshift and pVLTstop, indicates clonality in this region. In summary, these Illumina DNA-sequencing data show CRISPR/Cas9 edited viruses can be efficiently isolated by repeated plaque purifications, yielding highly pure mutant viruses.

In contrast to the viruses with large deletions in the VLT locus (Figure 4A), expression of VLT RNAs, RNA 60, RNA 61 and RNA 63 was not affected in VZV pVLTΔ1aa, pVLTshift and pVLTstop viruses compared to VZV pVLTwt (Figure 5H). Concordant with the data obtained for the viruses with large deletions, growth data for pVLT/pVLT-ORF63(+) viruses (pVLTwt and pVLTΔ1aa) and pVLT/pVLT-ORF63(−) viruses (pVLTshift and pVLTstop), showed that pVLT/pVLT-ORF63(−) viruses with indel mutations were significantly impaired in replication compared to pVLT/pVLT-ORF63(+) viruses at 72 hpi (Figure 5I). Furthermore, titration of cells infected with either the pVLTwt or pVLTstop virus at 72 hpi showed a significant reduction in virus titer for the pVLT/pVLT-ORF63(−) virus (Figure 5J). Thus, the loss of pVLT/pVLT-ORF63 coding potential impairs lytic VZV infection in ARPE-19 cells.

3.5. Generation of VZV pVLT/pVLT-ORF63 Mutant Viruses Using BAC Mutagenesis

The currently available anti-pVLT antibody recognizes the first 20 amino acids of pVLT [9], thereby prohibiting formal confirmation of the absence of pVLT/pVLT-ORF63 in the aforementioned VLT mutant viruses: the large deletions in the first set of VLT mutants (Figure 3) disturb the antibody binding site but potentially encode a N-terminal altered variant of pVLT/pVLT-ORF63, whereas in the set with VLT mutants generated by sgRNA VLT-9 (Table 1 and Figure 5) the stop codon is located 14 amino acids downstream of the antibody binding site. BAC-based mutagenesis is widely used to generate mutant herpesvirus genomes, and its power resides in studying viral protein functions at nucleotide-level precision. Thus, to create a mutant virus in which the absence of pVLT/pVLT-ORF63 could be confirmed and to substantiate our findings, we performed complementary BAC-based mutagenesis. We constructed a recombinant virus from the parental Oka genome cloned into a bacterial artificial chromosome (pOka-BAC) [20] in which the pVLT/pVLT-ORF63 start codon (ATG) is replaced for an isoleucine encoding codon (ATA), and its corresponding repaired virus, hereafter named VLTM1I and VLTM1IR, respectively (Figure 6A). We confirmed expression of VLT RNA in ARPE-19 cells infected with VZV VLTM1I or VLTM1IR by RT-PCR and subsequent Sanger sequencing (Figure 6A) and showed by cDNA qPCR that relative expression levels of VLT RNAs, RNA 60, RNA 61 and RNA 63 were similar between VLTM1I and VLTM1IR viruses (Figure 6B). Western blotting confirmed the absence of both pVLT and pVLT-ORF63 in VZV VLTM1I-infected cells (Figure 6C), whereas VLTM1I and VLTM1IR expressed pORF63 at comparable levels (Figure 6D). Unlike the CRISPR/Cas9-based pVLT/pVLT-ORF63 mutant viruses, infection of ARPE-19 cells with VLTM1I and VLTM1IR resulted in comparable frequencies of infected cells and cell-associated virus titers (Figure 6E,F). However, VLTM1I produced significantly smaller infectious foci compared to VLTM1IR at 6 dpi (Figure 6G,H). Possibly, the genetic background of CRISPR/Cas9-derived (EMC-1) and BAC-derived (pOka) viruses influences the rate of cell-to-cell spread, syncytia formation or induction of cell death. Indeed, we observed a marked difference in infectious focus morphology between CRISPR/Cas9-derived (EMC-1 based) and BAC-derived (pOka based) viruses (Figure S1). Together these data show that while pVLT and pVLT-ORF63 are dispensable for lytic VZV replication in ARPE-19 cells, their absence may reduce the efficiency of VZV spread in epithelial cells.

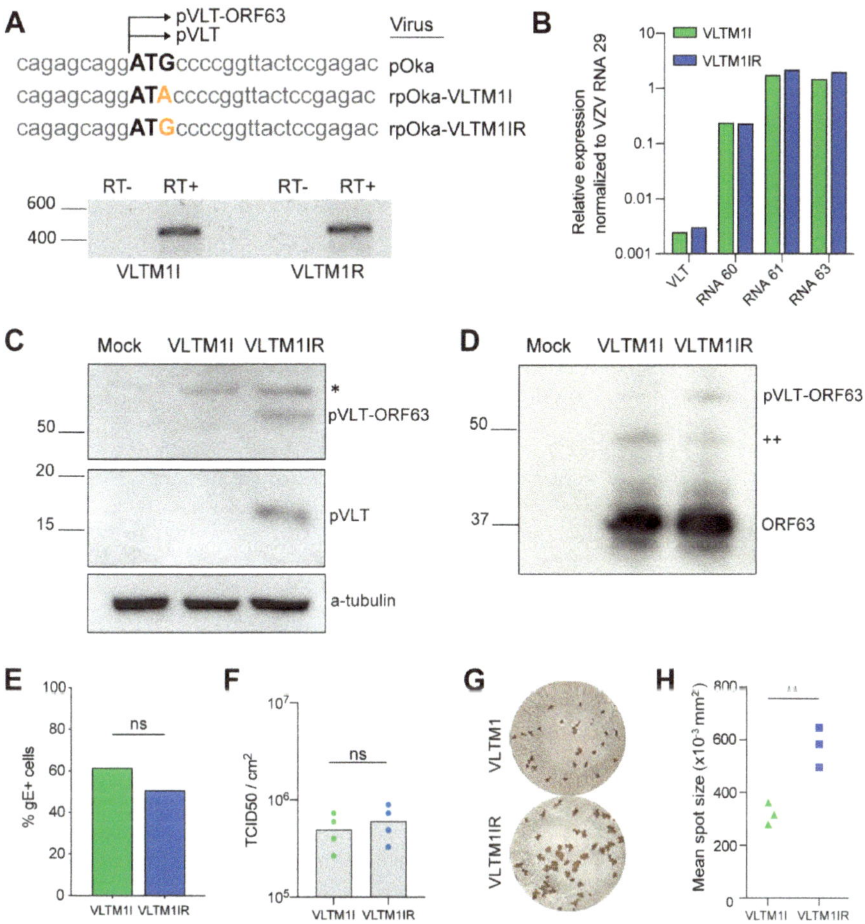

Figure 6. Generation of VZV pVLT/pVLT-ORF63 mutant viruses using BAC mutagenesis. (**A**) Top: Nucleotide sequence of VZV pOka, rpOka-VLTM1I and rpOka-VLTM1IR of the pVLT/pVLT-ORF63 ATG start codon in black, mutations in orange and its flanking regions and bottom: RT-PCR on RNA isolated from ARPE-19 cells infected with VZV VLTM1I or VLTM1IR using primers located in VLT exon 1 and exon 5. RT-/RT+: reverse transcriptase omitted or added in cDNA synthesis. (**B**) Relative expression level of VLT RNAs, RNA 60, RNA 61 and RNA 63 in ARPE-19 cells asynchronously infected with VLTM1I or VLTM1IR. Expression levels were normalized to VZV RNA29 to compensate for differences in infection. (**C**,**D**) Western blots on lysates of mock-, VZV VLTM1I- or VZV VLTM1IR-infected ARPE-19 cells stained with anti-pVLT, anti-alpha-tubulin (**C**) and anti-pORF63 (**D**) antibodies. Bands for pVLT and pVLT-ORF63 are indicated, * signifies an nonspecific band detected in all virus-infected cells, ++ indicates pORF63-N+ [10]. (**E**) Percentage VZV glycoprotein E (gE)–positive cells by flow cytometry of VZV VLTM1I and VLTM1IR in ARPE-19 cells at 72 hpi. Representative data for $n = 2$ independent experiments, $n = 3$ technical replicates. ns, not significant by unpaired Student's *t*-test. (**F**) Cell-associated VZV titers of monolayers infected with VLTM1I or VLTM1IR virus at 72 hpi as measured by TCID50 per cm2, ns: not significant by unpaired Student's *t*-test. (**G**,**H**) Infectious focus assay of VZV VLTM1I and VLTM1IR-infected ARPE-19 cells at 6 days post-infection ($n = 2$ independent experiments, 4 replicates each). (**G**) Representative images of infected wells stained for VZV gE by immunohistochemistry and (**H**) Mean spot size (×103 mm^2) of VLTM1I (green) and VLTM1IR (blue) viruses per well, ** $p < 0.01$ by unpaired Student's *t*-test.

4. Discussion

VZV latency in sensory neurons is defined by selective expression of VLT and VLT-ORF63 fusion transcripts [9,10]. VLT and VLT-ORF63 RNAs are unique as they are the only gene products that are present throughout the entire viral life cycle—during both lytic and latent phases. VLT and VLT63-1 encode for proteins that are expressed during lytic infection: pVLT, a protein of unknown function, and pVLT-ORF63, likely a key regulator in the transition from latency to lytic infection [10]. Here, we studied the role of pVLT and/or pVLT-ORF63 during lytic VZV infection in epithelial cells using complementary CRISPR/Cas9- and BAC-based mutagenesis. We demonstrated that CRISPR/Cas9 editing of non-essential VZV genes leads to functional protein disruption, but that editing of the VLT region during lytic VZV infection is inefficient and favors the selection of pVLT/pVLT-ORF63 coding virus mutants. Additionally, VZV mutants lacking pVLT and pVLT-ORF63 coding potential, generated by either CRISPR/Cas9- or BAC-based mutagenesis, were mildly impaired in virus replication in ARPE-19 cells. Collectively, these data indicate that pVLT and/or pVLT-ORF63 are dispensable for lytic VZV replication, but the presence of these viral proteins promotes efficient VZV infection in ARPE-19 epithelial cells.

Our study reports on locus-specific VZV genome editing using CRISPR/Cas9 technology and subsequent purification of mutant virus isolates. Non-essential VZV genes ORF8 and ORF11 could be efficiently targeted in lytically VZV-infected cells and produced infectious mutant progeny virus. The VLT locus, however, proved to be difficult to target, and no genome editing was observed when using 11 of 12 single sgRNAs targeting the VLT locus. It is unclear what causes this resistance of the VLT region to CRISPR/Cas9 genome editing, as targeting essential herpesvirus genes is possible and typically results in the outgrowth of escape mutants that carry in-frame deletions [14]. Many cellular factors could influence the binding of the SpCas9-sgRNA complex to mammalian genomes, such as chromatin accessibility, nucleosome positioning and certain chromatin modifications [13,35,36]. Although herpesviral genomes are immediately chromatinized upon nuclear entry, viral proteins actively reduce histone occupancy leading to an open chromatin structure of the viral genome during lytic infection. This, combined with the efficient editing of the VLT locus using two sgRNAs, suggests that impaired chromatin accessibility or binding of the SpCas9-sgRNA complex to the viral genome does not contribute to the relative resistance of the genomic VLT region. Alternatively, the VLT region may contain unknown essential regulatory elements that require perfect repair by the cellular DNA damage repair machinery in order to produce infectious progeny.

The CRISPR/Cas9 system provides the opportunity to edit clinical isolates [37], enabling studies of gene function in different genetic backgrounds, and to perform in vitro mutagenesis of viral genomes. Recovered genomic sequences are driven by selection (i.e., pressure during replication in permissive cell types) and, therefore, provide valuable insights into the importance of specific loci or favored mutation types. Most recovered virus isolates generated with the double sgRNA approach contained an upstream ATG start codon in-frame with the remainder of the pVLT/pVLT-ORF63 open reading frame, which suggests that the presence of N-terminally altered versions of these proteins is favored over their absence. Similarly, using CRISPR/Cas9 editing with sgRNA VLT-9, we were only able to obtain pVLT/pVLT-ORF63(-) viruses when rescued on cells constitutively expressing pVLT, suggesting that the presence of pVLT in trans provides an important survival advantage to mutant viruses. Indeed, we observed that pVLT/pVLT-ORF63(-) VZV was viable, but attenuated in growth compared to pVLT/pVLT-ORF63(+) VZV.

Cell cultures are oversimplified representations of the host environment, and thus requirements for survival in vivo can substantially differ from those in vitro. Indeed, the phenotype of single-ORF deletion VZV mutants can be markedly different between cell culture and skin tissue culture or xenograft models [38–42], as illustrated by skin-specific growth defects of ORF7-deletion viruses and combined T-cell- and skin-specific growth defects of ORF67-deletion viruses [11,43]. Therefore, future studies in these skin explant and xenograft models are of interest to determine the role of pVLT/pVLT-ORF63 in organ-specific lytic VZV replication in vivo. Similarly, the recent development of human neuron models to study VZV latency and reactivation in vitro [10,44,45] enables future studies to determine the role of pVLT and pVLT-ORF63 in the establishment, maintenance and egress from latency. The latter is of particular interest, because we have recently shown that ectopic expression of VLT63-1—encoding pVLT-ORF63—induces broad viral gene expression from latent VZV genomes [10].

Currently available antivirals and vaccines that counteract and prevent VZV disease, leave the latent VZV burden intact. CRISPR/Cas9, besides being a versatile tool for research, also holds promising therapeutic potential for the treatment of persistent virus infections, as illustrated by the application to excise the integrated SIV genome in vivo [46,47]. Additionally, several studies showed successful targeting of herpesvirus genomes with either CRISPR/Cas9 or other programmable endonucleases, such as reducing latent EBV [14,48], latent KSHV [49], quiescent and latent HSV-1 genomes in vitro and in vivo [13,50] and, importantly, reducing herpes keratitis occurrence in vivo [23]. In this study, we showed that in vitro editing of VZV genomes during lytic infection is feasible. Although the VLT locus was relatively resilient to CRISPR/Cas9-based genome editing using single sgRNAs, we identified one potent VLT-specific sgRNA (sgRNA VLT-9) and showed efficient editing of the VZV VLT genomic region when two sgRNAs were combined. This highlights the potential for therapeutic use in humans, where the VLT locus is the only site in the VZV genome that is transcriptionally active during the latent phase, suggesting a relatively open chromatin structure that could allow for SpCas9-sgRNA binding, cleavage and potential disruption of the viral genome.

In conclusion, this study describes the use of CRISPR/Cas9-based genome editing to generate VZV mutants. Specifically, we applied this method to investigate the role of pVLT and pVLT-ORF63 during lytic VZV infection in vitro. Our data provide valuable insight into the effectiveness of VLT-specific sgRNAs or combinations thereof that can be used to optimize the inactivation of latent VZV genomes in in vitro neuronal latency models and could ultimately be applied in the human host.

Supplementary Materials: The following are available online at https://www.mdpi.com/article/10.3390/v13112289/s1, Figure S1: Comparison of infectious foci produced by VZV EMC-1 and VZV pOka derived viruses, Table S1: sgRNA sequences, Table S2: primers used in this study, Table S3: Illumina sequencing of VZV pVLT mutants.

Author Contributions: Conceptualization, R.J.L., G.M.G.M.V., T.S. and W.J.D.O.; Software, D.P.D.; Validation, S.E.B., T.S. and W.J.D.O.; Formal Analysis, S.E.B., D.P.D., T.S. and W.J.D.O.; Investigation, S.E.B., D.P.D., C.M.E.S., L.A.A., T.S. and W.J.D.O.; Writing—Original Draft Preparation, S.E.B., T.S. and W.J.D.O.; Writing—Review and Editing, S.E.B., G.M.G.M.V., T.S. and W.J.D.O.; Visualization, S.E.B., T.S. and W.J.D.O.; Supervision, G.M.G.M.V. and W.J.D.O.; Funding Acquisition, S.E.B., G.M.G.M.V., T.S. and W.J.D.O. All authors have read and agreed to the published version of the manuscript.

Funding: S.E.B. was in part supported by the Nederlandse Organisatie voor Wetenschappelijk Onderzoek (NWO) project 022.005.032. G.M.G.M.V. and W.J.D.O. are supported by the National Institute Of Allergy And Infectious Diseases of the National Institutes of Health under Award Number R01AI151290; The content is solely the responsibility of the authors and does not necessarily represent the official views of the National Institutes of Health. D.P.D. is supported by a DZIF (Deutsches Zentrum für Infektionsforschung) Professorship. T.S. received funding from the Tokyo Biochemical Research Foundation, the Takeda Science Foundation, and the Japan Society for the Promotion of Science (JSPS KAKENHI JP24590551) and the Ministry of Education, Culture, Sports, Science and Technology (MEXT KAKENHI JP21H02741).

Data Availability Statement: All sequencing datasets generated as part of this study are available at the European Nucleotide Archive under the accession: PRJEB47381.

Acknowledgments: We thank Stipan Jonjic for providing the mouse monoclonal anti-VZV ORF8 antibody, Yasuko Mori (Kobe University) for the pOka-BAC genome and Gregory Smith (Northwestern University) for GS1783 *E. coli*.

Conflicts of Interest: The authors declare no conflict of interest.

References

1. Gershon, A.A.; Breuer, J.; Cohen, J.I.; Cohrs, R.J.; Gershon, M.D.; Gilden, D.; Grose, C.; Hambleton, S.; Kennedy, P.G.E.; Oxman, M.N.; et al. Varicella zoster virus infection. *Nat. Rev. Dis. Primers* **2015**, *1*, 15016. [CrossRef] [PubMed]
2. Gilden, D.H.; Vafai, A.; Shtram, Y.; Becker, Y.; Devlin, M.; Wellish, M. Varicella-zoster virus DNA in human sensory ganglia. *Nature* **1983**, *306*, 478–480. [CrossRef]
3. Depledge, D.; Sadaoka, T.; Ouwendijk, W. Molecular Aspects of Varicella-Zoster Virus Latency. *Viruses* **2018**, *10*, 349. [CrossRef] [PubMed]
4. Gilden, D.H.; Kleinschmidt-DeMasters, B.K.; LaGuardia, J.J.; Mahalingam, R.; Cohrs, R.J. Neurologic complications of the reactivation of varicella-zoster virus. *N. Engl. J. Med.* **2000**, *342*, 635–645. [CrossRef] [PubMed]
5. Johnson, R.W.; Rice, A.S.C. Clinical practice. Postherpetic Neuralgia. *N. Engl. J. Med.* **2014**, *371*, 1526–1533. [CrossRef]
6. Cunningham, A.L.; Lal, H.; Kovac, M.; Chlibek, R.; Hwang, S.-J.; Díez-Domingo, J.; Godeaux, O.; Levin, M.J.; McElhaney, J.E.; Puig-Barberà, J.; et al. Efficacy of the Herpes Zoster Subunit Vaccine in Adults 70 Years of Age or Older. *N. Engl. J. Med.* **2016**, *375*, 1019–1032. [CrossRef]
7. Braspenning, S.E.; Sadaoka, T.; Breuer, J.; Verjans, G.M.G.M.; Ouwendijk, W.J.D.; Depledge, D.P. Decoding the Architecture of the Varicella-Zoster Virus Transcriptome. *Mbio* **2020**, *11*, e01568-20. [CrossRef]
8. Ouwendijk, W.J.D.; Dekker, L.J.M.; van den Ham, H.-J.; Lenac Rovis, T.; Haefner, E.S.; Jonjic, S.; Haas, J.; Luider, T.M.; Verjans, G.M.G.M. Analysis of Virus and Host Proteomes During Productive HSV-1 and VZV Infection in Human Epithelial Cells. *Front. Microbiol.* **2020**, *11*, 1179. [CrossRef]
9. Depledge, D.P.; Ouwendijk, W.J.D.; Sadaoka, T.; Braspenning, S.E.; Mori, Y.; Cohrs, R.J.; Verjans, G.M.G.M.; Breuer, J. A spliced latency-associated VZV transcript maps antisense to the viral transactivator gene 61. *Nat. Commun.* **2018**, *9*, 1167. [CrossRef]
10. Ouwendijk, W.J.D.; Depledge, D.P.; Rajbhandari, L.; Lenac Rovis, T.; Jonjic, S.; Breuer, J.; Venkatesan, A.; Verjans, G.M.G.M.; Sadaoka, T. Varicella-zoster virus VLT-ORF63 fusion transcript induces broad viral gene expression during reactivation from neuronal latency. *Nat. Commun.* **2020**, *11*, 6324. [CrossRef]
11. Zhang, Z.; Selariu, A.; Warden, C.; Huang, G.; Huang, Y.; Zaccheus, O.; Cheng, T.; Xia, N.; Zhu, H. Genome-wide mutagenesis reveals that ORF7 is a novel VZV skin-tropic factor. *PLoS Pathog.* **2010**, *6*, e1000971. [CrossRef]
12. Cohen, J.I.; Nguyen, H. Varicella-zoster virus ORF61 deletion mutants replicate in cell culture, but a mutant with stop codons in ORF61 reverts to wild-type virus. *Virology* **1998**, *246*, 306–316. [CrossRef] [PubMed]
13. Oh, H.S.; Neuhausser, W.M.; Eggan, P.; Angelova, M.; Kirchner, R.; Eggan, K.C.; Knipe, D.M. Herpesviral lytic gene functions render the viral genome susceptible to novel editing by CRISPR/Cas9. *Elife* **2019**, *8*, e51662. [CrossRef] [PubMed]
14. Van Diemen, F.R.; Kruse, E.M.; Hooykaas, M.J.G.; Bruggeling, C.E.; Schürch, A.C.; van Ham, P.M.; Imhof, S.M.; Nijhuis, M.; Wiertz, E.J.H.J.; Lebbink, R.J. CRISPR/Cas9-Mediated Genome Editing of Herpesviruses Limits Productive and Latent Infections. *PLoS Pathog.* **2016**, *12*, e1005701. [CrossRef]
15. Sadaoka, T.; Yanagi, T.; Yamanishi, K.; Mori, Y. Characterization of the Varicella-Zoster Virus ORF50 Gene, Which Encodes Glycoprotein M. *J. Virol.* **2010**, *84*, 3488–3502. [CrossRef]
16. Lenac Rovis, T.; Bailer, S.M.; Pothineni, V.R.; Ouwendijk, W.J.D.; Simic, H.; Babic, M.; Miklic, K.; Malic, S.; Verweij, M.C.; Baiker, A.; et al. Comprehensive Analysis of Varicella-Zoster Virus Proteins Using a New Monoclonal Antibody Collection. *J. Virol.* **2013**, *87*, 6943–6954. [CrossRef]
17. Ouwendijk, W.J.D.; Geluk, A.; Smits, S.L.; Getu, S.; Osterhaus, A.D.M.E.; Verjans, G.M.G.M. Functional Characterization of Ocular-Derived Human Alphaherpesvirus Cross-Reactive CD4 T Cells. *J. Immunol.* **2014**, *192*, 3730–3739. [CrossRef] [PubMed]
18. Li, H.; Handsaker, B.; Wysoker, A.; Fennell, T.; Ruan, J.; Homer, N.; Marth, G.; Abecasis, G.; Durbin, R. 1000 Genome Project Data Processing Subgroup The Sequence Alignment/Map format and SAMtools. *Bioinformatics* **2009**, *25*, 2078–2079. [CrossRef]
19. Quinlan, A.R.; Hall, I.M. BEDTools: A flexible suite of utilities for comparing genomic features. *Bioinformatics* **2010**, *26*, 841–842. [CrossRef]
20. Nagaike, K.; Mori, Y.; Gomi, Y.; Yoshii, H.; Takahashi, M.; Wagner, M.; Koszinowski, U.; Yamanishi, K. Cloning of the varicella-zoster virus genome as an infectious bacterial artificial chromosome in Escherichia coli. *Vaccine* **2004**, *22*, 4069–4074. [CrossRef]
21. Hobom, U.; Brune, W.; Messerle, M.; Hahn, G.; Koszinowski, U.H. Fast screening procedures for random transposon libraries of cloned herpesvirus genomes: Mutational analysis of human cytomegalovirus envelope glycoprotein genes. *J. Virol.* **2000**, *74*, 7720–7729. [CrossRef] [PubMed]
22. Koshizuka, T.; Ota, M.; Yamanishi, K.; Mori, Y. Characterization of varicella-zoster virus-encoded ORF0 gene—Comparison of parental and vaccine strains. *Virology* **2010**, *405*, 280–288. [CrossRef]
23. Yin, D.; Ling, S.; Wang, D.; Dai, Y.; Jiang, H.; Zhou, X.; Paludan, S.R.; Hong, J.; Cai, Y. Targeting herpes simplex virus with CRISPR-Cas9 cures herpetic stromal keratitis in mice. *Nat. Biotechnol.* **2021**, *39*, 567–577. [CrossRef]
24. Sadaoka, T.; Serada, S.; Kato, J.; Hayashi, M.; Gomi, Y.; Naka, T.; Yamanishi, K.; Mori, Y. Varicella-Zoster Virus ORF49 Functions in the Efficient Production of Progeny Virus through Its Interaction with Essential Tegument Protein ORF44. *J. Virol.* **2014**, *88*, 188–201. [CrossRef] [PubMed]
25. Ross, J.; Williams, M.; Cohen, J.I. Disruption of the varicella-zoster virus dUTPase and the adjacent ORF9A gene results in impaired growth and reduced syncytia formation in vitro. *Virology* **1997**, *234*, 186–195. [CrossRef] [PubMed]
26. Che, X.; Reichelt, M.; Sommer, M.H.; Rajamani, J.; Zerboni, L.; Arvin, A.M. Functions of the ORF9-to-ORF12 gene cluster in varicella-zoster virus replication and in the pathogenesis of skin infection. *J. Virol.* **2008**, *82*, 5825–5834. [CrossRef] [PubMed]

27. Doench, J.G.; Fusi, N.; Sullender, M.; Hegde, M.; Vaimberg, E.W.; Donovan, K.F.; Smith, I.; Tothova, Z.; Wilen, C.; Orchard, R.; et al. Optimized sgRNA design to maximize activity and minimize off-target effects of CRISPR-Cas9. *Nat. Biotechnol.* **2016**, *34*, 184–191. [CrossRef]
28. Zhang, Z.; Rowe, J.; Wang, W.; Sommer, M.; Arvin, A.; Moffat, J.; Zhu, H. Genetic Analysis of Varicella-Zoster Virus ORF0 to ORF4 by Use of a Novel Luciferase Bacterial Artificial Chromosome System. *J. Virol.* **2007**, *81*, 9024–9033. [CrossRef] [PubMed]
29. Kallimasioti-Pazi, E.M.; Thelakkad Chathoth, K.; Taylor, G.C.; Meynert, A.; Ballinger, T.; Kelder, M.J.E.; Lalevée, S.; Sanli, I.; Feil, R.; Wood, A.J. Heterochromatin delays CRISPR-Cas9 mutagenesis but does not influence the outcome of mutagenic DNA repair. *PLoS Biol.* **2018**, *16*, e2005595. [CrossRef]
30. Knight, S.C.; Xie, L.; Deng, W.; Guglielmi, B.; Witkowsky, L.B.; Bosanac, L.; Zhang, E.T.; El Beheiry, M.; Masson, J.-B.; Dahan, M.; et al. Dynamics of CRISPR-Cas9 genome interrogation in living cells. *Science* **2015**, *350*, 823–826. [CrossRef]
31. Mandal, P.K.; Ferreira, L.M.R.; Collins, R.; Meissner, T.B.; Boutwell, C.L.; Friesen, M.; Vrbanac, V.; Garrison, B.S.; Stortchevoi, A.; Bryder, D.; et al. Efficient Ablation of Genes in Human Hematopoietic Stem and Effector Cells using CRISPR/Cas9. *Cell Stem Cell* **2014**, *15*, 643–652. [CrossRef]
32. Zhou, J.; Wang, J.; Shen, B.; Chen, L.; Su, Y.; Yang, J.; Zhang, W.; Tian, X.; Huang, X. Dual sgRNAs facilitate CRISPR/Cas9-mediated mouse genome targeting. *FEBS J.* **2014**, *281*, 1717–1725. [CrossRef] [PubMed]
33. Carpenter, J.E.; Henderson, E.P.; Grose, C. Enumeration of an extremely high particle-to-PFU ratio for Varicella-zoster virus. *J. Virol.* **2009**, *83*, 6917–6921. [CrossRef]
34. Sadaoka, T.; Schwartz, C.L.; Rajbhandari, L.; Venkatesan, A.; Cohen, J.I. Human Embryonic Stem Cell-Derived Neurons Are Highly Permissive for Varicella-Zoster Virus Lytic Infection. *J. Virol.* **2018**, *92*, e01108-17. [CrossRef] [PubMed]
35. Horlbeck, M.A.; Witkowsky, L.B.; Guglielmi, B.; Replogle, J.M.; Gilbert, L.A.; Villalta, J.E.; Torigoe, S.E.; Tjian, R.; Weissman, J.S. Nucleosomes impede Cas9 access to DNA in vivo and in vitro. *Elife* **2016**, *5*, e12677. [CrossRef] [PubMed]
36. Isaac, R.S.; Jiang, F.; Doudna, J.A.; Lim, W.A.; Narlikar, G.J.; Almeida, R. Nucleosome breathing and remodeling constrain CRISPR-Cas9 function. *Elife* **2016**, *5*, e13450. [CrossRef]
37. Finnen, R.L.; Banfield, B.W. CRISPR/Cas9 Mutagenesis of UL21 in Multiple Strains of Herpes Simplex Virus Reveals Differential Requirements for pUL21 in Viral Replication. *Viruses* **2018**, *10*, 258. [CrossRef]
38. Taylor, S.L.; Moffat, J.F. Replication of Varicella-Zoster Virus in Human Skin Organ Culture. *J. Virol.* **2005**, *79*, 11501–11506. [CrossRef]
39. Lloyd, M.G.; Smith, N.A.; Tighe, M.; Travis, K.L.; Liu, D.; Upadhyaya, P.K.; Kinchington, P.R.; Chan, G.C.; Moffat, J.F. A Novel Human Skin Tissue Model To Study Varicella-Zoster Virus and Human Cytomegalovirus. *J. Virol.* **2020**, *94*, e01082-20. [CrossRef]
40. François, S.; Sen, N.; Mitton, B.; Xiao, X.; Sakamoto, K.M.; Arvin, A. Varicella-Zoster Virus Activates CREB, and Inhibition of the pCREB-p300/CBP Interaction Inhibits Viral Replication In Vitro and Skin Pathogenesis In Vivo. *J. Virol.* **2016**, *90*, 8686–8697. [CrossRef]
41. Moffat, J.F.; Zerboni, L.; Sommer, M.H.; Heineman, T.C.; Cohen, J.I.; Kaneshima, H.; Arvin, A.M. The ORF47 and ORF66 putative protein kinases of varicella-zoster virus determine tropism for human T cells and skin in the SCID-hu mouse. *Proc. Natl. Acad. Sci. USA* **1998**, *95*, 11969–11974. [CrossRef]
42. Sato, B.; Ito, H.; Hinchliffe, S.; Sommer, M.H.; Zerboni, L.; Arvin, A.M. Mutational Analysis of Open Reading Frames 62 and 71, Encoding the Varicella-Zoster Virus Immediate-Early Transactivating Protein, IE62, and Effects on Replication In Vitro and in Skin Xenografts in the SCID-hu Mouse In Vivo. *J. Virol.* **2003**, *77*, 5607–5620. [CrossRef]
43. Moffat, J.; Ito, H.; Sommer, M.; Taylor, S.; Arvin, A.M. Glycoprotein I of Varicella-Zoster Virus Is Required for Viral Replication in Skin and T Cells. *J. Virol.* **2002**, *76*, 8468–8471. [CrossRef] [PubMed]
44. Markus, A.; Lebenthal-Loinger, I.; Yang, I.H.; Kinchington, P.R.; Goldstein, R.S. An In Vitro Model of Latency and Reactivation of Varicella Zoster Virus in Human Stem Cell-Derived Neurons. *PLoS Pathog.* **2015**, *11*, e1004885. [CrossRef] [PubMed]
45. Sadaoka, T.; Depledge, D.P.; Rajbhandari, L.; Venkatesan, A.; Breuer, J.; Cohen, J.I. In vitro system using human neurons demonstrates that varicella-zoster vaccine virus is impaired for reactivation, but not latency. *Proc. Natl. Acad. Sci. USA* **2016**, *113*, E2403–E2412. [CrossRef] [PubMed]
46. Mancuso, P.; Chen, C.; Kaminski, R.; Gordon, J.; Liao, S.; Robinson, J.A.; Smith, M.D.; Liu, H.; Sariyer, I.K.; Sariyer, R.; et al. CRISPR based editing of SIV proviral DNA in ART treated non-human primates. *Nat. Commun.* **2020**, *11*, 6065. [CrossRef]
47. Lebbink, R.J.; de Jong, D.C.M.; Wolters, F.; Kruse, E.M.; van Ham, P.M.; Wiertz, E.J.H.J.; Nijhuis, M. A combinational CRISPR/Cas9 gene-editing approach can halt HIV replication and prevent viral escape. *Sci. Rep.* **2017**, *7*, 41968. [CrossRef]
48. Wang, J.; Quake, S.R. RNA-guided endonuclease provides a therapeutic strategy to cure latent herpesviridae infection. *Proc. Natl. Acad. Sci. USA* **2014**, *111*, 13157–13162. [CrossRef]
49. Tso, F.Y.; West, J.T.; Wood, C. Reduction of Kaposi's Sarcoma-Associated Herpesvirus Latency Using CRISPR-Cas9 To Edit the Latency-Associated Nuclear Antigen Gene. *J. Virol.* **2019**, *93*, e02183-18. [CrossRef]
50. Aubert, M.; Strongin, D.E.; Roychoudhury, P.; Loprieno, M.A.; Haick, A.K.; Klouser, L.M.; Stensland, L.; Huang, M.; Makhsous, N.; Tait, A.; et al. Gene editing and elimination of latent herpes simplex virus in vivo. *Nat. Commun.* **2020**, *11*, 4148. [CrossRef] [PubMed]

viruses

MDPI

Article

Meningitis Caused by the Live Varicella Vaccine Virus: Metagenomic Next Generation Sequencing, Immunology Exome Sequencing and Cytokine Multiplex Profiling

Prashanth S. Ramachandran [1], Michael R. Wilson [1], Gaud Catho [2], Geraldine Blanchard-Rohner [3], Nicoline Schiess [4], Randall J. Cohrs [5], David Boutolleau [6], Sonia Burrel [6], Tetsushi Yoshikawa [7], Anne Wapniarski [1], Ethan H. Heusel [8], John E. Carpenter [8], Wallen Jackson [8], Bradley A. Ford [9] and Charles Grose [8,*]

1 Department of Neurology, Weill Institute for Neurosciences, University of California San Francisco, San Francisco, CA 94110, USA; Prashanth.Ramachandran@ucsf.edu (P.S.R.); Michael.Wilson@ucsf.edu (M.R.W.); awapniarski@gmail.com (A.W.)
2 Division of Pediatric Infectious Diseases, Geneva University Hospitals, Faculty of Medicine, University of Geneva, 1205 Geneva, Switzerland; Gaud.Catho@hcuge.ch
3 Pediatric Immunology and Vaccinology Unit, Division of General Pediatrics, Department of Pediatrics, Gynecology and Obstetrics, Geneva University Hospitals, University of Geneva, 1205 Geneva, Switzerland; Geraldine.BlanchardRohner@hcuge.ch
4 Department of Neurology, Johns Hopkins School of Medicine, Baltimore, MD 21205, USA; nschies1@jhmi.edu
5 Department of Neurology, University of Colorado School of Medicine, Aurora, CO 80045, USA; randall.cohrs@cuanschutz.edu
6 Virology Department, National Reference Center for Herpesviruses, Pitie-Salpetriere Hospital, Sorbonne University, 75013 Paris, France; david.boutolleau@aphp.fr (D.B.); sonia.burrel@aphp.fr (S.B.)
7 Department of Pediatrics, Fujita Health University School of Medicine, Aichi, Toyoake 470-1192, Japan; tetsushi@fujita-hu.ac.jp
8 Division of Infectious Diseases/Virology, Department of Pediatrics, University of Iowa, Iowa City, IA 52242, USA; ethan-heusel@uiowa.edu (E.H.H.); john-carpenter@uiowa.edu (J.E.C.); wallen-jackson@uiowa.edu (W.J.)
9 Department of Pathology, University of Iowa, Iowa City, IA 52242, USA; bradley-ford@uiowa.edu
* Correspondence: charles-grose@uiowa.edu

Citation: Ramachandran, P.S.; Wilson, M.R.; Catho, G.; Blanchard-Rohner, G.; Schiess, N.; Cohrs, R.J.; Boutolleau, D.; Burrel, S.; Yoshikawa, T.; Wapniarski, A.; et al. Meningitis Caused by the Live Varicella Vaccine Virus: Metagenomic Next Generation Sequencing, Immunology Exome Sequencing and Cytokine Multiplex Profiling. *Viruses* **2021**, *13*, 2286. https://doi.org/10.3390/v13112286

Academic Editor: Thomas Stamminger

Received: 8 October 2021
Accepted: 5 November 2021
Published: 16 November 2021

Abstract: Varicella vaccine meningitis is an uncommon delayed adverse event of vaccination. Varicella vaccine meningitis has been diagnosed in 12 children, of whom 3 were immunocompromised. We now report two additional cases of vaccine meningitis in twice-immunized immunocompetent children and we perform further testing on a prior third case. We used three methods to diagnose or investigate cases of varicella vaccine meningitis, none of which have been used previously on this disease. These include metagenomic next-generation sequencing and cytokine multiplex profiling of cerebrospinal fluid and immunology exome analysis of white blood cells. In one new case, the diagnosis was confirmed by metagenomic next-generation sequencing of cerebrospinal fluid. Both varicella vaccine virus and human herpesvirus 7 DNA were detected. We performed cytokine multiplex profiling on the cerebrospinal fluid of two cases and found ten elevated biomarkers: interferon gamma, interleukins IL-1RA, IL-6, IL-8, IL-10, IL-17F, chemokines CXCL-9, CXCL-10, CCL-2, and G-CSF. In a second new case, we performed immunology exome sequencing on a panel of 356 genes, but no errors were found. After a review of all 14 cases, we concluded that (i) there is no common explanation for this adverse event, but (ii) ingestion of an oral corticosteroid burst 3–4 weeks before onset of vaccine meningitis may be a risk factor in some cases.

Keywords: varicella-zoster virus; Oka strain; human herpesvirus 6; human herpesvirus 7; corticosteroids; serious adverse event; herpes zoster; IL-6; IL-10; innate immunity

1. Introduction

Twenty years ago, virologists at the Food and Drug Administration (FDA) carried out an analysis of antibody titers of 4631 children collected over the 4 years after varicella vaccination. They interpreted the results as showing frequent asymptomatic reactivation of latent varicella vaccine virus in immunized children with low serum antibody titers after vaccination [1]. Subsequent epidemiologic studies have clearly shown that symptomatic reactivation as herpes zoster can occur after varicella vaccination, but the prevalence is less than that after wild-type varicella-zoster virus (VZV) infection [2]. Nevertheless, eight cases of varicella vaccine meningitis, a consequence of herpes zoster, have been reported in children who have received one varicella vaccination [3,4]. Even more recently, four cases of varicella vaccine meningitis in twice-immunized children have been reported [5]. We now report two new cases of varicella vaccine meningitis that occurred in twice-immunized, previously healthy boys. The first case is notable because identification of the virus as varicella vaccine virus was made by metagenomic next-generation sequencing (mNGS) of cerebrospinal fluid (CSF). Thus, this report provides proof-of-principle that mNGS can be used to differentiate vaccine-type from wild-type VZV strains.

The second case is notable because an immunology exome analysis was performed in an attempt to define risk factors for varicella vaccine meningitis in apparently healthy children. Additionally, we propose that we have found a previously unrecognized risk factor—namely, bursts of oral corticosteroids. Re-assessment of all 14 cases of varicella vaccine meningitis revealed that 1 of the new cases as well as a previously published varicella vaccine meningitis case had received a burst of oral corticosteroids 3–4 weeks before onset of varicella vaccine meningitis [6]. As a final goal of defining the CNS immunology environment during the meningitis event, we carried out cytokine multiplex profiling on CSF samples from one of the current cases and a prior case of varicella vaccine meningitis.

2. Materials and Methods

2.1. Patients and Ethics Statement

Two new patients are described. Furthermore, we performed additional testing on an archived sample from an earlier varicella meningitis patient and we reviewed the histories of all published cases to define risk factors. In order to distinguish each patient separately, we have chosen to continue the nomenclature that was used in an earlier review that summarized data from 12 patients with varicella vaccine meningitis [5]. Therefore, the two new cases will be called case 13 and case 14. The family of case 13 lived in the United States; the family of case 14 lived in both the United States and Switzerland. Both children received their varicella vaccinations in the United States. These studies were approved by the Institutional Review Boards at the respective medical centers.

2.2. Next Generation Sequencing of Cerebrospinal Fluid

VZV genotyping and a search for co-infection was performed on a CSF sample from case 13, using mNGS [7,8]. Total nucleic acid was extracted from 90 μL of CSF using the Zymo Quick-DNA/RNA MagBead (Zymo Cat. No. R2130) via the Agilent Bravo liquid handling robot (Santa Clara, CA, USA). The nucleic acid was then divided, with half undergoing DNAse treatment to isolate RNA for RNA sequencing (RNA-seq) and the remainder being used for DNA sequencing (DNA-Seq).

RNA-Seq libraries were prepared using the New England Biolabs' NEBNext Ultra II RNA library preparation kit (NEB Cat No. E7770, Ipswich, MA, USA), DNA libraries were prepared using the New England Biolabs' NEBNext® Ultra™ II DNA Library preparation kit (NEB Cat No. E7645), both using the Echo Labcyte 525 and Agilent Bravo robots with a previously described protocol. Host ribosomal RNA depletion was performed using the Qiagen QIAseq FastSelect RNA removal kit (Qiagen Cat No. 333180, Hilden, Germany) at 1:100 dilution. Pooled libraries were size selected using Ampure beads. Final libraries were then sequenced on an Illumina Novaseq 6000 (San Diego, CA, USA) using 146 base pair paired-end sequencing. The mNGS workflow events are also illustrated in Figure 1.

A

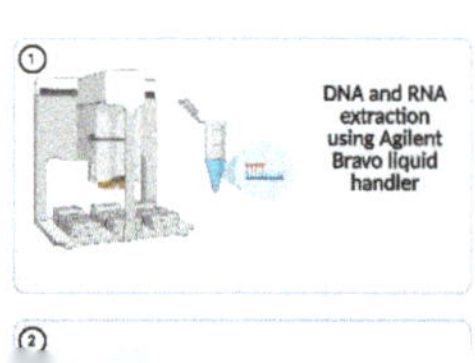

B

2.3. Sanger Sequencing

VZV genotyping on a CSF sample from case 14 was carried out as follows. After DNA extraction on the automated platform EMAG (BioMerieux, Marcy-l'Étoile, France), a duplex real-time PCR targeting the SNP108111 in ORF62 was performed on the LightCycler 480 (Roche Diagnostics, Basel, Switzerland) to discriminate vaccine from wild-type VZV. Thereafter, VZV genotyping was performed using conventional PCR and Sanger sequencing, targeting vaccine-associated SNPs in ORF1 (560, 685, and 703), ORF38 (69349), and ORF62 (105705, 106262, and 108111), as recommended by the Centers for Disease Control [9,10].

2.4. Immunology Exome Analysis

Immunology exome sequencing was performed on a panel of genes involved in immunologic and autoinflammatory disorders, by a method that has been previously described [11]. Exome capture was initiated by collecting genomic DNA extracted from white blood cells from case 14, using the SureSelect Human ALL Exon kit (Agilent Technologies, Santa Clara, CA, USA). High-throughput sequencing was performed in a HiSeq2000. The number of targeted genes was increased from 92 to 356. The full name of each gene as well as its genetic phenotype are listed in the catalog of Online Mendelian Inheritance in Man (OMIM) [12] or in HUGO (Human Genome Organization dictionary).

2.5. Measurement of Viral Antibody

VZV antibody in serum samples from case 13 was titrated by immunofluorescence (FAMA; fluorescent antibody to membrane antigen) on VZV-infected live cells [13]. Antibody titers to human herpes viruses (HHV) 6 and 7 were measured by immunofluorescence of fixed HHV6- and HHV7-infected cells [14]. Cell lines used in this report are available from the European Collection of Authenticated Cell Cultures and include MeWo cells (no. 93082609) and Molt-3 cells (90021901).

2.6. Measurement of Chemokines and Interleukins

CSF samples were stored frozen after the initial series of routine laboratory tests were performed. Cytokines in the CSF from cases 12 and 13 were analyzed by a bead-based multiplex fluorescence assay [15]. The Procartaplex Human Cytokine Magnetic 35-plex Bead Panel (Product #LHC6005M, Invitrogen/Thermo Fisher Scientific, Waltham, MA, USA) simultaneously evaluates 35 analytes (EGF, Eotaxin, FGF-basic, G-CSF, GM-CSF, HGF, IFNa, IFNγ, IL-1a, IL-1β, IL-1RA, IL-2, IL-2R, IL-3, IL-4, IL-5, IL-6, IL-7, IL-8, IL-9, IL-10, IL-12, IL-13, IL-17A, IL-17F, IL-22, IP-10, MIG, MIP-1α, MIP-1β, MCP-1, RANTES, TNFα, and VEGF). This assay was run on the Luminex® FLEXMAP 3D® instrument operated with xPONENT Software V4.2 (both from Luminex Corp., Austin, TX, USA). The cytokine profiling was carried out at the Bursky Center for Human Immunology and Immunotherapy Programs at the Washington University Immunomonitoring Laboratory, St. Louis, MO, USA.

3. Results

3.1. Varicella Vaccine Meningitis Case 13

Three days before admission, this previously healthy 7-year-old boy developed a headache, followed a few hours later by low grade fever. Two days before admission, his headaches worsened, and he began to vomit. Vomiting continued over the next 24 h whenever his parents attempted to feed him. When the vomiting persisted into the second day, he was admitted to a local hospital for intravenous (IV) hydration. His past history included febrile seizures at age 13 months and age 26 months; these were not further evaluated. He had received all recommended childhood immunizations, including Varivax at age 14 months and ProQuad at age 49 months.

After spending one night at the local hospital, he was transferred to a tertiary medical center. On admission, the child was in obvious pain. His physical examination did not disclose any rashes or abnormalities in major organ systems. Although the neurology examination was difficult to perform, he had mild nuchal rigidity but no focal neurological abnormalities. His non-contrast head CT imaging was normal.

His initial laboratory testing included the following values: ESR = 14 mm; CRP $\leq$ 0.5 mg/L; white blood cell count = 4000/mm^3 with 1790 neutrophils, 1130 lymphocytes, and 380 eosinophils. His hemoglobin was 12.2 gm/dL; AST/ALT = 20/30 units/L and his urinalysis was normal. His cerebrospinal fluid (CSF) was hazy with a glucose of 50 mg/dL, total protein of 60 mg/dL, and nucleated cell count = 544 cells/mm^3 with mainly lymphocytes. (A table of normal values is included in Table S1.) A Biofire FilmArray multiplex meningitis/encephalitis PCR (bioMerieux) was positive for both VZV and human herpesvirus 6 (HHV6). All Biofire VZV and HHV 6 reactions were repeated for verification.

Because the preliminary diagnosis was VZV meningitis, treatment with acyclovir IV at 30 mg/kg/day was initiated. The IV acyclovir was stopped after 4 days, when the child was markedly improved and his headaches resolved. He was discharged on a regimen of valacyclovir (500 mg three times daily) for 10 more days. The child was seen several times during the 18 months post discharge. His recovery from meningitis was complete.

3.2. Metagenomic Next Generation Sequencing of Cerebrospinal Fluid

Because a diagnosis of dual CNS infection with VZV and HHV6 was unexpected, we had an initial concern about a false positive result. In addition, the Biofire Filmarray panel cannot distinguish wild-type VZV from vaccine-type VZV. For the above two reasons, we pursued mNGS (Figure 1). Depth of sequencing for DNA-seq was 42,353,554 paired-end reads and 17,809,888 paired-end reads for RNA-seq. DNA-seq demonstrated 16.9rPM (716 reads) aligning to the VZV genome. RNA-seq had 1.6rPM (26 reads) aligning to VZV. The combination of both DNA and RNA reads allowed for 26.9% coverage of the VZV genome and definitive identification of the vaccine strain (GenBank AB097932), based on established criteria (Figure 1). Of great significance, one read aligning to ORF0 included the stop codon (560, *130R), which is mutated in the VZV vaccine strain [16]. More than 200 wild-type and vaccine-type VZV strains have been completely sequenced and the only other non-vaccine virus that has a similar ORF0 mutation is the highly passaged (>90) laboratory strain VZV Ellen (GenBank JQ972913) [17]. In addition, reads aligning with ORF62 contained additional SNPs found in the vaccine strain at nucleotides 105331, 105356, 105705, 107136, and 107165, including an extremely important fixed allele 105,705 [18]. Altogether, therefore, the mNGS data documenting a vaccine-type virus fulfilled the diagnostic criteria outlined by the Centers for Disease Control [9,10]. In addition, see Figure S1 for information about locations of the SNPs.

Further analysis of the mNGS data, however, failed to confirm the positive HHV6 result in the Biofire assay of the child's CSF, as described above. Instead, DNA-seq revealed two reads of 85 bp length aligning to HHV7 (Figure 2). These two reads only matched to HHV7 and no other virus when processed through BLAST in the NCBI database (GenBank U43400.1).

3.3. Viral Antibody Studies

VZV-specific IgM and IgG antibody titers detected by immunofluorescence on patient 13 were positive at titers of 1:4 and 1:64, respectively. Because the Biofire meningitis panel had a positive HHV6 signal and the mNGS assay detected HHV7 sequences but no HHV6 sequences, we also performed testing for both HHV6 and HHV7 antibodies. The boy had detectable IgG antibodies to both HHV6 and HHV7 at a titer of 1:16.

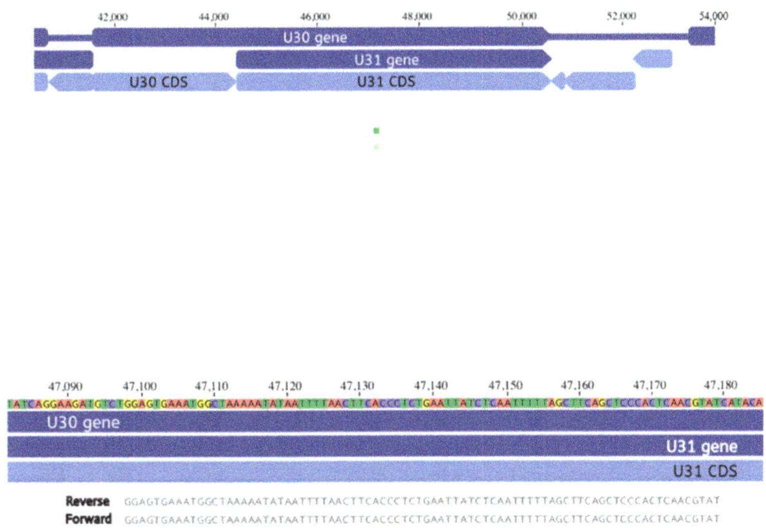

Figure 2. Identification of human herpesvirus 7 DNA in CSF by mNGS. The HHV7 nucleotide sequence includes 144,861 bp. The genome encodes ~100 open reading frames (ORFs). U30 includes 2816 bp; it is a capsid assembly protein. U31 includes 6179 bp, including the read from the mNGS of patient 13; it is a tegument protein. U = ORF located in the unique portion of the genome.

3.4. Corticosteroid Burst Treatment Prior to Varicella Vaccine Meningitis

Patient 13 had reported an erythematous rash about 4 cm in diameter over the right knee about 20 days before admission to the local hospital for meningitis. When seen at a medical clinic, poison ivy was diagnosed and oral prednisolone was prescribed at a daily dosage of 20 mg for 7 days (Table 1). Because case 13 had received a corticosteroid burst, we re-assessed the available clinical data from the 12 previous cases of varicella vaccine meningitis. We discovered that case 12 had also received a short course of prednisone before onset of her neurological disease. The timing was similar to case 13, but the dosage (mg/kg/day) was much lower (Table 1). She had taken prednisone because of an exacerbation of asthma. Obviously, ingestion of oral corticosteroids may lead to a transient immunocompromised state in an otherwise healthy child with more severe VZV disease. For these reasons, we decided to investigate the cytokine inflammatory profiles in the CSF samples from both patients who had received the corticosteroid bursts.

Table 1. History of corticosteroid burst before varicella meningitis.

Category	Patient 13	Patient 12
Gender	Male	Female
1st Vaccine	1 Year	1 Year
2nd Vaccine	4 Years	5 Years
Meningitis	7 Years	14 Years
Head Image	Normal	Normal
Prior Zoster	No	Yes
VZV type	Vaccine	Vaccine
Corticosteroids	Prednisolone	Prednisone
Time Before Meningitis	3 Weeks	4 Weeks
Daily Dosage	20 mg	20 mg

Table 1. *Cont.*

Category	Patient 13	Patient 12
Mg/Kg/Day	0.8 mg/kg	0.3 mg/kg
Duration	7 Days	5 days
Reason for Steroids	Poison Ivy	Asthma
Follow-up	Recovered	Recovered

3.5. Cytokine Profiling in the Cerebrospinal Fluid

An archived CSF sample from case 12 had been stored frozen [6]. CSF samples from cases 12 and 13 were examined in a 35-well cytokine multiplex assay (Table 2). The 10 most elevated cytokines in our patients' CSF samples included interferon gamma, interleukins IL-1RA; IL-6, IL-8, IL-10, and IL-17F; chemokines CCL2, CXCL9, and CXCL10; as well as G-CSF. Of note, IL-8 is elevated in CSF of adult cases of severe herpes zoster [19]. Granulocyte colony-stimulating factor (G-CSF) was first shown to be elevated in acute viral meningitis many years ago [20]. A table of normal values is included in Table S1.

Table 2. Cytokine levels in the cerebrospinal fluid.

Cytokine	PT. 13	PT. 12	Cytokine	PT. 13	PT. 12
IL-1A	3.19	1.02	IFNa	104.44	38.79
IL-1RA	184.63	1620.65	IFNg	75.76	865.85
IL-1b	6.14	3.70	MIP1a	39.27	117.88
IL-2	8.74	12.63	CXCL10	1178.58	507.83
IL-2R	104.40	79.92	CCL11	3.91	2.86
IL-3	10.42	11.61	Rantes	115.95	30.76
IL-4	16.42	23.54	GM-CSF	1.74	3.21
IL-5	43.76	11.55	TNFa	75.76	1.07
IL-6	90.57	31,816.58	HGF	93.61	187.21
IL-7	10.22	11.99	MIP-1b	91.13	156.86
IL-8	4034.30	11,604.79	CCL2	616.79	3025.46
IL-9	8.83	6.50	FGF2	28.21	15.03
IL-10	75.39	846.40	VEGF	2.35	4.36
IL-12	72.32	39.31	G-CSF	55.79	171.57
IL-13	15.48	18.67	CXCL9	81.16	273.94
IL-15	56.41	26.00	EGF	48.25	42.61
IL-17A	7.22	3.67			
IL-17F	348.68	72.26			
IL-22	22.12	70.20			

Note: The full name for each abbreviation is given in the catalog of Online Mendelian Inheritance in Man (OMIM^{R}) or HUGO (Human Genome Organization dictionary).

3.6. Case 14 and Immunology Exome Sequencing

A previously healthy 12-year-old boy presented to the emergency department with a 1-day history of fever, headache, phonophobia and photophobia. His immunizations were up to date, including varicella vaccinations (Varivax) at 14 months and 30 months. Because the child had neck stiffness, a lumbar puncture was performed. Laboratory analysis of CSF revealed 83 white cells/mm3 (88% lymphocytes). The CSF was tested by the meningitis panel and found to be positive for VZV and negative for enterovirus, herpes simplex virus, and HHV6, as well as bacteria. The child was hospitalized, and acyclovir IV (30 mg/kg/day) was initiated. No radiologic studies were ordered. On day 5, he was discharged home; his acyclovir was switched to oral administration for a total treatment course of 14 days. The virus in the CSF was confirmed to be the VZV vaccine strain by Sanger sequencing, as described in methods by the Centers for Disease Control [9,10]. Specifically, the fixed alleles found in the vaccine strain are included in Table 1 [10]. Because we were aware of results from case 13, we also performed a PCR analysis for HHV-7 DNA on the CSF; that test was negative.

Patient 14 also had an extensive immunology evaluation, which showed normal serum immunoglobulin IgG, IgA, and IgM levels and normal serum levels of vaccine antibodies for all tested antigens (measles, varicella, tetanus, and diphtheria). Anti-VZV antibodies were detected using a commercial VZV ELISA kit. An extended immunophenotyping showed normal subpopulations of B, T, and NK cells. Finally, genetic analysis was performed by immunology exome sequencing, which included a panel of 356 genes of primary immunodeficiency (Table 4). However, no errors in innate immune genes or other rare primary immunodeficiencies were discovered; specifically, there were no errors in RNA polymerase III genes (POLR3A and POLR3C) [21]. RNA polymerase III is a sensor of foreign DNA in the cytosol; errors in RNA polymerase III genes have been found in some people with severe wild-type varicella meningoencephalitis [21]. Furthermore, a frequently cited gene involved in susceptibility to HSV1 meningoencephalitis (TLR3) was intact [22]. The child had undergone a complete recovery when seen at 1 and 12 months after his hospitalization.

Table 3. Immunology exome sequencing of a panel of 356 genes.

ACD	C8B	CFP	FADD	IL10RA	LPIN2	NSMCE3	RSGRP1	SPINK5	TRAC
ACP5	C9	CFTR	FAS	IL10RB	LRBA	OAS1	RBCK1	SPPL2A	TRAF3IP2
ADA	CARD11	CHD7	FASLG	IL12B	LYST	ORAI1	RC3H1	SRP54	TREX1
ADA2	CARD14	CIB1	FAT4	IL12RB1	MAGT1	OTULIN	RECQL4	STAT1	TRIM22
ADAM17	CARD9	CIITA	FCGR3A	IL17F	MALT1	PARN	RELA	STAT2	TRNT1
ADAR	CARMIL2	CLPB	FCHO1	IL17RA	MAP3K14	PAX1	RFX5	STAT3	TTC37
AICDA	CASP10	COPA	FERMT3	IL17RC	MBL2	PEPD	RFXANK	STAT5B	TTC7A
AIRE	CASP8	CORO1A	FNIP1	IL1RN	MCM4	PGM3	RFXAP	STIM1	TYK2
AK2	CCBE1	CSF2RA	FOXN1	IL21	MEFV	PIK3CD	RHOH	STING1	UNC13D
ALPI	CD19	CSF2RB	FOXP3	IL21R	MOGS	PIK3CG	RIPK1	STK4	UNC93B1
AP1S3	CD247	CSF3R	FPR1	IL2RA	MPO	PIK3R1	RNASEH2A	STX11	UNG
AP3B1	CD27	CTC1	G6PC3	IL2RB	MRTFA	PLCG2	RNASEH2B	STXBP2	USB1
AP3D1	CD3D	CTLA4	G6PD	IL2RG	MSN	PMS2	RNASEH2C	TAP1	USP18
ARPC1B	CD3E	CTPS1	GATA1	IL36RN	MTHFD1	PNP	RNF168	TAP2	VPS13B
ATM	CD3G	CTSC	GATA2	IL6R	MVK	POLA1	RNF31	TAPBP	VPS45
ATP6AP1	CD4	CXCR4	GFI1	IL6ST	MYD88	POLD1	RORC	TAZ	WAS
B2M	CD40	CYBA	GINS1	IL7R	MYO5B	POLE	RPSA	TBK1	WDR1
BACH2	CD40LG	CYBB	GUCY2C	INO80	MYSM1	POLR3A	RTEL1	TBX1	WIPF1
BCL10	CD46	CYBC1	HAVCR2	IRAK4	NBN	POLR3C	SAMD9	TCF3	XIAP
BCL11B	CD55	DBR1	HAX1	IRF3	MCF1	POMP	SAMD9L	TCN2	ZAP70
BLM	CD59	DCLRE1B	HELLS	IRF7	NCF2	PRF1	SAMHD1	TERT	ZBTB24
BLNK	CD70	DCLRE1C	HPS1	IRF8	NCF4	PRKCD	SBDS	TET2	ZNF341
BLOC1S6	CD79A	DEF6	HPS4	IRF9	NCKAP1L	PRKDC	SEC61A1	TFRC	CR2
BTK	CD79B	DKC1	HPS6	ISG15	NCSTN	PSENEN	SERPING1	TGFB1	

Table 4. Immunology exome sequencing of a panel of 356 genes.

C1QA	CD81	DNAJC21	HTRA2	ITCH	NFE2L2	PSMA3	SGPL1	TICAM1
C1QB	CD8A	DNASE1L3	ICOS	ITGB2	NFKB1	PSMB10	SH2D1A	TINF2
C1QC	CDC42	DNASE2	IFIH1	ITK	NFKB2	PSMB4	SKIV2L	TLR3
C1R	CDCA7	DNMT3B	IFNAR1	ITPKB	NFKBIA	PSMB8	SLC29A3	TLR7
C1S	CEBPE	DOCK2	IFNGR1	IVNS1ABP	NHEJ1	PSMB9	SLC35C1	TMC6
C2	CFB	DOCK8	IFNGR2	JAGN1	NHP2	PSTPIP1	SLC37A4	TMC8
C3	CFD	EFL1	IGHM	JAK3	NLRC4	PTEN	SLC39A7	TNFAIP3
C4A	CFH	ELANE	IGKC	KRAS	NLRP1	PTPN2	SLC46A1	TNFRSF11A
C4B	CFHR1	EPG5	IGLL1	LAMTOR2	NLRP12	PTPRC	SLC7A7	TNFRSF13C
C5	CFHR3	ERBIN	IKBKB	LAT	NLRP3	RAB27A	SMARCAL1	TNFRSF1A
C6	CFHR4	ERCC6L2	IKBKG	LCK	NOD2	RAC2	SMARCD2	TNFRSF9
C7	CFHR5	EXTL3	IKZF1	LIG1	NOP10	RAG1	SOCS1	TOP2B
C8A	CFI	F12	IL10	LIG4	NRAS	RAG2	SP110	TPP2

Note: The full name for each abbreviation is given in the catalog of Online Mendelian Inheritance in Man (OMIM^{R}).

4. Discussion

Varicella vaccine meningitis presumably represents the clinical manifestations of herpes zoster of the trigeminal ganglion, whereby afferent fibers carry reactivated virus to the meninges [23,24]. An alternative source of virus is the superior cervical ganglion, which also houses latent virus and also innervates the dura [25]. This serious adverse event is rare. Only one case of varicella vaccine meningitis had been described before 2006 [26], when the CDC switched their recommendation from one to two varicella vaccine doses for children (Figure 3). With inclusion of the 2 new cases reported herein, the current tally is 14 cases (Figure 3). This tally includes eight children who only received one varicella vaccination (Table 5) [3,4]. There are now six case reports of varicella vaccine meningitis in twice-immunized immunocompetent adolescents [6,27,28]. There is also no pattern in the timing between meningitis and a second varicella vaccination (Table 5). We have recently reviewed the epidemiology of wild-type varicella meningitis in immunocompetent children [5]. Among the published cases, about one-half are preceded by clinical herpes zoster and one-half are not preceded by herpes zoster. Of interest, the majority of cases of wild-type varicella meningitis have occurred in adolescents, a possible suggestion of waning immunity since a bout of chickenpox in early childhood. The adolescents have recovered with no permanent neurological sequelae.

4.1. Corticosteroids and VZV Reactivation

In 2021, a group from Taiwan surveyed the medical records of 4.5 million children in their national database and reported an association between oral corticosteroid bursts and increased severe adverse events in children, including gastrointestinal bleeding, sepsis, pneumonia, and glaucoma [29]. Case 13 was diagnosed with poison ivy on the right knee 3 weeks before his bout of meningitis. After a visit to a local clinic, he was treated with prednisolone 20 mg daily for 7 days (Table 1). Based on this observation, we reviewed all prior case reports of varicella vaccine meningitis in immunocompetent children who had received either one or two vaccinations. Reports of meningitis in the once-immunized children contained no information about corticosteroid prescriptions. Among the three prior cases of varicella vaccine meningitis in immunocompetent children who had received two varicella vaccinations, the report of case 12 mentioned a prior corticosteroid burst.

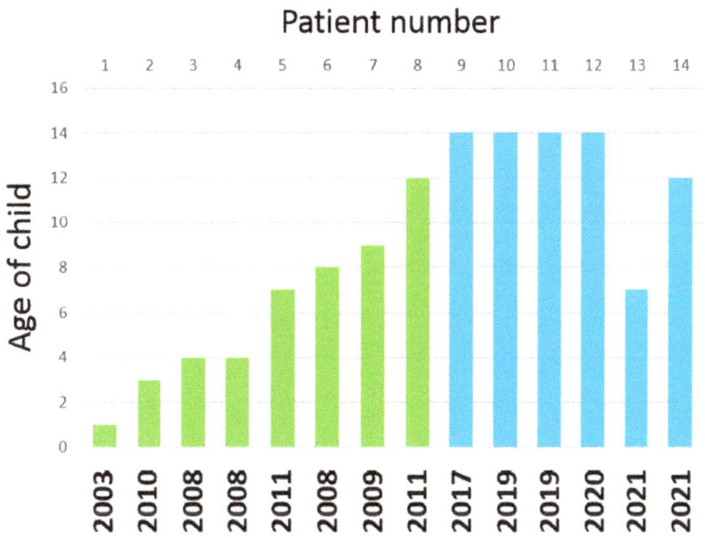

Figure 3. Fourteen cases of varicella vaccine meningitis. The cases are arranged first by whether the child had one varicella vaccination (green bar) or two varicella vaccinations (blue bar). Then, the height of each bar graph indicates the age of the patients when they developed meningitis. The year under each bar indicates the year in which the case was first reported in the medical literature. Data about the first 12 cases are taken from Reference 5.

Table 5. Interval between varicella vaccination and varicella meningitis.

Case	Age	Vaccine	Interval-1	Interval-2
1	1.2 year	1	11 weeks	–
2	3.5 years	1	20 months	–
3	4 years	1	18 months	–
4	4 years	1	32 months	–
5	7 years	1	6 years	–
6	8 years	1	7 years	–
7	9 years	1	8 years	–
8	12 years	1	11 years	–
9	14 years	2	13 years	2 years
10	14 years	2	13 years	10 years
11	14 years	2	13 years	4 years
12	14 years	2	13 years	9 years
13	7 years	2	6 years	2 years
14	12 years	2	11 years	9 years

Note: Interval-1 is the time between meningitis and the first varicella vaccination; interval-2 is the time between meningitis and the second varicella vaccination, among the group who had received two vaccinations. Age indicates the age when meningitis was diagnosed. Dash (–) means none.

There has been extensive reporting on adverse consequences in people who received corticosteroids prior to or during VZV infection, dating back almost 70 years, long before acyclovir was available [30,31]. Corticosteroids increased the likelihood of disseminated varicella and subsequent death; corticosteroids also facilitated VZV reactivation to cause herpes zoster [32]. Prednisone dosages usually associated with these severe consequences ranged from 0.5 to 2 mg/kg/day. Based on these publications, treatment with a prednisolone burst at a relatively high dosage likely allowed latent vaccine virus to reactivate in case 13. Further, we again point out the serological data that showed frequent subclinical reactivations in healthy children who had received varicella vaccination [1]. If any vaccinated child received a steroid burst during a subclinical reactivation, perhaps that child might develop clinical herpes zoster and even varicella vaccine meningitis.

4.2. Specificity of mNGS and the Biofire Filmarray Encephalitis Panel

In an analysis of specificity of the Biofire meningitis/encephalitis panel by the Biofire Company, they reported 1/7 false-positive results for VZV [33]. With regard to VZV specificity, we point out the overall importance of mNGS methodology for finalizing the primary diagnosis from an extremely small CSF sample (90 μL). Indeed, mNGS has the capability to assess genomic data in a sample that falls outside targeted PCR primer regions, allowing for better strain identification. Furthermore, any SNPs that may be present at prime sites leading to PCR failure could be captured by mNGS.

The data about HHV6 and HHV7 are a true conundrum. When patient 13 was hospitalized, we initially considered a positive HHV6 result in CSF by Biofire assay to be a false positive [34,35]. When reviewing the mNGS data, however, a new hypothesis emerged. We observed two 85 base-pair reads specific for HHV7, not HHV6 (Figure 2). When considered along with that result, we now think that there are two possible explanations for the positive PCR for HHV6 on the Biofire panel: (i) The Biofire result was a true positive, but the amount of HHV6 DNA detectable by the Biofire assay was below the limits of detection by mNGS and (ii) the Biofire result for HHV6 was a false positive in that primers for HHV6 in the Biofire assay amplified conserved sequences in the closely related HHV7 genome. In an analysis of specificity in the original publications about the Biofire assay by the Biofire company, they observed that HHV6 had the highest number of false positives, but they did not mention the possibility of HHV7 amplification by the HHV6 primers in the Biofire kit. The HHV7 U31 gene was discovered in the mNGS analysis, but that gene is unlikely to be a target in the Biofire kit because there is only a 46% identity to its HHV6 homolog [36]. However, the Biofire kit could be recognizing a HHV7 gene such as U77, which has 75% identity to its HHV6 homolog [36].

We also note that HHV7 was codetected in the CSF of patient 13 but not patient 14. The severity of meningitis was much greater in patient 13 than patient 14, as documented by the decision not to perform any CNS imaging on patient 14. A strikingly similar result was found earlier in a large virological analysis of childhood meningoencephalitis—namely, HHV7 was codetected in 2 of 5 cases of Epstein–Barr virus meningoencephalitis [37]. This pattern fits into a hypothesis that HHV7 codetection may be a marker for increased severity of viral meningoencephalitis.

4.3. Cytokine Profiles in CSF after Neurotropic Virus Infection

The cytokine profiles from cases 12 and 13 were very revealing when compared with the cytokine profiles in CSF samples from patients with enterovirus, HSV1, HSV2, and VZV CNS infections (Figure 4). The concentrations in the figure represent the highest values for each cytokine within each study group [38–46]. We selected IL-6, IL-8, IL-10, and CXCL-10 (IP-10) for inclusion in the figure because there were several papers that included these cytokines. As is apparent from reviewing the table, there is a commonality to the cytokine profiles in the CSF during both enterovirus and alpha herpesvirus CNS infections. Of note for VZV, extremely high IL-6 levels are also seen in a human skin organ culture model for VZV infection [47]. Because lumbar punctures are not performed in healthy children, there are minimal data on normal cytokine levels in the cerebrospinal fluids.

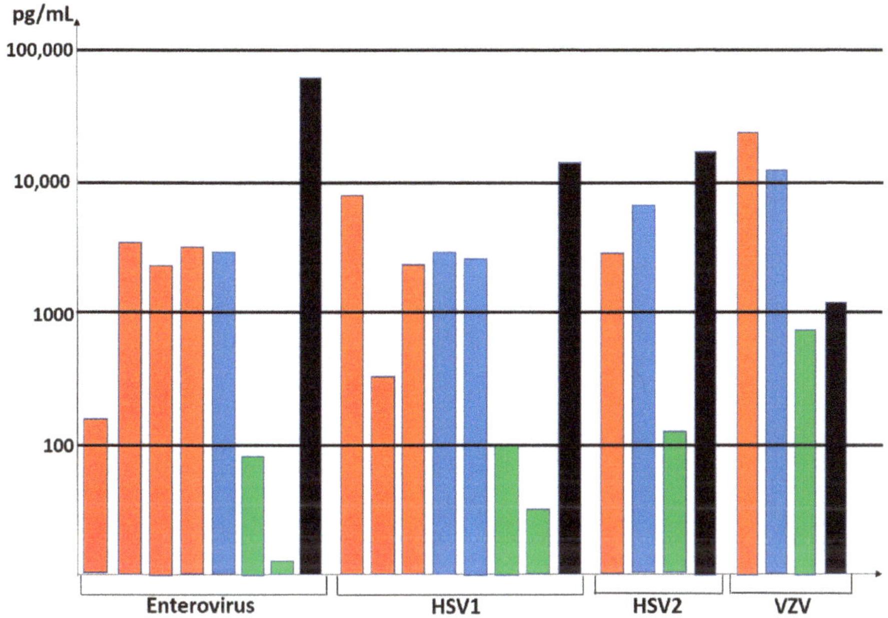

Figure 4. Cytokine levels in CSF from patients with HSV1, HSV2, enterovirus, and VZV infections in the central nervous system. Cytokine data about HSV1, HSV2, and enterovirus were obtained from References 38–46; data about VZV are from the current report. Bars: red = IL-6; blue = IL-8; green = IL-10; black = CXCL10.

4.4. Elevated CXCL10 in CSF

The importance of elevated CXCL10 in the CSF deserves further discussion. Increased CXCL10 production in the dorsal root ganglion (DRG) has been shown to be a manifestation of an antiviral host response to peripheral neuronal infection by wild-type VZV [48]. Increased CXCL10 has also been found in the CSF of patients with wild-type VZV infection [49]. Furthermore, increased CXC10 production has been documented in the DRG following reactivation of the closely related simian varicella virus [50]. Thus, the inflammatory response to varicella vaccine virus infection in the CNS closely mimics the response to wild-type virus infection in both the CNS and the peripheral nervous system [49].

5. Conclusions

To provide perspective, universal varicella vaccination has been the national public health policy in the United States for decades [51,52]. Over 200 million doses have administered, and the safety profile of the live attenuated varicella vaccine is excellent [53,54]. Although impaired for reactivation from latency, the vaccine virus certainly does enter ganglia and reactivate [55]. An original hypothesis put forward by Hope-Simpson was the need for periodic subclinical VZV reactivations to maintain VZV immunity and delay clinical herpes zoster [56]. An overview of VZV evolution also suggests that VZV reactivation is a mechanism by which the virus achieves fitness throughout the lifetime of its human host [57]. Therefore, it seems likely that periodic reactivations, usually subclinical but occasionally clinical, are required after live varicella vaccination to maintain immunity within immunized children into early adulthood [1,2]. We encourage continued varicella vaccination of young children [58].

Supplementary Materials: The following are available online at https://www.mdpi.com/article/10.3390/v13112286/s1, Figure S1: Additional data about mNGS reads, Table S1: Normal values for blood and cerebrospinal fluid tests.

Author Contributions: Conceptualization, M.R.W., R.J.C. and C.G.; methodology and investigation, P.S.R., M.R.W., D.B., S.B., T.Y., A.W., G.B.-R., E.H.H., J.E.C., W.J., B.A.F. and C.G.; validation of genomic sequence data, P.S.R., M.R.W., J.E.C., C.G. and R.J.C.; patient management, G.C., G.B.-R., N.S. and C.G.; writing—original draft preparation, all authors.; writing—review and editing, P.S.R., M.R.W., D.B., B.A.F. and C.G. All authors have read and agreed to the published version of the manuscript.

Funding: P.S.R. was supported by an American Academy of Neurology Clinical Research Training Scholarship (P0534134); M.R.W. was supported by NINDS grant NS096117 and the Rachleff Family Foundation; C.G. was supported by NIAID grant AI153817.

Institutional Review Board Statement: The study was conducted according to the guidelines of the Declaration of Helsinki, and the study was approved by the Institutional Review Board of the University of California San Francisco, Study title: Pathogen and autoantibody discovery in unexplained diseases of the nervous system, IRB number13-12236, dated 12 August 2019.

Informed Consent Statement: Informed consent was obtained from all subjects involved in the study.

Acknowledgments: R.J.C. was a collaborating member of the research team supported by NIH grant AI153817 at the University of Iowa; he died on 30 July 2021.

Conflicts of Interest: The authors declare no conflict of interest.

References

1. Krause, P.R.; Klinman, D.M. Varicella vaccination: Evidence for frequent reactivation of the vaccine strain in healthy children. *Nat. Med.* **2000**, *6*, 451–454. [CrossRef]
2. Harpaz, R.; Leung, J.W. The Epidemiology of Herpes Zoster in the United States during the Era of Varicella and Herpes Zoster Vaccines: Changing Patterns among Children. *Clin. Infect. Dis.* **2019**, *69*, 345–347. [CrossRef]
3. Pahud, B.A.; Glaser, C.A.; Dekker, C.L.; Arvin, A.M.; Schmid, D.S. Varicella zoster disease of the central nervous system: Epidemiological, clinical, and laboratory features 10 years after the introduction of the varicella vaccine. *J. Infect. Dis.* **2011**, *203*, 316–323. [CrossRef]
4. Horien, C.; Grose, C. Neurovirulence of varicella and the live attenuated varicella vaccine virus. *Semin. Pediatr. Neurol.* **2012**, *19*, 124–129. [CrossRef] [PubMed]
5. Heusel, E.H.; Grose, C. Twelve Children with Varicella Vaccine Meningitis: Neuropathogenesis of Reactivated Live Attenuated Varicella Vaccine Virus. *Viruses* **2020**, *12*, 1078. [CrossRef]
6. Ramachandran, V.; Elliott, S.C.; Rogers, K.L.; Cohrs, R.J.; Weinberger, M.; Jackson, W.; Carpenter, J.E.; Grose, C.; Bonthius, D.J. Varicella Vaccine Meningitis as a Complication of Herpes Zoster in Twice-Immunized Immunocompetent Adolescents. *J. Child Neurol.* **2020**, *35*, 889–895. [CrossRef]
7. Wilson, M.R.; Sample, H.A.; Zorn, K.C.; Arevalo, S.; Yu, G.; Neuhaus, J.; Federman, S.; Stryke, D.; Briggs, B.; Langelier, C.; et al. Clinical Metagenomic Sequencing for Diagnosis of Meningitis and Encephalitis. *N. Engl. J. Med.* **2019**, *380*, 2327–2340. [CrossRef] [PubMed]
8. Wilson, M.R.; O'Donovan, B.D.; Gelfand, J.M.; Sample, H.A.; Chow, F.C.; Betjemann, J.P.; Shah, M.P.; Richie, M.B.; Gorman, M.P.; Hajj-Ali, R.A.; et al. Chronic Meningitis Investigated via Metagenomic Next-Generation Sequencing. *JAMA Neurol.* **2018**, *75*, 947–955. [CrossRef]
9. Quinlivan, M.L.; Jensen, N.J.; Radford, K.W.; Schmid, D.S. Novel genetic variation identified at fixed loci in ORF62 of the Oka varicella vaccine and in a case of vaccine-associated herpes zoster. *J. Clin. Microbiol.* **2012**, *50*, 1533–1538. [CrossRef] [PubMed]
10. Schmid, D.S. *Varicella-zoster* virus vaccine: Molecular genetics. *Curr. Top. Microbiol. Immunol.* **2010**, *342*, 323–340. [CrossRef]
11. Baleydier, F.; Ranza, E.; Schappi, M.; Rougemont, A.L.; Merlini, L.; Ansari, M.; Blanchard-Rohner, G. Activated Phosphoinositide 3 Kinase Delta Syndrome (APDS): A Primary Immunodeficiency Mimicking Lymphoma. *J. Pediatr. Hematol. Oncol.* **2019**, *41*, e521–e524. [CrossRef]
12. Amberger, J.S.; Bocchini, C.A.; Schiettecatte, F.; Scott, A.F.; Hamosh, A. OMIM.org: Online Mendelian Inheritance in Man (OMIM(R)), an online catalog of human genes and genetic disorders. *Nucleic Acids Res.* **2015**, *43*, D789–D798. [CrossRef]
13. Williams, V.; Gershon, A.; Brunell, P.A. Serologic response to *Varicella-zoster* membrane antigens measured by direct immunofluorescence. *J. Infect. Dis.* **1974**, *130*, 669–672. [CrossRef]
14. Yoshikawa, T.; Asano, Y.; Kobayashi, I.; Nakashima, T.; Yazaki, T.; Suga, S.; Ozaki, T.; Wyatt, L.S.; Frenkel, N. Seroepidemiology of human herpesvirus 7 in healthy children and adults in Japan. *J. Med. Virol.* **1993**, *41*, 319–323. [CrossRef] [PubMed]
15. Cook, D.B.; McLucas, B.C.; Montoya, L.A.; Brotski, C.M.; Das, S.; Miholits, M.; Sebata, T.H. Multiplexing protein and gene level measurements on a single Luminex platform. *Methods* **2019**, *158*, 27–32. [CrossRef]
16. Gomi, Y.; Sunamachi, H.; Mori, Y.; Nagaike, K.; Takahashi, M.; Yamanishi, K. Comparison of the complete DNA sequences of the Oka varicella vaccine and its parental virus. *J. Virol.* **2002**, *76*, 11447–11459. [CrossRef]

17. Peters, G.A.; Tyler, S.D.; Carpenter, J.E.; Jackson, W.; Mori, Y.; Arvin, A.M.; Grose, C. The attenuated genotype of *Varicella-Zoster* virus includes an ORF0 transitional stop codon mutation. *J. Virol.* **2012**, *86*, 10695–10703. [CrossRef]
18. Depledge, D.P.; Yamanishi, K.; Gomi, Y.; Gershon, A.A.; Breuer, J. Deep Sequencing of Distinct Preparations of the Live Attenuated *Varicella-Zoster* Virus Vaccine Reveals a Conserved Core of Attenuating Single-Nucleotide Polymorphisms. *J. Virol.* **2016**, *90*, 8698–8704. [CrossRef]
19. Kotani, N.; Kudo, R.; Sakurai, Y.; Sawamura, D.; Sessler, D.I.; Okada, H.; Nakayama, H.; Yamagata, T.; Yasujima, M.; Matsuki, A. Cerebrospinal fluid interleukin 8 concentrations and the subsequent development of postherpetic neuralgia. *Am. J. Med.* **2004**, *116*, 318–324. [CrossRef]
20. Fukushima, K.; Ishiguro, A.; Shimbo, T. Transient elevation of granulocyte colony-stimulating factor levels in cerebrospinal fluid at the initial stage of aseptic meningitis in children. *Pediatr. Res.* **1995**, *37*, 160–164. [CrossRef]
21. Carter-Timofte, M.E.; Paludan, S.R.; Mogensen, T.H. RNA Polymerase III as a Gatekeeper to Prevent Severe VZV Infections. *Trends Mol. Med.* **2018**, *24*, 904–915. [CrossRef]
22. Zhang, S.Y.; Jouanguy, E.; Ugolini, S.; Smahi, A.; Elain, G.; Romero, P.; Segal, D.; Sancho-Shimizu, V.; Lorenzo, L.; Puel, A.; et al. TLR3 deficiency in patients with herpes simplex encephalitis. *Science* **2007**, *317*, 1522–1527. [CrossRef] [PubMed]
23. Esiri, M.M.; Tomlinson, A.H. Herpes Zoster. Demonstration of virus in trigeminal nerve and ganglion by immunofluorescence and electron microscopy. *J. Neurol. Sci.* **1972**, *15*, 35–48. [CrossRef]
24. LaGuardia, J.J.; Cohrs, R.J.; Gilden, D.H. Prevalence of *Varicella-Zoster* virus DNA in dissociated human trigeminal ganglion neurons and nonneuronal cells. *J. Virol.* **1999**, *73*, 8571–8577. [CrossRef] [PubMed]
25. Uddman, R.; Edvinsson, L.; Hara, H. Axonal tracing of autonomic nerve fibers to the superficial temporal artery in the rat. *Cell Tissue Res.* **1989**, *256*, 559–565. [CrossRef]
26. Levin, M.J.; Dahl, K.M.; Weinberg, A.; Giller, R.; Patel, A.; Krause, P.R. Development of resistance to acyclovir during chronic infection with the Oka vaccine strain of *Varicella-Zoster* virus, in an immunosuppressed child. *J. Infect. Dis.* **2003**, *188*, 954–959. [CrossRef]
27. Chen, Y.C.; James, A.; Kung, E.; Madhavan, V. A Case of Herpes Zoster and Meningitis in a Twice-Vaccinated Healthy Adolescent. *J. Pediatr. Infect. Dis.* **2017**, *12*, 142–144. [CrossRef]
28. Harrington, W.E.; Mato, S.; Burroughs, L.; Carpenter, P.A.; Gershon, A.; Schmid, D.S.; Englund, J.A. Vaccine Oka Varicella Meningitis in Two Adolescents. *Pediatrics* **2019**, *144*, e20191522. [CrossRef] [PubMed]
29. Yao, T.C.; Wang, J.Y.; Chang, S.M.; Chang, Y.C.; Tsai, Y.F.; Wu, A.C.; Huang, J.L.; Tsai, H.J. Association of Oral Corticosteroid Bursts with Severe Adverse Events in Children. *JAMA Pediatr.* **2021**, *175*, 723–729. [CrossRef]
30. Shee, J.C.; Fehrsen, P. Reactivation of varicella virus by cortisone therapy. *Br. Med. J.* **1953**, *2*, 82. [CrossRef]
31. Good, R.A.; Smith, R.T.; Vernier, R.L. Serious untoward reactions to therapy with cortisone and adrenocorticotropin in pediatric practice. II. *Pediatrics* **1957**, *19*, 272–284.
32. Price, N.B.; Grose, C. Corticosteroids Contribute to Serious Adverse Events Following Live Attenuated Varicella Vaccination and Live Attenuated Zoster Vaccination. *Vaccines* **2021**, *9*, 23. [CrossRef]
33. Leber, A.L.; Everhart, K.; Balada-Llasat, J.M.; Cullison, J.; Daly, J.; Holt, S.; Lephart, P.; Salimnia, H.; Schreckenberger, P.C.; DesJarlais, S.; et al. Multicenter Evaluation of BioFire FilmArray Meningitis/Encephalitis Panel for Detection of Bacteria, Viruses, and Yeast in Cerebrospinal Fluid Specimens. *J. Clin. Microbiol.* **2016**, *54*, 2251–2261. [CrossRef]
34. Pandey, U.; Greninger, A.L.; Levin, G.R.; Jerome, K.R.; Anand, V.C.; Dien Bard, J. Pathogen or Bystander: Clinical Significance of Detecting Human Herpesvirus 6 in Pediatric Cerebrospinal Fluid. *J. Clin. Microbiol.* **2020**, *58*, 5. [CrossRef] [PubMed]
35. Green, D.A.; Pereira, M.; Miko, B.; Radmard, S.; Whittier, S.; Thakur, K. Clinical Significance of Human Herpesvirus 6 Positivity on the FilmArray Meningitis/Encephalitis Panel. *Clin. Infect. Dis.* **2018**, *67*, 1125–1128. [CrossRef] [PubMed]
36. Nicholas, J. Determination and analysis of the complete nucleotide sequence of human herpesvirus. *J. Virol.* **1996**, *70*, 5975–5989. [CrossRef]
37. Parisi, S.G.; Basso, M.; Del Vecchio, C.; Andreis, S.; Franchin, E.; Bello, F.D.; Pagni, S.; Biasolo, M.A.; Manganelli, R.; Barzon, L.; et al. Virological testing of cerebrospinal fluid in children aged less than 14 years with a suspected central nervous system infection: A retrospective study on 304 consecutive children from January 2012 to May 2015. *Eur. J. Paediatr. Neurol.* **2016**, *20*, 588–596. [CrossRef]
38. Maric, L.S.; Lepej, S.Z.; Gorenec, L.; Grgic, I.; Trkulja, V.; Rode, O.D.; Roglic, S.; Grmoja, T.; Barisic, N.; Tesovic, G. Chemokines CXCL10, CXCL11, and CXCL13 in acute disseminated encephalomyelitis, non-polio enterovirus aseptic meningitis, and neuroborreliosis: CXCL10 as initial discriminator in diagnostic algorithm? *Neurol. Sci.* **2018**, *39*, 471–479. [CrossRef] [PubMed]
39. Lind, L.; Studahl, M.; Persson Berg, L.; Eriksson, K. CXCL11 production in cerebrospinal fluid distinguishes herpes simplex meningitis from herpes simplex encephalitis. *J. Neuroinflamm.* **2017**, *14*, 134. [CrossRef] [PubMed]
40. Park, S.E.; Song, D.; Shin, K.; Nam, S.O.; Ko, A.; Kong, J.; Kim, Y.M.; Yeon, G.M.; Lee, Y.J. Prospective research of human parechovirus and cytokines in cerebrospinal fluid of young children less than one year with sepsis-like illness: Comparison with enterovirus. *J. Clin. Virol.* **2019**, *119*, 11–16. [CrossRef]
41. Wang, S.M.; Lei, H.Y.; Su, L.Y.; Wu, J.M.; Yu, C.K.; Wang, J.R.; Liu, C.C. Cerebrospinal fluid cytokines in enterovirus 71 brain stem encephalitis and echovirus meningitis infections of varying severity. *Clin. Microbiol. Infect.* **2007**, *13*, 677–682. [CrossRef]

42. Xu, Y.; Li, S.; Cai, C.; Liu, J.; Wang, Y.; Jiang, Y.; Du, L.; Chen, Z. Characterization of inflammatory cytokine profiles in cerebrospinal fluid of hand, foot, and mouth disease children with enterovirus 71-related encephalitis in Hangzhou, Zhejiang, China. *Medicine* **2019**, *98*, e18464. [CrossRef] [PubMed]
43. Lin, T.Y.; Hsia, S.H.; Huang, Y.C.; Wu, C.T.; Chang, L.Y. Proinflammatory cytokine reactions in enterovirus 71 infections of the central nervous system. *Clin. Infect. Dis.* **2003**, *36*, 269–274. [CrossRef]
44. Kamei, S.; Taira, N.; Ishihara, M.; Sekizawa, T.; Morita, A.; Miki, K.; Shiota, H.; Kanno, A.; Suzuki, Y.; Mizutani, T.; et al. Prognostic value of cerebrospinal fluid cytokine changes in herpes simplex virus encephalitis. *Cytokine* **2009**, *46*, 187–193. [CrossRef]
45. Rosler, A.; Pohl, M.; Braune, H.J.; Oertel, W.H.; Gemsa, D.; Sprenger, H. Time course of chemokines in the cerebrospinal fluid and serum during herpes simplex type 1 encephalitis. *J. Neurol. Sci.* **1998**, *157*, 82–89. [CrossRef]
46. Aurelius, E.; Andersson, B.; Forsgren, M.; Skoldenberg, B.; Strannegard, O. Cytokines and other markers of intrathecal immune response in patients with herpes simplex encephalitis. *J. Infect. Dis.* **1994**, *170*, 678–681. [CrossRef]
47. Jarosinski, K.W.; Carpenter, J.E.; Buckingham, E.M.; Jackson, W.; Knudtson, K.; Moffat, J.F.; Kita, H.; Grose, C. Cellular Stress Response to *Varicella-Zoster* Virus Infection of Human Skin Includes Highly Elevated Interleukin-6 Expression. *Open Forum Infect. Dis.* **2018**, *5*, ofy118. [CrossRef]
48. Steain, M.; Gowrishankar, K.; Rodriguez, M.; Slobedman, B.; Abendroth, A. Upregulation of CXCL10 in human dorsal root ganglia during experimental and natural *Varicella-Zoster* virus infection. *J. Virol.* **2011**, *85*, 626–631. [CrossRef]
49. Lind, L.; Eriksson, K.; Grahn, A. Chemokines and matrix metalloproteinases in cerebrospinal fluid of patients with central nervous system complications caused by *Varicella-Zoster* virus. *J. Neuroinflamm.* **2019**, *16*, 42. [CrossRef]
50. Ouwendijk, W.J.; Abendroth, A.; Traina-Dorge, V.; Getu, S.; Steain, M.; Wellish, M.; Andeweg, A.C.; Osterhaus, A.D.; Gilden, D.; Verjans, G.M.; et al. T-cell infiltration correlates with CXCL10 expression in ganglia of cynomolgus macaques with reactivated simian varicella virus. *J. Virol.* **2013**, *87*, 2979–2982. [CrossRef]
51. Gershon, A.A.; Arvin, A.M.; Levin, M.J.; Seward, J.F.; Schmid, D.S. Varicella vaccine in the United States: A decade of prevention and the way forward. *J. Infect. Dis.* **2008**, *197* (Suppl. 2), S39–S40. [CrossRef]
52. Gershon, A.A.; Breuer, J.; Cohen, J.I.; Cohrs, R.J.; Gershon, M.D.; Gilden, D.; Grose, C.; Hambleton, S.; Kennedy, P.G.E.; Oxman, M.N.; et al. Varicella zoster virus infection. *Nat. Rev. Primers* **2015**, *1*, 15016. [CrossRef] [PubMed]
53. Leung, J.; Marin, M. Update on trends in varicella mortality during the varicella vaccine era-United States, 1990–2016. *Hum. Vaccin Immunother.* **2018**, *14*, 2460–2463. [CrossRef]
54. Woodward, M.; Marko, A.; Galea, S.; Eagel, B.; Straus, W. Varicella Virus Vaccine Live: A 22-Year Review of Postmarketing Safety Data. In *Open Forum Infectious Diseases*; Oxford University Press: Oxford, UK, 2019; Volume 6. [CrossRef]
55. Sadaoka, T.; Depledge, D.P.; Rajbhandari, L.; Venkatesan, A.; Breuer, J.; Cohen, J.I. In vitro system using human neurons demonstrates that *Varicella-Zoster* vaccine virus is impaired for reactivation, but not latency. *Proc. Natl. Acad. Sci. USA* **2016**, *113*, E2403–E2412. [CrossRef]
56. Hope-Simpson, R.E. The Nature of Herpes Zoster: A Long-Term Study and a New Hypothesis. *Proc. R. Soc. Med.* **1965**, *58*, 9–20. [CrossRef] [PubMed]
57. Grose, C. Pangaea and the Out-of-Africa Model of *Varicella-Zoster* Virus Evolution and Phylogeography. *J. Virol.* **2012**, *86*, 9558–9565. [CrossRef] [PubMed]
58. Takahashi, M. Effectiveness of live varicella vaccine. *Expert Opin. Biol. Ther.* **2004**, *4*, 199–216. [CrossRef]

viruses

Review

Impact of Cultured Neuron Models on α-Herpesvirus Latency Research

Angus C. Wilson

Department of Microbiology, School of Medicine, New York University, New York, NY 10012, USA; angus.wilson@nyulangone.org; Tel.: +1-212-263-0206

Abstract: A signature trait of neurotropic α-herpesviruses (α-HV) is their ability to establish stable non-productive infections of peripheral neurons termed latency. This specialized gene expression program is the foundation of an evolutionarily successful strategy to ensure lifelong persistence in the host. Various physiological stresses can induce reactivation in a subset of latently-infected neurons allowing a new cycle of viral productive cycle gene expression and synthesis of infectious virus. Recurring reactivation events ensure transmission of the virus to new hosts and contributes to pathogenesis. Efforts to define the molecular basis of α-HV latency and reactivation have been notoriously difficult because the neurons harboring latent virus in humans and in experimentally infected live-animal models, are rare and largely inaccessible to study. Increasingly, researchers are turning to cultured neuron infection models as simpler experimental platforms from which to explore latency and reactivation at the molecular level. In this review, I reflect on the strengths and weaknesses of existing neuronal models and briefly summarize the important mechanistic insights these models have provided. I also discuss areas where prioritization will help to ensure continued progress and integration.

Keywords: latency; reactivation; HSV-1; HSV-2; VZV; neurons; cell culture

Citation: Wilson, A.C. Impact of Cultured Neuron Models on α-Herpesvirus Latency Research. *Viruses* **2022**, *14*, 1209. https://doi.org/10.3390/v14061209

Academic Editors: Charles Grose, Ravi Mahalingam and Joel Rovnak

Received: 1 April 2022
Accepted: 27 May 2022
Published: 2 June 2022

1. Prologue

The 'fireside chats' hosted by the late Dr. Randall (Randy) J. Cohrs (1952–2021) were a unique element of the annual Colorado Alpha-Herpesvirus Latency Society (CALS) gatherings. With his signature good humor and faux disorganization, Randy would read aloud suggestions for topics of discussion written on scraps of paper by anonymous meeting participants (see Figure 1). Often these talking points touched on fundamental questions already circulating within the α-herpesviruses (α-HV) latency field. How do we operationally define latency? Why is vaccine development against the human simplex viruses proving to be so difficult? How might neurotrophic viruses contribute to Alzheimer's disease (AD) and other devastating neurodegenerative disorders? What are the strengths and limitations of the current model systems? With an eye to the future, Randy made sure the voices of trainees and newly independent investigators were granted equal attention. From his perspective, reaching definitive and actionable answers during these convivial sessions was less important than simulating an open dialogue between friends and colleagues.

Randy championed the idea that the scientific enterprise thrives when there is a free exchange of ideas, a viewpoint shared by his long-standing colleague Dr. Donald (Don) H. Gilden (1937–2016). Together Randy and Don established CALS as the embodiment of this philosophy and quickly recruited a diverse family of loyal attendees from all over the world. Randy's sudden passing in July 2021 was a profound shock and he is sorely missed. In tribute to his substantial contributions to both the substance and practice of virology, this article reflects on a theme that was often touched on during the fireside chats, namely the potential of cultured neuron infection models to bring about major advances in our understanding of α HV latency. By continuing these amicable discussions, I hope to inspire the research community to redouble efforts to embrace and improve on the existing models and to unashamedly draw on new strategies and technologies from other areas of biomedical research. As a field, we need to ensure the new knowledge these models and methodologies will undoubtedly provide is quickly integrated into our shared understanding of the replication biology and pathophysiology of these important viruses.

Figure 1. Randy Cohrs in his element reading out discussion topic suggestions in front of a roaring fire on the last night of CALS 2017. Photograph generously provided with permission by CALS.

2. Part 1: The Ins and Outs of In Vitro Models

The first tissue and cell culture systems (commonly referred to as in vitro, meaning 'within the glass') were introduced in the early 1900s and have been used extensively to this day [1]. Techniques to culture neurons were developed almost immediately and have become increasingly sophisticated with respect to the diversity of neuronal subtypes and different culture environments [2]. This review is focused on neurotrophic α-herpesviruses (α-HV), viruses of clinical and veterinary importance that persist for long periods in the peripheral nervous system through a specialized mode of infection known as latency. There are three α-HV that latently infect humans: herpes simplex virus type 1 (HSV-1), herpes simplex virus type 2 (HSV-2), and varicella zoster virus (VZV), and nearly every person on the planet is eventually colonized by at least one of these viruses. Although live-animal infection models have been used extensively to study latency as well as reactivation, which is the transition back into productive replication, the field is seeing a clear shift towards the use of in vitro models. This has led to major advances in our understanding of the intricate relationship between these viruses and the host neuron and are beginning to reveal the molecular details of how neuronal stress responses alter this relationship allowing the virus to reactivate.

3. Impact of Cell Culture Models in Modern Biology

Building on discoveries made in yeast, fruit flies, plants and other genetically tractable organisms, the now widespread use of cultured cells as experimental models, has allowed generations of researchers to tease out the molecular details of biological processes in ways that are very challenging or impossible at the organismal level. One has only to look at the extent of our current understanding of gene expression, genome replication, cell division, macromolecular trafficking, and metabolic homeostasis to appreciate the bountiful fruits of this endeavor [3]. As the sophistication of cell culture models increases their potential as experimental platforms has expanded with in vitro models being used to explore higher-level biological processes such as pluripotency, circadian rhythms, aging and senescence, neuronal networking, and of course, host-pathogen interactions [4–8].

It is probably fair to say that a large fraction of what we know about the replication of HSV-1, HSV-2, and VZV has come from studies using infected cultured cells [9]. Often the cells used have been highly-transformed cell lines such as Vero, HEK293 and HeLa cells that have only a passing resemblance to the cells of normal tissues. These are favored because they are highly-permissive to viral replication, easy to grow and increasingly easy to manipulate. This strategy works because researchers can demonstrate that the in vitro model recapitulates the biological processes of interest, without necessarily reproducing every aspect of what might happen in vivo, meaning in the living human or animal surrogate. This tacit acknowledgement has been critical to the advancement of biomedical research, rendering a wide spectrum of biological processes, including disease states, accessible to descriptive and manipulative studies, and in many cases to some level of biochemical reconstitution. In the case of detrimental traits, the in vitro modeling of processes that lead to disease frequently accelerates the identification and refinement of pharmacological interventions. Once molecular connections are established in cell culture, they can be revisited in live-animal models or when possible, in humans. This transfer of findings from in vitro to in vivo is becoming easier with the advent of rapid genome engineering methods and the growing use of integrative approaches such as systems biology and data-driven epidemiology.

4. Neuronal Latency Models

In the last 40 years, a variety of in vitro models have been used to study the mechanisms of α-HV latency and reactivation and are eloquently described in other reviews [9–12]. The different models can be organized into three broad categories depending on the neuronal cells they are built upon. First are the models that use primary neurons harvested from fully or partially dissociated peripheral ganglia isolated from rats, mice, or chickens [13–17]. Second are neuron like cells differentiated from transformed cell lines including neuroblastomas of rat or human origin [18–21], and third are human neurons generated by directed differentiation of either induced pluripotent cells (iPSC) or embryonic stem cell (ESC) lines [22–24]. Advantages of primary neurons is that these offer a non-transformed genetic background, and have undergone terminal differentiation and some degree of specialization in vivo. The chief disadvantage is they cannot be amplified and must be continuously isolated from freshly dissected ganglia, which is tedious and expensive. Of note, neurons isolated from embryonic or prenatal animals adapt better to being placed into cell culture but as a consequence, exhibit a less mature neuronal phenotype. As discussed below there is evidence this alters their capacity to establish and maintain α-HV latency.

Neurons generated by controlled differentiation in vitro are easier to obtain in bulk but if derived from an immortal cell line are not considered primary neurons, or if generated from dedifferentiated precursors such as primary fibroblasts, require time-consuming differentiation protocols that in some cases involve an intermediary neural stem cell stage. Fortunately, it is often possible to freeze down intermediary stages and later complete the differentiation as needed thereby accelerating the speed at which studies can be conducted. For completeness it is worth mentioning that a few studies have used non-neuronal cells to establish non-productive infections resembling latency, either through the use of culture conditions such as heat stress that are inhibitory to productive replication, or by infecting with mutant viruses lacking immediate-early genes [25–29]. Although quicker and less expensive to generate than *bona fide* neuronal cultures, none of these non-neuronal models have gained traction, most likely due to concerns about the in vivo relevance of any findings.

The majority of published α-HV latency studies using in vitro models are focused on HSV-1, rather than HSV-2 or VZV, which are equally important human disease agents, if not more so. In the case of VZV, this reflects a poorly understood block to viral replication in cells from rats, mice and hamsters although interestingly, this barrier is reduced in cells from guinea pigs [30,31]. The availability of human neurons generated by guided differentiation of iPSC or ESC should go a long way to redressing this imbalance [24,32].

Careful comparisons of HSV-1 and HSV-2 in murine primary neuron cultures has revealed striking differences in the capacity of sensory and sympathetic neurons to support both produce replication and latency by each virus [33]. Comparative studies will eventually map the viral and host determinants of this selectivity, providing valuable insight into virus-host interactions. Although studied extensively, neurotropic α-herpesviruses of veterinarian importance such as pseudorabies virus (PRV) and bovine herpesvirus 1 (BHV-1) have similarly lagged in terms of in vitro latency studies, presumably because the natural hosts can be accessed experimentally, albeit with some important practical caveats [34,35]. The same rationale applies to simian varicella virus (SVV), a relative of VZV for which there is an animal model albeit with significant financial and ethical limitations [36]. We should expect to see more in vitro latency studies with all of these viruses as cultured neuron models are accepted as legitimate experimental platforms and their usage becomes routine in molecular virology laboratories. Direct virus-to-virus comparisons within the same or similar neuronal models will provide useful information about shared requirements for neuronal factors or processes but might also yield some interesting surprises.

Since 2010, substantial advances have been made using HSV-1 latency/reactivation models based on primary neurons isolated from the superior cervical ganglia (SCG) of prenatal rats and neonatal mice; adaptions of models pioneered by Christine Wilcox and Eugene Johnson in the late 1980's [17,37–39]. Primary SCG cultures are attractive because they can be maintained for long periods using nerve growth factor (NGF) and are considered to be more homogeneous than neurons prepared from other ganglia. Similar models have been established using neurons from other rat peripheral ganglia including dorsal root ganglion (DRG) [38], trigeminal ganglion (TG) [40], geniculate ganglion [41], and vestibular ganglion [42]. Primary neuron infection models have also been established using SCG-, DRG- and TG-derived neurons from both pre-natal and post-natal mice [14,15]. These neurons exhibit different growth factor requirements as well as different capacities to support HSV-1 and HSV-2 latency [15,43,44].

Peripheral ganglia are complex tissues and in addition to the neuronal cell bodies, include large numbers of satellite glial cells (SGCs) and infiltrating immune cells. Typically, these non-neuronal cells (accounting for almost 90% of the cell mass of a ganglion) are removed through the combined action of plating and exposure to mitotic poisons such as 5-fluorouracil and aphidicolin that kill proliferating cells and spare the post-mitotic neurons, which in principle should make up the bulk of the surviving cultures. However, a recent single cell RNA sequencing analysis of rat SCG cultures prepared in this manner found a surprising number of cells displaying transcriptomic signatures of fibroblasts and satellite glia were still present following HSV-1 infection [45]. Conceivably, these cells did not divide during the five or so days of culture establishment and thus like the post-mitotic neurons, which comprised 60–70% of the cultures, were not eliminated. Nonetheless, neither the latency-associated transcript (LAT), a signature of HSV-1 latency in human ganglia, nor viral productive cycle transcripts were detected in the non-neuronal cells suggesting that few if any were infected. Thus, although other cells may be present, neurons appear to be the predominant source of viral genomes and viral gene products.

An area of lingering contention is the use of antivirals during the initial exposure of neuron cultures to infectious virus. This is to prevent unwanted productive replication by the input virus. Even if this happens in just few cells, the resulting superinfection can overwhelm the entire culture before latency is established. While antivirals are not required to establish latency *per se*, they greatly increase the efficiency with which latent cultures can prepared and maintained. Infection in the presence of antivirals also allows for higher infectious doses (typically 1–3 plaque forming units per neuron), which may improve quantitative measurements by lowering the background signal due to uninfected cells. Most infection protocols use acyclovir (ACV), an acyclic guanine analog that that blocks viral DNA replication by inhibiting the HSV-1 DNA polymerase and by acting as a chain-terminator when incorporated into nascent genomes [46]. In some models ACV is supplemented with interferon (IFN) to achieve tighter control [23]. Whether exposure to ACV results in more damage to the DNA of the persistent genomes than is observed with in vivo models is unknown [47]. Methods are available to address this important question, although as will be discussed later we currently lack the tools to distinguish genomes that can actually participate in reactivation events from those that cannot. Careful quantitative studies in mouse models and in humans indicate that individual neurons can carry many copies of the HSV-1 or VZV genome [48–50], but as yet there is very little information on how many genomes are competent to engage in either partial (abortive) or complete programs of productive cycle gene expression. Elegant studies using isogenic reporter viruses expressing different fluorescent proteins have convincingly shown that during productive infection only a handful of HSV-1 or PRV genomes are transcribed in each infected cell indicating there is some as yet uncharacterized limitation imposed by the host cell [51–53]. Studies are needed to determine whether a similar bottleneck applies to genomes undergoing reactivation in neurons.

Although neurons are highly responsive to IFN, they make very little of their own when infected by viruses [54,55]. Instead, the evidence suggests that IFN produced by other cells acts on the infected neurons to control the virus through an unorthodox autophagic response [55]. Potential sources of IFN include infected non-neuronal cells at the site of primary infection [56] as well as infiltrating immune cells within the ganglia [57]. Along these lines, a recent study showed that exposure of murine SCG neurons to type I IFN did not alter the ability of incoming HSV-1 to establish latency but instead limited the potential for reactivation [58]. This was correlated with the entrapment of viral genomes within promyelocytic leukemia nuclear bodies (PML-NBs) that were formed in response to IFN treatment. Rescue can be achieved by depletion of the host PML protein using RNA interference, but this alone was not sufficient to induce reactivation, indicating that other layers of control such as repressive chromatin or transcription factor availability are also important [58].

Developing methods to consistently establish latency-like infections with wild type virus and without recourse to antiviral treatments is the ambition of many laboratories and this may be easier in some neuronal types than others. In the LUHMES model of HSV-1 latency for example, ACV treatment has been shortened to just the first 48 h [18]. It is not clear if modifying protocols to include only brief exposure to an antiviral will be sufficient to eliminate the concerns discussed above. This observation may also be telling us there is a critical window of opportunity at the beginning of the establishment period when the virus can potentially escape intrinsic neuronal control and warrants further study. Neuron age and the accompanying changes in neurotrophic support requirements may also be an important factor in determining whether antivirals are effective. This age-dependence may reflect known changes in epigenetic control mechanisms and differential growth factor receptor expression as neurons commit to a fully differentiated state [59]. There are indications that primary neurons isolated from adult mice are better able to control HSV-1 or HSV-2 than neurons from the equivalent ganglia or pre-natal or neonatal animals and most excitingly, that latent infections capable of reactivation can be established in adult neurons without the use of ACV or IFN [15] A drawback of extending the maturation period is that it adds to the difficulty and expense of generating bulk cultures and further lengthens experimental turnaround times.

Another strategy to establish latency without antivirals is the use of specialized culture devices to infect neurons via the distal axons but not by the cell body or neurites [16,24,60]. This mimics the physical separation of axon terminals located near the sites of primary infection (often in surface tissues such as the skin, wet mucosa, and cornea) from the neuronal cell body located a substantial distance away in the ganglion [61]. While the viral capsid containing the genome is actively transported to the nuclear pores via microtubule networks, is unclear if the viral transcription factor VP16, which is released from the tegument during uncoating, has the capacity to reach the nucleus. As such a plausible hypothesis is that axonal infection allows the heterochromatinization of incoming genomes in the absence of viral tegument factors that would otherwise antagonize this process to sustain productive replication [61,62]. Unfortunately, there are practical limitations to using axonal infections as a means to routinely establish latency without recourse to antivirals. Diffusion barriers made from cloning cylinders and silicone grease have been used to isolate the axons from the cell bodies of dissociated chick ganglia, but in practice it is difficult to achieve and maintain complete fluidic isolation [16]. Preformed microfluidic devices [63] offer a simpler alternative and have been used successfully to establish latent VZV infections of human ESC-derived neurons that could be subsequently reactivated by sequestering NGF with an antibody, but for reasons that remain unclear, not with HDAC or STAT3 inhibitors [24].

A drawback of chamber devices is that only a relatively small numbers of neurons (at most a few hundred) can extend axons through microgrooves separating the two changes and this limited scale is not conducive to many experimental applications. Regardless of these practical constraints, low inoculum axonal infection of rat SCG neuron cultures with PRV resulted in a latency-like infection that could be sustained for more than 20 days without antivirals and could still be reactivated by superinfection with non-replicating UV-inactivated virus [60,64]. Exposure of the cell bodies to viral particles lacking functional genomes (either UV-inactivated or purified light particles) prevented functional PRV introduced via the axons from establishing a latency-like infection [64]. Infection of neonatal rat DRG neuron cultures with HSV-1 recombinants defective for replication and spread in the absence of antivirals found that viral immediate-early (IE) promoter activity was effectively extinguished within 6–8 days but that viral genomes were retained and could reinitiate IE transcription upon withdrawal of NGF or treatment with the deacetylase inhibitor trichostatin A [65]. This suggests a model in which transcriptional silencing is either established quickly yielding a low copy number latency-like infection or silencing takes place more slowly and permits some replication but is ultimately abortive and again resolves as a latent infection. This is supported by recent studies using a new reporter virus (HSV-1 Stayput-GFP) that is fully competent for DNA replication and gene expression but cannot spread from neuron-to-neuron [66]. With this virus, latent infections of murine primary SCG and TG neurons can be readily established and maintained for extended periods in the absence of antiviral treatments. Cultures can then be efficiently reactivated via the two-step mechanism using a novel combination of LY294002, forskolin, and heat shock.

5. Part 2: What Have We Learned?

The use of in vitro latency/reactivation models has yielded a wealth of information on the molecular, and to some extent physical, properties of neurons that allow α-HV to establish the persistent infections that we recognize as latency. A picture is emerging in which intracellular pathways that maintain homeostasis within neurons are also required to maintain the viral latency program. This is consistent with the long-standing idea that latency is imposed on the virus by the neuronal environment but can be antagonized by viral immediate-proteins [67]. This places the spotlight on the first few hours if not minutes after a neuron becomes infected—largely uncharted waters from in terms of direct observation. Maintenance of latency is also an active process requiring continuous intracellular signaling. As the mechanisms responsible for the silencing of productive cycle genes come into focus it will become apparent how these vary between neurons. Ultimately, this will help to explain the neuron subtype preferences of each α-HV and also why reactivation is so asynchronous.

6. Maintenance of Latency

In vitro latency/reactivation models, especially those based on rodent primary neurons, have provided a wealth of information on the signaling pathways required to sustain α-HV latency. This knowledge has emerged through manipulations that can be difficult to use in live-animal models such as small molecule inhibitors that would be toxic or difficult administer [13,14,68]. Likewise, lentiviruses have proven to be effective genetic tools to deliver short-hairpin RNAs to deplete critical neuronal proteins require for organismal viability or to express proteins of interest without inadvertently inducing reactivation [58,68–71]. During development, the survival of sympathetic and sensory neurons is dependent on a signaling cascade initiated by the binding of neurotrophic growth factor (NGF) to the high affinity receptor TrkA [72]. Studies by Christine Wilcox and Eugene Johnson in the late 1980s showed that removal of NGF was an effective inducer of HSV-1 reactivation in primary neuron infection models they had developed [17,37,38]. Subsequent studies in similar rat SCG-derived neuron cultures demonstrated that continuous NGF signaling is required to maintain HSV-1 latency and that transient cessation of translation using a 3-h puromycin pulse triggers reactivation [13,70]. As a result, the cellular serine-threonine kinase AKT (also called protein kinase B) has emerged as a critical nexus (see Figure 2) that couples the control of viral latency to important physiological parameters such as nutrient availability, growth factor signaling, and genome integrity [69]. The three isoforms that constitute AKT help to control the activity of mTORC1, a multiprotein complex with kinase activity that regulates cap-dependent mRNA translation. It is not clear is protein synthesis is required to maintain levels of labile repressive factors or to avoid triggering a stress-response that feeds into the pathways discussed in the next section.

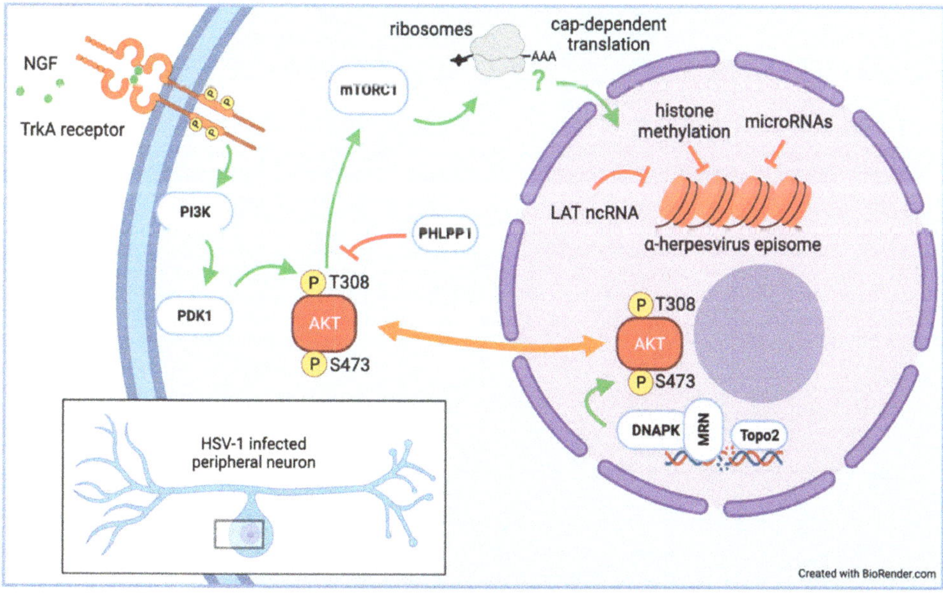

Figure 2. Host kinase AKT integrates NGF signaling with maintenance of chromosomal integrity to maintain HSV-1 latency. Studies in rat SCG neurons have shown that the AKT integrates external signaling from NGF interacting with its cognate receptor TrkA and intracellular DNA damage signaling initiating at topoisomerase 2 (Topo2)-induced DNA breaks. By shuttling between the cytoplasm and nucleus, AKT is kept in an active state through simultaneous phosphorylation of threonine-308 ($T308^P$) by PDK1 in the cytoplasm and of serine 473 ($S473^P$) by DNAPK in the nucleus. The AKT-mTORC1 axis ensures continuous cap-dependent protein synthesis, which is required to maintain the HSV-1 genome in a transcriptionally repressed state. Silencing of viral productive cycle genes involves the combined activities of repressive facultative heterochromatin chromatin, microRNAs, and potentially by lncRNAs such as the viral latency-associated transcript (LAT).

Table 1. Inducers of α-herpesvirus reactivation in neuronal latency models.

Inducer	Molecular Target(s)	Latency Model	Refs.
Nerve growth factor (NGF) depletion	TrkA receptor tyrosine kinase	rat SCG, hESC neurons	[13,17,24,38]
dexamethasone	Glucocorticoid receptor (GR)	murine SCG	[14]
LY294002, Wortmannin	phosphatidylinositol 3-kinase (PI3-K)	rat SCG, murine SCG, human LUHMES	[13,14,18]
sodium butyrate (NaB), trichostatin A (TSA)	(histone) deacetylases (HDACs)	hESC neurons, murine DRG	[23,43]
AKT inhibitor VIII	allosteric AKT inhibitor	rat SCG	[70]
rapamycin, PP242	mTORC1-selective inhibitor (cap-dependent translation)	rat SCG	[70]
puromycin	ribosome (global translation inhibitor)	rat SCG	[70]
hypoxia	4E-BP hyperphosphorylation by mTORC1 (cap-dependent translation)	rat SCG	[70]
mirin	Mre11 nuclease activity of MRN complex	rat SCG	[45,69]
forskolin	adenylate cyclase	murine SCG, ND-PC12	[68,73]
8-Bromo-cAMP	membrane-permeable cAMP derivative	murine SCG	[68]
shRNA depletion of NGF signaling or DNA damage response factors	e.g., PDK1, raptor, Ku80, TOP2β, Gadd45β, Gadd45γ	rat SCG	[13,45,69,70]
bleomycin	radiomimetic, generates DNA breaks	rat SCG	[69]
etoposide	forms ternary complex with DNA & topoisomerase II (generates dsDNA breaks)	rat SCG	[69]
HSF1A	activates heat shock factor 1 (HSF-1)	rat SCG	[45]
HSV superinfection	transactivates viral promoters	rat SCG, chick eTGE	[16,60]
capsaicin	vanilloid receptor-1 (VR-1), Ca2+ flux	rat DRG	[40]
hexamethylene bisacetamide (HMBA)	broad spectrum kinase inhibitor	chick eTGE	[16]

7. Mechanisms of Reactivation

In humans, natural α-HV infections are characterized by periodic reactivation events often linked to environmental and physiological insults that include sunburn, tissue damage, nerve resection, hormonal changes, psychological stress, toxin exposure, and the response to other infectious agents [74]. The identification of specific treatments that induce reactivation in various neuronal culture models has been instructive in terms of exposing the underlying circuitry that unites a seemingly amorphous collection of triggers. The advantage of cell-permeable inducers (see Table 1) is that they can be applied to cultures in a consistent manner, typically by supplementing the culture medium using chemically-synthesized compounds obtained from a commercial source. This standardization should enhance cross-lab reproducibility. Additionally, many inducers act on defined molecular targets that function within well-studied cellular processes such as intracellular signal transduction pathways, homeostatic regulation of protein synthesis and DNA damage response relays.

Interruption of NGF signaling using the PI3 kinase inhibitor LY294002 has been used extensively to reactivate HSV-1 in both rat and murine SCG models [13,14]. This has led to a two-step or bi-phasic model (see Figure 3) to account for the transition from a transcriptionally repressed state to full expression of the productive cycle genes [71]. Viral produce cycle gene expression begins with a synchronous wave of all productive cycle viral transcripts regardless of kinetic class termed Phase I that peaks at 18–20 h post-stimulus. There is evidence that Phase I may not require new viral protein synthesis or transactivation by VP16. Instead, neuronal factors including the stress-response kinases dual leucine-zipper kinase (DLK) and c-Jun N-terminal kinase (JNK) are required [14]. Phosphorylation of serine-10 on histone H (H3S10P) by JNK results in the so-called 'methyl/phospho switch', a curious combinatorial mark that allows transcription by RNA polymerase II (RNAPII) in the presence of the otherwise repressive histone H3 lysine-9 trimethyl (H3K9me3) histone modification [75]. Despite detectable late gene expression, there is no evidence that any viral DNA replication occurs in Phase I and viral promoters remain associated with heterochromatin, raising the possibility that genomes that are 'animated' in this manner but can easily return to the previous latent state. Lastly, it is apparent from highly sensitive RNA FISH that initiation of phase I does not occur simultaneously in all latently infected neurons [45].

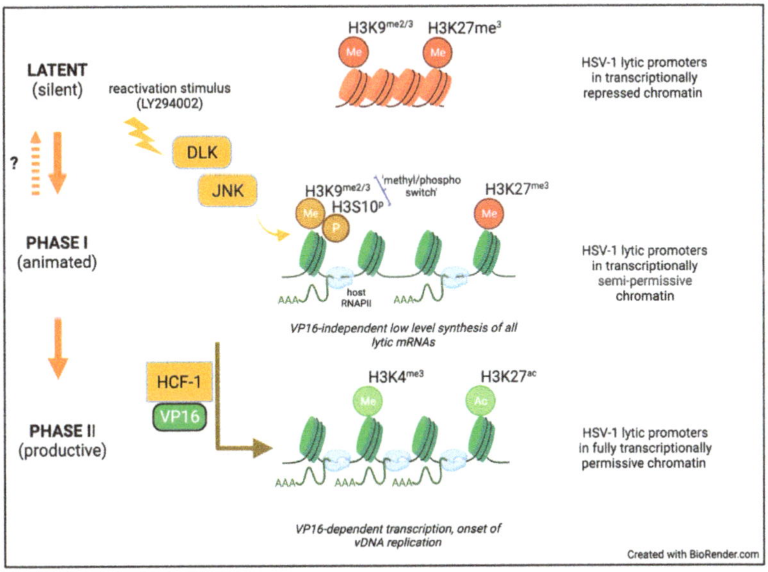

Figure 3. Biphasic model of HSV-1 reactivation. Studies using the rat and mouse SCG neuron infection models have found that HSV-1 reactivation proceeds through two mechanistically distinct steps referred to as Phase I and Phase II. Stresses such as interruption of growth factor signaling activates dual leucine zipper kinase (DLK) which in turn activate JNK resulting in phosphorylation of neuronal transcription factors and the repressive chromatin associated with the viral genome. JNK targets include serine-10 of histone H3 ($H3S10^{P}$), which is adjacent to methylated lysine-9 ($H3K9^{me3}$), a mark of repressive chromatin. This combinatorial mark ('methyl/phospho switch') renders chromatin permissive to RNAPII transcription, allowing widescale but low-level expression of viral productive cycle mRNAs. This transient animation of the viral transcriptome is termed Phase I. Although not yet test directly, dephosphorylation could potentially return animated genomes to their original transcriptionally silent state. The viral regulator VP16 is synthesized during Phase I but accumulates in the cytoplasm and does not influence viral gene expression. However, in a few neurons, VP16 is transported into the nucleus together with the coactivator HCF-1 and together recruit additional chromatin modifiers that remove repressive modifications and likely add activating marks such as methylation on lysine-4 of histone ($H3K4^{me3}$) and/or acetylation of lysine 27 ($H3K27^{ac}$) rendering the chromatin fully permissive for transcription. This VP16-dependent step is termed Phase II and results in higher levels of productive cycle gene transcription and permits the viral DNA genome amplification, which is not detected in Phase I.

The second wave of viral lytic gene expression, Phase II, occurs approximately 48 h post stimulus. This is characterized by higher levels of viral gene transcription, onset of viral DNA replication and ultimately production of infectious virus. In contrast to Phase I, the repressive heterochromatic marks are removed and it seems likely that active euchromatic marks such as di or tri-methylated histone H3 lysine-4 (H3K4me2/3) are installed, consistent with robust viral gene transcription. As has been extensively described for acute replication in non-neuronal cells, viral protein synthesis, VP16-mediated transactivation, and viral DNA replication are all required for Phase II. Recent single-molecule RNA FISH analyses support the notion that Phase I and Phase II are sequential events and that individual genomes transcribe multiple viral genes during Phase I [45]. The biphasic program is not unique to reactivation in response to LY294002 but can be observed with mirin a small molecule inhibitor of the MRN complex which results in topoisomerase 2-mediated DNA breaks [45,69], as well as in response to increased neuronal excitation [68]. Neuronal hyperexcitability is associated with prolonged stress conditions and IL1-β release, and can be mimicked in vitro using the natural bicyclic compound forskolin, which acts within the neuron to activate adenylate cyclase. The accompanying rise in intracellular cAMP levels also results in DLK activation. Importantly ex vivo (meaning 'outside of the living body') reactivation in explanted TGs from latently-infected mice, also begins with a DLK-dependent but histone demethylase-independent wave of viral gene transcription that is essentially indistinguishable from Phase I as defined in SCG neuron cultures [76]. It is notable that in the explant model, viral productive cycle mRNAs can be detected by 5 h rather than 18 h consistent with known differences in the kinetics of DLK activation in response to the extreme trauma associated with axotomy [77].

It is important acknowledge that other reactivation pathways have been proposed based on in vivo models but have yet to tested directly in the context of the in vitro models considered here. The best characterized invokes the selective expression of the viral transcription factor VP16 (encoded by UL48) after exposure to hyperthemic stress in the mouse ocular infection model [78]. VP16 is recruited to enhancer elements associated with the five viral immediate early genes which include ICP0, a chromatin modifier, and ICP4, a sequence-specific transcription factor required for early and late gene expression, and thus should be sufficient to overcome epigenetic silencing. Indeed, expression of VP16 from an adenoviral vector is a potent inducer of reactivation in the rat SCG model without the need for another stimulus [79]. The capacity to reactivate in response to hyperthermal stress maps to the UL48 promoter, which contains sequence motifs recognized by several cellular transcription factors involved in various stress responses [80]. Whether DLK/JNK are required for VP16-induced reactivation is not known. It is possible that stress-kinases are themselves activated by hyperthermic stress and facilitate RNAPII transcription from VP16-responsive viral promoters. Although ectopic VP16 is sufficient to induce reactivation in vitro, the levels of expression are probably much higher than in vivo and might elicit a DLK/JNK stress response. Finally, another potentially distinct in vivo reactivation stimulus is trans-corneal iontophoresis of adrenaline/epinephrine in the rabbit ocular model [81,82]. As with changes in core body temperature it is not entirely clear how exposure to secreted hormones alters intra-neuronal signaling and whether this reactivation pathway also feeds through the DLK and JNK.

The ability to reactivate is an essential property of latency [83], and models based on the neuroblastoma line SH-SY5Y [12] or immortalized HD10.6 cells [20], have encountered difficulties achieving robust reactivation. While this is frustrating given the potential benefits of these more scalable platforms, there might be value in understanding why latent virus is not fully responsive. Does poor reactivation reflect a defect in the neuronal response to the reactivation stimulus itself or some aspect of the chromatin-associated with genomes? Monitoring DLK/JNK activation, Phase I transcription and viral DNA replication in Phase II will help to distinguish defects in signaling from downstream transcriptional events.

8. Part 3: The Future

As the use of in vitro α-HV latency models increases it seems a good time for the community as a whole to consider priorities. How many different neuronal models do we actually need? What features should to be developed or refined to best tackle the major gaps in our knowledge? How do we apply the findings from neuronal models to live-animal models and if possible, to humans? The sections below will discuss various opportunities for consolidation or broader development.

9. Seeking Consistency

A shared aspiration of research scientists is to achieve consistent results that can be replicated and expanded upon by others [84]. With the proliferation of neuronal infection models there is a risk that poorly understood differences in these models will hamper progress by generating findings that do not carry over from one model to the next. Do contradictory findings reflect weaknesses of the models, differences in experimental design or unrecognized complexity in the biology? How much does it matter if a model does not perfectly mirror everything that happens in experimental animals or humans? Regardless of these uncertainties, the ability to replicate key findings in more than one model system—which can be both time consuming and expensive—needs to be recognized as a strength by reviewers of manuscripts and grants. Too often efforts in this direction are dismissed as showing a 'lack of innovation' rather than evidence of experimental rigor.

As a field we need to consider the strengths and weaknesses of consolidation versus diversification. Consolidation means that a small number of neuron sources and/or infection protocols gain widespread usage. A benefit is that favored models become more extensively characterized and the likelihood that new findings are independently validated will increase. The downside is that favored models might unwittingly omit important features of latency in humans or experimentally tractable small-animal models such as mice and rabbits. This may not be so detrimental to mechanistic studies, but could impede progress in the development and testing of therapeutics. Diversification might mean that the number of different models could grow to a point where almost every laboratory uses its own model to the exclusion of others. As a result, generalizable conclusions may be obscured by the idiosyncrasies of individual models. This could impact the consolidation of knowledge and reduce the impetus towards further validation through animal or human studies.

So, what can be done to ensure consistency? This is where a community approach might be fruitful. Investigators could enter into collaborations in which they deliberately try to validate or refute key findings in different models without the need to necessarily break new ground. This of course might draw the criticism mentioned above. Cross-model comparisons could be simplified by the sharing of reference viruses. These might represent the most frequently used laboratory strains or engineered reporter viruses than can be used to establish benchmarks in terms of infectivity, replication efficiency, ease of maintenance in a non-productive infection state and lastly, their ability to reactivate in response to defined stimuli. The importance of strain background is well understood in animal infection models and is likely to hold true in cultured neurons. RNA sequencing of productively infected SH-SY5Y neuroblastoma cells has shown that different HSV-1 strains impact the neuronal transcriptome in different ways and can manifest as different effects on cell morphology, cell-cell interactions and the relative expression of viral proteins [85]. With respect to latency, careful comparisons in the human LUHMES model found a profound difference between HSV-1 strain 17syn+ and KOS(M) in terms of reactivation in response to the phosphatidylinositol 3-kinase (PI3-K) inhibitor LY294002 [86]. Understanding the molecular basis for these differences could be instructive but also potentially distracting in terms of model characterization and data sharing.

10. Seeking Scalability

As discussed above, a major challenge to primary neuron models is the practical difficulty and expense of generating infected cultures in sufficient bulk for many biochemical applications. Informative techniques such as PAR-CLIP, ribosome profiling (Ribo-seq), and proximity-labeling all benefit from larger quantities of sample and also require several biological replicates. The same is true for chromatin immunoprecipitation (ChIP), especially when combined with next generation sequencing (ChIP-seq). ChIP-seq has been used to good effect in studies of acute HSV-1 infections [87], non-productive HSV-1 infections in non-neuronal models [88], profiling of insulator factors associated with latent HSV-1 in murine ganglia [89]. The power of this approach is clear from studies of latency and reactivation of lymphotropic γ-herpesviruses, where tens of millions of latently infected cells can be readily obtained [90]. ChIP-seq protocols require the isolation of cross-linked chromatin prior to fragmentation and antibody-mediated capture of the proteins of interest and it is notoriously difficult to achieve consistent chromatin preparations when working with small numbers of cells. Models based on the differentiation of proliferating cells lines such as LUHMES or on proliferating iPSC or ESC precursors offer tremendous promise assuming that sizeable cultures of fully differentiated neurons can be generated on a consistent basis. Alternatively, investigators may opt for targeted nuclease strategies requiring less material and with a better signal-to-noise ratio than standard ChIP-seq [91,92].

11. Next Generation Models

It is likely we will see more studies using α-HV latency models based on genetically-isogenic populations of human neurons exhibiting transcriptomes resembling those of *bona fide* sensory or sympathetic neurons. Because of the perceived value in regenerative medicine, directed reprogramming techniques capable of converting untransformed cells such as primary fibroblasts into functional neurons are becoming more efficient and more nuanced in terms of the neuronal subtypes generated, and do not require cells to pass through an embryonic phenotype [93–96]. However, there are still significant challenges relating to the length of time required to generate usable cultures and unresolved difficulties in achieving desirable uniformity in sufficient quantity. Single-cell RNA-sequencing of more than a hundred iPSC-derived sensory neuron cultures found substantial variability in the gene expression profiles between cultures, more that is detected in postmortem DRG from human donors [97]. Thus, despite the fact that iPSC-derived sensory neurons express classic neuronal markers and exhibit the expected morphology and electrophysiological properties, they are still transcriptionally distinct from the mature neurons encountered by virus during natural infections. The extent to which these differences matter in terms of virus-host interactions needs to be determined but should be kept in mind by those embarking on human in vitro models. Some differentiation protocols generate neurons that more closely resemble the CNS and may serve as excellent models to study the consequences of spread from the PNS into the CNS resulting in acute but very severe outcomes such as herpes simplex encephalitis (HSE) or longer-term pathologies such as AD and AD-related dementias [98].

Recent studies are also beginning to explore the consequences of switching from traditional two-dimensional (2D) cultures into three-dimensional (3D) structures termed organoids [99,100]. Aggregates provide more points of neuron-to-neuron contact in addition to more contacts between neurons and the extracellular matrix. Work in other areas has found that organoid models are better predictors of drug responses in vivo than conventional 2D monolayers. As witnessed by recent efforts to model SARS-CoV-2 infections using liquid-air interface cultures that mimic the epithelial layers of the airway and lungs, organoids are increasingly valued in studies of host-pathogen interactions [101].

Ultimately it might possible to develop mixed cultures composed of neurons and satellite glial cells (SGCs). In peripheral ganglia, the SGCs are found tightly wrapped around the neuronal cell body or soma and are encased within a sheath of connective tissue [102]. The activity of the neurons and SGCs is reciprocally modulated via gap junctions acting in concert with secreted proteins and small-molecules [103]. At a minimum, these interactions could enhance the ability of neurons to enforce α-HV latency by reducing the stresses associated with in vitro culture. Latently-infected ganglia also contain infiltrating T-cells, macrophages and keratinocytes, which again are absent from most cultured neuronal models [104]. These immune cells are known to be important contributors to neuroinflammation and neuropathic pain [105]. There is clear evidence from mice that infiltrating cytolytic T-cells suppress productive replication of HSV-1 through a noncytolytic involving secretion of α-interferon and granzyme B, which degrades ICP4, an essential viral transcription factor [106,107]. As already discussed, it may be difficult to achieve sufficient scale in these complex but perhaps more realistic models.

12. Looking to the Horizon

With the growing sophistication of in vitro latency studies, we can expect rapid progress in several key areas. For example, there are still substantial gaps in our knowledge of the chromatin associated with viral genomes, especially during the establishment period. Precision studies are needed to determine if there are additional changes in chromatin composition or post-translational modification over time that can explain why reactivation is less efficient after sustained periods. Studies using the new spread-deficient reporter found that although genome copy number and LAT levels remained relatively constant, the proportion of neurons undergoing reactivation at 30 days post-infection was much reduced compared to 8 or 16 days [66].

With the ability to infect a higher proportion of neuron cultures than occurs by natural infection in vivo, it has become much easier to ask if the transcriptome and proteome of host neurons is changed due to the presence of latent virus and/or the accumulation of latency-associated RNAs. The establishment process has been especially difficult to study in live-animal models because of the spatial and temporal disconnect between primary infections at peripheral sites such as the eye or footpad and the neuronal nuclei situated in the peripheral ganglia. In animals, colonization of TGs is highly asynchronous with new infections and superinfections taking place over several days [80]. As such it is uncertain if incoming genomes undergo limited replication before they are incorporated into heterochromatin and whether the act of replication influences this process in any way.

Although we have a relatively complete picture of the stepwise transition from latency into productive replication, it is still unclear how JNK and other stress kinases are actually targeted to the chromatin surrounding viral promoters. Conceivably this requires neuronal factors, which may themselves be regulated by stress signals. The specific details are important because they might lead to prophylactic strategies to limit reactivation in humans during periods of immunological vulnerability. Another open question is why viral DNA replication is not detected until Phase II, even though many viral mRNA and proteins are synthesized in Phase I. In vitro models will allow systematic analysis of the relative abundance and localization of the largely virus-encoded DNA replication machinery, perhaps revealing unforeseen differences in the control of viral DNA replication in post-mitotic neurons compared to proliferating cell types.

Finally, α-HV latency is characterized by the sustained expression of the latency-associated transcripts (LATs) or the VZV latency-associated transcripts (VLTs) [108,109]. By analogy to the cellular functions of long non-coding RNAs (lncRNAs), these may act *in cis* to influence the chromatin of the viral genomes they are transcribed from and again, in vitro models will be instrumental in identifying any molecular targets [110]. The same is true for host and viral microRNAs which are strongly implicated in dampening productive cycle gene expression [111,112]. Use of human rather than rodent neurons will be beneficial if perfect nucleotide complementarity is critical.

The organization of latent α-HV genomes in terms of the distribution of factors and modifications reflective of facultative heterochromatin and/or euchromatin as well as differences in topology and subnuclear location remain very much a black box. Cryptic heterogeneity may help to explain why only a small subset of latently-infected neurons reactivate in vitro even though the entire culture is exposed to the same stimulus. Does this reflect neuron to neuron differences in the numbers of reactivation-competent genomes or some other variables? Indeed, it is still unclear if multiple genomes within an individual nucleus initiate productive cycle gene transcription or whether a single responsive genome is sufficient.

The availability of sensitive single-cell imaging techniques including the ability to localize and characterize nascent RNA will help to resolve these important questions. Single-cell RNA sequencing (scRNA-seq) has emerged as a highly effective tool to identify and characterize cells by their individual transcriptomic signature and can yield dynamic information through RNA velocity modeling, metabolic labeling and other refinements [113,114]. Use of scRNA-seq to study α-HV latency has been hampered by an inability to reliably detect low abundance viral RNAs or even detect abundant latency products such as the stable LAT introns and mature microRNAs that lack poly(A) tails. Fortunately, sensitivity has greatly improved with the introduction of newer chemistries such as Chromium v3 from 10× Genomics [115]. Using this approach, a recent study was able to detect sufficient numbers of HSV-1 productive cycle transcripts in rat SCG neurons treated with LY294002 to distinguish neurons undergoing viral reactivation from either uninfected neurons or latently-infected neurons that had not responded to the stimulus [45]. This analysis identified a small number of host mRNAs that were upregulated by the stress stimulus but interestingly, only in virus-infected cells. Prominent among these were mRNAs encoding components of the heat shock response pathway and the Gadd45 protein family. Depletion of Gadd45b, and to a lesser extent, Gadd45g, increased the frequency of reactivation. Surprisingly, this was insensitive to JNK inhibition hinting at a mechanism that is less reliant on the methyl/phospho switch. In support of a suppressive role, direct expression of Gadd45b from a lentiviral vector reduced expression of both ICP4 and viral late transcripts in response to loss of NGF signaling (LY294002) or dsDNA break repair (mirin). Taken together these new studies identify Gadd45b as an intrinsic restriction factor that limits reactivation by suppressing viral late gene expression.

Interestingly, single neuron immunofluorescence imaging (shown schematically in Figure 4) revealed that the Gadd45b protein is found throughout the nucleoplasm and cytoplasm in most neurons but in two discrete subpopulations is either excluded from the nucleus of neurons engaged in new viral DNA synthesis or forms discrete nuclear puncta in infected neurons where DNA synthesis is absent. It is tempting to imagine these puncta correspond to locations of viral genomes but this needs to be addressed experimentally. Nonetheless, differential localization of the host Gadd45b protein provides a useful marker to distinguish between successful and abortive reactivation events. The presence of multiple neuronal subtypes may be a confounding factor that contributes to the heterogeneity evident in both establishment of latency and subsequent reactivation. With the exception of the Gadd45b study, this has not been investigated extensively in vitro. Moreover, because these are low-frequency events, they are essentially invisible in bulk analyses and require simultaneous detection of viral and host markers at the level of individual neurons.

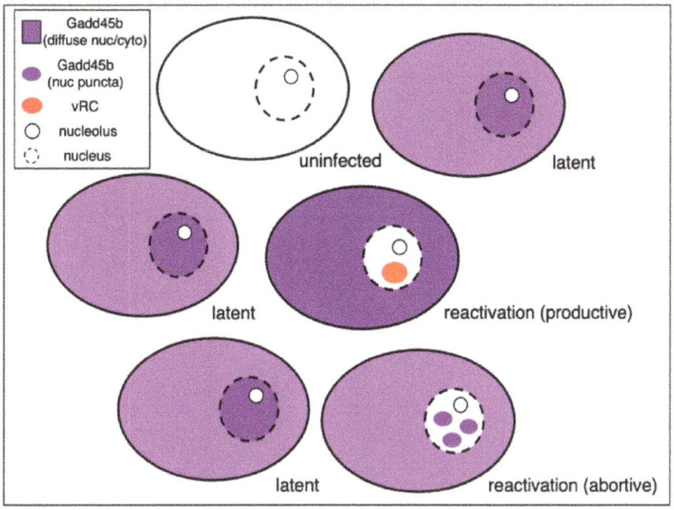

Figure 4. Subcellular localization of the Gadd45b protein acts as a marker of HSV-1 reactivation heterogeneity within a neuronal population. Schematic showing the different patterns of Gadd45b protein (purple) accumulation within the cell bodies of latently-infected rat SCG neurons treated with reactivation inducer LY294002. Protein levels are elevated only in latently infected neurons and is distributed throughout the cytoplasm and nucleoplasm. In a small number of neurons, Gadd45b is excluded from the nucleus and these neurons contain EdU-positive foci (red) corresponding to sites of active viral DNA replication (vRC) a marker of Phase II. These individual neurons are considered to be engaged in successful reactivation. In another rare population, Gadd45b accumulates as discrete puncta with the nucleus. These neurons are always EdU-negative and likely have not entered into Phase II and represent unsuccessful or abortive reactivation events. For full details see Hu et al., 2021 [45].

13. Closing Thoughts

The number of researchers using neuronal infection models to study α-HV latency and reactivation is at an all-time high and seems poised to continue increasing. This surge in popularity is propelled by the wealth of mechanistic insights that have emerged in the last few years and a growing appreciation for the potential of in vitro models to address long-standing mechanistic questions that are out of reach otherwise. Technological advances have played a major role in expanding the experimental possibilities. Innovations include the use of fluorescent reporter viruses that provide readouts in real time, a growing arsenal of chemical inhibitors with clearly defined molecular targets, the ease by which lentiviral transduction and RNA interference can be used to deplete host and viral gene products, and a wealth of exquisitely sensitive assay tools. It is encouraging to see many researchers embracing emerging technologies such as organoid culture, super-resolution imaging and single molecule detection. As often happens in science, technical innovations expand the sorts of biological questions that can be asked. Increasingly, studies are shifting from population-level (bulk) measurements to assays that trace processes or events in individual cells. In time, this precision will almost certainly extend to the level of individual viral genomes, something that would have been inconceivable not so long ago. Finally, there is growing societal and political pressure to 'refine, reduce, and replace' the use of animals in biomedical research and some investigators may be drawn to human iPSC and ESC-based models for this reason alone [116]. With so much to look forward to, our thoughts return to Randy Cohrs and the satisfaction he would have found in harnessing these exciting technologies to solve the puzzles in virology that interested him so deeply.

Funding: Work in my laboratory is supported by National Institutes of Health (NIH) grants AI130618 and AI147163.

Institutional Review Board Statement: Not applicable.

Informed Consent Statement: Not applicable.

Data Availability Statement: Not applicable.

Acknowledgments: My thanks to Ian Mohr: Tony Huang, and Anna Cliffe for their constructive comments on the manuscript and to Maria Nagel for access to photographs from the CALS collection. I am indebted to Dan Depledge, Andrea Bertke, and Moses Chao along with the members of our labs, and the many participants from CALS and other gatherings for insightful questions and observations that have shaped my thinking. Any errors, omissions, or misunderstandings are my own.

Conflicts of Interest: The author declares no conflict of interest. The sponsors had no role in the conceptualization or writing of the manuscript.

References

1. Taylor, M.W. A History of Cell Culture. In *Viruses and Man: A History of Interactions*; Taylor, M.W., Ed.; Springer International Publishing: Cham, Switzerland, 2014; pp. 41–52. ISBN 978-3-319-07758-1.
2. Millet, L.J.; Gillette, M.U. Over a Century of Neuron Culture: From the Hanging Drop to Microfluidic Devices. *Yale J. Biol. Med.* **2012**, *85*, 501–521. [PubMed]
3. Alberts, B.; Heald, R.; Johnson, A.; Morgan, D.; Raff, M.; Roberts, K.; Walter, P. *Molecular Biology of the Cell*, 7th ed.; W. W. Norton & Company: New York, NY, USA, 2022; ISBN 978-0-393-88482-1.
4. Keller, J.M.; Frega, M. Past, Present, and Future of Neuronal Models In Vitro. In *In Vitro Neuronal Networks*; Advances in Neurobiology; Springer: Cham, Switzerland, 2019; Volume 22, pp. 3–17. [CrossRef]
5. Ramos-Ibeas, P.; Nichols, J.; Alberio, R. States and Origins of Mammalian Embryonic Pluripotency In Vivo and in a Dish. *Curr. Top. Dev. Biol.* **2017**, *128*, 151–179. [CrossRef] [PubMed]
6. Hu, J.L.; Todhunter, M.E.; LaBarge, M.A.; Gartner, Z.J. Opportunities for Organoids as New Models of Aging. *J. Cell Biol.* **2018**, *217*, 39–50. [CrossRef] [PubMed]
7. Matsu-ura, T.; Dovzhenok, A.; Aihara, E.; Rood, J.; Le, H.; Ren, Y.; Rosselot, A.E.; Zhang, T.; Lee, C.; Obrietan, K.; et al. Intercellular Coupling of the Cell Cycle and Circadian Clock in Adult Stem Cell Culture. *Mol. Cell* **2016**, *64*, 900–912. [CrossRef]
8. Park, A.J.; Wright, M.A.; Roach, E.J.; Khursigara, C.M. Imaging Host–Pathogen Interactions Using Epithelial and Bacterial Cell Infection Models. *J. Cell Sci.* **2021**, *134*, jcs250647. [CrossRef]
9. Knipe, D.M.; Heldwein, E.E.; Mohr, I.J.; Sodroski, C.N. Herpes Simplex Viruses: Mechanisms of Lytic and Latent Infection. In *Field's Virology: DNA Viruses*; Lippincott Williams & Wilkins: Philadelphia, PA, USA, 2021; pp. 235–296. ISBN 9781975112578.
10. Thellman, N.M.; Triezenberg, S.J. Herpes Simplex Virus Establishment, Maintenance, and Reactivation: In Vitro Modeling of Latency. *Pathogens* **2017**, *6*, 28. [CrossRef]
11. Koyuncu, O.O.; MacGibeny, M.A.; Enquist, L.W. Latent versus Productive Infection: The Alpha Herpesvirus Switch. *Future Virol.* **2018**, *13*, 431–443. [CrossRef]
12. Goldstein, R.S.; Kinchington, P.R. Varicella Zoster Virus Neuronal Latency and Reactivation Modeled In Vitro. In *Current Topics in Microbiology and Immunology*; Springer: Berlin/Heidelberg, Germany, 2021; pp. 1–32.
13. Camarena, V.; Kobayashi, M.; Kim, J.Y.; Roehm, P.C.; Perez, R.; Gardner, J.; Wilson, A.C.; Mohr, I.; Chao, M.V. Nature and Duration of Growth Factor Signaling through Receptor Tyrosine Kinases Regulates HSV-1 Latency in Neurons. *Cell Host Microbe* **2010**, *8*, 320–330. [CrossRef]
14. Cliffe, A.R.; Arbuckle, J.H.; Vogel, J.L.; Geden, M.J.; Rothbart, S.B.; Cusack, C.L.; Strahl, B.D.; Kristie, T.M.; Deshmukh, M. Neuronal Stress Pathway Mediating a Histone Methyl/Phospho Switch Is Required for Herpes Simplex Virus Reactivation. *Cell Host Microbe* **2015**, *18*, 649–658. [CrossRef]
15. Yanez, A.; Harrell, T.; Sriranganathan, H.; Ives, A.; Bertke, A.S. Neurotrophic Factors NGF, GDNF and NTN Selectively Modulate HSV1 and HSV2 Lytic Infection and Reactivation in Primary Adult Sensory and Autonomic Neurons. *Pathogens* **2017**, *6*, 5. [CrossRef]
16. Hafezi, W.; Lorentzen, E.U.; Eing, B.R.; Müller, M.; King, N.J.C.; Klupp, B.; Mettenleiter, T.C.; Kühn, J.E. Entry of Herpes Simplex Virus Type 1 (HSV-1) into the Distal Axons of Trigeminal Neurons Favors the Onset of Nonproductive, Silent Infection. *PLoS Pathog.* **2012**, *8*, e1002679. [CrossRef] [PubMed]
17. Wilcox, C.L.; Johnson, E.M. Characterization of Nerve Growth Factor-Dependent Herpes Simplex Virus Latency in Neurons In Vitro. *J. Virol.* **1988**, *62*, 393–399. [CrossRef] [PubMed]
18. Edwards, T.G.; Bloom, D.C. Lund Human Mesencephalic (LUHMES) Neuronal Cell Line Supports Herpes Simplex Virus 1 Latency In Vitro. *J. Virol.* **2019**, *93*, 419. [CrossRef] [PubMed]
19. Shipley, M.M.; Mangold, C.A.; Kuny, C.V.; Szpara, M.L. Differentiated Human SH-SY5Y Cells Provide a Reductionist Model of Herpes Simplex Virus 1 Neurotropism. *J. Virol.* **2017**, *91*, e00958-17. [CrossRef] [PubMed]

20. Thellman, N.M.; Botting, C.; Madaj, Z.; Triezenberg, S.J. An Immortalized Human Dorsal Root Ganglia Cell Line Provides a Novel Context to Study Herpes Simplex Virus Type-1 Latency and Reactivation. *J. Virol.* **2017**, *91*, e00080-17. [CrossRef] [PubMed]
21. Danaher, R.J.; Jacob, R.J.; Miller, C.S. Establishment of a Quiescent Herpes Simplex Virus Type 1 Infection in Neurally-Differentiated PC12 Cells. *J. Neurovirol.* **1999**, *5*, 258–267. [CrossRef] [PubMed]
22. Markus, A.; Lebenthal-Loinger, I.; Yang, I.H.; Kinchington, P.R.; Goldstein, R.S. An in Vitro Model of Latency and Reactivation of Varicella Zoster Virus in Human Stem Cell-Derived Neurons. *PLoS Pathog.* **2015**, *11*, e1004885. [CrossRef]
23. Pourchet, A.; Modrek, A.S.; Placantonakis, D.G.; Mohr, I.; Wilson, A.C. Modeling HSV-1 Latency in Human Embryonic Stem Cell-Derived Neurons. *Pathogens* **2017**, *6*, 24. [CrossRef]
24. Sadaoka, T.; Depledge, D.P.; Rajbhandari, L.; Venkatesan, A.; Breuer, J.; Cohen, J.I. In Vitro System Using Human Neurons Demonstrates That Varicella-Zoster Vaccine Virus Is Impaired for Reactivation, but Not Latency. *Proc. Natl. Acad. Sci. USA* **2016**, *113*, E2403–E2412. [CrossRef]
25. McMahon, R.; Walsh, D. Efficient Quiescent Infection of Normal Human Diploid Fibroblasts with Wild-Type Herpes Simplex Virus Type 1. *J. Virol.* **2008**, *82*, 10218–10230. [CrossRef]
26. Hammer, S.M.; Richter, B.S.; Hirsch, M.S. Activation and Suppression of Herpes Simplex Virus in a Human T Lymphoid Cell Line. *J. Immunol.* **1981**, *127*, 144–148. [PubMed]
27. Cohen, E.M.; Avital, N.; Shamay, M.; Kobiler, O. Abortive Herpes Simplex Virus Infection of Nonneuronal Cells Results in Quiescent Viral Genomes That Can Reactivate. *Proc. Natl. Acad. Sci. USA* **2020**, *117*, 635–640. [CrossRef] [PubMed]
28. Vahlne, A.; Lycke, E. Herpes Simplex Virus Infection of Mouse Neuroblastoma Cells. *Proc. Soc. Exp. Biol. Med.* **1977**, *156*, 82–87. [CrossRef] [PubMed]
29. Russell, J.; Preston, C.M. An *In Vitro* Latency System for Herpes Simplex Virus Type 2. *J. Gen. Virol.* **1986**, *67*, 397–403. [CrossRef]
30. Laemmle, L.; Goldstein, R.S.; Kinchington, P.R. Modeling Varicella Zoster Virus Persistence and Reactivation—Closer to Resolving a Perplexing Persistent State. *Front. Microbiol.* **2019**, *10*, 1634. [CrossRef]
31. Mahalingam, R.; Gershon, A.; Gershon, M.; Cohen, J.I.; Arvin, A.; Zerboni, L.; Zhu, H.; Gray, W.; Messaoudi, I.; Traina-Dorge, V. Current In Vivo Models of Varicella-Zoster Virus Neurotropism. *Viruses* **2019**, *11*, 502. [CrossRef]
32. Sadaoka, T.; Schwartz, C.L.; Rajbhandari, L.; Venkatesan, A.; Cohen, J.I. Human Embryonic Stem Cell Derived Neurons Are Highly Permissive for Varicella-Zoster Virus Lytic Infection. *J. Virol.* **2017**, *92*, e01108-17. [CrossRef]
33. Ives, A.M.; Bertke, A.S. Stress Hormones Epinephrine and Corticosterone Selectively Modulate Herpes Simplex Virus 1 (HSV-1) and HSV-2 Productive Infections in Adult Sympathetic, but Not Sensory, Neurons. *J. Virol.* **2017**, *91*, e00582-17. [CrossRef]
34. Pomeranz, L.E.; Reynolds, A.E.; Hengartner, C.J. Molecular Biology of Pseudorabies Virus: Impact on Neurovirology and Veterinary Medicine. *Microbiol. Mol. Biol. R.* **2005**, *69*, 462–500. [CrossRef]
35. Ostler, J.B.; Sawant, L.; Harrison, K.; Jones, C. Regulation of Neurotropic Herpesvirus Productive Infection and Latency-Reactivation Cycle by Glucocorticoid Receptor and Stress-Induced Transcription Factors. *Vitam. Horm.* **2021**, *117*, 101–132. [CrossRef]
36. Sorel, O.; Messaoudi, I. Varicella Virus-Host Interactions During Latency and Reactivation: Lessons From Simian Varicella Virus. *Front. Microbiol.* **2018**, *9*, 3170. [CrossRef]
37. Wilcox, C.L.; Johnson, E.M. Nerve Growth Factor Deprivation Results in the Reactivation of Latent Herpes Simplex Virus in vitro. *J. Virol.* **1987**, *61*, 2311–2315. [CrossRef] [PubMed]
38. Wilcox, C.L.; Smith, R.L.; Freed, C.R.; Johnson, E.M. Nerve Growth Factor-Dependence of Herpes Simplex Virus Latency in Peripheral Sympathetic and Sensory Neurons in vitro. *J. Neurosci.* **1990**, *10*, 1268–1275. [CrossRef] [PubMed]
39. Colgin, M.A.; Smith, R.L.; Wilcox, C.L. Inducible Cyclic AMP Early Repressor Produces Reactivation of Latent Herpes Simplex Virus Type 1 in Neurons In Vitro. *J. Virol.* **2001**, *75*, 2912–2920. [CrossRef]
40. Hunsperger, E.A.; Wilcox, C.L. Capsaicin-Induced Reactivation of Latent Herpes Simplex Virus Type 1 in Sensory Neurons in Culture. *J. Gen. Virol.* **2003**, *84*, 1071–1078. [CrossRef]
41. Kuhn, M.A.; Nayak, S.; Camarena, V.; Gardner, J.; Wilson, A.; Mohr, I.; Chao, M.V.; Roehm, P.C. A Cell Culture Model of Facial Palsy Resulting from Reactivation of Latent Herpes Simplex Type 1. *Otol. Neurotol.* **2012**, *33*, 87–92. [CrossRef] [PubMed]
42. Roehm, P.C.; Camarena, V.; Nayak, S.; Gardner, J.B.; Wilson, A.; Mohr, I.; Chao, M.V. Cultured Vestibular Ganglion Neurons Demonstrate Latent HSV1 Reactivation. *Laryngoscope* **2011**, *121*, 2268–2275. [CrossRef] [PubMed]
43. Bertke, A.S.; Swanson, S.M.; Chen, J.; Imai, Y.; Kinchington, P.R.; Margolis, T.P. A5-Positive Primary Sensory Neurons Are Nonpermissive for Productive Infection with Herpes Simplex Virus 1 In Vitro. *J. Virol.* **2011**, *85*, 6669–6677. [CrossRef] [PubMed]
44. Margolis, T.P.; Imai, Y.; Yang, L.; Vallas, V.; Krause, P.R. Herpes Simplex Virus Type 2 (HSV-2) Establishes Latent Infection in a Different Population of Ganglionic Neurons than HSV-1: Role of Latency-Associated Transcripts. *J. Virol.* **2007**, *81*, 1872–1878. [CrossRef]
45. Hu, H.; Srinivas, K.P.; Wang, S.; Chao, M.V.; Lionnet, T.; Mohr, I.; Wilson, A.C.; Depledge, D.P.; Huang, T.T. Single-cell Transcriptomics Identifies Gadd45b as a Regulator of Herpesvirus-reactivating Neurons. *EMBO Rep.* **2022**, *23*, e53533. [CrossRef]
46. Elion, G.B. Mechanism of Action and Selectivity of Acyclovir. *Am. J. Med.* **1982**, *73*, 7–13. [CrossRef]
47. Millhouse, S.; Su, Y.-H.; Zhang, X.; Wang, X.; Song, B.P.; Zhu, L.; Oppenheim, E.; Fraser, N.W.; Block, T.M. Evidence That Herpes Simplex Virus DNA Derived from Quiescently Infected Cells *In Vitro*, and Latently Infected Cells *In Vivo*, Is Physically Damaged. *J. Neurovirol.* **2010**, *16*, 384–398. [CrossRef] [PubMed]

48. Sawtell, N.M. Comprehensive Quantification of Herpes Simplex Virus Latency at the Single-Cell Level. *J. Virol.* **1997**, *71*, 5423–5431. [CrossRef] [PubMed]
49. Wang, K.; Lau, T.Y.; Morales, M.; Mont, E.K.; Straus, S.E. Laser-Capture Microdissection: Refining Estimates of the Quantity and Distribution of Latent Herpes Simplex Virus 1 and Varicella-Zoster Virus DNA in Human Trigeminal Ganglia at the Single-Cell Level. *J. Virol.* **2005**, *79*, 14079–14087. [CrossRef]
50. Catez, F.; Picard, C.; Held, K.; Gross, S.; Rousseau, A.; Theil, D.; Sawtell, N.M.; Labetoulle, M.; Lomonte, P. HSV-1 Genome Subnuclear Positioning and Associations with Host-Cell PML-NBs and Centromeres Regulate LAT Locus Transcription during Latency in Neurons. *PLoS Pathog.* **2012**, *8*, e1002852. [CrossRef]
51. Kobiler, O.; Lipman, Y.; Therkelsen, K.; Daubechies, I.; Enquist, L.W. Herpesviruses Carrying a Brainbow Cassette Reveal Replication and Expression of Limited Numbers of Incoming Genomes. *Nat. Commun.* **2010**, *1*, 146. [CrossRef]
52. Taylor, T.J.; McNamee, E.E.; Day, C.; Knipe, D.M. Herpes Simplex Virus Replication Compartments Can Form by Coalescence of Smaller Compartments. *Virology* **2003**, *309*, 232–247. [CrossRef]
53. Cohen, E.M.; Kobiler, O. Gene Expression Correlates with the Number of Herpes Viral Genomes Initiating Infection in Single Cells. *PLoS Pathog.* **2016**, *12*, e1006082. [CrossRef]
54. Yordy, B.; Iijima, N.; Huttner, A.; Leib, D.A.; Iwasaki, A. A Neuron-Specific Role for Autophagy in Antiviral Defense against Herpes Simplex Virus. *Cell Host Microbe* **2012**, *12*, 334–345. [CrossRef]
55. Katzenell, S.; Leib, D.A. Herpes Simplex Virus and Interferon Signaling Induce Novel Autophagic Clusters in Sensory Neurons. *J. Virol.* **2016**, *90*, 4706–4719. [CrossRef]
56. Laval, K.; Cleemput, J.V.; Vernejoul, J.B.; Enquist, L.W. Alphaherpesvirus Infection of Mice Primes PNS Neurons to an Inflammatory State Regulated by TLR2 and Type I IFN Signaling. *PLoS Pathog.* **2019**, *15*, e1008087. [CrossRef] [PubMed]
57. Divito, S.; Cherpes, T.L.; Hendricks, R.L. A Triple Entente: Virus, Neurons, and CD8+ T Cells Maintain HSV-1 Latency. *Immunol. Res.* **2006**, *36*, 119–126. [CrossRef]
58. Suzich, J.B.; Cuddy, S.R.; Baidas, H.; Dochnal, S.; Ke, E.; Schinlever, A.R.; Babnis, A.; Boutell, C.; Cliffe, A.R. PML-NB-Dependent Type I Interferon Memory Results in a Restricted Form of HSV Latency. *EMBO Rep.* **2021**, *22*, e52547. [CrossRef] [PubMed]
59. Luo, W.; Wickramasinghe, S.R.; Savitt, J.M.; Griffin, J.W.; Dawson, T.M.; Ginty, D.D. A Hierarchical NGF Signaling Cascade Controls Ret-Dependent and Ret-Independent Events during Development of Nonpeptidergic DRG Neurons. *Neuron* **2007**, *54*, 739–754. [CrossRef]
60. Koyuncu, O.O.; Song, R.; Greco, T.M.; Cristea, I.M.; Enquist, L.W. The Number of Alphaherpesvirus Particles Infecting Axons and the Axonal Protein Repertoire Determines the Outcome of Neuronal Infection. *mBio* **2015**, *6*, e00276-15. [CrossRef]
61. Roizman, B.; Sears, A.E. An Inquiry into the Mechanisms of Herpes Simplex Virus Latency. *Annu. Rev. Microbiol.* **1987**, *41*, 543–571. [CrossRef]
62. Roizman, B.; Whitley, R.J. An Inquiry into the Molecular Basis of HSV Latency and Reactivation. *Annu. Rev. Microbiol.* **2013**, *67*, 355–374. [CrossRef]
63. Taylor, A.M.; Blurton-Jones, M.; Rhee, S.W.; Cribbs, D.H.; Cotman, C.W.; Jeon, N.L. A Microfluidic Culture Platform for CNS Axonal Injury, Regeneration and Transport. *Nat. Methods* **2005**, *2*, 599–605. [CrossRef]
64. Koyuncu, O.O.; MacGibeny, M.A.; Hogue, I.B.; Enquist, L.W. Compartmented Neuronal Cultures Reveal Two Distinct Mechanisms for Alpha Herpesvirus Escape from Genome Silencing. *PLoS Pathog.* **2017**, *13*, e1006608. [CrossRef]
65. Arthur, J.L.; Scarpini, C.G.; Connor, V.; Lachmann, R.H.; Tolkovsky, A.M.; Efstathiou, S. Herpes Simplex Virus Type 1 Promoter Activity during Latency Establishment, Maintenance, and Reactivation in Primary Dorsal Root Neurons In Vitro. *J. Virol.* **2001**, *75*, 3885–3895. [CrossRef]
66. Dochnal, S.; Merchant, H.Y.; Schinlever, A.R.; Babnis, A.; Depledge, D.P.; Wilson, A.C.; Cliffe, A.R. DLK-Dependent Biphasic Reactivation of Herpes Simplex Virus Latency Established in the Absence of Antivirals. *J. Virol.* **2022**, e00508-22. [CrossRef] [PubMed]
67. Efstathiou, S.; Preston, C.M. Towards an Understanding of the Molecular Basis of Herpes Simplex Virus Latency. *Virus Res.* **2005**, *111*, 108–119. [CrossRef]
68. Cuddy, S.R.; Schinlever, A.R.; Dochnal, S.; Seegren, P.V.; Suzich, J.; Kundu, P.; Downs, T.K.; Farah, M.; Desai, B.N.; Boutell, C.; et al. Neuronal Hyperexcitability Is a DLK-Dependent Trigger of Herpes Simplex Virus Reactivation That Can Be Induced by IL-1. *eLife* **2020**, *9*, e58037. [CrossRef] [PubMed]
69. Hu, H.-L.; Shiflett, L.A.; Kobayashi, M.; Chao, M.V.; Wilson, A.C.; Mohr, I.; Huang, T.T. TOP2β-Dependent Nuclear DNA Damage Shapes Extracellular Growth Factor Responses via Dynamic AKT Phosphorylation to Control Virus Latency. *Mol. Cell* **2019**, *74*, 466–480.e4. [CrossRef] [PubMed]
70. Kobayashi, M.; Wilson, A.C.; Chao, M.V.; Mohr, I. Control of Viral Latency in Neurons by Axonal MTOR Signaling and the 4E-BP Translation Repressor. *Genes Dev.* **2012**, *26*, 1527–1532. [CrossRef] [PubMed]
71. Kim, J.Y.; Mandarino, A.; Chao, M.V.; Mohr, I.; Wilson, A.C. Transient Reversal of Episome Silencing Precedes VP16-Dependent Transcription during Reactivation of Latent HSV-1 in Neurons. *PLoS Pathog.* **2012**, *8*, e1002540. [CrossRef] [PubMed]
72. Glebova, N.O.; Ginty, D.D. Growth and Survival Signals Controlling Sympathetic Nervous System Development. *Annu. Rev. Neurosci.* **2005**, *28*, 191–222. [CrossRef]

73. Danaher, R.J.; Jacob, R.J.; Miller, C.S. Herpesvirus Quiescence in Neuronal Cells. V: Forskolin-Responsiveness of the Herpes Simplex Virus Type 1 Alpha0 Promoter and Contribution of the Putative CAMP Response Element. *J. Neurovirol.* **2003**, *9*, 489–497. [CrossRef]
74. Suzich, J.B.; Cliffe, A.R. Strength in Diversity: Understanding the Pathways to Herpes Simplex Virus Reactivation. *Virology* **2018**, *522*, 81–91. [CrossRef]
75. Sabbattini, P.; Sjoberg, M.; Nikic, S.; Frangini, A.; Holmqvist, P.-H.; Kunowska, N.; Carroll, T.; Brookes, E.; Arthur, S.J.; Pombo, A.; et al. An H3K9/S10 Methyl-Phospho Switch Modulates Polycomb and Pol II Binding at Repressed Genes during Differentiation. *Mol. Biol. Cell* **2014**, *25*, 904–915. [CrossRef]
76. Whitford, A.L.; Clinton, C.A.; Kennedy, E.B.L.; Dochnal, S.A.; Suzich, J.B.; Cliffe, A.R. Ex Vivo Herpes Simplex Virus Reactivation Involves a Dual Leucine Zipper Kinase-Dependent Wave of Lytic Gene Expression That Is Independent of Histone Demethylase Activity and Viral Genome Synthesis. *J. Virol.* **2022**, e00475-22. [CrossRef] [PubMed]
77. Adib, E.A.; Smithson, L.J.; Collins, C.A. An Axonal Stress Response Pathway: Degenerative and Regenerative Signaling by DLK. *Curr. Opin. Neurobiol.* **2018**, *53*, 110–119. [CrossRef] [PubMed]
78. Thompson, R.L.; Preston, C.M.; Sawtell, N.M. De Novo Synthesis of VP16 Coordinates the Exit from HSV Latency In Vivo. *PLoS Pathog.* **2009**, *5*, e1000352. [CrossRef] [PubMed]
79. Linderman, J.A.; Kobayashi, M.; Rayannavar, V.; Fak, J.J.; Darnell, R.B.; Chao, M.V.; Wilson, A.C.; Mohr, I. Immune Escape via a Transient Gene Expression Program Enables Productive Replication of a Latent Pathogen. *Cell Rep.* **2017**, *18*, 1312–1323. [CrossRef]
80. Sawtell, N.M.; Thompson, R.L. De Novo Herpes Simplex Virus VP16 Expression Gates a Dynamic Programmatic Transition and Sets the Latent/Lytic Balance during Acute Infection in Trigeminal Ganglia. *PLoS Pathog.* **2016**, *12*, e1005877. [CrossRef]
81. Rivera, L.; Beuerman, R.W.; Hill, J.M. Corneal Nerves Contain Intra Axonal HSV-1 after Virus Reactivation by Epinephrine Iontophoresis. *Curr. Eye Res.* **2009**, *7*, 1001–1008. [CrossRef]
82. Rootman, D.S.; Haruta, Y.; Hill, J.M.; Kaufman, H.E. Corneal Nerves Are Necessary for Adrenergic Reactivation of Ocular Herpes. *Investig. Ophth. Vis. Sci.* **1988**, *29*, 351–356.
83. Sawtell, N.M.; Thompson, R.L. Herpes Simplex Virus and the Lexicon of Latency and Reactivation: A Call for Defining Terms and Building an Integrated Collective Framework. *F1000Research* **2016**, *5*, 2038. [CrossRef]
84. Samuel, S.; König-Ries, B. Understanding Experiments and Research Practices for Reproducibility: An Exploratory Study. *PeerJ* **2021**, *9*, e11140. [CrossRef]
85. Mangold, C.A.; Rathbun, M.M.; Renner, D.W.; Kuny, C.V.; Szpara, M.L. Viral Infection of Human Neurons Triggers Strain Specific Differences in Host Neuronal and Viral Transcriptomes. *PLoS Pathog.* **2021**, *17*, e1009441. [CrossRef]
86. Grams, T.R.; Edwards, T.G.; Bloom, D.C. Herpes Simplex Virus 1 Strains 17syn+ and KOS(M) Differ Greatly in Their Ability To Reactivate from Human Neurons In Vitro. *J. Virol.* **2020**, *94*. [CrossRef] [PubMed]
87. Lang, F.; Li, X.; Vladimirova, O.; Hu, B.; Chen, G.; Xiao, Y.; Singh, V.; Lu, D.; Li, L.; Han, H.; et al. CTCF Interacts with the Lytic HSV-1 Genome to Promote Viral Transcription. *Sci. Rep.* **2017**, *7*, 39861. [CrossRef] [PubMed]
88. Ferenczy, M.W.; DeLuca, N.A. Epigenetic Modulation of Gene Expression from Quiescent Herpes Simplex Virus Genomes. *J. Virol.* **2009**, *83*, 8514–8524. [CrossRef] [PubMed]
89. Washington, S.D.; Singh, P.; Johns, R.N.; Edwards, T.G.; Mariani, M.; Frietze, S.; Bloom, D.C.; Neumann, D.M. The CCCTC Binding Factor, CTRL2, Modulates Heterochromatin Deposition and the Establishment of HSV-1 Latency In Vivo. *J. Virol.* **2019**, *93*, e00415-19. [CrossRef] [PubMed]
90. Günther, T.; Fröhlich, J.; Herrde, C.; Ohno, S.; Burkhardt, L.; Adler, H.; Grundhoff, A. A Comparative Epigenome Analysis of Gammaherpesviruses Suggests Cis-Acting Sequence Features as Critical Mediators of Rapid Polycomb Recruitment. *PLoS Pathog.* **2019**, *15*, e1007838. [CrossRef] [PubMed]
91. Meers, M.P.; Bryson, T.D.; Henikoff, J.G.; Henikoff, S. Improved CUT&RUN Chromatin Profiling Tools. *eLife* **2019**, *8*, e46314. [CrossRef]
92. Kaya-Okur, H.S.; Wu, S.J.; Codomo, C.A.; Pledger, E.S.; Bryson, T.D.; Henikoff, J.G.; Ahmad, K.; Henikoff, S. CUT&Tag for Efficient Epigenomic Profiling of Small Samples and Single Cells. *Nat. Commun.* **2019**, *10*, 1930. [CrossRef]
93. Ng, A.H.M.; Khoshakhlagh, P.; Arias, J.E.R.; Pasquini, G.; Wang, K.; Swiersy, A.; Shipman, S.L.; Appleton, E.; Kiaee, K.; Kohman, R.E.; et al. A Comprehensive Library of Human Transcription Factors for Cell Fate Engineering. *Nat. Biotechnol.* **2021**, *39*, 510–519. [CrossRef]
94. Herdy, J.; Schafer, S.; Kim, Y.; Ansari, Z.; Zangwill, D.; Ku, M.; Paquola, A.; Lee, H.; Mertens, J.; Gage, F.H. Chemical Modulation of Transcriptionally Enriched Signaling Pathways to Optimize the Conversion of Fibroblasts into Neurons. *eLife* **2019**, *8*, 1502. [CrossRef]
95. Zhu, S.; Stanslowsky, N.; Fernández-Trillo, J.; Mamo, T.M.; Yu, P.; Kalmbach, N.; Ritter, B.; Eggenschwiler, R.; Ouwendijk, W.J.D.; Mzinza, D.; et al. Generation of HiPSC-Derived Low Threshold Mechanoreceptors Containing Axonal Termini Resembling Bulbous Sensory Nerve Endings and Expressing Piezo1 and Piezo2. *Stem Cell Res.* **2021**, *56*, 102535. [CrossRef]
96. Lampert, A.; Bennett, D.L.; McDermott, L.A.; Neureiter, A.; Eberhardt, E.; Winner, B.; Zenke, M. Human Sensory Neurons Derived from Pluripotent Stem Cells for Disease Modelling and Personalized Medicine. *Neurobiol. Pain* **2020**, *8*, 100055. [CrossRef] [PubMed]

97. Schwartzentruber, J.; Foskolou, S.; Kilpinen, H.; Rodrigues, J.; Alasoo, K.; Knights, A.J.; Patel, M.; Goncalves, A.; Ferreira, R.; Benn, C.L.; et al. Molecular and Functional Variation in IPSC-Derived Sensory Neurons. *Nat. Genet.* **2018**, *50*, 54–61. [CrossRef] [PubMed]
98. Bergström, P.; Trybala, E.; Eriksson, C.E.; Johansson, M.; Satir, T.M.; Widéhn, S.; Fruhwürth, S.; Michno, W.; Nazir, F.H.; Hanrieder, J.; et al. Herpes Simplex Virus 1 and 2 Infections during Differentiation of Human Cortical Neurons. *Viruses* **2021**, *13*, 2072. [CrossRef] [PubMed]
99. D'Aiuto, L.; Naciri, J.; Radio, N.; Tekur, S.; Clayton, D.; Apodaca, G.; Maio, R.D.; Zhi, Y.; Dimitrion, P.; Piazza, P.; et al. Generation of Three-Dimensional Human Neuronal Cultures: Application to Modeling CNS Viral Infections. *Stem Cell Res. Ther.* **2018**, *9*, 134. [CrossRef]
100. D'Aiuto, L.; Bloom, D.C.; Naciri, J.N.; Smith, A.; Edwards, T.G.; McClain, L.; Callio, J.A.; Jessup, M.; Wood, J.; Chowdari, K.; et al. Modeling HSV-1 Infections in Human CNS Neuronal Cells Using Two-Dimensional and Three- Dimensional Cultures Derived from Induced Pluripotent Stem Cells. *J. Virol.* **2019**, *93*, e00111-19. [CrossRef]
101. Blutt, S.E.; Estes, M.K. Organoid Models for Infectious Disease. *Annu. Rev. Med.* **2021**, *73*, 167–182. [CrossRef]
102. Pannese, E. The Structure of the Perineuronal Sheath of Satellite Glial Cells (SGCs) in Sensory Ganglia. *Neuron Glia Biol.* **2010**, *6*, 3–10. [CrossRef]
103. Huang, L.M.; Gu, Y.; Chen, Y. Communication between Neuronal Somata and Satellite Glial Cells in Sensory Ganglia. *Glia* **2013**, *61*, 1571–1581. [CrossRef]
104. Ji, R.-R.; Chamessian, A.; Zhang, Y.-Q. Pain Regulation by Non-Neuronal Cells and Inflammation. *Science* **2016**, *354*, 572–577. [CrossRef]
105. Koyuncu, O.O.; Hogue, I.B.; Enquist, L.W. Virus Infections in the Nervous System. *Cell Host Microbe* **2013**, *13*, 379–393. [CrossRef]
106. Knickelbein, J.E.; Khanna, K.M.; Yee, M.B.; Baty, C.J.; Kinchington, P.R.; Hendricks, R.L. Noncytotoxic lytic granule-mediated CD8+ T cell inhibition of HSV-1 reactivation from neuronal latency. *Science* **2008**, *322*, 268–271. [CrossRef] [PubMed]
107. Cherpes, T.L.; Busch, J.L.; Sheridan, B.S.; Harvey, S.A.K.; Hendricks, R.L. Medroxyprogesterone Acetate Inhibits CD8+ T Cell Viral-Specific Effector Function and Induces Herpes Simplex Virus Type 1 Reactivation. *J. Immunol.* **2008**, *181*, 969–975. [CrossRef] [PubMed]
108. Depledge, D.P.; Ouwendijk, W.J.D.; Sadaoka, T.; Braspenning, S.E.; Mori, Y.; Cohrs, R.J.; Verjans, G.M.G.M.; Breuer, J. A Spliced Latency-Associated VZV Transcript Maps Antisense to the Viral Transactivator Gene 61. *Nat. Commun.* **2018**, *9*, 1581. [CrossRef] [PubMed]
109. Phelan, D.; Barrozo, E.R.; Bloom, D.C. HSV1 Latent Transcription and Non-Coding RNA: A Critical Retrospective. *J. Neuroimmunol.* **2017**, *308*, 65–101. [CrossRef]
110. Ali, T.; Grote, P. Beyond the RNA-Dependent Function of LncRNA Genes. *eLife* **2020**, *9*, e60583. [CrossRef]
111. Sun, B.; Yang, X.; Hou, F.; Yu, X.; Wang, Q.; Oh, H.S.; Raja, P.; Pesola, J.M.; Vanni, E.A.H.; McCarron, S.; et al. Regulation of Host and Virus Genes by Neuronal MiR-138 Favours Herpes Simplex Virus 1 Latency. *Nat. Microbiol.* **2021**, *6*, 682–696. [CrossRef]
112. Flores, O.; Nakayama, S.; Whisnant, A.W.; Javanbakht, H.; Cullen, B.R.; Bloom, D.C. Mutational Inactivation of HSV-1 MicroRNAs Identifies Viral MRNA Targets and Reveals Phenotypic Effects in Culture. *J. Virol.* **2013**, *87*, 6589–6603. [CrossRef]
113. Erhard, F.; Baptista, M.A.P.; Krammer, T.; Hennig, T.; Lange, M.; Arampatzi, P.; Jürges, C.S.; Theis, F.J.; Saliba, A.-E.; Dölken, L. ScSLAM-Seq Reveals Core Features of Transcription Dynamics in Single Cells. *Nature* **2019**, *571*, 419–423. [CrossRef]
114. Hein, M.Y.; Weissman, J.S. Functional Single-Cell Genomics of Human Cytomegalovirus Infection. *Nat. Biotechnol.* **2022**, *40*, 391–401. [CrossRef]
115. Wang, X.; He, Y.; Zhang, Q.; Ren, X.; Zhang, Z. Direct Comparative Analyses of 10X Genomics Chromium and Smart-Seq2. *Genom. Proteom. Bioinform.* **2021**, *19*, 253–266. [CrossRef]
116. Aerts, L.; Miccoli, B.; Delahanty, A.; Witters, H.; Verstraelen, S.; Strooper, B.D.; Braeken, D.; Verstreken, P. Do We Still Need Animals? Surveying the Role of Animal-free Models in Alzheimer's and Parkinson's Disease Research. *EMBO J.* **2022**, *41*, e110002. [CrossRef] [PubMed]

viruses

Review

The Biology of Varicella-Zoster Virus Replication in the Skin

Cristina Tommasi [1,*] and Judith Breuer [2,*]

[1] School of Cellular and Molecular Medicine, University of Bristol, Bristol BS8 1TD, UK
[2] Department of Infection, Institute of Child Health, University College London, London WC1N 1EH, UK
* Correspondence: cristina.tommasi@bristol.ac.uk (C.T.); j.breuer@ucl.ac.uk (J.B.)

Abstract: The replication of varicella-zoster virus (VZV) in skin is critical to its pathogenesis and spread. Primary infection causes chickenpox, which is characterised by centrally distributed skin blistering lesions that are rich in infectious virus. Cell-free virus in the cutaneous blistering lesions not only spreads to cause further cases, but infects sensory nerve endings, leading to the establishment of lifelong latency in sensory and autonomic ganglia. The reactivation of virus to cause herpes zoster is again characterised by localised painful skin blistering rash containing infectious virus. The development of *in vitro* and *in vivo* models of VZV skin replication has revealed aspects of VZV replication and pathogenesis in this important target organ and improved our understanding of the vaccine strain vOKa attenuation. In this review, we outline the current knowledge on VZV interaction with host signalling pathways, the viral association with proteins associated with epidermal terminal differentiation, and how these interconnect with the VZV life cycle to facilitate viral replication and shedding.

Keywords: varicella-zoster virus; skin; epidermis; keratinocytes; epidermal differentiation; cutaneous blistering lesions; keratins; autophagy; immunity; vaccine

Citation: Tommasi, C.; Breuer, J. The Biology of Varicella-Zoster Virus Replication in the Skin. *Viruses* **2022**, *14*, 982. https://doi.org/10.3390/v14050982

Academic Editors: Charles Grose, Ravi Mahalingam and Joel Rovnak

Received: 29 March 2022
Accepted: 4 May 2022
Published: 6 May 2022

1. Introduction

Varicella-zoster virus (VZV) is a member of the human alphaherpesvirus family. Primary VZV infection causes chickenpox (varicella) in susceptible individuals, which typically manifests in immunocompetent individuals as a skin blistering rash preceded by flu-like symptoms and is usually self-limiting. In immunocompromised individuals, pregnant women as well as healthy adults, VZV infection may cause serious illness and even be lethal, often as a consequence of bacterial superinfections. A vaccine for VZV exists, which is based on the live attenuated parental VZV Oka strain [1,2]. Because the vaccine virus is live, it is not suitable for individuals with compromised immune systems. In addition, it can sometimes cause rashes, establishes latency in sensory nerves and reactivates in a small number of cases to cause herpes zoster (HZ). Consequently, research for alternative VZV vaccines, and therefore for a better understanding of VZV life cycle, is warranted.

VZV is transmitted between hosts by contact with the cutaneous blistering lesions containing the highly infectious cell-free virus or through inhalation of the aerosolized virions released from the skin lesions and to some extent the respiratory tract. During primary infection, VZV initially infects cells of the mucosa lining the upper respiratory tract where it is initially detected by dendritic cells (DCs) and transported to lymphoid tissues, primarily the tonsils, where it infects skin homing markers expressing T-lymphocytes that then deliver it to the skin [3,4]. The virus replicates in the skin epidermis before being released as cell-free virus from the blistering lesions. While replicating in the epidermis, the virus can gain access to the terminal endings of local sensory nerves and travel retrogradely through the axon to the cell bodies where it establishes latency. When reactivated, VZV travels within the axon in anterograde manner to reach the innervated dermatome where it causes herpes zoster (HZ) or shingles, characterised by a localised painful vesicular rash [4,5].

The tropism of VZV for the skin is well recognised, however our understanding of the molecular mechanisms of VZV replication in the epidermal tissue of the skin are still incomplete. VZV being a human-restricted virus with limited infectivity for other species, has limited the use of animal models for the study of viral natural history and pathogenicity. Most research into VZV cellular interactions has been in human embryonic lung fibroblasts (HELF), human lung MRC-5 fibroblasts and the human melanoma MeWo cell line. While these cell types are relatively easy to infect with VZV, none of them is representative of the human epidermis, which is the actual site of much VZV replication and shedding. An *in vivo* VZV infection model that closely mimics VZV infection of human skin *in vivo* was developed by the Arvin group and is based on the infection with VZV of xenografts of human fetal skin implanted into SCID-hu mice [3,6,7]. VZV infection of hTERT and human primary keratinocytes, organotypic human skin culture and human skin explants, also provides better *in vitro* representation of VZV infection of the epidermis [8–13]. The calcium-switch model of VZV infection, which allows analysis of VZV during keratinocyte differentiation provides a useful 2D *in vitro* model of VZV infection in human epidermis [8,9].

This review will explore what is understood to date about the mechanisms of VZV replication in the epidermis of the skin, as discovered in the very few existing human skin models of VZV infection.

2. VZV Life Cycle

During primary infection, VZV infects epithelial cells of the upper respiratory tract mucosa. It is believed that at this location the virus encounters and infects DCs, which then travel to the tonsils and other regional lymphoid tissues, where they transfer the virus to T lymphocytes [14,15]. It is not clear how T cells get infected by DCs, however it is well recognised that T lymphocytes play a crucial role in the dissemination of the virus from the initial sites of infection to the rest of the body, particularly the skin, during the viraemic stage of infection. VZV-infected T cells are mainly memory $CD4^+$ T cells expressing activation and skin-homing markers, such as CC-chemokine receptor 4 (CCR4) and cutaneous leukocyte antigen (CLA), but VZV also infects and activates naïve T cells [3,16]. It has been demonstrated *in vitro* that natural killer (NK) cells can also be infected by VZV and consequently increase the expression of skin-homing proteins [17]. However, their role in VZV skin pathogenesis *in vivo* and whether they contribute to the virus dissemination to T cells has not yet been elucidated. The incubation period during VZV infection is about 10–21 days, which is the time between VZV infection and the manifestation of symptoms, including the skin vesicular rash. Studies in SCID-hu mice models indicate that T cells reach the skin within days after infection, but it appears that the virus is able to evade initial antiviral responses through its interaction with interferon (IFN)-mediated innate immunity and apoptotic pathways (reviewed in [4]). These cause the virus to remain undetected for several days and therefore have enough time to replicate and produce the infectious virions-filled cutaneous blistering lesions that are needed for spread to new hosts [18,19]. During replication in the skin, VZV infects the terminal endings of sensory ganglia, which innervate the skin dermis and epidermis. It is believed that the intraepidermal nerve fibres encounter VZV at the level of the cutaneous lesions and are thereby infected by the cell-free virus [20,21]. Infection of neurons by cell-free virus allows the establishment of latency, whereas infection with cell-associated virus results in lytic infection [20,22,23]. VZV then travels to the neuronal cell bodies through retrograde axonal transport and there remains latent [24,25]. An additional possible method of neurons infection may involve the direct transmission of the virus from VZV-infected T cells [16] to neuronal cell bodies, but the precise mechanisms for this are still unclear [26]. Once reactivated, VZV travels back to the skin causing HZ, which is characterised by a painful blistering rash confined to the skin area innervated by the ganglion where reactivation occurred [5].

3. VZV and Skin

3.1. The Skin

The skin represents the outermost organ of the human body and encompasses the following tissues from the innermost to the outermost: hypodermis, dermis and epidermis. The dermis and epidermis represent the skin connective and epithelial tissues, respectively. While VZV replicates both the in the dermis and the epidermis, the latter represents the major site of viral replication, as well as the location where blistering lesions laden with infectious VZV virions form, guaranteeing virus transmission to new susceptible hosts.

The keratinocytes, which are the main cell type in the epidermis, are subject to cycles of proliferation, differentiation and death that allow preservation of epidermal homeostasis as well as of epidermal barrier function [27–29]. They maintain their proliferative capacity in the basal epidermal layer and when starting to differentiate they cease dividing and move to occupy the suprabasal spinous, granular, upper-granular and cornified layers (Figure 1). In the suprabasal layers, the keratinocytes express markers of terminal differentiation, until they ultimately loose nuclei and organelles and become corneocytes. Corneocytes are essentially bundles of filaments surrounded by cross-linked proteins and insoluble lipids and are ultimately shed out as dead cells [27,29,30]. Among the proteins that keratinocytes express during the distinct stages of differentiation, there are keratins, which are usually organised in heterodimers to constitute the keratin intermediate filaments (KIFs) that extend from the cell nucleus to the cell-cell desmosomes junctions. The basal keratinocytes typically express keratin 5 (K5) and 14 (K14). With differentiation, K5 and 14 are replaced by K1 and K10 and other "specialised" keratins such as K9 in the palms and soles' skin and K2 in thickened skin areas [31,32]. Keratin 15 (K15), which in the hair follicles (HFs) is associated with bulge stem cells, is also expressed by undifferentiated keratinocytes and lost when keratinocytes differentiate [32]. Other proteins that mark epidermal terminal differentiation include, but are not limited to, involucrin (IVL), loricrin, filaggrin (FLG), trichohyalin (TCHH) and small proline-rich proteins (SPRRs) [27].

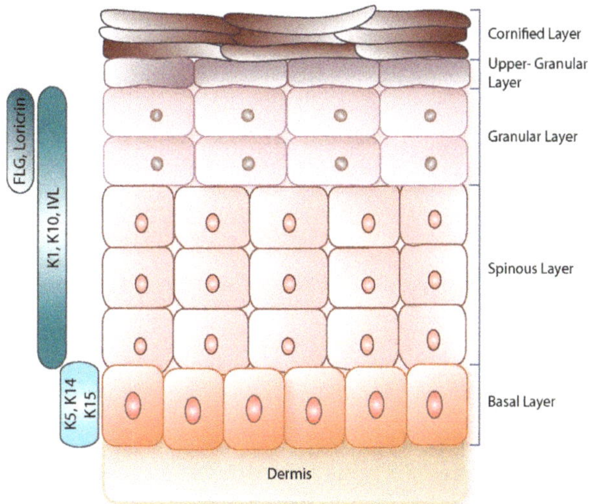

Figure 1. Schematic of the human epidermis. The human epidermis is a pluristratified tissue made up of basal, spinous, granular, upper-granular and cornified layers, in ascending order. The keratinocytes in the basal layer express keratins K5 and K14, as well as K15, which are then substituted by markers of epidermal differentiation such as K1, K10, IVL in the spinous layer. With progression of differentiation, other markers such as FLG and loricrin are expressed in the granular layer. The keratinocytes at the level of the cornified layer are called corneocytes as they are devoid of organelles and nuclei and are finally sloughed off.

3.2. VZV Skin Tropism

The earliest histological evidence of VZV infection from skin biopsies of both chickenpox and HZ is observed in the HFs [33,34]. This was also detected in skin xenografts on SCID-hu mice, where 24 h post systemic injection of VZV-infected T cells, a high number of T cells was observed in the HFs' skin cells, as well as in basal epidermal keratinocytes [18]. It is likely that VZV-carrying T lymphocytes reach the skin cells surrounding the HFs bulb through the extensive microvasculature which wraps around, possibly by diapedesis, however this has not been formally demonstrated [4]. The mechanisms whereby the virus is then transferred from T cells to HFs' skin cells are not known. Based on mathematical modelling applied to cutaneous blistering lesions following VZV vaccination, it has been inferred that each skin lesion arises from no more than three virions [35]. It could be hypothesised therefore that one virion (three at most) is transferred from a T cell to a skin cell and from there replicates to generate one blister. It is also unclear which are the exact skin epithelial cells targeted by the virus at the level of the HFs and how the virus is then transferred from the HFs to the interfollicular basal keratinocytes.

The virus replicates, spreads among cells and is highly cell-associated in the interfollicular basal keratinocytes, but becomes infectious only when keratinocytes differentiate into the upper epidermal layers, where it accumulates as infectious cell-free virions in the cutaneous blistering lesions [4,8,20,36]. Evidence from VZV infection of human keratinocytes in the calcium-switch model shows an increase in viral copy number occurring with epidermal differentiation, and this in turn is associated with increased expression of VZV genes classed as early or late in the replication cycle, with the loss of immediate early gene expression. Specifically, VZV immediate early genes are generally expressed in the basal epidermal layer, whereas late genes, such as the glycoproteins gC and gE, tend to be expressed in the suprabasal epidermal layers [8]. This provides some indication that VZV depends upon epidermal terminal differentiation to fully mature into cell-free and infectious virions that can be released by the skin lesions for infection of new hosts. The "physical" barrier represented by epidermal terminal differentiation of keratinocytes, as differentiated keratinocytes are more resistant to infection, may contribute to blisters remaining discrete lesions in the skin. However, a major contribution to this state is likely to come from the induction of IFNα and IFNβ, as well as immunity regulators, such as phosphorylated signal transducer and activator of transcription 1 (pSTAT1) and NF-κB in bystander epidermal cells, that also contain viral spread in the skin [4,18].

From studies conducted mainly in MeWo cells and lung fibroblasts, it has been determined that VZV enters the cell through initial interaction with cell surface heparan sulfate proteoglycans (HSPGs) [37], followed by binding to additional receptors, namely the insulin-degrading enzyme (IDE) and the mannose 6-phosphate receptor (MPR) [37,38]. The binding to these receptors is mediated by a number of viral glycoproteins. Specifically, as the MPR interacts with mannose-6-phopsphate groups, it can be bound by the glycoproteins gB, gI, gH, gE [36,39], whereas the IDE receptor can be bound by gE [38,40]. However, the role of gE-IDE in mediating cell-to-cell spread is still debated as it has been reported by later studies that IDE binds a precursor rather than a mature form of the gE protein and that this interaction occurs in the cell cytoplasm [41]. The roles of these receptors for VZV ingress into keratinocytes and for viral cell-to-cell spread in the epidermis are still unclear. It has been shown that the HSPGs and the MPR receptors are expressed in the basal epidermal layer and lost in the suprabasal layers [36,42]. However, viral glycoproteins are usually expressed only when infected keratinocytes differentiate [8], which would limit their interaction with MPR receptors at the level of the basal undifferentiated keratinocytes. Hence, further investigation in skin models of VZV infection will be necessary to determine whether these are the sole receptors needed by VZV for infection and to evaluate how their differential expression throughout the diverse epidermal layers is used by VZV in its life cycle. MPRs have also been implicated in the generation of cell-free VZV virions in the epidermal suprabasal layers, because MPRs also line the vesicles that transport newly enveloped virions to the plasma membrane and late endosomes [36,37,43]. In this study, the transport of virions to late endosomes was interpreted to result in virion degradation in the basal epidermal layer. In contrast, the absence of MPRs in the upper epidermal layers would prevent their transit to late endosomes, therefore allowing their maturation into cell-free infectious virions [36]. However, further studies are required to understand how cell-free VZV virions are generated in the epidermis and to understand whether autophagy (which is discussed in paragraph 4.2) may play a role in this process.

3.3. *Viral Components of VZV Replication in the Skin*

Early work using skin grafted into SCID-hu mice [3,6,7] showed differences between the behaviour of VZV in cells monolayers, even in the case of skin-derived cells, and differentiated epidermis. Using deletion mutagenesis in SCID-hu mice skin xenografts, a number of VZV proteins which appear to be dispensable for VZV infection in cell monolayers, were identified as essential to VZV pathogenesis *in vivo* (reviewed in [4]). Examples include the viral glycoprotein gC (encoded by ORF14), gM (encoded by ORF50), the regulatory protein ORF10 and the kinase ORF47, [6,7,44–46]. Further viral proteins that are essential to VZV skin pathogenesis *in vivo* encompass proteins encoded by the ORF9–12 gene cluster and the transcriptional transactivators IE62 (encoded by ORF62 and ORF71) and IE63 (encoded by ORF63 and ORF70), although for IE63 the relevance in VZV infection *in vivo* is dependent on the phosphorylation status [4,45,47]. The glycoproteins that are essential in skin pathogenesis *in vivo* encompass the glycoprotein gE (encoded by ORF68), which is essential for VZV replication and cell–cell spread in skin [48,49], and its partner gI [50], as well as glycoproteins gB (encoded by ORF31) and gH (encoded by ORF 37), which, in a complex with gL (encoded by ORF60), regulate both viral ingress into the cell and cell–cell fusion to generate syncytia [51–55]. Syncytia production is a hallmark of VZV infection and it is caused by the fusion of infected with uninfected cells to form polykaryocytes [4,52,53]. The fusion process is triggered by the interaction of gB and gH/gL, which for this purpose localise at the surface of infected cells, with receptors on the cell membranes of neighbouring cells. First, a fusion pore between adjacent cells membranes is formed, then the membranes fully fuse and combine their cytoplasm and nuclei to generate the multinucleated structure typical of syncytia (reviewed in [56]). The cytoplasmic domain of gB is indispensable for the fusion process, particularly the immunoreceptor tyrosine-based inhibition motif (ITIM) [52] and the lysine cluster downstream of the ITIM [57]. It is important to note that syncytia formation relies on a fine balance of cell-cell fusion; too little fusion would not allow sufficient cell-to-cell spread, whereas too much fusion would be detrimental for the virions' assembly and replication. The process of fusion therefore needs to be tightly controlled by the virus, probably through manipulation of host proteins, such as the phosphatase calcineurin [56].

4. VZV Interaction with Host Epidermal Pathways

4.1. *VZV Interplay with Epidermal Terminal Differentiation*

Like other skin epitheliotropic viruses, e.g., cutaneous human papilloma viruses (HPVs) [58,59], VZV life cycle and epidermal terminal differentiation are tightly interconnected. VZV profoundly alters gene and protein expression in the host keratinocytes, particularly in pathways that are unique to the epidermis, namely epidermal differentiation and epidermal barrier function pathways [8,9] (Figure 2).

In *in vitro* models of VZV infection of human epidermis, it was demonstrated that VZV induces the degradation of the suprabasal epidermal differentiation marker K10, as well as the downregulation of a number of corneodesmosomal proteins [8,9]. As K10, together with K1, forms the core of the intermediate filament network of the suprabasal keratinocytes and is a crucial component of the epidermal barrier function [60–62], its degradation is probably caused by VZV to facilitate blistering lesion formation. Indeed, the disruption of keratins is often associated with skin blistering conditions [63].

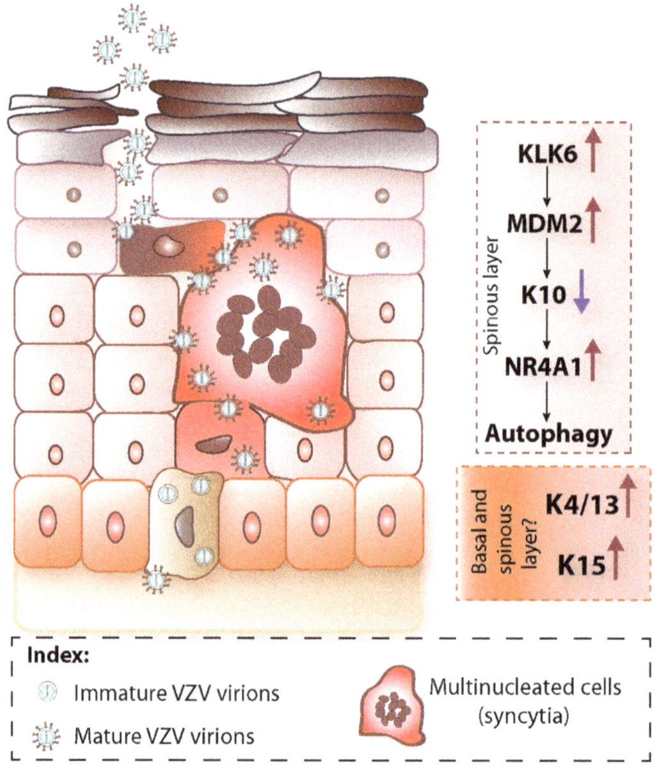

Figure 2. Schematic of VZV interplay with epidermal terminal differentiation. Left model of VZV-infected epidermis: VZV infects epidermal basal keratinocytes where starts to replicate. The newly formed virions in the basal keratinocytes are immature and do not express glycoproteins. In the suprabasal layers VZV virions mature together with epidermal differentiation and express glycoproteins. They also spread cell-to-cell via syncytia formation. At the uppermost epidermal layers mature and cell-free virions are released. Right panel: summary of the major changes induced by VZV to epidermal host pathways. They include in the spinous layer the downregulation of keratin K10 due to its degradation and caused by MDM2, which is upregulated by increased levels of KLK6. K10 downregulation has a structural effect on blister and syncytia formation, but also a signalling effect by upregulating NR4A1, which in turn induces activation of autophagy pathways. Keratin K15, which is normally expressed in the basal layer, is upregulated and its expression is found also in differentiated keratinocytes. K4 and K13 are also upregulated in differentiated keratinocytes during VZV infection.

VZV keratinocytes infection is also characterised by upregulation of a number of proteases, most notably kallikreins [8]. Kallikreins are serine proteases, whose most recognised role in skin is mediating epidermal desquamation through the degradation of corneodesmosomal proteins [64,65]. Kallikrein 5 (KLK5) and 7, whose role in epidermal desquamation is very well established, are highly upregulated in VZV-infected and differentiated keratinocytes, concomitant with downregulation of some of their known targets, notably desmoglein-1 (DSG1), desmocollin-1 (DSC1) [8,66]. It can be hypothesised that, through the upregulation of KLK5 and KLK7 levels, VZV exploits the process of epidermal desquamation for its shedding from the cornified envelope. Notably, it was observed that although KLK5 and KLK7 are highly upregulated in VZV keratinocytes infection, the most upregulated kallikreins are KLK6, 12 and 13, which are not strictly associated with skin desquamation [8]. In fact, an important role for KLK6 in VZV epidermal infection appears to be to upregulate cytoplasmic levels of the murine double minute 2 (MDM2) protein, allowing MDM2 to bind and ubiquitinate K10 and target it for proteosome degradation [9].

The formation of syncytia structures is well documented in MeWo cells [56], and multinucleated cells are also observed in VZV-infected keratinocytes, which have been differentiated with calcium-switch [9], as well as in skin from the SCID-hu mouse model [52]. However, the molecular mechanisms by which VZV overcomes host keratinocyte barriers to induce fusion have not yet been elucidated and warrant further investigation. We observed that in infected keratinocytes MDM2 interacts with K10 in nascent syncytial structures. The interaction is absent in larger syncytia, by which stage K10 is already lost, suggesting that K10 degradation may also contribute to syncytia expansion [9]. Indeed, the disruption of KIFs would lead to a disruption of cell–cell junctions, which may facilitate cell–cell fusion. Consistent with this, VZV infection of keratinocytes is characterised by an extensive downregulation of adhesion proteins [8].

K10 not only serves a structural function in VZV infection of the epidermis. In line with increasing studies reporting involvement of keratins in gene expression and signalling pathways [67,68], K10 degradation was found to also have a signalling role in VZV infection of epidermis, by upregulating the expression of the transcription factor nuclear receptor subfamily 4 group A member 1 (NR4A1) [9,69]. NR4A1 expression is non-nuclear in VZV-infected keratinocytes [9], which has been associated with a non-transcriptional role of NR4A1, for example in autophagy pathways [70]. In the case of VZV skin infection, cytoplasmic NR4A1 induces autophagy activation [9].

In the transcriptional signature of VZV-infected keratinocytes, *KRT4/13* (which are mucosal keratins, [32]) and *KRT15* genes are also upregulated [8]. Recent work showing the upregulation of K17 and WNT5a expression [71], together with K15 [8], seems to suggest an involvement of wound healing pathways in VZV infection, but the exact mechanisms and which precise characters are involved are unknown.

4.2. VZV Interplay with Epidermal Autophagy

Macroautophagy, hereafter called "autophagy", is a process involving the degradation and recycling of proteins and cellular components, which is activated in response to cellular stress, e.g., following microbes' infection, to guarantee cell homeostasis. The autophagy events are mediated by a number of proteins called autophagy-related genes (ATGs), as well as not ATG proteins [72] and involve the formation of a autophagophore which surrounds the to-be-degraded cargo, then matures into a double-membrane autophagosome, which eventually fuses with the lysosome to generate the autolysosome, the place where degradation of the cargo occurs (reviewed in [73,74]). In some cases, autophagosomes may fuse with late endosomes to form amphisomes, before fusion with lysosomes [73,75]. The termination of the different stages of autophagosome maturation, including autophagosome/amphisome fusion with the lysosome, is known as autophagic flux. Autophagy is constitutively operative in the epidermal granular layer [76], where it has a pivotal role in epidermal homeostasis by contributing to epidermal terminal differentiation and barrier function, as also proven by a number of inflammation and skin barrier diseases characterised by defects in autophagy ([77] and reviewed in [78]). Akinduro et al. [76] showed that a type of autophagic nuclear degradation called nucleophagy occurs in keratinocytes at the level of the granular layer, demonstrating that nucleophagy plays a role in the process of nuclei loss during late terminal differentiation.

In VZV-infected and differentiated keratinocytes and skin explants, NR4A1 has been found to be crucial in the induction of autophagy mechanisms, as demonstrated by the decreased expression of key autophagy markers, such as LC3, when NR4A1 was knocked down [9]. Notably, LC3, which in infected cells is widely expressed in the cytoplasm, following NR4A1 downregulation, gets re-localised around the cell nuclei [9], which is a characteristic feature of LC3 in terminally differentiated keratinocytes [76].

The analyses of VZV-infected fibroblasts and MeWo cells, as well as keratinocytes and skin explants, indicate that autophagy has a proviral effect during VZV skin infection and suggest that the virus exploits autophagy events for completion of its life cycle [9,79,80]. It was shown, mostly in MRC-5 fibroblasts and MeWo cells, that, unlike HSV-1, in VZV infection the autophagic flux is not interrupted [81,82]. In fact, no genes with autophagy inhibition function have been so far identified in the VZV genome, whereas the HSV-1 genome expresses at least two autophagy inhibitors genes, ICP34.5 and US11 [83,84]. Notwithstanding, a recent investigation showed in VZV-infected MRC-5 fibroblasts that the late stages of the autophagy flux are inhibited [85]. In keeping with this, in VZV-infected and differentiated keratinocytes lysosomal proteins, such as lysosomal associated membrane protein 2 (LAMP2), and other autolysosome trafficking proteins, such as vesicle-associated membrane protein 7 (VAMP7), are downregulated [9]. It has also been observed, in fibroblasts and MeWo cells, that VZV virions localize in single membranes vesicles, presumably amphisomes, as characterised by both the autophagosome protein LC3 and the endosomal protein Rab11 [86]. Given that the synthesis and maturation of some VZV glycoproteins, such as gE, are reduced when autophagy is inhibited [80], it could be hypothesised that the autophagy machinery is involved in the secondary envelopment of VZV virions or, as hypothesised by Grose et al. [86], in the virions exocytosis pathway, through employment of amphisomes' vesicles. More future studies will be needed to clarify the molecular mechanisms of autophagy in VZV infection of the skin, to understand whether the virus co-opts amphisomes structures for the completion of its life cycle, also with respect to the specifics of the diverse epidermal layers that the virus navigates through. Many aspects are still unclear and in some cases contradictory. For example, Chen et al. [36] showed that MPR expression is reduced in the suprabasal epidermal layers, yet amphisomes express MPR proteins as they are the result of the fusion of autophagosomes and late endosomes [75].

4.3. VZV Interplay with Epidermal Innate Immunity

The major innate immunity cells involved in the reaction to VZV infection, including DCs, Langerhans and NK cells, have been recently reviewed in [87]. Here, we focus on discussing how VZV restrains the innate immune responses that involve cytokines and antiviral proteins produced by keratinocytes upon infection. In studies of VZV-infected skin in the SCID-hu mouse model, it was shown that VZV averts the degradation of inhibitor of NF-κBα (IκBα) and NF-κB is segregated in the cytoplasm of infected cells [88]. Concomitantly, infected keratinocytes also display activation of STAT3 (and therefore increase levels of pSTAT3) which in turn up-regulates the expression of survivin, which blocks host cells' apoptosis, thereby guaranteeing the virus survival in the host cell [89]. In addition, in the infected keratinocytes, VZV is able to block, through ORF61, the antiviral cellular mechanism induced by IFNs and represented by the protein promyelocytic leukemia protein- nuclear bodies (PML-NB) formation, which entraps nascent virions in the nucleus, thereby preventing them from leaving the nucleus to complete their life cycle [90]. Finally, in fibroblasts and melanoma cells, it was demonstrated that VZV IE62 mediates the block of IFN regulatory factor 3 (IRF3) phosphorylation to prevent IFNβ production [91] and VZV IE63 blocks eukaryotic initiation factor 2 α (eIF-2α) phosphorylation, which is normally a consequence of IFNα activation to inhibit protein translation [92].

5. VZV Vaccine and Skin

The varicella vaccine was generated by Takahashi and colleagues in 1974 [2] by serial passaging, first in human embryonic fibroblasts and then in in guinea pig embryo fibroblasts, of VZV virus isolated from a boy with chickenpox [1,2]. The VZV strain used for the vaccine generation is called Oka, consequently the parental and vaccine strains have been termed pOka and vOka, respectively. This live attenuated vOka vaccine was licensed for the prevention of chickenpox initially in the US in 1995 and since then in several mainly high income countries [93]. Since its introduction in the US, the vaccine has drastically reduced the number of varicella clinical cases and deaths. Moreover, the administration of a second dose of vaccine has significantly decreased breakthrough infections [94]. The vaccine is safe [95] and effective in preventing chickenpox or reducing the severity of it, as well as in building a long-last immunity against the virus [93]. However, being live, the vaccine is not suitable for administration to immunocompromised individuals [96]. The vOka vaccine, like wild-type VZV, stimulates the generation of antibodies as well as cell-mediated immunity (CMI). CMI is crucial for clearance of the virus once an individual has been infected, as the virus is highly cell-associated, as well as for the prevention of virus reactivation and HZ. However, CMI declines with age and this is believed to increase the risk of HZ in older age groups [94]. For this reason, the same vOka strain has been formulated at higher concentration for the prevention of HZ in older adults, as a therapeutic vaccine [97]. A newer glycoprotein E subunit vaccine adjuvanted with $AS01_B$ adjuvant has recently been shown to be extremely effective in preventing HZ and its commonest sequela, post herpetic neuralgia [98].

Moffat et al. [7] demonstrated that vOka strain of VZV is attenuated for growth compared to pOka in the SCID-hu mouse skin xenograft model, although not in tissue culture monolayers [7,99]. No difference was found in the level of vOka titer recovered from infected T cells in the SCID-hu mouse model [3], indicating that the mechanism of vOka attenuation is likely to be reduced skin replication. Additionally, vOka appears to be impaired for reactivation [23,95]. Because the vOka strain has been produced by serial passaging of pOka in cells, it has accumulated a variety of genetic mutations, with 137 single-nucleotide polymorphisms (SNPs) shared by all the vOka vaccines and additional SNPs specific for certain commercial vaccines preparations, as identified by deep sequencing [100–103]. Only a few SNPs are at near fixation for the vaccine allele, with four present in the ORF62 gene and one in ORF0 [100,104,105]. ORF62 encodes the VZV transactivator IE62, which regulates the transcription of immediate early genes and is crucial in VZV skin pathogenesis [4]. ORF62 is the gene that in the vOka strain contains the major number of SNPs compared to pOka, of which four are highly fixed, with two of them causing R958G and S628G substitutions in two highly conserved regions of the ORF62 gene [106]. It also appears that vOka IE62 induces a decreased level of gene transcription compared to pOka IE62 [107]. All in all, the current available evidence suggests that IE62 may play an important role in the skin attenuation of the VZV vaccine and studies aimed at elucidating these mechanisms are underway.

Understanding the molecular mechanisms causing attenuation in skin of the existing VZV vaccine will shed light on pivotal events of VZV skin replication that could be targeted by new antiviral drugs, as well as pave the way for the generation of novel vaccines that could be directed to a wider number of recipients, including immunocompromised children.

6. Conclusions and Future Perspectives

Work on models that reproduce the architecture of human skin or at very least recapitulate keratinocyte differentiation has provided important insights into how VZV interacts with host epidermal pathways. In particular, the pivotal role of the specialised skin keratinocyte cell in the VZV replication cycle is likely to shed light on VZV tropism and pathogenesis. These mechanisms are just beginning to be unveiled and this work is also complicated by VZV infection being highly interconnected with the host epidermal biology, which itself is characterised by molecular pathways that have often been only partially elucidated. An example is epidermal autophagy, whose pathways have only quite recently begun to be uncovered [76,78]. An important tool to untangle such complexity and that could provide important insights into VZV infection at different stages of keratinocyte differentiation, is single cell RNA-sequencing, which has been already used successfully to identify cells at different stages of differentiation from human skin tissues [108,109]. This approach will allow to address the questions that remain unanswered concerning the role of epidermal differentiation in the VZV life cycle, particularly which differentiation pathways, together with K10 degradation, are implicated in VZV skin replication and blistering lesion formation. It will be important to elucidate the role of K15 and whether wound healing pathways are activated, at which stage of infection and for which purpose. Future studies will also need to address the molecular mechanisms and functions of autophagy and other cell stress processes in the context of VZV replication in the epidermis. Elucidation of these represents new druggable pathways for antiviral therapies and vaccine targets. For example, we have already shown that the small-molecule inhibitor nutlin-3, which is known to block MDM2 function [110], is able to restore K10 expression in VZV-infected keratinocytes and significantly reduce VZV growth in keratinocytes monolayers and skin explants [9]. A similar approach could be taken using existing inhibitors of autophagy.

Finally, a better understanding of VZV pathogenesis in the skin could help to develop new vaccines that are not based on the live attenuated virus. The existing vaccine vOka, being attenuated for growth in skin, is actually itself an important tool for the study of VZV skin infection and the comparative analysis with the parental strain pOka can further the understanding of the molecular mechanisms of VZV replication in skin. Moreover, a better knowledge of VZV skin pathogenesis will probably prove to be very valuable to further understand the pathogenesis of other skin epitheliotropic viruses that may share common mechanisms of replication in skin.

Author Contributions: All authors contributed to the writing and editing of the manuscript. All authors have read and agreed to the published version of the manuscript.

Funding: This research received no external funding.

Conflicts of Interest: The authors declare no conflict of interest.

References

1. Breuer, J. Vaccination to prevent varicella and shingles. *J. Clin. Pathol.* **2001**, *54*, 743–747. [CrossRef] [PubMed]
2. Takahashi, M.; Otsuka, T.; Okuno, Y.; Asano, Y.; Yazaki, T.; Isomura, S. Live vaccine used to prevent the spread of varicella in children in hospital. *Lancet* **1974**, *304*, 1288–1290. [CrossRef]
3. Moffat, J.F.; Stein, M.D.; Kaneshima, H.; Arvin, A.M. Tropism of varicella-zoster virus for human CD4+ and CD8+ T lymphocytes and epidermal cells in SCID-hu mice. *J. Virol.* **1995**, *69*, 5236–5242. [CrossRef] [PubMed]
4. Zerboni, L.; Sen, N.; Oliver, S.L.; Arvin, A.M. Molecular mechanisms of varicella zoster virus pathogenesis. *Nat. Rev. Microbiol.* **2014**, *12*, 197–210. [CrossRef] [PubMed]
5. Arvin, A.M. Varicella-zoster virus. *Clin. Microbiol. Rev.* **1996**, *9*, 361–381. [CrossRef]
6. Moffat, J.F.; Zerboni, L.; Sommer, M.H.; Heineman, T.C.; Cohen, J.I.; Kaneshima, H.; Arvin, A.M. The ORF47 and ORF66 putative protein kinases of varicella-zoster virus determine tropism for human T cells and skin in the SCID-hu mouse. *Proc. Natl. Acad. Sci. USA* **1998**, *95*, 11969–11974. [CrossRef]
7. Moffat, J.F.; Zerboni, L.; Kinchington, P.R.; Grose, C.; Kaneshima, H.; Arvin, A.M. Attenuation of the vaccine Oka strain of varicella-zoster virus and role of glycoprotein C in alphaherpesvirus virulence demonstrated in the SCID-hu mouse. *J. Virol.* **1998**, *72*, 965–974. [CrossRef]
8. Jones, M.; Dry, I.R.; Frampton, D.; Singh, M.; Kanda, R.K.; Yee, M.B.; Kellam, P.; Hollinshead, M.; Kinchington, P.R.; O'Toole, E.A.; et al. RNA-seq Analysis of Host and Viral Gene Expression Highlights Interaction between Varicella Zoster Virus and Keratinocyte Differentiation. *PLoS Pathog.* **2014**, *10*, e1003896. [CrossRef]
9. Tommasi, C.; Rogerson, C.; Depledge, D.P.; Jones, M.; Naeem, A.S.; Venturini, C.; Frampton, D.; Tutill, H.J.; Way, B.; Breuer, J.; et al. Kallikrein-Mediated Cytokeratin 10 Degradation Is Required for Varicella Zoster Virus Propagation in Skin. *J. Investig. Dermatol.* **2020**, *140*, 774–784.e11. [CrossRef]
10. Jarosinski, K.W.; Carpenter, J.E.; Buckingham, E.M.; Jackson, W.; Knudtson, K.; Moffat, J.F.; Kita, H.; Grose, C. Cellular Stress Response to Varicella-Zoster Virus Infection of Human Skin Includes Highly Elevated Interleukin-6 Expression. *Open Forum Infect. Dis.* **2018**, *5*, ofy118. [CrossRef]
11. Lloyd, M.G.; Smith, N.A.; Tighe, M.; Travis, K.L.; Liu, D.; Upadhyaya, P.K.; Kinchington, P.R.; Chan, G.C.; Moffat, J.F. A Novel Human Skin Tissue Model To Study Varicella-Zoster Virus and Human Cytomegalovirus. *J. Virol.* **2020**, *94*, e01082-20. [CrossRef] [PubMed]
12. Taylor, S.L.; Moffat, J.F. Replication of varicella-zoster virus in human skin organ culture. *J. Virol.* **2005**, *79*, 11501–11506. [CrossRef] [PubMed]
13. Rowe, J.; Greenblatt, R.J.; Liu, D.; Moffat, J.F. Compounds that target host cell proteins prevent varicella-zoster virus replication in culture, ex vivo, and in SCID-Hu mice. *Antivir. Res.* **2010**, *86*, 276–285. [CrossRef] [PubMed]
14. Abendroth, A.; Morrow, G.; Cunningham, A.L.; Slobedman, B. Varicella-zoster virus infection of human dendritic cells and transmission to T cells: Implications for virus dissemination in the host. *J. Virol.* **2001**, *75*, 6183–6192. [CrossRef] [PubMed]
15. Morrow, G.; Slobedman, B.; Cunningham, A.L.; Abendroth, A. Varicella-Zoster Virus Productively Infects Mature Dendritic Cells and Alters Their Immune Function. *J. Virol.* **2003**, *77*, 4950–4959. [CrossRef]

16. Ku, C.C.; Padilla, J.A.; Grose, C.; Butcher, E.C.; Arvin, A.M. Tropism of varicella-zoster virus for human tonsillar CD4(+) T lymphocytes that express activation, memory, and skin homing markers. *J. Virol.* **2002**, *76*, 11425–11433. [CrossRef]
17. Campbell, T.M.; McSharry, B.P.; Steain, M.; Ashhurst, T.M.; Slobedman, B.; Abendroth, A. Varicella zoster virus productively infects human natural killer cells and manipulates phenotype. *PLoS Pathog.* **2018**, *14*, e1006999. [CrossRef]
18. Ku, C.C.; Zerboni, L.; Ito, H.; Graham, B.S.; Wallace, M.; Arvin, A.M. Varicella-zoster virus transfer to skin by T Cells and modulation of viral replication by epidermal cell interferon-alpha. *J. Exp. Med.* **2004**, *200*, 917–925. [CrossRef]
19. Abendroth, A.; Arvin, A.M. Immune evasion as a pathogenic mechanism of varicella zoster virus. *Semin. Immunol.* **2001**, *13*, 27–39. [CrossRef]
20. Gershon, M.D.; Gershon, A.A. VZV infection of keratinocytes: Production of cell-free infectious virions in vivo. *Curr. Top. Microbiol. Immunol.* **2010**, *342*, 173–188.
21. Annunziato, P.W.; Lungu, O.; Panagiotidis, C.; Zhang, J.H.; Silvers, D.N.; Gershon, A.A.; Silverstein, S.J. Varicella-Zoster Virus Proteins in Skin Lesions: Implications for a Novel Role of ORF29p in Chickenpox. *J. Virol.* **2000**, *74*, 2005–2010. [CrossRef] [PubMed]
22. Gershon, A.A.; Chen, J.; Gershon, M.D. A model of lytic, latent, and reactivating varicella-zoster virus infections in isolated enteric neurons. *J. Infect. Dis.* **2008**, *197* (Suppl. 2), S61–S65. [CrossRef] [PubMed]
23. Sadaoka, T.; Depledge, D.P.; Rajbhandari, L.; Venkatesan, A.; Breuer, J.; Cohen, J.I. In vitro system using human neurons demonstrates that varicella-zoster vaccine virus is impaired for reactivation, but not latency. *Proc. Natl. Acad. Sci. USA* **2016**, *113*, E2403–E2412. [CrossRef] [PubMed]
24. Markus, A.; Grigoryan, S.; Sloutskin, A.; Yee, M.B.; Zhu, H.; Yang, I.H.; Thakor, N.V.; Sarid, R.; Kinchington, P.R.; Goldstein, R.S. Varicella-zoster virus (VZV) infection of neurons derived from human embryonic stem cells: Direct demonstration of axonal infection, transport of VZV, and productive neuronal infection. *J. Virol.* **2011**, *85*, 6220–6233. [CrossRef]
25. Grigoryan, S.; Kinchington, P.R.; Yang, I.H.; Selariu, A.; Zhu, H.; Yee, M.; Goldstein, R.S. Retrograde axonal transport of VZV: Kinetic studies in hESC-derived neurons. *J. Neurovirol.* **2012**, *18*, 462–470. [CrossRef]
26. Zerboni, L.; Ku, C.C.; Jones, C.D.; Zehnder, J.L.; Arvin, A.M. Varicella-zoster virus infection of human dorsal root ganglia in vivo. *Proc. Natl. Acad. Sci. USA* **2005**, *102*, 6490–6495. [CrossRef]
27. Candi, E.; Schmidt, R.; Melino, G. The cornified envelope: A model of cell death in the skin. *Nat. Rev. Mol. Cell Biol.* **2005**, *6*, 328–640. [CrossRef]
28. Blanpain, C.; Fuchs, E. Epidermal homeostasis: A balancing act of stem cells in the skin. *Nat. Rev. Mol. Cell Biol.* **2009**, *10*, 207–217. [CrossRef]
29. Proksch, E.; Brandner, J.M.; Jensen, J.M. The skin: An indispensable barrier. *Exp. Dermatol.* **2008**, *17*, 1063–1072. [CrossRef]
30. Eckhart, L.; Lippens, S.; Tschachler, E.; Declercq, W. Cell death by cornification. Biochim. *Biophys. Acta* **2013**, *1833*, 3471–3480. [CrossRef]
31. Fuchs, E.; Cleveland, D.W. A Structural Scaffolding of Intermediate Filaments in Health and Disease. *Science* **1998**, *279*, 514–519. [CrossRef] [PubMed]
32. Moll, R.; Divo, M.; Langbein, L. The human keratins: Biology and pathology. *Histochem. Cell Biol.* **2008**, *129*, 705–733. [CrossRef] [PubMed]
33. Iwasaki, T.; Muraki, R.; Kasahara, T.; Sato, Y.; Sata, T.; Kurata, T. Pathway of viral spread in herpes zoster: Detection of the protein encoded by open reading frame 63 of varicella-zoster virus in biopsy specimens. *Arch. Virol. Suppl.* **2001**, *17*, 109–119.
34. Muraki, R.; Iwasaki, T.; Sata, T.; Sato, Y.; Kurata, T. Hair follicle involvement in herpes zoster: Pathway of viral spread from ganglia to skin. *Virchows Arch. Int. J. Pathol.* **1996**, *428*, 275–280. [CrossRef]
35. Weinert, L.A.; Depledge, D.P.; Kundu, S.; Gershon, A.A.; Nichols, R.A.; Balloux, F.; Welch, J.J.; Breuer, J. Rates of vaccine evolution show strong effects of latency: Implications for varicella zoster virus epidemiology. *Mol. Biol. Evol.* **2015**, *32*, 1020–1028. [CrossRef] [PubMed]
36. Chen, J.J.; Zhu, Z.; Gershon, A.A.; Gershon, M.D. Mannose 6-phosphate receptor dependence of varicella zoster virus infection in vitro and in the epidermis during varicella and zoster. *Cell* **2004**, *119*, 915–926. [CrossRef]
37. Zhu, Z.; Gershon, M.D.; Ambron, R.; Gabel, C.; Gershon, A.A. Infection of cells by varicella zoster virus: Inhibition of viral entry by mannose 6-phosphate and heparin. *Proc. Natl. Acad. Sci. USA* **1995**, *92*, 3546–3550. [CrossRef]
38. Li, Q.; Krogmann, T.; Ali, M.A.; Tang, W.J.; Cohen, J.I. The amino terminus of varicella-zoster virus (VZV) glycoprotein E is required for binding to insulin-degrading enzyme, a VZV receptor. *J. Virol.* **2007**, *81*, 8525–8532. [CrossRef]
39. Gabel, C.A.; Dubey, L.; Steinberg, S.P.; Sherman, D.; Gershon, M.D.; Gershon, A.A. Varicella-zoster virus glycoprotein oligosaccharides are phosphorylated during posttranslational maturation. *J. Virol.* **1989**, *63*, 4264–4276. [CrossRef]
40. Ali, M.A.; Li, Q.; Fischer, E.R.; Cohen, J.I. The insulin degrading enzyme binding domain of varicella-zoster virus (VZV) glycoprotein E is important for cell-to-cell spread and VZV infectivity, while a glycoprotein I binding domain is essential for infection. *Virology* **2009**, *386*, 270–279. [CrossRef]
41. Carpenter, J.E.; Jackson, W.; de Souza, G.A.; Haarr, L.; Grose, C. Insulin-degrading enzyme binds to the nonglycosylated precursor of varicella-zoster virus gE protein found in the endoplasmic reticulum. *J. Virol.* **2010**, *84*, 847–855. [CrossRef] [PubMed]
42. Jung, J.Y.; Oh, J.H.; Kim, Y.K.; Shin, M.H.; Lee, D.; Chung, J.H. Acute UV Irradiation Increases Heparan Sulfate Proteoglycan Levels in Human Skin. *J. Korean Med. Sci.* **2012**, *27*, 300–306. [CrossRef] [PubMed]

43. Gershon, A.A.; Sherman, D.L.; Zhu, Z.; Gabel, C.A.; Ambron, R.T.; Gershon, M.D. Intracellular transport of newly synthesized varicella-zoster virus: Final envelopment in the trans-Golgi network. *J. Virol.* **1994**, *68*, 6372–6390. [CrossRef] [PubMed]
44. Besser, J.; Ikoma, M.; Fabel, K.; Sommer, M.H.; Zerboni, L.; Grose, C.; Arvin, A.M. Differential Requirement for Cell Fusion and Virion Formation in the Pathogenesis of Varicella-Zoster Virus Infection in Skin and T Cells. *J. Virol.* **2004**, *78*, 13293–13305. [CrossRef] [PubMed]
45. Che, X.; Reichelt, M.; Sommer, M.H.; Rajamani, J.; Zerboni, L.; Arvin, A.M. Functions of the ORF9-to-ORF12 gene cluster in varicella-zoster virus replication and in the pathogenesis of skin infection. *J. Virol.* **2008**, *82*, 5825–5834. [CrossRef]
46. Zerboni, L.; Sung, P.; Sommer, M.; Arvin, A. The C-terminus of varicella-zoster virus glycoprotein M contains trafficking motifs that mediate skin virulence in the SCID-human model of VZV pathogenesis. *Virology* **2018**, *523*, 110–120. [CrossRef]
47. Baiker, A.; Bagowski, C.; Ito, H.; Sommer, M.; Zerboni, L.; Fabel, K.; Hay, J.; Ruyechan, W.; Arvin, A.M. The immediate-early 63 protein of Varicella-Zoster virus: Analysis of functional domains required for replication in vitro and for T-cell and skin tropism in the SCIDhu model in vivo. *J. Virol.* **2004**, *78*, 1181–1194. [CrossRef]
48. Berarducci, B.; Ikoma, M.; Stamatis, S.; Sommer, M.; Grose, C.; Arvin, A.M. Essential functions of the unique N-terminal region of the varicella-zoster virus glycoprotein E ectodomain in viral replication and in the pathogenesis of skin infection. *J. Virol.* **2006**, *80*, 9481–9496. [CrossRef]
49. Berarducci, B.; Rajamani, J.; Zerboni, L.; Che, X.; Sommer, M.; Arvin, A.M. Functions of the unique N-terminal region of glycoprotein E in the pathogenesis of varicella-zoster virus infection. *Proc. Natl. Acad. Sci. USA* **2010**, *107*, 282–287. [CrossRef]
50. Oliver, S.L.; Sommer, M.H.; Reichelt, M.; Rajamani, J.; Vlaycheva-Beisheim, L.; Stamatis, S.; Cheng, J.; Jones, C.; Zehnder, J.; Arvin, A.M. Mutagenesis of varicella-zoster virus glycoprotein I (gI) identifies a cysteine residue critical for gE/gI heterodimer formation, gI structure, and virulence in skin cells. *J. Virol.* **2011**, *85*, 4095–4110. [CrossRef]
51. Oliver, S.L.; Sommer, M.; Zerboni, L.; Rajamani, J.; Grose, C.; Arvin, A.M. Mutagenesis of varicella-zoster virus glycoprotein B: Putative fusion loop residues are essential for viral replication, and the furin cleavage motif contributes to pathogenesis in skin tissue in vivo. *J. Virol.* **2009**, *83*, 7495–7506. [CrossRef] [PubMed]
52. Oliver, S.L.; Brady, J.J.; Sommer, M.H.; Reichelt, M.; Sung, P.; Blau, H.M.; Arvin, A.M. An immunoreceptor tyrosine-based inhibition motif in varicella-zoster virus glycoprotein B regulates cell fusion and skin pathogenesis. *Proc. Natl. Acad. Sci. USA* **2013**, *110*, 1911–1916. [CrossRef] [PubMed]
53. Yang, E.; Arvin, A.M.; Oliver, S.L. The cytoplasmic domain of varicella-zoster virus glycoprotein H regulates syncytia formation and skin pathogenesis. *PLoS Pathog.* **2014**, *10*, e1004173. [CrossRef] [PubMed]
54. Vleck, S.E.; Oliver, S.L.; Reichelt, M.; Rajamani, J.; Zerboni, L.; Jones, C.; Zehnder, J.; Grose, C.; Arvin, A.M. Anti-glycoprotein H antibody impairs the pathogenicity of varicella-zoster virus in skin xenografts in the SCID mouse model. *J. Virol.* **2010**, *84*, 141–152. [CrossRef] [PubMed]
55. Vleck, S.E.; Oliver, S.L.; Brady, J.J.; Blau, H.M.; Rajamani, J.; Sommer, M.H.; Arvin, A.M. Structure-function analysis of varicella-zoster virus glycoprotein H identifies domain-specific roles for fusion and skin tropism. *Proc. Natl. Acad. Sci. USA* **2011**, *108*, 18412–18417. [CrossRef]
56. Oliver, S.L.; Zhou, M.; Arvin, A.M. Varicella-zoster virus: Molecular controls of cell fusion-dependent pathogenesis. *Biochem. Soc. Trans.* **2020**, *48*, 2415–2435. [CrossRef]
57. Yang, E.; Arvin, A.M.; Oliver, S.L. The Glycoprotein B Cytoplasmic Domain Lysine Cluster Is Critical for Varicella-Zoster Virus Cell-Cell Fusion Regulation and Infection. *J. Virol.* **2017**, *91*, e01707–e01716. [CrossRef]
58. Haller, K.; Stubenrauch, F.; Pfister, H. Differentiation-dependent transcription of the epidermodysplasia verruciformis-associated human papillomavirus type 5 in benign lesions. *Virology* **1995**, *214*, 245–255. [CrossRef]
59. Kajitani, N.; Satsuka, A.; Kawate, A.; Sakai, H. Productive Lifecycle of Human Papillomaviruses that Depends Upon Squamous Epithelial Differentiation. *Front. Microbiol.* **2012**, *3*, 152. [CrossRef]
60. Steinert, P.M. Analysis of the mechanism of assembly of mouse keratin 1/keratin 10 intermediate filaments in vitro suggests that intermediate filaments are built from multiple oligomeric units rather than a unique tetrameric building block. *J. Struct. Biol.* **1991**, *107*, 175–188. [CrossRef]
61. Steinert, P.M. Organization of coiled-coil molecules in native mouse keratin 1/keratin 10 intermediate filaments: Evidence for alternating rows of antiparallel in-register and antiparallel staggered molecules. *J. Struct. Biol.* **1991**, *107*, 157–174. [CrossRef]
62. Steinert, P.M.; Marekov, L.N. The proteins elafin, filaggrin, keratin intermediate filaments, loricrin, and small proline-rich proteins 1 and 2 are isodipeptide cross-linked components of the human epidermal cornified cell envelope. *J. Biol. Chem.* **1995**, *270*, 17702–17711. [CrossRef] [PubMed]
63. Uitto, J.; Richard, G.; McGrath, J.A. Diseases of epidermal keratins and their linker proteins. *Exp. Cell Res.* **2007**, *313*, 1995–2009. [CrossRef] [PubMed]
64. Simon, M.; Jonca, N.; Guerrin, M.; Haftek, M.; Bernard, D.; Caubet, C.; Egelrud, T.; Schmidt, R.; Serre, G. Refined characterization of corneodesmosin proteolysis during terminal differentiation of human epidermis and its relationship to desquamation. *J. Biol. Chem.* **2001**, *276*, 20292–20299. [CrossRef]
65. Borgoño, C.A.; Michael, I.P.; Komatsu, N.; Jayakumar, A.; Kapadia, R.; Clayman, G.L.; Sotiropoulou, G.; Diamandis, E.P. A potential role for multiple tissue kallikrein serine proteases in epidermal desquamation. *J. Biol. Chem.* **2007**, *282*, 3640–3652. [CrossRef]

66. Caubet, C.; Jonca, N.; Brattsand, M.; Guerrin, M.; Bernard, D.; Schmidt, R.; Egelrud, T.; Simon, M.; Serre, G. Degradation of corneodesmosome proteins by two serine proteases of the kallikrein family, SCTE/KLK5/hK5 and SCCE/KLK7/hK7. *J. Investig. Dermatol.* **2004**, *122*, 1235–1244. [CrossRef]
67. Hobbs, R.P.; DePianto, D.J.; Jacob, J.T.; Han, M.C.; Chung, B.M.; Batazzi, A.S.; Poll, B.G.; Guo, Y.; Han, J.; Ong, S.; et al. Keratin-dependent regulation of Aire and gene expression in skin tumor keratinocytes. *Nat. Genet.* **2015**, *47*, 933–938. [CrossRef]
68. Nair, R.R.; Hsu, J.; Jacob, J.T.; Pineda, C.M.; Hobbs, R.P.; Coulombe, P.A. A role for keratin 17 during DNA damage response and tumor initiation. *Proc. Natl. Acad. Sci. USA* **2021**, *118*, e2020150118. [CrossRef]
69. Niu, G.; Lu, L.; Gan, J.; Zhang, D.; Liu, J.; Huang, G. Dual roles of orphan nuclear receptor TR3/Nur77/NGFI-B in mediating cell survival and apoptosis. *Int. Rev. Cell Mol. Biol.* **2014**, *313*, 219–258.
70. Pawlak, A.; Strzadala, L.; Kalas, W. Non-genomic effects of the NR4A1/Nur77/TR3/NGFIB orphan nuclear receptor. *Steroids* **2015**, *95*, 1–6. [CrossRef]
71. Zerboni, L.; Sung, P.; Lee, G.; Arvin, A. Age-Associated Differences in Infection of Human Skin in the SCID Mouse Model of Varicella-Zoster Virus Pathogenesis. *J. Virol.* **2018**, *92*, e00002–e00018. [CrossRef]
72. Mizushima, N.; Noda, T.; Yoshimori, T.; Tanaka, Y.; Ishii, T.; George, M.D.; Klionsky, D.J.; Ohsumi, M.; Ohsumi, Y. A protein conjugation system essential for autophagy. *Nature* **1998**, *395*, 395–398. [CrossRef] [PubMed]
73. Lamb, C.A.; Yoshimori, T.; Tooze, S.A. The autophagosome: Origins unknown, biogenesis complex. *Nat. Rev. Mol. Cell Biol.* **2013**, *14*, 759–774. [CrossRef] [PubMed]
74. Lamb, C.A.; Dooley, H.C.; Tooze, S.A. Endocytosis and autophagy: Shared machinery for degradation. *BioEssays News Rev. Mol. Cell. Dev. Biol.* **2013**, *35*, 34–45. [CrossRef] [PubMed]
75. Berg, T.O.; Fengsrud, M.; Strømhaug, P.E.; Berg, T.; Seglen, P.O. Isolation and characterization of rat liver amphisomes. Evidence for fusion of autophagosomes with both early and late endosomes. *J. Biol. Chem.* **1998**, *273*, 21883–21892. [CrossRef]
76. Akinduro, O.; Sully, K.; Patel, A.; Robinson, D.J.; Chikh, A.; McPhail, G.; Braun, K.M.; Philpott, M.P.; Harwood, C.A.; Byrne, C.; et al. Constitutive Autophagy and Nucleophagy during Epidermal Differentiation. *J. Investig. Dermatol.* **2016**, *136*, 1460–1470. [CrossRef]
77. Aymard, E.; Barruche, V.; Naves, T.; Bordes, S.; Closs, B.; Verdier, M.; Ratinaud, M.-H. Autophagy in human keratinocytes: An early step of the differentiation? *Exp. Dermatol.* **2011**, *20*, 263–268. [CrossRef]
78. Liu, C.; Gu, L.; Ding, J.; Meng, Q.; Li, N.; Dai, G.; Li, Q.; Wu, X. Autophagy in skin barrier and immune-related skin diseases. *J. Dermatol.* **2021**, *48*, 1827–1837. [CrossRef]
79. Takahashi, M.N.; Jackson, W.; Laird, D.T.; Culp, T.D.; Grose, C.; Haynes, J.I.; Benetti, L. Varicella-zoster virus infection induces autophagy in both cultured cells and human skin vesicles. *J. Virol.* **2009**, *83*, 5466–5476. [CrossRef]
80. Buckingham, E.M.; Carpenter, J.E.; Jackson, W.; Grose, C. Autophagy and the effects of its inhibition on varicella-zoster virus glycoprotein biosynthesis and infectivity. *J. Virol.* **2014**, *88*, 890–902. [CrossRef]
81. Carpenter, J.E.; Jackson, W.; Benetti, L.; Grose, C. Autophagosome formation during varicella-zoster virus infection following endoplasmic reticulum stress and the unfolded protein response. *J. Virol.* **2011**, *85*, 9414–9424. [CrossRef] [PubMed]
82. Buckingham, E.M.; Carpenter, J.E.; Jackson, W.; Zerboni, L.; Arvin, A.M.; Grose, C. Autophagic flux without a block differentiates varicella-zoster virus infection from herpes simplex virus infection. *Proc. Natl. Acad. Sci. USA* **2015**, *112*, 256–261. [CrossRef] [PubMed]
83. Orvedahl, A.; Alexander, D.; Tallóczy, Z.; Sun, Q.; Wei, Y.; Zhang, W.; Burns, D.; Leib, D.A.; Levine, B. HSV-1 ICP34.5 confers neurovirulence by targeting the Beclin 1 autophagy protein. *Cell Host Microbe* **2007**, *1*, 23–35. [CrossRef] [PubMed]
84. Lussignol, M.; Queval, C.; Bernet-Camard, M.F.; Cotte-Laffitte, J.; Beau, I.; Codogno, P.; Esclatine, A. The herpes simplex virus 1 Us11 protein inhibits autophagy through its interaction with the protein kinase PKR. *J. Virol.* **2013**, *87*, 859–871. [CrossRef] [PubMed]
85. Graybill, C.; Morgan, M.J.; Levin, M.J.; Lee, K.S. Varicella-zoster virus inhibits autophagosome-lysosome fusion and the degradation stage of mTOR-mediated autophagic flux. *Virology* **2018**, *522*, 220–227. [CrossRef]
86. Grose, C.; Buckingham, E.M.; Carpenter, J.E.; Kunkel, J.P. Varicella-Zoster Virus Infectious Cycle: ER Stress, Autophagic Flux, and Amphisome-Mediated Trafficking. *Pathogens* **2016**, *5*, 67. [CrossRef]
87. Gerada, C.; Campbell, T.M.; Kennedy, J.J.; McSharry, B.P.; Steain, M.; Slobedman, B.; Abendroth, A. Manipulation of the Innate Immune Response by Varicella Zoster Virus. *Front. Immunol.* **2020**, *11*, 1. [CrossRef]
88. Jones, J.O.; Arvin, A.M. Inhibition of the NF-kappaB pathway by varicella-zoster virus in vitro and in human epidermal cells in vivo. *J. Virol.* **2006**, *80*, 5113–5124. [CrossRef]
89. Sen, N.; Che, X.; Rajamani, J.; Zerboni, L.; Sung, P.; Ptacek, J.; Arvin, A.M. Signal transducer and activator of transcription 3 (STAT3) and survivin induction by varicella-zoster virus promote replication and skin pathogenesis. *Proc. Natl. Acad. Sci. USA* **2012**, *109*, 600–605. [CrossRef] [PubMed]
90. Wang, L.; Oliver, S.L.; Sommer, M.; Rajamani, J.; Reichelt, M.; Arvin, A.M. Disruption of PML nuclear bodies is mediated by ORF61 SUMO-interacting motifs and required for varicella-zoster virus pathogenesis in skin. *PLoS Pathog.* **2011**, *7*, e1002157. [CrossRef]
91. Sen, N.; Sommer, M.; Che, X.; White, K.; Ruyechan, W.T.; Arvin, A.M. Varicella-zoster virus immediate-early protein 62 blocks interferon regulatory factor 3 (IRF3) phosphorylation at key serine residues: A novel mechanism of IRF3 inhibition among herpesviruses. *J. Virol.* **2010**, *84*, 9240–9253. [CrossRef] [PubMed]

92. Ambagala, A.P.N.; Cohen, J.I. Varicella-Zoster virus IE63, a major viral latency protein, is required to inhibit the alpha interferon-induced antiviral response. *J. Virol.* **2007**, *81*, 7844–7851. [CrossRef] [PubMed]
93. Flatt, A.; Breuer, J. Varicella vaccines. *Br. Med. Bull.* **2012**, *103*, 115–127. [CrossRef] [PubMed]
94. Gershon, A.A.; Gershon, M.D.; Shapiro, E.D. Live Attenuated Varicella Vaccine: Prevention of Varicella and of Zoster. *J. Infect. Dis.* **2021**, *224*, S387–S397. [CrossRef] [PubMed]
95. Woodward, M.; Marko, A.; Galea, S.; Eagel, B.; Straus, W. Varicella Virus Vaccine Live: A 22-Year Review of Postmarketing Safety Data. *Open Forum Infect. Dis.* **2019**, *6*, ofz295. [CrossRef]
96. Gershon, A.A. Varicella Vaccine: Rare Serious Problems—but the Benefits Still Outweigh the Risks. *J. Infect. Dis.* **2003**, *188*, 945–947. [CrossRef]
97. Oxman, M.N.; Levin, M.J.; Johnson, G.R.; Schmader, K.E.; Straus, S.E.; Gelb, L.D.; Arbeit, R.D.; Simberkoff, M.S.; Gershon, A.A.; Davis, L.E.; et al. A vaccine to prevent herpes zoster and postherpetic neuralgia in older adults. *N. Engl. J. Med.* **2005**, *352*, 2271–2284. [CrossRef]
98. Lal, H.; Cunningham, A.L.; Godeaux, O.; Chlibek, R.; Diez-Domingo, J.; Hwang, S.J.; Levin, M.J.; McElhaney, J.E.; Poder, A.; Puig-Barberà, J.; et al. Efficacy of an Adjuvanted Herpes Zoster Subunit Vaccine in Older Adults. *N. Engl. J. Med.* **2015**, *372*, 2087–2096. [CrossRef]
99. Zerboni, L.; Hinchliffe, S.; Sommer, M.H.; Ito, H.; Besser, J.; Stamatis, S.; Cheng, J.; Distefano, D.; Kraiouchkine, N.; Shaw, A.; et al. Analysis of varicella zoster virus attenuation by evaluation of chimeric parent Oka/vaccine Oka recombinant viruses in skin xenografts in the SCIDhu mouse model. *Virology* **2005**, *332*, 337–346. [CrossRef]
100. Depledge, D.P.; Kundu, S.; Jensen, N.J.; Gray, E.R.; Jones, M.; Steinberg, S.; Gershon, A.; Kinchington, P.R.; Schmid, D.S.; Balloux, F.; et al. Deep Sequencing of Viral Genomes Provides Insight into the Evolution and Pathogenesis of Varicella Zoster Virus and Its Vaccine in Humans. *Mol. Biol. Evol.* **2014**, *31*, 397–409. [CrossRef]
101. Depledge, D.P.; Yamanishi, K.; Gomi, Y.; Gershon, A.A.; Breuer, J. Deep Sequencing of Distinct Preparations of the Live Attenuated Varicella-Zoster Virus Vaccine Reveals a Conserved Core of Attenuating Single-Nucleotide Polymorphisms. *J. Virol.* **2016**, *90*, 8698–8704. [CrossRef] [PubMed]
102. Gomi, Y.; Sunamachi, H.; Mori, Y.; Nagaike, K.; Takahashi, M.; Yamanishi, K. Comparison of the complete DNA sequences of the Oka varicella vaccine and its parental virus. *J. Virol.* **2002**, *76*, 11447–11459. [CrossRef] [PubMed]
103. Quinlivan, M.; Gershon, A.A.; Steinberg, S.P.; Breuer, J. An evaluation of single nucleotide polymorphisms used to differentiate vaccine and wild type strains of varicella-zoster virus. *J. Med. Virol.* **2005**, *75*, 174–180. [CrossRef]
104. Quinlivan, M.L.; Jensen, N.J.; Radford, K.W.; Schmid, D.S. Novel genetic variation identified at fixed loci in ORF62 of the Oka varicella vaccine and in a case of vaccine-associated herpes zoster. *J. Clin. Microbiol.* **2012**, *50*, 1533–1538. [CrossRef] [PubMed]
105. Peters, G.A.; Tyler, S.D.; Carpenter, J.E.; Jackson, W.; Mori, Y.; Arvin, A.M.; Grose, C. The attenuated genotype of varicella-zoster virus includes an ORF0 transitional stop codon mutation. *J. Virol.* **2012**, *86*, 10695–10703. [CrossRef] [PubMed]
106. Jeon, J.S.; Won, Y.H.; Kim, I.K.; Ahn, J.H.; Shin, O.S.; Kim, J.H.; Lee, C.H. Analysis of single nucleotide polymorphism among Varicella-Zoster Virus and identification of vaccine-specific sites. *Virology* **2016**, *496*, 277–286. [CrossRef]
107. Ko, H.; Lee, G.M.; Shin, O.S.; Song, M.J.; Lee, C.H.; Kim, Y.E.; Ahn, J.-H. Analysis of IE62 mutations found in Varicella-Zoster virus vaccine strains for transactivation activity. *J. Microbiol. Seoul Korea* **2018**, *56*, 441–448. [CrossRef]
108. Cheng, J.B.; Sedgewick, A.J.; Finnegan, A.I.; Harirchian, P.; Lee, J.; Kwon, S.; Fassett, M.S.; Golovato, J.; Gray, M.; Ghadially, R.; et al. Transcriptional Programming of Normal and Inflamed Human Epidermis at Single-Cell Resolution. *Cell Rep.* **2018**, *25*, 871–883. [CrossRef]
109. Wang, S.; Drummond, M.L.; Guerrero-Juarez, C.F.; Tarapore, E.; MacLean, A.L.; Stabell, A.R.; Wu, S.C.; Gutierrez, G.; That, B.T.; Benavente, C.A.; et al. Single cell transcriptomics of human epidermis identifies basal stem cell transition states. *Nat. Commun.* **2020**, *11*, 4239. [CrossRef]
110. Vassilev, L.T.; Vu, B.T.; Graves, B.; Carvajal, D.; Podlaski, F.; Filipovic, Z.; Kong, N.; Kammlott, U.; Lukacs, C.; Klein, C.; et al. In Vivo Activation of the p53 Pathway by Small-Molecule Antagonists of MDM2. *Science* **2004**, *303*, 844–848. [CrossRef]

viruses

Review

Antiviral Drug Discovery for the Treatment of COVID-19 Infections

Teresa I. Ng [1,*], Ivan Correia [2], Jane Seagal [3,†], David A. DeGoey [4], Michael R. Schrimpf [4], David J. Hardee [4], Elizabeth L. Noey [5] and Warren M. Kati [1]

1 Virology Drug Discovery, AbbVie Inc., North Chicago, IL 60064, USA; warren.kati@abbvie.com
2 Department of Cell and Protein Sciences, Drug Discovery Science and Technology, AbbVie Inc., Worcester, MA 01605, USA; ivan.correia@abbvie.com
3 Department of Biologics Discovery, Drug Discovery Science and Technology, AbbVie Inc., Worcester, MA 01605, USA; jseagal@alivamab.com
4 Department of Centralized Medicinal Chemistry, Drug Discovery Science and Technology, AbbVie Inc., North Chicago, IL 60064, USA; david.degoey@abbvie.com (D.A.D.); michael.schrimpf@abbvie.com (M.R.S.); david.hardee@abbvie.com (D.J.H.)
5 Department of Structural Biology, Drug Discovery Science and Technology, AbbVie Inc., North Chicago, IL 60064, USA; elizabeth.noey@abbvie.com
* Correspondence: teresa.ng@abbvie.com
† Current address: AlivaMab Discovery Services, San Diego, CA 92121, USA.

Citation: Ng, T.I.; Correia, I.; Seagal, J.; DeGoey, D.A.; Schrimpf, M.R.; Hardee, D.J.; Noey, E.L.; Kati, W.M. Antiviral Drug Discovery for the Treatment of COVID-19 Infections. *Viruses* **2022**, *14*, 961. https://doi.org/10.3390/v14050961

Academic Editors: Charles Grose, Ravi Mahalingam and Joel Rovnak

Received: 28 March 2022
Accepted: 29 April 2022
Published: 4 May 2022

Abstract: The coronavirus disease 2019 (COVID-19) pandemic is caused by the severe acute respiratory syndrome coronavirus 2 (SARS-CoV-2), a recently emerged human coronavirus. COVID-19 vaccines have proven to be successful in protecting the vaccinated from infection, reducing the severity of disease, and deterring the transmission of infection. However, COVID-19 vaccination faces many challenges, such as the decline in vaccine induced immunity over time, and the decrease in potency against some SARS-CoV-2 variants including the recently emerged Omicron variant, resulting in breakthrough infections. The challenges that COVID-19 vaccination is facing highlight the importance of the discovery of antivirals to serve as another means to tackle the pandemic. To date, neutralizing antibodies that block viral entry by targeting the viral spike protein make up the largest class of antivirals that has received US FDA emergency use authorization (EUA) for COVID-19 treatment. In addition to the spike protein, other key targets for the discovery of direct-acting antivirals include viral enzymes that are essential for SARS-CoV-2 replication, such as RNA-dependent RNA polymerase and proteases, as judged by US FDA approval for remdesivir, and EUA for Paxlovid (nirmatrelvir + ritonavir) for treating COVID-19 infections. This review presents an overview of the current status and future direction of antiviral drug discovery for treating SARS-CoV-2 infections, covering important antiviral targets such as the viral spike protein, non-structural protein (nsp) 3 papain-like protease, nsp5 main protease, and the nsp12/nsp7/nsp8 RNA-dependent RNA polymerase complex.

Keywords: antiviral; COVID-19; SARS-CoV-2; drug discovery; coronavirus; spike protein; Mpro; RdRp; PLpro

1. Introduction

Severe acute respiratory syndrome coronavirus 2 (SARS-CoV-2), a novel human coronavirus that emerged in late 2019, is the etiological agent of coronavirus disease 2019 (COVID-19) [1,2]. The COVID-19 pandemic has presented unprecedented challenges to health care, economics, and societies globally. At the time of writing (February of 2022), SARS-CoV-2 infection has exceeded 400 million cases, resulting in approximately 5.8 million deaths worldwide [3]. SARS-CoV-2 is a member of the betacoronavirus genus, the same genus as the two highly pathogenic human betacoronaviruses known as SARS-CoV and MERS-CoV, which were responsible for the deadly outbreaks in 2002 and 2012, respectively [4,5].

1.1. Medical Countermeasures for COVID-19

To combat the COVID-19 pandemic, there has been immense effort in the discovery and development of medical countermeasures including vaccines and drug treatments for COVID-19. COVID-19 vaccine development achieved scientific breakthroughs, delivering several vaccine candidates that received the US Food and Drug Administration (FDA) emergency use authorization (EUA) in less than a year after the start of the COVID-19 pandemic. COVID-19 vaccines have significantly reduced morbidity and mortality in the vaccinated population [6]. However, lessons learned from the use of COVID-19 vaccines since the first vaccine approval in December 2020 indicate that COVID-19 vaccination alone cannot effectively address the current pandemic, and additional treatment options are required to end this global challenge. COVID-19 vaccination faces many hurdles to control COVID-19 infections worldwide: limited availability that results in inequality in global access, logistic challenges to distribute vaccines requiring special storage conditions to remote areas, a requirement for 2 doses for some vaccines to establish satisfactory immune-protection [7], waning immunity in as little as 6 months after completion of the initial vaccine administration thus requiring boosting [8–10], and emergence of SARS-CoV-2 variants resistant to the immunity induced by vaccines [11–14]. Breakthrough infections in vaccinated populations with the recently emerged SARS-CoV-2 Omicron variant exposed some of these limitations of COVID-19 vaccines and highlight the need for other medical treatments such as drug therapy, especially those that are broad spectrum and can be administered orally, to complement the use of vaccines. The strategies for COVID-19 drug discovery can be divided into two categories, targeting host factors or viral proteins that are important for the life cycle and/or pathogenesis of SARS-CoV-2 infections. This review focuses on the discovery of COVID-19 drugs that directly act against viral proteins. Direct-acting antiviral therapeutics have a good track record for treating viral diseases, such as those caused by human immunodeficiency virus (HIV), hepatitis C virus (HCV), hepatitis B virus (HBV), herpesviruses, and influenza virus. In addition, some of the recently developed COVID-19 direct-acting antivirals have also demonstrated efficacy in clinical settings.

1.2. Representative Viral Targets for COVID-19 Antiviral Intervention

Several SARS-CoV-2-encoded proteins have been identified as promising molecular targets for antiviral intervention due to their essential roles in the viral life cycle [15–17]. The entry of SARS-CoV-2 is mediated by the binding of the viral spike (S) protein to the host cell receptor angiotensin-converting enzyme 2 (ACE2) [18,19]. After entry, SARS-CoV-2 viral RNA is translated by the host to produce two polyproteins from two overlapping open reading frames (ORFs), ORF1a and ORF1b. The polyproteins are then proteolytically cleaved by two virally encoded cysteine proteases, the non-structural protein (nsp) 3 papain-like protease (PLpro) and the nsp5 main protease (Mpro, also known as 3CLpro) to yield 16 individual nsps [20]. A subset of these nsps associate to form a replication–transcription complex that mediates RNA synthesis, capping and proofreading. The nsp12 RNA-dependent RNA polymerase (RdRp) is a key viral enzyme that mediates viral replication and transcription. In short, the S protein, PLpro, Mpro, and RdRp represent prime targets for SARS-CoV-2 antiviral drug discovery. Not surprisingly, SARS-CoV-2 antivirals that have received US FDA formal approval or EUA for COVID-19 treatment encompass inhibitors targeting many of these viral proteins (Table 1).

Table 1. Antiviral drugs for the treatment of COVID-19 infections in the US.

COVID-19 Drug	Viral Target	Drug Modality	Delivery	Approval Status	Discovery Approach
Sotrovimab	Spike	Biologic	IV	EUA [1]	Developed for SARS-CoV-2
Bebtelovimab	Spike	Biologic	IV	EUA [1]	Developed for SARS-CoV-2
Tixagevimab + Cilgavimab	Spike	Biologic	IM	EUA [2]	Developed for SARS-CoV-2
Bamlanivimab + Etesevimab	Spike	Biologic	IV	EUA [1,3]	Developed for SARS-CoV-2
Casirivimab + Imdevimab	Spike	Biologic	IV/SubQ	EUA [1,3]	Developed for SARS-CoV-2
Remdesivir	RdRp	Small molecule	IV	Approved	Repurposed Ebola inhibitor
Molnupiravir	RdRp	Small molecule	Oral	EUA	Repurposed VEEV inhibitor
Paxlovid (Nirmatrelvir + Ritonavir)	Mpro	Small molecule	Oral	EUA	Nirmatrelvir designed for SARS-CoV-2; ritonavir used as a PK enhancer

[1] For post-exposure treatment of COVID-19. [2] For pre-exposure prophylaxis of COVID-19 in special populations. [3] Use limited by the FDA in January 2022 to treat COVID-19 due to the Omicron variant. IV: intravenous; EUA: emergency use authorization; IM: intramuscular; SubQ: subcutaneous; RdRp: RNA-dependent RNA polymerase; VEEV: Venezuelan equine encephalitis virus; Mpro: main protease; PK: pharmacokinetic.

1.3. Strategies of Antiviral Drug Discovery for COVID-19

Different strategies have been used in antiviral drug discovery for COVID-19, such as drug repurposing of approved or investigational drugs beyond their original indications, high-throughput screening, computer-aided virtual screening, and structure-based drug discovery. At the onset of the COVID-19 pandemic, there was great interest in drug repurposing as an expedited means to identify COVID-19 drugs [21,22]. There was indeed some success in drug repurposing for COVID-19 treatment. Remdesivir (RDV), which was originally developed for the treatment of Ebola virus infection, was found to be active against SARS-CoV-2 [23,24] and has successfully been developed into a COVID-19 drug [25]. Molnupiravir was originally discovered for Venezuelan equine encephalitis virus (VEEV) infection, but was later found to have antiviral activity against a number of respiratory viruses, including influenza and, most recently, SARS-CoV-2 [26,27]. When the COVID-19 pandemic started, the development of molnupiravir was quickly switched from influenza to COVID-19, and it received US FDA EUA to treat COVID-19 infections in December 2021 [26,27]. In addition to drug repurposing, novel molecules including a number of anti-S neutralizing monoclonal antibodies (mAbs) and an Mpro inhibitor have also been discovered and developed as COVID-19 drugs (Table 1). Antiviral drugs used for the treatment of COVID-19 infections provide protection from infection or improvement in recovery, but all COVID-19 antivirals available to date have some limitations that may make them not suitable for use in the general population. For example, molnupiravir is not recommended for use in pregnancy because it may cause fetal harm [28]. Another example is Paxlovid (ritonavir-boosted nirmatrelvir), which has the potential for complex drug–drug interactions with concomitant medications because (1) ritonavir is a CYP3A inhibitor, and may therefore increase plasma concentrations of drugs that are predominantly metabolized by CYP3A, and (2) ritonavir and nirmatrelvir are substrates of CYP3A; drugs that induce CYP3A may thus decrease ritonavir and nirmatrelvir drug levels in plasma, reducing Paxlovid therapeutic potency [29,30]. Drug resistance of emerging SARS-CoV-2 variants is also a concern as exemplified by the action of the FDA in January 2022 to limit the use of the anti-S mAb cocktail of bamlanivimab and etesevimab, as well as the mAb cocktail of casirivimab and imdevimab in COVID-19 patients, as both of these mAb cocktails are not active against the Omicron variant, which was circulating in the US at a very high frequency at that time [31]. Furthermore, patients receiving COVID-19 drugs that require IV administration (e.g., most anti-S mAbs and RDV) need access to medical facilities capable of delivering these drugs. Due to these limitations, there is an urgent need to discover novel and improved antiviral drugs with convenience of administration, improved safety and drug property profiles, and broad-spectrum coronavirus coverage to combat the current COVID-19 crisis as well as emerging coronaviruses with spillover and pandemic risk. This review provides an up-to-date overview on the discovery of antiviral drugs for COVID-19, covering the functions of important antiviral targets such as the viral S protein, Mpro, RdRp, and PLpro, and the different inhibitors against these targets.

2. Spike Protein (S Protein)

Coronaviruses are enveloped viruses and their successful entry into the host target cells requires completion of two steps [32–34]. The first is binding to a host cell receptor and the second is fusion of the viral envelope with the host cell membrane, which releases viral genome into the cytoplasm, enabling viral replication. Both steps are mediated by the S protein, a heavily glycosylated class I fusion protein that is present on the envelope of the virus as a trimer [35]. The importance of the viral S protein to host cell entry makes it an ideal target for antibody (Ab)-based therapeutics [36–38].

2.1. Structural Organization of Spike

Each monomer in the S trimer consists of the receptor-binding S1 and a membrane-anchored S2 subunit that contains the membrane fusion machinery (Figure 1a) [19,32,34,39,40]. The receptor-binding S1 subunit has an N-terminal domain (NTD) and a receptor-binding domain (RBD), also called the C-terminal binding domain (CTD). The RBD is composed of a core region and a receptor-binding motif (RBM) that the virus uses to interact with the host cell receptor. The RBM of the S protein is prone to mutations, while the core region is more conserved [41]. The membrane-anchored S2 subunit has the fusion peptide (FP), a domain rich in hydrophobic residues, which is inserted into the host cell membrane. The S2 domain also contains two heptad repeats, HR1 and HR2, that form a six-helix bundle to complete the fusion process and delivery of the viral genome into the cytoplasm [40]. Like many other viral fusion proteins, the SARS-CoV-2 S protein is covered by a glycan shield protecting the S protein from host immune recognition. Both SARS-CoV and SARS-CoV-2 S proteins present a different glycosylation pattern from that of the HIV-1 envelope protein, showing a larger presence of complex N-glycans relative to oligomannose type [35]. The SARS-CoV-2 S protein has 22 predicted N-glycosylation sites per protomer plus at least two predicted O-glycosylation sites. The RBD is less protected by glycans and is therefore more immunogenic. Natural Ab responses are mostly directed toward the RBD and, as described further below, many mutations arise in this domain to escape Ab neutralization [42].

2.2. Interaction of Spike with Receptor and Mechanism of Viral Entry

SARS-CoV-2 and SARS-CoV share the same host cell receptor, ACE2, whereas for MERS-CoV, the receptor is dipeptidyl peptidase 4 (DPP4) [43–45]. The RBM has a high tolerance for mutations and both SARS-CoV and SARS-CoV-2 bind to the receptor in a similar manner despite low sequence similarity. In addition to depending on ACE2 for host cell entry, both SARS-CoV and SARS-CoV-2 depend on entry activation by host cell proteases (cathepsin L, TMPRSS2) acting at the S1/S2 boundary and an S2 site upstream of the fusion peptide called S2′ of the S protein (Figure 1a) [46]. The SARS-CoV-2 S protein also has a furin cleavable site that potentiates SARS-CoV-2 infectivity (Figure 1a) [47]. Of interest, although the interaction of the S protein with relevant host receptor plays a critical role in tissue tropism, other "background genes", including nucleocapsid and replicase as well as accessory genes, may also impact tissue tropism [48].

Cryo-EM and single-particle reconstruction have provided structural information on the trimeric S protein [34,49]. In the closed pre-fusion conformation, all three RBDs lie flat ("down" state) and make the RBMs inaccessible for biological interactions, while in the open pre-fusion conformation, one or more RBDs are lifted ("up" state) to expose the corresponding RBMs, enabling S protein/ACE2 interactions (Figure 1b) [49].

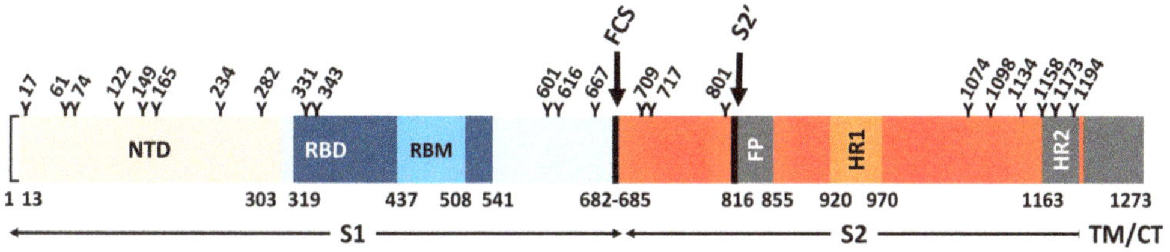

(a)

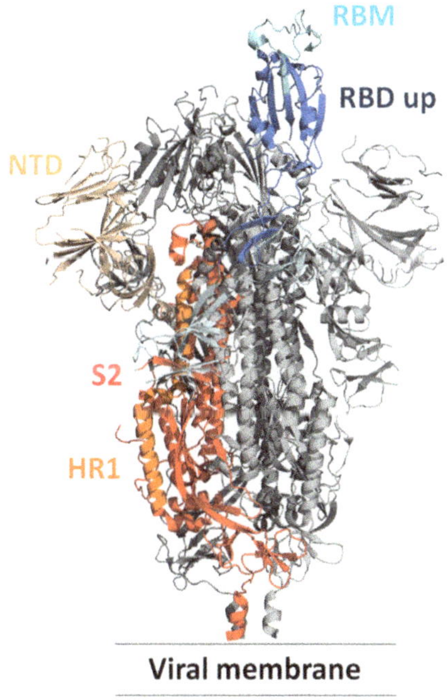

(b)

Figure 1. (**a**) Domain architecture of the SARS-CoV-2 spike protein, comprising the N-terminal domain (NTD), the receptor-binding domain (RBD), the receptor-binding motif (RBM), the furin cleavage site (FCS), the S2′ cleavage site, the fusion peptide (FP), and heptad repeats 1 and 2 (HR1 and HR2), as they relate to the S1 and S2 subunits, as well as the transmembrane domain (TM) and the cytoplasmic tail (CT). Glycosylation sites are marked at the top of the figure. (**b**) Side view of the pre-fusion structure of the SARS-CoV-2 spike protein (PDB ID 6VYB [39]) with a single RBD in the "up" state and exposing the RBM. The two RBD "down" protomers are shown in gray and the RBD "up" protomer is shown in color corresponding to the schematic in (**a**).

As depicted in Figure 2, S protein binding to ACE2 destabilizes the pre-fusion trimer, resulting in shedding of the S1 subunit, and significant conformational change in the membrane-bound S2 subunit. The S2 subunit acquires an elongated shape, enabling the insertion of the fusion peptide into the host cell membrane (intermediate conformation). The intermediate conformation is unstable and rapidly transitions to a stable post-fusion conformation in which HR1 and HR2 form a six-helix bundle. During this transition, viral and host-cell membranes are brought to a close proximity, resulting in the membrane fusion [34,49,50].

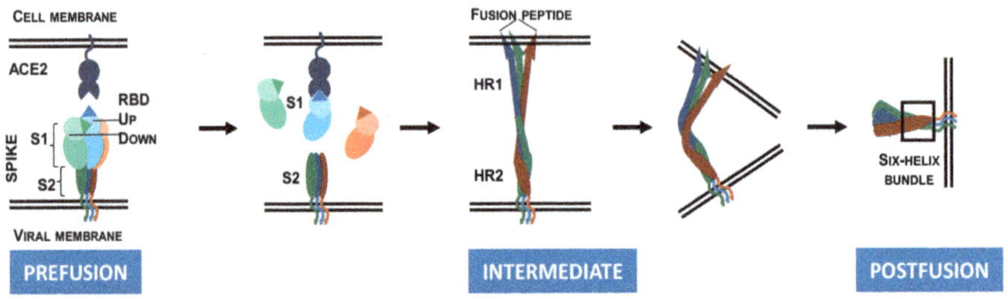

Figure 2. Conformational changes in the SARS-CoV-2 spike ectodomain during membrane fusion. Pre-fusion conformation: Spike protein with two RBDs in the "down" state and one RBD in the "up" state with the RBM exposed and available for binding to the ACE2 receptor. The spike protein/ACE2 interactions induce shedding of the S1 subunits. Intermediate conformation: The S2 subunits become elongated and reach out to the host cell membrane, enabling insertion of the fusion peptide. Post-fusion conformation: HR2 forms a six-helix bundle with HR1 inducing fusion of the viral membrane with the host cell membrane.

2.3. Anti-SARS-CoV-2 Antibodies Recognizing the Spike RBD

Monoclonal antibodies that target SARS-CoV-2 have been isolated from convalescent COVID-19 patients, SARS-CoV patients, as well as immunized wild-type and transgenic mice. Three-dimensional structures are available for mAbs either in complex with the ECD or the RBD of the S protein, and most mAbs target immunodominant epitopes in the RBM though several of them target the core RBD, the NTD and the S2 subunit [51–53].

Antibodies targeting the RBD of the S protein can be assigned to four functional classes based on binding epitope determined by cryo-EM or high-resolution X-ray crystallography [54]. Class 1 and 2 Abs directly block ACE2, whereas class 3 and 4 do not; class 1 and 4 only bind to the "up" RBDs, whereas class 2 and 3 bind to RBDs regardless of their "up" and "down" states [33,55].

- Class 1 mAbs are ACE2 blocking and only bind to the "up" RBDs and prevent viral entry into the cell. Many are VH3-53 or VH3-66 Abs. Examples include etesevimab (LY-CoV016), casirivimab (REGN10933) and tixagevimab (AZD8895).
- Class 2 mAbs are ACE2 blocking and bind both the "up" and "down" RBDs and contact adjacent RBDs. Shedding of S1 is reported when the RBD is captured in the "up" state and premature conversion to the post-fusion state which prevents fusion of the viral membrane with the host cell membrane. Examples include bamlanivimab (LY-CoV555) and cilgavimab (AZD1061).
- Class 3 mAbs do not block ACE2 and bind both the "up" and "down" RBDs. Contact with adjacent RBDs limits movement and can lock the RBD in a closed conformation. Examples of Class 3 Abs include sotrovimab (VIR-7831), bebtelovimab (LY-CoV1404) and imdevimab (REGN10987).
- Class 4 mAbs do not block ACE2 and bind only the "up" RBDs. Shedding of S1 is reported when the RBD is captured in the "up" state. Examples include C1C-A3, CR3022 and S304.

There are multiple mechanisms for how Abs neutralize and clear viruses. Antibodies can bind the RBD and block (directly or indirectly) binding to receptor thus preventing viral entry into the host cells [36]. Some Class 2 mAbs induce premature shedding of the S1 domain [56], thus inducing the post-fusion state and preventing fusion of the host-viral membranes. Bivalent crosslinking of the S proteins can result in steric hindrance or aggregate virions and neutralize viral entry. Finally, the Fc portion of Abs can interact with Fc gamma receptors found on myeloid and natural killer cells. These interactions are important for viral clearance brought about by engaging different receptors and inducing either Ab-dependent cell-mediated phagocytosis (ADCP), Ab-dependent cellular cytotoxicity

(ADCC) or activation of the complement pathway [52]. For effective antiviral protection, a cocktail of mAbs with more than one mechanism of action might be required.

2.4. Escape Mutations

RNA viruses replicating via an RdRp have high rates of mutation in nature, which presents challenges for designing effective vaccines or Ab therapeutics [57,58].

Multiple structures of the S protein in complex with ACE2 are available; both SARS-CoV and SARS-CoV-2 bind to the receptor in a similar manner despite low sequence similarity in their RBM, indicating a high tolerance for mutations [18,19,43,59–62]. SARS-CoV-2 variants of concern (VOCs) have developed resistance to neutralizing Abs, including some clinical Abs [63]. The B.1.351 (Beta) VOC was found to have the largest magnitude of immune evasion upon acquiring the E484K and K417N mutations (Table 2) [64], whereas B.1.617.2 (Delta) quickly outcompeted all other circulating variants through acquisition of mutations (L452R and T478K) that enhanced transmission and pathogenicity, as well as eroded neutralizing Ab responses. The recently emerged SARS-CoV-2 Omicron variant harbors 37 amino acid substitutions in the S protein, 15 of which are in the RBD and 10 in the RBM [42,65]. Despite a significant number of mutations around the ACE2-binding site, Omicron binds to ACE2 with enhanced affinity relative to the Wuhan-Hu-1 strain and is the dominant strain circulating around the globe at the time of writing (February of 2022).

Out of the eight mAbs currently authorized by the US FDA (Table 1), six (bamlanivimab, etesevimab, casirivimab, imdevimab, cilgavimab, and tixagevimab) directly block binding of the S protein to ACE2 [42]. These mAbs are frequently used in combination to bring about maximum coverage. The second class of mAbs, represented by sotrovimab, do not block ACE2 binding, but neutralize SARS-CoV-2 by targeting non-RBM epitopes shared across many sarbecoviruses, including SARS-CoV. Comparison of the in vitro neutralizing activity of therapeutic mAbs from these two groups against Wuhan Hu 1 S and Omicron S proteins using VSV pseudoviruses revealed that the RBM-specific mAbs lost their neutralizing activity except for the cilgavimab and tixagevimab cocktail, where there was a ~200-fold reduction in potency [14,42]. Sotrovimab was found to have 3-fold reduced potency against Omicron likely brought about by the G339D mutation (not shown in Table 2). Bebtelovimab was granted EUA by the US FDA in February 2022. Bebtelovimab binds and potently neutralizes all currently known VOCs of SARS-CoV-2 including the Omicron variant [66]. As shown in Table 2, the binding epitope of bebtelovimab is very similar to imdevimab, and the structural location of the epitope is closer to the canonical Class 3 binder, VIR-S309 [66].

2.5. Conclusions and Future Directions

The RBD is highly immunogenic but is prone to accumulate mutations. Under strong immune selection, escape mutants rapidly arise as we observe with the Omicron variant [14,42]. Further, as we prepare for emerging coronaviruses with "pandemic potential", there is limited opportunity for having Abs with cross-reactivity to SARS-CoV-2 and other coronaviruses due to the high diversity in RBD sequences. The membrane-anchored S2 subunit, which contains the membrane fusion machinery exhibits a higher level of protein sequence conservation across coronavirus S proteins. Several teams have identified a class of S2-targeting Abs with broad reactivity towards several human betacoronaviruses from distinct subgenera, and characterized their antiviral activity, epitope and in vivo protective efficacy [52,56,68–70]. S2-specific mAbs can prove to be very useful, and their binding could inhibit conformational changes necessary for membrane fusion to occur. S2-specific mAbs, however, may not be sufficiently potent in viral neutralization, and enhanced Fc effector function might be necessary to achieve in vivo efficacy [52]. These mAbs can be used in combination with clinically proven neutralizing anti-SARS-CoV-2 therapies to achieve a broad neutralization spectrum across all SARS-CoV-2 variants.

Table 2. Spike RBD sequence of SARS-CoV-2 Wuhan-Hu-1 with highlighted footprints of ACE2 (pale blue), variants of concern Alpha, Beta, Gamma, Delta and Omicron with positions of mutations (peach), and 8 mAbs with binding epitopes on the RBM (pale green) [42,66,67]. (AA#: amino acid numbering of the spike protein.).

AA#	ACE2	Alpha	Beta	Gamma	Delta (+)	Omicron	Etesevimab	Bamlanivimab	Cilgavimab	Tixagevimab	Casirivimab	Imdevimab	Regdanvimab	Bebtelovimab
417	K		N	T	N	N								
438	S													
439	N													
440	N					K								
441	L													
442	D													
443	S													
444	K													
445	V													
446	G					S								
447	G													
448	N													
449	Y													
450	N													
451	Y													
452	L				R									
453	Y													
454	R													
455	L													
456	F													
457	R													
458	K													
459	S													
460	N													
461-469														
470	T													
471	E													
472	I													
473	Y													
474	Q													
475	A													
476	G													
477	S					N								
478	T				K	K								
479	P													
480	C													
481	N													

Table 2. *Cont.*

482	G													
483	V													
484	E		K	K		A								
485	G													
486	F													
487	N													
488	C													
489	Y													
490	F													
491	P													
492	L													
493	Q					K								
494	S													
495	Y													
496	G					S								
497	F													
498	Q					R								
499	P													
500	T													
501	N	Y	Y	Y		Y								
502	G													
503	V													
504	G													
505	Y					H								

3. Main Protease (Mpro)

3.1. Structural Organization and Function of Mpro

The SARS-CoV-2 Mpro is a 33.8-kDa protein which is responsible for proteolytic cleavage of viral polyproteins and is essential for viral replication. X-ray analysis has revealed that two Mpro proteins associate to form a dimer which is required for catalytic activity, although it has been shown that Mpro exists as a mixture of monomer and dimer in solution [71,72]. Mpro is composed of three domains, with domains I and II forming the catalytic active site including the catalytic dyad, Cys145 and His41, and substrate-binding subsites, S1, S2, S4 and S1′ (Figure 3). Mpro recognizes and cleaves the polyproteins at 11 sites with highly conserved sequences characterized by the P1 (substrate residue N-terminal to the cleavage site) glutamine residue (Figure 4) which forms a hydrogen bond between the amide side chain carbonyl and the conserved S1 residue His163. Mpro is an attractive target for the development of small molecule antiviral therapeutics for the current pandemic as well as broad-spectrum coronavirus antivirals [71,72]. This great interest stems from the recognition that Mpro has a highly conserved active site across coronaviruses and there is no human cysteine protease with similar substrate specificity, suggesting Mpro inhibitors could be developed into broad-spectrum anti-coronavirus antivirals with high selectivity.

3.2. Discovery of Mpro Inhibitors

During the early stages of the COVID-19 pandemic in 2020, drug discovery organizations worldwide scoured their collections for existing drugs or candidates that could be repurposed to battle the novel coronavirus [21,22,75]. Many of the leading candidates for COVID-19 treatment originated in programs targeting the human coronaviruses SARS-CoV and MERS-CoV, in addition to unrelated rhinoviruses, enteroviruses, HIV and HCV.

SARS-CoV-2 Mpro inhibitors can be classified as either *covalent*, which bear an electrophilic "warhead" that traps the catalytic cysteine to form a complex where the inhibitor is linked to the enzyme, or *non-covalent*, which do not form such an adduct (Figure 5). Many of the former are peptidomimetics that maintain a P1 glutamine-mimetic fragment to provide specificity and improve affinity, as well as a lipophilic P2 substituent. Commonly used warheads include aldehydes (or masked aldehydes), activated ketones, Michael acceptors and nitriles. Formation of the inhibitor–enzyme adduct is typically reversible and is highly stabilized by extensive hydrogen-bonding and electronic effects that mimic the transition state for protease cleavage of the enzyme substrate. In contrast, non-covalent inhibitor designs rely on hydrogen bonding, van der Waals forces, and hydrophobic interactions to create reversible binding that blocks the Mpro active site and may or may not involve the catalytic residues. It has been hypothesized that these types of drugs will have lower potential for off-target toxicities due to their lack of reactive functionalities but may suffer from weaker inhibitory activity or shorter duration of action. Non-covalent scaffolds have also served as starting points for the design of covalent inhibitors by addition of reactive groups at appropriate positions.

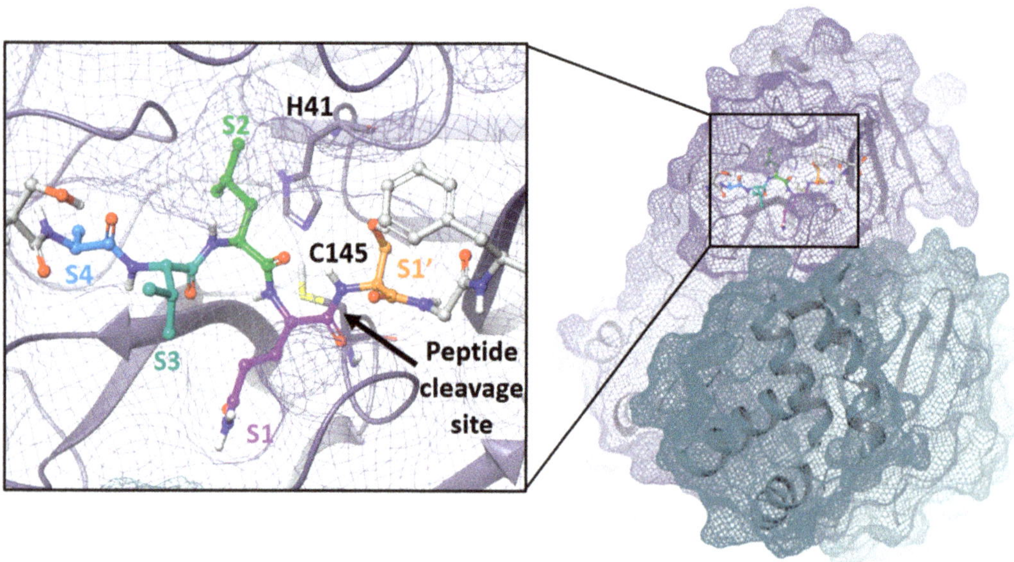

Figure 3. Mpro dimer and active site with peptide substrate bound. Image produced from PDB ID 7N89 with C145A mutant modeled back to cysteine [73].

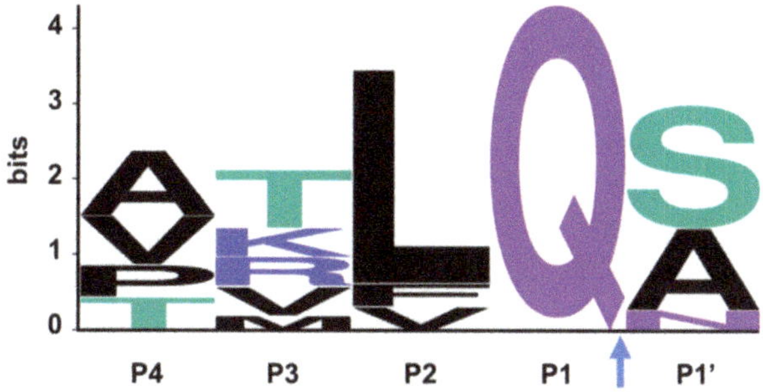

Figure 4. The consensus recognition sequence cleaved by SARS-CoV-2 (Uniprot code P0DTD1) Mpro. The cleavage site is marked by the blue arrow. Image generated by WebLogo [74].

The Ugi class of non-covalent inhibitors originated in a high-throughput screen of the NIH sample library in search of novel SARS-CoV Mpro inhibitors in the early 2000s. The simple dipeptide structure is easily assembled using the Ugi multicomponent coupling methodology, which allowed for rapid SAR exploration. The lead compound ML 188 (**1**) was later shown to have similar and perhaps improved activity against SARS-CoV-2 Mpro (IC_{50} = 2.5 µM). The benzotriazole class was also discovered by screens targeting SARS-CoV [76,77]. In response to the COVID-19 pandemic, work on this series was refocused on SARS-CoV-2, leading to improved analogs such as CCF981 (**2**) that exhibited several orders of magnitude improvement in Mpro potency (IC_{50} = 68 nM) with submicromolar antiviral activity in infected Vero E6 cells (CPE EC_{50} = 497 nM). Perampanel, an anti-epileptic agent, was found to be a weak Mpro inhibitor through a repurposing effort that evaluated approximately 2000 known drugs [78]. Structure-guided optimization afforded a perampanel analog (**3**), which has a SARS-CoV-2 Mpro inhibitory potency of IC_{50} = 0.17 µM and antiviral EC_{50} = ~1 µM in a plaque assay in Vero E6 cells with low cytotoxicity. The isoquinoline-containing Mpro inhibitor (**4**) is a lead compound discovered through a unique crowd-sourced initiative called the COVID-19 Moonshot program. Scientists worldwide were able to contribute designs through a web portal, and donations enabled the synthesis and testing of compounds [79]. Data and x-ray structures provided by partners including the Diamond Light Source and the Weizmann Institute were made publicly available in real time. Lead compound **4** is a potent inhibitor of SARS-CoV-2 Mpro (IC_{50} = 37 nM) and demonstrates antiviral activity against several circulating variants in HeLa-ACE2 cells [80]. The consortium is hoping to advance this compound or an analog to clinical development. Shionogi has reported on S-217622 (**5**), an orally delivered Mpro inhibitor with activity against SARS-CoV-2 variants (EC_{50} = 23.9–61.7 nM, 293T-ACE2-TMPRSS2 cells), that has completed Phase 2a studies [81,82].

GC376 (**6**) is the prodrug form of a peptidomimetic drug GC373 (**7**), which was originally developed for feline coronavirus (feline infectious peritonitis virus) infections [83]. Anivive, a biotech company focused on animal healthcare, is reportedly working with partners to expand use of the compound to treat COVID-19 infections in humans [84]. The chemical structure of GC376 features a gamma-lactam as a P1 glutamine bioisostere and an aldehyde-bisulfite adduct as a latent covalent warhead. X-ray structures establish that the isobutyl sidechain occupies the lipophilic S2 site and that key hydrogen-bonding interactions are made between the inhibitor and backbone amides of Glu166, His164 and Gln189. The active form **7** exhibits moderate potency against SARS-CoV-2 Mpro (IC_{50} = 0.40 µM), similar to its level of activity against FIPV Mpro (IC_{50} = 0.49 µM) but weaker than observed with SARS-CoV (IC_{50} = 0.070 µM). Plaque assays in infected Vero E6 cells demonstrate that both compounds **6** and **7** have antiviral activity (EC_{50} = 0.92 and 1.5 µM, respectively). In a virus yield reduction assay, both compounds decreased SARS-CoV-2 titers by approximately three orders of magnitude [83].

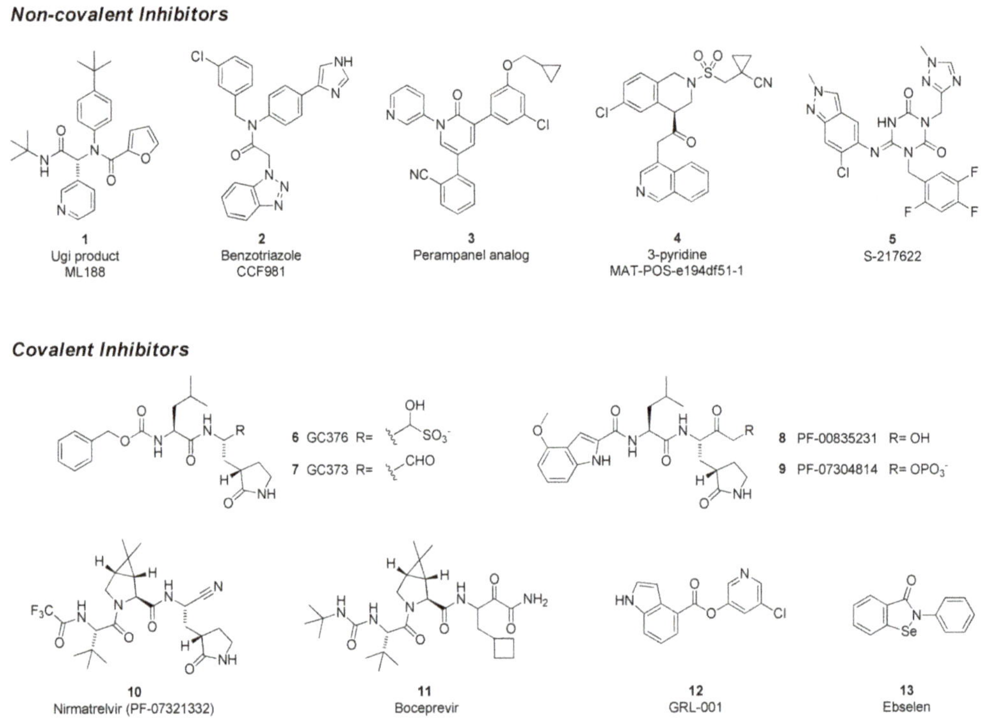

Figure 5. Chemical structures of representative SARS-CoV-2 Mpro inhibitors.

GC376 is the prototype of a rapidly expanding class of covalent peptidomimetic Mpro inhibitors. Many publications describe related compounds, with the warhead and P3–P4 regions most frequently targeted for modification. At the beginning of the COVID-19 pandemic, Pfizer advanced PF-00835231 (**8**) and its phosphate prodrug form PF-07304814 (**9**) as intravenous (IV) drugs for early-stage infections [85,86]. The parent is potent in biochemical assays, with subnanomolar affinity for SARS-CoV-2 Mpro (K_i = 0.27 nM) and good activity across a broad panel of alpha- and beta-coronaviruses. Since the compound is a substrate for efflux transporters, the in vitro antiviral activity is highly dependent on the cell line and the concentration of added efflux pump inhibitor. PF-00835231 reduced viral titers in a dose-dependent fashion in SARS-CoV- and SARS-CoV-2-infected mouse models at unbound plasma C_{min} concentrations of approximately 0.7 fold of the in vitro EC_{90}. Despite the poor oral pharmacokinetic profile of PF-00835231, due to the urgency of the pandemic as well as acceptable efficacy and safety of the compound, the clinical trials of PF-00835231 (as an IV administered drug) were initiated in late 2020.

An IV drug that must be administered in a clinical setting is not optimal for treating COVID-19 in the early stages of an infection when an antiviral is most likely to have maximum benefit. Therefore, an oral drug is highly desirable, and Pfizer followed PF-00835231 with a second compound that meets this criterion. Nirmatrelvir (PF-07321332, **10**) is the active component of Paxlovid, a two-drug combination that was recently granted EUA in the United States for the treatment of COVID-19 infections [87]. The second component is ritonavir, a CYP3A inhibitor that improves exposure of the antiviral compound. Structurally, nirmatrelvir contains the same glutamine-mimetic fragment as PF-00835231 but replaces the aldehyde warhead with a less reactive yet still reversible nitrile group. The remainder of the structure resembles boceprevir (**11**), an HCV protease inhibitor that has also been shown to inhibit SARS-CoV-2 Mpro. The bicyclic proline variation and trifluoroacetamide groups were necessary to improve permeability and oral absorption. Nirmatrelvir has slightly lower affinity for SARS-CoV-2 Mpro (K_i = 3.11 nM) than the earlier IV drug candidate **8** but maintains similar antiviral activity, with an EC_{50} = 78 nM in A549-ACE2 cells. Clinical data revealed high efficacy (up to 89% reduction in risk of hospitalization or death) in at-risk patients who began treatment within 5 days of infection [88]. Whereas GC376, nirmatrelvir and related compounds are reversible covalent inhibitors, GRL-001 (**12**) and ebselen (**13**) are representatives of a category of non-peptidic small molecules that react with the catalytic cysteine in an irreversible and time-dependent manner to produce an inactivated enzyme.

Several additional SARS-CoV-2 Mpro inhibitors have been developed for which clinical trials have been initiated or preclinical data have been presented, but their chemical structures have not been disclosed yet. Enanta has released preclinical data for EDP-235, an oral Mpro inhibitor with a biochemical IC_{50} of 5.8 nM and antiviral activity in primary human airway epithelial cells with an EC_{90} of 33 nM [89]. Cocrystal Pharma has announced plans to advance two Mpro inhibitors, CDI-988 and CDI-873, to clinical trials in 2022 [90]. Aligos has reported preclinical data for ALG-097111, which demonstrated a biochemical IC_{50} of 7 nM and antiviral EC_{50} of 200 nM (A549-ACE2 cells) [91].

3.3. Conclusions and Future Directions

The discovery and advancement of the first-generation SARS-CoV-2 Mpro inhibitors to clinical trials and the approval of Paxlovid for the treatment of COVID-19 infections are unprecedented in their speed, with Paxlovid demonstrating significant clinical benefit in reducing severe illness, hospitalization, and death. In the future, Mpro inhibitors that do not require pharmacokinetic boosting may enable broader utilization of these inhibitors with less risk for drug–drug interactions and use for pre- and post-exposure prophylaxis for high-risk patient populations. Previous experience with direct-acting antivirals against other viruses such as HIV and HCV suggests that combinations of drugs with different mechanisms of action may provide higher efficacy and prevent the emergence of resistance, an area of research that will become clearer as the first Mpro inhibitor sees more widespread use.

4. RNA-Dependent RNA Polymerase (RdRp)

4.1. Structural Organization and Function of RdRp

Faithful replication of the coronavirus genome is a complicated process that includes RNA synthesis, proofreading, template switching and 5′-capping, resulting in the formation of genome-length (+)-strand RNA for incorporation into newly formed virions as well as a variety of subgenomic (+)-strand RNAs that are translated into structural and accessory proteins [92]. The core SARS-CoV-2 RdRp consists of the 934 amino acid nsp12 protein, the 85 amino acid nsp7 protein and the 200 amino acid nsp8 protein. RNA synthesis occurs within the nsp12 protein but only when the nsp7 and nsp8 accessory proteins are complexed with it in a 1:1:2 stoichiometry [93,94]. Cryo-electron microscopy studies have shown that the polymerase domain of nsp12 resembles a human "right hand" consisting of fingers, thumb and palm subdomains. The two copies of nsp8 bind to nsp12 on opposing sides of the RNA-binding cleft. The C-terminal domain of one copy of nsp8 also binds to nsp7, whereas the C-terminal domain of the second nsp8 copy adopts a different fold and interacts with nsp12 directly. When long duplex RNA is present, the N-terminal domains of both nsp8 copies form long helical extensions that contact the RNA duplex product. These extensions likely act as sliding poles to enhance the processivity of the RdRp core complex and enable it to efficiently replicate the long genome of the coronaviruses (Figure 6) [94,95]. Additional proteins are thought to associate with the core RdRp complex to perform the helicase and proofreading functions.

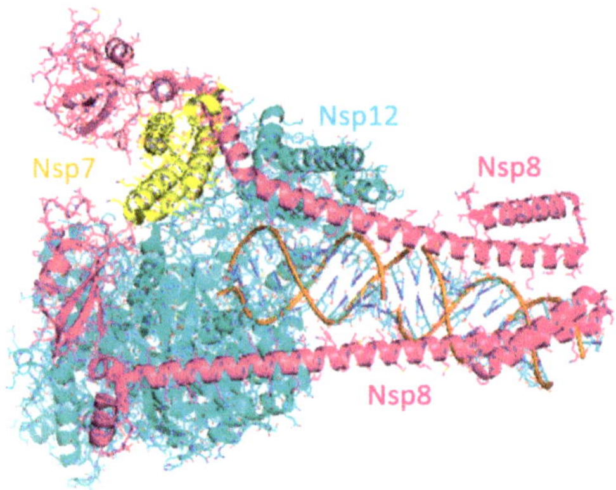

Figure 6. Structure of SARS-CoV-2 RdRp complex. Image produced from PDB ID 6YYT. (Nsp7: yellow; nsp8: magenta; nsp12: cyan.)

4.2. Discovery of RdRp Inhibitors

The SARS-CoV-2 RdRp was deemed to be an attractive drug target given the pharmaceutical industry's success in discovering drugs, most notably nucleoside and nucleotide analogs and their prodrugs, that inhibit the RdRps for other viruses such as HIV, HBV and HCV [96,97]. Nucleos(t)ide analogs often possess inhibitory activity against the RdRps from several viral families. Indeed, drug repurposing studies identified a group of nucleos(t)ide analogs that inhibit the SARS-CoV-2 RdRp and have subsequently been investigated for their potential to elicit therapeutic benefits in humans infected with the SARS-CoV-2 virus (Figure 7). These compounds are prodrugs that require transformation to the nucleotide monophosphate level before being metabolized by host cell enzymes to the corresponding nucleotide triphosphate which is the form that is recognized by SARS-CoV-2 RdRp. Accordingly, antiviral potencies for nucleos(t)ide analogs are dependent, at least in part, by the ability of the host cells to synthesize the triphosphate form. For example, RDV (**14**) inhibits replication of the SARS-CoV-2 virus with EC_{50} values of 1 μM in Vero E6 cells [23], 0.28 μM in Calu-3 cells, 0.115 μM in H549-ACE2 cells and 0.0099 μM in primary human airway epithelial cells [24].

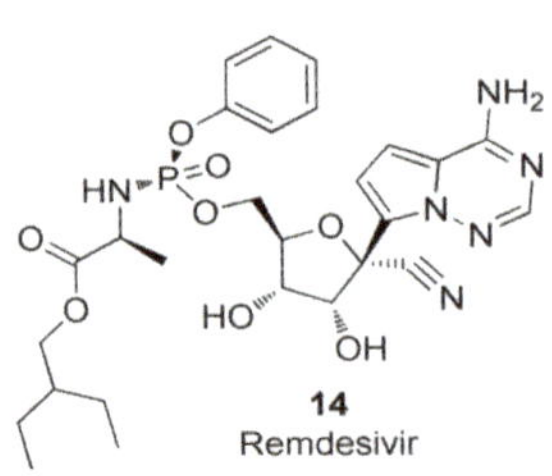

14
Remdesivir

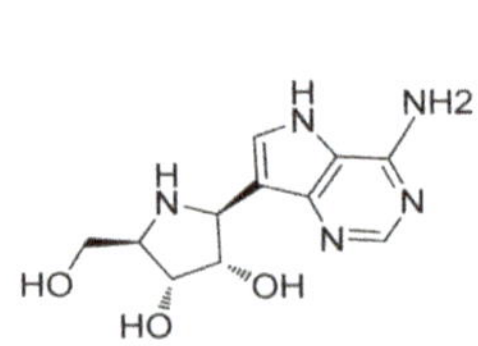

15
Galidesivir

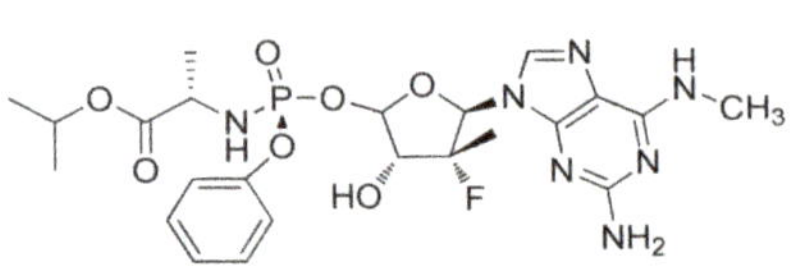

16
AT-527

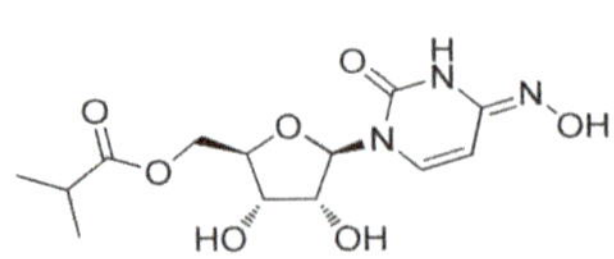

17
Molnupiravir

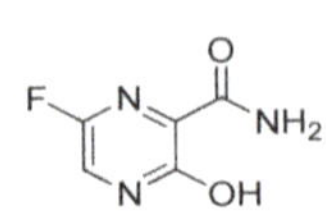

18
Favipiravir

Figure 7. Chemical structures of representative SARS-CoV-2 RdRp inhibitors.

Interestingly, although these compounds belong to the nucleos(t)ide analog class, several distinct inhibitory mechanisms of action are represented. Incorporation of the RDV nucleotide into RNA by SARS-CoV-2 RdRp stops RNA synthesis after three additional nucleotides are incorporated. Further translocation and nucleotide incorporation are blocked at this point due to a steric clash between the conserved S861 side chain and the 1′-cyano group of the incorporated RDV nucleotide [98,99]. Should this translocation blockade be overcome by high NTP levels leading to full-length (−)-strand RNA containing the RDV base in several locations, a second inhibitory mechanism may come into play that includes reduced efficiency of incorporation for UTP across from template RDV, as well as for incorporation of the next correct nucleotide in the primer strand [100]. Although mechanistic studies of SARS-CoV-2 RdRp with galidesivir (**15**) have not been reported, it has been shown that nucleotide incorporation into RNA by the RdRp of HCV results in RNA chain termination after two additional NTPs are incorporated [101]. Similarly, mechanistic studies are lacking for SARS-CoV-2 RdRp inhibition by AT-527 (**16**), whose base is converted into guanine in vivo to form the active triphosphate metabolite AT-9010 [102]. However, it should be noted that the uridine analog of AT-9010, sofosbuvir, induces immediate chain termination after incorporation into RNA by SARS-CoV-2 RdRp, likely mediated by strong steric hindrance introduced by its bulky 2′-methyl group [103,104]. The mechanism of action for molnupiravir (**17**) is complex and distinct from chain-terminating nucleotides. The 4-oxime moiety of molnupiravir's chemical structure can exist as an equilibrium of tautomeric forms—the keto form resembling uridine and the enol form resembling cytosine (Figure 8). Steady-state kinetic experiments demonstrated that molnupiravir triphosphate preferentially incorporates opposite template guanosine when incorporated into RNA by SARS-CoV-2 RdRp, thereby acting predominantly as a cytosine analog. Notably, neither this step nor incorporation of subsequent nucleotides seems to be appreciably inhibited [105,106]. In contrast, when molnupiravir's base is present in the template, it can resemble cytosine or uridine, leading to incorporation of GTP or ATP into the opposite strand, the latter resulting in a G-to-A transition mutation if one considers the sequence G→M (molnupiravir) then M→A. C-to-U transition mutations can occur if one considers several rounds of RNA synthesis and the sequence C→G; G→M; M→A; A→U [105]. Incorporation of a guanosine nucleotide opposite the template molnupiravir base does inhibit the RdRp, but the inhibition can be overcome by high NTP concentrations [105]. Thus, molnupiravir can modestly inhibit RNA synthesis, but it also acts as a viral genome mutagenic agent. Given that molnupiravir treatment of SARS-CoV-2 virus induces significant G-to-A and C-to-U transition mutations in the viral RNA, it seems that the mutagenic effect predominates, leading to an "error catastrophe" and the eventual failure to generate replication competent viral populations [107]. Similarly, favipiravir (**18**) treatment of SARS-CoV-2-infected cells results in a significant elevation of C-to-U and G-to-A transition mutations in the viral genome, consistent with its incorporation primarily as a guanosine analog but with the ability to efficiently serve as a template base for pairing with UTP, suggesting that induction of viral mutagenesis likely plays a role in the antiviral mechanism of action for this compound [108].

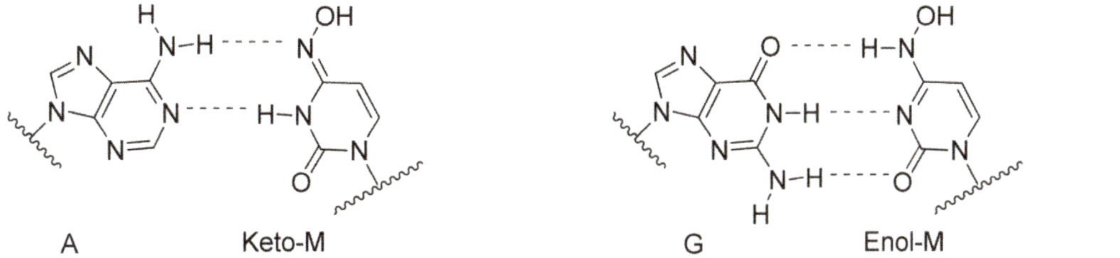

Figure 8. Keto-enol tautomeric equilibrium provides opportunities for molnupiravir (M) in the RNA template to form Watson–Crick hydrogen bonds with incoming ATP or GTP.

Remdesivir was approved by the US FDA in October, 2020 for the treatment of COVID-19 infections in hospitalized subjects. The approval of this intravenously administered drug was based on three clinical studies, the ACTT1 study [109] and two studies sponsored by Gilead [110,111], but omitted results from a study conducted in China [112]. Shortly after approval, the results from the much larger Solidarity trial were published [113]. A meta-analysis of four of these randomized controlled trials with 7334 patients concluded that subjects treated with RDV were more likely to demonstrate recovery and were associated with higher rates of hospital discharge, but there was no significant reduction in mean time to clinical improvement or mortality [114].

Molnupiravir exhibits an in vitro antiviral EC_{50} value of 0.3 µM against SARS-CoV-2 virus when tested in Vero cells [107] as well as broad-spectrum antiviral activity against seasonal coronaviruses [115]. In preclinical models, molnupiravir inhibits SARS-CoV-2 replication in the Syrian hamster model [116,117] and mice [107] and blocks SARS-CoV-2 transmission in ferrets [118]. Given its mechanism of action as a viral mutagenesis agent, molnuprivar has been closely scrutinized for its potential to elicit DNA mutagenesis in host cells and tissues. One lab has found that molnupiravir displays host mutational activity in an animal cell culture assay [119] although molnupiravir was negative in a 28 day transgenic rodent mutagenicity study [28]. In addition, this orally administered compound was safe and well tolerated in Phase 1 studies conducted in healthy human volunteers [27]. A Phase 3 clinical study of 1408 unvaccinated participants demonstrated that molnupiravir treatment for 5 days reduced the risk of hospitalization and death by 30% [120]. Molnupiravir received EUA by the US FDA in December, 2021 for treatment of mild-to-moderate COVID-19 infections. Molnupiravir is not authorized for use in patients younger than 18 years of age because it may affect bone growth and cartilage formation and it is also important to recognize the drug may cause fetal harm when administered to pregnant individuals [28].

Galidesivir inhibits replication of SARS-CoV and MERS-CoV viruses in Vero E6 cells with EC_{50} values of 57.7 and 68.4 µM, respectively [101]. Early administration of galidesivir in a COVID-19 animal model reduced the viral burden and pathology in lung tissue [121]. A small Phase I study in COVID-19 patients demonstrated that galidesivir was safe and generally well tolerated, but it did not show signs of significant clinical benefit. Consequently, the sponsor has discontinued plans to develop galidesivir for treatment of COVID-19 [121]. AT-527 was initially developed as a drug to treat HCV infections but its EC_{90} value of 0.47 µM in a virus yield reduction assay against SARS-CoV-2 in primary human airway epithelial cells suggested potential utility for treating COVID-19 [102]. Unfortunately, AT-527 failed to meet its primary goal of reducing SARS-CoV-2 RNA at various intervals in a Phase 2 clinical trial in subjects with mild or moderate COVID-19 in the outpatient setting, leading the sponsor to update its clinical development strategy [122]. The orally administered drug favipiravir, approved to treat novel influenza in Japan, inhibits SARS-CoV-2 replication and the generation of cytopathic effects in Vero E6 cells with EC_{50} values of 207 and 118 µM, respectively [108]. Russia has approved favipiravir for treating COVID-19 infections and several other countries such as Mexico, India and Malaysia have granted EUA for this indication. A meta-analysis on clinical studies that evaluated the efficacy and safety of favipiravir as a treatment for COVID-19 found that there was a significant clinical and radiological improvement following treatment with favipiravir in comparison to the standard of care but with no significant differences on viral clearance, oxygen support requirement and side effect profiles [123].

4.3. Conclusions and Future Directions

Drug repurposing studies initiated at the beginning of the COVID-19 pandemic identified at least two compounds from the nucleos(t)ide analog inhibitor class, remdesivir and molnupiravir, with sufficient antiviral activity against the SARS-CoV-2 virus to merit approval or EUA by the US FDA for treating COVID-19 infections in select populations. However, both drugs have their limitations and so research is currently underway to discover new RdRp inhibitors including those from the non-nucleoside inhibitor class with improved safety and efficacy properties. Non-nucleosides are compounds that bind directly to the viral RdRp without the need for chemical transformation to the nucleotide triphosphate form and are represented in drug therapies to treat HIV and cure HCV infections [124,125]. Non-nucleosides often possess high selectivity for their viral RdRps and so it is understandable that no potent non-nucleoside inhibitors of SARS-CoV-2 RdRp emerged from early drug repurposing studies. Discovering non-nucleoside inhibitors of SARS-CoV-2 RdRp will require screening and medicinal chemistry optimization—activities at the very early stages of the drug discovery process. Fortunately, recent advances in DNA encoded libraries [126] and virtual screening [127,128] offer the ability to rapidly screen billions of molecules and should facilitate that first step on the road towards the discovery of highly selective non-nucleoside SARS-CoV-2 RdRp inhibitors.

5. Papain-Like Protease (PLpro)

5.1. Structural Organization and Functions of PLpro

SARS-CoV-2 nsp3 is a large multidomain protein essential for viral replication [129]. PLpro, a ~36 kDa protein, is one of the domains of nsp3 and exists as a monomer in biological settings [130]. It has a ubiquitin-like (UBL) domain at the N-terminus, and the rest of the protein has the architecture of the ubiquitin-specific protease (USP) fold (Figure 9) [130,131]. The USP fold is topologically organized into three subdomains, thumb, palm and fingers, which together form a structure resembling a right hand [132]. The SARS-CoV-2 PLpro active site contains a canonical cysteine protease catalytic triad comprised of residues Cys111, His272, and Asp286 [130,131]. It cleaves at the N terminus of the LXGG motifs (Figure 10) between nsp1, nsp2, nsp3 and nsp4, liberating the nsp1, nsp2, and nsp3 from the viral polyproteins [130,133]. Therefore, PLpro plays an essential role in the processing and maturation of the SARS-CoV-2 polyproteins.

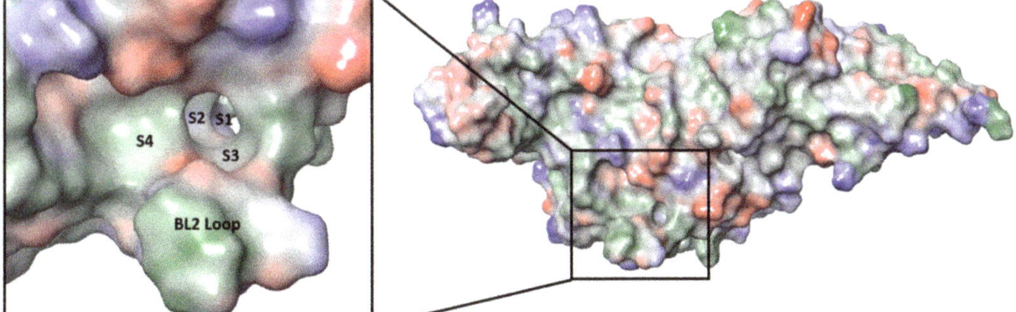

Figure 9. SARS-CoV-2 PLpro structure and substrate-binding site, with substrate-binding subsites of S1–S4 and the BL2 loop enlarged in the box. Image produced from PDB ID 6WUU. (UBL: ubiquitin-like domain; USP: ubiquitin-specific protease fold.)

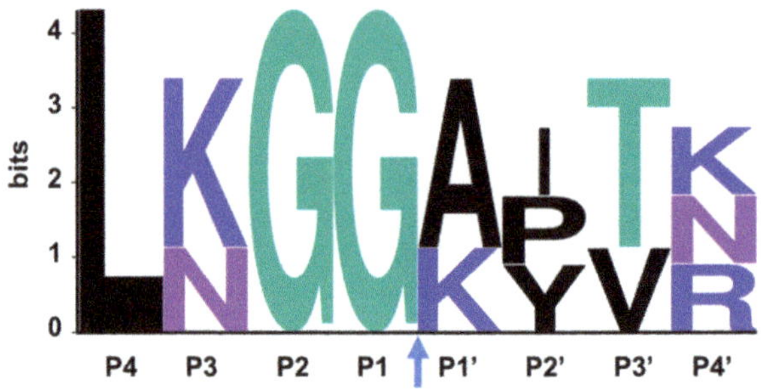

Figure 10. The consensus recognition sequence cleaved by SARS-CoV-2 (Uniprot code P0DTD1) PLpro. The cleavage site is marked by the blue arrow. Image generated by WebLogo.

PLpro is encoded by all coronaviruses, often in two copies, as denoted by PL1pro and PL2pro [134]. However, a number of coronaviruses, including SARS-CoV, SARS-CoV-2 and MERS-CoV, have only one copy of PLpro. The PLpro sequence between SARS-CoV-2 and SARS-CoV is highly conserved, with 83% sequence identity, and 90% sequence similarity. In contrast, the SARS-CoV-2 PLpro sequence is only 31% identical and 49% similar to that of MERS-CoV [135]. In fact, several SARS-CoV PLpro inhibitors are also known to be active against SARS-CoV-2 PLpro but not against the MERS-CoV enzyme [135,136]. In addition to the protease activity, PLpros from SARS-CoV, SARS-CoV-2 and MERS-CoV were shown in separate studies to possess deubiquitinating and deISGylating capabilities, despite the difference in the PLpro sequences among these three coronaviruses [136–138]. These additional enzymatic activities provide catalytic functions for cleaving ubiquitin (Ub) or ISG15 modifications from host proteins, blocking the induction of type I interferons and expression of cellular cytokines [138]. As a result, PLpro facilitates viral replication in host cells through antagonism of the host antiviral innate immune response [139,140]. Therefore, inhibition of PLpro activity can serve dual functions, controlling viral replication by targeting protease activity and restoring host immune response by targeting the deubiquitinating and deISGylating activities, making PLpro an excellent target for antiviral drug discovery.

5.2. Discovery of PLpro Inhibitors

While efforts to discover inhibitors for SARS-CoV-2 PLpro have been reported in the literature, there are currently no approved therapies or assets in clinical trials utilizing this mechanism. The majority of selective PLpro inhibitors target the substrate-binding pocket, although this region presents challenges for druggability (Figure 9). The S1 and S2 subsites are small and narrow to recognize the glycine residues found in PLpro substrates (Figure 10), limiting access to the active-site cysteine. The S3 and S4 regions provide a larger binding pocket but the pocket is bordered by the flexible BL2 loop.

The high conservation of the PLpro active site between SARS-CoV-2 and SARS-CoV has accelerated the study of non-covalent inhibitors by enabling repurposing of naphthalene-based inhibitors developed to target SARS-CoV (Figure 11). Compounds such as GRL-0617 (**19**) and **20** were originally reported by Ghosh [141,142] to inhibit SARS-CoV PLpro by binding to the S3/S4 pocket and were found to have modest activity against SARS-CoV-2 (**19**: IC_{50} = 1.6 μM; **20**: IC_{50} = 2.6 μM) [135,143,144]. Efforts to improve the activity of these first-generation compounds have often grown the inhibitors to target additional interactions. Xiong and coworkers improved potency by converting the naphthalene of GRL-0617 to a substituted 2-phenylthiophene in XR8-89 (**21**, IC_{50} = 0.11 μM) in order to reach further down the BL2 groove, while also appending an azetidine to the aniline to interact with Glu167 [143]. Tan and coworkers used compound **20** as a starting point and enhanced activity by attaching the piperazine urea in **22** (IC_{50} = 0.44 μM) to extend around the BL2 loop [144]. A high-throughput screen by Wang identified dibenzylamine hits such as Jun9-13-9 (**23**, IC_{50} = 6.7 μM) that during optimization converged with the earlier naphthalene inhibitors to yield more potent compounds such as Jun9-75-4 (**24**, IC_{50} = 0.62 μM) [145]. While activity gains have recently been made in this class of non-covalent PLpro inhibitors, further improvements in antiviral potency will be required to generate candidates with clinical efficacy.

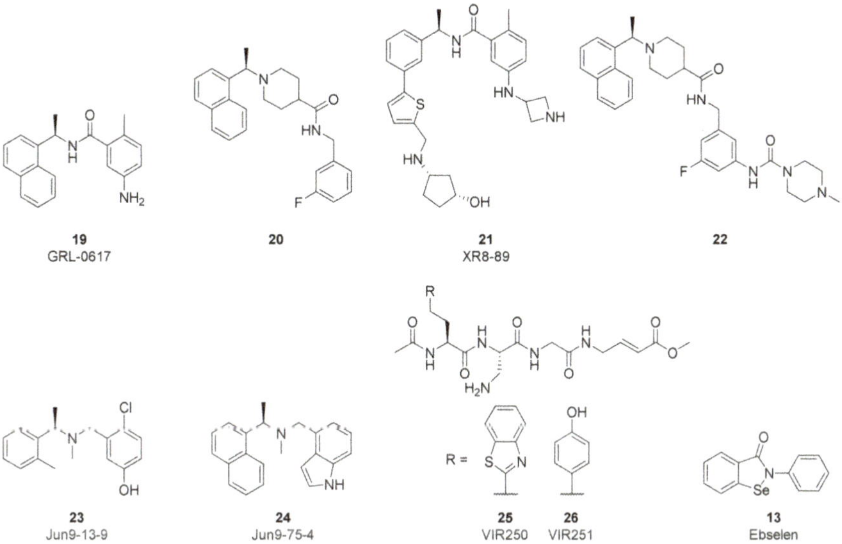

Figure 11. Chemical structures of representative SARS-CoV-2 PLpro inhibitors.

Efforts to produce covalent SARS-CoV-2 PLpro inhibitors have also been described. Olsen and coworkers used a combinatorial library of fluorogenic tetrapeptide substrates to identify the optimal peptide sequences for PLpro reactivity [146]. Replacement of the chromophore with an α,β-unsaturated ester warhead provided inhibitors VIR250 (**25**) and VIR251 (**26**). Co-crystal structures highlighted the key interactions with the protease and verified the covalent inhibition mechanism. Li and coworkers used **19** as the starting point for their covalent inhibitor and appended a sulfonium-tethered peptide to generate a peptide drug conjugate [147]. Both strategies incorporated the glycine residues of the PLpro substrates to navigate the S1 and S2 sites. Further advancements in covalent strategies will need to overcome the modest activity and challenging peptidic properties of these early compounds.

High-throughput screens of known bioactive compounds have led to reports that a variety of other classes of compounds inhibit SARS-CoV-2 PLpro. Organoselenium compounds such as Ebselen (**13**) were reported to inhibit PLpro [148], but others have found these act as non-selective modifiers of cysteine, including the cysteine at the active site of PLpro [149]. Validation studies of reported quinone and nucleoside hits also suggest these classes of compounds are non-specific inhibitors [150]. The limited antiviral activity and poor selectivity of these hits are significant challenges to advancing them into useful tools and highlight the importance of confirming PLpro screening results across multiple assays.

5.3. Conclusions and Future Directions

SARS-CoV-2 PLpro is a multifunctional enzyme with protease, deubiquitinase, and deISGylating activities, with the latter two activities involved in blocking the expression of type I interferons in the infected cells. In principle, targeting the activities of PLpro may not only inhibit viral replication but also block the viral-mediated evasion of host innate immunity. One recent report presented data that support this hypothesis: GRL-0617 could inhibit SARS-CoV-2 replication as well as maintain the antiviral interferon signaling in the infected cells [136]. The discovery of potent PLpro inhibitors has been challenging due to the apparent lack of well-defined binding pockets at its substrate-binding pocket. However, the growing number of structural studies conducted with PLpro to identify potential binding pockets and the increasing interest in optimizing known PLpro inhibitors based on structural data would certainly facilitate the understanding of the druggability of PLpro and advance the drug discovery for this antiviral target.

6. Summary

In this review, we discuss the current status and future strategies for the discovery of small- and large-molecule antiviral therapeutics for COVID-19 and, potentially, emerging coronaviruses. There is an increasing number of inhibitors reported to have anti-SARS-CoV-2 activity in vitro and in vivo. Several of these inhibitors [e.g., some Mpro or RdRp inhibitors, and at least two examples of anti-S mAbs (sotrovimab and bebtelovimab)] have also been shown to be efficacious for a reasonably wide spectrum of coronaviruses and/or COVID-19 variants [87,98,99,151]. This is not surprising due to the high protein sequence conservation among coronaviruses and variants at the catalytic sites and substrate-binding sites of Mpro and RdRp, as well as parts of the S protein that are important for viral entry. PLpro protein sequence is less conserved among coronaviruses when compared to those of Mpro and RdRp. The breadth of the antiviral spectrum of an optimized PLpro inhibitor remains to be determined. This review has focused on small molecules and mAbs that directly target viral proteins. In addition to these drug modalities, other therapeutic approaches (e.g., CRISPR, RNAi, or antisense oligonucleotide) may also prove to be useful for COVID-19 therapies.

Resistance generation during antiviral therapy can occur during treatment for chronic viral infections such as HIV and HCV. With the short duration of COVID-19 antiviral therapy, development of drug resistance to small molecule inhibitors may not be an issue. Should drug resistance become a concern, the availability of SARS-CoV-2 inhibitors targeting different viral proteins will allow combination therapy to minimize the generation of resistance.

Drug discovery is progressing at lightning speed to address the urgent need of the COVID-19 crisis. We bear witness to unprecedented partnerships among industry, academics and/or government to join forces in the discovery and development of COVID-19 drugs. In addition, the basic research, drug discovery and clinical development frameworks established by decades of antiviral research on human pathogenic viruses enable investigators to quickly apply prior knowledge and cutting-edge technology to discover treatments for COVID-19 infections. Recognizing the importance of antivirals, the US government has pledged to support a new initiative called the Antiviral Program for Pandemics by committing $3.2 billion in funding for research on COVID-19 antiviral drug development, as well as new drugs for viruses that could cause future pandemics [152]. With the availability of several validated and druggable targets for SARS-CoV-2, as well as the unprecedented partnerships and resources dedicated to COVID-19 drug discovery, we expect that a rich pipeline of COVID-19 drugs will emerge that will become important weapons to tackle COVID-19 and future coronaviruses with pandemic risk.

Author Contributions: All authors wrote, reviewed, and edited the manuscript. All authors have read and agreed to the published version of the manuscript.

Funding: The financial support for this research was provided by AbbVie.

Acknowledgments: We would like to thank Bernhard Sielaff for assistance in generating the figure for the domain architecture of the SARS-CoV-2 S protein, Charles Hutchins for assistance in generating the figure for the structure of the RdRp complex, and Morteza Khabiri for assistance in generating the figure for the consensus recognition sequence of PLpro.

Conflicts of Interest: The authors declare no conflict of interest.

Disclosures: T.I.N., I.C., D.A.D., M.R.S., D.J.H., E.L.N. and W.M.K. are employees of AbbVie. J.S. was an employee of AbbVie at the time of writing. AbbVie participated in the interpretation of data, review, and approval of the publication.

References

1. Wu, F.; Zhao, S.; Yu, B.; Chen, Y.-M.; Wang, W.; Song, Z.-G.; Hu, Y.; Tao, Z.-W.; Tian, J.-H.; Pei, Y.-Y.; et al. A new coronavirus associated with human respiratory disease in China. *Nature* **2020**, *579*, 265–269. [CrossRef] [PubMed]
2. Zhu, N.; Zhang, D.; Wang, W.; Li, X.; Yang, B.; Song, J.; Zhao, X.; Huang, B.; Shi, W.; Lu, R.; et al. A Novel Coronavirus from Patients with Pneumonia in China, 2019. *N. Engl. J. Med.* **2020**, *382*, 727–733. [CrossRef] [PubMed]
3. COVID-19 Dashboard by the Center for Systems Science and Engineering at Johns Hopkins University. Available online: https://coronavirus.jhu.edu/map.html (accessed on 21 March 2022).
4. Lee, N.; Hui, D.; Wu, A.; Chan, P.; Cameron, P.; Joynt, G.M.; Ahuja, A.; Yung, M.Y.; Leung, C.; To, K.; et al. A Major Outbreak of Severe Acute Respiratory Syndrome in Hong Kong. *N. Engl. J. Med.* **2003**, *348*, 1986–1994. [CrossRef] [PubMed]
5. Zaki, A.M.; Van Boheemen, S.; Bestebroer, T.M.; Osterhaus, A.D.M.E.; Fouchier, R.A.M. Isolation of a Novel Coronavirus from a Man with Pneumonia in Saudi Arabia. *N. Engl. J. Med.* **2012**, *367*, 1814–1820. [CrossRef] [PubMed]
6. Tregoning, J.S.; Flight, K.E.; Higham, S.L.; Wang, Z.; Pierce, B.F. Progress of the COVID-19 vaccine effort: Viruses, vaccines and variants versus efficacy, effectiveness and escape. *Nat. Rev. Immunol.* **2021**, *21*, 626–636. [CrossRef]
7. Arunachalam, P.S.; Scott, M.K.D.; Hagan, T.; Li, C.; Feng, Y.; Wimmers, F.; Grigoryan, L.; Trisal, M.; Edara, V.V.; Lai, L.; et al. Systems vaccinology of the BNT162b2 mRNA vaccine in humans. *Nature* **2021**, *596*, 410–416. [CrossRef]
8. Thomas, S.J.; Moreira, E.D., Jr.; Kitchin, N.; Absalon, J.; Gurtman, A.; Lockhart, S.; Perez, J.L.; Pérez Marc, G.; Polack, F.P.; Zerbini, C.; et al. Safety and Efficacy of the BNT162b2 mRNA COVID-19 Vaccine through 6 Months. *N. Engl. J. Med.* **2021**, *385*, 1761–1773. [CrossRef]
9. Goldberg, Y.; Mandel, M.; Bar-On, Y.M.; Bodenheimer, O.; Freedman, L.; Haas, E.J.; Milo, R.; Alroy-Preis, S.; Ash, N.; Huppert, A. Waning Immunity after the BNT162b2 Vaccine in Israel. *N. Engl. J. Med.* **2021**, *385*, e85. [CrossRef]
10. Pegu, A.; O'Connell, S.E.; Schmidt, S.D.; O'Dell, S.; Talana, C.A.; Lai, L.; Albert, J.; Anderson, E.; Bennett, H.; Corbett, K.S.; et al. Durability of mRNA-1273 vaccine–induced antibodies against SARS-CoV-2 variants. *Science* **2021**, *373*, 1372–1377. [CrossRef]
11. Accorsi, E.K.; Britton, A.; Fleming-Dutra, K.E.; Smith, Z.R.; Shang, N.; Derado, G.; Miller, J.; Schrag, S.J.; Verani, J.R. Association between 3 Doses of mRNA COVID-19 Vaccine and Symptomatic Infection Caused by the SARS-CoV-2 Omicron and Delta Variants. *JAMA* **2022**, *327*, 639. [CrossRef]
12. Ai, J.; Zhang, H.; Zhang, Y.; Lin, K.; Zhang, Y.; Wu, J.; Wan, Y.; Huang, Y.; Song, J.; Zhangfan, F.; et al. Omicron variant showed lower neutralizing sensitivity than other SARS-CoV-2 variants to immune sera elicited by vaccines after boost. *Emerg. Microbes Infect.* **2022**, *1*, 337–343. [CrossRef] [PubMed]
13. Lusvarghi, S.; Pollett, S.D.; Neerukonda, S.N.; Wang, W.; Wang, R.; Vassell, R.; Epsi, N.J.; Fries, A.C.; Agan, B.K.; Lindholm, D.A.; et al. SARS-CoV-2 Omicron neutralization by therapeutic antibodies, convalescent sera, and post-mRNA vaccine booster. *bioRxiv* **2021**. [CrossRef]
14. Cao, Y.; Wang, J.; Jian, F.; Xiao, T.; Song, W.; Yisimayi, A.; Huang, W.; Li, Q.; Wang, P.; An, R.; et al. Omicron escapes the majority of existing SARS-CoV-2 neutralizing antibodies. *Nature* **2021**, *602*, 657–663. [CrossRef]
15. Poduri, R.; Joshi, G.; Jagadeesh, G. Drugs targeting various stages of the SARS-CoV-2 life cycle: Exploring promising drugs for the treatment of COVID-19. *Cell. Signal.* **2020**, *74*, 109721. [CrossRef]
16. Mei, M.; Tan, X. Current Strategies of Antiviral Drug Discovery for COVID-19. *Front. Mol. Biosci.* **2021**, *8*, 671263. [CrossRef]
17. Gil, C.; Ginex, T.; Maestro, I.; Nozal, V.; Barrado-Gil, L.; Cuesta-Geijo, M.; Urquiza, J.; Ramírez, D.; Alonso, C.; Campillo, N.E.; et al. COVID-19: Drug Targets and Potential Treatments. *J. Med. Chem.* **2020**, *63*, 12359–12386. [CrossRef] [PubMed]
18. Shang, J.; Ye, G.; Shi, K.; Wan, Y.; Luo, C.; Aihara, H.; Geng, Q.; Auerbach, A.; Li, F. Structural basis of receptor recognition by SARS-CoV-2. *Nature* **2020**, *581*, 221–224. [CrossRef]
19. Lan, J.; Ge, J.; Yu, J.; Shan, S.; Zhou, H.; Fan, S.; Zhang, Q.; Shi, X.; Wang, Q.; Zhang, L.; et al. Structure of the SARS-CoV-2 spike receptor-binding domain bound to the ACE2 receptor. *Nature* **2020**, *581*, 215–220. [CrossRef]
20. Naqvi, A.A.T.; Fatima, K.; Mohammad, T.; Fatima, U.; Singh, I.K.; Singh, A.; Atif, S.M.; Hariprasad, G.; Hasan, G.M.; Hassan, I. Insights into SARS-CoV-2 genome, structure, evolution, pathogenesis and therapies: Structural genomics approach. *Biochim. et Biophys. Acta (BBA)–Mol. Basis Dis.* **2020**, *1866*, 165878. [CrossRef]
21. Ng, Y.L.; Salim, C.K.; Chu, J.J.H. Drug repurposing for COVID-19: Approaches, challenges and promising candidates. *Pharmacol. Ther.* **2021**, *228*, 107930. [CrossRef]
22. Bellera, C.L.; Llanos, M.; Gantner, M.E.; Rodriguez, S.; Gavernet, L.; Comini, M.; Talevi, A. Can drug repurposing strategies be the solution to the COVID-19 crisis? *Expert Opin. Drug Discov.* **2021**, *16*, 605–612. [CrossRef] [PubMed]
23. Riva, L.; Yuan, S.; Yin, X.; Martin-Sancho, L.; Matsunaga, N.; Pache, L.; Burgstaller-Muehlbacher, S.; De Jesus, P.D.; Teriete, P.; Hull, M.V.; et al. Discovery of SARS-CoV-2 antiviral drugs through large-scale compound repurposing. *Nature* **2020**, *586*, 113–119. [CrossRef] [PubMed]
24. Prescribing Information for Veklurytm (Remdesivir). Available online: https://www.gilead.com/-/media/files/pdfs/medicines/COVID-19/veklury/veklury_pi.pdf (accessed on 21 March 2022).

25. Pardo, J.; Shukla, A.M.; Chamarthi, G.; Gupte, A. The journey of remdesivir: From Ebola to COVID-19. *Drugs Context* **2020**, *9*, 1–9. [CrossRef] [PubMed]
26. Painter, G.R.; Natchus, M.G.; Cohen, O.; Holman, W.; Painter, W.P. Developing A Direct Acting, Orally Available Antiviral Agent in a Pandemic: The Evolution of Molnupiravir as a Potential Treatment for COVID-19. *Curr. Opin. Virol.* **2021**, *50*, 17–22. [CrossRef]
27. Painter, W.P.; Holman, W.; Bush, J.A.; Almazedi, F.; Malik, H.; Eraut, N.C.J.E.; Morin, M.J.; Szewczyk, L.J.; Painter, G.R. Human Safety, Tolerability, and Pharmacokinetics of Molnupiravir, a Novel Broad-Spectrum Oral Antiviral Agent with Activity against SARS-CoV-2. *Antimicrob. Agents Chemother.* **2021**, *65*, e02428-20. [CrossRef]
28. Fact Sheet for Healthcare Providers: Emergency Use Authorization for Molnupiravir. Available online: https://www.fda.gov/media/155054/download (accessed on 21 March 2022).
29. Fact Sheet for Healthcare Providers: Emergency Use Authorization for Paxlovid. Available online: https://www.covid19oralrx-patient.com/files/Final-Emergency-Use-Full-Prescribing-Info-HCP-Fact-Sheet-COVID-19-Oral-Antiviral.pdf (accessed on 21 March 2022).
30. NIH COVID-19 Treatment Guidelines. Ritonavir-Boosted Nirmatrelvir (Paxlovid). Available online: https://www.covid19treatmentguidelines.nih.gov/therapies/antiviral-therapy/ritonavir-boosted-nirmatrelvir--paxlovid-/ (accessed on 21 March 2022).
31. FDA. Coronavirus (COVID-19) Update: FDA Limits Use of Certain Monoclonal Antibodies to Treat COVID-19 Due to the Omicron Variant. Press Announcements. Available online: https://www.fda.gov/news-events/press-announcements/coronavirus-covid-19-update-fda-limits-use-certain-monoclonal-antibodies-treat-covid-19-due-omicron (accessed on 21 March 2022).
32. Tortorici, M.A.; Veesler, D. Structural insights into coronavirus entry. *Adv. Virus Res.* **2019**, *105*, 93–116. [CrossRef]
33. Finkelstein, M.; Mermelstein, A.; Miller, E.; Seth, P.; Stancofski, E.-S.; Fera, D. Structural Analysis of Neutralizing Epitopes of the SARS-CoV-2 Spike to Guide Therapy and Vaccine Design Strategies. *Viruses* **2021**, *13*, 134. [CrossRef]
34. Wrapp, D.; Wang, N.; Corbett, K.S.; Goldsmith, J.A.; Hsieh, C.-L.; Abiona, O.; Graham, B.S.; McLellan, J.S. Cryo-EM structure of the 2019-nCoV spike in the prefusion conformation. *Science* **2020**, *367*, 1260–1263. [CrossRef]
35. Casalino, L.; Gaieb, Z.; Goldsmith, J.A.; Hjorth, C.K.; Dommer, A.C.; Harbison, A.M.; Fogarty, C.A.; Barros, E.P.; Taylor, B.C.; McLellan, J.S.; et al. Beyond Shielding: The Roles of Glycans in the SARS-CoV-2 Spike Protein. *ACS Cent. Sci.* **2020**, *6*, 1722–1734. [CrossRef]
36. Baum, A.; Fulton, B.O.; Wloga, E.; Copin, R.; Pascal, K.E.; Russo, V.; Giordano, S.; Lanza, K.; Negron, N.; Ni, M.; et al. Antibody cocktail to SARS-CoV-2 spike protein prevents rapid mutational escape seen with individual antibodies. *Science* **2020**, *369*, 1014–1018. [CrossRef]
37. Hansen, J.; Baum, A.; Pascal, K.E.; Russo, V.; Giordano, S.; Wloga, E.; Fulton, B.O.; Yan, Y.; Koon, K.; Patel, K.; et al. Studies in humanized mice and convalescent humans yield a SARS-CoV-2 antibody cocktail. *Science* **2020**, *369*, 1010–1014. [CrossRef] [PubMed]
38. Rappazzo, C.G.; Tse, L.V.; Kaku, C.I.; Wrapp, D.; Sakharkar, M.; Huang, D.; Deveau, L.M.; Yockachonis, T.J.; Herbert, A.S.; Battles, M.B.; et al. Broad and potent activity against SARS-like viruses by an engineered human monoclonal antibody. *Science* **2021**, *371*, 823–829. [CrossRef] [PubMed]
39. Walls, A.C.; Park, Y.-J.; Tortorici, M.A.; Wall, A.; McGuire, A.T.; Veesler, D. Structure, Function, and Antigenicity of the SARS-CoV-2 Spike Glycoprotein. *Cell* **2020**, *181*, 281–292.e6. [CrossRef] [PubMed]
40. Cai, Y.; Zhang, J.; Xiao, T.; Peng, H.; Sterling, S.M.; Walsh, R.M.; Rawson, S.; Rits-Volloch, S.; Chen, B. Distinct conformational states of SARS-CoV-2 spike protein. *Science* **2020**, *369*, 1586–1592. [CrossRef]
41. Starr, T.N.; Greaney, A.J.; Hilton, S.K.; Ellis, D.; Crawford, K.H.D.; Dingens, A.S.; Navarro, M.J.; Bowen, J.E.; Tortorici, M.A.; Walls, A.C.; et al. Deep mutational scanning of SARS-CoV-2 receptor binding domain reveals constraints on folding and ACE2 binding. *Cell* **2020**, *182*, 1295–1310.e1220. [CrossRef]
42. Cameroni, E.; Saliba, C.; Bowen, J.E.; Rosen, L.E.; Culap, K.; Pinto, D.; De Marco, A.; Zepeda, S.K.; di Iulio, J.; Zatta, F.; et al. Broadly neutralizing antibodies overcome SARS-CoV-2 Omicron antigenic shift. *Nature* **2021**, *602*, 664–670. [CrossRef]
43. Li, F.; Li, W.; Farzan, M.; Harrison, S.C. Structure of SARS Coronavirus Spike Receptor-Binding Domain Complexed with Receptor. *Science* **2005**, *309*, 1864–1868. [CrossRef]
44. Wang, N.; Shi, X.; Jiang, L.; Zhang, S.; Wang, D.; Tong, P.; Guo, D.; Fu, L.; Cui, Y.; Liu, X.; et al. Structure of MERS-CoV spike receptor-binding domain complexed with human receptor DPP4. *Cell Res.* **2013**, *23*, 986–993. [CrossRef]
45. Li, W.; Moore, M.J.; Vasilieva, N.; Sui, J.; Wong, S.K.; Berne, M.A.; Somasundaran, M.; Sullivan, J.L.; Luzuriaga, K.; Greenough, T.C.; et al. Angiotensin-converting enzyme 2 is a functional receptor for the SARS coronavirus. *Nature* **2003**, *426*, 450–454. [CrossRef]
46. Millet, J.K.; Whittaker, G.R. Host cell proteases: Critical determinants of coronavirus tropism and pathogenesis. *Virus Res.* **2015**, *202*, 120–134. [CrossRef]
47. Peacock, T.P.; Goldhill, D.H.; Zhou, J.; Baillon, L.; Frise, R.; Swann, O.C.; Kugathasan, R.; Penn, R.; Brown, J.C.; Sanchez-David, R.Y.; et al. The furin cleavage site in the SARS-CoV-2 spike protein is required for transmission in ferrets. *Nat. Microbiol.* **2021**, *6*, 899–909. [CrossRef] [PubMed]
48. Weiss, S.R. Forty years with coronaviruses. *J. Exp. Med.* **2020**, *217*, e20200537. [CrossRef] [PubMed]
49. Ke, Z.; Oton, J.; Qu, K.; Cortese, M.; Zila, V.; McKeane, L.; Nakane, T.; Zivanov, J.; Neufeldt, C.J.; Cerikan, B.; et al. Structures and distributions of SARS-CoV-2 spike proteins on intact virions. *Nature* **2020**, *588*, 498–502. [CrossRef] [PubMed]

50. Lai, A.L.; Millet, J.K.; Daniel, S.; Freed, J.H.; Whittaker, G.R. The SARS-CoV Fusion Peptide Forms an Extended Bipartite Fusion Platform that Perturbs Membrane Order in a Calcium-Dependent Manner. *J. Mol. Biol.* **2017**, *429*, 3875–3892. [CrossRef]
51. Chi, X.; Yan, R.; Zhang, J.; Zhang, G.; Zhang, Y.; Hao, M.; Zhang, Z.; Fan, P.; Dong, Y.; Yang, Y.; et al. A neutralizing human antibody binds to the N-terminal domain of the Spike protein of SARS-CoV-2. *Science* **2020**, *369*, 650–655. [CrossRef]
52. Pinto, D.; Park, Y.-J.; Beltramello, M.; Walls, A.C.; Tortorici, M.A.; Bianchi, S.; Jaconi, S.; Culap, K.; Zatta, F.; De Marco, A.; et al. Cross-neutralization of SARS-CoV-2 by a human monoclonal SARS-CoV antibody. *Nature* **2020**, *583*, 290–295. [CrossRef]
53. Wang, C.; van Haperen, R.; Gutiérrez-Álvarez, J.; Li, W.; Okba, N.M.A.; Albulescu, I.; Widjaja, I.; van Dieren, B.; Fernandez-Delgado, R.; Sola, I.; et al. A conserved immunogenic and vulnerable site on the coronavirus spike protein delineated by cross-reactive monoclonal antibodies. *Nat. Commun.* **2021**, *12*, 1715. [CrossRef]
54. Barnes, C.O.; Jette, C.A.; Abernathy, M.E.; Dam, K.M.A.; Esswein, S.R.; Gristick, H.B.; Malyutin, A.G.; Sharaf, N.G.; Huey-Tubman, K.E.; Lee, Y.E.; et al. Structural classification of neutralizing antibodies against the SARS-CoV-2 spike receptor-binding domain suggests vaccine and therapeutic strategies. *bioRxiv* **2020**. [CrossRef]
55. Nabel, K.G.; Clark, S.A.; Shankar, S.; Pan, J.; Clark, L.E.; Yang, P.; Coscia, A.; McKay, L.G.A.; Varnum, H.H.; Brusic, V.; et al. Structural basis for continued antibody evasion by the SARS-CoV-2 receptor binding domain. *Science* **2022**, *375*, eabl6251. [CrossRef]
56. Li, W.; Chen, Y.; Prévost, J.; Ullah, I.; Lu, M.; Gong, S.Y.; Tauzin, A.; Gasser, R.; Vézina, D.; Anand, S.P.; et al. Structural basis and mode of action for two broadly neutralizing antibodies against SARS-CoV-2 emerging variants of concern. *Cell Rep.* **2022**, *38*, 110210. [CrossRef]
57. Guo, H.; Hu, B.-J.; Yang, X.-L.; Zeng, L.-P.; Li, B.; Ouyang, S.; Shi, Z.-L. Evolutionary Arms Race between Virus and Host Drives Genetic Diversity in Bat Severe Acute Respiratory Syndrome-Related Coronavirus Spike Genes. *J. Virol.* **2020**, *94*, e00902-20. [CrossRef] [PubMed]
58. Dolan, P.T.; Whitfield, Z.J.; Andino, R. Mechanisms and Concepts in RNA Virus Population Dynamics and Evolution. *Annu. Rev. Virol.* **2018**, *5*, 69–92. [CrossRef] [PubMed]
59. Lu, R.; Zhao, X.; Li, J.; Niu, P.; Yang, B.; Wu, H.; Wang, W.; Song, H.; Huang, B.; Zhu, N.; et al. Genomic characterisation and epidemiology of 2019 novel coronavirus: Implications for virus origins and receptor binding. *Lancet* **2020**, *395*, 565–574. [CrossRef]
60. Wang, Q.; Zhang, Y.; Wu, L.; Niu, S.; Song, C.; Zhang, Z.; Lu, G.; Qiao, C.; Hu, Y.; Yuen, K.Y.; et al. Structural and Functional Basis of SARS-CoV-2 Entry by Using Human ACE2. *Cell* **2020**, *181*, 894–904.e889. [CrossRef]
61. Yan, R.; Zhang, Y.; Li, Y.; Xia, L.; Guo, Y.; Zhou, Q. Structural basis for the recognition of SARS-CoV-2 by full-length human ACE2. *Science* **2020**, *367*, 1444–1448. [CrossRef]
62. Yi, C.; Sun, X.; Ye, J.; Ding, L.; Liu, M.; Yang, Z.; Lu, X.; Zhang, Y.; Ma, L.; Gu, W.; et al. Key residues of the receptor binding motif in the spike protein of SARS-CoV-2 that interact with ACE2 and neutralizing antibodies. *Cell. Mol. Immunol.* **2020**, *17*, 621–630. [CrossRef]
63. Thomson, E.C.; Rosen, L.E.; Shepherd, J.G.; Spreafico, R.; Filipe, A.D.S.; Wojcechowskyj, J.A.; Davis, C.; Piccoli, L.; Pascall, D.J.; Dillen, J.; et al. Circulating SARS-CoV-2 spike N439K variants maintain fitness while evading antibody-mediated immunity. *Cell* **2021**, *184*, 1171–1187.e20. [CrossRef]
64. Tegally, H.; Wilkinson, E.; Giovanetti, M.; Iranzadeh, A.; Fonseca, V.; Giandhari, J.; Doolabh, D.; Pillay, S.; San, E.J.; Msomi, N.; et al. Emergence and rapid spread of a new severe acute respiratory syndrome-related coronavirus 2 (SARS-CoV-2) lineage with multiple spike mutations in South Africa. *medRxiv* 2020. [CrossRef]
65. Sheward, D.J.; Pushparaj, P.; Das, H.; Kim, C.; Kim, S.; Hanke, L.; Dyrdak, R.; McInerney, G.M.; Albert, J.; Murrell, B.; et al. Structural basis of Omicron neutralization by affinity-matured public antibodies. *bioRxiv* **2022**. [CrossRef]
66. Westendorf, K.; Žentelis, S.; Foster, D.; Vaillancourt, P.; Wiggin, M.; Lovett, E.; van der Lee, R.; Hendle, J.; Pustilnik, A.; Sauder, J.M.; et al. LY-CoV1404 (bebtelovimab) potently neutralizes SARS-CoV-2 variants. *bioRxiv* **2022**. [CrossRef]
67. Jawad, B.; Adhikari, P.; Podgornik, R.; Ching, W.-Y. Key Interacting Residues between RBD of SARS-CoV-2 and ACE2 Receptor: Combination of Molecular Dynamics Simulation and Density Functional Calculation. *J. Chem. Inf. Model.* **2021**, *61*, 4425–4441. [CrossRef] [PubMed]
68. Hsieh, C.-L.; Werner, A.P.; Leist, S.R.; Stevens, L.J.; Falconer, E.; Goldsmith, J.A.; Chou, C.-W.; Abiona, O.M.; West, A.; Westendorf, K.; et al. Stabilized coronavirus spike stem elicits a broadly protective antibody. *Cell Rep.* **2021**, *37*, 109929. [CrossRef] [PubMed]
69. Zhou, P.; Yuan, M.; Song, G.; Beutler, N.; Shaabani, N.; Huang, D.; He, W.-T.; Zhu, X.; Callaghan, S.; Yong, P.; et al. A human antibody reveals a conserved site on beta-coronavirus spike proteins and confers protection against SARS-CoV-2 infection. *Sci. Transl. Med.* **2022**, *14*, eabi9215. [CrossRef] [PubMed]
70. Sauer, M.M.; Tortorici, M.A.; Park, Y.-J.; Walls, A.C.; Homad, L.; Acton, O.J.; Bowen, J.E.; Wang, C.; Xiong, X.; de van der Schueren, W.; et al. Structural basis for broad coronavirus neutralization. *Nat. Struct. Mol. Biol.* **2021**, *28*, 478–486. [CrossRef]
71. Zhang, L.; Lin, D.; Sun, X.; Curth, U.; Drosten, C.; Sauerhering, L.; Becker, S.; Rox, K.; Hilgenfeld, R. Crystal structure of SARS-CoV-2 main protease provides a basis for design of improved α-ketoamide inhibitors. *Science* **2020**, *368*, 409–412. [CrossRef]
72. Jin, Z.; Du, X.; Xu, Y.; Deng, Y.; Liu, M.; Zhao, Y.; Zhang, B.; Li, X.; Zhang, L.; Peng, C.; et al. Structure of Mpro from SARS-CoV-2 and discovery of its inhibitors. *Nature* **2020**, *582*, 289–293. [CrossRef]
73. Kneller, D.W.; Zhang, Q.; Coates, L.; Louis, J.M.; Kovalevsky, A. Michaelis-like complex of SARS-CoV-2 main protease visualized by room-temperature X-ray crystallography. *IUCrJ* **2021**, *8*, 973–979. [CrossRef]

74. Crooks, G.E.; Hon, G.; Chandonia, J.-M.; Brenner, S.E. WebLogo: A Sequence Logo Generator. *Genome Res.* **2004**, *14*, 1188–1190. [CrossRef]
75. Jang, W.D.; Jeon, S.; Kim, S.; Lee, S.Y. Drugs repurposed for COVID-19 by virtual screening of 6218 drugs and cell-based assay. *Proc. Natl. Acad. Sci. USA* **2021**, *118*, e2024302118. [CrossRef]
76. Han, S.H.; Goins, C.M.; Arya, T.; Shin, W.-J.; Maw, J.; Hooper, A.; Sonawane, D.P.; Porter, M.R.; Bannister, B.E.; Crouch, R.D.; et al. Structure-Based Optimization of ML300-Derived, Noncovalent Inhibitors Targeting the Severe Acute Respiratory Syndrome Coronavirus 3CL Protease (SARS-CoV-2 3CL^{pro}). *J. Med. Chem.* **2021**, *65*, 2880–2904. [CrossRef]
77. Jacobs, J.; Grum-Tokars, V.; Zhou, Y.; Turlington, M.; Saldanha, S.A.; Chase, P.; Eggler, A.; Dawson, E.S.; Baez-Santos, Y.M.; Tomar, S.; et al. Discovery, Synthesis, And Structure-Based Optimization of a Series of *N*-(*tert*-Butyl)-2-(*N*-arylamido)-2-(pyridin-3-yl) Acetamides (ML188) as Potent Noncovalent Small Molecule Inhibitors of the Severe Acute Respiratory Syndrome Coronavirus (SARS-CoV) 3CL Protease. *J. Med. Chem.* **2012**, *56*, 534–546. [CrossRef]
78. Zhang, C.-H.; Stone, E.A.; Deshmukh, M.; Ippolito, J.A.; Ghahremanpour, M.M.; Tirado-Rives, J.; Spasov, K.A.; Zhang, S.; Takeo, Y.; Kudalkar, S.N.; et al. Potent Noncovalent Inhibitors of the Main Protease of SARS-CoV-2 from Molecular Sculpting of the Drug Perampanel Guided by Free Energy Perturbation Calculations. *ACS Central Sci.* **2021**, *7*, 467–475. [CrossRef] [PubMed]
79. von Delft, F.; Calmiano, M.; Chodera, J.; Griffen, E.; Lee, A.; London, N.; Matviuk, T.; Perry, B.; Robinson, M.; von Delft, A. A white-knuckle ride of open COVID drug discovery. *Nature* **2021**, *594*, 330–332. [CrossRef] [PubMed]
80. Achdout, H.; Aimon, A.; Bar-David, E.; Barr, H.; Ben-Shmuel, A.; Bennett, J.; Boby, M.L.; Borden, B.; Bowman, G.R.; Brun, J.; et al. Open Science Discovery of Oral Non-Covalent SARS-CoV-2 Main Protease Inhibitor Therapeutics. *bioRxiv* **2021**. [CrossRef]
81. Shionogi. Shionogi R&D Day 2021. Available online: https://www.shionogi.com/content/dam/shionogi/global/investors/ir-library/presentation/2021/e_20210929_4.pdf (accessed on 21 March 2022).
82. Sasaki, M.; Tabata, K.; Kishimoto, M.; Itakura, Y.; Kobayashi, H.; Ariizumi, T.; Uemura, K.; Toba, S.; Kusakabe, S.; Maruyama, Y.; et al. Oral administration of S-217622, a SARS-CoV-2 main protease inhibitor, decreases viral load and accelerates recovery from clinical aspects of COVID-19. *bioRxiv* **2022**. [CrossRef]
83. Vuong, W.; Khan, M.B.; Fischer, C.; Arutyunova, E.; Lamer, T.; Shields, J.; Saffran, H.A.; McKay, R.T.; van Belkum, M.J.; Joyce, M.A.; et al. Feline coronavirus drug inhibits the main protease of SARS-CoV-2 and blocks virus replication. *Nat. Commun.* **2020**, *11*, 4282. [CrossRef]
84. Anivive. Anivive Repurposes Veterinary Drug GC376 for COVID-19 and Submits Pre-IND to FDA. Available online: https://www.prnewswire.com/news-releases/anivive-repurposes-veterinary-drug-gc376-for-covid-19-and-submits-pre-ind-to-fda-301065619.html (accessed on 21 March 2022).
85. Hoffman, R.L.; Kania, R.S.; Brothers, M.A.; Davies, J.F.; Ferre, R.A.; Gajiwala, K.S.; He, M.; Hogan, R.J.; Kozminski, K.; Li, L.Y.; et al. Discovery of Ketone-Based Covalent Inhibitors of Coronavirus 3CL Proteases for the Potential Therapeutic Treatment of COVID-19. *J. Med. Chem.* **2020**, *63*, 12725–12747. [CrossRef]
86. Boras, B.; Jones, R.M.; Anson, B.J.; Arenson, D.; Aschenbrenner, L.; Bakowski, M.A.; Beutler, N.; Binder, J.; Chen, E.; Eng, H.; et al. Preclinical characterization of an intravenous coronavirus 3CL protease inhibitor for the potential treatment of COVID19. *Nat. Commun.* **2021**, *12*, 6055. [CrossRef]
87. Owen, D.R.; Allerton, C.M.N.; Anderson, A.S.; Aschenbrenner, L.; Avery, M.; Berritt, S.; Boras, B.; Cardin, R.D.; Carlo, A.; Coffman, K.J.; et al. An oral SARS-CoV-2 M ^{pro} inhibitor clinical candidate for the treatment of COVID-19. *Science* **2021**, *374*, 1586–1593. [CrossRef]
88. Pfizer. Pfizer's Novel COVID-19 Oral Antiviral Treatment Candidate Reduced Risk of Hospitalization or Death by 89% in Interim Analysis of Phase 2/3 EPIC-HR Study. Available online: https://www.pfizer.com/news/press-release/press-release-detail/pfizers-novel-covid-19-oral-antiviral-treatment-candidate (accessed on 21 March 2022).
89. Enanta. Enanta Pharmaceuticals Presents New Data for Edp-235, Its Lead Oral Protease Inhibitor Designed for the Treatment of COVID-19, at The ISIRV–Who Virtual Conference 2021. Available online: https://www.enanta.com/investors/news-releases/press-release/2021/Enanta-Pharmaceuticals-Presents-New-Data-for-EDP-235-its-Lead-Oral-Protease-Inhibitor-Designed-for-the-Treatment-of-COVID-19-at-the-ISIRVWHO-Virtual-Conference-2021/default.aspx (accessed on 21 March 2022).
90. Cocrystal Pharam. Cocrystal Pharma Selects Two Lead Antiviral Drug Candidates for Its COVID-19 Oral Drug Program. Available online: https://www.cocrystalpharma.com/news/press-releases/detail/139/cocrystal-pharma-selects-two-lead-antiviral-drug-candidates (accessed on 21 March 2022).
91. Vandyck, K.; Abdelnabi, R.; Gupta, K.; Jochmans, D.; Jekle, A.; Deval, J.; Misner, D.; Bardiot, D.; Foo, C.S.; Liu, C.; et al. ALG-097111, a potent and selective SARS-CoV-2 3-chymotrypsin-like cysteine protease inhibitor exhibits in vivo efficacy in a Syrian Hamster model. *Biochem. Biophys. Res. Commun.* **2021**, *555*, 134–139. [CrossRef]
92. Snijder, E.J.; Decroly, E.; Ziebuhr, J. The Nonstructural Proteins Directing Coronavirus RNA Synthesis and Processing. *Adv. Virus Res.* **2016**, *96*, 59–126. [CrossRef] [PubMed]
93. Subissi, L.; Posthuma, C.C.; Collet, A.; Zevenhoven-Dobbe, J.C.; Gorbalenya, A.E.; Decroly, E.; Snijder, E.J.; Canard, B.; Imbert, I. One severe acute respiratory syndrome coronavirus protein complex integrates processive RNA polymerase and exonuclease activities. *Proc. Natl. Acad. Sci. USA* **2014**, *111*, E3900–E3909. [CrossRef] [PubMed]
94. Hillen, H.S. Structure and function of SARS-CoV-2 polymerase. *Curr. Opin. Virol.* **2021**, *48*, 82–90. [CrossRef]
95. Hillen, H.S.; Kokic, G.; Farnung, L.; Dienemann, C.; Tegunov, D.; Cramer, P. Structure of replicating SARS-CoV-2 polymerase. *Nature* **2020**, *584*, 154–156. [CrossRef] [PubMed]

96. Geraghty, R.; Aliota, M.; Bonnac, L. Broad-Spectrum Antiviral Strategies and Nucleoside Analogues. *Viruses* **2021**, *13*, 667. [CrossRef]
97. Sofia, M.J. *Nucleosides and Nucleotides for the Treatment of Viral Diseases. Annual Reports in Medicinal Chemistry*; Academic Press: Cambridge, MA, USA, 2014; Volume 49, pp. 221–247. ISSN 0065-7743. ISBN 9780128001677. [CrossRef]
98. Gordon, C.J.; Tchesnokov, E.P.; Feng, J.Y.; Porter, D.P.; Götte, M. The antiviral compound remdesivir potently inhibits RNA-dependent RNA polymerase from Middle East respiratory syndrome coronavirus. *J. Biol. Chem.* **2020**, *295*, 4773–4779. [CrossRef]
99. Gordon, C.J.; Tchesnokov, E.P.; Woolner, E.; Perry, J.K.; Feng, J.Y.; Porter, D.P.; Götte, M. Remdesivir is a direct-acting antiviral that inhibits RNA-dependent RNA polymerase from severe acute respiratory syndrome coronavirus 2 with high potency. *J. Biol. Chem.* **2020**, *295*, 6785–6797. [CrossRef]
100. Tchesnokov, E.P.; Gordon, C.J.; Woolner, E.; Kocinkova, D.; Perry, J.K.; Feng, J.Y.; Porter, D.P.; Götte, M. Template-dependent inhibition of coronavirus RNA-dependent RNA polymerase by remdesivir reveals a second mechanism of action. *J. Biol. Chem.* **2020**, *295*, 16156–16165. [CrossRef]
101. Warren, T.K.; Wells, J.; Panchal, R.G.; Stuthman, K.S.; Garza, N.L.; Van Tongeren, S.A.; Dong, L.; Retterer, C.J.; Eaton, B.P.; Pegoraro, G.; et al. Protection against filovirus diseases by a novel broad-spectrum nucleoside analogue BCX4430. *Nature* **2014**, *508*, 402–405. [CrossRef]
102. Good, S.S.; Westover, J.; Jung, K.H.; Zhou, X.-J.; Moussa, A.; La Colla, P.; Collu, G.; Canard, B.; Sommadossi, J.-P. AT-527, a Double Prodrug of a Guanosine Nucleotide Analog, Is a Potent Inhibitor of SARS-CoV-2 In Vitro and a Promising Oral Antiviral for Treatment of COVID-19. *Antimicrob. Agents Chemother.* **2021**, *65*, e02479-20. [CrossRef]
103. Jockusch, S.; Tao, C.; Li, X.; Chien, M.; Kumar, S.; Morozova, I.; Kalachikov, S.; Russo, J.J.; Ju, J. Sofosbuvir terminated RNA is more resistant to SARS-CoV-2 proofreader than RNA terminated by Remdesivir. *Sci. Rep.* **2020**, *10*, 16577. [CrossRef] [PubMed]
104. Yuan, C.; Goonetilleke, E.C.; Unarta, I.C.; Huang, X. Incorporation efficiency and inhibition mechanism of 2′-substituted nucleotide analogs against SARS-CoV-2 RNA-dependent RNA polymerase. *Phys. Chem. Chem. Phys.* **2021**, *23*, 20117–20128. [CrossRef] [PubMed]
105. Gordon, C.J.; Tchesnokov, E.P.; Schinazi, R.F.; Götte, M. Molnupiravir promotes SARS-CoV-2 mutagenesis via the RNA template. *J. Biol. Chem.* **2021**, *297*, 100770. [CrossRef]
106. Kabinger, F.; Stiller, C.; Schmitzová, J.; Dienemann, C.; Kokic, G.; Hillen, H.S.; Höbartner, C.; Cramer, P. Mechanism of molnupiravir-induced SARS-CoV-2 mutagenesis. *Nat. Struct. Mol. Biol.* **2021**, *28*, 740–746. [CrossRef] [PubMed]
107. Sheahan, T.P.; Sims, A.C.; Zhou, S.; Graham, R.L.; Pruijssers, A.J.; Agostini, M.L.; Leist, S.R.; Schäfer, A.; Dinnon, K.H., 3rd; Stevens, L.J.; et al. An orally bioavailable broad-spectrum antiviral inhibits SARS-CoV-2 in human airway epithelial cell cultures and multiple coronaviruses in mice. *Sci. Transl. Med.* **2020**, *12*, eabb5883. [CrossRef] [PubMed]
108. Shannon, A.; Selisko, B.; Le, N.-T.-T.; Huchting, J.; Touret, F.; Piorkowski, G.; Fattorini, V.; Ferron, F.; Decroly, E.; Meier, C.; et al. Rapid incorporation of Favipiravir by the fast and permissive viral RNA polymerase complex results in SARS-CoV-2 lethal mutagenesis. *Nat. Commun.* **2020**, *11*, 4682. [CrossRef] [PubMed]
109. Beigel, J.H.; Tomashek, K.M.; Dodd, L.E.; Mehta, A.K.; Zingman, B.S.; Kalil, A.C.; Hohmann, E.; Chu, H.Y.; Luetkemeyer, A.; Kline, S.; et al. Remdesivir for the Treatment of COVID-19—preliminary report. *N. Engl. J. Med.* **2020**, *383*, 1813–1826. [CrossRef]
110. Spinner, C.D.; Gottlieb, R.L.; Criner, G.J.; López, J.R.A.; Cattelan, A.M.; Viladomiu, A.S.; Ogbuagu, O.; Malhotra, P.; Mullane, K.M.; Castagna, A.; et al. Effect of Remdesivir vs Standard Care on Clinical Status at 11 Days in Patients with Moderate COVID-19: A Randomized Clinical Trial. *JAMA* **2020**, *324*, 1048–1057. [CrossRef]
111. Goldman, J.D.; Lye, D.C.; Hui, D.S.; Marks, K.M.; Bruno, R.; Montejano, R.; Spinner, C.D.; Galli, M.; Ahn, M.-Y.; Nahass, R.G.; et al. Remdesivir for 5 or 10 Days in Patients with Severe COVID-19. *N. Engl. J. Med.* **2020**, *383*, 1827–1837. [CrossRef]
112. Wang, Y.; Zhang, D.; Du, G.; Du, R.; Zhao, J.; Jin, Y.; Fu, S.; Gao, L.; Cheng, Z.; Lu, Q.; et al. Remdesivir in adults with severe COVID-19: A randomised, double-blind, placebo-controlled, multicentre trial. *Lancet* **2020**, *395*, 1569–1578. [CrossRef]
113. WHO Solidarity Trial Consortium. Repurposed Antiviral Drugs for COVID-19—Interim WHO Solidarity Trial Results. *N. Engl. J. Med.* **2021**, *384*, 497–511. [CrossRef] [PubMed]
114. Al-Abdouh, A.; Bizanti, A.; Barbarawi, M.; Jabri, A.; Kumar, A.; Fashanu, O.E.; Khan, S.U.; Zhao, D.; Antar, A.A.; Michos, E.D. Remdesivir for the treatment of COVID-19: A systematic review and meta-analysis of randomized controlled trials. *Contemp. Clin. Trials* **2021**, *101*, 106272. [CrossRef] [PubMed]
115. Wang, Y.; Li, P.; Solanki, K.; Li, Y.; Ma, Z.; Peppelenbosch, M.P.; Baig, M.S.; Pan, Q. Viral polymerase binding and broad-spectrum antiviral activity of molnupiravir against human seasonal coronaviruses. *Virology* **2021**, *564*, 33–38. [CrossRef] [PubMed]
116. Rosenke, K.; Hansen, F.; Schwarz, B.; Feldmann, F.; Haddock, E.; Rosenke, R.; Barbian, K.; Meade-White, K.; Okumura, A.; Leventhal, S.; et al. Orally delivered MK-4482 inhibits SARS-CoV-2 replication in the Syrian hamster model. *Nat. Commun.* **2021**, *12*, 2295. [CrossRef]
117. Abdelnabi, R.; Foo, C.S.; De Jonghe, S.; Maes, P.; Weynand, B.; Neyts, J. Molnupiravir Inhibits Replication of the Emerging SARS-CoV-2 Variants of Concern in a Hamster Infection Model. *J. Infect. Dis.* **2021**, *224*, 749–753. [CrossRef]
118. Cox, R.M.; Wolf, J.D.; Plemper, R.K. Therapeutically administered ribonucleoside analogue MK-4482/EIDD-2801 blocks SARS-CoV-2 transmission in ferrets. *Nat. Microbiol.* **2021**, *6*, 11–18. [CrossRef]
119. Zhou, S.; Hill, C.S.; Sarkar, S.; Tse, L.V.; Woodburn, B.M.D.; Schinazi, R.F.; Sheahan, T.P.; Baric, R.S.; Heise, M.T.; Swanstrom, R. β-d-N 4-hydroxycytidine inhibits SARS-CoV-2 through lethal mutagenesis but is also mutagenic to mammalian cells. *J. Infect. Dis.* **2021**, *224*, 415–419. [CrossRef]

120. Jayk Bernal, A.; Gomes da Silva, M.M.; Musungaie, D.B.; Kovalchuk, E.; Gonzalez, A.; Delos Reyes, V.; Martín-Quirós, A.; Caraco, Y.; Williams-Diaz, A.; Brown, M.L.; et al. Molnupiravir for Oral Treatment of COVID-19 in Nonhospitalized Patients. *N. Engl. J. Med.* **2022**, *386*, 509–520. [CrossRef]
121. BioCryst. BioCryst Provides Update on Galidesivir Program. Available online: https://ir.biocryst.com/news-releases/news-release-details/biocryst-provides-update-galidesivir-program (accessed on 21 March 2022).
122. Atea. Atea Pharmaceuticals Provides Update and Topline Results for Phase 2 MOONSONG Trial Evaluating AT-527 in the Outpatient Setting. Available online: https://ir.ateapharma.com/news-releases/news-release-details/atea-pharmaceuticals-provides-update-and-topline-results-phase-2 (accessed on 21 March 2022).
123. Shrestha, D.B.; Budhathoki, P.; Khadka, S.; Shah, P.B.; Pokharel, N.; Rashmi, P. Favipiravir versus other antiviral or standard of care for COVID-19 treatment: A rapid systematic review and meta-analysis. *Virol. J.* **2020**, *17*, 141. [CrossRef]
124. Sluis-Cremer, N. Future of nonnucleoside reverse transcriptase inhibitors. *Proc. Natl. Acad. Sci. USA* **2018**, *115*, 637–638. [CrossRef]
125. Kati, W.; Koev, G.; Irvin, M.; Beyer, J.; Liu, Y.; Krishnan, P.; Reisch, T.; Mondal, R.; Wagner, R.; Molla, A.; et al. *In Vitro* Activity and Resistance Profile of Dasabuvir, a Nonnucleoside Hepatitis C Virus Polymerase Inhibitor. *Antimicrob. Agents Chemother.* **2015**, *59*, 1505–1511. [CrossRef] [PubMed]
126. Gironda-Martínez, A.; Donckele, E.J.; Samain, F.; Neri, D. DNA-Encoded Chemical Libraries: A Comprehensive Review with Succesful Stories and Future Challenges. *ACS Pharmacol. Transl. Sci.* **2021**, *4*, 1265–1279. [CrossRef] [PubMed]
127. Gorgulla, C.; Boeszoermenyi, A.; Wang, Z.-F.; Fischer, P.D.; Coote, P.W.; Das, K.M.P.; Malets, Y.S.; Radchenko, D.S.; Moroz, Y.S.; Scott, D.A.; et al. An open-source drug discovery platform enables ultra-large virtual screens. *Nature* **2020**, *580*, 663–668. [CrossRef] [PubMed]
128. Gorgulla, C.; Das, K.M.P.; Leigh, K.E.; Cespugli, M.; Fischer, P.D.; Wang, Z.-F.; Tesseyre, G.; Pandita, S.; Shnapir, A.; Calderaio, A.; et al. A multi-pronged approach targeting SARS-CoV-2 proteins using ultra-large virtual screening. *iScience* **2021**, *24*, 102021. [CrossRef]
129. Lei, J.; Kusov, Y.; Hilgenfeld, R. Nsp3 of coronaviruses: Structures and functions of a large multi-domain protein. *Antivir. Res.* **2018**, *149*, 58–74. [CrossRef]
130. Báez-Santos, Y.M.; St. John, S.E.; Mesecar, A.D. The SARS-coronavirus papain-like protease: Structure, function and inhibition by designed antiviral compounds. *Antiviral. Res.* **2015**, *115*, 21–38. [CrossRef]
131. Gao, X.; Qin, B.; Chen, P.; Zhu, K.; Hou, P.; Wojdyla, J.A.; Wang, M.; Cui, S. Crystal structure of SARS-CoV-2 papain-like protease. *Acta Pharm. Sin. B* **2021**, *11*, 237–245. [CrossRef]
132. Ye, Y.; Akutsu, M.; Reyes-Turcu, F.; I Enchev, R.; Wilkinson, K.D.; Komander, D. Polyubiquitin binding and cross-reactivity in the USP domain deubiquitinase USP21. *EMBO Rep.* **2011**, *12*, 350–357. [CrossRef]
133. Harcourt, B.H.; Jukneliene, D.; Kanjanahaluethai, A.; Bechill, J.; Severson, K.M.; Smith, C.M.; Rota, P.A.; Baker, S.C. Identification of Severe Acute Respiratory Syndrome Coronavirus Replicase Products and Characterization of Papain-Like Protease Activity. *J. Virol.* **2004**, *78*, 13600–13612. [CrossRef]
134. Woo, P.C.Y.; Huang, Y.; Lau, S.K.P.; Yuen, K.-Y. Coronavirus Genomics and Bioinformatics Analysis. *Viruses* **2010**, *2*, 1804–1820. [CrossRef]
135. Osipiuk, J.; Azizi, S.-A.; Dvorkin, S.; Endres, M.; Jedrzejczak, R.; Jones, K.A.; Kang, S.; Kathayat, R.S.; Kim, Y.; Lisnyak, V.G.; et al. Structure of papain-like protease from SARS-CoV-2 and its complexes with non-covalent inhibitors. *Nat. Commun.* **2021**, *12*, 743. [CrossRef] [PubMed]
136. Shin, D.; Mukherjee, R.; Grewe, D.; Bojkova, D.; Baek, K.; Bhattacharya, A.; Schulz, L.; Widera, M.; Mehdipour, A.R.; Tascher, G.; et al. Papain-like protease regulates SARS-CoV-2 viral spread and innate immunity. *Nature* **2020**, *575*, 210–216. [CrossRef] [PubMed]
137. Lindner, H.A.; Fotouhi-Ardakani, N.; Lytvyn, V.; Lachance, P.; Sulea, T.; Meénard, R. The Papain-Like Protease from the Severe Acute Respiratory Syndrome Coronavirus Is a Deubiquitinating Enzyme. *J. Virol.* **2005**, *79*, 15199–15208. [CrossRef] [PubMed]
138. Mielech, A.M.; Kilianski, A.; Baez-Santos, Y.M.; Mesecar, A.D.; Baker, S.C. MERS-CoV papain-like protease has deISGylating and deubiquitinating activities. *Virology* **2014**, *450-451*, 64–70. [CrossRef]
139. Devaraj, S.G.; Wang, N.; Chen, Z.; Chen, Z.; Tseng, M.; Barretto, N.; Lin, R.; Peters, C.J.; Tseng, C.-T.K.; Baker, S.C.; et al. Regulation of IRF-3-dependent Innate Immunity by the Papain-like Protease Domain of the Severe Acute Respiratory Syndrome Coronavirus. *J. Biol. Chem.* **2007**, *282*, 32208–32221. [CrossRef]
140. Békés, M.; van der Heden van Noort, G.J.; Ekkebus, R.; Ovaa, H.; Huang, T.T.; Lima, C.D. Recognition of Lys48-Linked Di-ubiquitin and Deubiquitinating Activities of the SARS Coronavirus Papain-like Protease. *Mol. Cell* **2016**, *62*, 572–585. [CrossRef]
141. Ghosh, A.K.; Takayama, J.; Aubin, Y.; Ratia, K.; Chaudhuri, R.; Baez, Y.; Sleeman, K.; Coughlin, M.; Nichols, D.B.; Mulhearn, D.C.; et al. Structure-Based Design, Synthesis, and Biological Evaluation of a Series of Novel and Reversible Inhibitors for the Severe Acute Respiratory Syndrome—Coronavirus Papain-Like Protease. *J. Med. Chem.* **2009**, *52*, 5228–5240. [CrossRef]
142. Ghosh, A.K.; Takayama, J.; Rao, K.V.; Ratia, K.; Chaudhuri, R.; Mulhearn, D.C.; Lee, H.; Nichols, D.B.; Baliji, S.; Baker, S.C.; et al. Severe Acute Respiratory Syndrome Coronavirus Papain-like Novel Protease Inhibitors: Design, Synthesis, Protein−Ligand X-ray Structure and Biological Evaluation. *J. Med. Chem.* **2010**, *53*, 4968–4979. [CrossRef]
143. Shen, Z.; Ratia, K.; Cooper, L.; Kong, D.; Lee, H.; Kwon, Y.; Li, Y.; Alqarni, S.; Huang, F.; Dubrovskyi, O.; et al. Design of SARS-CoV-2 PLpro Inhibitors for COVID-19 Antiviral Therapy Leveraging Binding Cooperativity. *J. Med. Chem.* **2021**, *65*, 2940–2955. [CrossRef]

144. Shan, H.; Liu, J.; Shen, J.; Dai, J.; Xu, G.; Lu, K.; Han, C.; Wang, Y.; Xu, X.; Tong, Y.; et al. Development of potent and selective inhibitors targeting the papain-like protease of SARS-CoV-2. *Cell Chem. Biol.* **2021**, *28*, 855–865.e9. [CrossRef]
145. Ma, C.; Sacco, M.D.; Xia, Z.; Lambrinidis, G.; Townsend, J.A.; Hu, Y.; Meng, X.; Szeto, T.; Ba, M.; Zhang, X.; et al. Discovery of SARS-CoV-2 Papain-like Protease Inhibitors through a Combination of High-Throughput Screening and a FlipGFP-Based Reporter Assay. *ACS Central Sci.* **2021**, *7*, 1245–1260. [CrossRef] [PubMed]
146. Rut, W.; Lv, Z.; Zmudzinski, M.; Patchett, S.; Nayak, D.; Snipas, S.J.; El Oualid, F.; Huang, T.T.; Bekes, M.; Drag, M.; et al. Activity profiling and crystal structures of inhibitor-bound SARS-CoV-2 papain-like protease: A framework for anti–COVID-19 drug design. *Sci. Adv.* **2020**, *6*, eabd4596. [CrossRef] [PubMed]
147. Liu, N.; Zhang, Y.; Lei, Y.; Wang, R.; Zhan, M.; Liu, J.; An, Y.; Zhou, Y.; Zhan, J.; Yin, F.; et al. Design and Evaluation of a Novel Peptide–Drug Conjugate Covalently Targeting SARS-CoV-2 Papain-like Protease. *J. Med. Chem.* **2022**, *65*, 876–884. [CrossRef] [PubMed]
148. Weglarz-Tomczak, E.; Tomczak, J.M.; Talma, M.; Burda-Grabowska, M.; Giurg, M.; Brul, S. Identification of ebselen and its analogues as potent covalent inhibitors of papain-like protease from SARS-CoV-2. *Sci. Rep.* **2021**, *11*, 3640. [CrossRef]
149. Ma, C.; Hu, Y.; Townsend, J.A.; Lagarias, P.I.; Marty, M.T.; Kolocouris, A.; Wang, J. Ebselen, Disulfiram, Carmofur, PX-12, Tideglusib, and Shikonin Are Nonspecific Promiscuous SARS-CoV-2 Main Protease Inhibitors. *ACS Pharmacol. Transl. Sci.* **2020**, *3*, 1265–1277. [CrossRef]
150. Ma, C.; Wang, J. Validation and Invalidation of SARS-CoV-2 Papain-like Protease Inhibitors. *ACS Pharmacol. Transl. Sci.* **2022**, *5*, 102–109. [CrossRef]
151. Cathcart, A.L.; Havenar-Daughton, C.; Lempp, F.A.; Ma, D.; Schmid, M.A.; Agostini, M.L.; Guarino, B.; Rosen, L.E.; Tucker, H.; Dillen, J.; et al. The dual function monoclonal antibodies VIR-7831 and VIR-7832 demonstrate potent in vitro and in vivo activity against SARS-CoV-2. *bioRxiv* **2021**. [CrossRef]
152. NIH. Biden Administration to Invest $3 Billion from American Rescue Plan as Part of COVID-19 Antiviral Development Strategy. Available online: https://www.niaid.nih.gov/news-events/biden-administration-invest-3-billion-american-rescue-plan-part-covid-19-antiviral (accessed on 21 March 2022).

viruses

MDPI

Review

Rational Design of a Skin- and Neuro-Attenuated Live Varicella Vaccine: A Review and Future Perspectives

Wei Wang [1], Dequan Pan [1], Tong Cheng [1,*] and Hua Zhu [2,*]

[1] State Key Laboratory of Molecular Vaccinology and Molecular Diagnostics, National Institute of Diagnostics and Vaccine Development in Infectious Diseases, School of Life Sciences, School of Public Health, Xiamen University, Xiamen 361102, China; lukewang@xmu.edu.cn (W.W.); dequanpan@stu.xmu.edu.cn (D.P.)

[2] Department of Microbiology and Molecular Genetics, New Jersey Medical School, Rutgers University, Newark, NJ 070101, USA

* Correspondence: tcheng@xmu.edu.cn (T.C.); zhuhu@njms.rutgers.edu (H.Z.)

Abstract: Primary varicella-zoster virus (VZV) infection causes varicella, which remains a prominent public health concern in children. Current varicella vaccines adopt the live-attenuated Oka strain, vOka, which retains the ability to infect neurons, establish latency and reactivate, leading to vaccine-associated zoster in some vaccinees. Therefore, it is necessary to develop a safer next-generation varicella vaccine to help reduce vaccine hesitancy. This paper reviews the discovery and identification of the skin- and neuro-tropic factor, the open reading frame 7 (*ORF7*) of VZV, as well as the development of a skin- and neuro-attenuated live varicella vaccine comprising an *ORF7*-deficient mutant, v7D. This work could provide insights into the research of novel virus vaccines based on functional genomics and reverse genetics.

Keywords: varicella-zoster virus; varicella; VZV; neuro-attenuated; ORF7; vaccine

Citation: Wang, W.; Pan, D.; Cheng, T.; Zhu, H. Rational Design of a Skin- and Neuro-Attenuated Live Varicella Vaccine: A Review and Future Perspectives. *Viruses* **2022**, *14*, 848. https://doi.org/10.3390/v14050848

Academic Editors: Charles Grose, Ravi Mahalingam and Joel Rovnak

Received: 24 March 2022
Accepted: 18 April 2022
Published: 20 April 2022

1. Introduction

Varicella zoster virus (VZV) is a neurotropic human α-herpesvirus that causes varicella (or chickenpox) during the primary infection, and then establishes lifelong latency in neurons of the dorsal root, cranial, enteric and autonomic ganglia [1]. Years later, the dormant VZV can reactivate to cause herpes zoster (HZ, or shingles) [2]. Varicella is a common and highly contagious skin disease in children, but its symptoms are usually mild and self-limiting. HZ is an acute infectious disease commonly seen in the elderly and in individuals with human immunodeficiency virus (HIV) infection, cancer, or those treated with corticosteroids and other immunosuppressive agents. The HZ incidence rate increases significantly with age. HZ is characterized by painful, unilateral clustered vesicular rashes, and is sometimes complicated by post-herpetic neuralgia (PHN), a refractory chronic pain syndrome persisting for months or longer after the HZ rash has resolved [2]. In addition, VZV infection may cause serious, life-threatening complications, and even death, in newborns, adults and immunocompromised or immunodeficient individuals [3].

2. Current Status of Varicella Vaccine

Vaccination is a cost-effective way to prevent varicella. Currently, only live-attenuated vaccines are available for protection against varicella, and the most widely used varicella vaccines employ the Oka vaccine strain of VZV (vOka). vOka was developed by Michiaki Takahashi in 1974 through serial passages of the wild-type parent Oka (pOka) strain of VZV in cell cultures [4,5]. This method resulted in a mixture of virus haplotypes, but the mechanism of vOka attenuation is still not well-characterized [6,7]. The vOka varicella vaccine has been introduced into routine childhood immunization programs in many countries [8]. Over the last few decades, studies have shown that vOka varicella vaccines are generally safe and effective in preventing varicella-related morbidity and mortality in healthy children worldwide [9–12]. Furthermore, varicella vaccination results in a decreased incidence of HZ in children, and provides protection for susceptible adults or immunosuppressed patients who are at high risk of fatal varicella [12,13].

However, there have been concerns raised about the use of vOka varicella vaccines following the analysis of adverse events (AEs) reported in post-market surveillance [14–22]. One notable AE is vOka-mediated HZ, which is associated with vOka's ability to infect human neurons, establish latency and subsequently reactivate [23,24]. Although lower HZ incidences have been reported in vaccinated children than in unvaccinated ones by several studies [18,25,26], vOka is associated with a substantial proportion of post-vaccination HZ cases. For example, in three safety surveillance studies of vOka-Varivax from Merck, polymerase chain reaction (PCR) analysis identified vOka in 40.7% (48/118) [14], 35.2% (57/162) [15] and 47.1% (8/17) [16], respectively, of all clinical specimens collected from the reported post-vaccination HZ cases; the remaining specimens were positive for wild-type VZV, tested negative, or were considered inadequate for testing. Furthermore, there have been a few cases of vOka-mediated severe meningitis which occurred following the onset of HZ in both once- and twice-immunized children [22]. The vOka varicella vaccine, which is recommended for children aged 12 months to 12 years, has been on the market for only about 30 years, while HZ mainly occurs in adults over 50 years old. Therefore, it will take decades to confirm whether varicella vaccination causes a rise in the incidence of vOka-mediated HZ among the older population. More importantly, it is desirable to develop a safer varicella vaccine that will help to improve the acceptance of varicella vaccination, in order to better establish herd immunity against varicella.

To date, concerns over the safety of vOka have led to many studies that have evaluated the use of inactivated viruses or recombinant viral subunits as vaccines against VZV infection [27,28]. Vaccine candidates of both types were generally safe, immunogenic and effective for the prevention of varicella in healthy or immunocompromised adults, but extensive efforts are still required to further evaluate their use in healthy children. On the other hand, advances in reverse genetics of VZV have enabled the identification and manipulation of virulence factors [29–31], thus providing an opportunity for rational design of a novel live-attenuated varicella vaccine.

3. Identification of *ORF7* as a Skin- and Neuro-Tropic Factor of VZV

VZV induces pathological abnormalities, mainly in human skin and nerve tissues, during infection. Characterizing the crucial genes related to the skin- and neuro-tropisms of VZV will aid in the understanding of VZV pathogenesis, and it will also facilitate the development of improved vaccines. However, VZV is strictly human-specific, and has a highly cell-associated nature in cultured cells. These qualities make it challenging to study VZV gene functions and distinguish VZV from the other α-herpesviruses, which can infect a wide range of host species and generate large amounts of cell-free particles, leading to a much earlier and clearer understanding of their replication and pathogenesis compared to VZV. Recently, thanks to advances in molecular virology and reverse genetics, VZV gene function in specific cells and tissues has been characterized extensively by mutagenesis of the VZV genome, through the use of human tissues cultured ex vivo or implanted in severe combined immunodeficient (SCID) mice in vivo (Figure 1) [32–34].

We aimed to identify the key viral virulence and tissue-tropic factor(s) for the rational design of a next-generation, safer, live-attenuated VZV vaccine. For this purpose, we used bacterial artificial chromosome (BAC) technology to develop an efficient method for generating recombinant VZV mutants [31,35]. To better monitor the growth of mutant viruses in vitro and in vivo, a BAC clone of the wild-type VZV pOka strain, carrying both green fluorescence protein (GFP) and luciferase (luc) reporter genes, was constructed, and PCR methods were utilized to replace each open reading frame (ORF) in the BAC clone with a selectable marker (e.g., the *galK* gene), via homologous recombination in *E. coli* [31,32]. The resulting BACs were isolated and transfected into human melanoma cells. A total of 26 single-ORF deletion mutants expressing GFP and luc were reconstituted, and then screened in ex vivo human skin organ cultures (SOCs) [32]. In addition to previously reported skin-tropic genes such as *ORF10* [36], *ORF14* [37] and *ORF47* [38], we identified *ORF7* as a novel gene essential for VZV replication and spread in human skin [32]. Subsequently, to identify any potential neurotropic factors of VZV, we screened 18 ORF deletion mutants, that can grow normally in either epithelial cells or fibroblasts, in differentiated neuroblastoma (SH-SY5Y) cells and human embryonic stem cell (hESC)-derived neurons cultured in vitro [33]. Among them, only the *ORF7* deletion mutant showed a severe growth defect in these neuronal cell models, and its neuro-attenuated phenotype was further confirmed in human fetal dorsal root ganglia (DRGs), both ex vivo and in vivo [33]. Therefore, we identified *ORF7* as the first, and currently only, known full-length neurotropic gene of VZV. Altogether, our work identified *ORF7* as a skin- and neurotropic factor of VZV, and we found that while retaining replication competence in other cell lines, an *ORF7*-deficient mutant that is defective in spread in human skin and neurons would be a potential live-attenuated vaccine candidate against VZV infection.

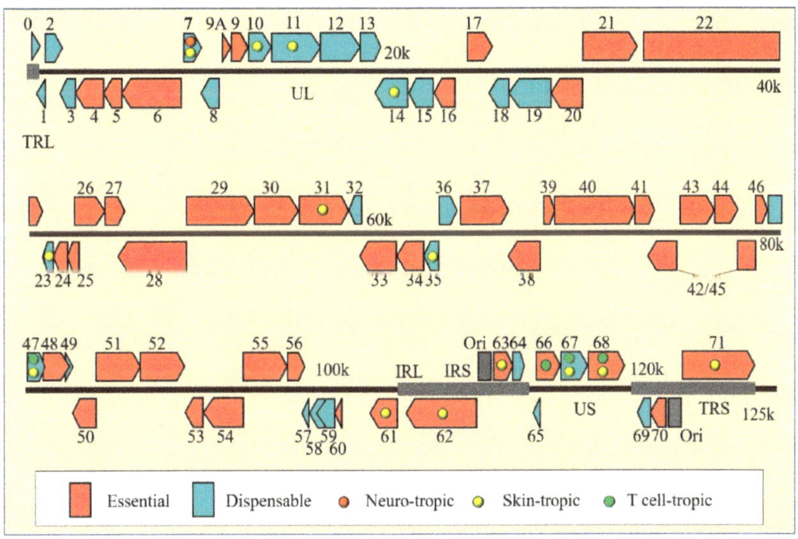

Figure 1. Genome-wide map of VZV genes required for virus replication and tissue tropism. Genomic arrangement of the open reading frames (ORFs) is based on the complete genome sequence of VZV pOka strain. VZV ORFs are colored according to their essentiality for virus growth in human melanoma cells, epithelial cells or fibroblasts in vitro. Essential and dispensable ORFs with functional domains that determine VZV tropisms are labeled with differently colored filled circles, depending on their specific function in human dorsal root ganglia (DRG), skin and T cells in vivo. The image is adapted with modifications from reference [32].

4. Development of an *ORF7*-Deficient, Live-Attenuated Varicella Vaccine, v7D

To construct the *ORF7*-deficient VZV vaccine virus, we employed a stop-codon mutation strategy to minimize changes to the original genome sequence of the wild-type virus [39]. Firstly, the 11 bp region downstream of the ATG start codon of *ORF7* was replaced with a three-frame stop-codon cassette in the BAC clone of the VZV wild-type pOka strain, for the abrogation of *ORF7* expression. Then, the loxP-flanked BAC vector containing the GFP expression cassette was excised from the viral genome by co-transfecting the purified BAC clone with a Cre expression vector into human fibroblasts, thus reconstituting the *ORF7*-deficient candidate vaccine virus, v7D. Compared with the original wild-type strain of VZV-pOka, there was no difference in the genome sequence of v7D, except for the stop-codon mutation within *ORF7* and a single 34 bp loxP site in the noncoding region between *ORF60* and *ORF61*. Whole-genome sequencing of serially passaged v7D showed that its genome was stable for at least 25 passages in cell culture.

To thoroughly confirm the skin- and neuro-attenuation phenotype of v7D, a variety of cell and animal models currently available for evaluation of VZV infection and tissue tropism were used [39]. Similar to the reported *ORF7* deletion mutant [32,33], the lack of *ORF7* expression prevents the replication and cell-to-cell spread of v7D in human skin and neuronal cells, both in vitro and in SCID-hu mice in vivo. In contrast, wild-type VZV and vOka established a robust lytic infection in these cells. In addition, v7D failed to infect DRGs in the guinea pig and cotton rat models of VZV infection, and showed no intrathalamic neurovirulence or repeated-dose systemic toxicity in non-human primates. These preclinical toxicity evaluations show that v7D is highly skin- and neuro-attenuated, and should have a reduced risk of vaccine-associated complications, thus meeting the critical safety criteria for a next-generation live-attenuated varicella vaccine.

Despite the attenuation phenotype, v7D exhibited wild-type growth in MRC-5 human fetal lung fibroblasts, allowing for the production of v7D as a live vaccine [39]. Furthermore, similar to vOka, v7D retains full lymphotropism, which may help the replication and dissemination of vaccine viruses in vivo, and can functionally activate human dendritic cells (DCs) to process and present viral antigens to T cells in vitro [39]. These properties indicate that v7D has the ability to function as a live vaccine and stimulate human adaptive immunities against VZV infection.

VZV only infects humans, and thus VZV vaccine studies are limited to the use of common experimental animal models, which lack virus–host interactions under natural infection scenarios, to obtain immunogenicity data as indirect clues to VZV vaccine efficacy in humans. For preclinical immunogenicity evaluation of v7D, small animal models, including mice, rats, guinea pigs and rabbits, as well as non-human primate models, were used [39]. Three subcutaneous immunizations with the same doses of v7D and vOka induced similar and long-lasting VZV-specific antibody and T-cell responses, suggesting that the immunogenicity of v7D was non-inferior to that of vOka in small animals. v7D was also immunogenic in non-human primates. These preclinical toxicity and immunogenicity data have supported the initiation of a first-in-human phase I trial of the v7D-based, live-attenuated varicella vaccine in China (NO. ChiCTR1900022284).

5. Future Directions

Our preclinical data suggest that v7D is a promising, safer, live vaccine candidate against varicella. Nevertheless, many questions remain and should be addressed in future research.

5.1. Safety of the v7D Varicella Vaccine in Humans

Through future clinical observations, the v7D vaccine is expected to show several safety advantages compared to the vOka vaccine, including lower frequency of both varicella-like rashes and vaccine-associated HZ cases. However, fully confirming the long-term safety of v7D will require decades of observation for possible reversion to wild-type virulence, latency and reactivation of the vaccine virus, and revealing all AEs associated with v7D vaccination.

The possibility of reversion to a wild-type phenotype always raises concerns over the use of live-attenuated vaccines. However, like all herpesviruses, the VZV DNA genome is highly stable, and we have not observed the occurrence of wild-type reversion of v7D during serial cell passaging and long-term in vivo experiments. An early study showed that an *ORF61* mutant with stop-codon mutation reverted to wild-type virus in cell culture, but this virus retained the coding sequence of a RING finger domain of pORF61, which could have contributed to its wild-type reversion [40]. For v7D, the stop-codon mutation abrogated the expression of the entire *ORF7*, which could be beneficial to the stability of the virus genome and its attenuation phenotype. Nevertheless, it is necessary to continue to evaluate the risk of wild-type reversion of v7D in subsequent studies.

In addition, we do not know whether v7D would establish neuronal latency, although v7D cannot establish an initial lytic infection in human neurons and is much less likely to reactivate from possible latency, if at all. Therefore, it will be an important research question to determine whether v7D establishes latency and expresses VZV latency-associated RNA transcripts (VLT) in human neurons, based on the in vitro and in vivo models that are available to study VZV latency.

5.2. Immunogenicity and Efficacy of the v7D Varicella Vaccine in Humans

v7D has distinct properties of attenuation that may enable it to induce different patterns of immune response, compared to that of vOka, in humans. Therefore, with reference to an early study of vOka [41], the dynamics of the immunity induced by v7D of different doses should be investigated in clinical trials to determine an appropriate dose for its clinical use. Furthermore, since a routine two-dose varicella vaccination schedule for children that has been applied in several countries has been shown to provide better protection against varicella, further reducing breakthrough cases and the risk of outbreaks [42,43], it would be necessary to initiate further research on multiple dosing of the v7D vaccine to obtain a satisfactory protective efficacy. In addition, anti-VZV serum antibody is always used as a reliable indicator for serological confirmation of vaccine-induced VZV-specific immunity in humans, but there is still a lack of cell-mediated immunity (CMI) analysis in this field. Future clinical research is also necessary to investigate CMI to v7D vaccination and to reveal its role in the efficacy of varicella vaccine.

5.3. Mechanisms of v7D Attenuation and Vaccine-Induced Protective Immunity

v7D is a pure and genetically defined VZV strain that has a replication defect in both human skin and neuronal cells due to the abrogation of *ORF7* expression. *ORF7* encodes a virion tegument protein that localizes to the Golgi apparatus in the infected cells [44], the secondary envelopment site for VZV virus particles. Studies have shown that the deletion of *ORF7* does not prevent viral entry, viral genome replication, viral protein expression or retrograde transport of virus particles from axon terminals to somata, but substantially affects syncytia formation, secondary envelopment and subsequent cell-to-cell spreading of progeny viruses in differentiated skin and neuronal cells in vitro [39,44,45]. Therefore, pORF7 seems to work in conjunction with other viral or cellular proteins and acts as an essential part to regulate cell-to-cell fusion and secondary envelopment of VZV virus particles in infected human skin and neurons. Previously, we found that pORF7 interacts with pORF53 and they co-localize in the trans-Golgi network (TGN) in VZV-infected cells [46]. Most recently, the structure of pUL51 in complex with pUL7, the pORF7 and pORF53 homologues from herpes simplex virus-1 (HSV-1), has suggested a conserved role for pORF7 and its homologues in promoting membrane scission during cytoplasmic envelopment of nascent virions [47]. Further evidence is needed to validate and thoroughly characterize these mechanisms.

In addition, to overcome the limitation of VZV's strict human specificity, future research on the mechanisms of v7D attenuation and protective immunity induced by v7D vaccine could be complemented by work in non-human primate models of simian varicella virus, which recapitulates the clinical features of VZV infection in humans [48].

6. Conclusions

Varicella remains an important public health concern worldwide. The vaccine virus vOka retains neurovirulence and may cause vaccine-associated HZ in some vaccinees. Thus, a safer neuro-attenuated vaccine is needed to improve vaccine acceptance and aid in establishing herd immunity against varicella, together with vOka vaccines. Through a genome-wide mutagenesis screen, we identified *ORF7* as a skin- and neuro-tropic factor of VZV, and we generated an *ORF7*-deficient, pOka-derived live virus vaccine candidate, v7D. Compared to vOka, v7D has a reduced risk of vaccine-associated AE in humans, and shows similar immunogenicity in animal models. Future research should focus on testing the safety and effectiveness of the v7D vaccine, determining its immune procedure and dosage through clinical trials, and understanding the mechanism of v7D attenuation and protective immune response induced by this vaccine virus. Together, this knowledge could help guide the future use of v7D as a safe viral vector for the prevention and possible treatment of infectious diseases.

Author Contributions: The manuscript was written by W.W. and D.P. and edited by T.C. and H.Z. All authors have read and agreed to the published version of the manuscript.

Funding: This research was supported by grants from the National Natural Science Foundation of China (Nos. 81871648 and 82171833).

Acknowledgments: We thank Janine Zeng for critically reading the manuscript.

Conflicts of Interest: The authors declare no conflict of interest.

References

1. Cohen, J.I.; Straus, S.E.; Arvin, A.M. *Fields Virology*, 5th ed.; Lippincott Williams & Wilkins: Philadelphia, PA, USA, 2007.
2. Gilden, D.H.; Kleinschmidt-DeMasters, B.K.; LaGuardia, J.J.; Mahalingam, R.; Cohrs, R.J. Neurologic complications of the reactivation of varicella-zoster virus. *N. Engl. J. Med.* **2000**, *342*, 635–645. [CrossRef] [PubMed]
3. Gnann, J.W., Jr. Varicella-zoster virus: Atypical presentations and unusual complications. *J. Infect. Dis.* **2002**, *186* (Suppl. 1), S91–S98. [CrossRef] [PubMed]
4. Takahashi, M.; Otsuka, T.; Okuno, Y.; Asano, Y.; Yazaki, T.; Isomura, S. Live vaccine used to prevent the spread of varicella in children in hospital. *Lancet* **1974**, *304*, 1288–1290. [CrossRef]
5. Takahashi, M.; Okuno, Y.; Otsuka, T.; Osame, J.; Takamizawa, A. Development of a live attenuated varicella vaccine. *Biken J.* **1975**, *18*, 25–33. [PubMed]
6. Depledge, D.P.; Kundu, S.; Jensen, N.J.; Gray, E.R.; Jones, M.; Steinberg, S.; Gershon, A.; Kinchington, P.R.; Schmid, D.S.; Balloux, F.; et al. Deep sequencing of viral genomes provides insight into the evolution and pathogenesis of varicella zoster virus and its vaccine in humans. *Mol. Biol. Evol.* **2014**, *31*, 397–409. [CrossRef] [PubMed]
7. Depledge, D.P.; Yamanishi, K.; Gomi, Y.; Gershon, A.A.; Breuer, J. Deep Sequencing of Distinct Preparations of the Live Attenuated Varicella-Zoster Virus Vaccine Reveals a Conserved Core of Attenuating Single-Nucleotide Polymorphisms. *J. Virol.* **2016**, *90*, 8698–8704. [CrossRef]
8. Quinlivan, M.; Breuer, J. Clinical and molecular aspects of the live attenuated Oka varicella vaccine. *Rev. Med. Virol.* **2014**, *24*, 254–273. [CrossRef]
9. Wolfson, L.J.; Daniels, V.J.; Altland, A.; Black, W.; Huang, W.; Ou, W. The Impact of Varicella Vaccination on the Incidence of Varicella and Herpes Zoster in the United States: Updated Evidence From Observational Databases, 1991–2016. *Clin. Infect. Dis.* **2020**, *70*, 995–1002. [CrossRef]
10. Morino, S.; Tanaka-Taya, K.; Satoh, H.; Arai, S.; Takahashi, T.; Sunagawa, T.; Oishi, K. Descriptive epidemiology of varicella based on national surveillance data before and after the introduction of routine varicella vaccination with two doses in Japan, 2000–2017. *Vaccine* **2018**, *36*, 5977–5982. [CrossRef]
11. Woodward, M.; Marko, A.; Galea, S.; Eagel, B.; Straus, W. Varicella Virus Vaccine Live: A 22-Year Review of Postmarketing Safety Data. In *Open Forum Infectious Diseases*; Oxford University Press: Oxford, UK, 2019; Volume 6. [CrossRef]
12. Gershon, A.A.; Gershon, M.D.; Shapiro, E.D. Live Attenuated Varicella Vaccine: Prevention of Varicella and of Zoster. *J. Infect. Dis.* **2021**, *224*, S387–S397. [CrossRef]
13. Harpaz, R. Do varicella vaccination programs change the epidemiology of herpes zoster? A comprehensive review, with focus on the United States. *Expert Rev. Vaccines* **2019**, *18*, 793–811. [CrossRef] [PubMed]
14. Chaves, S.S.; Haber, P.; Walton, K.; Wise, R.P.; Izurieta, H.S.; Schmid, D.S.; Seward, J.F. Safety of varicella vaccine after licensure in the United States: Experience from reports to the vaccine adverse event reporting system, 1995–2005. *J. Infect. Dis.* **2008**, *197* (Suppl. 2), S170–S177. [CrossRef] [PubMed]
15. Galea, S.A.; Sweet, A.; Beninger, P.; Steinberg, S.P.; LaRussa, P.S.; Gershon, A.A.; Sharrar, R.G. The safety profile of varicella vaccine: A 10-year review. *J. Infect. Dis.* **2008**, *197* (Suppl. 2), S165–S169. [CrossRef] [PubMed]
16. Goulleret, N.; Mauvisseau, E.; Essevaz-Roulet, M.; Quinlivan, M.; Breuer, J. Safety profile of live varicella virus vaccine (Oka/Merck): Five-year results of the European Varicella Zoster Virus Identification Program (EU VZVIP). *Vaccine* **2010**, *28*, 5878–5882. [CrossRef]
17. Yoshikawa, T.; Ando, Y.; Nakagawa, T.; Gomi, Y. Safety profile of the varicella vaccine (Oka vaccine strain) based on reported cases from 2005 to 2015 in Japan. *Vaccine* **2016**, *34*, 4943–4947. [CrossRef]
18. Weinmann, S.; Chun, C.; Schmid, D.S.; Roberts, M.; Vandermeer, M.; Riedlinger, K.; Bialek, S.R.; Marin, M. Incidence and clinical characteristics of herpes zoster among children in the varicella vaccine era, 2005–2009. *J. Infect. Dis.* **2013**, *208*, 1859–1868. [CrossRef]

19. Sharrar, R.G.; LaRussa, P.; Galea, S.A.; Steinberg, S.P.; Sweet, A.R.; Keatley, R.M.; Wells, M.E.; Stephenson, W.P.; Gershon, A.A. The postmarketing safety profile of varicella vaccine. *Vaccine* **2000**, *19*, 916–923. [CrossRef]
20. Wise, R.P.; Salive, M.E.; Braun, M.M.; Mootrey, G.T.; Seward, J.F.; Rider, L.G.; Krause, P.R. Postlicensure safety surveillance for varicella vaccine. *JAMA* **2000**, *284*, 1271–1279. [CrossRef]
21. Krause, P.R.; Klinman, D.M. Varicella vaccination: Evidence for frequent reactivation of the vaccine strain in healthy children. *Nat. Med.* **2000**, *6*, 451–454. [CrossRef]
22. Heusel, E.H.; Grose, C. Twelve Children with Varicella Vaccine Meningitis: Neuropathogenesis of Reactivated Live Attenuated Varicella Vaccine Virus. *Viruses* **2020**, *12*, 1078. [CrossRef]
23. Zerboni, L.; Ku, C.-C.; Jones, C.D.; Zehnder, J.L.; Arvin, A.M. Varicella-zoster virus infection of human dorsal root ganglia in vivo. *Proc. Natl. Acad. Sci. USA* **2005**, *102*, 6490–6495. [CrossRef] [PubMed]
24. Sadaoka, T.; Depledge, D.P.; Rajbhandari, L.; Venkatesan, A.; Breuer, J.; Cohen, J.I. In vitro system using human neurons demonstrates that varicella-zoster vaccine virus is impaired for reactivation, but not latency. *Proc. Natl. Acad. Sci. USA* **2016**, *113*, E2403–E2412. [CrossRef] [PubMed]
25. Weinmann, S.; Naleway, A.L.; Koppolu, P.; Baxter, R.; Belongia, E.A.; Hambidge, S.J.; Irving, S.A.; Jackson, M.L.; Klein, N.P.; Lewin, B.; et al. Incidence of Herpes Zoster Among Children: 2003–2014. *Pediatrics* **2019**, *144*, e20182917. [CrossRef] [PubMed]
26. Civen, R.; Chaves, S.S.; Jumaan, A.; Wu, H.; Mascola, L.; Gargiullo, P.; Seward, J.F. The incidence and clinical characteristics of herpes zoster among children and adolescents after implementation of varicella vaccination. *Pediatr. Infect. Dis. J.* **2009**, *28*, 954–959. [CrossRef]
27. L'Huillier, A.G.; Hirzel, C.; Ferreira, V.H.; Ierullo, M.; Ku, T.; Selzner, N.; Schiff, J.; Juvet, S.; Miao, C.; Schmid, D.S.; et al. Evaluation of Recombinant Herpes Zoster Vaccine for Primary Immunization of Varicella-seronegative Transplant Recipients. *Transplantation* **2021**, *105*, 2316–2323. [CrossRef]
28. Levin, M.J.; Weinberg, A. Immune Responses to Varicella-Zoster Virus Vaccines. In *Current Topics in Microbiology and Immunology*; Springer: Berlin/Heidelberg, Germany, 2022. [CrossRef]
29. Cohen, J.I.; Seidel, K.E. Generation of varicella-zoster virus (VZV) and viral mutants from cosmid DNAs: VZV thymidylate synthetase is not essential for replication in vitro. *Proc. Natl. Acad. Sci. USA* **1993**, *90*, 7376–7380. [CrossRef]
30. Niizuma, T.; Zerboni, L.; Sommer, M.H.; Ito, H.; Hinchliffe, S.; Arvin, A.M. Construction of varicella-zoster virus recombinants from parent Oka cosmids and demonstration that ORF65 protein is dispensable for infection of human skin and T cells in the SCID-hu mouse model. *J. Virol.* **2003**, *77*, 6062–6065. [CrossRef]
31. Zhang, Z.; Huang, Y.; Zhu, H. A highly efficient protocol of generating and analyzing VZV ORF deletion mutants based on a newly developed luciferase VZV BAC system. *J. Virol. Methods* **2008**, *148*, 197–204. [CrossRef]
32. Zhang, Z.; Selariu, A.; Warden, C.; Huang, G.; Huang, Y.; Zaccheus, O.; Cheng, T.; Xia, N.; Zhu, H. Genome-wide mutagenesis reveals that ORF7 is a novel VZV skin-tropic factor. *PLoS Pathog.* **2010**, *6*, e1000971. [CrossRef]
33. Selariu, A.; Cheng, T.; Tang, Q.; Silver, B.; Yang, L.; Liu, C.; Ye, X.; Markus, A.; Goldstein, R.S.; Cruz-Cosme, R.S.; et al. ORF7 of varicella-zoster virus is a neurotropic factor. *J. Virol.* **2012**, *86*, 8614–8624. [CrossRef]
34. Zerboni, L.; Sen, N.; Oliver, S.L.; Arvin, A.M. Molecular mechanisms of varicella zoster virus pathogenesis. *Nat. Rev. Microbiol.* **2014**, *12*, 197–210. [CrossRef]
35. Zhang, Z.; Rowe, J.; Wang, W.; Sommer, M.; Arvin, A.; Moffat, J.; Zhu, H. Genetic analysis of varicella-zoster virus ORF0 to ORF4 by use of a novel luciferase bacterial artificial chromosome system. *J. Virol.* **2007**, *81*, 9024–9033. [CrossRef] [PubMed]
36. Che, X.; Zerboni, L.; Sommer, M.H.; Arvin, A.M. Varicella-zoster virus open reading frame 10 is a virulence determinant in skin cells but not in T cells in vivo. *J. Virol.* **2006**, *80*, 3238–3248. [CrossRef] [PubMed]
37. Moffat, J.F.; Zerboni, L.; Kinchington, P.R.; Grose, C.; Kaneshima, H.; Arvin, A.M. Attenuation of the vaccine Oka strain of varicella-zoster virus and role of glycoprotein C in alphaherpesvirus virulence demonstrated in the SCID-hu mouse. *J. Virol.* **1998**, *72*, 965–974. [CrossRef] [PubMed]
38. Moffat, J.F.; Zerboni, L.; Sommer, M.H.; Heineman, T.C.; Cohen, J.I.; Kaneshima, H.; Arvin, A.M. The ORF47 and ORF66 putative protein kinases of varicella-zoster virus determine tropism for human T cells and skin in the SCID-hu mouse. *Proc. Natl. Acad. Sci. USA* **1998**, *95*, 11969–11974. [CrossRef] [PubMed]
39. Wang, W.; Pan, D.; Fu, W.; Ye, X.; Han, J.; Yang, L.; Jia, J.; Liu, J.; Zhu, R.; Zhang, Y.; et al. Development of a skin- and neuro-attenuated live vaccine for varicella. *Nat. Commun.* **2022**, *13*, 824. [CrossRef]
40. Cohen, J.I.; Nguyen, H. Varicella-zoster virus ORF61 deletion mutants replicate in cell culture, but a mutant with stop codons in ORF61 reverts to wild-type virus. *Virology* **1998**, *246*, 306–316. [CrossRef]
41. Weibel, R.E.; Kuter, B.J.; Neff, B.J.; Rothenberger, C.A.; Fitzgerald, A.J.; Connor, K.A.; Morton, D.; Mclean, A.A.; Scolnick, E.M. Live Oka/Merck Varicella Vaccine in Healthy-Children—Further Clinical and Laboratory Assessment. *JAMA* **1985**, *254*, 2435–2439. [CrossRef]
42. Perella, D.; Wang, C.; Civen, R.; Viner, K.; Kuguru, K.; Daskalaki, I.; Schmid, D.S.; Lopez, A.S.; Tseng, H.F.; Newbern, E.C.; et al. Varicella Vaccine Effectiveness in Preventing Community Transmission in the 2-Dose Era. *Pediatrics* **2016**, *137*, e20152802. [CrossRef]
43. Leung, J.; Lopez, A.S.; Blostein, J.; Thayer, N.; Zipprich, J.; Clayton, A.; Buttery, V.; Andersen, J.; Thomas, C.A.; Del Rosario, M.; et al. Impact of the US Two-dose Varicella Vaccination Program on the Epidemiology of Varicella Outbreaks: Data from Nine States, 2005–2012. *Pediatr. Infect. Dis. J.* **2015**, *34*, 1105–1109. [CrossRef]

44. Jiang, H.-F.; Wang, W.; Jiang, X.; Zeng, W.-B.; Shen, Z.-Z.; Song, Y.-G.; Yang, H.; Liu, X.-J.; Dong, X.; Zhou, J.; et al. ORF7 of Varicella-Zoster Virus Is Required for Viral Cytoplasmic Envelopment in Differentiated Neuronal Cells. *J. Virol.* **2017**, *91*, e00127-17. [CrossRef] [PubMed]
45. Grigoryan, S.; Kinchington, P.R.; Yang, I.H.; Selariu, A.; Zhu, H.; Yee, M.; Goldstein, R.S. Retrograde axonal transport of VZV: Kinetic studies in hESC-derived neurons. *J. Neurovirol.* **2012**, *18*, 462–470. [CrossRef]
46. Wang, W.; Fu, W.; Pan, D.; Cai, L.; Ye, J.; Liu, J.; Liu, C.; Que, Y.; Xia, N.; Zhu, H.; et al. Varicella-zoster virus ORF7 interacts with ORF53 and plays a role in its trans-Golgi network localization. *Virol. Sin.* **2017**, *32*, 387–395. [CrossRef] [PubMed]
47. Butt, B.G.; Owen, D.J.; Jeffries, C.M.; Ivanova, L.; Hill, C.H.; Houghton, J.W.; Ahmed, M.F.; Antrobus, R.; Svergun, D.I.; Welch, J.J.; et al. Insights into herpesvirus assembly from the structure of the pUL7:pUL51 complex. *eLife* **2020**, *9*, e53789. [CrossRef] [PubMed]
48. Mahalingam, R.; Messaoudi, I.; Gilden, D. Simian varicella virus pathogenesis. *Curr. Top. Microbiol. Immunol.* **2010**, *342*, 309–321. [CrossRef]

viruses

Review

Comparative Analysis of the Simian Varicella Virus and Varicella Zoster Virus Genomes

Wayne L. Gray

Biology Department, University of Mississippi, Oxford, MS 38677, USA; wlgray@olemiss.edu

Abstract: Varicella zoster virus (VZV) and simian varicella virus (SVV) cause varicella (chickenpox) in children and nonhuman primates, respectively. After resolution of acute disease, the viruses establish latent infection in neural ganglia, after which they may reactivate to cause a secondary disease, such as herpes zoster. SVV infection of nonhuman primates provides a model to investigate VZV pathogenesis and antiviral strategies. The VZV and SVV genomes are similar in size and structure and share 70–75% DNA homology. SVV and VZV DNAs are co-linear in gene arrangement with the exception of the left end of the viral genomes. Viral gene expression is regulated into immediate early, early, and late transcription during in vitro and in vivo infection. During viral latency, VZV and SVV gene expression is limited to transcription of a viral latency-associated transcript (VLT). VZV and SVV are closely related alphaherpesviruses that likely arose from an ancestral varicella virus that evolved through cospeciation into species-specific viruses.

Keywords: varicella zoster virus; simian varicella virus; herpesvirus

Citation: Gray, W.L. Comparative Analysis of the Simian Varicella Virus and Varicella Zoster Virus Genomes. *Viruses* **2022**, *14*, 844. https://doi.org/10.3390/v14050844

Academic Editors: Charles Grose, Ravi Mahalingam and Joel Rovnak

Received: 29 March 2022
Accepted: 15 April 2022
Published: 19 April 2022

1. Introduction

Varicella zoster virus (VZV) and simian varicella virus (SVV) are herpesviruses that cause erythematous diseases in humans and Old World monkeys, respectively. VZV (*Human herpesvirus 3*) produces varicella (chickenpox) in children, characterized by vesicular skin lesions over the entire body [1]. After resolution of the primary disease, VZV establishes latent infection in neural ganglia throughout the neuraxis. Subsequently, the virus may be induced to reactivate to cause herpes zoster (shingles) with vesicular skin lesions associated with one or a few dermatomes [1].

SVV (*Cercopithecine herpesvirus 9*) causes a varicella-like disease in cercopithecoid monkeys, including patas and vervet monkeys, and several macaque species [2]. Infected monkeys exhibit fever, malaise, and a whole-body vesicular skin rash. Like VZV, SVV establishes latent infection in neural ganglia [3]. Stress or immunosuppression may induce viral reactivation although unlike VZV and for unclear reasons, SVV reactivation disease may be systemic and not associated with dermatomal spread [4]. Sporadic epizootics of simian varicella in facilities housing nonhuman primates have sometimes been associated with severe disease with high morbidity and mortality, while other outbreaks have exhibited milder disease [2,5].

Studies of VZV pathogenesis and development and evaluation of antiviral strategies are limited since VZV is exclusively a human virus. Based on the clinical and pathological similarities of simian and human varicella, SVV infection of nonhuman primates is a useful experimental model to study mechanisms of VZV pathogenesis and to develop and evaluate antiviral chemotherapies and vaccines [2].

SVV and VZV are species-specific herpesviruses. SVV replicates optimally in cell cultures of simian origin, such as African green monkey (Vero) cells, while VZV grows best in human cell culture, such as melanoma (Mewo) cells. Experimental VZV infection does not induce a varicella-like illness in nonhuman primates. Likewise, there are no reported cases of SVV causing disease in humans.

SVV and VZV are morphologically similar herpesviruses with an icosahedral capsid enclosing the viral double-stranded DNA genomes and surrounded by a viral envelope. The viruses are antigenically related. Immune serum from SVV infected monkeys immunoprecipitates cross-reacting VZV antigens and also neutralizes VZV in plaque-reduction neutralization tests (PRNT) [6,7]. Conversely, VZV immune serum from zoster patients immunoprecipitates SVV antigens and neutralizes SVV. In addition, experimental immunization of nonhuman primates with VZV induces immune protection against simian varicella following SVV challenge [8].

SVV and VZV also share extensive genetic relatedness. While the viral DNAs have distinct restriction endonuclease profiles, they share 70–75% DNA homology as determined by DNA hybridization assays using conditions of varying stringency as well as DNA sequence analysis [9,10]. Considering the close antigenic and genetic relatedness between SVV and VZV, the basis of the species-specificity of the viruses is unclear. This review compares the genetic characteristics of SVV and VZV genomes, their gene expression, and their evolutionary relationship.

2. Size, Genetic Content, and Structure of the SVV and VZV Genomes

The SVV and VZV genomes are the smallest in size among the alphaherpesviruses. The complete DNA sequence of the VZV prototype Dumas strain is 124,884 base pairs (bp) in length [11] although the DNA sequences of other VZV isolates vary with a size range of 124,770 to 125,945 bp. The size of the SVV genome is slightly smaller with 124,785 bp (deltaherpesvirus isolate) [10].

The guanosine + cytosine (G + C) content of SVV DNA was determined to be 40.8% by buoyant density analysis on cesium chloride gradients and confirmed to be 40.4% by DNA sequence analysis [10,12]. The VZV genome has a slightly higher G + C % content of 46% [11].

SVV and VZV DNA exist in an extrachromosomal configuration in infected cells. The viral DNAs may be in a circular form as shown by PCR utilizing primers from the termini of the viral genomes [13,14]. An unpaired 3′ nucleotide at the end of viral DNA may facilitate DNA circulation. The viral DNAs may also exist in concatameric head-to-tail forms, which are generated through the rolling circle mechanism of DNA replication as occurs in herpesvirus DNA synthesis [13]. SVV and VZV DNA also exist in a circular, extrachromosomal form in neurons of latently infected ganglia.

Consistent with other herpesviruses, the SVV and VZV genomes include direct and inverted repeat sequences [11,15]. Electron microscopy of denatured and then re-annealed SVV DNA revealed molecular structures consisting of a S 7.2 kb double-stranded stem with a ≈5.2 kb single-stranded loop [16]. This stem and loop structure was linked to a ≈100 bp single-stranded DNA stretch. Subsequent DNA sequence analysis confirmed that SVV DNA consists of a 20 kb short (S) component that includes a 5.2 kb unique (US) sequence bracketed by 7.5 kp internal and terminal inverted repeat (IRS and TRS) sequences [10]. This S component is covalently linked to a 104.7 kb long (L) component, which includes a 104.0 kb unique long sequence (UL) bracketed by 65 bp inverted repeat sequences (TRL and IRL).

This structure of SVV DNA is similar to the organization of the VZV genome (Figure 1, Table 1). The sizes of the various components of the SVV and VZV genomes are comparable. For example, the inverted repeat sequences of SVV DNA S component (TRS/IRS) at 7557 bp are slightly longer than that found in the VZV genome by 238 bp, while the SVV US sequence is 328 bp shorter than the VZV US.

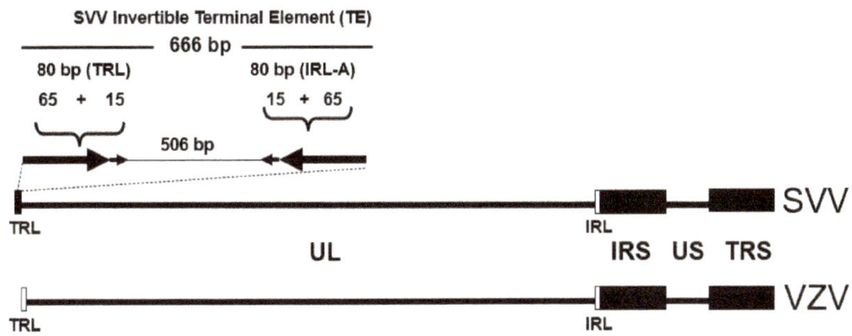

Figure 1. The SVV and VZV genomes consist of a short (S) component linked to a long (L) component. The S includes a unique sequence (US) flanked by internal and terminal inverted repeats (IRS, TRS). The L component includes a unique long (UL) segment bracketed by internal and terminal repeats (IRL, TRL). The SVV genome has an invertible terminal element (TE) that does not exist in VZV DNA.

Table 1. Comparison of the SVV and VZV genomes.

	Size (Base Pairs)		G + C %	
	SVV [1]	VZV [2]	SVV [1]	VZV [2]
TRS/IRS	7557	7319.5	65.0%	59.0%
US	4904	5232	39.1%	42.8%
Total S	20,018	19,871	58.6%	54.7%
TRL/IRL	65	88.5	69.3%	68.4%
UL	104,036	104,836	38.3%	44.3%
TE [3]	666	-	53.8%	-
Total L	104,767	105,013	38.3%	44.3%
Total Genome	124,785	124,884	40.4%	46.0%

[1] data from Gray et al. (2001) [10]; [2] data derived from Davison and Scott (1986) [11]; [3] terminal element-SVV genome left end.

The S and the L components of the SVV and VZV genomes can invert relative to each other, permitting the viral genomes to exist in two equimolar isomeric forms. This finding is based upon the existence of 0.5 molar DNA fragments in the restriction endonuclease profiles of the DNAs [16].

The SVV and VZV genomes include tandem direct repeat sequences [10,11]. SVV DNA has four of these repeat elements, R1, R2, R3, and R4, which are distributed along the SVV genome (Figure 1). The R1, located between ORFs 7 and 8, includes an A + T-rich 37 bp sequence that is repeated three times. The R2 present within the glycoprotein C gene (ORF 14) consists of two G + C-rich 83 bp sequences with an intervening 13 bp sequence. The R3 is the largest and most complex repeat element comprising seventeen 9 bp sequences and eight 12 base sequences and is within ORF 22. The R4 is a 16 bp G + C-rich sequence located between ORFs 62 and 63. VZV DNA also has repeat elements that correspond in location with the SVV R2, R3, and R4 although they vary in length and sequence composition [11]. The VZV genome has an R1 element that does not correspond in location with the SVV R1 but rather is positioned within VZV ORF 11. While direct repeat elements are conserved within herpesvirus genomes, their function is not understood.

The left terminus of the SVV and VZV genomes includes the most structural diversity between the viral DNAs. The left end of SVV genome includes a 666 bp terminal element that is not present within VZV DNA (Figure 1) [17]. This terminal element includes a 506 bp unique sequence surrounded by two 80 bp inverted repeat sequences. The inverted repeat sequences include the 65 bp IRL and TRL inverted repeats that bracket the UL component of SVV DNA. The terminal element is conserved among SVV isolates but may vary in length since some SVV isolates have an additional 46 bp repeat within the unique sequence. While VZV DNA does not contain a comparable terminal element, this component is found in the genomes of other *Varicelloviruses*, including pseudorabies herpesvirus (PRV) and equine herpesviruses types 1 and 4 (EHV-1, EHV-4) [17].

3. Gene Organization of the SVV and VZV Genomes

The SVV genome encodes 73 identified open reading frames (ORFs), of which three (ORFs 69, 70, and 71) are duplicated within the IRS and TRS (Figure 2). VZV DNA includes 72 ORFs with the same three duplicated genes. Viral ORFs along the SVV and VZV genomes are co-linear with respect to gene location. With only two exceptions, each SVV gene has a corresponding VZV homologue (Table S1). The homology based upon predicted amino acid sequence ranges from 75% for glycoprotein B (ORF 31) to 27% for ORF 1.

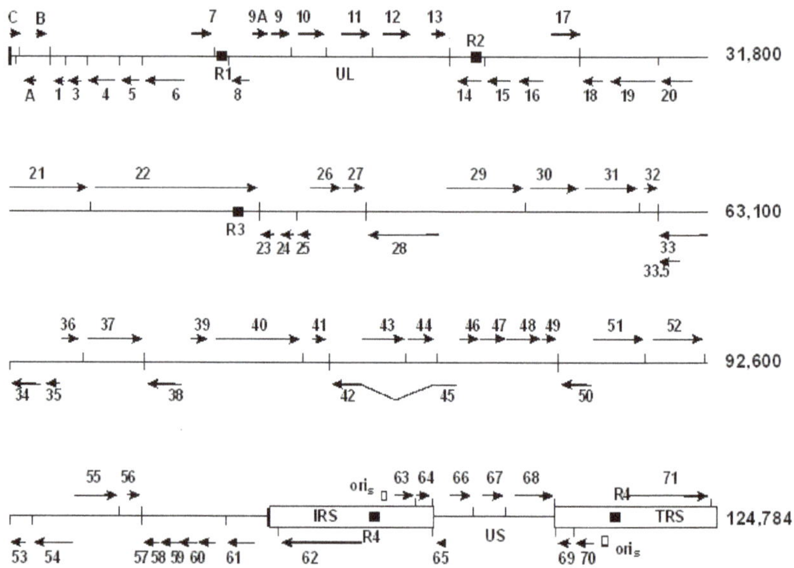

Figure 2. The SVV genome includes 73 ORFs, each of which are designated as arrows indicating the site on each DNA strand. The R1, R2, R3, and R4 direct repeats are indicated as boxes. The VZV genome has a similar gene map except for the left end of the viral DNA. VZV ORFs have the same nomenclature as their SVV homologs, while unique SVV ORFs are designated with letters (A, B, C). Vertical lines indicate poly A sites. Reprinted from Current Topics in Microbiology and Immunology [15] with permission from the publisher.

The major differences in genetic content of SVV and VZV DNA occur at the left ends of the viral genomes (Figure 3). SVV DNA does not include a homologous gene to the VZV ORF 2, which is expressed as a 31 kD phosphoprotein within membranes of VZV infected cells but not within viral particles [18]. The VZV ORF 2 is nonessential, as deletion mutants replicate efficiently in cell culture and establish latency in ganglia of latently infected cotton rats.

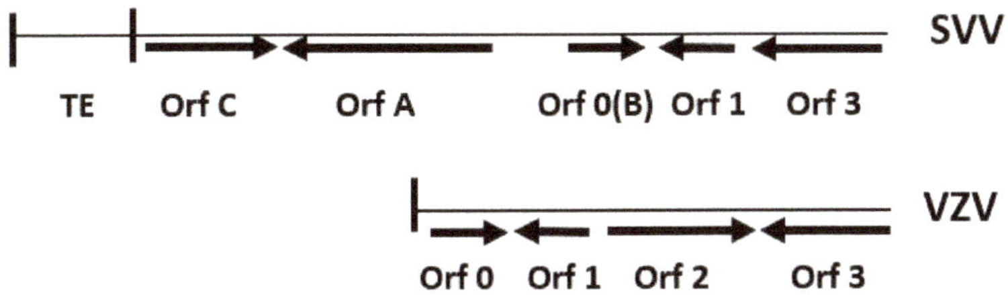

Figure 3. The left ends of the SVV and VZV genomes. SVV and VZV ORFs are designated by arrows indicating the site on each DNA strand. SVV DNA includes a terminal element (TE) but does not have an ORF 2. VZV DNA does not include a TE, ORF A, or ORF C. Vertical lines indicate TRL sequences.

The SVV left end includes an ORF A that is not found within the VZV genome [19]. The SVV ORF A is a truncated paralog of the SVV ORF 4 and ortholog of VZV ORF 4. The VZV ORF 4 is essential for in vitro replication and encodes an immediate early (IE) protein that transactivates VZV promoters [20–22]. Based upon extensive homology to VZV ORF 4, the SVV ORF 4 is also proposed as a viral transactivator protein. However, in an experimental study, the SVV ORF A did not transactivate SVV promoters [23]. This lack of transactivation may be due to the truncated SVV ORF A lacking the highly acidic amino-terminus and nuclear localization signal that is critical for VZV ORF 4 transactivation of reporter promoters [22]. The SVV ORF A promoter is stimulated by the SVV IE ORF 62 protein, and the 1.0 kb transcript is expressed in infected Vero cells [23]. An SVV ORF A deletion mutant replicates efficiently in Vero cells, indicating the gene is not essential for in vitro replication [23]. However, the ORF A is conserved within various SVV isolates and is expressed in tissues of SVV-infected monkeys, indicating its likely importance in viral pathogenesis [19].

The SVV ORF B, as originally identified on the SVV genome, is now known to be a homolog of the VZV ORF 0 (or ORF S/L), sharing 35% amino acid identity [19]. The VZV ORF 0 is essential for efficient replication in cell culture and in vivo and encodes a transmembrane protein that may play a role in viral DNA cleavage and packaging of concatemeric VZV DNA [20,24].

The SVV ORF C is a more recently identified gene located at the extreme left terminus of the SVV genome [25]. The SVV ORF C is a duplicated paralog of the SVV ORF 0 and encodes a 123 amino acid putative transmembrane protein. While the VZV genome does not have a corresponding gene to the SVV ORF C, it does have the homologous VZV ORF 0 [24]. The homology between the SVV ORF C, SVV ORF 0, and VZV ORF 0 is concentrated within the protein central region and the carboxy terminal hydrophobic domain (Figure 4).

The SVV ORF C, SVV ORF 0, and VZV ORF 0 are orthologs of the herpes simplex virus (HSV) unique long ORF 56 (UL56) gene family, which is conserved among alpha-herpesviruses [25]. These UL56 orthologs include PPXY motifs, which are a consensus sequence for interaction with tryptophan-tryptophan (W-W) domains of cellular ubiquitin ligases that promote the vacuolar protein sorting pathway to induce budding of enveloped viruses from infected cells [26,27].

```
VZV Orf 0                                                                            PVTLTSSPSMDDVAT PI PYL   33
                                                                                     |++ ++   |+   +|  |  |
SVV Orf C   MSAATHRPST PPPGVYVLFK RPRSSPNAFQ PAGIVCVSPERH VFIPLSSSALSS LQS IAP PTP   61

VZV Orf 0   PTYAE    A VADAPPPYRS          RESLV FS                         PP LF       PH   62
            | ||     |+|   || |            ++   |                          || +        |
SVV Orf C   PTYDRTC A I ASNPPTYEHAPPPTYEQACAF GS NSSLP  D LR IEPRTEPRTD LPPAYT S  D T H   123
            ||  +|     ++ |    |     |  ||             |+     ++ +     |    | ||| ||  |  +|
SVV Orf 0   PTVERF     ILS S TELTLSSFS PI TY              NAEYM  N VYLPTYEEAIQ D LPPAYRSRE TL   56

VZV Orf 0   VENGTTQQSYDC  LDCAYDG IHR  LQLALF R IRK       CC VPA  FL I  LFGI        LT  LTAVVVA   115
               +     ++    |  +  |+   +      +     | ||+     |  |    ++ +     |          +  |
SVV Orf C   ATHNVPMETL QC CNSAHSA VWLQDV I VAR IRR F I PPCL VLG  LV V VAAI  I LL  L I S LNPH   181
            +|          +   | |           +  +   +  +  +|    +|    ||   ++   + |+| |+ +++   |
SVV Orf 0   STLTTDTGCMD CI CTGLYK       I HER  LTSCVRYC  VPAMFVLFAVLT  LTAV I LV  I L A AI PDV   114
```

Figure 4. Amino acid alignment of the C-terminal region of the SVV and VZV UL56 homologs including SVV ORFs C and 0 (B) and VZV ORF 0. Vertical lines indicate amino acid identity, and + indicates amino acid similarity. PPRX motifs are in bold. Predicted transmembrane domains are underlined.

4. Transcription of the SVV and VZV Genomes

Initial transcript mapping of the SVV and VZV genomes during lytic infection employed Northern blot analysis and labeled restriction endonuclease DNA fragment probes [28,29]. These studies revealed similarities in SVV and VZV transcription profiles and confirmed that all regions of the viral genome are transcriptionally active during acute infection in cell culture. Subsequent gene array analyses confirmed transcription of each SVV and VZV ORF during in vitro infection [30,31].

Most recently, RNA sequencing approaches (RNA-Seq, dRNA-Seq) have provided comprehensive analyses of the SVV and VZV transcriptomes [32,33]. The transcriptional start sites (TSS) and cleavage and polyadenylation sites (CPAS) for each of the SVV and VZV protein coding genes have been identified. A complex transcriptional pattern has revealed alternative TSS usage, read-through transcription, and RNA splicing and the existence of various transcript isoforms for individual viral ORFs. While the transcriptional profiles for the SVV and VZV genomes were found to be quite similar, differences in the number and sizes of transcript isomers for specific SVV and VZV genes were identified.

5. Regulation of Gene Expression of the SVV and VZV Genomes

Similar to other herpesviruses, the SVV and VZV genomes are coordinately expressed into kinetic stages. While the cell-associated nature of SVV and VZV complicates the generation of high-titer viral stocks needed for synchronous in vitro infection, studies confirm that the viruses express IE, early, and late gene products.

IE gene expression does not require prior viral protein synthesis and generates regulatory proteins that transactivate the promoters of viral genes. SVV and VZV DNAs have been predicted to express four IE genes, ORFs 4, 61, 62, and 63, based upon homology to the HSV-1 IE genes. The IE62, encoded by ORFs 62 and 71, is considered as the major viral transactivator of IE, early, and late gene promoters [34]. The SVV IE62 transactivates SVV ORF 28/29 early gene expression by interacting with a 16 bp palindrome sequence within the ORF 28/29 early promoter that includes a cellular upstream stimulatory factor-binding site (USF) [35]. The VZV IE62 also interacts with viral promoters and transcription factors to transactivate viral gene expression [34,36].

The SVV and VZV ORF 61 encode a homolog to the HSV-1 ICP0 transactivator protein. The SVV IE61 transactivates its own promoter as well as IE, early, and late gene promoters [37]. The protein includes a ring finger motif and a nuclear localization sequence that is required for transactivation. The VZV IE61 also has transactivation properties, which are dependent on cell type and transfection conditions [38]. The SVV and VZV ORF 61 are dispensable for viral replication in cell culture [37,39].

Studies of the SVV and VZV ORF 63 gene products have shown differential effects on the regulation of viral promoters. While the SVV 63 protein did not transactivate the ORF 21 early gene promoter by itself, the SVV 63 upregulated SVV 62 transactivation of the ORF 21 promoter in neuronal cells [40]. In contrast, the SVV 63 down-regulated SVV 62 transactivation of the ORF 21 promoter in Vero and Mewo cells. The SVV ORF 63 is not essential for in vitro replication, but its expression stimulates viral growth [41]. The VZV ORF 63 protein represses ORF 62 IE transcription, activates the thymidine kinase early gene expression, and has no effect on expression of late glycoprotein genes [42].

The SVV ORF 4 protein only weakly transactivated an early promoter but stimulated IE62 transactivation of the same promoter [23]. As indicated above, the SVV ORF A truncated protein homolog of ORF 4 does not transactivate SVV promoters [23]. The VZV ORF 4 is an IE gene that is essential for viral replication, can transactivate viral gene expression, and can augment IE62 stimulation of viral promoters [21,36].

Herpesvirus early gene products include enzymes involved in DNA synthesis. Several SVV and VZV early genes have been characterized including the deoxyuridine nucleotidohydrolase (dUTPase, ORF 8), uracil glycosylase (ORF 59), and thymidine kinase (ORF 36), which express enzymes involved in viral DNA replication and repair [43–46].

Herpesvirus late genes encode structural proteins including the viral capsid and envelope glycoproteins. Several VZV ORFs encode proteins involved in the structure and assembly process of viral capsid formation [47]. While SVV capsid assembly has not been analyzed, the SVV and VZV major capsid proteins encoded by ORF 40 are similar in size (156 kD) and share over 70% amino acid identity. SVV and VZV express at least nine glycoproteins, including gB, gC, gE, gH, gI, gK, gL, gN, and gM, which play a critical role in viral attachment, penetration, and cell-to-cell spread [48–50]. The SVV and VZV genomes do not encode a homolog to the HSV-1 glycoprotein D.

The development of RNA-Seq and dRNA-Seq methodology has permitted low titer, cell-free virus synchronous infection to analyze the kinetics of VZV transcription and gene expression [33]. VZV-infected culture cells were infected in the presence of cycloheximide (CHX) or phosphonoacetic acid (PAA) to identify IE or early transcripts, respectively. Late mRNAs were identified later in infection without metabolic inhibitors. VZV ORFs 4, 61, and 63 were highly expressed as IE genes in CHX-infected cells. ORF 0 was also expressed as an IE gene. Interestingly, VZV ORF 62 was not expressed under IE conditions but rather expressed late during infection, suggesting that a virion tegument derived ORF 62 protein is responsible for transactivation of IE transcription. Early gene expression was dependent on IE proteins, and 53 early transcripts were identified. Late gene expression included 28 leaky late (LL) and 41 true late (TL) transcripts, which were expressed at low levels prior to DNA synthesis or after DNA replication, respectively. To date, a similar kinetic analysis for SVV gene expression has not been reported.

6. Expression of SVV and VZV Viral Latency Associated Transcripts (VLTs)

Following primary varicella infection, SVV and VZV establish life-long latency within cells of neural ganglia [1,51]. Induction of viral reactivation may occur to cause secondary disease, such as herpes zoster [4]. An understanding of the molecular basis of viral latency and reactivation is needed for development of antiviral strategies against herpes zoster and postherpetic neuralgia.

SVV gene expression in neural ganglia of latently infected monkeys is restricted to a viral latency-associated transcript (VLT) [52]. This transcript was initially detected at low abundance by reverse-transcriptase PCR (RT-PCR) in trigeminal, cervical, and lumbar neural ganglia but not in lung or liver tissues during SVV latency. The SVV VLT was localized to the genomic UL/IRS junction and oriented antisense to the SVV ORF 61 (HSV-1 ICP0 homolog), which encodes an IE transactivator protein (Figure 5).

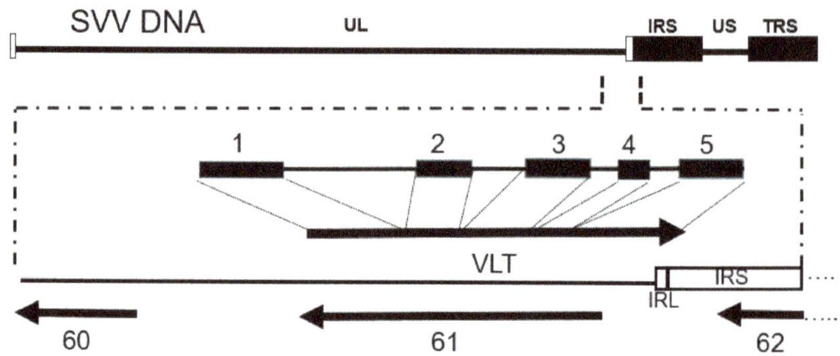

Figure 5. The SVV latency associated transcript (VLT) is a spliced RNA consisting of five exons (blocks) and introns (lines) mapping to the UL/IRS junction of SVV DNA and antisense to ORF 61. Multiple other isoforms are also transcribed from this locus during lytic infection. From Braspenning et al. [32].

Subsequently, RNA-Seq was employed to reveal a VZV VLT in human trigeminal ganglia [53]. The VZV VLT, like the SVV VLT, was determined to map antisense to the viral ORF 61. While multiple spliced VLT isoforms were generated during lytic infection, a single VLT isoform was detected in latently infected ganglia. The VLT was associated with a 136 amino acid protein expressed with late kinetics in cultured cells and in skin biopsy derived from a shingles patient but not in human ganglia.

RNA-Seq analysis has also provided a detailed characterization of the SVV VLT [32]. Similar to the VZV VLT, during lytic infection of culture cells, the SVV VLT locus is expressed as multiple spliced transcript isoforms with alternative transcription start sites. Some of these isoforms are predicted to be translated as peptides with homology to the VZV VLT protein. However, in ganglia derived from latently infected monkeys, a single spliced VLT isoform was detected at low abundance. This VLT was detected within neurons but not non-neuronal cells of latently infected ganglia [32].

Expression of the SVV and VZV VLT oriented antisense to the ORF 61 and HSV-1 ICP0 homolog is consistent with gene expression during latency of other alphaherpesviruses [52]. HSV-1, HSV-2, EHV-1, EHV-4, PRV, bovine herpesvirus 1 (BHV-1), and feline herpesvirus 1 (FHV-1) each express a transcript antisense to an ICP0 homolog in neural ganglia of latently infected animals [32].

The role of the SVV and VZV VLT in viral latency and reactivation is not understood. A SVV ORF 61 deletion mutant, with a corresponding disrupted VLT, produced clinical varicella in rhesus macaques and established latent infection in neural ganglia [54]. In addition, a VZV ORF 61 deletion mutant was able to establish latent infection in neural ganglia of infected cotton rats [39]. These results indicate that the SVV and VZV VLT is not essential for the establishment of viral latency. It is possible that the VLT may play a role in maintenance of the latent state or induction of reactivation by binding of the VLT to the complementary ORF 61 promoter and inhibiting IE61 expression and transactivation of other viral promoters. Such a hypothesis is contradicted by studies demonstrating that HSV-1 reactivation in mice is not mediated by VLT antisense regulation of the ICP0 gene [55]. Finally, the SVV and VZV VLT may promote neuronal cell survival and enhance viral latency by inhibiting apoptosis, as has been proposed for HSV-1 and BHV-1 VLT [56,57]. Hopefully, further studies will elucidate the mechanisms by which the SVV and VZV VLT promote viral latency and/or reactivation.

7. Evolutionary Relationship between SVV and VZV

A primordial herpesvirus infecting ancestral vertebrates gave rise to the *Herpesviridae* family an estimated 400 million years ago (mya) [58,59]. Figure 6 shows a phylogenetic tree of the *Alphaherpesvirinae* assembled based upon alignment of amino acid sequences for six conserved genes among alpha-, beta-, and gammaherpesviruses [59]. Based upon this analysis, the alphaherpesviruses diverged from the beta and gamma herpesviruses around 190 to 200 mya [59]. Then, around 70 mya, the alphaherpesviruses evolved into the *simplexvirus* (ultimately HSV-1, HSV-1, herpesvirus B (HVB), and simian agent 8 (SA8)), and *varicellovirus* (VZV, EHV-1, EHV-1 PRV, FHV-2) genera [58].

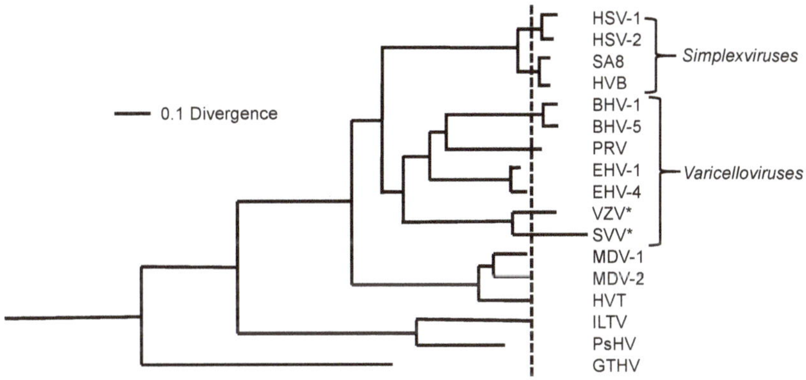

Figure 6. Phylogenetic tree of the alphaherpesviruses constructed based on amino acid alignment of six conserved herpesvirus genes [59]. SVV and VZV are denoted with an asterisk. Abbreviations: MDV-1, MDV-2, Marek's disease virus; HVT, herpesvirus of turkeys; ILTV, infectious laryngotracheitis virus; PsHV, psittacid herpesvirus 1; GTHV, green turtle herpesvirus. Other abbreviations are indicated in the text. Reprinted from *Virus Research* [59] with permission from the publisher.

An ancestral varicella virus may have arisen in an early African primate around 70 mya [58]. Fossil and phylogenetic evidence indicate that cercopithecoid monkeys and hominoids (great apes) diverged around 25 to 30 mya. Based upon the hypothesis of cospeciation, SVV and VZV likely evolved from this ancestral varicella virus, which diverged as closely related, species-specific viruses for Old World monkeys or humans, respectively.

VZV evolution continues as evident by the current existence of five major phylogeographical clades, which are differentiated by single-nucleotide polymorphisms (SNP) [60]. Clades 1, 3, and 4 include VZV isolates predominately derived from Europe and North America. Clade 2 and clade 5 include VZV isolates of Asian and Indian origin, respectively. While the clades are distinguished by specific SNPs, each of the VZV isolates share at least 99.8% DNA sequence identity. The clades represent clusters of VZV strains presumed to have evolved from a common ancestor.

The SVV deltaherpesvirus (DHV) is currently the only SVV isolate that has been completely sequenced and annotated [10]. DHV was derived from an infected patas monkey during a 1973 outbreak at a primate facility in Louisiana, U.S.A. GenBank also includes an unannotated DNA sequence of SVV isolated from an infected African Green monkey in China. Additional SVV isolates have been derived from various species of Old World monkeys and from outbreaks at various geographical locations [5]. Future genetic analyses of these SVV isolates will reveal further insight regarding the diversity of SVV isolates and the coevolution of SVV and VZV.

Supplementary Materials: The following are available online at https://www.mdpi.com/article/10.3390/v14050844/s1, Table S1. Comparison of the SVV and VZV genomes.

Funding: The manuscript was supported by NIH grants AI52373 and AI37571.

Acknowledgments: Thanks to Louie Perotti and William Marshall for their help with preparing the manuscript.

Conflicts of Interest: The author declare no conflict of interest.

References

1. Kennedy, P.G.E.; Gershon, A.A. Clinical features of varicella-zoster virus infection. *Viruses* **2018**, *10*, 609. [CrossRef] [PubMed]
2. Gray, W.L. Simian varicella: A model for human varicella-zoster virus infections. *Rev. Med. Virol.* **2004**, *14*, 363–381. [CrossRef] [PubMed]
3. Mahalingam, R.; Smith, D.; Wellish, M.; Wolf, W.; Dueland, A.N.; Cohrs, R.; Soike, K.; Gilden, D. Simian varicella virus DNA in dorsal root ganglia. *Proc. Natl. Acad. Sci. USA* **1991**, *88*, 2750–2752. [CrossRef]
4. Mahalingam, R.; Traina-Dorge, V.; Wellish, M.; Deharo, E.; Singletary, M.L.; Ribka, E.P.; Sanford, R.; Gilden, D. Latent simian varicella virus reactivates in monkeys treated with tacrolimus with or without exposure to irradiation. *J. Neurovirol.* **2010**, *16*, 342–354. [CrossRef] [PubMed]
5. Gray, W.L. Simian varicella in old world monkeys. *Comp. Med.* **2008**, *58*, 22–30.
6. Fletcher, T.M., 3rd; Gray, W.L. Simian varicella virus: Characterization of virion and infected cell polypeptides and the antigenic cross-reactivity with varicella-zoster virus. *J. Gen. Virol.* **1992**, *73*, 1209–1215. [CrossRef] [PubMed]
7. Felsenfeld, A.D.; Schmidt, N.J. Antigenic relationships among several simian varicella-like viruses and varicella-zoster virus. *Infect. Immun.* **1977**, *15*, 807–812. [CrossRef]
8. Felsenfeld, A.D.; Schmidt, N.J. Varicella-zoster virus immunizes patas monkeys against simian varicella-like disease. *J. Gen. Virol.* **1979**, *42*, 171–178. [CrossRef]
9. Gray, W.L.; Oakes, J.E. Simian varicella virus DNA shares homology with human varicella-zoster virus DNA. *Virology* **1984**, *136*, 241–246. [CrossRef]
10. Gray, W.L.; Starnes, B.; White, M.W.; Mahalingam, R. The DNA sequence of the simian varicella virus genome. *Virology* **2001**, *284*, 123–130. [CrossRef]
11. Davison, A.J.; Scott, J.E. The complete DNA sequence of varicella-zoster virus. *J. Gen. Virol.* **1986**, *67*, 1759–1816. [CrossRef] [PubMed]
12. Clarke, P.; Rabkin, S.D.; Inman, M.; Mahalingam, R.; Cohrs, R.; Wellish, M.; Gilden, D. Molecular analysis of simian varicella virus DNA. *Virology* **1992**, *190*, 597–605. [CrossRef]
13. Clarke, P.; Beer, T.; Gilden, D.H. Configuration and terminal sequences of the simian varicella virus genome. *Virology* **1995**, *207*, 154–159. [CrossRef] [PubMed]
14. Clarke, P.; Beer, T.; Cohrs, R.; Gilden, D. Configuration of latent varicella-zoster virus dna. *J. Virol.* **1995**, *69*, 8151–8154. [CrossRef]
15. Gray, W.L. Simian varicella virus: Molecular virology. *Curr. Top. Microbiol. Immunol.* **2010**, *342*, 291–308. [PubMed]
16. Gray, W.L.; Pumphrey, C.Y.; Ruyechan, W.T.; Fletcher, T.M. The simian varicella virus and varicella zoster virus genomes are similar in size and structure. *Virology* **1992**, *186*, 562–572. [CrossRef]
17. Mahalingam, R.; Gray, W.L. The simian varicella virus genome contains an invertible 665 base pair terminal element that is absent in the varicella zoster virus genome. *Virology* **2007**, *366*, 387–393. [CrossRef]
18. Sato, H.; Pesnicak, L.; Cohen, J.I. Varicella-Zoster virus open reading frame 2 encodes a membrane phosphoprotein that is dispensable for viral replication and for establishment of latency. *J. Virol.* **2002**, *76*, 3575–3578. [CrossRef]
19. Mahalingam, R.; White, T.; Wellish, M.; Gilden, D.H.; Soike, K.; Gray, W.L. Sequence analysis of the leftward end of simian varicella virus (EcoRI-I fragment) reveals the presense of an 8-bp repeat flanking the unique long segment and an 881-bp open-reading frame that is absent in the varicella zoster virus genome. *Virology* **2000**, *274*, 420–428. [CrossRef]
20. Zhang, Z.; Rowe, J.; Wang, W.; Sommer, M.H.; Arvin, A.M.; Moffat, J.; Zhu, H. Genetic analysis of varicella-zoster virus ORF0 to ORF4 by use of a novel luciferase bacterial artificial chromosome system. *J. Virol.* **2007**, *81*, 9024–9033. [CrossRef]
21. Defechereux, P.; Debrus, S.; Baudoux, L.; Rentier, B.; Piette, J. Varicella-zoster virus open reading frame 4 encodes an immediate-early protein with posttranscriptional regulatory properties. *J. Virol.* **1997**, *71*, 7073–7079. [CrossRef]
22. Moriuchi, M.; Moriuchi, H.; Debrus, S.; Piette, S.; Cohen, J.I. The acidic amino-terminal region of varicella-zoster virus open reading frame 4 protein is required for transactivation and can functionally replace the corresponding region of herpes simplex virus ICP27. *Virology* **1995**, *208*, 378–382. [CrossRef]
23. Gray, W.L. The simian varicella virus ORF A is expressed in infected cells, but is non-essential for replication in cell culture. *Arch. Virol.* **2012**, *157*, 1803–1806. [CrossRef] [PubMed]
24. Kaufer, B.B.; Smejkal, B.; Osterrieder, N. The vericella-zoster virus ORFS/L (ORF0) gene is required for effecient viral replication and contains an element involved in DNA cleavage. *J. Virol.* **2010**, *84*, 11661–11669. [CrossRef]
25. Davison, A.J. Herpesvirus systematics. *Vet. Microbiol.* **2010**, *143*, 52–69. [CrossRef]
26. Martin-Serrano, J.; Eastman, S.W.; Chung, W.; Bieniasz, P.D. HECT ubiquitin ligases link viral and cellular PPXY motifs to the vacuolar protein-sorting pathway. *J. Cell Biol.* **2005**, *168*, 89–101. [CrossRef]
27. Wills, J.W.; Cameron, C.E.; Wilson, C.B.; Xiang, Y.; Bennett, R.P.; Leis, J. An assembly domain of the Rous sarcoma virus gag protein required late in budding. *J. Virol.* **1994**, *68*, 6605–6618. [CrossRef]

28. Gray, W.L.; Gusick, N.; Fletcher, T.M.; Pumphrey, C.Y. Characterization and mapping of simian varicella virus transcripts. *J. Gen. Virol.* **1993**, *74*, 1639–1643. [CrossRef] [PubMed]
29. Ostrove, J.M.; Reinhold, W.; Fan, C.M.; Zorn, S.; Hay, J.; Straus, S.E. Transcription mapping of the varicella-zoster virus genome. *J. Virol.* **1985**, *56*, 600–606. [CrossRef]
30. Deitch, S.B.; Gilden, D.H.; Wellish, M.; Smith, J.; Cohrs, R.; Mahalingam, R. Array analysis of simian varicella virus gene transcription in productively infected cells in tissue culture. *J. Virol.* **2005**, *79*, 5315–5325. [CrossRef] [PubMed]
31. Cohrs, R.; Hurley, M.P.; Gilden, D.H. Array analysis of viral gene transcription during lytic infection of cells in tissue culture with varicella-zoster virus. *J. Virol.* **2003**, *77*, 11718–11732. [CrossRef] [PubMed]
32. Braspenning, S.E.; Verjans, G.M.; Mehraban, T.; Messaoudi, I.; Depledge, D.P.; Ouwendijk, W.J.D. The architecture of the simian varicella virus transcriptome. *PLoS Pathog.* **2021**, *17*, e1010084. [CrossRef]
33. Braspenning, S.E.; Sadaoka, T.; Breuer, J.; Verjans, G.M.; Ouwendijk, W.J.D.; Depledge, D.P. Decoding the architecture of the varicella-zoster virus transcriptome. *mBio* **2020**, *11*, e01568-20. [CrossRef] [PubMed]
34. Perera, L.P.; Mosca, J.D.; Sadeghi-Zadeh, M.; Ruyechan, W.T.; Hay, J. The varicella-zoster virus immediate early protein, IE62 can positively regulate its cognate promoter. *Virology* **1992**, *191*, 346–354. [CrossRef]
35. Ou, Y.; Gray, W.L. Simian varicella virus gene 28 and 29 promoters share a common upstream stimulatory factor-binding site and are induced by IE62 transactivation. *J. Gen. Virol.* **2006**, *87*, 1501–1508. [CrossRef]
36. Inchaupse, G.; Nagpal, S.; Ostrove, J.M. Mapping of two varicella-zoster virus-encoded genes that activate the expression of viral early and late genes. *Virology* **1989**, *173*, 700–709.
37. Gray, W.L.; Davis, K.; Ou, Y.; Ashburn, C.; Ward, T.M. Simian varicella virus gene 61 encodes a viral transactivator but is non-essential for in vitro replication. *Arch. Virol.* **2007**, *152*, 553–563. [CrossRef]
38. Moriuchi, H.; Moriuchi, M.; Straus, S.E.; Cohen, J.I. Varicella-zoster virus (VZV) open reading frame 61 protein transactivates VZV gene promoters and enhances the infectivity of VZV DNA. *J. Virol.* **1993**, *67*, 4290–4295. [CrossRef]
39. Sato, H.; Pesnicak, L.; Cohen, J.I. Use of a rodent model to show that varicella-zoster virus ORF61 is dispensable for establishment of latency. *J. Med. Virol.* **2003**, *70*, S79–S81. [CrossRef]
40. Mahalingam, R.; Gilden, D.; Wellish, M.; Pugazhenthi, S. Transactivation of the simian varicella virus (SVV) open reading frame (ORF) 21 promoter by SVV ORF 62 is upregulated in neuronal cells but downregulated in non-neuronal cells by SVV ORF 63 protein. *Virology* **2006**, *345*, 244–250. [CrossRef]
41. Brazeau, E.; Wellish, M.; Kaufer, B.B.; Tischer, B.K.; Gray, W.L.; Zhou, F.; Osterrieder, N. Simian varicella virus open reading frame 63/70 expression is required for efficient virus replication in culture. *J. Neurovirol.* **2011**, *17*, 274–280. [CrossRef] [PubMed]
42. Jackers, P.; Defechereux, P.; Baudoux, L.; Lambert, C.; Massaer, M.; Merville-Louis, M.P.; Rentier, B.; Piette, J. Characterization of regulatory functions of the varicella-zoster virus gene 63-encoded protein. *J. Virol.* **1992**, *66*, 3899–3903. [CrossRef]
43. Ward, T.M.; Williams, M.V.; Traina-Dorge, V.; Gray, W.L. The simian varicella virus uracil DNA glycosylase and dUTPase genes are expressed in vivo, but are non-essential for replication in cell culture. *Virus Res.* **2009**, *142*, 78–84. [CrossRef]
44. Pumphrey, C.Y.; Gray, W.L. Identification and analysis of the simian varicella virus thymidine kinase gene. *Arch. Virol.* **1996**, *141*, 43–55. [CrossRef]
45. Ashburn, C.; Gray, W.L. Identification and characterization of the simian varicella virus uracil DNA glycosylase. *Arch. Virol.* **1999**, *144*, 2161–2172. [CrossRef] [PubMed]
46. Roberts, G.B.; Fyfe, J.A.; McKee, S.A.; Rahim, S.G.; Daluge, S.M.; Almond, M.R.; Rideout, J.L.; Koszalka, G.W.; Krenitsky, T.A. Varicella-zoster virus thymidine kinase characterization and substrate specificity. *Biochem. Pharmacol.* **1993**, *46*, 2209–2218. [CrossRef]
47. Chaudhuri, V.; Sommer, M.; Rajamani, J.; Zerboni, L.; Arvin, A.M. Functions of varicella-zoster virus ORF23 capsid protein in viral replication and the pathogenesis of skin infection. *J. Virol.* **2008**, *82*, 10231–10246. [CrossRef]
48. Grose, C.; Carpenter, J.E. Overview of varicella-zoster virus glycoproteins gC, gH, and gL. *Curr. Top. Microbiol. Immunol.* **2010**, *342*, 113–128.
49. Pumphrey, C.Y.; Gray, W.L. DNA sequence and transcriptional analysis of the simian varicella virus glycoprotein B gene. *J. Gen. Virol.* **1994**, *75*, 3219–3227. [CrossRef]
50. Pumphrey, C.Y.; Gray, W.L. DNA sequence of the simian varicella virus (SVV) gH gene and analysis of the SVV and varicella zoster virus gH transcripts. *Virus Res.* **1995**, *38*, 55–70. [CrossRef]
51. Mahalingam, R.; Clarke, P.; Wellish, M.; Dueland, A.N.; Soike, K.F.; Gilden, D.H.; Cohrs, R. Prevalence and distribution of latent simian varicella virus DNA in monkey ganglia. *Virology* **1992**, *188*, 193–197. [CrossRef]
52. Ou, Y.; Davis, K.A.; Traina-Dorge, V.; Gray, W.L. Simian varicella virus expresses a latency-associated transcript that is antisense to open reading frame 61 (ICP0) mRNA in neural ganglia of latently infected monkeys. *J. Virol.* **2007**, *81*, 8149–8156. [CrossRef] [PubMed]
53. Depledge, D.P.; Ouwendijk, W.J.D.; Sadaoka, T.; Braspenning, S.E.; Mori, Y.; Cohrs, R.J.; Verjans, G.M.; Breuer, J. A spliced latency-associated VZV transcript maps antisense to the viral transactivator gene 61. *Nat. Commun.* **2018**, *9*, 1167. [CrossRef] [PubMed]
54. Meyer, C.; Kerns, A.; Haberthur, K.; Dewane, J.; Walker, J.; Gray, W.L.; Messaoudi, I. Attenuation of the adaptive immune response in rhesus macaques infected with simian varicella virus lacking open reading frame 61. *J. Virol.* **2013**, *87*, 2151–2163. [CrossRef] [PubMed]

55. Perng, G.C.; Ghiasi, H.; Slanina, S.M.; Nesburn, A.B.; Wechsler, S.L. The spontaneous reactivation function of the herpes simplex virus type 1 LAT gene resides completely within the first 1.5 kilobases of the 8.3-kilobase primary transcript. *J. Virol.* **1996**, *70*, 976–984. [CrossRef]
56. Perng, G.C.; Ciacci-Zanella, J.; Stone, M.; Henderson, G.; Yukht, A.; Slanina, S.M.; Hofman, F.M.; Ghiasi, H.; Nesburn, A.B.; Wechsler, S.L. Virus-induced neuronal apoptosis blocked by the herpes simplex virus latency-associated transcript. *Science* **2000**, *287*, 1500–1503. [CrossRef]
57. Ciacci-Zanella, J.; Stone, M.; Henderson, G.; Jones, C. The latency-related gene of bovine herpesvirus 1 inhibits programmed cell death. *J. Virol.* **1999**, *73*, 9734–9740. [CrossRef]
58. Grose, C. Pangea and the out-of-Africa model of varicella-zoster virus evolution and phylogeography. *J. Virol.* **2012**, *86*, 9558–9565. [CrossRef]
59. McGeoch, D.J.; Rixon, F.J.; Davison, A.J. Topics in herpesvirus genomics and evolution. *Virus Res.* **2006**, *117*, 90–104. [CrossRef]
60. Breuer, J.; Grose, C.; Norberg, P.; Tipples, G.; Schmid, D.S. A proposal for common nomenclature for viral clades that form the species varicella-zoster virus: Summary of VZV Nomenclature Meeting 2008, Barts and the London School of Medicine and Dentistry, 24–25 July 2008. *J. Gen. Virol.* **2010**, *91*, 821–828. [CrossRef]

viruses

MDPI

Review

Distinguishing Features Common to Dual Fatal Herpes Simplex Virus Infections That Occur in Both a Pregnant Woman and Her Newborn Infant

Nathan B. Price [1,*] and Kelly E. Wood [2]

[1] Division of Infectious Diseases, Department of Pediatrics, University of Arizona, Tucson, AZ 85719, USA
[2] Newborn Nursery Service and Hospitalist Service, Division of General Pediatrics and Adolescent Medicine, Stead Family Department of Pediatrics, University of Iowa, Iowa, IA 52242, USA; kelly-wood@uiowa.edu
* Correspondence: pricen@peds.arizona.edu

Abstract: Deaths from herpes simplex virus type 1 (HSV-1) and herpes simplex virus type 2 (HSV-2) are rare. A major exception is perinatally acquired HSV-1 or HSV-2 infection where the neonatal death rate is substantial. Fatal HSV infection also occurs occasionally in pregnant women. The goal of this review is to enumerate the reports that describe dual deaths of both a pregnant woman and her newborn from a herpesvirus infection. A total of 15 reports were found in the medical literature, of which five described pregnant women with HSV encephalitis and 10 described women with disseminated HSV infection. When the virus was typed, most cases of dual mother/newborn deaths were caused by HSV-2. Of interest, in two situations caused by HSV-1, the pregnant woman probably acquired her primary HSV-1 infection from one of her children and not by sexual transmission. Complete genomic sequencing was performed on one set of HSV-1 isolates collected from mother (blood) and newborn (blood and skin). The mother's strain and the newborn's skin strain were 98.9% identical. When the newborn's two strains were compared, they were 97.4% identical. Only one mother was tested by the HerpeSelect IgG antibody kit. During the nine days of her undiagnosed disseminated infection preceding her death, her serology was negative. In summary, although dual mother/newborn deaths from HSV infection are rare, they continue to be reported as recently as 2017.

Keywords: herpes simplex virus type 1; herpes simplex virus type 2; neonatal herpesvirus infection; HSV UL6; herpesvirus encephalitis; herpesvirus hepatitis; acyclovir; HerpeSelect test; viral sequencing; HELLP syndrome

Citation: Price, N.B.; Wood, K.E. Distinguishing Features Common to Dual Fatal Herpes Simplex Virus Infections That Occur in Both a Pregnant Woman and Her Newborn Infant. *Viruses* **2021**, *13*, 2542. https://doi.org/10.3390/v13122542

Academic Editors: Charles Grose, Ravi Mahalingam and Joel Rovnak

Received: 29 November 2021
Accepted: 16 December 2021
Published: 18 December 2021

1. Introduction

There are two human herpes simplex viruses, called herpes simplex virus type 1 (HSV-1) and herpes simplex virus type 2 (HSV-2) [1]. Their structures are shown after examination by electron microscopy (EM) (Figure 1). HSV-1 is usually acquired during childhood, as a disease presenting with gingivostomatitis, while HSV-2 is often acquired as a sexually transmitted disease in adolescence and early adulthood [2]. However, over the past 30 years, physicians at several major metropolitan areas, including the University of Washington Medical Center in Seattle, have observed that HSV-1 has become a more common cause of genital disease than HSV-2 [1]. As with all herpesviruses, after primary infection both HSV-1 and HSV-2 enter a latent state within the dorsal root ganglia. Both viruses periodically reactivate throughout the life of an infected individual [3–5]. Death from a primary infection or reactivation with either HSV-1 or HSV-2 is exceedingly rare in immunocompetent children and adults [6]. However, when an otherwise healthy woman contracts a primary infection during pregnancy, the likelihood of death rises considerably [7,8]. Furthermore, there is also a concurrent risk that the fetus or newborn infant becomes infected [9]. What is extraordinarily unusual is for both a pregnant woman and her newborn child to die after a primary herpesvirus infection. This rare scenario in medicine is the subject of this review. The goal of the review is to search for further insight into pathogenesis of a herpesvirus infection when both mother and infant die from the same infection.

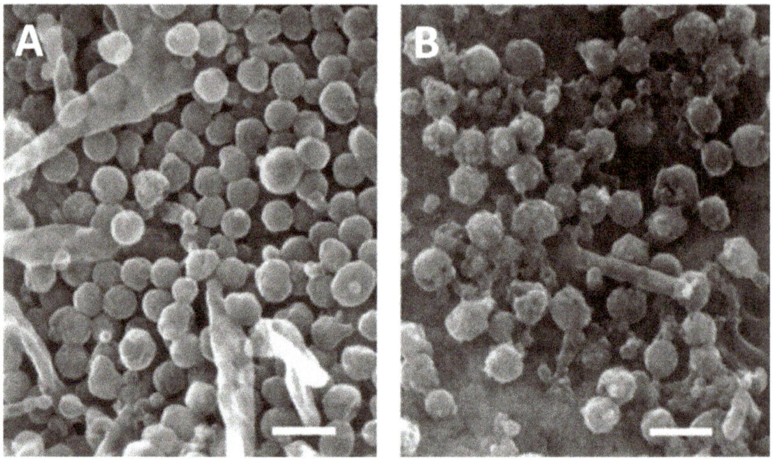

Figure 1. Scanning electron micrographs of herpes viruses. (**A**). HSV-1. (**B**). HSV-2. The appearance of complete enveloped HSV-1 particles is more uniform than HSV-2 particles on the surface of an infected cultured cell. Micrographs obtained from the archives of the University of Iowa Central Microscopy Research Facility. Bar, 400 nanometers.

2. Evolution of the Viruses

HSV-1 is an ancient virus that has coevolved with primates throughout 60 million years of evolution in Africa, finally coevolving with ancestors to *Homo sapiens* into contemporary times [10,11]. By contrast, HSV-2, also commonly called genital herpes, evolved around 1 million years ago when a herpes virus from the great ape species bonobo was acquired by *Homo erectus* [12]. The mechanism for this zoonotic spread is likely the same as for human acquisition of HIV-1, namely *Homo erectus* hunted and ate the herpesvirus-infected bonobo for food in East Africa, while contemporary human inhabitants living in West Africa around 100 years ago hunted and ate the retrovirus-infected chimpanzee for food [13]. In other words, Lucy Australopithecus, who lived 3 million years ago in the country now known as Ethiopia would likely have had an HSV-1 infection passed onto her from her parents or siblings, but she could not have had HSV-2 infection, since HSV-2 had not yet arisen [14,15]. However, by the time *Homo sapiens* migrated out of Africa into the Middle East 70,000 years ago, some would have been infected with both HSV-1 and HSV-2. Thus, HSV-1 and HSV-2 infections are found in nearly all human populations distributed around the world. As HSV-1 infected humankind spread into Asia and Europe over past millennia, a sufficient number of polymorphisms have arisen in the genomes, that current strains of HSV-1 can be segregated into Asian and European groupings or clades [16]. There does not appear to be any differences in virulence between the geographic clades of viruses. A similar pattern of geographic genome variation is seen within varicella-zoster virus (VZV) clades [17]. Likewise, there is no difference in virulence between the different geographic VZV clades.

3. Past Published Cases of Dual Infection with Dual Deaths

As noted in the Introduction, deaths following HSV infection are rare. After a search of the medical literature, we have found 15 reports of dual deaths of a pregnant woman (Table 1) and her newborn infant (Table 2). We have divided the cases into two groups. The first group includes pregnant women who developed HSV encephalitis (cases 1–5) and the second group includes pregnant women with a disseminated HSV infection (cases 6–15).

Vignettes of each case are included below. We recognize that the documentation of HSV infection in some of the earlier cases (especially the newborn) is less than that in the later cases.

Table 1. Fatal Herpes in Mothers.

Case #	Year	Age	Serotype	Findings at Death	Day of Death	Reference
1	1966	31	unk [a]	Encephalitis	10	[18]
2	1975	28	unk	Encephalitis	11	[19]
3	1979	20	unk	Encephalitis	14	[20]
4	1979	22	unk	Encephalitis	3	[20]
5	1979	17	unk	Encephalitis	14	[20]
6	1974	23	1	Liver necrosis	0.5	[21]
7	1980	21	2	Liver necrosis	12	[22,23] [b]
8	1982	33	unk	Liver necrosis	14	[7]
9	1985	18	unk	Liver necrosis	9	[24,25] [b]
10	1992	20	2	Liver necrosis	4	[26]
11	1996	30	2	Liver necrosis	9	[27]
12	1996	21	2	Liver necrosis	8	[8]
13	2002	27	2	Liver necrosis	18	[28]
14	2017	18	2	Liver necrosis	9	[29]
15	2017	41	1	Liver necrosis	7	[30,31] [b]

Note: [a], unk, serotype not known; [b], same case reported by two different author groups.

Table 2. Fatal Herpes in Fetuses and Newborns.

Case #	Year	Serotype	Maternal Infection	GA (Weeks)	Age at Death (Days)	Antiviral
1	1966	Unk	encephalitis	28	IUFD	No
2	1975	Unk	encephalitis	28	0	No
3	1979	Unk	encephalitis	28	0	No
4	1979	Unk	encephalitis	16	0	No
5	1979	Unk	encephalitis	24	0	No
6	1974	1	disseminated	28	IUFD	No
7	1980	2	disseminated	36	10	No
8	1982	Unk	disseminated	25	IUFD	No
9	1985	Unk	disseminated	34	2	No
10	1992	2	disseminated	34	14	Yes
11	1996	2	disseminated	32	11	Yes
12	1996	2	disseminated	13	IUFD	No
13	2002	2	disseminated	27	1	No
14	2017	2	disseminated	26	4	No
15	2017	1	disseminated	30	5	Yes

Abbreviations: GA: gestational age; Unk: serotesting not performed; IUFD: Intrauterine fetal death.

3.1. *Case 1, Minnesota, USA*

A 31-year-old woman developed runny nose, headache, hallucinations, confusion, and fever over the course of a week [18]. She was admitted at 7 months of gestation and had no oral HSV lesions on examination. Testing of the cerebrospinal fluid (CSF) showed lymphocytic pleocytosis and an electroencephalogram (EEG) was abnormal. She developed seizures, became comatose and died on the 10th day of hospitalization. Immediately post-mortem, a fetus was delivered by cesarean section (C-section), but the fetus was not viable. Maternal autopsy showed viral cytopathic effect in the brain, and brain tissue culture grew HSV which was not serotyped. Fetal autopsy studies were not done.

3.2. *Case 2, Massachusetts, USA*

A 28-year-old woman developed malaise, slurred speech, amnesia, and unexplained falls over the course of 10 days [19]. After she developed fever and seizures, she was admitted at 28 weeks of gestation. CSF examination showed pleocytosis. She was given empiric antibiotics, but she developed additional seizures and became comatose. She progressively worsened and died 3 weeks after initial onset of symptoms. The day after her hospitalization, a fetus was delivered spontaneously and died within 3 h of birth. Maternal autopsy showed brain edema and necrosis. Culture from the temporal lobe grew HSV, but no typing was done. Infant autopsy showed hyaline membrane disease of prematurity, but no viral cultures were performed.

3.3. Case 3, Bogota, Columbia

A 20-year-old woman was admitted with seizures and treated for eclampsia at 28 weeks of gestation [20]. The next day she delivered a dead fetus. Subsequently, the woman developed headache, fever, hallucinations, bizarre behavior and delirium. CSF examination showed lymphocytic pleocytosis. She was started on empiric therapy for tuberculosis. She progressively worsened, became comatose, had worsening seizures and died about 2 weeks after admission. Autopsy showed necrotizing encephalitis involving bilateral temporal lobes with viral inclusion bodies in the neurons and glia suggestive of an HSV infection. No fetal autopsy was done.

3.4. Case 4, Bogota, Columbia

A 22-year-old woman was admitted to a psychiatric hospital because of 2 months of bizarre behavior [20]. She was diagnosed with and treated for schizophrenia. At 16 weeks of gestation, she had a spontaneous abortion and was admitted to the hospital for curettage. She then developed hypotension, bradycardia and became unresponsive. She was treated for septic shock but died 3 days after admission. Autopsy showed bitemporal subacute necrotizing encephalitis with viral inclusion bodies in the neurons. Electron microscopy showed abundant herpesvirus-like particles. No fetal autopsy was done.

3.5. Case 5, Bogota, Columbia

A 17-year-old young woman developed bizarre behavior at 24 weeks of gestation [20]. A week later she was admitted to the hospital because of coma. CSF examination showed normal cell count, protein, and glucose. Labor was induced and she delivered a dead fetus. Subsequently, the woman developed fevers, seizures and had an abnormal EEG consistent with temporal lobe involvement. Right temporal lobe biopsy showed edema, necrotizing encephalitis and viral inclusion bodies suggestive of HSV infection. She was treated with steroids but continued to worsen and died 2 weeks after admission. Autopsy findings in the mother were similar to the biopsy findings; no fetal autopsy was done.

3.6. Case 6, Texas, USA

A 23-year-old woman developed fever, sore throat, malaise and shortness of breath at 28 weeks of gestation [21]. She was noted to have an erythematous posterior pharynx with ulcerated lesions. Her 2-year-old child at home was recovering from herpes gingivostomatitis at the time. The mother developed worsening respiratory distress and was admitted to the hospital. No fetal heart tones were heard, and the uterus measured the same as it had 2 weeks prior. She had leukocytosis, anemia and proteinuria. She developed hypoxia, hypotension and died within 16 h of hospitalization from cardiopulmonary arrest. Autopsy showed hypopharyngeal ulcers as well as necrosis of the liver and adrenal glands with Cowdrey Type A inclusions. EM showed viral particles consistent with HSV. Tissue cultures grew HSV-1. The fetus appeared normal for gestation on gross examination, but viral cultures were not done.

3.7. Case 7, Missouri, USA

A 21-year-old woman at 36 weeks of gestation fell several times at home [22,23]. Over the course of the next 5 days, she developed abdominal pain and fever not responsive to oral antibiotics. She was admitted to the hospital and found to have ruptured membranes with purulent fluid and a cervical lesion with a white pseudomembrane. The mother was treated with antibiotics and the fetus was delivered by C-section within hours of maternal admission. The mother developed multiorgan failure of the liver and kidneys, in addition to encephalopathy. A cervical smear was consistent with HSV infection; a liver biopsy showed necrosis and grew HSV-2. Adenine arabinoside was started on hospital day 9, but she continued to deteriorate and died 12 days after admission. Autopsy showed liver necrosis. The newborn died at 10 days of life and autopsy showed hepatoadrenal necrosis. No cultures were done.

3.8. Case 8, Louisiana, USA

A 33-year-old woman at 24 weeks of gestation developed fever, malaise, nausea, vomiting and abdominal pain over 4 days [7]. She was admitted to the hospital and found to have ulcerative cervicitis. Cultures of lesions later grew HSV, type not noted. She was started on empiric antibiotics, developed hepatitis and underwent exploratory laparotomy; she was found to have many small white nodules on the liver. Antibiotics were started. She continued to deteriorate and had spontaneous contractions; the fetus failed to progress and a stillborn infant was delivered by hysterotomy. Liver biopsy showed necrosis and Cowdrey Type A inclusions. Her condition continued to deteriorate and she developed bleeding, encephalopathy, seizures, respiratory and renal failure; she died on the 14th hospital day. Autopsy showed liver necrosis with Cowdrey type A inclusions and EM inclusions consistent with herpesvirus. She also had pulmonary embolism, cerebral edema and cerebellar tonsillar herniation. Tissue cultures were negative. No autopsy of the fetus was done.

3.9. Case 9, Michigan, USA

An 18-year-old woman at 34 weeks of gestation developed fever, malaise and otalgia over 4 days [24,25]. She had no mucocutaneous HSV lesions on exam. She was admitted with hepatitis and started on empiric antibiotics. Her hepatitis worsened and she developed coagulopathy. The infant was delivered by C-section 5 days after admission. The mother developed seizures and died 4 days after the infant was delivered. Autopsy showed liver necrosis. Liver, rectal and throat cultures grew HSV, type not specified. HSV antibodies were negative on admission and positive on hospital day 8. The infant developed hyaline membrane disease of prematurity and intraventricular hemorrhage and died 2 days after delivery. The infant's culture results for HSV were negative.

3.10. Case 10, Hamburg, Germany

A 20-year-old woman was treated for 2 weeks with antibiotics for pyelonephritis at one hospital 1 [26]. She developed hepatitis, coagulopathy and was transferred to a different hospital at 34 weeks of gestation. Liver biopsy done at the prior facility showed necrosis and she was started on acyclovir. The mother's condition worsened and the fetus showed signs of distress; a C-section was done 24 h after transfer. The mother had some improvement for about 12 h, then developed seizures, encephalopathy and respiratory failure. She died 4 days after transfer. Autopsy showed pyelitis of the kidney and liver necrosis. Culture from the liver grew HSV-2. The infant was treated with acyclovir but died of encephalitis at 14 days of age. Intrathecal antibodies were positive for HSV-2.

3.11. Case 11, South Carolina, USA

A 30-year-old woman at 31 weeks of gestation was admitted with a 2-day history of fever, cough and abdominal tenderness [27]. She was started on antibiotics for pyelonephritis. She developed hypoxia, pulmonary infiltrates, coagulopathy and hepatitis. Fetal distress was noted and the infant was delivered by C-section on hospital day 5. One of four maternal blood cultures grew Group B Streptococcus. She developed respiratory and renal failure and died on hospital day 9. Autopsy showed liver necrosis, and immunostaining showed Cowdry type A inclusions positive for HSV. Liver tissue cultures grew HSV 2. Inclusions were also seen in lung and spleen tissue. The infant developed hyaline membrane disease and was started on acyclovir on day 6 after the possibility of HSV infection of the mother was raised. Despite this therapy, the infant died at 11 days of life. Cultures from the infant's blood, urine, throat and CSF grew HSV-2. Autopsy was not done.

3.12. Case 12, Texas, USA

A 21-year-old woman at 12 weeks of gestation had fever, myalgia, urinary frequency and urgency for 3 days not responsive to nitrofurantoin [8]. She was admitted and started on empiric antibiotics. She had no mucocutaneous lesions. She developed cough, respiratory distress and lung infiltrates on chest film, followed by hepatitis and coagulopathy. She worsened and died 8 days after admission. Autopsy of the woman showed necrotizing bronchopneumonia with Cowdry type A inclusions in the lung that stained positive for HSV. Liver necrosis was seen with positive HSV staining, and cultures grew HSV 2. Uterine necrosis was seen with few cells suspicious for viral inclusions, but uterine tissue cultures were negative for HSV. A fetal autopsy was not performed.

3.13. Case 13, California, USA

A 21-year-old woman at 27 weeks of gestation presented with fever and vaginal discharge [28]. She was found to have a right lower lobe pneumonia and leukopenia and did not respond to 4 days of empiric antibiotics. She was transferred to a higher-level medical facility. She did not have any mucocutaneous lesions. She developed hepatitis, coagulopathy and within 24 h of transfer she developed hypotension and respiratory failure. Fetus had appeared healthy, but labor was induced due to risk of ongoing sepsis to the fetus. C-section was not attempted due to maternal condition and the fetus was delivered stillborn 9 h later. The mother's condition continued to worsen, and she developed skin vesicles on her thigh, forearm and back. She was started on acyclovir and cultures of the lesions grew HSV-2. She developed anasarca and then died after repeated episodes of ventricular tachycardia 18 days after admission. Autopsy showed liver and lung necrosis with immunohistochemical staining positive for HSV. Autopsy of the fetus was not done.

3.14. Case 14, Ohio, USA

An 18-year-old woman at 26 weeks of gestation developed shortness of breath and fever [29]. She was admitted and found to have hepatitis and coagulopathy. A fetus was delivered by C-section and mother continued to deteriorate; she was transferred for liver transplant evaluation. She developed renal and respiratory failure and was started on acyclovir on hospital day 3. She died of multiorgan failure on hospital day 9. A blood HSV-2 polymerase chain reaction (PCR) test and a herpes IgM test were positive. The newborn died of HSV-2 sepsis at 4 days of life.

3.15. Case 15, Iowa, USA

This dual-case where both the pregnant mother and her infant died from HSV-1 infection has been described in two reports [30,31]. These are summarized in the section below.

4. A Recent Case of Fatal Dual Infection

The mother in Case 15 was a 41-year-old female in the 29th week of gestation during her sixth pregnancy [30]. She had two living children and three prior miscarriages. She was admitted to a hospital because of a 2-day history of fever and cough (Figure 2). During an earlier clinic visit, she had been diagnosed with a urinary tract infection. After an Infectious Disease consultation, she was treated with ceftriaxone and oseltamivir. Her liver function tests on admission included aspartate aminotransferase (AST) 90 units and alanine aminotransferase (ALT) 61 units, and her creatinine was 0.7 mg/dL. Within 3 days, her AST and ALT levels were 5652 and 1559 and her creatinine was 3.2. Her platelet count had fallen from 172,000 to 139,000/mm^3. She became disoriented. The main diagnoses that were considered included HELLP syndrome (hemolysis, elevated liver enzymes and lowered platelets) and acute fatty liver of pregnancy [32]. She was transferred on an emergent basis to a referral hospital, where a C-section was performed around 2 AM. An order was placed for a HerpeSelect IgG antibody test (Focus Diagnostics, Cypress, CA, USA) after the C-section and the results showed no detectable antibody to either HSV-1 or HSV-2. Her hepatic function and renal function continued to worsen throughout the day, and she died around 11 PM.

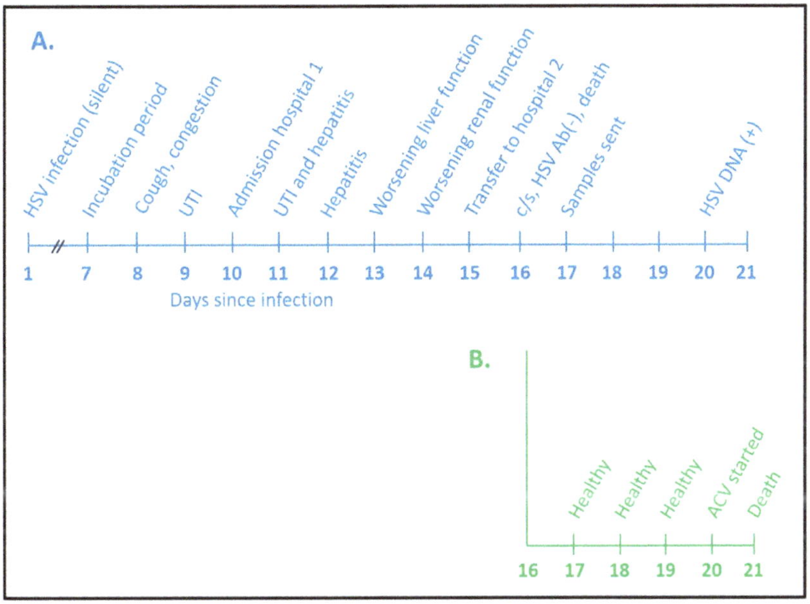

Figure 2. Timeline of hospitalization of pregnant woman and her newborn infant. For this timeline, a typical incubation period for primary HSV-1 infection was considered to be 7 days [6]. The diagram documents a day-by-day rapid progression to death of a pregnant woman (**A**, blue) and her newborn infant (**B**, green) after disseminated HSV-1 infection in the woman. UTI, urinary tract infection; c/s, cesarean section; ACV, acyclovir.

A complete autopsy was performed on the following day. Numerous additional samples were collected and sent, mainly to outside diagnostic facilities. These tests included several PCR assays. The results were returned 3 days later. A PCR assay for HSV-1 was positive in a blood sample. Subsequent immunohistochemical analysis of the liver and adrenal glands showed multiple foci of infection by HSV-1, but no foci of HSV-1 infection were found in the kidney. In Figure 2, we have placed the time of HSV-1 infection for the mother as one week before she began to develop symptoms of illness (incubation period); she died 16 days after contracting the infection.

The transfer note from the obstetrics staff to the neonatology staff cited severe preeclampsia as the reason for the emergency C-section [31]. The newborn underwent a physical examination by neonatologists shortly after the C-section; although premature, the examination and weight (1.6 kg) were normal for 30 weeks of gestation (Figure 2). The blood indices and ALT test were also normal. A similar normal examination was recorded during the first 3 days of life. However, when the deceased mother's positive HSV-1 PCR results were returned on day of life 4, the infant had begun having apneic spells; he was immediately placed on IV acyclovir (60 mg/kg/day). The neonate died on day 5 of life of disseminated HSV-1 infection. A blood PCR test collected on day 4 subsequently was found to be positive for HSV-1. The infant never developed a skin rash. An autopsy was not performed on the neonate.

Since the newborn had a normal physical examination and a normal ALT test on the first day of life, we speculate that the newborn may have contracted HSV-1 infection during the emergent C-section. An alternative possibility is that the fetus was infected via placental transfer a few days or a few hours before the C-section. Overall, however, intrauterine HSV infection is an uncommon event [1,9].

5. Pathogenesis of Fatal Herpes Infection in Infected Pregnant Women

After a review of the 15 reports, we have delineated two very different modes of pathogenesis leading to death of both pregnant mother and newborn by HSV infection. The first is HSV encephalitis in the pregnant women. Several reports about the pathogenesis of HSV encephalitis are in the medical literature and this illness will not be further discussed [33,34]. The second is a disseminated HSV infection in the mother, sometimes called herpes sepsis [35]. Furthermore, because of this review, we in turn have distinguished two forms of disseminated HSV infection in the pregnant woman. With regard to HSV-1, in both reported cases, we speculate that the mother acquired a primary HSV-1 infection from one of her young children with or without gingivostomatis, probably through exchange of oral secretions laden with HSV-1. Mother #15 was the oldest mother in Table 1 to die from disseminated herpes. She was a homemaker who also ran a private childcare center in her home. We speculate that she contracted a primary HSV-1 infection from either one of her own children or from one of the children in childcare. This event probably occurred around 2 weeks before she was admitted to hospital. Thus, this case closely resembles case #6, where the mother probably acquired HSV-1 from her child. This mode of HSV-1 transmission from an HSV-1 seropositive child to an HSV-1-seronegative mother is strikingly similar to a common mode of transmission when a CMV-seronegative pregnant women acquires a primary CMV infection from her CMV-seropositive child and subsequently delivers a newborn with a congenital CMV infection [36]. Since the seroprevalence of HSV-1 antibody in American adolescents is only 31% [37], a majority of young pregnant women are at risk of a primary HSV-1 infection.

A second possible mode of HSV-1 acquisition by mother #15 was sexual; under this scenario, her early UTI may have been a misdiagnosed HSV-1 primary genital infection. With regard to HSV-2, all the dual deaths listed in Tables 1 and 2 followed dissemination from a primary or recurrent genital infection [38,39]. One of the maternal organs most often damaged by disseminated HSV infection is the liver (Figure 3). Liver necrosis caused by extensive HSV-2 replication was documented in all the autopsies from cases 6 to 15 (Table 1).

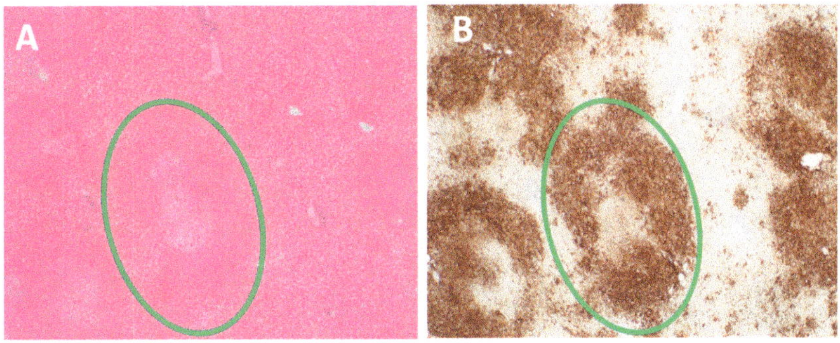

Figure 3. Pathology in liver infected with HSV. Inspection of the surface of the liver demonstrated numerous nodules (2–5 mm in diameter). (**A**). Hematoxylin and eosin stain of an HSV-infected liver section. Viral cytopathology is not easily distinguished by routine staining (green circle). (**B**). Immunohistochemistry. Viral cytopathology is easily distinguished after special immunostaining of an HSV-infected liver with anti-HSV antibody. The dark brown color indicates necrotic viral foci (green circle). Micrographs obtained from the archives of the University of Iowa Department of Pathology. Magnification, 20×.

6. Herpesvirus Diagnostic Testing

For diagnosis of a disseminated herpesvirus infection, a PCR-based HSV test of blood is preferred; for diagnosis of herpesvirus encephalitis, a PCR-based HSV test of cerebrospinal fluid is preferred, and for diagnosis of a herpesvirus genital infection, a PCR-based HSV test of swabs collected from the genital area is preferred [1]. Herpesvirus serology is not recommended because the results can be confusing. This report illustrates that a pregnant woman can contract a primary HSV-1 infection, develop HSV-1 viremia, transmit the virus to her fetus, undergo C-section, and then die from an HSV-1 sepsis syndrome before her immune system produces HSV-1 specific IgG antibody detectable by the HerpeSelect Test. This conclusion was not readily apparent from the medical literature, where older reviews of herpesvirus infection during pregnancy include tables where many of the included maternal cases have detectable HSV-1 antibody [8]. Our dual case report indicates that there may be a delay in virus specific antibody responses in pregnant women when tested by the HerpeSelect assay. The HerpeSelect test can discriminate HSV-1 and HSV-2 antibodies because the test contains only one viral protein as antigen (gG protein); the gG protein between HSV-1 and HSV-2 does not share epitopes.

Virologists have already titrated sera collected periodically from patients with primary HSV-1 infection and patients with primary HSV-2 infection [40]. The ages of the HSV-1 patients ranged from 2 to 27 years, while the ages of the HSV-2 patients ranged from 29 to 43 years. HSV-1 specific IgG antibody was detected first between 5 to 10 days, while HSV-2 specific IgG antibody was detected by 15 days. Thus, IgG antibody to HSV-1 or HSV-2 was detectable in assays that used whole virus antigen by 13–15 days after primary infection, rather than just the HSV gG protein. This age range also fits perfectly with testing for appearance of IgG antibody to other viruses, such as measles, mumps, rubella and COVID-19 (Figure 4) [41,42].

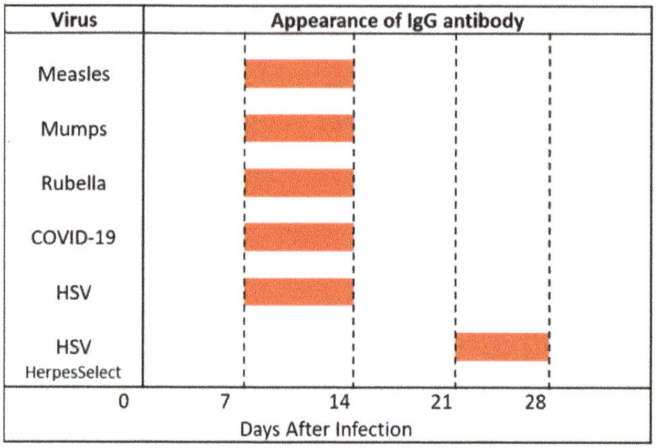

Figure 4. Timeline of appearance of antiviral IgG antibodies in human serum samples. The table shows the timeline by weeks post-infection when specific human IgG antibodies against several common viral pathogens usually are detectable (red rectangles). Data from references [40–43].

The sensitivity of the HerpeSelect test has been assessed in over 100 patients [43]. In patients with culture-documented first episodes of genital herpes, the median time from onset of symptoms to seroconversion by HerpeSelect ELISA was 25 days for patients with primary genital HSV-1 infection and 21 days for those with primary HSV-2 infection (Figure 4). Thus, the HerpeSelect IgG test appears highly specific but less sensitive than older HSV testing methods for detection of the earliest IgG antibodies, which are directed against HSV-1 proteins called gB and Gd [44].

7. Genomic Analyses of the HSV-1 Strains

Extensive sequencing was performed on the viral isolates collected from dual case 15. HSV-1 was isolated from the blood of the mother and from the skin (around the mouth and nose) and blood of the newborn. Complete genomic sequencing was performed on each of the three strains at the PennState Huck Institute of the Life Sciences [31]. The entire HSV-1 genome contains 152 kilobases and encodes over 80 genes. The mother's strain and the infant's skin strain were 98.9% identical and the mother's strain and the infant's blood strain were 96.8% identical. When the baby's two strains were compared, they were 97.4% identical. The majority of the genetic variations were found in the noncoding regions of the genes and in intergenic regions.

Minor variants were found in the open reading frames within each of the HSV-1 strains. Minor variants are alternative alleles that exist in extremely low frequency in any population of a single virus strain. A total of 39 minor variants were detected in the maternal strain, 42 minor variants in the neonate's skin strain and 28 minor variants in the infant's blood strain [31]. By identification of these minor variants, it was apparent that some minor variants in the mother's HSV-1 strain were found in both of the infant's two strains. Thus, there did not appear to be a bottleneck event during transmission of virus from pregnant mother to infant, presumably at the time of the emergency C-section. However, of more interest, there were 23 minor variants in the infant's two HSV-1 strains that were not present in the maternal strain [31]. This result indicated that HSV-1 genetic mutations had occurred in the infant's HSV- population within only 6 days of life.

The single most important nonsynonymous minor variant that occurred in theinfant's two HSV1 strains was found in an HSV-1 gene known as unique-long 6 (UL6) (Figure 5). The HSV1 structural protein is called the portal protein because it is attached to the viral capsid, where it facilitates entry and packaging of newly synthesized viral DNA genomes from the nucleus into the capsid itself [45]. After the viral DNA is packaged within a capsid, the capsid exits the nucleus and enters the cytoplasm, where the capsid undergoes envelopment before finally being released from the infected cell. Since no experiments have yet been performed with the UL6 variant, we do not know if the mutation documented in the HSV-1 portal protein of the infant hinders or accelerates encapsidation of viral genomes. The GenBank accession numbers for the maternal strain, infant's skin strain, and infant's blood strain are listed, respectively: MK 952185, MK 952183, and MK 952184.

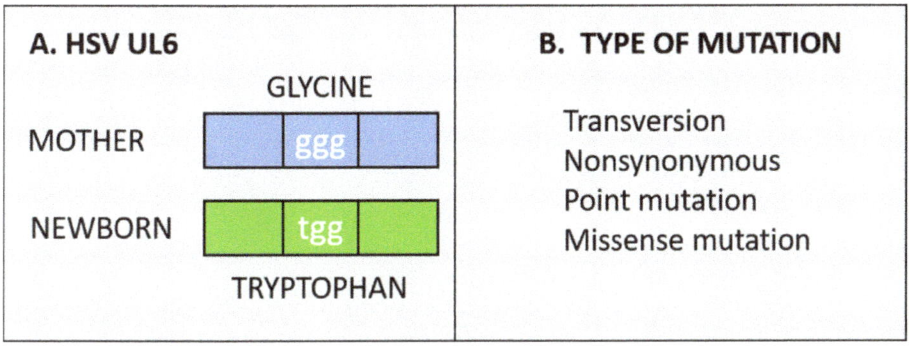

Figure 5. Mutation in codon 486 of HSV UL6 portal protein. (**A**). Diagram of UL6. (**B**). Names of the UL6 mutation. This missense mutation in one open reading frame of the HSV genome was a minor variant detected in the blood of an HSV-infected newborn infant who lived for only 6 days. The variant arose in the infant because it was not detected in the sequencing of the maternal HSV-1 strain.

8. Conclusions

Our survey of the literature has uncovered 15 reports of dual mother/newborn deaths from HSV infection, more than found in other reviews (Tables 1 and 2). The fact that HSV encephalitis may occur more commonly in pregnant women has not been widely discussed. Nevertheless, dual deaths following maternal HSV encephalitis seem to be a disease that has largely disappeared, probably due to the frequent use of intravenous acyclovir treatment in any person suspected of having HSV encephalitis [46].

In sharp contrast, dual mother/newborn deaths following primary HSV-1 infections in a pregnant woman are still being reported. As shown in case #15, the symptoms of a primary HSV-1 infection in the mother can closely resemble other noninfectious illnesses of pregnant women. A negative HerpeSelect IgG titer may be a clue that the pregnant woman potentially has a primary HSV-1 infection. Obviously an HSV-1 DNA PCR test must be performed on a throat swab and a blood sample to confirm this diagnosis. We speculate that both of our HSV-1 cases may have been caused by HSV-1 transmission from a child to a pregnant mother. There were no documented dual deaths from an HSV-1 primary maternal genital infection in the above 15 reports. Whether the six dual-mother/newborn deaths from a documented disseminated HSV-2 infection were caused by a primary HSV-2 infection or the first clinical presentation of a previously acquired but asymptomatic HSV-2 infection cannot be determined by the limited virological testing preformed on the six pregnant women in the above case reports.

Dual deaths of a pregnant woman and her newborn from disseminated HSV infection remain a rare event, with only 10 cases reported in the USA since 1966. In contrast, 68 newborns diagnosed with perinatal HSV infection died in 2017 in the USA [47,48]. This number represents a doubling of the rate of neonatal deaths due to HSV infection in 1995. Based on this review, we reach a counter-intuitive conclusion that the more sick the mother is with disseminated HSV infection (in the absence of rash), the greater the likelihood of an incorrect diagnosis, such as HELLP syndrome or acute fatty liver of pregnancy [32]. There are two pathways by which to reduce the number of deaths from herpesvirus infections: (i) upgrade clinical microbiology laboratories at community hospitals or (ii) develop a herpesvirus vaccine. Currently many community hospitals do not have the technology to perform same-day PCR analyses on a variety of clinical samples. On the other hand, there is considerable interest in the development of a vaccine against both HSV-1 and HSV-2 [49]. Promising experimental data about immunization within a mouse model system have shown that HSV-2 vaccination of a pregnant mouse can prevent HSV-2 infection in her newborn pups [50].

Author Contributions: N.B.P. and K.E.W.performed a literature search and then wrote the review article. All authors have read and agreed to the published version of the manuscript.

Funding: This research received no external funding.

Acknowledgments: We thank Charles Grose (University of Iowa) for helpful comments during the preparation of this review article.

Conflicts of Interest: The authors declare no conflict of interest.

References

1. Melvin, A.J.; Mohan, K.M.; Vora, S.B.; Selke, S.; Sullivan, E.; Wald, A. Neonatal herpes simplex virus infection: Epidemiology and outcomes in the modern era. *J. Pediatric Infect. Dis. Soc.* **2021**. [CrossRef]
2. Nahmias, A.J.; Roizman, B. Infection with herpes-simplex viruses 1 and 2. *N. Engl. J. Med.* **1973**, *289*, 667–674. [CrossRef]
3. Stevens, J.G. Latent herpes simplex virus and the nervous system. *Curr. Top. Microbiol. Immunol.* **1975**, *70*, 31–50. [PubMed]
4. Bloom, D.C. Alphaherpesvirus Latency: A Dynamic State of Transcription and Reactivation. *Adv. Virus Res.* **2016**, *94*, 53–80.
5. Sawtell, N.M.; Thompson, R.L. Alphaherpesvirus Latency and Reactivation with a Focus on Herpes Simplex Virus. *Curr. Issues Mol. Biol.* **2021**, *41*, 267–356. [CrossRef]
6. Juretic, M. Natural history of herpetic infection. *Helv. Paediatr. Acta* **1966**, *21*, 356–368.
7. Wertheim, R.A.; Brooks, B.J., Jr.; Rodriguez, F.H., Jr.; Lesesne, H.R.; Jennette, J.C. Fatal herpetic hepatitis in pregnancy. *Obstet. Gynecol.* **1983**, *62* (Suppl. 3), 38s–42s. [PubMed]
8. Young, E.J.; Chafizadeh, E.; Oliveira, V.L.; Genta, R.M. Disseminated herpesvirus infection during pregnancy. *Clin. Infect. Dis.* **1996**, *22*, 51–58. [CrossRef] [PubMed]
9. Kimberlin, D.W. Herpes simplex virus infections in neonates and early childhood. *Semin. Pediatr. Infect. Dis.* **2005**, *16*, 271–281. [CrossRef]
10. McGeoch, D.J.; Davison, A.J. The molecular evolutionary history of the herpesviruses. In *Origin and Evolution of Viruses*; Domingo, E., Webster, R., Holland, J., Eds.; Academic Press: New York, NY, USA, 1999; pp. 441–465.
11. Davison, A.J. Herpesvirus systematics. *Vet. Microbiol.* **2010**, *143*, 52–69. [CrossRef] [PubMed]
12. Wertheim, J.O.; Hostager, R.; Ryu, D.; Merkel, K.; Angedakin, S.; Arandjelovic, M.; Ayimisin, E.A.; Babweteera, F.; Bessone, M.; Brun-Jeffery, K.J.; et al. Discovery of Novel Herpes Simplexviruses in Wild Gorillas, Bonobos, and Chimpanzees Supports Zoonotic Origin of HSV-2. *Mol. Biol. Evol.* **2021**, *38*, 2818–2830. [CrossRef] [PubMed]
13. Sharp, P.M.; Hahn, B.H. Origins of HIV and the AIDS pandemic. *Cold Spring Harb. Perspect. Med.* **2011**, *1*, a006841. [CrossRef]
14. Johanson, D.C. *Lucy's Legacy: The Quest for Human Origins*; Random House: New York, NY, USA, 2009.
15. Grose, C.; Johanson, D.C. Transmission of Cytomegalovirus, Epstein-Barr Virus, and Herpes Simplex Virus Infections: From the Lucy Australopithecus Epoch to Modern-Day Netherlands. *J. Pediatr.* **2016**, *170*, 9–10. [CrossRef]
16. Szpara, M.L.; Gatherer, D.; Ochoa, A.; Greenbaum, B.; Dolan, A.; Bowden, R.J.; Enquist, L.W.; Legendre, M.; Davison, A.J. Evolution and diversity in human herpes simplex virus genomes. *J. Virol.* **2014**, *88*, 1209–1227. [CrossRef] [PubMed]
17. Grose, C. Pangaea and the Out-of-Africa Model of Varicella-Zoster Virus Evolution and Phylogeography. *J. Virol.* **2012**, *86*, 9558–9565. [CrossRef] [PubMed]
18. Rawls, W.E.; Dyck, P.J.; Klass, D.W.; Greer, H.D., 3rd; Herrmann, E.C., Jr. Encephalitis associated with herpes simplex virus. *Ann. Intern. Med.* **1966**, *64*, 104–115. [CrossRef]
19. Jewett, J.F. Committee on Maternal Welfare Herpes simplex encephalitis. *N. Engl. J. Med.* **1975**, *292*, 531. [CrossRef]
20. Roman-Campos, G.; Navarro de Roman, L.I.; Toro, G.; Vergara, I. Herpes encephalitis in pregnancy. *Am. J. Obstet. Gynecol.* **1979**, *135*, 158–159. [CrossRef]
21. Goyette, R.E.; Donowho, E.M., Jr.; Hieger, L.R.; Plunkett, G.D. Fulminant herpesvirus hominis hepatitis during pregnancy. *Obstet. Gynecol.* **1974**, *43*, 191–195.
22. Kobbermann, T.; Clark, L.; Griffin, W.T. Maternal death secondary to disseminated herpesvirus hominis. *Am. J. Obstet. Gynecol.* **1980**, *137*, 742–743. [CrossRef]
23. Hamory, B.H.; Luger, A.; Kobberman, T. Herpesvirus hominis hepatitis of mother and newborn infant. *South. Med. J.* **1981**, *74*, 992–995. [CrossRef]
24. Goyert, G.L.; Bottoms, S.F.; Sokol, R.J. Anicteric presentation of fatal herpetic hepatitis in pregnancy. *Obstet. Gynecol.* **1985**, *65*, 585–588. [PubMed]

25. Jacques, S.M.; Qureshi, F. Herpes simplex virus hepatitis in pregnancy: A clinicopathologic study of three cases. *Hum. Pathol.* **1992**, *23*, 183–187. [CrossRef]
26. Wolf, H.; Kuhler, O.; Henke, P.; Klose, G. Liver dystrophy in disseminated herpes simplex infection in pregnancy. *Geburtshilfe Frauenheilkd.* **1992**, *52*, 123–125. [CrossRef]
27. Gelven, P.L.; Gruber, K.K.; Swiger, F.K.; Cina, S.J.; Harley, R.A. Fatal disseminated herpes simplex in pregnancy with maternal and neonatal death. *South. Med. J.* **1996**, *89*, 732–734. [CrossRef] [PubMed]
28. Frederick, D.M.; Bland, D.; Gollin, Y. Fatal disseminated herpes simplex virus infection in a previously healthy pregnant woman. A case report. *J. Reprod. Med.* **2002**, *47*, 591–596. [PubMed]
29. Natu, A.; Iuppa, G.; Packer, C.D. Herpes Simplex Virus Hepatitis: A Presentation of Multi-Institutional Cases to Promote Early Diagnosis and Management of the Disease. *Case Rep. Hepatol.* **2017**, *2017*, 3180984. [CrossRef]
30. Masadeh, M.; Shen, H.; Lee, Y.; Gunderson, A.; Brown, K.; Bellizzi, A.; Tanaka, T. A fatal case of herpes simplex virus hepatitis in a pregnant patient. *Intractable Rare Dis. Res.* **2017**, *6*, 124–127. [CrossRef] [PubMed]
31. Shipley, M.M.; Renner, D.W.; Pandey, U.; Ford, B.; Bloom, D.C.; Grose, C.; Szpara, M.L. Personalized viral genomic investigation of herpes simplex virus 1 perinatal viremic transmission with dual fatality. *Cold Spring Harb. Mol. Case Stud.* **2019**, *5*, a004382. [CrossRef] [PubMed]
32. Rath, W.; Tsikouras, P.; Stelzl, P. HELLP Syndrome or Acute Fatty Liver of Pregnancy: A Differential Diagnostic Challenge: Common Features and Differences. *Geburtshilfe Frauenheilkd.* **2020**, *80*, 499–507. [CrossRef]
33. Whitley, R.J. Herpes simplex encephalitis: Adolescents and adults. *Antiviral. Res.* **2006**, *71*, 141–148. [CrossRef]
34. Kennedy, P.G.; Steiner, I. Recent issues in herpes simplex encephalitis. *J. Neurovirol.* **2013**, *19*, 346–350. [CrossRef] [PubMed]
35. Young, E.J.; Killam, A.P.; Greene, J.F., Jr. Disseminated herpesvirus infection. Association with primary genital herpes in pregnancy. *JAMA* **1976**, *235*, 2731–2733. [CrossRef] [PubMed]
36. Pass, R.F.; Hutto, S.C.; Reynolds, D.W.; Polhill, R.B. Increased frequency of cytomegalovirus infection in children in group day care. *Pediatrics* **1984**, *74*, 121–126. [CrossRef] [PubMed]
37. Xu, F.; Lee, F.K.; Morrow, R.A.; Sternberg, M.R.; Luther, K.E.; Dubin, G.; Markowitz, L.E. Seroprevalence of herpes simplex virus type 1 in children in the United States. *J. Pediatr.* **2007**, *151*, 374–377. [CrossRef] [PubMed]
38. Vontver, L.A.; Hickok, D.E.; Brown, Z.; Reid, L.; Corey, L. Recurrent genital herpes simplex virus infection in pregnancy: Infant outcome and frequency of asymptomatic recurrences. *Am. J. Obstet. Gynecol.* **1982**, *143*, 75–84. [CrossRef]
39. Corey, L.; Wald, A. Maternal and neonatal herpes simplex virus infections. *N. Engl. J. Med.* **2009**, *361*, 1376–1385. [CrossRef] [PubMed]
40. Kurtz, J.B. Specific IgG and IgM antibody responses in herpes-simplex-virus infections. *J. Med. Microbiol.* **1974**, *7*, 333–341. [CrossRef] [PubMed]
41. Evans, A.S. *Viral Infections of Humans*, 1st ed.; Plenum Publishing: New York, NY, USA, 1976.
42. Mulligan, M.J.; Lyke, K.E.; Kitchin, N.; Absalon, J.; Gurtman, A.; Lockhart, S.; Neuzil, K.; Raabe, V.; Bailey, R.; Swanson, K.A.; et al. Phase I/II study of COVID-19 RNA vaccine BNT162b1 in adults. *Nature* **2020**, *586*, 589–593. [CrossRef]
43. Ashley-Morrow, R.; Krantz, E.; Wald, A. Time course of seroconversion by HerpeSelect ELISA after acquisition of genital herpes simplex virus type 1 (HSV-1) or HSV-2. *Sex. Transm. Dis.* **2003**, *30*, 310–314. [CrossRef]
44. Kuhn, J.E.; Dunkler, G.; Munk, K.; Braun, R.W. Analysis of the IgM and IgG antibody response against herpes simplex virus type 1 (HSV-1) structural and nonstructural proteins. *J Med Virol* **1987**, *23*, 135–150. [CrossRef] [PubMed]
45. Newcomb, W.W.; Juhas, R.M.; Thomsen, D.R.; Homa, F.L.; Burch, A.D.; Weller, S.K.; Brown, J.C. The UL6 gene product forms the portal for entry of DNA into the herpes simplex virus capsid. *J. Virol.* **2001**, *75*, 10923–10932. [CrossRef]
46. Dodd, K.C.; Michael, B.D.; Ziso, B.; Williams, B.; Borrow, R.; Krishnan, A.; Solomon, T. Herpes simplex virus encephalitis in pregnancy—A case report and review of reported patients in the literature. *BMC Res. Notes* **2015**, *8*, 118. [CrossRef] [PubMed]
47. Donda, K.; Sharma, M.; Amponsah, J.K.; Bhatt, P.; Okaikoi, M.; Chaudhari, R.; Dapaah-Siakwan, F. Trends in the incidence, mortality, and cost of neonatal herpes simplex virus hospitalizations in the United States from 2003 to 2014. *J. Perinatol.* **2019**, *39*, 697–707. [CrossRef] [PubMed]
48. Slutsker, J.S.; Schillinger, J.A. Assessing the Burden of Infant Deaths Due to Herpes Simplex Virus, Human Immunodeficiency Virus, and Congenital Syphilis: United States, 1995 to 2017. *Sex. Transm. Dis.* **2021**, *48*, S4–S10. [CrossRef] [PubMed]
49. Johnston, C.; Gottlieb, S.L.; Wald, A. Status of vaccine research and development of vaccines for herpes simplex virus. *Vaccine* **2016**, *34*, 2948–2952. [CrossRef] [PubMed]
50. Patel, C.D.; Backes, I.M.; Taylor, S.A.; Jiang, Y.; Marchant, A.; Pesola, J.M.; Coen, D.M.; Knipe, D.M.; Ackerman, M.E.; Leib, D.A. Maternal immunization confers protection against neonatal herpes simplex mortality and behavioral morbidity. *Sci. Transl. Med.* **2019**, *11*, eaau6039. [CrossRef] [PubMed]

viruses

MDPI

Conference Report

The 21st Annual Meeting of the Rocky Mountain Virology Association

Laura A. St Clair [1], Ali L. Brehm [1], Shelby Cagle [2], Tillie Dunham [1], Jonathan Faris [3], Paul Gendler [4], Monica E. Graham [5], Sandra L. Quackenbush [2], Joel Rovnak [2] and Rushika Perera [1,*]

1 Center for Vector-Borne Infectious Diseases, Department of Microbiology, Immunology, and Pathology, Colorado State University, Fort Collins, CO 80523, USA; stclairl@colostate.edu (L.A.S.C.); Ali.Brehm@colostate.edu (A.L.B.); Tillie.Dunham@rams.colostate.edu (T.D.)
2 Department of Microbiology, Immunology, and Pathology, Colorado State University, Fort Collins, CO 80523, USA; Shelby.Cagle@colostate.edu (S.C.); Sandra.Quackenbush@colostate.edu (S.L.Q.); Joel.Rovnak@colostate.edu (J.R.)
3 Department of Chemical and Biological Engineering, University of Colorado Boulder, Boulder, CO 80309, USA; Jonathan.Faris@colorado.edu
4 Department of Molecular, Cellular, and Development Biology, University of Colorado Boulder, Boulder, CO 80309, USA; Paul.Gendler@colorado.edu
5 Department of Immunology & Microbiology, University of Colorado Anschutz Medical Campus, Aurora, CO 80045, USA; monica.graham@cuanschutz.edu
* Correspondence: Rushika.Perera@colostate.edu

Abstract: Nestled within the Rocky Mountain National Forest, 114 scientists and students gathered at Colorado State University's Mountain Campus for this year's 21st annual Rocky Mountain National Virology Association meeting. This 3-day retreat consisted of 31 talks and 30 poster presentations discussing advances in research pertaining to viral and prion diseases. The keynote address provided a timely discussion on zoonotic coronaviruses, lessons learned, and the path forward towards predicting, preparing, and preventing future viral disease outbreaks. Other invited speakers discussed advances in SARS-CoV-2 surveillance, molecular interactions involved in flavivirus genome assembly, evaluation of ethnomedicines for their efficacy against infectious diseases, multi-omic analyses to define risk factors associated with long COVID, the role that interferon lambda plays in control of viral pathogenesis, cell-fusion-dependent pathogenesis of varicella zoster virus, and advances in the development of a vaccine platform against prion diseases. On behalf of the Rocky Mountain Virology Association, this report summarizes select presentations.

Keywords: flavivirus; SARS-CoV-2; pandemic preparedness; host–virus interactions; prion; vaccines; interferon lambda; virus

Citation: St Clair, L.A.; Brehm, A.L.; Cagle, S.; Dunham, T.; Faris, J.; Gendler, P.; Graham, M.E.; Quackenbush, S.L.; Rovnak, J.; Perera, R. The 21st Annual Meeting of the Rocky Mountain Virology Association. *Viruses* **2021**, *13*, 2392. https://doi.org/10.3390/v13122392

Academic Editor: Eric O. Freed

Received: 23 November 2021
Accepted: 27 November 2021
Published: 29 November 2021

1. Introduction

In 2000, the Rocky Mountain Virology Club (RMVC) was formed to create an environment where regional scientists could showcase their recent advances in virology, graduate students could interact with senior scientists, and productive collaborations could flourish among graduate students, postdocs, junior investigators, and senior scientists across our region. In 2009, we added prion biologists and investigators to our group, and in 2010, RMVC became the Rocky Mountain Virology Association (RMVA). For most of our gatherings, we have held the annual RMVA meeting at the Colorado State University Mountain Campus located ~50 miles west of the main CSU campus in Fort Collins, CO. In 2020, our annual gathering was held virtually, like many other conferences, due to the ongoing coronavirus pandemic. We were determined to ensure that this year's meeting could once again be held in person. To that end, we required that all attendees provide proof of vaccination and a negative COVID-19 test taken with 48 h of the conference. We also wore masks and practiced social distancing as much as possible. Happily, we report that there have been no reported instances of COVID-19 infections amongst our participants following this year's gathering. Taking these safety precautions allowed all of us to enjoy a much-needed, secluded retreat in the beautiful Colorado Rocky Mountains, and to reconnect (or form new connections) with friends, colleagues, and mentors over our shared love of viruses and prions.

This year's three-day gathering was attended by 114 participants made up of undergraduate students, graduate students, postdocs, research scientists, and junior and senior faculty members (Figure 1). We enjoyed 31 oral presentations and 30 poster presentations, including the keynote address provided by Dr. Ralph Baric and talks from our invited speakers: Drs. Nicole Ehrhart, Joyce Jose, Rupika Delgoda, James Heath, Helen Lazear, Stefan Oliver, and Holger Wille. Talks ranged from advances made in SARS-CoV-2 surveillance, detection, and vaccine development; characterization and analyses of potential emerging zoonoses; molecular determinants of viral replication and pathogenesis; and advances in prion biology that have culminated in advanced understanding of prion disease dissemination and development of a prion vaccine platform. For a bit of levity, students also participated in the "Longest Coherent Sentence" contest, using a word cloud comprised of all the words from the abstracts presented at this year's conference. The winner, Brynn Lauterbach of Colorado State University, took home the prize with her 109-word sentence (Figure 2).

Figure 1. Attendees at the 21st Annual Rocky Mountain Virology Association Conference.

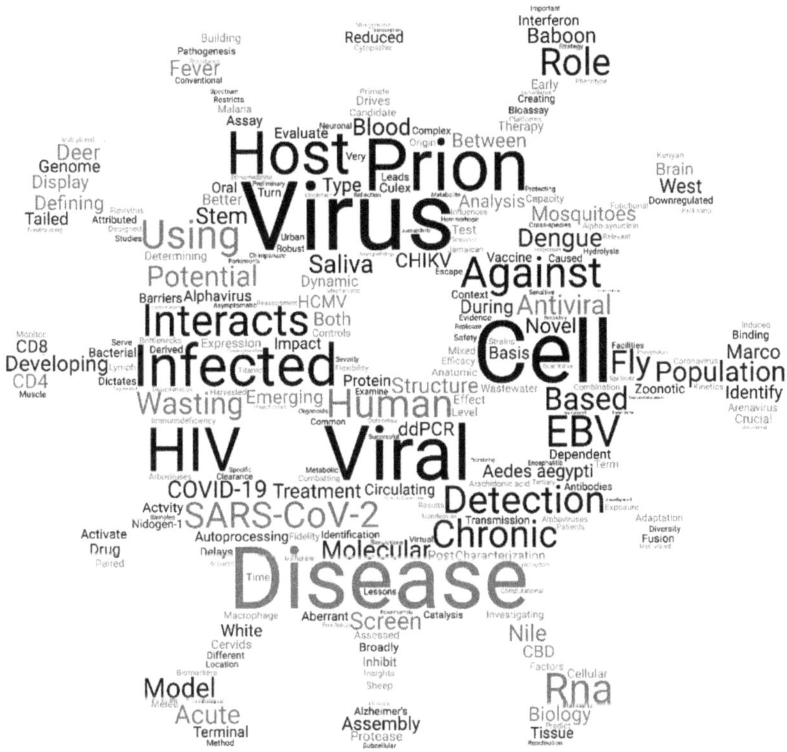

Figure 2. The Longest Coherent Sentence Word Cloud and Award. Brynn Lauterbach of Colorado State University won with her sentence of 109 words: "Surveillance of exposure, asymptomatic transmission, dissemination, activity, progression, and antiviral drug resistance of important emerging cross-species viral disease of humans, urban *Aedes aegypti* and *Culex* mosquitoes, Scandinavian sheep, white tailed deer, Jamaican baboons, Kenyan chimpanzees, cervids, and specific relevant primates in different locations impacts the detection of arbovirus, flavivirus, arenavirus, and alphaviruses and is crucial in determining the level of host spillover and developing a cellular or RNA vaccine, stem cell-derived therapy, circulating neutralizing antibodies, and creating complex and sensitive quantitative immunological bioassays, novel computational model simulations, and ddPCR in experimentally assessing and characterizing expression of cellular factors, biomarkers, and susceptibility of samples against a spectrum of combination treatments." The word cloud was generated using the WordArt.com word cloud generator, and image licensing was provided to Laura A. St. Clair.

However, this year's gathering was also bittersweet for many of us in attendance, as our dear friend, colleague, founding member, and president of RMVA, Dr. Randall J. Cohrs (Figure 3), passed away suddenly this past July. To say that Randy was a cornerstone of the RMVA conference and within our virology community is an understatement, and his absence was assuredly felt at this year's conference. Professor Emerita Dr. Kathryn V. Holmes of the University of Colorado School of Medicine, who was unable to join us physically, provided a poignant video tribute to Randy that highlighted his years of service and dedication to the virology community. Randy loved a good story, but especially those that unraveled the inner workings of viral and prion diseases. He poured that love into the formation of innumerable global collaborations, and in the formation of regional, national, and international conferences where knowledge can be shared, and training and mentorship could be extended to students and junior scientists. His kindness, generosity, friendship, mentorship, and perpetual childlike joy for science are greatly missed, but were a blessing to all of us that knew him. To honor his legacy, the Randall J. Cohrs Awards Fund was established this year to support the top student and postdoc poster and presentation awards and to support the Randall J. Cohrs Keynote address. We would also like to dedicate this year's conference report to his memory. Selected abstracts are presented below.

Figure 3. Dr. Randall J. Cohrs. Photo permitted for use by the Cohrs family.

2. Summary of Scientific Sessions

2.1. *Keynote Speaker*

Dr. Ralph Baric (Department of Microbiology and Immunology, William R. Kenan, Jr. Distinguished Professor, Department of Epidemiology, University of North Carolina School of Medicine, Chapel Hill, NC, USA) presented the keynote address and discussed emerging coronavirus pathogenesis and prevention. Emerging viruses constantly arise from zoonotic reservoirs to threaten global health and economic stability, illustrated by the ongoing SARS-CoV2 COVID-19 pandemic that has infected over 200 million people and caused over 4.6 million deaths worldwide. Zoonotic coronaviruses include rich pools of strains poised for human emergence, as evidenced by their ability to use human receptors for docking and entry, replicate efficiently in primary human airway epithelial cells, and cause disease in transgenic mice. Using well-developed reverse genetic and animal models of human disease, his group explores the evolutionary and pathogenic potential of SARS-CoV-2 and other emerging coronaviruses that are poised to cause acute and chronic disease, discuss mechanisms of severe acute and chronic lung injury, and explore new strategies for antiviral control and vaccine design. An underlying theme is that human globalization, behavior, and activities will increase the probability of epidemic and pandemic viral disease outbreaks in the 21st century, requiring new investments and strategies for preparedness at the national and international levels.

2.2. *Prions and Protein Aggregation*

Brianne Coleman (Department of Microbiology, Immunology, and Pathology, Colorado State University, Fort Collins, CO, USA) updated on her work on detecting prions in specific blood cell populations harvested from pre-clinical white-tailed deer infected with chronic wasting disease. Chronic wasting disease (CWD) is a terminal, infectious prion disease endemic within captive and free- ranging cervid populations across North America, Korea, and Scandinavia. Big-game hunting (1 in 36 Americans) and subsequent human consumption of contaminated meat (7000–15,000 CWD+ cervids/year; rising 20%/year) intensifies the need to characterize prion peripheralization in cervids. It has been established that prion infectivity is present in the blood of prion-infected animals, including deer and humans. Yet to be revealed is the longitudinal profile of specific blood cell subsets associated with prion infection. Using her white-tailed deer model, she has established a reliable, consistent method for isolating blood cell populations throughout the disease course—from minutes post-inoculation to terminal disease. Throughout the CWD disease course, she has isolated platelets, polymorphonuclear cells (PMNs), total peripheral blood mononuclear cells (PBMCs), and specific blood cell subsets, including CD4, CD8, and CD14. All of these cell subsets are being analyzed for the presence of amyloid seeding activity (prions) with real-time quaking-induced conversion (RT-QuIC) assays. To date, she has analyzed total PBMC and specific blood cell subsets with RT-QuIC and has identified the presence of amyloid seeding activity near the time of tonsil biopsy positivity. Longitudinal analysis of these cell populations is ongoing. Results from these studies will identify the role specific blood cell subsets play in establishing and maintaining CWD prion infections and may offer early prion antemortem diagnosis. All animal studies were performed following guidelines and protocols approved by the Institutional Animal Care and Use Committee of Colorado State University.

Vincenzo Gilberto, along with Stephanie McGrath and Julie A. Moreno (Department of Environmental and Radiological Health Sciences and Department of Clinical Sciences, Colorado State University, Fort Collins, CO, USA), presented his work on neuronal treatment of models of Alzheimer's disease using CBD and trazodone. Alzheimer's disease (AD) is one of the most common forms of neurodegeneration and is typically characterized by either the presence of Amyloid-Beta plaques (Aβ) or Tau Tangles (NFT). Even though there is no current cure for AD, it is well known that targeting of signaling pathways involved in reactive oxygen species (ROS) or unfolded protein response (UPR) improves behavioral deficits, glial inflammation, and neuronal toxicity. This research aims to utilize specific strains from the model organism *C. elegans* that have been genetically modified to contain two common misfolded proteins found to aggregate and accumulate in AD patients' brains, amyloid-β, and the hyperphosphorylation of tau (P-tau). They hypothesize that combinational drug stacking of compounds that target both ROS production and UPR will improve the behavioral qualities associated with AD and will also extend the lifespan of these neurodegenerative nematode models to that of their control counterparts. To address this hypothesis, they have used CBD and trazodone to inhibit ROS and the UPR, respectively. Using both one-way and two-way ANOVA, their previous experiments have revealed that neurodegenerative *C. elegans* motility is significantly worse compared to their control counterparts and that early-stage exposure to trazodone significantly improved the motility of the neurodegenerative nematodes. The next steps for this project include, but are not limited to, isolated and combinational drug treatments of both trazodone and CBD to measure their motility and life span. Once optimal doses are identified, they aim to implement a late-stage rescue of neurodegenerative *C. elegans* with combinational drug stacking. All animal studies were performed following guidelines and protocols approved by the Institutional Animal Care and Use Committee of Colorado State University.

Caitlyn Kraft, along with C. Hoover, N. Denkers, and Candace Mathiason (Department of Microbiology, Immunology, Colorado State University, Fort Collins, CO, USA), presented her work on detection of chronic wasting disease prions in muscle tissue. Prion diseases affecting animals that are regularly consumed by humans pose a significant public health risk—the most notable example being bovine spongiform encephalopathy, or Mad Cow Disease, which spread to humans through consumption of infected muscle. Chronic wasting disease (CWD) affects moose, elk, and deer, all of which are commonly hunted and consumed by humans. While clear evidence of CWD transmission to humans has not been demonstrated, establishing whether CWD prions are present in muscle is of high interest. Currently, conventional assays, such as immunohistochemistry (IHC) and enzyme-linked immunosorbent assay (ELISA), are unable to detect the low concentration of prions likely to be found outside neural or lymphatic tissues. Here, they combined two prion amplification assays—protein misfolding cyclic amplification (PMCA) and real-time quaking-induced conversion (RT-QuIC)—to demonstrate the presence of prion seeding activity in the muscle of CWD-positive white-tailed deer. Hamstring muscle samples from 24 CWD-infected deer were subjected to five rounds of PMCA, followed by a readout of the amplification product with RT-QuIC. They demonstrated that 16 of the 24 samples (66.67%) contained prion seeding activity. To compare the sensitivity of PMCA and RT-QuIC, a subset of muscle samples were examined with IHC and standard RT-QuIC. Zero (0) of the four samples assessed with IHC and one of the four samples assessed with RT-QuIC tested positive. They concluded that the combination of prion amplification assays allowed for the detection of low prion concentrations present in the muscle of CWD-positive deer. The consequence of consumption of these low levels of CWD prions remains unknown. All animal studies were performed following guidelines and protocols approved by the Institutional Animal Care and Use Committee of Colorado State University.

Casey P. McDermott [1], along with Collin Bantle [1], Dev Aldaz [2], Savannah Rocha [2], Katriana Popichak [2], Ken Olson [2], and Ronald Tjalkens [1] ([1] Toxicology Program, Dept. of Environmental and Radiological Health Sciences, [2] Dept. of Microbiology, Immunology, and Pathology, Colorado State University, Fort Collins, CO, USA), examined how alpha-synuclein overexpression delays neuropathology in a viral model of Parkinson's disease. Parkinson's disease (PD) is a multifaceted movement disorder caused by the interaction of multiple factors, including age, environmental exposures, and genetic predisposition. The hallmarks of PD include glial reactivity, neuronal loss in the Substantia Nigra (the movement center of the brain), loss of dopamine in the striatum, and alpha synuclein aggregation. Infection with Western Equine Encephalitic Virus (WEEV) in animals and humans has previously been shown to cause neurological symptoms similar to those of PD. In this study, they examined how WEEV could propagate neuropathology in transgenic mice expressing a variant of human alpha-synuclein (A53T) associated with early onset of PD. Normally, these transgenic mice do not show any motor deficits, α-syn aggregation, or neuronal loss until 16 months of age. They postulate that infection with WEEV would accelerate aggregation of A53T alpha-synuclein at as early as 4 months of age (1 month post infection). To test this hypothesis, mice were intranasally infected with WEEV at 3 months of age and examined for neuropathological changes at 1 and 3 months post-infection. Cellular markers of neuropathology were assessed in multiple brain regions by immunohistochemistry and immunofluorescence staining, including glial reactivity, protein aggregation, and neuronal loss. At 1 month post-infection, they saw extensive astrogliosis in both the Substantia Nigra and the striatum. There was a significant increase in microgliosis in the Nigra as well; however, there was no loss of dopaminergic neurons. Total α-syn was also measured in all regions, and was only found to be significant in the cortex. Three months post-infection, microgliosis and astrogliosis had ceased, and they began to see loss of TH neurons. Alpha-synuclein aggregation also began to rise in the SN and the cortex. This supports the field in such that α-syn could have an antiviral application, even in a form that should be pre-disposed to misfolding.

Julianna Sun [1,2], along with Jifeng Bian[1], Sehun Kim[1], Jenna Crowell [1], Bailey Huser [1], EmmaKate Raisley [1], and Glenn C. Telling[1,2] ([1] Prion Research Center and the Department of Microbiology, Immunology, and Pathology, Colorado State University, [2] Program in Cell and Molecular Biology, Colorado State University, Fort Collins, CO, USA), presented her work detailing the breadth, properties, and zoonotic potential of emergent prion disease strains in Scandinavian cervids. The replicative properties of prions, which are extraordinary proteinaceous infectious agents causing lethal neurodegenerative diseases in humans and animals, challenge fundamental concepts of inheritance and infection. During this process, PrPSc, a conformationally altered counterpart of host-encoded PrPC, templates its infective conformation on PrPC, resulting in exponential prion accumulation. Remarkably, prions share the property of heritable strain diversity with nucleic-acid-containing pathogens. Transmission of bovine spongiform encephalopathy to humans epitomizes the unpredictable effects of strains on the prion host range and the devastating effects of prion zoonoses. Chronic wasting disease (CWD), a burgeoning and ineradicable disease of wild and captive North American (NA) cervids, raises similar concerns. The discovery of CWD in Norway in 2016 raised questions about its relationship to the NA epidemic. Their finding that emergent Norwegian cases are, in fact, caused by novel prion strains with unpredictable adaptive potential increases uncertainties about CWD zoonosis. The development of genetically modified mice has been key for advancing our understanding of the molecular parameters controlling prion transmission, including the effects of species-specific variation in PrP primary structures. While deer express glutamine (Q) at PrP residue 226, elk express aspartate (E). To investigate the role of residue 226 on strain selection, they developed gene-targeted mice in which murine PrP was replaced by elk or deer PrP, referred to as GtE or GtQ, respectively. They showed that residue 226 controls the selective propagation of distinct CWD strains and that the influences of residue 226 differ between Norwegian and NA CWD strains. Additional CWD cases have been diagnosed in Norway, Sweden, and Finland. Here, they described the results of studies that illustrate additional unprecedented strain properties of these CWD isolates and described approaches to addressing the risks they pose to sympatric species and humans exposed to CWD. This work was supported by NINDS grants 1R01NS109376 and 1R01NS121682. All animal studies were performed following guidelines and protocols approved by the Institutional Animal Care and Use Committee of Colorado State University.

Holger Wille (Centre for Prions and Protein Folding Diseases and Department of Biochemistry, University of Alberta, Canada) presented his and his colleagues' efforts to design prion disease vaccines. Chronic wasting disease (CWD) is a prion disease that affects captive and free-ranging cervids in many U.S. states and Canadian provinces, including Colorado and neighboring states. CWD is spreading uncontrollably through environmental contamination and animal-to-animal contacts. The prion diseases are caused by the misfolding of the cellular prion protein (PrPC), which adopts a β-sheet rich conformation when it is converted into the infectious state (PrPSc). Prion protein peptides have been used in previous attempts to create prion vaccines, but in most cases, the structures of the antigens were poorly controlled, resulting in a lack of vaccine efficacy. Here, he presented their approach to designing structure-based vaccines that present specific antigenic sequences in a structurally controlled format, thereby creating a prophylactic prion vaccine. The prion domain of the fungal HET-s protein is unrelated to the mammalian prion protein, but is thought to adopt a similar, β-sheet rich conformation. This innocuous scaffold protein was engineered as a vaccine carrier to express PrPSc surface epitopes in a structurally controlled manner. They selected seven discontinuous amino acids, which were predicted to be surface exposed in PrPSc, and inserted them into equivalent positions in the scaffold protein. The vaccine was produced in *E. coli*, purified, refolded, controlled for its structural fidelity, and used to immunize a transgenic mouse model (TgP101L mice) of a genetic prion disease in humans: Gerstmann–Sträussler–Scheinker disease. Their structure-based vaccine targeting PrPSc produced an immune response that was specific for the infectious conformer only and did not react with PrPC. This specificity for PrPSc was observed in all animals that were immunized with the vaccine, while animals that received the unmodified scaffold protein produced only unspecific immune responses. Moreover, unimmunized TgP101L mice developed disease symptoms at ~177 days of age (±17 days), animals that were immunized with the unmodified scaffold protein started to fall ill at ~166 days of age (±26 days), while TgP101L mice that were immunized with the prion vaccine remained free of symptoms until ~448 days of age (±39 days). Therefore, their prion vaccine extended the health span of the TgP101L mice by >250%. Additional vaccine efficacy trials in other prion disease models are currently underway. Furthermore, a prophylactic vaccine efficacy trial in CWD-challenged white-tailed deer will be initiated shortly in collaboration with Candace Mathiason and her team at Colorado State University. The project was supported by awards from the Alberta Prion Research Institute, the Alberta Livestock and Meat Agency, and the Creutzfeldt-Jakob Disease Foundation. All animal studies were performed following guidelines and protocols approved by the Institutional Animal Care and Use Committee of the University of Alberta. All studies using human subjects or tissue samples have been approved by the Institutional Review Board of the University of Alberta.

2.3. Arboviruses

Stephanie E. Ander [1], along with Kathryn S. Carpentier [1] and Thomas E. Morrison [1] ([1] Department of Immunology and Microbiology, University of Colorado School of Medicine), presented on their work on the magnitude and duration of vertebrate viremia. These parameters are critical determinants of arbovirus transmission, geographic spread, and disease severity. However, mechanisms determining arboviral viremia levels are poorly defined. Their laboratory previously defined roles for liver Kupffer cells and scavenger receptor SR-A6 (MARCO) in mediating control of arthritogenic alphavirus viremia. To determine whether this is specific for arthritogenic alphaviruses or functions more broadly, they evaluated the clearance of Eastern (EEEV) and Venezuelan (VEEV) equine encephalitis viruses using recombinant virus particles composed of a chimeric SINV-EEEV or -VEEV genome encapsulated by EEEV or VEEV structural proteins. Following intravenous inoculation of mice, they found that $EEEV^{FL93}$ and enzootic $VEEV^{PIXV}$ particles are cleared from circulation by $Cd169^{+}$ phagocytic cells, while epizootic $VEEV^{TrD}$ particles are resistant and persist at >3 hpi within the bloodstream. Furthermore, $EEEV^{FL93}$ clearance is dictated by the presence of a basic patch on the E2 glycoprotein. Alanine substitution of these residues not only produces clearance-resistant $EEEV^{FL93}$ particles, but has also been shown to enhance viremia and viral pathogenesis following subcutaneous inoculation. While similar, the specific mechanisms mediating vascular clearance of arthritogenic and encephalitic alphaviruses are distinct: $EEEV^{FL93}$ and $VEEV^{PIXV}$ vascular clearances have slower kinetics and are MARCO-independent. Collectively, their findings suggest that phagocytic cells control the magnitude and duration of viremia following infection with a broad group of alphaviruses. Ongoing studies are aimed at defining the specific phagocytic cells and receptors that mediate clearance of circulating $EEEV^{FL93}$ and $VEEV^{PIXV}$ and the molecular features of EEEV and VEEV particles that promote or evade vascular clearance. All animal studies were performed following guidelines and protocols approved by the Institutional Animal Care and Use Committee of The University of Colorado Anschutz Medical Campus.

Anna Burnet (Department of Microbiology, Immunology, and Pathology, University of Colorado Anschutz, Boulder, CO, USA) discussed her efforts in defining a role for heme in reactivation of EBV from latency in B cells during acute malaria. The connection between Epstein–Barr virus (EBV) and *Plasmodium falciparum* malaria in their role in the development of EBV-associated Burkitt lymphoma in Sub-Saharan Africa has long been accepted by the scientific community, yet the mechanisms involved remain unknown. EBV-positive Burkitt lymphoma is a non-Hodgkin's lymphoma with peak incidence at 6 years of age. A major hallmark of malarial disease is hemolysis of red blood cells, which causes the release of heme in large quantities. She found that hemin treatment of latently infected EBV-positive B cells in culture results in viral reactivation from latency by activating transcription of immediate early genes BZLF1 and BRLF1, as well as degradation of Bach2 protein by 24 h. Hemin treatment additionally increases CD138 expression, demonstrating that cells are being driven to shift to plasma cells. She hypothesized that heme released during acute malaria infection drives differentiation of latently infected EBV-positive memory B cells via specific heme-binding to Bach2, an important regulator of B cell proliferation and differentiation. Importantly, Bach2 is a repressor of Blimp-1, a protein essential for B-cell terminal differentiation. Furthermore, she hypothesized that the binding interaction between heme and Bach2 can allow for an increase in Blimp-1, which, in turn, can induce lytic replication by initiating transcription of immediate early viral genes BZLF1 and BRLF1. No animal or human studies were performed.

Jasmine Donkoh (Department of Microbiology, Immunology and Pathology, Colorado State University, Fort Collins, CO, USA) presented on her work studying the effect of dengue virus infection on macrophage gene expression and phenotype. Dengue virus (DENV) is the most prevalent arthropod-borne flavivirus in the world, causing dengue hemorrhagic fever and dengue shock syndrome. Macrophages are a site of DENV replication, and the imbalance between macrophage pro-inflammatory and anti-inflammatory phenotypes during DENV infection leads to severe disease outcomes. To investigate the effect of macrophage phenotype on DENV serotype 2 (DENV2) replication, she treated THP-1 macrophages with either interferon gamma (IFNγ) or interleukin 4 (IL-4) to polarize them towards pro-inflammatory or anti-inflammatory phenotypes, respectively, prior to infection. DENV2 showed preferential replication in cells treated with IL-4 compared to cells treated with IFNγ or non-treated cells. DENV2 infection upregulates expression of the anti-inflammatory cytokine, interleukin 10 (IL-10), and the inflammatory chemokine, CXCL10, in all phenotypes. DENV2 infection of macrophages also leads to induced expression of the host mediator complex protein, cyclin dependent kinase 8 (CDK8). CDK8 is a transcriptional co-factor that regulates expression of IFNγ-stimulated genes and certain cytokines/chemokines. Expression of CDK8 increases coincidently with DENV2 replication throughout a 36-h time course. Treatment with Senexin B, a CDK8 inhibitor, decreased DENV2 mRNA and infectious particles in THP-1 macrophages. She also found that inhibition of CDK8 activity with Senexin B increases IL-10, but decreases CXCL10 gene expression, regardless of macrophage phenotype. This suggests dependence upon CDK8 activity for virus induction of CXCL10 and IL-10. She hypothesizes that CDK8 plays a key role in regulating a macrophage's ability to transcribe genes that control polarization and antiviral immunity. No animal or human studies were performed.

Tillie Dunham, along with Mark Stenglein (Department of Microbiology, Immunology and Pathology, Colorado State University, Fort Collins, CO, USA), discussed vertical transmission of insect-specific viruses in *Aedes Aegypti* mosquitos. *Aedes Aegypti* mosquitoes are a major vector of pathogenic arboviruses, including the dengue and zika viruses. In this project, she quantified the vertical transmission efficiency of insect-specific viruses that are common in wild mosquitoes. These viruses cannot be transmitted to vertebrates, but may impact the biology of their host. Viruses with efficient vertical transmission have an inherent ability to spread through host populations and so could have important applications in biocontrol. She first reared mosquitoes from eggs, crossed them, and sampled adults and offspring from those crosses. Then, she performed RNA extraction and RT and qPCR in order to screen for viruses in each sample. She found that three viruses, Verdadero virus, Renna virus, and *Aedes* anphevirus, exhibited efficient biparent vertical transmission. This property likely underlies their success in the wild and their persistent infection of colonized mosquitoes and means that these could be good candidates for gene delivery through mosquito populations. No animal or human studies were performed.

Emily Fitzmeyer, along with Emily N. Gallichotte, Nicole R. Sexton, and Gregory D. Ebel (Department of Microbiology, Immunology, and Pathology, Colorado State University, Fort Collins, CO, USA), presented her efforts using barcoded West Nile virus to examine the impact of tissue and cellular bottlenecks on virus populations in *Culex* mosquitoes. Due to error-prone replication, RNA viruses exist within hosts as genetically and phenotypically complex swarms. For arboviruses such as West Nile virus (WNV), these swarms fluctuate as they encounter bottlenecks and different selective environments when cycling between hosts. Bottlenecks alter the composition of virus populations by stochastically reducing overall diversity, which can impact virus fitness and transmission potential. Using a molecularly barcoded WNV (bcWNV) allows them to quantitatively measure population bottlenecks in two common WNV mosquito vectors: *Culex tarsalis* and *Culex quinquefasciatus*. This approach allows them to examine changes in virus population structure as the population moves through the midgut, salivary glands, and saliva—the main bottlenecks associated with infection, dissemination, and transmission. Additionally, it is not currently known which cell populations in the midgut and salivary glands become infected and what role intracellular bottlenecks might play in shaping virus populations. Previous work with barcoded viruses shows that single cells may be infected with several unique WNV genomes (i.e., become polyinfected). They therefore will use single-cell sorting and sequencing of bcWNV-infected mosquito tissues to determine the number of viral barcodes present in both whole tissues and specific cell populations within these tissues. While it is known that there are bottlenecks at the tissue level, they predict that interactions at the cellular level are mainly responsible for altering the diversity of the total virus population. From this work, they will establish how both whole-tissue and intracellular bottlenecks alter virus populations within mosquitos. This work was supported by NIH grant R01 AI067380. No animal or human studies were performed.

Monica Graham [1], Benjamin Akiyama [2], and David Beckham [1] ([1] Department of Immunology and Microbiology, University of Colorado Anschutz Medical Campus, [2] Department of Biochemistry and Molecular Genetics, University of Colorado Anschutz Medical Campus, Aurora, CO, USA) discussed the role of secondary and tertiary structures within the 3′ untranslated region (UTR) of Zika virus (ZIKV) in viral cytopathic effects. The dumbbell-1 (DB-1) structure is one of the least-studied RNA structures in the 3′UTR. Previous studies suggest that DB-1 in ZIKV and other flaviviruses is important for genome replication and cytopathic effect (CPE). However, these studies have not investigated how the highly conserved DB-1 structure is related to its function. To investigate the structure–function relationship of DB-1, they created two mutant infectious ZIKV clones: TL.PK, which disrupts a crucial tertiary fold, and p.2.5′, which disrupts the conserved secondary structure. In A549 cells, viral genome replication is modestly reduced, but the CPE of both mutants is significantly abrogated. They investigated sub-genomic flaviviral RNA (sfRNA) formation with their TL.PK and p.2.5′ mutants during A549 infection and found that their mutants produced less sfRNA compared to ZIKV-WT. The two mutants also did not produce the smallest sfRNA species seen in ZIKV-WT, sfRNA3. To investigate the decreased CPE, they assayed mutant-infected cells for cell viability and caspase induction. They found that cell viability was significantly increased in TL.PK- and p.2.5′-infected cells compared to ZIKV-WT. It was also determined that TL.PK- and p.2.5′-infected cells had decreased levels of activated caspases 1 and 3 compared to ZIKV-WT. Overall, these data suggest that the DB-1 structure plays an important role in ZIKV CPE and sfRNA formation during mammalian cell infection. Their current hypothesis is that the phenotypic change in CPE is a result of altered sfRNA production during infection. No animal or human studies were performed.

Shannon N. Hinsdale, along with Kathryn S. Carpentier and Thomas E. Morrison (Department of Immunology and Microbiology, University of Colorado School of Medicine, Aurora, Colorado, USA), discussed the development of an shRNA-based screen to identify ISGs with antiviral activity against type I interferon sensitive alphaviruses. Alphaviruses are positive-sense RNA arboviruses that can cause severe human disease. Alphavirus infection elicits and is sensitive to type I IFN. Type I IFN signaling transcriptionally activates antiviral interferon-stimulated genes (ISGs). However, specific ISGs that control alphavirus infection remain poorly characterized. Here, Ross River virus (RRV) is used as a representative alphavirus to identify ISGs that control alphavirus replication. RRV DC5692 is an attenuated strain of RRV with enhanced sensitivity to type I IFN both in vitro and in vivo compared with RRV T48. This phenotype was attributed to six nonsynonymous nucleotide changes in nsP1. They developed a flow-cytometry-based ISG knockdown screen by utilizing lentiviruses that co-express GFP and shRNA targeting ISGs and an mKate-expressing RRV encoding the six nsP1 mutations to identify ISGs that control alphavirus replication. Murine embryonic fibroblasts (MEFs) were chosen for the screen after comparing transduction efficiency in L929 cells and MEFs via flow cytometry. Virus MOI (FFU/cell) and type I IFN treatment conditions were optimized by pre-treating WT or transduced MEFs with a range of type I IFN concentrations and infecting with a range of MOIs. An MOI of 3 was determined to be sufficient for infection of ~80% of cells and a concentration of 7.5 IU/mL type I IFN suppressed the number of mKate-positive cells by more than 10-fold. Using these conditions, they anticipate that this screen will identify specific ISGs that control alphavirus replication and are antagonized by determinants in nsP1, thus improving the understanding of alphavirus pathogenesis.

Helen Lazear (Department of Microbiology and Immunology, University of North Carolina Chapel Hill, Chapel Hill, NC, USA) presented her lab's work investigating how interferon lambda (IFN-λ, type III IFN) restricts viral infection at anatomic barriers.

Crossing anatomic barriers is a key step in viral infection, and the ability to cross external epithelial barriers (such as the respiratory and gastrointestinal tracts or the skin) and internal barriers (such as the blood–brain barrier or placenta) contributes to viral tissue tropism and transmission mechanisms. As a mosquito-borne virus, Zika virus (ZIKV) must first surmount the external barrier of the skin, and its ability to cross the placental barrier allows ZIKV to cause birth defects. Among the specialized immune mechanisms that protect anatomic barriers, IFN-λ provides front-line protection by activating a local antiviral response similar to IFN-αβ (type I IFN), but with reduced inflammation. IFN-λ is constitutively secreted from human placental cells and reduces ZIKV transplacental transmission in mice, but the mechanism by which it restricts congenital infection is unknown. They generated mouse pregnancies that lacked either maternal or fetal IFN-λ responsiveness and found that IFN-λ signals exclusively to maternal tissues to exert antiviral effects against ZIKV. Interestingly, infection earlier in pregnancy (E7 rather than E9) resulted in IFN-λ-mediated pathology: IFN-λ-responsive dams had higher rates of fetal resorption than non-responsive dams. Maternal IFN-λ-mediated pathology was also elicited by poly(I:C) treatment, which was similarly dependent on gestational stage. These findings identify an unexpected effect of IFN-λ signaling specifically in maternal (rather than placental or fetal) tissues and highlight the complexity of immune signaling at the maternal–fetal interface, where disparate outcomes can result from signaling at different gestational stages. While IFN-λ does not restrict replication of ZIKV in peripheral tissues, they found a skin-specific role for IFN-λ in restricting ZIKV dissemination. To further investigate the antiviral effects of IFN-λ in the skin, they used a herpes simplex virus (HSV-1) skin infection model. They found that mice lacking IFN-λ signaling developed more severe skin lesions compared to WT mice, identifying a protective role for IFN-λ in the skin. Altogether, this work highlights the importance of IFN-λ-mediated immunity at both external and internal barriers for controlling viral pathogenesis. This work was supported by R01 AI139512. All animal studies were performed following guidelines and protocols approved by the Institutional Animal Care and Use Committee of the University of North Carolina Chapel Hill.

Frances S. Li, along with Kathryn S. Carpentier and Thomas E. Morrison (Department of Immunology and Microbiology, University of Colorado School of Medicine, Aurora, CO, USA), discussed defining MARCO–virus interactions that are important for alphavirus clearance from the circulation. Arboviruses, such as mosquito-borne alphaviruses, are major public health concerns, and the capacity of an arbovirus to be transmitted in a human–mosquito–human transmission cycle has fueled disease outbreaks worldwide. Major determinants of arbovirus transmission, geographic spread, and pathogenesis are the magnitude and duration of viremia in the vertebrate host. Previously, they determined that multiple arthritogenic alphaviruses, including chikungunya (CHIKV) and Ross River (RRV) viruses, are cleared efficiently from murine circulation by scavenger receptor A6 (MARCO) expressed on liver macrophages. Here, they find that MARCO-dependent clearance of CHIKV is contingent on the presence of specific biochemical features of the virion surface E2 and E1 glycoproteins: a lysine (K) residue at E2-200 (K200), a negative charge at E2-208, and a positive charge at E1-61. Similar distinct electrostatic requirements were also observed for RRV clearance, where a lysine residue is required at E2-251 (K251), a basic histidine (H) residue at E2-232 (H232), and an acidic aspartic (D) residue at E2-246 (D246). Utilizing an in vitro cell culture system, they uncovered that ectopic expression of MARCO promoted internalization of CHIKV and RRV particles, and this effect was dependent on the scavenger receptor cysteine-rich (SRCR) domain of MARCO, which displayed distinct electrostatic distributions. Collectively, these findings suggest that CHIKV and RRV particles ionically interact with MARCO SRCR domain via unique charge characteristics surrounding CHIKV E2 K200 and RRV E2 K251. Ultimately, this project may provide new insight into the molecular mechanisms that dictate arthritogenic alphavirus transmission, dissemination, and pathogenesis in vertebrate hosts. This work was supported by the following funding sources: AI123348, AI148144, and AI140567. All mice experiments were performed at the University of Colorado, Anschutz Medical Campus and adhere to the Institutional Animal Care and Use Committee (IACUC) guidelines.

Cormac J. Lucas, along with Ryan M. Sheridan, Bennett J Davenport, Jay R. Hesselberth, and Thomas E. Morrison (Department of Immunology and Microbiology, University of Colorado Anschutz Medical Campus; RNA Bioscience Initiative, University of Colorado Anschutz Medical Campus, Aurora, CO, USA), discussed how Chikungunya virus interacts with floor and Marco+ LECs in the draining lymph node early after infection. Pathogenic Chikungunya virus (CHIKV) strains evade B cell responses to establish persistent infection and induce infiltration of the draining lymph node (dLN) with inflammatory iNOS-expressing myeloid cells that impair normal B cell responses and disrupt normal lymph node organization. The LN cells that interact with pathogenic CHIKV strains to initiate disruption of normal lymphoid tissue function remain to be identified. In prior studies, they found that accumulation of CHIKV in the dLN is dependent on expression of the scavenger receptor MARCO. Using single-cell RNA sequencing of murine LN non-hematopoietic cells at two early timepoints, 8 and 24 h post-infection, they found that CHIKV RNA accumulates predominantly in two subsets of lymphatic endothelial cells (LECs), floor LECs, and MARCO+ LECs, with increasing MARCO expression correlating with higher viral RNA counts. Moreover, both of these LEC subsets lack expression of Mxra8, a CHIKV cell entry receptor, suggesting that MARCO facilitates CHIKV internalization in LECs. Decreased expression of host genes, increased subgenomic viral RNA counts in CHIKV+ cells, and the disappearance of the floor LECs by 24 h suggest that floor and MARCO+ LECs may support viral replication. A subset of LN cells upregulate innate immune genes at 8 h, but the LN displays a dominant ISG response by 24 h. Further work will investigate the susceptibility and permissiveness of LECs to CHIKV infection in vitro and identify innate immune signaling pathways regulated by MARCO in LECs to better understand the contribution of these cells to immune responses initiated in the dLN. All animal studies were performed following guidelines and protocols approved by the Institutional Animal Care and Use Committee of The University of Colorado Anschutz Medical Campus.

Gabriela Ramirez [1], along with Paul S. Soma [1], Nunya Chotiwan [1], Nurul Islam [1], Barbara Graham [1], Austin J. Mejia [2], Laura St Clair [1], John T. Belisle [1], Elizabeth A. McGraw [2], and Rushika Perera [1] ([1] Department of Microbiology, Immunology and Pathology, Colorado State University, Fort Collins, CO; [2] Biology Department and The Center for Infectious Disease Dynamics, Pennsylvania State University, University Park, PA 16802), discussed her work investigating the metabolic differences that define virus–host interactions between arboviruses and *Aedes aegypti* mosquitoes. The metabolic landscape, also referred to as the metabolome, plays a critical role in the replication of arboviruses in *Ae. aegypti* mosquitoes. As obligate parasites, viruses must strike a balance between commensalism and competition within their hosts for effective viral entry, replication, and transmission. Insecticides, ecological conditions, geographic distribution, age of vector, and endosymboints (such as *Wolbachia*) also alter metabolic conditions in mosquitoes and influence vector competence. To investigate which host resources are required and/or altered by arbovirus infection, they used high-resolution mass-spectrometry-based metabolomics approaches combined with loss-of-function analyses in *Ae. Aegypti* infected with Zika (ZIKV), dengue (DENV), and Chikungunya (CHIKV) viruses. Intriguingly, they found that multiple lipid classes are altered. Major components of cellular membranes, such as phosphoglycerolipids, were increased in parallel with active viral replication in the midgut. Certain sphingolipids were also altered at early times post-infection. These are particularly interesting, as they are well-characterized as potent bioactive signaling molecules in mammals. However, their role in mosquito biology remains unknown. Additionally, they were able to identify unique features that metabolically distinguish ZIKV, DENV, and CHIKV infection with the *Ae. aegypti* vector. These data lend insight into how unique metabolic biosignatures induced by these three viruses can inform on tolerance of co-infection. Progress in these studies was presented. No animal or human studies were performed.

Molly Ring, along with Anna-Sophia Leon, Ashley Janich, Paula Lado, Nguyen C., Pugh G., and Brian Foy (Center for Vector-Borne Infectious Diseases, Department of Microbiology, Immunology, and Pathology, Colorado State University, Fort Collins, CO, USA), presented her work on the efficacy of ivermectin and convectional insecticides against wild-type West Nile virus vectors from urban and rural Larimer County. Colorado has some of the highest West Nile virus (WNV) case numbers in the United States each year, and perennially has counties with some of the highest WNV disease incidence. Current WNV control is limited to applying mosquito larvicides to water sources and spraying insecticides to control adult mosquito vectors in residential areas when risk of WNV transmission is high. The latter has limited proof of efficacy, can be poorly targeted, has environmental toxicity concerns, and may be ineffective against mosquito populations if they are already resistant to similar insecticides applied to residential and agricultural areas. Wild birds are WNV reservoirs because they infect local mosquito vectors that feed upon them, and these vectors subsequently transmit WNV to humans. They are developing an alternative WNV control strategy that treats birds with the drug ivermectin (IVM) to kill mosquitoes that blood feed upon them to reduce WNV transmission risk to humans. They investigated the susceptibility of local rural and urban wild-type *Culex tarsalis* to the adulticide permethrin using insecticide bioassays and to IVM by blood feeding them on IVM-treated chickens and observing their survivorship in semi-field mesocosms they constructed. Bioassays determined that the lab strain was susceptible to permethrin, and the rural strain was deemed possibly resistant, but future testing is needed to confirm this. There was significantly lower survivorship in mosquitoes that fed on IVM-treated birds compared to the control group. These preliminary data suggest that they can potentially implement this control method in the field to help reduce WNV transmission. Funding for this project was provided by NIH grant R01AI148633. All animal studies were performed following guidelines and protocols approved by the Institutional Animal Care and Use Committee of Colorado State University.

Paul S. Soma [1], along with Rebekah C. Gullberg [1], Barbara G. Andre [1], Kimberly Anderson [1], Stephanie Mills [1], Elena Lian [1], Kristen Krieger [1], Lionel Gresh [2], Raquel Burger-Calderon [5], Amber Hopf-Jannasch [3], Angel Balmaseda [4], Barry Beaty [1], Eva Harris [5], Carol D. Blair, [1] and Rushika Perera [1] ([1] Dept. of Microbiology, Immunology and Pathology, Colorado State University, Fort Collins, CO, USA; [2] Sustainable Sciences Institute, Managua, Nicaragua; [3] Metabolite Profiling Facility, Bindley Bioscience Center, Purdue University, West Lafayette, IN, USA; [4] Laboratorio Nacional de Virología, Centro Nacional de Diagnóstico y Referencia, Ministry of Health, Managua, Nicaragua; [5] Division of Infectious Diseases and Vaccinology, School of Public Health, University of California, Berkeley, CA, USA), shared their research on the identification of metabolic biomarkers that predict dengue disease severity. Dengue viruses (DENVs) place over 2 billion people at risk of infection each year, rendering them the most aggressive arboviruses worldwide. There are four serotypes of DENVs. Infection with one serotype does not cross-protect from infection with other serotypes. No treatment options exist due to complications in disease pathology mediated by the immune response. Using a liquid chromatography–high-resolution mass spectrometry (LC-MS) untargeted metabolomics approach, they identified a small metabolite biosignature panel in acute-phase pediatric patient serum that is associated with dengue disease severity and pathogenesis. This metabolite panel has high potential for early prediction of dengue disease severity and improved clinical management. Additionally, metabolite biosignatures for Zika virus and Chikungunya infections have been established. Untargeted LC-MS analysis provides putative identification of detected metabolites through chromatographic retention time (RT) and accurate mass (*m/z*) measurements of metabolite ions (e.g., $[\text{Metabolite+H}]^+$). Putatively identified metabolites must be further validated using tandem mass spectrometry (MS/MS) methods and/or RT matching with synthetic standards. Collision-induced dissociation (CID) of chromatographically resolved serum metabolites (LC-MS/MS analysis) was performed in a 'TopN' data-dependent manner with an inclusion list for metabolites of particular interest. CID of metabolite parent ions generated product ions that provided metabolite structural information. Experimental CID spectra for putatively identified metabolites were searched against MS/MS spectral databases (mzCloud, METLIN, HMDB) or compared to CID spectra from synthetic standards when available. Confident identification of metabolite biomarkers of dengue, Zika, and Chikungunya diseases aids in elucidating relevant metabolomic pathways and lays a foundation for simplified point-of-care prognosis. All human studies were performed following guidelines and protocols approved by the Institutional Review Board of Colorado State University.

Laura St. Clair [1], along with Carissa Drake [2], Michael Spedding [3], David Priestman [2], Frances Platt [2], and Rushika Perera [1] ([1] Center for Vector-borne Infectious Diseases, Department of Microbiology, Immunology and Pathology, Colorado State University, Fort Collins, CO, USA; [2] Department of Pharmocology, University of Oxford, Oxford, UK; [3] Spedding Research Solutions SAS, Le Vesinet, France), presented on her work analyzing the effect of inhibition of glucosylceramide hydrolysis during DENV2 infection in human hepatoma cells. Sphingolipids (SLs) are potent bioactive signaling molecules involved in nearly all major biological responses. Previously, her colleagues and others showed that there are significant alterations of SL metabolism during infection with dengue viruses (DENVs) in both human and mosquito hosts. Her colleagues demonstrated that disruption of de novo ceramide biosynthesis in the mosquito reduces DENV2 genome replication and infectious virus release. They also demonstrated that there are significant differences in the SLs found in human sera between the febrile, defervescent, and convalescent phases of dengue disease. Others have shown that dysregulation of SLs is evident in vascular leakage and thrombocytopenia. However, it remains unclear how viral gene products might alter SLs during infection or what role specific SLs play in the viral lifecycle. In an siRNA-mediated screen of enzymes within the SL pathway, she discovered that losses of function of the two enzymes responsible for hydrolysis of glucosylceramide (GluCer) have opposite effects on DENV2 replication and release. Using chemical modulators, she confirmed that both the inhibition of non-lysosomal beta-glucocerebrosidase (GBA2) and upregulation of the lysosomal GBA1 enzyme reduce viral replication and release. These data suggest that subcellular location of GluCer hydrolysis is critical for the DENV2 lifecycle. She systematically investigated how specific parts of the DENV lifecycle are affected by these enzymes. Additionally, she completed analyses to determine their activity over time and determined how they are modulated during infection. Here, she presented an update on the progress of these studies. No animal or human studies were performed.

Elisa Thrasher, along with Emma Harris and Rebekah Kading (Department of Microbiology, Immunology, and Pathology, Colorado State University), investigates the dynamic bacterial community of *Aedes aegypti*. Arbovirus (arthropod-borne virus) transmission by *Aedes aegypti* mosquitoes is a worldwide public health crisis that requires in-depth research for novel methods to disrupt viral transmission. The mosquito midgut harbors a multitude of bacterial endosymbionts that directly interact with invading viruses entering through an infectious blood meal. Various bacterial species have been identified as constitutive members that are naturally present in field-collected and lab-reared *Ae. aegypti* microbiota and have been shown to provide immunity to entomopathogens and vulnerability to arboviruses, bacterial translocation, and antibiotic resistance. Understanding the microbial composition and its relationship to the mosquito is an essential step in developing efficient paratransgenic techniques that disrupt viral transmission. This study investigated the composition and plasticity of the microbiota of adult female *Aedes aegypti* mosquitoes. They collected female *Ae. aegypti* mosquitoes (n = 8–10) and surface sterilized, lysed, and plated them onto non-selective agar using time-course sampling methods (at emergence, 1, 2, and 3 weeks post-emergence). A total of 104 of the 232 nonselective agar plates developed observable bacterial colonies, from which DNA was extracted. Unique phenotypic colonies were propagated at a biologically relevant temperature of 28 °C and identified via Sanger sequencing of the 16s ribosomal RNA gene. Future works to examine shifts in bacterial composition through application of antibiotics as a means of selective pressure would, when taken together with these data, provide a larger, novel snapshot of the various bacterial species within the midgut that could be targeted for genetic manipulation.

Brian C. Ware, together with Bennett J. Davenport and Thomas E. Morrison (Department of Immunology and Microbiology, University of Colorado School of Medicine, Aurora, CO, USA), presented work on how CHIKV-infected cells display reduced MHC Class I expression and a reduced capacity to activate CD8 T cells. Arthritogenic alphaviruses, including Chikungunya (CHIKV), Mayaro (MAYV), and Ross River (RRV) viruses, are re-emerging global health threats with no approved vaccines or therapies. Infection with these viruses causes debilitating pain and inflammation in musculoskeletal tissue for months to years due partially to the persistence of viral RNA and antigens. Prior studies revealed that arthritogenic alphavirus infection evades the B cell response to establish persistence. However, mechanisms by which these infections evade CD8 T cell responses, another critical arm of antiviral adaptive immunity, are undefined. Indeed, anti-CHIKV CD8 T cells are present in lymphoid tissues and sites of infection, and these cells can kill peptide-loaded target cells. Remarkably, genetic deletion of CD8 T cells does not alter CHIKV burdens in joint-associated tissues during acute or chronic infection, leading them to hypothesize that infected cells escape surveillance by recruited CD8 T cells. To investigate this idea, CHIKV-infected joint cells, including fibroblasts and myeloid cells, were assessed for MHC-I cell surface expression. They found marked loss of MHC-I expression coincident with high levels of viral replication. Furthermore, in vitro infected fibroblasts failed to activate CD8 T cells above the bystander background, suggesting that CHIKV infection suppresses the antigen presentation capacity of infected cells. Lastly, mutating the nonstructural protein 2 of CHIKV ameliorates the reduction in MHC-I expression, suggesting a role for nsP2 in the evasion of the CD8 T cell response and establishment of persistent infection. These results highlight a role for CHIKV nsP2 in evading the CD8 T cell response and contributing to CHIKV persistence. No animal or human studies were performed.

Joseph Westrich (Department of Microbiology, Immunology, and Pathology, Colorado State University) presented about the longitudinal viral progression and immunological responses to Bluetongue Virus 17 in experimentally infected sheep. Bluetongue virus (BTV) is an economically important arthropod-borne pathogen that infects ruminant species worldwide. The severity of BTV infections range from asymptomatic to lethal, with the most severe cases succumbing to disease within one week. Animals that survive the infection often require months to fully recover. The immune response to BTV infection is thought to contribute to the propagation of disease in addition to being critical in the ultimate resolution of infection. Although BTV has been recognized since the 1800s, much of the cellular and cytokine response remains poorly understood due to limited reagent availability for the natural host species. To gain a greater understanding of the role the immune response plays in BTV infection, he infected a cohort of sheep with BTV-17 and longitudinally monitored each for clinical disease, viremia, and specific immunological parameters. Immune cells and cytokines were evaluated with traditional flow cytometry, RNA flow cytometry, RT-qPCR, and/or fluorescent-based antibody arrays. All BTV-inoculated sheep exhibited clinical signs characteristic of BTV disease. Circulating virus was observed as early as 3 days post=inoculation (dpi) and remained detectable for the remainder of the study (24 dpi). A distinct pan-leukopenia was observed between 8 and 14 dpi that rebounded to mock-inoculated control levels at 17 dpi. He observed increased expression of pro-inflammatory cytokines after 8 dpi—notably, the pro-inflammatory cytokine CXCL10. Taken together, he established a sheep model of BTV infection and successfully monitored the longitudinal immunological response and viral progression using a combination of traditional methods and cutting-edge technology.

2.4. Viral Detection and Other Novel Techniques

Noelia Altina [1], along with David Maranon [1], John Anderson [1], and Jeff Wilusz [1] ([1] Microbiology, Immunology, Pathology Department, Colorado State University, Fort Collins, CO, USA), investigated fundamental knowledge gaps in the RNA biology of

SARS-CoV-2. Specifically, they are interested in the molecular mechanisms underlying SARS-CoV-2 mRNA capping, cap-proximal adenosine methylation, and the impact of viral RNA–protein interactions on host cell RNA biology. First, to characterize the putative role of the viral Nsp12 protein as the guanylyltransferase involved in SARS-CoV-2 mRNA capping, they successfully expressed and purified recombinant protein. They are currently performing biochemical assays and mutagenesis studies to gain insight into enzymatic mechanisms and potential drug targets. Second, they obtained evidence suggesting that the A residue at the 5′ end of SARS-CoV-2 mRNAs likely contains an 'm6Am' methylation modification. This is significant because all cellular mRNAs that initiate at an A residue contain a similar m6Am modification. Thus, they hypothesize that SARS-CoV-2 is modifying its mRNAs to prevent their detection as 'non-self'. They are currently testing this hypothesis and will identify the mechanism of SARS-CoV-2 m6Am RNA modification. Finally, two cellular proteins (hnRNPA1 and PTBP) that regulate alternative splicing bind to the abundant viral mRNAs in related coronaviruses and become mis-localized to the cytoplasm during infection. Therefore, they hypothesize that SARS-CoV-2 mRNAs can sequester these splicing factors, leading to significant changes in alternative mRNA splicing. RT-PCR analyses of the splicing patterns across select exons of hnRNPA1/PTBP-targeted pre-mRNAs confirm this hypothesis and suggest a new mechanism for the impact of SARS-CoV-2 infection on cellular RNA biology. These studies collectively focus on attractive targets for developing novel broad-spectrum anti-coronavirus drugs. No animal or human studies were performed.

Ali L. Brehm, along with Mark Stenglein (Department of Microbiology, Immunology, and Pathology, Colorado State University, Fort Collins, CO, USA), presented their work on creating a homozygous fly line utilizing non-lethal genotyping. *Drosophila melanogaster* is used widely as a model organism due to its high number of orthologous genes with humans. New lines of flies are created regularly that contain and express genes of interest through multiple generations of inbreeding. Inbreeding can lead to the maintenance of deleterious mutations and creates stocks of *D. melanogaster* that are considered weak, and which may be more susceptible to pathogens that create mass die-offs. Using non-lethal DNA extraction and PCR, a fly line that was homozygous for an endogenous viral element (EVE) that was 95% similar to the RNA-dependent RNA polymerase gene of galbut virus was created. Galbut virus is a double-stranded RNA multipartite partitivirus that is of particular interest because it is found in 100% of wild populations of *D. melanogaster* and has the potential to be used in cross-species biocontrol efforts, but despite being found in all populations, only 60% of individuals are infected. More information regarding *D. melanogaster* resistance to galbut virus is needed to determine if galbut virus EVEs may limit its use in biocontrol. Non-lethal genotyping has the potential to allow the revival of significant fly lines that are weak due to inbreeding depression, as was the case in the original galbut EVE fly line. Funding for this project was provided by NSF IOS 2048214. No animal or human studies were performed.

Luke Davis, along with Philip Fox, Kaitlyn Wagner, Liddy Gordon, and Mark Zabel (Department of Microbiology, Immunology, and Pathology, Colorado State University, Fort Collins, CO, USA), presented an update on CSU's efforts to develop a high-throughput pooling and ddPCR-based strategy for COVID-19 screening of asymptomatic populations. Screening of asymptomatic populations for the presence of SARS-CoV-2 is an effective way to permit in-person classes and gatherings while still ensuring the health and safety of those involved. However, the costs of periodic screening and the materials and reagents required can be prohibitive of such a strategy, especially during a pandemic, when common reagents become difficult to obtain. They developed a paired-pooling strategy that was implemented for a droplet digital PCR assay to detect SARS-CoV-2 in saliva, which reduced costs (<10 USD/sample) and reagent consumption while supporting high-throughput (>2000 samples/day). This assay was implemented in an asymptomatic screening program at CSU, and in the Fall 2020 and Spring 2021 semesters, over 150,000 samples were screened. The centerpiece of this strategy is the paired-pooling approach, in which each sample is added to a set of two pools such that no two pools contain more than one common sample. With this approach, one positive sample can be identified out of 64 total samples without requiring a rerun of all the samples within the positive pools. As the percentage of positive samples increases, the number of required reruns grows with the square of the number of positive samples, so this approach works best for screening asymptomatic populations with a low percentage of positive samples. The paired-pooling approach can be implemented with different assays and specimen types and represents an effective way to perform asymptomatic screening while keeping costs and material use low. All human studies were performed following guidelines and protocols approved by the Institutional Review Board of Colorado State University.

Nicole Ehrhart [1,2], along with Greg Ebel [2] and Kristy Pabilonia [2] ([1] Columbine Health Systems Center for Healthy Aging and Department of Clinical Sciences, [2] Department of Microbiology, Immunology, and Pathology, Colorado State University, Fort Collins, CO, USA), presented her group's surveillance efforts of staff at nursing homes and assisted living facilities. Nursing homes and assisted living facilities, collectively known as long-term care facilities (LTCFs), provide care for some of the most vulnerable populations in society. Communal living spaces and the need for daily contact between staff and residents creates an opportunistic environment for the spread of any infectious disease. In particular, the morbidity and mortality within LTCFs throughout the COVID-19 pandemic has demonstrated the extreme vulnerability of both LTCF residents and staff to transmissible viral illness. To detect pre- and asymptomatic infections among LTCF staff, they performed weekly SARS-CoV-2 surveillance testing of staff and residents at 31 LTCFs over 30–60 consecutive weeks. They collected infection and outbreak patterns, genomic sequencing data, serum neutralizing antibody data, and the success or failure of various infection mitigation strategies. They archived one of the largest known collections of longitudinal diagnostic specimens. They concluded that: (1) LTCF staff represent a disproportionally high percentage of SARS-CoV-2 infections as compared with non-healthcare community members regardless of whether or not LTCF staff have direct patient contact, (2) LTCF staff who tested positive were most likely to be infected from exposure within their workplace and not the community, (3) early identification of infected staff via regular surveillance testing was effective at mitigating outbreaks, and (4) the testing cadence and type was highly correlated with mitigation efficacy. Infectious disease response agility and future pandemic preparedness will require more proactive strategies to prevent the high morbidity and mortality within LTFCs in the future.

Emmanuel Ijezie and Elizabeth Fortunato (Department of Biological Sciences University of Idaho, Moscow, ID, USA) presented their research investigating the role that Nidogen-1 plays in aberrant neurodevelopment of HCMV-infected brain organoids. The ability to generate three-dimensional brain organoids in vitro has provided an ethical system to model and study viral–host interactions within a developing human brain. Congenital human cytomegalovirus (HCMV) is the leading cause of viral-induced birth defects in humans. These birth defects include microcephaly, sensorineural hearing loss, and cognitive impairment. The process in which the developing fetus incurs these neurological defects is still poorly understood. To elucidate some of these mechanisms, their lab has adapted a brain organoid development protocol that utilizes HCMV-infected induced pluripotent stem cells to generate in vitro brain organoid structures that model the first and second trimester in fetal brain development. Early in development, brain organoids generate neural rosette structures that model neural tube formation in vertebrates. These neural rosettes further develop into structures with multiple neural cell layers. Both infected and uninfected brain organoids were collected, frozen, and sectioned for qualitative and quantitative histological and immunofluorescence (IF) protein analysis. Histological analysis of infected brain organoids revealed that fewer neural rosettes were formed. IF analysis of neural rosettes showed that nidogen-1 protein expression in the basal lamina was greatly reduced in infected neural rosettes compared to uninfected neural rosettes. IF analysis also indicated that expression of neural progenitor cell protein markers Pax-6 and Nestin was downregulated in infected neural rosettes compared to uninfected neural rosettes. IF analysis of dissociated brain organoids detected viral proteins IE1 and PP71 early during organoid development. Infected cortical structures also displayed decreased B-tubulin III expression, indicating a delay in neurogenesis. In conclusion, the progression of neural development is disrupted in neural rosettes within infected brain organoids. This disruption in the development and downregulation of cellular proteins correlates with the reanimation of viral proteins within these brain organoids. Understanding the role of HCMV in downregulating neural developmental processes in brain organoids could elucidate the mechanisms behind the disease phenotypes seen in congenitally infected newborns. This study was funded by NIH RO1 AI051463 and AI139503 to E.A.F., NIH P20 RR016454. No animal or human studies were performed.

Tanya Jolly [1], along with Luke Davis [1], August Luc [1], Jim Huang [1], Susan De Long [2], Philip Fox [1], and Carol Wilusz [1] ([1] Wastewater Testing Laboratory/Molecular Quantification Core, Dept. of Microbiology, Immunology, and Pathology, and [2] Department of Civil & Environmental Engineering, Colorado State University), discussed a robust and reproducible assay for monitoring SARS-CoV2 in wastewater from treatment plants and college dormitories. Wastewater-based epidemiology represents an unbiased, cost-effective, and non-invasive means of monitoring a population for infectious agents. SARS-CoV2 infects the human gastrointestinal tract and is shed in feces prior to establishing itself in the respiratory system, from which it can be disseminated to others and causes most of its symptoms. This means that circulating virus can often be detected in the sewer prior to cases arriving at the clinic. They developed a molecular assay to detect SARS-CoV2 in wastewater that is sensitive, rapid, and resistant to inhibition. Since August 2020, their facility has received up to 100 samples a week of wastewater from university dorms and wastewater treatment plants across the Colorado Front Range. Their protocol allows concentration of viral particles via ultrafiltration, extraction of viral RNA, and quantification of the SARS-CoV2 genome copy number using probe-based multiplex droplet digital PCR. Data are reported to stakeholders within 24 h of sample receipt and inform public health messaging and decision making. Variant strains can be detected by further analysis through sequencing. Acknowledgements: Their assay was developed in collaboration with GT Molecular Inc., the Colorado Department of Public Health and Environment, the Colorado Wastewater Collaborative with funding from the Colorado State University Office of the Vice President for Research, and Metro Water Recovery.

Lexi Keene, along with Tillie Dunham and Mark Stenglein (Department of Microbiology, Immunology, and Pathology, Colorado State University, Fort Collins, CO, USA), discussed her testing efforts involved in building a better fly trap for catching wild flies. *Drosophila melanogaster* have been utilized since the early 1900s, and these insects have contributed substantially to the understanding of the natural world. Most *Drosophila* research has relied on lab-reared strains that are inbred and long removed from their geno-/phenotypes. Studying wild flies offers a way to extend the relevance of this laboratory model. Her PhD research is focused on studying viruses that naturally infect *D. melanogaster*, which can reveal principles of host–virus interactions that can further the study of disease in animals and plants. To accomplish this, she needed a method to efficiently trap live flies. Traps baited with banana and yeast have been an effective way to capture wild flies; however, the bait decomposes rapidly, making it difficult for flies to lay eggs and for adults to be retrieved. Thus, the ability to capture wild flies needs to be through the use of food that is stable over time and during transportation while also providing a place for the flies to reproduce. Three types of food were tested for longevity in different environments. Cornmeal food proved the most stable. Additionally, four additives were tested to determine which would attract flies. The flies were overwhelmingly attracted to food that had a banana component and, surprisingly, were almost repulsed by the addition of ethyl isovalerate, which had been identified as the key attractant in marula fruit, the purported ancestral food source of *D. melanogaster*. She is continuing to optimize the bait composition and the efficacy of these traps in real-world environments. The work was supported by NSF grant #2048214 and qCMB T32: T32GM132057. All animal studies were performed following guidelines and protocols approved by the Institutional Animal Care and Use Committee of Colorado State University.

Stefan L. Oliver (Department of Pediatrics—Infectious Diseases, Stanford University School of Medicine, California, USA) presented his lab's work studying the cell-fusion-dependent pathogenesis of varicella-zoster virus (VZV). VZV is an alphaherpesvirus of medical importance and is the causative agent of chicken pox (varicella) and shingles (zoster). Although considered benign diseases, both varicella and zoster can produce complications. Zoster is painful, potentially leading to protracted post-herpetic neuralgia. VZV has also been linked to stroke and has been related to giant cell arteritis in some cases. A hallmark of VZV pathology is the formation of multinucleated cells driven by cell–cell fusion (abbreviated as cell fusion) mediated by the VZV glycoproteins gB, gH, and gL, which constitute the fusion complex of herpesviruses, including VZV. These evolutionarily conserved glycoproteins are required for virion entry into cells, with the gH-gL heterodimer priming the trimeric gB fusogen to complete the membrane fusion process. The adaptive immune response can target the fusion complex, producing antibodies that bind to both the gB trimer and the gH-gL heterodimer. An Achilles' heel for VZV and a new functional domain for gB were revealed by cryo-EM and X-ray crystallography derived near atomic resolution structures of neutralizing antibodies bound to these critical glycoproteins. Expression of gB, gH, and gL during VZV infection and trafficking to the cell surface enables cell fusion, which is inhibited by these neutralizing antibodies. Moreover, recent evidence supports the concept that cellular processes are required for regulating cell fusion induced by gB/gH-gL. Mutations within the carboxyl domains of either gB or gH have profound effects on fusion regulation and dramatically restrict the ability of VZV to replicate in human skin. This loss of regulation modifies the transcriptome of VZV-infected cells. Furthermore, cellular proteins have significant effects on the regulation of gB/gH-gL-mediated cell fusion and the replication of VZV, exemplified by the cellular phosphatase calcineurin. His presentation focused on the current state-of-the-art knowledge about the molecular controls of cell-fusion-dependent pathogenesis caused by VZV. This research was supported by a Stanford Bio-X Interdisciplinary Initiatives Seed Grant and the National Institutes of Health through grants P41-GM103832, R01-GM079429, R01-AI102546, R37-AI20459, and S10-OD021600. All animal studies were performed following guidelines and protocols approved by the Institutional Animal Care and Use Committee of Stanford University.

Carol Wilusz [1], along with Phil Fox [1], Laura Bankers [2], Shannon R. Matzinger [2], Brian Erly [2], Joshua Goldmann-Torres [3], Mazdak Arabi [4], and Susan De Long [4] ([1] Department of Microbiology, Immunology, and Pathology, Colorado State University, [2] Colorado Department of Public Health and Environment, [3] Metro Water Recovery, [4] Department of Civil and Environmental Engineering, Colorado State University, Fort Collins, CO, USA), presented on their efforts to establish wastewater-based epidemiology (WBE) at Colorado State University. WBE has been established as a viable, valuable, and cost-effective means of monitoring infectious disease within a community. SARS-CoV-2 is shed in feces as early as one to two days post-infection, and viral RNA is detectable for several days in wastewater. Thus, detection and quantification of its genome can be used to evaluate trends in infection rate and inform public health actions. Over the last year, they have monitored wastewater from 21 treatment plants across the state of Colorado for SARS-CoV-2 and correlated the data to the clinical caseload in the population they represent through census tracts. The viral RNA load over time closely correlates with and slightly precedes increases in reported clinical cases. The ability to track the virus through wastewater is more valuable than ever as the epidemiology of SARS-CoV-2 changes with increased vaccination, new variants, and changing testing and infection patterns. Moreover, the isolated RNA can be used to detect the presence of novel and known variants of the virus. For example, mutations associated with the Alpha and Delta variants have been detected via mutation-specific PCR and/or whole-genome sequencing prior to clinical cases being observed within the Colorado population.

2.5. Antivirals, Treatments, and Vaccines

Corey L. Campbell [1,*], along with Trey K. Snell [1], Susi Bennett [1], John Wyckoff [2], Darragh Heaslip [2], Emma K. Harris [1], Daniel Hartman [1], Elena Lian [1], Brian H. Bird [3], Mark D. Stenglein [1], Richard A. Bowen [1], and Rebekah C. Kading [1] ([1] Arthropod-borne & Infectious Diseases Laboratory, Department of Microbiology, Immunology, and Pathology, Colorado State University, Fort Collins, CO, USA. [2] Department of Neurology, University of Colorado School of Medicine, Boulder, CO, USA. [3] College of Veterinary Medicine and Biomedical Sciences, Colorado State University, Fort Collins, CO, USA), presented her work on human vaccine candidate development against Rift Valley fever virus (RVFV). RVFV is a mosquito-borne pathogen with significant human and veterinary health consequences that periodically emerges in epizootics. RVF causes fetal loss and death in ruminants, and in humans, it can lead to liver and renal disease, delayed-onset encephalitis, retinitis, and, in some cases, severe hemorrhagic fever. A live attenuated vaccine candidate, DDVax, was developed by the deletion of the virulence factors NSs and NSm from a clinical isolate, ZH501, and has proven safe and immunogenic in rodents, pregnant sheep, and non-human primates. Deletion of NSm also severely restricted mosquito midgut infection and inhibited vector-borne transmission. To demonstrate environmental safety, this study investigated the replication, dissemination, and transmission efficiency of DDVax in mosquitoes following oral exposure compared to RVFV strains MP-12 and ZH501. Infection and dissemination profiles were also measured in mosquitoes 7 days after feeding on goats inoculated with DDvax or MP-12. Hypothesis: DDVax should infect mosquitoes at significantly lower rates than other RVF strains and, due to lack of NSm, be transmission incompetent. Exposure of *Ae. aegypti* and *Cx. tarsalis* to 6–8 log10 plaque-forming units (PFU)/mL DDVax by artificial bloodmeal resulted in significantly reduced DDVax infection rates in mosquito bodies compared to controls. Plaque assays indicated negligible transmission of infectious DDVax in *Cx. tarsalis* saliva (1/140 sampled) and none in *Ae aegypti* saliva (0/120). Serum from goats inoculated with DDVax or MP-12 did not harbor detectable infectious virus according to the plaque assay at 1, 2, or 3 days post-inoculation; infectious virus was, however, recovered from mosquito bodies that fed on goats vaccinated with MP-12 (13.8% and 4.6%, respectively), but strikingly, DDvax-positive mosquito bodies were greatly reduced (4% and 0%, respectively). Furthermore, DDVax, unlike MP-12, did not disseminate to legs/wings in any of the goat-fed mosquitoes. Collectively, these results are consistent with a beneficial environmental safety profile. All animal studies were performed following guidelines and protocols approved by the Institutional Animal Care and Use Committee of Colorado State University.

Chaoping Chen [1], along with Liangqun Huang [1] and Megan Gish [1] ([1] Biochemistry and Molecular Biology Department, Colorado State University, Fort Collins, CO, USA), presented on a main protease (M^{PRO}) of SARS-CoV-2 that is indispensable for viral replication. It is initially synthesized as part of polyprotein precursors whose autoproteolysis liberates free mature M^{pro}. They herein studied autoprocessing of fusion precursors with the mature M^{pro}, along with a few flanking amino acids (to keep the cleavage sites) sandwiched between tags in transfected mammalian cells. Mutating the conserved Gln of C-terminal P1 residue to Glu abolished C-terminal autoprocessing, while it had no impact on N-terminal autoprocessing. Interestingly, mutating the conserved N-terminal P1 Gln to Glu did not ablate N-terminal processing, indicating distinct catalysis kinetics at N- and C-terminal autoproteolysis. Various upstream fusion tags (sGST, GST, CD63, Nsp4) also showed diverse N-terminal autoprocessing efficiencies, suggesting regulation of M^{pro} precursor catalysis in the polyprotein context. Furthermore, mutation QtoE at the N-terminal P1 position altered precursor catalysis and outcomes, as demonstrated by different susceptibilities to several preclinical M^{pro} inhibitors. N-terminal processing of precursors with Gln at P1 was more resistant to drug inhibition than those with Glu at P1 in the GST-fusion context. Additionally, the mature M^{pro} released from Q precursors was self-degradation prone and could be suppressed by boceprevir, calpain inhibitor II, and GC376, whereas those released from E precursors were self-degradation resistant and irresponsive to drug treatment. Many amino acid variations at the N-terminal P1 position were well tolerated with various processing efficiencies, but a few containing β-branched side chains blocked N-terminal processing. These results collectively revealed substrate diversity and catalysis plasticity of fusion precursors, providing insights into drug discovery targeting the M^{pro}. No animal or human studies were performed.

Rupika Delgoda [1], along with S. Elmegerhi [2], W. Irvine [1], D. Picking [1], S. Francis [1], and Rushika Perera [2] ([1] Natural Products Institute, University of the West Indies, Mona, Jamaica; [2] Center for Vector-Borne Infectious Diseases, Colorado State University, Fort Collins, CO, USA), presented their work on evaluating the potential of Jamaican ethnomedicines for their efficacy against infectious diseases. Targeting novel chemical scaffoldings entrapped in nature's resources has yielded valuable pharmaceuticals for current medical use. Despite modern combinatorial capacities, bioprospecting of microbial, terrestrial, and marine organisms remains a thriving endeavor, due largely to their unique molecular offerings. Probing biodiversity with a history of human use has also had rewarding results, often curtailing the protracted discovery pathway. Given the high cultural reliance on ethnomedicines in the Caribbean, they delved into popular practices in Jamaica in search of bioactive molecules against infectious diseases. Community-based surveys revealed that 73% of persons rely on medicinal plants for primary care, with a large portion (78%) using them to treat respiratory tract ailments. Of the 51 plant families in use, Fabaceae, Lamiaceae, Asteraceae, Malvaceae, and Piperaceae are the most commonly used, hosting multitudes of natural bioactive alkaloid, polyphenol, terpene, and flavonoid products. They highlighted plants prepared following ethnomedical extraction protocols, tested against the SARS-COV-2, and using cytoprotection and plaque assays on Vero cells. Several plant extracts showed promising results in reducing cell death due to virus infection and reducing infectious virus release. Taken together with knowledge on anti-inflammatory activity, as well as the impact on drug-metabolizing cytochrome P450 enzymes for combination therapy, a useful appraisal has emerged for these ethnomedicines as infectious diseases therapeutics. No animal or human studies were performed.

Suad Elmegerhi [1,*], Elena Lian [1], Gabriela Ramirez [1], Carley McAlister [1], Laura St. Clair [1], Camryn S Guenther [1], Rupika Delgoda [2], and Rushika Perera [1] ([1] Department of Microbiology, Immunology, and Pathology, Colorado State University, Fort Collins, CO; [2] Natural Products Institute, University of the West Indies, Mona, Jamaica) discussed their lab's efforts in developing platforms to test antivirals against SARS-CoV-2. Despite the great efforts by the international community to develop effective vaccine candidates to mitigate the COVID-19 outbreak, many new variants and new cases are still on the rise. Therefore, there is still a need for effective and readily available antiviral therapies to help reduce disease severity and escalating cases. Re-purposing FDA-approved therapeutics or compounds in the preliminary stages of approval paves a quicker path to medical treatment and is currently a focus for antiviral development. Performing more targeted testing of compounds that interfere with the general coronavirus life cycle is another route being pursued. Some compounds that include natural products have proven antiviral properties against many viruses and have been used to treat rhinovirus and influenza virus infections. In this study, they tested the antiviral activity of different drugs and investigational compounds, natural products, peptides, enzyme inhibitors, and disinfectant-treated textile materials against SARS-CoV-2. The platforms developed for antiviral testing and results were presented. No animal or human studies were performed.

Camryn S. Guenther, along with Elena Lian, Laura St. Clair, and Rushika Perera (Center for Vector-Borne Infectious Diseases, Department of Microbiology, Immunology and Pathology, Colorado State University, Fort Collins, CO, USA), discussed her investigations into the roles of phospholipase A2 and arachidonic acid during DENV2 infection. Dengue viruses are transmitted by *Aedes aegypti* mosquitoes and cause over 400 million infections annually. These viruses hijack host lipid pathways in the cell to support their replication. A pathway of interest is arachidonic acid (AA) metabolism. AA and lysophospholipids are released by cytosolic and secreted phospholipase A2 (PLA2) enzymes from cell membrane phospholipids. This process is called the Lands cycle. AA is also a well-known precursor of antiviral inflammatory mediators. Previous studies by their laboratory have already shown that reduction of AA through a knockdown of AA-producing PLA2 isoforms (2A, 4A, and 4C) decreased infectious viral titer during a DENV2 infection. The goal of the current study was to identify why AA plays such an important role during the viral life cycle. First, they have investigated which steps of the viral lifecycle are impacted by the reduction of AA through an siRNA-mediated knockdown of PLA2 (2A, 4A, and 4C). Second, they conducted experiments to determine if exogeneous AA addition could recover infectious viral titer in the siRNA-treated cells. It is their hypothesis that PLA2 and AA may have dual roles during virus infection, both in supporting viral replication and creating the antiviral inflammatory precursor molecules. Progress on this work was presented. No animal or human studies were performed.

Jim Heath (Institute for Systems Biology, Seattle, WA, USA) presented on research from his lab about quantitative analysis of post-acute sequelae in COVID-19 patients. Post-acute sequelae from COVID-19 (PASC) represent a potential global crisis. However, quantifiable risk factors for distinct PASC and their biological associations are poorly resolved. He discussed the ISB/Swedish INCOV study in which his group investigated several hundred COVID-19 patients and healthy donors from two cohorts using clinical data, patient-reported symptoms, viral load, anti-SARS-CoV-2 antibodies and autoantibodies, and multi-omic analyses of blood plasma and circulating immune cells. They resolved four factors that are significantly associated with PASC: SARS-CoV-2 RNAemia near the time of initial COVID-19 diagnosis, Epstein–Barr virus (EBV) viremia during acute disease, bystander activation of cytomegalovirus (CMV), and specific autoantibodies at convalescence. They also identified potential therapeutic strategies for a subset of patients suffering from PASC. Systematic analysis of symptom-associated immunological signatures revealed coordinated polarization of innate and adaptive immunity that was independently associated with acute infection severity and PASC. Their analyses suggest that the heterogeneous symptoms of long COVID arise from multiple independent sources. All human studies were performed following guidelines and protocols approved by the Institutional Review Board of the Institute of Systems Biology.

Sven Heinz, along with Samuel J. Roth and Christopher Benner (Dept. of Medicine, Division of Endocrinology and Metabolism, University of California San Diego), examined how transcriptome phenotyping identifies viruses that inhibit host cell transcription termination. Cells that recognize viral-pathogen-associated molecular patterns rapidly activate production of antiviral proteins. Some viruses, such as herpes simplex virus 1 and influenza A virus, subvert the host cell antiviral response in part by globally inhibiting transcription termination. This is caused by viral proteins that inhibit transcript cleavage and polyadenylation, leading to "reading through" of RNA polymerase II past the end of genes. The resulting unnaturally long readthrough transcripts, which include those coding for antiviral proteins, remain in the nucleus and are not translated into protein. Applying a bioinformatics approach to detecting readthrough transcription in public RNA-seq datasets from cells infected with various viruses, they found that Zika, Sindbis, and Rift Valley Fever virus (RVFV) inhibit transcription termination. They confirmed this finding for RVFV, where they showed that expressing RVFV NSs protein alone is sufficient to induce global transcriptional readthrough. Their findings suggest that genome-wide inhibition of transcription termination by viral proteins may represent a broader viral strategy to undermine the host cell antiviral response. No animal or human studies were performed.

Ryan Jeep [1], along with Christian Sanders [1], Megan Gish [1], Lillian Huang [1], and Chaoping Chen [1] ([1] Department of Biochemistry and Molecular Biology, Colorado State University, Fort Collins, CO, USA), discussed their research on HIV protease inhibitor (PI) resistance. PI resistance compromises treatment efficacy and prognosis of combination antiretroviral therapy. The currently available PIs target the active site of mature protease (PR); however, resistance-associated mutations (RAMs) identified in patients experiencing drug resistance map to various regions of protease. When introduced into *E. coli* for protein purification and in vitro characterization, many RAMs do not display drug resistance at the magnitudes observed in patients. They previously reported that precursor autoprocessing is context-dependent, as its activity and outcomes can be modulated by sequences upstream of p6*-PR. Notably, the mature PRs released from select fusion precursors were self-degradation prone, resembling those purified from *E. coli*. In contrast, precursors with the 26aa maltose-binding protein signal peptide (SigP) at the N-terminus released mature PRs resistant to self-degradation resembling those found in viral particles. Therefore, SigP-containing fusion precursors likely represent a more biologically relevant "context" and are expected to manifest the clinically observed drug resistance. To investigate this context, they engineered a panel of fusion precursor constructs containing either the wild-type or an indinavir-resistance-associated double mutation (77I82T) in the PR region. Both wild-type and 77I82T mutant precursors showed similar responses to PI when released from constructs without SigP. However, the 77I82T mutant displayed a significant resistance profile when the N-terminal SigP was present in the precursors, which matches their viral infectivity analysis. These results support their context-dependent hypothesis and suggest that SigP-containing precursors better recapitulate PI response observed in a viral setting. This platform provides a unique insight for studying drug resistance. No animal or human studies were performed.

Jeffrey Kim (Department of Microbiology, Immunology, and Pathology, College of Veterinary Medicine and Biomedical Sciences, Colorado State University, Fort Collins, CO, USA) provided an update on combination antiretroviral therapy treatment for feline immunodeficiency virus (FIV). FIV is a naturally occurring retrovirus that causes progressive immune dysfunction in cats. Although effective treatments, including a combination antiretroviral therapy (cART), have been developed for human immunodeficiency virus, there is no therapy for FIV. This study aimed to assess a cART of Doulutegravir, Tenofovir, and Emtricitabine as a treatment for FIV in domestic cats. Eighteen 6- to 10-month-old cats were used with female intact and male neutered cats equally distributed among cART, placebo, and control groups. He inoculated twelve cats with FIV in addition to six naïve cats as controls and quantified the blood viral and proviral loads via digital droplet PCR pre-FIV inoculation and in weeks 0–24. Beginning in week 4, cats were injected with cART ($n = 6$) or a 15% kleptose placebo ($n = 6$), the cART vehicle, daily for the FIV-infected cats. Infection peaked at week 2 with mean viral loads of 6.43×10^6 and 12.4×10^7 copies/mL among cART and placebo cats, respectively; mean proviral loads were 5.43×10^4 and 4.94×10^4 copies/10^6 cells among cART and placebo cats, respectively. Viral and proviral loads steadily declined afterwards and were not statistically significant. Hematologic analysis demonstrated marked neutropenia in placebo cats compared to cART ($p = 0.05$), with the lowest at week 6 with mean absolute values/uL of 651 and 1340, respectively. The mean neutrophil values in cART cats normalized by week 16, but placebo cats remained neutropenic. No clinical signs associated with neutropenia were observed. These results suggest that this cART protocol does not reduce FIV viral and proviral loads, but may impact the hematologic impacts of FIV infection. All animal studies were performed following guidelines and protocols approved by the Institutional Animal Care and Use Committee of Colorado State University.

Mani Kuan [1], L. Caruso [2], A. Zavala [1], P. Rana [1], J. O'Dowd [1], I. Tempera [2], and E. Fortunato E [1] (Department of Biological Sciences, University of Idaho [1] and The Wistar Institute [2]) discussed how HCMV pp71 interaction with host DNA leads to downregulation of the basement membrane protein Nidogen 1. Nidogen 1 (NID1) is an important basement membrane protein. Their previous work found that HCMV infection rapidly induced chromosome 1 breaks and downregulated expression of NID1 both at the transcriptional and post-translational levels. NID1 downregulation promoted HCMV dissemination. They determined several viral proteins and some of their mechanisms involved in regulating

NID1. Here, they investigated gene expression of NID1. Since regulation started by 24 hpi, they screened the most prominent viral proteins present at this timepoint, including incoming virus tegument and rapidly expressed immediate early (IE) proteins. They screened pp65, pp71, UL35, IE1, and IE2. They used an adenovirus (Ad) expression system to introduce each of these HCMV proteins into human foreskin fibroblasts. Using RT-qPCR, they found that pp71 downregulated NID1 gene transcription. Surprisingly, WF28-71, a fibroblast clone that expresses small amounts of pp71, suppressed NID1 transcript levels as efficiently as during HCMV infection and, therefore, resulted in a dramatically reduced steady-state (ss) level of NID1 protein. CCCTC-binding factor (CTCF) is a transcription factor that contributes with various regulatory roles in gene expression. There are CTCF binding sites located directly adjacent to the chromosome 1 break sites and the NID1 promoter. They suspected that pp71 might be interacting with CTCF to influence NID1 transcription. Chromatin immunoprecipitation found both pp71 and CTCF bound at these sites during HCMV infection. Knockdown CTCF fibroblasts infected with Ad-pp71 recovered the NID1 transcription and ss level of NID1 protein. pp71 is a key player in HCMV's efforts to eliminate cellular NID1 and affects this paradigm via direct binding to the host cell genomic DNA and interaction with CTCF. No animal or human studies were performed.

Gabriela Samayoa Reyes (University of Colorado, Anschutz Medical Campus, Denver, CO) presented characterization of EBV in saliva of HIV+ mothers in a Kenyan population and presented evidence of mixed EBV infections and both methylated and unmethylated viral genomes. HIV infection is associated with EBV shedding in saliva, suggesting an increased risk of EBV transmission to infants born to mothers with HIV. She investigated the following questions: (i) Is HIV status a risk factor for EBV shedding in postpartum mothers? (ii) Is there a difference in EBV strains shed between HIV- positive (HIV+) and HIV-negative (HIV−) individuals? (iii) Is the viral DNA detected by qPCR representative of cell-associated infection or attributable to virions? Samples were from a cohort study in which HIV+ and HIV− pregnant women from Western Kenya were followed through delivery and the postpartum period. DNA was extracted from saliva and EBV load was measured by qPCR. She observed a significant relationship between EBV shedding and HIV status; HIV− mothers had a greater proportion of detectable EBV in their saliva compared to HIV+ mothers (p-value: 0.005). Of the mothers that were EBV shedders, those with HIV had significantly greater log EBV copies/mL of saliva (p-value: 0.015). Overall, she found that 47.3% of the EBV-positive samples were infected with EBV type 1, 23% with EBV type 2, and 24.3% were co-infected with no difference in frequency of EBV type attributable to the HIV− or HIV+ mothers. She analyzed the EBV genome methylation patterns using bisulfite sequencing to assess whether they were detecting cell-associated DNA (methylated) vs. virion-derived DNA (unmethylated). Interestingly, she observed two major groups that could be segregated based on methylation patterns and viral load. One group had high methylation and low viral load; the other was an unmethylated group associated with a higher EBV viral load, indicative of virion shedding. The presence of methylation indicates that a variable proportion of EBV DNA in saliva derives from cellular sources. In summary, HIV+ mothers that shed EBV have a higher viral load compared to HIV− mothers. All human studies were performed following guidelines and protocols approved by the Institutional Review Board of the University of Colorado, Anschutz Medical Campus.

Author Contributions: L.A.S.C., A.L.B., S.C., T.D., J.F., P.G. and M.E.G. assembled the abstracts. L.A.S.C. wrote the abstract, introduction, and figure legends. L.A.S.C., S.L.Q., J.R. and R.P. oversaw the assembly of the final version. All authors have read and agreed to the published version of the manuscript.

Funding: Funding for this conference was supported, in part, by the National Institute of Allergy and Infectious Diseases of the National Institutes of Health under award number R13 AI164787-01. All content is solely the responsibility of the organizers and presenters and does not necessarily represent the official views of the National Institutes of Health; nor does the mention of trade names, commercial practices, or organizations imply endorsement by the U.S. Government.

Acknowledgments: Additional financial support for the conference was provided by donations from Charles Grose, Charles H. Calisher, Sandra L. Quackenbush, Rushika Perera, Randall J. Cohrs, Sara Sawyer, Wendy Maury, Jeff Wilusz, Jaquelin Dudley, Luis Rodriguez, Dan Konet, John Parker, and Suzanne Vernon. The conference was sponsored by Agilent Technologies and Darwin Biosciences. Funding to establish the Randall J. Cohrs Awards fund was provided by the generous support of: John Beckham, Sonja Best, Carol Blair, Charles H. Calisher, Nels Elde, Sonia Flores, Lee Fortunato, Anne and Michael Gershon, James Goodrich, Charles Grose, Sven Heinz, Kathryn Holmes, Dan Konet, Geoff Letchworth, Ian Lipkin, Candi Lostroh, Ravi Mahalingam, Wendy Maury, Iok Ng, Stefan Oliver, Olve Peersen, Rushika Perera, Huntington Potter, Sandra Quackenbush, Joel Rovnak, Mario Santiago, Sara Sawyer, Mark Stenglein, Kenneth Tyler, Derrick Waller, Susan Weiss, and Mark Zabel. Funding for the Richard A. Bessen Scholarship to support graduate and postdoc trainees as well as an annual guest speaker in prion biology was supported by donations from Robert and Amy Bessen.

Conflicts of Interest: The authors declare no conflict of interest.

MDPI
St. Alban-Anlage 66
4052 Basel
Switzerland
Tel. +41 61 683 77 34
Fax +41 61 302 89 18
www.mdpi.com

Viruses Editorial Office
E-mail: viruses@mdpi.com
www.mdpi.com/journal/viruses

www.ingramcontent.com/pod-product-compliance
Ingram Content Group UK Ltd.
Pitfield, Milton Keynes, MK11 3LW, UK
UKHW060121010826
14089UKWH00006B/101

9 783036 552590